Modeling in Geomechanics

Modeling in Geomechanics

Musharraf Zaman
University of Oklahoma, USA

Giancarlo Gioda
Politecnico di Milano, Italy

John Booker
University of Sydney, Australia

JOHN WILEY & SONS, LTD

Chichester • Weinheim • New York • Brisbane • Singapore • Toronto

Copyright © 2000 by John Wiley & Sons Ltd,
        Baffins Lane, Chichester,
        West Sussex PO19 1UD, England

        National 01243 779777
        International (+44) 1243 779777

e-mail (for orders and customer service enquiries): cs-books@wiley.co.uk

Visit our Home Page on http://www.wiley.co.uk
                        or
                http://www.wiley.com

Other Wiley Editorial Offices

John Wiley & Sons, Inc., 605 Third Avenue,
New York, NY 10158-0012, USA

Wiley-VCH Verlag GmbH, Pappelallee 3,
D-69469 Weinheim, Germany

Jacaranda Wiley Ltd, 33 Park Road, Milton,
Queensland 4064, Australia

John Wiley & Sons (Asia) Pte Ltd, 2 Clementi Loop #02-01,
Jin Xing Distripark, Singapore 0512

John Wiley & Sons (Canada) Ltd, 22 Worcester Road,
Rexdale, Ontario M9W 1L1, Canada

Library of Congress Cataloging-in-Publication Data
Modeling in geomechanics/[edited by] J.R. Booker, G. Gioda, M. Zaman.
    p. cm.
  Includes bibliographical references and index.
  ISBN 0-471-49218-3
  1. Engineering geology – Mathematical models.   2. Soil mechanics – Mathematical
models. I. Booker, John R.   II. Gioda, G. (Giancarlo)   III. Zaman, Musharraf.

TA705 .M62 2000
624.1′5136′05118 – dc21

                                                                00-27332

British Library Cataloguing in Publication Data

A catalogue record for this book is available from the British Library

ISBN 0 471 49218 3

Typeset in 10/12pt Times Roman by Laser Words, Madras, India
Printed and bound in Great Britain by Bookcraft (Bath) Ltd
This book is printed on acid-free paper responsibly manufactured from sustainable
forestry, in which at least two trees are planted for each one used for paper production.

Contents

23 Modeling fluid flow and rock deformation in fractured porous media

24 Models for uncertainty and its propagation with applications to geomechanics

Preface

Geomechanics is an interdisciplinary field that involves study of natural and man-made systems with emphasis on the mechanics of various interacting phenomena. It comprises aspects of numerous engineering and scientific disciplines such as civil, mechanical, hydraulic, materials, and geological engineering with appropriate bases in mathematics, mechanics and physics.

Geomechanics covers both fundamental and practical aspects, and recognizes the need for a rational process of simplification through integration of theory, experiments and verification toward the development of procedures for solutions of complex practical problems. Geomechanics with its emphasis on both basic and applied aspects, and through its interdisciplinary nature, is capable of participation in and integration of emerging scientific and technological developments so as to contribute to a wide range of areas. Among the topics in the continuing developments are constitutive models for materials, interfaces and joints to characterize stress-deformation, damage, fracture, failure and instability (lique-faction); laboratory testing; computational procedures; verification of constitutive and computer models; flow and mass transport in porous media; coupled flow and consolidation; uncertainty and its propagation; and improved analysis, design and construction of geotechnical systems such as structures and foundations, dams, caverns, tunnels, boreholes, and offshore structures.

In recent years, with the extraordinary growth of computing resources and power, the progress in the development of new theories and techniques for the analysis of geomechanics problems has far surpassed their use by practitioners. This has led to a gap between our ability to deal with complex, interdisciplinary problems in geomechanics and the impact of these developments in engineering practice. The increasing gap between theory and practice is not only due to the natural hesitancy of many practicing engineers in adopting new tools for their everyday work, but also due to the difficulties inherent in the process of representing an actual engineering problem by a satisfactory mathematical (numerical or analytical) model. The "modeling process" is often not a straightforward task for most practicing engineers. It requires both experience and an adequate level of knowledge in a specific field of engineering as well as an adequate background of the fundamental basis of various solution techniques.

This book aims at reducing the gap between theory and practice through the contributions of many well accomplished and well regarded researchers and practitioners from various branches of geomechanics including numerical and analytical methods, modeling and testing. It is intended to provide help to researchers and practicing engineers in the modeling, analysis and testing processes of geomechanical problems and to broaden the horizons of researchers toward applications of the theories and models. Wherever possible, efforts are made to illustrate relevant aspects of innovative numerical and analytical techniques as well as their application to a variety of engineering problems of practical significance.

The chapters in the book are grouped into three parts with some overlap between them: (a) Computational Procedures; (b) Constitutive Modeling and Testing; and (c) Modeling and Simulation.

The part on *Computational Procedures* includes four chapters. Two of these chapters address coupled processes in geomechanics: one related to a common problem in geotechnical engineering, namely, consolidation, and the other on the integration of boundary element method with other methods. The integrated (finite element and boundary element) method is shown to be an effective tool for analyzing complex geomechanics problems such as underground structures including shaft-cavern interaction. The asynchronous parallel algorithm can be adopted by users to analyze consolidation problems using modern computers with parallel computing capabilities. The illustrative computational strategies for soil plasticity models presented in this part can be readily implemented in a finite element-type code. The algorithm presented for numerical estimation of material parameters can be a useful tool for many geomechanical problems.

The part on *Constitutive Modeling and Testing* includes a diverse group of constitutive models ranging from classical plasticity to gradiant and Cosserat theories to the recent novel models based on the "Disturbed State Concept." Micromechanics-based models as well as modeling of interfaces and joints are also addressed. One of the distinguishing features of the contributions in this part is that for all the constitutive models presented, relevant applications are also included. The range of applications varies from simulation of laboratory testing of concrete and rocks to flow of granular materials to liquefaction. Some chapters in this part also outline steps for the evaluation of model parameters and implementation of constitutive models in computer codes for solution of boundary value problems. The chapter on laboratory testing provides an informative review of the commonly used laboratory tests and procedure for evaluation of model parameters. A companion chapter provides a review of non-destructive testing techniques for material characterization including advantages and shortcomings of various methods.

The third part, *Modeling and Simulation*, comprises a major portion of the book, and includes numerous examples illustrating how numerical and analytical methods can be used effectively. The modeling and simulation aspects and the practical boundary value problems addressed in this part include: seepage, consolidation, mass transport with emphasis on earth dams, three-dimensional hydraulic fracturing, rigid pavements including vehicle-pavement interaction, dynamic soil-structure interaction during seismic events, contact problems with emphasis on rigid circular plates, flow through fractured-porous rocks, anisotropic poroelasticity with applications to borehole problems, uncertainty in geomechanics and its propagation, impact of contaminant on landfills, slope stability, and foundations for offshore structures. A number of examples highlight the importance of coupling (e.g., fluid-solid, fracture-matrix, interfaces and joints, and soil-structure) and discusses some novel ways to include them in numerical modeling and simulation. Some of the chapters are written by practitioners with the goal of increasing the impact of numerical and analytical methods in solving real-life problems in geomechanics and reducing the gap between theory and practice.

In summary, this book is a collection of manuscripts prepared by well known engineers and scientists and practitioners. An attempt has been made in each part to summarize the relevant developments achieved in recent years and to provide examples on how these developments can be applied in modeling practical engineering problems. This book is expected to be a useful resource for the researcher, student and practitioner.

This book is one of the last pieces of work by Professor John Booker, who passed away while preparation of this book was in progress. Professor Booker was one of the worlds finest researchers in theoretical geomechanics. He produced an enormous body of published research, including papers in soil mechanics, foundation engineering, and environmental geomechanics. He was a Fellow of the Australian Academy of Science.

Dedication

This book is dedicated to Chandra S. Desai, Regents' Professor, Department of Civil Engineering and Engineering Mechanics, University of Arizona, in appreciation of his pioneering contributions in research, teaching, and professional service in a wide range of engineering disciplines, particularly geomechanics. Professor Desai has been involved in interdisciplinary research in areas such as geomechanics, solid and structural mechanics, structural dynamics and earthquake analysis, flow and deformation in porous media, mass transport, geoenvironmental engineering, tillage and raindrop analysis, materials for construction in space, multi-component track and pavement systems, natural hazards such as landslides, and electronic packaging. Many of his contributions in constitutive modeling of engineering materials and interfaces, development of innovative laboratory test devices, and computational finite element procedures have been pioneering and have provided alternative directions for research, analysis, testing and design. He has authored or co-authored and edited or co-edited 19 books and proceedings, 16 book chapters, and over 250 technical papers. One of the important attributes of his work has been the integration of theory, laboratory testing, computer analysis, and solution of practical and field problems – an approach that is challenging and often difficult. For relating research to practical problems, he has collaborated with various government and private agencies. It has been because of this integration that his work has a wide and lasting impact on both fundamental aspects and practice in engineering.

Professor Desai has received a number of recognitions and awards e.g., Meritorious Civilian Service Award by U.S. Army Corps of Engineers; Alexander von Humboldt U.S. Senior Scientist Prize by the German Government; Outstanding Contributions Medal by the International Association for Computer Methods and Advances in Geomechanics; and Outstanding Contributions in Mechanics Medal by the Czech Academy of Sciences. He has been recognized by The University of Arizona with the prestigious title of Regents' Professor, and by two high level teaching awards.

Professor Desai has made important and lasting contributions to the profession. He has served on various committees of the ASCE, TRB, ASTM, ISRM, and ASME, and as editorial board member of ten technical journals. He pioneered the successful series of international conferences on Computer Methods and Advances in Geomechanics. He also pioneered the series of conferences on Constitutive Laws for Engineering Materials: Theory and Applications; their scope includes a wide range of engineering disciplines. He is the founding President of the International Association for Computer Methods and Advances in Geomechanics, and Chief Editor of the *International Journal for Numerical and Analytical Methods in Geomechanics*.

Acknowledgements

The editors would like express their thanks to the authors for their contributions to this book. The review process involved modification of a number of chapters for which the assistance of the authors in this endeavor was invaluable. A number of contributed chapters could not be included in this book because of the late submission and the limited nature of the topics. The editors are nevertheless thankful to those individuals who made efforts to prepare those contributions.

The preparation of this book took significantly longer than originally expected for various reasons. The editors are thankful to the contributing authors for their patience and understanding.

The editors are thankful to their family members, Afroza Zaman, Jessica Zaman, Ashiq Zaman, for their support and understanding that made this book possible. Professor John Booker passed away while preparation of this book was in progress. It was a great loss for the geomechanics community. We would like to thank his family members for their valuable support in completing this project.

The editors gratefully acknowledge the School of Civil Engineering and Environmental Science at the University of Oklahoma for the secretarial support. Many professional colleagues and friends provided assistance in reviewing the manuscripts and other support at various stages of this project. We are thankful to them for their assistance and cooperation.

Musharraf Zaman
The University of Oklahoma, Norman, Oklahoma, U.S.A.

Giancarlo Gioda
Politecnico di Milano, Milan, Italy

John R. Booker (Deceased)
University of Sydney, Sydney, Australia

List of Contributors

Younane Abousleiman
School of Engineering and Architecture,
Lebanese American University, Byblos,
Lebanon

Mao Bai
Rock Mechanics Institute, The University of
Oklahoma, Norman, OK 73019, USA

Gernot Beer
Institute for Structural Analysis, University of
Technology, Graz, Austria

John R. Booker
School of Civil and Mining Engineering, The
University of Sydney, Sydney, NSW 2006,
Australia

René de Borst
Delft University of Technology, Faculty of
Civil Engineering, PO Box 5048, NL-2600 GA
Delft, The Netherlands

Bruce J. Carter
Cornell University, Ithaca, NY, USA

O. Cazacu
Department of Aerospace Engineering,
Mechanics & Engineering Science, University
of Florida, Gainesville, Florida, USA

A. H. C. Chan
School of Civil Engineering, University of
Brimingham, UK

Ching S. Chang
Department of Civil and Environmental
Engineering, University of Massachusetts,
Amherst, MA 01003, USA

John T. Christian
Consulting Engineer,
23 Fredana Road, Waban, MA 02168, USA

Annamaria Cividini
Department of Structural Engineering,
Politecnico di Milano, Piazza Leonardo da
Vinci 32, 20133 Milano, Italy

N. D. Cristescu
Department of Aerospace Engineering,
Mechanics & Engineering Science, University
of Florida, Gainesville, Florida, USA

Lizheng Cui
Rock Mechanics Institute, The University of
Oklahoma, Norman, OK 73019, USA

J. Desroches
Schlumberger Well Services, Houston, Texas,
USA

Eric Drumm
Department of Civil and Environmental
Engineering, University of Tennessee,
Knoxville, TN 37996, USA

Kenneth Fishman
McMahon & Mann Consulting Engineer,
Buffalo, New York 14214, USA

Roger Ghanem
The John Hopkins University, Baltimore, MD
21218, USA

Giancarlo Gioda
Department of Structural Engineering,
Politecnico di Milano, Piazza Leonardo da
Vinci 32, 20133 Milano, Italy

Arend E. Groen
Delft University of Technology, Faculty of
Civil Engineering, PO Box 5048, NL-2600 GA
Delft, The Netherlands

A. R. Ingraffea
Cornell University, Ithaca, NY, USA

Takashi Ito
Department of Civil Engineering, Toyota
College of Technology, 2-1 Elsei-cho, Toyota,
Aichi 471, Japan

Noboru Kikuchi
Department of Mechanical Engineering and
Applied Mechanics, The University of
Michigan, Ann Arbor, MI 48109-2125, USA

T. Kundu
Department of Civil Engineering and
Engineering Mechanics, Univeristy of Arizona,
Tucson, AZ 85721, USA

R. Matteazzi
Dipartimento di Costruzioni e Trasporti,
Universita' di Padova, Via Marzolo 9, I-35153
Padova, Italy

Kyran D. Mish
Lawrence Livermore National Laboratory,
Livermore, CA 94550, USA

Hans-B Mühlhaus
CSIRO, Division of Exploration & Mining,
PO Box 437, Nedlunds WA 6009, Australia

Kanthasamy K. Muraleetharan
School of Civil Engineering & Environmental
Science, Univeristy of Oklahoma, Norman, OK
73019, USA

James D. Murff
Exxon Production Research Co., P.O. Box
2189, Houston, Texas 77001, USA

M. Pastor
CETA/CEDEX Madrid, Spain

R. Kerry Rowe
The University of Western Ontario,
London, Ontario, Canada

Adel S. Saada
Department of Civil Engineering,
Case Western Reserve University, Cleveland,
Ohio 44106, USA

Bernard A. Schrefler
Dipartimento di Costruzioni e Trasporti,
Universita' di Padova, Via Marzolo 9,
I-35131 Padova, Italy

A. P. S. Selvadurai
Department of Civil Engineering and Applied
Mechanics, McGill University, 817 Sherbrooke
Street West, Montreal, Quebec, Canada H3A
2K6

Bert Sluys
Delft University of Technology, Faculty of
Civil Engineering, PO Box 5048, NL-2600 GA
Delft, The Netherlands

Kenjiro Terada
Department of Naval Architecture and Ocean
Engineering, The University of Tokyo, 7-3-1
Hongo, Bunkyo-ku, Tokyo 113, Japan

Laurent Vulliet
Swiss Federal Institute of Technology (EPFL),
Ecublens, CH 1015 Lausanne, Switzerland

X. Wang
Research Institute of Engineering Mechanics,
Dalian University of Technology, Dalian,
116023, China

P. A. Wawrzynek
Cornell University, Ithaca, NY, USA

G. Wilje Wathugala
Louisiana State University, Baton Rouge,
Louisiana, USA

Changfu Wei
School of Civil Engineering & Environmental
Science, Univeristy of Oklahoma, Norman, OK
73019, USA

Luther W. White
Department of Mathematics and Institute for
Reservoir Characterization, The University of
Oklahoma, Norman, Oklahoma 73019, USA

Musharraf Zaman
School of Civil Engineering & Environmental
Science, University of Oklahoma, Norman, OK
73019, USA

O. C. Zienkiewicz
Institute of Numerical Methods in Engineering,
University of Wales, Swansea, UK

X. Zheng
School of Civil and Mining Engineering,
The University of Sydney, Sydney, NSW
2006, Australia

PART 1

Computational Procedures

1

Boundary Element and Coupled Methods in Geomechanics

Gernot Beer
University of Technology, Graz, Austria

1.1 Introduction

Boundary Element Methods (BEM) offer some significant advantages for the modeling of problems in geomechanics because they can deal easily with infinite and semi-infinite domains, and require a discretization effort which is an order of magnitude smaller than that required for the Finite Element Method (FEM). The classical BEM assumes that the domain is homogeneous, isotropic, unjointed and elastic, and this has placed restrictions on the applicability of the method to problems in geomechanics. However, with the multi-region rigid-plastic BEM as presented later in this chapter, some of these difficulties have been overcome.

With this method it is still very difficult or impossible to model other features which are essential in geomechanics: ground support (e.g. rock anchors), sequential excavation and backfill, creep and swelling, viscoplastic behavior, etc. Here a combination with the FEM seems to be the 'best of both worlds'.

The chapter will give an overview of the state-of-the-art in the application of BEM and coupled methods to practical problems in geomechanics. First, the theoretical background of the BEM and the multi-region rigid-plastic BEM will be briefly reviewed. It will be shown that the latter is better suited for the modeling of joints and faults than the FEM. Next, various methods for coupling finite and boundary element discretization will be discussed.

The main part of the chapter will be devoted to the presentation of industrial applications of BEMs and coupled models in the field of rock mechanics and tunneling.

1.2 The Boundary Element Method

The method has been explained in various textbooks (for example, see Beer and Watson, 1992), but its main features are reviewed here because it is still not very well known. The main difference from the FEM is that, instead of local shape functions defined for each finite element, global shape functions are defined for a region. The important property of these special shape functions is that they satisfy the governing differential equations exactly. For elasticity, a fundamental solution of the differential equation (for example, the solution of a concentrated load in an infinite domain) can be employed. Because shape functions are defined globally, there is no need to discretize the continuum itself, as is needed in the FEM. Only the surface(s) where boundary conditions are to be satisfied need to be discretized. In the case of modeling underground excavations, for example, only the surfaces of the excavation have to be divided into boundary elements.

Bettis' reciprocal theorem can be employed to enforce the satisfaction of the boundary conditions. At a point P the following integral equation can be written:

$$\mathbf{c}(P)\mathbf{u}(P) = \int_S \mathbf{T}(P, Q)\mathbf{u}(Q)\,dS + \int_S \mathbf{U}(P, Q)\mathbf{t}(Q)\,dS \tag{1.1}$$

In Eq. (1.1), Q is a point on surface S (also the integration variable), $\mathbf{u}(Q)$ is the displacement vector and $\mathbf{t}(Q)$ the traction vector at Q, $\mathbf{T}(P, Q)$ is a matrix of fundamental solutions for the tractions and $\mathbf{U}(P, Q)$ is a matrix of fundamental solutions for displacements at Q due to a unit point load at P. $\mathbf{c}(P)$ is a matrix which depends upon the tangents to the surface at point P.

The integral equations can be changed into algebraic equations by dividing the surface S into boundary elements, over which the variation of tractions and displacements are approximated by linear or parabolic shape functions (Figure 1.1).

Infinite boundary elements are used to model surfaces which are not well defined, such as the extent of a half-space, the excavation boundary of very long tunnels, etc. These elements have been described in detail by Beer and Watson (1989). The geometry of the element and the variation of the tractions and displacements inside the element are described by:

$$\mathbf{x}(\xi, \eta) = \sum N_j(\xi, \eta)\mathbf{x}_j \tag{1.2}$$

$$\mathbf{t}(\xi, \eta) = \sum N_j(\xi, \eta)\mathbf{t}_j \tag{1.3}$$

$$\mathbf{u}(\xi, \eta) = \sum N_j(\xi, \eta)\mathbf{u}_j \tag{1.4}$$

Here $\mathbf{x}_j$, $\mathbf{t}_j$, $\mathbf{u}_j$ are the cartesian coordinates of node j and the values of tractions and displacements at node j, respectively.

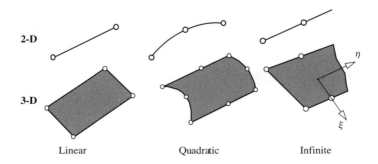

Figure 1.1   Isoparametric boundary elements

N_j are either linear or parabolic shape functions of the intrinsic coordinates ξ, η. For the infinite boundary elements, special shape functions are used for describing the geometry, and it is assumed that the displacements and tractions decay either in the infinite direction or remain constant. After introducing the isoparametric elements, the following equations are obtained for point P:

$$\mathbf{c}(P)\mathbf{u}(P) + \sum_{b=1}^{p}\sum_{j=1}^{n(b)} \Delta\mathbf{T}_j^b u_j^b = \sum_{b=1}^{p}\sum_{j=1}^{n(b)} \Delta\mathbf{U}_j^b t_j^b \tag{1.5}$$

where

$$\Delta\mathbf{T}_j^b = \int_{S_b} \mathbf{T}(P, Q)N_j \, dS_Q \tag{1.6}$$

$$\Delta\mathbf{U}_j^b = \int_{S_b} \mathbf{U}(P, Q)N_j \, dS_Q \tag{1.7}$$

In Eq. (1.5), p is the number of elements and $n(b)$ is the number of nodes of element b. The integrations over boundary elements are carried out numerically using Gauss quadrature.

Using the point collocation method, we conveniently choose the collocation points P to be the nodal points of the boundary elements $(P_1, P_2 \ldots)$ to obtain a square system of simultaneous equations:

$$[\mathbf{A}]\{\mathbf{u}\} = [\mathbf{B}]\{\mathbf{t}\} \tag{1.8}$$

where $[\mathbf{A}]$ and $[\mathbf{B}]$ are unsymmetric and fully populated coefficient matrices, and $\{\mathbf{u}\}$ and $\{\mathbf{t}\}$ are vectors containing the components of displacements and tractions on all the nodes of the boundary elements mesh. If the tractions on all the nodes at the boundary are known, then the displacements can be computed from Eq. (1.8). For a problem where tractions are known on some nodes and displacements on others, the system of equations can be rearranged so that unknown quantities are on the left-hand side and known ones on the right-hand side.

1.2.1 Multiple regions

The method as explained so far can only deal with elastic, unjointed, homogeneous material, and seems not ideally suited for solving problems in geomechanics. We will show here how the method can be extended to model piecewise inhomogeneous, jointed material.

The idea is to divide the continuum to be analysed into subvolumes (regions). For each region we can write the integral equation separately. For the two region problem depicted in Figure 1.2, we have, for example, for region **I**:

$$[\mathbf{A}]^{\mathbf{I}}\{\mathbf{u}\}^{\mathbf{I}} = [\mathbf{B}]^{\mathbf{I}}\{\mathbf{t}\}^{\mathbf{I}} \tag{1.9}$$

and for region **II**

$$[\mathbf{A}]^{\mathbf{II}}\{\mathbf{u}\}^{\mathbf{II}} = [\mathbf{B}]^{\mathbf{II}}\{\mathbf{t}\}^{\mathbf{II}} \tag{1.10}$$

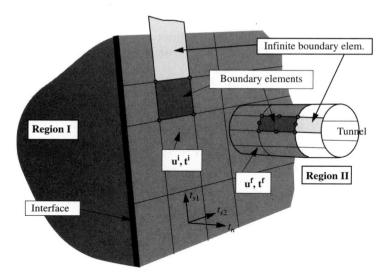

Figure 1.2   Problem with two regions

At all the nodes of each region, except at the interface between regions, either the tractions or the displacements are known. For the nodes on the interface, both displacements and tractions are unknown. By partitioning the vector of displacements and tractions, Eq. (1.10) can be rearranged as follows:

$$[\mathbf{A}^f - \mathbf{B}^i] \begin{bmatrix} \mathbf{u}^f \\ \mathbf{t}^i \end{bmatrix} = [\mathbf{B}^f - \mathbf{A}^i] \begin{bmatrix} \mathbf{t}^f \\ \mathbf{u}^i \end{bmatrix} \tag{1.11}$$

where $\mathbf{u}^f$, $\mathbf{t}^f$ are vectors containing the displacements and tractions of the nodes not on the interface ('free' nodes) and $\mathbf{u}^i$, $\mathbf{t}^i$ are the corresponding vectors for the nodes on the interface. Assuming for the moment that all tractions applied to the 'free' nodes are known, then we can write the following system of equations:

$$[\mathbf{A}^a]^\mathbf{I} \begin{bmatrix} \mathbf{u}^f \\ \mathbf{t}^i \end{bmatrix} = \{\mathbf{b^I}\} - [\mathbf{A}^i]^\mathbf{I}\{\mathbf{u}^i\}^\mathbf{I} \tag{1.12}$$

where

$$\{\mathbf{b^I}\} = [\mathbf{B}^f]^\mathbf{I}\{\mathbf{t}^f\}^\mathbf{I} \tag{1.13}$$

$$[\mathbf{A}^a]^\mathbf{I} = [\mathbf{A}^f - \mathbf{B}^i]^\mathbf{I} \tag{1.14}$$

Equation (1.12) gives a relationship between the displacements at the free nodes, the tractions at the interface and the displacements at the interface. If all displacements at the interface are zero, then the solution for the displacements at the free nodes $\{\mathbf{u}^{f0}\}$ and the tractions at the interface nodes $\{\mathbf{t}^{i0}\}$ can be obtained. Similarly, we may obtain results for the case where we give each node at the interface a unit displacement in turn. In this case, we obtain a matrix of solutions one for each unit displacement value. The matrix of solutions can be partitioned up into two parts, one containing the displacements at the free nodes $[\mathbf{U}^f]$ and the other containing the tractions at the interface nodes $[\mathbf{K}^i]$. The latter can be considered a 'stiffness matrix' for the interface. For each boundary region we may now write a relationship between the tractions and the displacements at the interface only. For example, for region **I** we have

$$\{\mathbf{t}^i\}^{\mathbf{I}} = \{\mathbf{t}^{i0}\}^{\mathbf{I}} + [\mathbf{K}^i]^{\mathbf{I}}\{\mathbf{u}^i\}^{\mathbf{I}} \tag{1.15}$$

The scheme just explained is similar to condensation used in the finite element method. It has the advantage that when we deal with interface behavior, we need only consider the degrees of freedom at the interface.

If no slip and separation occurs at the interface and no external tractions are applied at the interface, then we can write the equations of compatibility and equilibrium at the interface as

$$\{\mathbf{u}^i\}^{\mathbf{I}} = \{\mathbf{u}^i\}^{\mathbf{II}} \tag{1.16}$$

$$\{\mathbf{t}^i\}^{\mathbf{I}} + \{\mathbf{t}^i\}^{\mathbf{II}} = 0 \tag{1.17}$$

Using these conditions, the following system of equations at the interface can be obtained:

$$\{\mathbf{t}^{i0}\}^{\mathbf{I}} + \{\mathbf{t}^{i0}\}^{\mathbf{II}} + [[\mathbf{K}^i]^{\mathbf{I}} + [\mathbf{K}^i]^{\mathbf{II}}]\{\mathbf{u}^i\} = 0 \tag{1.18}$$

which can be solved for the displacements at the interface. Once the displacements at the interface have been determined, the tractions can be computed using Eq. (1.15). The displacements at the free nodes can finally be computed by

$$\{\mathbf{u}^f\}^{\mathbf{I}} = [\mathbf{U}^i]^{\mathbf{I}}\{\mathbf{u}^i\}^{\mathbf{I}} \tag{1.19}$$

1.2.2 Modeling of joints and faults

With the multi-region method it is relatively straightforward to model slip or separation, because in contrast to the FEM the nodal unknowns at the interface are tractions rather than forces. The conditions for the onset of slip and separation can be written as

$$F_S = |\tau| + \sigma_n \tan(\phi) - c \tag{1.20}$$

and

$$F_{\mathbf{T}} = \sigma_n - T \tag{1.21}$$

In the above, $|\tau|$ is the resultant shear stress acting on the interface, σ_n is the stress acting normal to the interface, ϕ is the angle of friction, c the cohesion and T the tensile strength.

The values of τ and σ_n are computed from the interface tractions as follows:

$$\tau = \sqrt{t_{s1}^2 + t_{s2}^2} \tag{1.22}$$

$$\sigma_n = t_n \tag{1.23}$$

where t_n, t_{s1}, t_{s2} are the traction components in a local coordinate system with directions normal and tangential to the fault plane (Figure 1.2). These are computed by

$$\mathbf{t}^{\text{local}} = \begin{bmatrix} t_n \\ t_{s1} \\ t_{s2} \end{bmatrix} = \mathbf{T}^T \begin{bmatrix} t_x \\ t_y \\ t_z \end{bmatrix} \tag{1.24}$$

where **T** is the transformation matrix. Equation (1.15) can be transformed so that it relates the local interface tractions with local interface displacements:

$$\{t^{i,\text{local}}\}^{\mathbf{I}} = [\mathbf{T}]^T \{t^{i0}\}^{\mathbf{I}} + [\mathbf{T}]^T [\mathbf{K}^i]^{\mathbf{I}} [\mathbf{T}] \{u^{i,\text{local}}\}^{\mathbf{I}} \qquad (1.25)$$

where $\{u^{i,\text{local}}\}$ is the vector containing the local components of displacements at the interface nodes.

The procedure of modeling joint behavior is as follows:

- An analysis is made assuming the interface between the regions does not slip or separate.
- With the traction values obtained from Eq. (1.25) we check the yield conditions at each node.
- At a particular node if both F_S, F_T are less than zero, then no action need to be taken.
- If $F_S > 0$, then the degrees of freedom in tangential directions are disconnected and the excessive shear stresses are applied in the opposite direction. This changes the way in which the system of Eq. (1.18) is assembled.
- If $F_T > 0$, then all degrees of freedom are disconnected from that node and the excessive normal stresses are applied in the opposite direction.

The iteration process proceeds until either all yield conditions are satisfied at all nodes that is $F \leq 0$, or a situation arises where two boundary element regions are completely separated. Each iteration involves an assembly and a new solution of the system of equations. Details of the numerical implementation are shown in Beer (1993). The method is termed the *rigid-plastic joint model* because, unlike the methods used in the FEM, no elastic slip and dilation is allowed to occur prior to joint failure.

1.2.3   Test example

The procedure just outlined is tested and compared with the FEM on a simple example. The test example used by Day and Potts (1994) to demonstrate the problems with oscillatory behavior that sometimes occurs with joint finite elements is that of an elastic block on a rigid foundation. Figure 1.3 shows the dimensions of the problem and the loading.

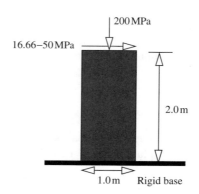

Figure 1.3   Test example to show oscillatory behavior of joint finite elements

The block is loaded by a normal stress of 200 MPa and a shear stress at the top which is increased until toppling occurs.

The theory predicts that tension will start to occur at the left edge of the interface when the shear stress reaches 16.66 MPa. At 50 MPa, toppling will occur. The theoretical shear stress distribution for this problem is not known, but the solution will be somewhere between the solution for a infinitely long rigid block on flexible foundations (constant distribution) and for an infinitely long flexible block on a rigid foundation (parabolic distribution).

Figure 1.4 shows the distribution of shear stress at the interface obtained with the FEM for an applied shear stress of 16.66 MPa and a ratio of joint stiffness (k) to elastic modulus of the block (E) of 1000. For this joint property oscillatory behavior was observed.

The oscillatory behavior of the interface stresses was found to be not only dependent on the stiffness assigned to the joint element, but also on the size of the finite element adjacent to the joint. If the joint stiffness was reduced or the dimension of the finite element adjacent to the joint was reduced, then the oscillatory stress distribution disappeared. For details, see Beer (1997). In contrast to the FEM, the rigid-plastic BEM does not require elements inside the block; it is not necessary to specify a joint stiffness, and the results of the analysis do not depend upon the discretization. Furthermore, there are much fewer iterations required to model the toppling of the block (Figure 1.5).

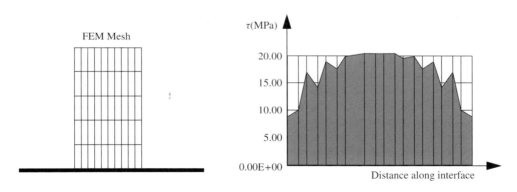

Figure 1.4   Distribution of shear stress at the interface for k/E = 1000, FEM model

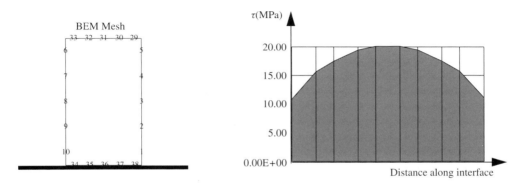

Figure 1.5   Distribution of shear stress at interface, BEM model

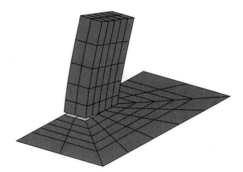

Figure 1.6   Displaced shape of block (exaggerated) after three iterations.

Figure 1.6 shows the results of a three-dimensional analysis of a block on an elastic halfspace, where half of the block and the halfspace is discretized into three-dimensional (surface) boundary elements and symmetry conditions are applied. For the case of an applied shear stress of 50 Mpa (toppling is predicted for this case), the displaced shape is shown after three iterations. The opening of the interface can be clearly seen.

1.2.4   Practical applications

Two applications in geomechanics of the rigid-plastic multi-region BEM will be shown. The first one is the three-dimensional analysis of a tunnel approaching a fault, the other deals with the determination of the stresses in a fault zone.

Tunnel approaching fault

This analysis was carried out as part of a project which dealt with the detection of faults and changes in rock stiffness using *in situ* displacement measurements. It has been found (Schubert and Budil, 1995) for example that when a tunnel is approaching a fault, the direction of the displacement vector changes, and from this change the presence of geological features may be determined. To give meaningful results the analysis had to be three-dimensional, and a model was required which used a small amount of human and computer resources. The boundary element mesh used is shown in Figure 1.7, and consists of two regions made up of eight node isoparametric boundary elements with a quadratic shape function, and six node infinite boundary elements with $1/r$ type decay of displacements in the infinite direction at the interface. For the infinitely long tunnel, infinite boundary elements with constant displacements were used to simulate plane strain behavior.

Only half of the tunnel and interface are discretized with symmetry conditions assumed. For this type of model, it is assumed that the nonlinear behavior can only occur in the fault plane, the rock mass itself is assumed to be elastic. One of the results of the analysis is shown in Figure 1.8 as contours of slip at the interface for a particular position of the tunnel.

The time required to make the mesh for this problem was considerably shorter than the time required for making an equivalent mesh with finite elements. The computing time for one excavation stage was 10 minutes on a 166 MHz Pentium PC. The results of the analysis were found to compare well with observed displacement directions.

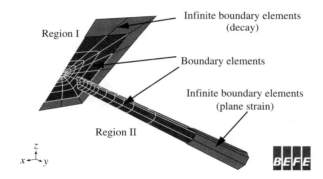

Region I

Infinite boundary elements
(decay)

Boundary elements

Infinite boundary elements
(plane strain)

Region II

Figure 1.7   Boundary element mesh for tunnel approaching fault

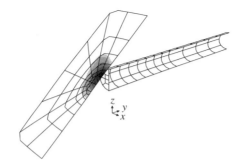

Figure 1.8   Contours of slip on interface (darker areas = higher values of slip)

Determination of stresses in a fault zone

The second example is concerned with the computation of the state of stress in a soft zone prior to the excavation of a tunnel. The example is taken from an actual tunnel site in Austria, where an accident occurred when the tunnel was driven through a soft zone of ground just before it reached harder rock. The purpose of the analysis was to confirm with a numerical simulation the proposition put forward by the experts investigating the accident that the virgin stress in the disturbed zone had been significantly altered by the difference in stiffness of the two zones and the presence of faults. This modification in the stress was thought to be caused by arching as the soft zone 'creeps' under self-weight (Figure 1.9).

For the analysis a boundary element discretization with two regions (one finite and one semi-infinite) was used. The mesh is shown in Figure 1.10.

The material properties were chosen as follows:

- Undisturbed zone: E $= 150\,000$ MPa$v = 0.3$
- Fault zone: E $= 50\,000$ MPa$v = 0.3$, $\gamma = 3000$ kg/m^3
- Fault: Angle of friction $= 5°$, Cohesion $= 0.5$ MPa

To simulate the arching effect, gravity was applied to the finite region. The analysis was elastic only, and the 'creep' effect modeled only by assigning a lower stiffness to the finite

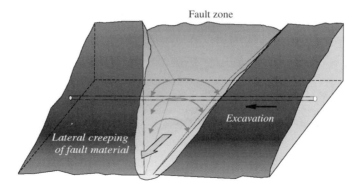

Fault zone

Excavation

Lateral creeping of fault material

Figure 1.9   Schematic of the arching effect being modeled

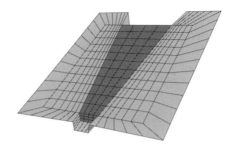

Figure 1.10   Boundary element discretization

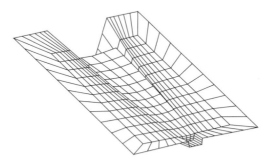

Figure 1.11   Displaced shape of fault zone after three iterations

region. Figures 1.11 and 1.12 show some results of the BEM analysis after convergence had been achieved after only three iterations. The displaced shape and stress vectors on a result plane are shown, and the slip in the fault plane and the arching of the stresses can be clearly seen. Further results are presented by Schubert and Riedmüller (1995).

The total computing time required for this non-linear three-dimensional problem was 14 minutes on a Pentium PC.

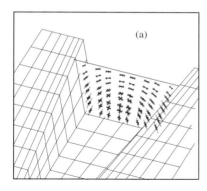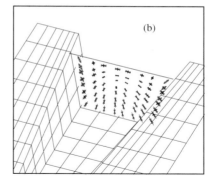

Figure 1.12 Stress trajectories plotted on a result plane inside the finite region after (a) three iterations and (b) one

Conclusions

In this section it has been shown that the multi-region boundary element method can be applied very efficiently to the analysis of jointed and faulted rock. It was found that run times are of an order of magnitude smaller than for the finite element method. For the last example (determination of stresses in a fault zone), a comparison was made with a three-dimensional finite element analysis. Although the results were comparable, the FEM took about 10 times longer to compute the results than the BEM analysis.

It has also been demonstrated here that joints and faults can be modeled in a much better way with the rigid-plastic multi-region BEM than with joint finite elements. There are fewer parameters one has to worry about, and the convergence is much faster.

The BEM, however, still has some disadvantages which are difficult to overcome at present:

- It cannot deal easily with nonlinear phenomena such as creep, plasticity and swelling.
- Sequential excavation, backfill and installation of the shotcrete lining cannot be modeled as easily as with the FEM.
- Steel arches, rock anchors and rock bolts cannot be modeled.

Here a combination of the FEM and BEM seems to give the 'best of both worlds', namely the flexibility of the FEM and the efficiency of the BEM.

1.3 The Coupled Boundary Element/Finite Element Method

In the coupled BE/FE method, the problem domain is divided up into one finite element region and one or more boundary element regions. Nonlinear effects and sequential excavation and construction are confined to the finite element region. Figure 1.13 shows two examples of coupled meshes: in the mesh on the left, only the shotcrete lining has been discretized into finite elements, and this would allow the placing of the shotcrete and its nonlinear behavior to be modeled. For the mesh at the right, not only the shotcrete shell but also part of the ground and anchors have been modeled with finite elements. This

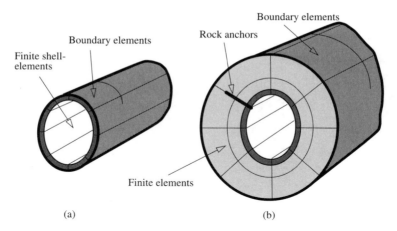

Figure 1.13   Possibilities of modeling tunnel construction with finite and boundary elements

mesh would, for example, allow the modeling of nonlinear behavior of the ground and the effect of ground support to be modeled.

The coupling theory explained here is similar to that already explained for the multi-region BEM. The only difference is that for the finite element region we work with nodal point forces **F** instead of tractions **t**. Therefore, before we can apply the equilibrium conditions, we must either convert the nodal point forces to tractions, or *vice versa*.

Using the principle of virtual displacements, we can obtain a vector of nodal point forces at node i, $\mathbf{F}_i^e$ for an element **e** which is loaded by a distributed load **t** on one side as (see Figure 1.14):

$$\mathbf{F}_i^e = \int_S N_i \mathbf{t} \, dS \tag{1.26}$$

In Eq. (1.26), N_i are the isoparametric shape functions and **t** is a vector of tractions acting on the boundary of element **e**. Expressing these tractions in terms of nodal point tractions (Eq. (1.3)), we obtain

$$\mathbf{F}_i^e = \sum \int_S N_i N_j \mathbf{t}_j^e \, dS \tag{1.27}$$

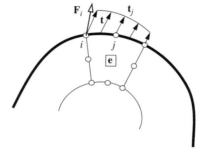

Figure 1.14   Relationship between nodal point forces and nodal tractions

The vector of forces at all nodes of element $\mathbf{F^e}$ may be expressed in terms of nodal point tractions $\mathbf{t^e}$ by

$$\mathbf{F^e} = \mathbf{M^e t^e} \tag{1.28}$$

where $\mathbf{M^e}$ is a matrix of integrals of shape function products whose coefficients are given by

$$\mathbf{M_{ij}^e} = \int_S N_i N_j \, dS \tag{1.29}$$

Alternatively, the vector of tractions at all the element nodes may be expressed as

$$\mathbf{t^e} = (\mathbf{M^e})^{-1} \mathbf{F^e} \tag{1.30}$$

There are two possible ways in which coupling can be achieved. One is to treat the FEM region in the same way as a boundary element subregion (unsymmetric coupling); the other one is to treat the BEM subregions as a 'superelement' in the standard assembly process (symmetric coupling).

1.3.1 Unsymmetric coupling

For the finite element region the relationship between all nodal point forces and displacements for each element can be written as

$$\mathbf{F^e} = \mathbf{K^e u^e} \tag{1.31}$$

This can be converted into a relationship between tractions and displacements, as with the BEM:

$$\mathbf{t^e} = (\mathbf{M^e})^{-1} \mathbf{K^e u^e} \tag{1.32}$$

The modified 'stiffness' matrix for the element, i.e. one that relates nodal tractions instead of forces to displacements, can be written as

$$\mathbf{\hat{K}^e} = (\mathbf{M^e})^{-1} \mathbf{K^e} \tag{1.33}$$

This 'stiffness' matrix can now be assembled with the boundary element regions, as explained in Section 1.2.1. For efficiency reasons, however, it may be necessary to first condense out all the nodes of the finite element region which are not on the interface, and then do the conversion of nodal forces to tractions.

With this method of coupling the analysis proceeds with the following steps:

(1) The stiffness matrix for the finite element region is assembled.

(2) For efficiency reasons, we eliminate the internal degrees of freedom so that we are left with a relationship between forces and displacements on the interface with boundary element regions only.

(3) The stiffness matrix and the force vector are converted as explained previously.

(4) The analysis proceeds as explained for the multi-region method, with the finite element region treated in the same way as the boundary element regions.

The method just outlined has the advantage that there is no need to 'symmetrize' the boundary element matrices as explained by Beer and Watson (1992), and that joints and faults can be still treated within the boundary element regions.

1.3.2   Symmetric coupling

Here we treat each boundary element region as a superelement and assemble it into the finite element equations in the usual way.

For each region, we modify Eq. (1.15) so that nodal point forces appear on the left-hand side:

$$\{\mathbf{F}^i\}^{\mathbf{I}} = [\mathbf{M}]\{\mathbf{t}^{i0}\}^{\mathbf{I}} + [\mathbf{M}][\mathbf{K}^i]^{\mathbf{I}}\{\mathbf{u}^i\}^{\mathbf{I}} \qquad (1.34)$$

Here $[\mathbf{M}]$ is a matrix assembled from element contributions $\mathbf{M}^e$.

The resulting stiffness matrix of the boundary element region **I** is unfortunately unsymmetric because $[\mathbf{K}^i]$ is. Since most finite element solvers can only deal with symmetric matrices, it is convenient to try to make the stiffness matrix symmetric. This can be achieved by taking an average value of the off-diagonal coefficients. There is some theoretical justification for this (see Beer and Watson (1992) for a full description), but this process may introduce some error. It should be noted that the unsymmetry of the coefficient matrices in the BEM are only due to the discretization and that if linear boundary elements are used it is less pronounced than if parabolic elements are used.

1.3.3   Test example

A test example is shown here to demonstrate that minimal error is introduced by the coupling process. It is a circular excavation in an infinite domain with a virgin compressive stress of 1 Mpa applied in the horizontal direction, and of 2 MPa in the vertical direction. Two different coupled meshes have been used, and they are shown in Figure 1.15, together with the displaced shapes. Parabolic boundary elements coupled with finite elements were used.

One axis of symmetry (y-axis) was assumed. Both symmetric and unsymmetric coupling was applied. The distribution of stress along a horizontal line from the centre of the circle is plotted in Figure 1.16 for the two different coupled meshes. It can be seen that the results of the analysis using mesh 1, which were computed as internal points,

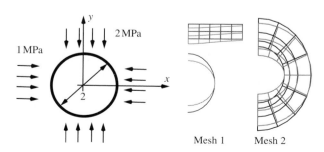

Figure 1.15   Test example, different discretization used and displaced shape computed

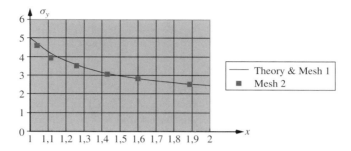

Figure 1.16   Distribution of vertical stress: comparison of results

are indistinguishable from the theoretical solution, whereas the results of mesh 2, which were computed at Gauss points of finite elements, show a small error. For this case, the difference between symmetric and unsymmetric coupling was negligible.

1.3.4   Case study – analysis of CERN caverns

The CERN organization in Geneva is planning to extend the facilities of the existing Large Electron Positron Collider (LEP) to install a new particle collider machine with considerably higher collision energy, the Large Hadron Collider (LHC). To accommodate the space requirements of the new LHC, additional caverns have to be built. At Point 5 a new detector, the Compact Muon Solenoid, will be installed. This requires the construction of two large caverns (experimental cavern UXC55 and service cavern USC55) with spans of up to 26 m and more than 30 m height, two large new shafts and a number of auxiliary tunnels in a geological sequence of marls and sandstones (Molasse) underneath a major aquifer. The large caverns are separated by a 7 m thick concrete pillar. Rock support and watertight secondary linings requiring minimum maintenance effort had to be designed. The design of the secondary lining structure implies a particular technical challenge due to the geometrical complexity of the structure and the load cases to be considered, which include excavation, rock mass creep and swelling. A series of analyses involving three-dimensional models using only boundary elements, as well as models with finite elements coupled with boundary elements, were carried out.

Boundary element analysis

The first analysis was carried out on a global model which simulated all caverns (existing and new) using only boundary element discretization and assuming elastic/homogeneous rock mass behavior. Figure 1.17 shows the discretization into quadratic (eight node) boundary elements and quadratic (six mode) infinite boundary elements.

  The results of these calculations give an indication of the global stress distribution in the rock mass, and allow us to assess and verify geometrical simplifications that might be justifiably introduced in the coupled models involving the extremely complex load cases swelling and rock mass creep. For post-processing and the display of results, stresses and displacements can be plotted either on the excavation surfaces or on additional specified planes in the rock mass. The global analysis was also used to determine locations where the sections for simplified plane strain analyses could be defined.

As an example, the distribution of the maximum compressive stress is shown in Figure 1.17 on two planes inside the rock mass (only the tip of the second plane is visible). This analysis, although fast in terms of mesh preparation and computing time (approximately two hours on a 32 MB Pentium 166 notebook) cannot deal with sequential concrete pillar construction, sequential cavern excavation and rock mass creep or swelling. For these cases, the coupled BE/FE analysis was found to be an alternative to a full 3-D analysis, giving some important advantages.

Coupled nonlinear analyses

Three-dimensional coupled analyses were carried out to study the effects in terms of sectional forces induced in the secondary lining structure by load cases such as swelling of marl layers, rock mass creep, lining weight, lining concrete creep, and lining temperature changes. For concrete design purposes, related sectional forces are derived. Preliminary analyses have shown that the large caverns can be modeled separately using symmetry conditions to simulate the influence of the caverns on each other.

The meshes in Figure 1.18 consist of finite and infinite boundary elements which model the rock mass, and finite elements which model the secondary concrete lining structure and the part of the rock mass where nonlinear material behavior is expected to occur. The thickness of the secondary linings is 800 mm for the large experimental cavern and 500 mm for the slightly smaller service cavern. The finite elements of the concrete lining are connected to the boundary elements representing the rock mass by joint elements which have a very low shear stiffness assigned so as to simulate the slip between the rock and the lining. Comparison calculations have shown that these interfaces are essential

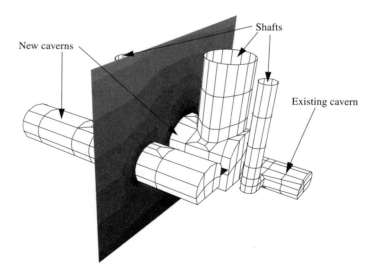

Figure 1.17   Boundary element mesh of caverns of the Hadron collider and contours of maximum compressive stress

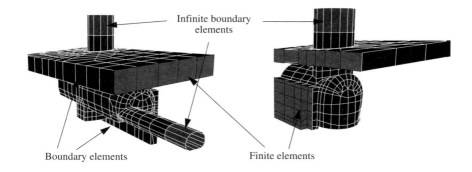

Figure 1.18   Coupled meshes for service cavern (left) and experimental cavern (right)

for the proper rock mass lining interaction simulation, especially in areas of distinct three-dimensional structural behavior, such as the shaft intersections.

Swelling and rock mass creep are assumed to be confined to a zone of weak marl, which is approximately 7 m thick and located 4 m above the crown of the caverns. Therefore, this additional zone of finite elements has been introduced above the cavern crowns to apply swelling and rock mass creep loads. Swelling is simulated by a temperature equivalent approach. Rock mass creep is simulated by using a real creep model, which is dependent on the actual local stress levels in the model. Therefore, it was necessary to introduce to the rock mass in the model the *in situ* stress field conditions. It has to be mentioned that the *in situ* stress conditions are not necessary for load cases such as swelling or lining weight, because the lining structure will be installed only after the excavations are completed and stabilized by appropriate rock support. Therefore, such load cases are creating sectional forces in the lining which are independent from the actual stress conditions in the rock mass.

After running the model and for the post-processing of results, the finite elements representing the concrete lining were degenerated to shell elements. This allows the computation and display of sectional forces in a local co-ordinate system, and is enhancing the evaluation procedure to determine the amount and extent of necessary concrete rein-forcements to be installed in the secondary lining structure. Figure 1.19 shows a result of such an analysis in the form of the computed internal sectional forces in the concrete shell for the load case swelling.

Note that, if one were to represent such a model by finite elements only, the whole rock mass would have to be filled up with solid elements. These elements would need to be extended up to a distance sufficiently far away from the excavation so that the influence of artificially applied boundary conditions on the results is negligible. This would require a FEM model of a size in the order of 50 000 to 100 000 elements. In the coupled BEM/FEM case, this large space extending from the actual area of interest to the area of negligible disturbance is represented just by the boundary element region. The mesh for the larger of the two caverns had less than 1000 boundary and finite elements, and took approximately 20 minutes to run on a 32 MB Pentium Notebook.

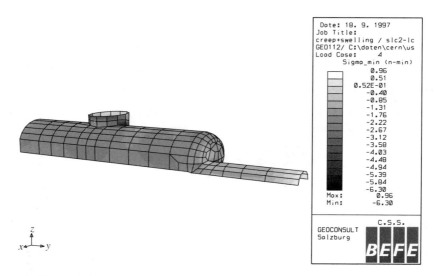

Figure 1.19   Normal forces in concrete shell due to load case swelling

Conclusions

It has been shown that the coupled boundary element finite element method can be used very effectively for the modeling of problems in geomechanics. Even difficult reinforcement design problems related to complex rock mass lining interactions at geometrically difficult locations such as shaft/cavern/pillar intersections are possible to solve in a relatively short time. In the case of the design of the underground structures for the CERN LHC project, the complex rock mass lining interaction required the used of three-dimensional calculation models. Time and cost savings were possible due to the use of meshes, which were considerably smaller since the ground itself can be represented by surface instead of volume discretization. Smaller meshes have definite advantages not only due to shorter computing time, but also in the pre-processing stage due to less efforts for mesh generation, as well as in the post-processing stage because of the shorter time needed for data evaluation. In addition, due to the fact that there is no subdivision of the bulk rock mass into finite elements, errors related to improper discretization are easier to control especially in three-dimensional models.

1.4   Summary and Outlook

We have attempted in this chapter to show that the 'Cinderellas' of numerical methods, the boundary element and coupled methods, can be applied very well to the analysis of problems in geomechanics. They have been shown not only to be very efficient in terms of computer and human resources, but also – contrary to common perception – it has been shown that they are able to handle discontinuous nonlinear behavior well, in some cases better than the FEM. By coupling the FEM and BEM, the best of both worlds can be obtained, and the benefits of each method, that is the ease with which the BEM can deal with infinite and semi-infinite domains, and the ease with which the finite element method can deal with nonlinear material behavior, sequential excavation/construction and ground support can be fully exploited.

A brief review of the theoretical background of the BEM and the coupled BEM/FEM has been presented here. This was followed by test examples which serve not only to demonstrate how the methods work, but also show the accuracy which can be obtained. Large scale industrial applications are presented to demonstrate that the methods are not only of academic interest, but also result in significantly lower computing times and efforts in mesh generation.

Only applications in statics have been presented here, but there has been substantial progress also in the dynamic, transient analysis of problems in geomechanics using the BEM, especially dealing with the initiation and propagation of cracks due to the detonation of an explosive charge (Tabatabai-Stocker and Beer, 1999).

The author believes that their potential is not being fully exploited at the moment, and it is hoped that this chapter can make a small contribution to the awareness of this potential in the geomechanics community. Many applications involving unbounded domains with unbounded fault planes exist where the BEM, together with the special infinite boundary elements, can be used very effectively. This could include, for example, the analysis of plate tectonics for earthquake prediction, the formation and propagation of cracks, and many others.

Acknowledgement

Thanks are due to Ms. Tabatabai-Stocker for the review of the manuscript.

References

Beer, G. An efficient numerical method for modelling initiation and propagation of cracks along material interfaces. *Int. J. Numer. Meth. Engng.*, **36**, (1993) 3579–3594.

Beer, G. Efficient discretisation methods for the numerical simulation in geomechanics. *IUTAM Symposium on Discretisation Methods*, Vienna, (1997).

Beer, G. and Watson, J. O. Infinite boundary elements. *Int. J. Numer. Meth. Engng.*, **28**, (1989) 1233–1247.

Beer, G. and Watson, J. O. *Introduction to Finite and Boundary Element Methods for Engineers.* Chichester: Wiley, (1992).

Day, R. A. and Potts, D. M. Zero thickness interface elements – numerical stability and application. *Int. J. Numer. Anal. Methods Geomech.*, **18**, (1994) 689–708.

Schubert, W. and Budil, A. The importance of longitudinal deformation in tunnel excavation. *ISRM Conference*, Tokyo, (1995).

Schubert, W. and Riedmüller, G. *Geotechnische Nachlese eines Verbruches – Erkenntnisse und Impulse*, 10 Christian Veder Kolloquium, TU Graz, (April 1995).

Tabatabai-Stocker, B. and Beer, G. A boundary element method for modelling cracks along material interfaces in transient dynamics. *Comm. Numer. Meth. Engng.* (accepted 1999).

2

Computational Strategies for Standard Soil Plasticity Models

René de Borst and **Arend E. Groen**
Delft University of Technology, The Netherlands

2.1 Introduction

Since the early developments in the 1950s and 1960s by Drucker and Prager and by the Cambridge school, plasticity theory has become an established framework for modeling the inelastic behavior of soils. Nowadays, it has earned a broad interest and acceptance (Drucker *et al.*, 1957; Roscoe and Burland, 1968; Matsuoka and Nakai, 1974; Lade, 1977; Desai, 1980). The Cam-Clay model (Roscoe and Schofield, 1963) has become widely accepted as a constitutive model that pairs a relative simplicity – and therefore a limited number of parameters – to a description of the essential mechanical properties of clays. For sand under monotonic loading, double-hardening models (Lade, 1977) have become popular. In these models, a shear yield surface is complemented by a cap, which limits the hydrostatic stresses and enables capturing the inelastic straining under pure compaction. Typically, these sand models are furnished with a non-associative flow rule, a hardening curve with a hyperbolic shape, and elastic moduli that are stress-level dependent.

Both classes of constitutive models – Cam-Clay models and double-hardening models – have a yield surface that is dependent on the hydrostatic stress level and are equipped with stress-level dependent elastic moduli. Furthermore, the Cam-Clay model possesses a non-associative hardening rule, while double-hardening models typically are furnished with a non-associative flow rule. In either case, this leads to a non-symmetric tangential matrix for the incremental relation between stress rate and strain rate. Because of this lack of symmetry, and because of the highly nonlinear nature of both classes of constitutive models, an explicit integration will only be conditionally stable. Nonetheless, most integration procedures utilize an explicit procedure (Britto and Gunn, 1989; Desai *et al.*, 1991; Hicks, 1995). Since the early 1990s, a few attempts have been made to apply implicit integration to plasticity models for soils (Borja, 1991; Hofstetter *et al.*, 1993; Simo *et al.*, 1993). Such an integration procedure allows for much larger loading steps. Moreover, the resulting algorithm is, in general, amenable to exact linearization, which ensures quadratic convergence when a Newton–Raphson strategy is applied on structural level.

This contribution is ordered as follows. After a brief introduction into plasticity theory, we shall formulate a fully implicit Euler backward return-mapping algorithm originally proposed by Borja (1991) which incorporates the use of stress invariants as primary variables and a secant update for the elastic properties. The algorithm will be applied for the integration of the Modified Cam-Clay model and a double-hardening model for sand. The numerical performance and robustness will be demonstrated by simulation of common laboratory tests and realistic boundary value problems, i.e. the simulation of a guided pipe-jacking in soft clay and an analysis of the Leaning Tower of Pisa.

2.2   A Note on Stress and Strain Definitions

In almost all soil mechanics applications, only compressive stresses are present. Therefore, it would be convenient to use a sign convention in which compressive stresses are positive (Figure 2.1(a)). On the other hand, in a general purpose finite element code, tensile stresses are usually regarded positive (Figure 2.1(b)). As the objective is to demonstrate how elasto-plastic soil models can be conveniently implemented into a general-purpose finite element code, it is assumed that tensile stresses are positive.

The stress tensor $\boldsymbol{\sigma}$ and the strain tensor $\boldsymbol{\varepsilon}$ are introduced in a vector format as

$$\boldsymbol{\sigma} = [\sigma_{xx}, \sigma_{yy}, \sigma_{zz}, \sigma_{xy}, \sigma_{yz}, \sigma_{zx}]^{\mathrm{T}} \tag{2.1}$$

$$\boldsymbol{\varepsilon} = [\varepsilon_{xx}, \varepsilon_{yy}, \varepsilon_{zz}, 2\varepsilon_{xy}, 2\varepsilon_{yz}, 2\varepsilon_{zx}]^{\mathrm{T}} \tag{2.2}$$

in which the engineering shear strains have been utilized instead of the tensorial shear strains. The hydrostatic pressure and the volumetric strain are defined as

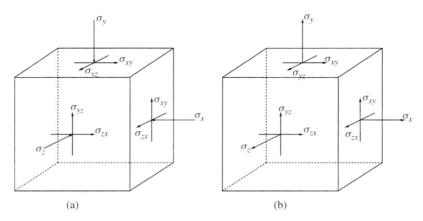

(a)   (b)

Figure 2.1   Possible sign conventions for the stress tensor. (a) Sign convention normally utilized in soil mechanics. Compression is regarded positive. (b) Sign convention normally utilized in finite element codes and utilized in this contribution. Tension is regarded positive

$$p = \tfrac{1}{3}(\sigma_{xx} + \sigma_{yy} + \sigma_{zz}) \tag{2.3}$$

$$\varepsilon_v = \varepsilon_{xx} + \varepsilon_{yy} + \varepsilon_{zz} \tag{2.4}$$

in which p is the hydrostatic tensile component and ε_v is the volumetric strain. Upon introduction of the projection vector $\boldsymbol{\pi} = [1, 1, 1, 0, 0, 0]^{\mathrm{T}}$, one can define the deviatoric stresses ($\boldsymbol{\xi}$) and strains ($\boldsymbol{\gamma}$),

$$\boldsymbol{\xi} = \boldsymbol{\sigma} - p\boldsymbol{\pi} \tag{2.5}$$

$$\boldsymbol{\gamma} = \boldsymbol{\varepsilon} - \tfrac{1}{3}\varepsilon_v \boldsymbol{\pi} \tag{2.6}$$

The effective deviatoric stress (q) is defined according to

$$q = \sqrt{\tfrac{3}{2}\boldsymbol{\xi}^{\mathrm{T}}\boldsymbol{R}\boldsymbol{\xi}}; \quad \boldsymbol{R} = \mathrm{diag}[1, 1, 1, 2, 2, 2] \tag{2.7}$$

The Lode angle θ is defined as

$$\cos 3\theta = -\frac{27}{2}\frac{J_3}{q^3}; \quad J_3 = \xi_{xx}\xi_{yy}\xi_{zz} + \xi_{xy}\xi_{yz}\xi_{zx} - 2(\xi_{xx}\xi_{yz}^2 + \xi_{yy}\xi_{zx}^2 + \xi_{zz}\xi_{xy}^2) \tag{2.8}$$

2.3 Computational Plasticity

A fundamental notion in plasticity theory is the existence of a yield function that bounds the elastic domain. Plasticity can only occur if the stresses $\boldsymbol{\sigma}$ satisfy the general yield criterion

$$f(\boldsymbol{\sigma}, \kappa) = 0 \tag{2.9}$$

where the hardening variable κ controls the amount of hardening or softening. Plastic behavior is characterized by the presence of irreversible strains upon load removal. These are introduced via the usual decomposition of the strain rate vector $\dot{\boldsymbol{\varepsilon}}$ in an elastic, reversible contribution $\dot{\boldsymbol{\varepsilon}}^e$ and a plastic, irreversible contribution $\dot{\boldsymbol{\varepsilon}}^p$:

$$\dot{\boldsymbol{\varepsilon}} = \dot{\boldsymbol{\varepsilon}}^e + \dot{\boldsymbol{\varepsilon}}^p \tag{2.10}$$

The elastic strain rate is related to the stress rate by an elastic constitutive law

$$\dot{\boldsymbol{\sigma}} = \boldsymbol{E}^{\mathrm{t}}(\boldsymbol{\varepsilon}^e)\dot{\boldsymbol{\varepsilon}}^e \tag{2.11}$$

It is assumed that the tangential elastic stiffness matrix $\boldsymbol{E}^{\mathrm{t}}(\boldsymbol{\varepsilon}^e)$ is solely dependent on the elastic strain field $\boldsymbol{\varepsilon}^e$. The assumption of a non-associative flow rule yields

$$\dot{\boldsymbol{\varepsilon}}^p = \dot{\lambda}\frac{\partial g(\boldsymbol{\sigma}, \kappa)}{\partial \boldsymbol{\sigma}} \tag{2.12}$$

where $g(\boldsymbol{\sigma}, \kappa)$ is the plastic potential function and $\dot{\lambda}$ is the (rate of the) plastic multiplier. The hardening variable κ is defined in a rate form as

$$\dot{\kappa} = \dot{\kappa}(\boldsymbol{\sigma}, \dot{\boldsymbol{\varepsilon}}^p) \tag{2.13}$$

on the basis of which a number of different measures for $\dot{\kappa}$ can be postulated. This contribution is restricted to strain hardening

$$\dot{\kappa} = \dot{\kappa}(\dot{\varepsilon}^{\mathrm{p}}) \tag{2.14}$$

which is integrated along the loading path to give

$$\kappa = \int_0^t \dot{\kappa}\,d\tau \tag{2.15}$$

Loading/unloading is conveniently established in Kuhn–Tucker form

$$\dot{\lambda} \geq 0; \quad f \leq 0; \quad \dot{\lambda}f = 0 \tag{2.16}$$

Neither the yield function in Eq. (2.9) nor the plastic potential $g(\sigma, \kappa)$ in Eq. (2.12) need to be C_1 continuous. In fact, the assumption of only piecewise C_1 continuous yield functions and plastic potentials allows for an independent description of different irreversible phenomena within one material. For example, tensile failure and compressive crushing can be treated independently for concrete (Feenstra, 1993), or for masonry (Lourenço, 1996). In this contribution, piecewise C_1 continuous yield functions and plastic potentials will be adopted for modeling shear failure and irrecoverable compression independently. Before elaborating the details of the constitutive modeling, a possible description of plasticity with only piecewise C_1 continuous yield functions and plastic potentials will be presented.

Assume that the yield function is described by a set of C_1 continuous yield functions

$$f_i(\sigma, \kappa_i) \leq 0 \tag{2.17}$$

in which κ_i is the hardening variable representative for plastic flow on that C_1 continuous yield surface. The additive decomposition of the strain rate given by Eq. (2.10) is generalized to (Koiter, 1953)

$$\dot{\varepsilon} = \dot{\varepsilon}^{\mathrm{e}} + \sum_{i=1}^{i=n} \dot{\varepsilon}_i^{\mathrm{p}} \tag{2.18}$$

with

$$\dot{\varepsilon}_i^{\mathrm{p}} = \dot{\lambda}_i \frac{\partial g_i(\sigma, \kappa_i)}{\partial \sigma} \tag{2.19}$$

$g_i(\sigma, \kappa_i)$ is a C_1 continuous plastic potential, which governs the flow direction on each C_1 continuous yield function f_i. Like in single surface plasticity, κ_i is integrated along the loading path

$$\kappa_i = \int_0^t \dot{\kappa}_i(\dot{\varepsilon}_i^{\mathrm{p}})\,d\tau \tag{2.20}$$

The Kuhn–Tucker conditions defined by Eq. (2.16) are generalized to

$$\dot{\lambda}_i \geq 0; \quad f_i \leq 0; \quad \dot{\lambda}_i f_i = 0 \tag{2.21}$$

2.3.1   Return-mapping via invariants with use of elastic secant moduli

The integration of the rate equations resulting from flow theory can be regarded as follows. At stage n the stresses, the total and elastic strains and the hardening parameter(s) are known. The objective is to determine the update of these parameters

$$\{\sigma_n, \varepsilon_n, \varepsilon^e_n, \kappa_{i,n}\} \rightarrow \{\sigma_{n+1}, \varepsilon_{n+1}, \varepsilon^e_{n+1}, \kappa_{i,n+1}\} \tag{2.22}$$

in which the subscript $n+1$ refers to the new, yet unknown state. In the remainder, the subscript $n+1$ will be dropped where convenient. Within the framework of a finite element code based on the displacement method, the total strains are updated according to

$$\varepsilon_{n+1} = \varepsilon_n + \Delta\varepsilon \tag{2.23}$$

in which the incremental strain $\Delta\varepsilon$ is determined from the strain-displacement relation $\Delta\varepsilon = B\,\Delta\mathbf{a}$, with $\Delta\mathbf{a}$ the nodal displacement increments for the respective element and B the strain-displacement operator in a typical integration (sampling) point for the respective element*. Henceforth, an integration algorithm based on Eq. (2.23) will be referred to as 'strain-driven'. What remains is to determine $\{\sigma_{n+1}, \varepsilon^e_{n+1}, \kappa_{i,n+1}\}$ with the constraint given by the update of ε_{n+1} in Eq. (2.23).

The additive decomposition of the strain rate in Eq. (2.18) is now generalized to strain increments

$$\Delta\varepsilon = \Delta\varepsilon^e + \sum_{i=1}^{i=n} \Delta\varepsilon^p_i \tag{2.24}$$

The elastic contribution to the strain increment is determined by the elastic constitutive law

$$\sigma_{n+1} = \sigma_n + \overline{E}(\varepsilon^e_n, \varepsilon^e_{n+1})\Delta\varepsilon^e \tag{2.25}$$

in which $\overline{E}$ is the *secant* elastic stiffness matrix. The plastic contribution to the strain increment is determined by the flow rule

$$\Delta\varepsilon^p_i = \Delta\lambda_i \frac{\partial g_i(\sigma_{n+1}, \kappa_{i,n+1})}{\partial\sigma_{n+1}} \tag{2.26}$$

in which the hardening variable $\kappa_{i,n+1}$ at the end of a step is determined by generalization of the integral in Eq. (2.20) for a finite increment

$$\kappa_{i,n+1} = \kappa_{i,n} + \Delta\kappa_i(\Delta\varepsilon^p_i) \tag{2.27}$$

It has been shown by Simo *et al.* (1988) that an Euler backward integration algorithm is equivalent to a constrained optimization problem governed by discrete Kuhn–Tucker conditions:

$$\Delta\lambda_i \geq 0; \quad f_i(\sigma_{n+1}, \kappa_{i,n+1}) \leq 0; \quad \Delta\lambda_i f_i(\sigma_{n+1}, \kappa_{i,n+1}) = 0 \tag{2.28}$$

---

* Applicable to standard isoparametric elements. For $\overline{B}$-elements (Hughes, 1980) the strain-displacement relation takes a slightly different form. The strain increment $\Delta\varepsilon$ is then determined as $\Delta\varepsilon = \overline{B}\,\Delta\mathbf{a}$.

Algorithmically, it is assumed that initially all deformations are elastic,

$$\varepsilon_{\text{trial}}^{\text{e}} = \varepsilon_n^{\text{e}} + \Delta\varepsilon; \quad \Delta\lambda_{i,\text{trial}} = 0 \tag{2.29}$$

which results in the elastic 'trial' stress

$$\sigma_{\text{trial}} = \sigma_n + \overline{E}(\varepsilon_n^{\text{e}}, \varepsilon_{\text{trial}}^{\text{e}})\Delta\varepsilon \tag{2.30}$$

and in an estimate for the hardening variable(s)

$$\Delta\kappa_{i,\text{trial}} = 0 \rightarrow \kappa_{i,\text{trial}} = \kappa_{i,n} \tag{2.31}$$

Violation of any of the yield functions $f_i(\sigma_{\text{trial}}, \kappa_{i,\text{trial}}) \leq 0$ determines which flow system is active. Then, Eqs. (2.24)–(2.28) are solved for these active flow systems under condition of Eq. (2.23) with the requirement that the yield function representative for the active flow system is enforced rigorously

$$f_i(\sigma_{n+1}, \kappa_{1,n+1}) = 0 \tag{2.32}$$

and with initial conditions defined by Eqs. (2.29)–(2.31). If none of the yield functions is violated then the trial stress state in Eq. (2.30) lies in the elastic domain and the update in Eq. (2.22) is given by the initial state in Eqs. (2.29)–(2.31) and condition in Eq. (2.23).

Rigorous enforcement of the active yield function(s) given by Eq. (2.32) combined with the initial state defined by Eqs. (2.29)–(2.31) can be interpreted as an elastic 'trial' stress being 'mapped back' onto the yield surface(s) representative for the active flow system(s). Therefore, the name 'Euler backward return-mapping' has been coined. It has been shown in different studies (e.g. Ortiz and Popov, 1985; and Simo and Taylor, 1986) that the implicit Euler backward return-mapping algorithm is unconditionally stable and accurate for J_2-plasticity. However, even when the yield surface is highly distorted, the Euler Backward algorithm is stable (Ortiz and Popov, 1985) and accurate (de Borst and Feenstra, 1990; Schellekens and de Borst, 1990).

For convenience in soil mechanics, the stresses and strains are decomposed into volumetric and deviatoric contributions. The additive decomposition in Eq. (2.24) is separated into a volumetric and a deviatoric part

$$\Delta\varepsilon_v = \Delta\varepsilon_v^{\text{e}} + \sum_{i=1}^{i=n} \Delta\varepsilon_{v,i}^{\text{p}}; \quad \Delta\gamma = \Delta\gamma^{\text{e}} + \sum_{i=1}^{i=n} \Delta\gamma_i^{\text{p}} \tag{2.33}$$

The volumetric and deviatoric plastic strain increments are determined by

$$\Delta\varepsilon_{v,i}^{\text{p}} = \Delta\lambda_i \frac{\partial g_i}{\partial p}; \quad \Delta\gamma_i^{\text{p}} = \Delta\lambda_i \frac{\partial g_i}{\partial \xi} \tag{2.34}$$

in which it is assumed that the plastic potential g_i is solely a function of the first and the second stress invariants and of the hardening variable, i.e.

$$g_i = g_i(p, q, \kappa_i) \tag{2.35}$$

Equation (2.35) covers a wide range of plastic potentials that are applicable to soils, including for example the Modified Cam-Clay model (Roscoe and Burland, 1968). Note that still $f_i = f_i(p, q, \cos 3\theta, \kappa_i)$. With the assumption of isotropy, the volumetric and deviatoric elastic strain increment are determined by

$$p = p_n + \overline{K} \Delta \varepsilon_v^e; \quad \boldsymbol{\xi} = \boldsymbol{\xi}_n + 2\overline{G} \boldsymbol{R}^{-1} \Delta \boldsymbol{\gamma}^e \tag{2.36}$$

in which $\overline{K}$ and $\overline{G}$ are the *secant* elastic bulk modulus and shear modulus, respectively. For computational convenience, it is assumed that both $\overline{K}$ and $\overline{G}$ are solely dependent on the elastic volumetric strain ε_v^e. This restriction yields the following form for the elastic secant moduli

$$\overline{K} = \overline{K} \left(\varepsilon_{v,n}^e, \varepsilon_{v,n}^e + \Delta \varepsilon_v - \sum_{i=1}^{i=n} \Delta \lambda_i \frac{\partial g_i}{\partial p} \right) \tag{2.37}$$

$$\overline{G} = \overline{G} \left(\varepsilon_{v,n}^e, \varepsilon_{v,n}^e + \Delta \varepsilon_v - \sum_{i=1}^{i=n} \Delta \lambda_i \frac{\partial g_i}{\partial p} \right) \tag{2.38}$$

Combination of Eqs. (2.33), (2.34) and (2.36) allows the evolution of the hydrostatic pressure to be written as

$$p = p_n + \overline{K} \left[\Delta \varepsilon_v - \sum_{i=1}^{i=n} \Delta \lambda_i \frac{\partial g_i}{\partial p} \right] \tag{2.39}$$

Similarly, combination of Eqs. (2.33), (2.34) and (2.36) gives for the deviatoric part

$$\boldsymbol{\xi} = \boldsymbol{\xi}_n + 2\overline{G} \boldsymbol{R}^{-1} \left[\Delta \boldsymbol{\gamma} - \sum_{i=1}^{i=n} \Delta \lambda_i \frac{\partial g_i}{\partial \boldsymbol{\xi}} \right] \tag{2.40}$$

The assumption for the plastic potential in Eq. (2.35) leads to the following convenient form for the plastic flow direction in the deviatoric plane

$$\frac{\partial g_i}{\partial \boldsymbol{\xi}} = m_i \boldsymbol{R} \boldsymbol{\xi} \tag{2.41}$$

in which $m_i = m_i(p, q, \kappa_i)$ is a scalar function. Equations (2.40) and (2.41) now yield an explicit expression for the deviatoric stress update

$$\boldsymbol{\xi} = \left[1 + 2\overline{G} \sum_{i=1}^{i=n} \Delta \lambda_i m_i \right]^{-1} [\boldsymbol{\xi}_n + 2\overline{G} \boldsymbol{R}^{-1} \Delta \boldsymbol{\gamma}] \tag{2.42}$$

The actual integration algorithm as proposed by Borja (1991) can now be written in the following way:

$$\boldsymbol{r} = \boldsymbol{r}(\boldsymbol{a}(\boldsymbol{\varepsilon}), \boldsymbol{\xi}(\boldsymbol{a}, \boldsymbol{\varepsilon}), \boldsymbol{\varepsilon})) \tag{2.43}$$

with $\boldsymbol{a} = [p, q, \cos 3\theta, \overline{G}, \Delta \kappa_i, \Delta \lambda_i]^{\mathrm{T}}$. The set of residuals $\boldsymbol{r}$ is defined by

$$\boldsymbol{r}^{\mathrm{T}} = [\boldsymbol{r}_1, \boldsymbol{r}_2, \boldsymbol{r}_3, \boldsymbol{r}_4]^{\mathrm{T}} \tag{2.44}$$

In r_1, the update of the stress invariants is collected

$$
r_1 = \begin{bmatrix}
p - p\left(\Delta\varepsilon_v - \sum_{i=1}^{i=n} \Delta\lambda_i \dfrac{\partial g_i}{\partial p}\right) \\[2ex]
q - \sqrt{\tfrac{3}{2}\boldsymbol{\xi}^{\mathrm{T}}\boldsymbol{R}\boldsymbol{\xi}} \\[2ex]
\cos 3\theta + \dfrac{27}{2}\dfrac{\xi_{xx}\xi_{yy}\xi_{zz} + \xi_{xy}\xi_{yz}\xi_{zx} - 2(\xi_{xx}\xi_{yz}^2 + \xi_{yy}\xi_{zx}^2 + \xi_{zz}\xi_{xy}^2)}{q^3}
\end{bmatrix}
\tag{2.45}
$$

in which $\boldsymbol{\xi}$ is determined from the explicit update presented in Eq. (2.42). r_2 contains the evolution of the secant shear modulus

$$
r_2 = \overline{G} - \overline{G}\left(\Delta\varepsilon_v - \sum_{i=1}^{i=n} \Delta\lambda_i \dfrac{\partial g_i}{\partial p}\right)
\tag{2.46}
$$

r_3 contains the evolution of the increment of the hardening variables

$$
r_3 = \Delta\kappa_i - \Delta\kappa_i\left(\Delta\lambda_i \dfrac{\partial g_i}{\partial p}, \, \Delta\lambda_i m_i \boldsymbol{R}\boldsymbol{\xi}\right)
\tag{2.47}
$$

and r_4 represents the active/inactive flow systems

$$
r_4 = f_i(p, q, \cos 3\theta, \Delta\kappa_i) \quad \text{or} \quad r_4 = \Delta\lambda_i
\tag{2.48}
$$

in which $r_4 = f_i$ when the representative flow system is active and $r_4 = \Delta\lambda_i$ when the representative flow system is inactive. The determination of which flow system is active will be addressed at the end of this section.

Under the condition that ε is constant, the set in Eq. (2.43) can be solved with a Newton–Raphson iterative scheme

$$
\delta\boldsymbol{a} = -\left[\dfrac{\partial \boldsymbol{r}}{\partial \boldsymbol{a}} + \dfrac{\partial \boldsymbol{r}}{\partial \boldsymbol{\xi}}\dfrac{\partial \boldsymbol{\xi}}{\partial \boldsymbol{a}}\right]^{-1} \delta\boldsymbol{r}
\tag{2.49}
$$

To complete the return-mapping algorithm, it is assumed that the initial conditions for the solution of system of Eqs. (2.45)–(2.48) and the deviatoric stress update in Eq. (2.42) are defined by the elastic 'trial' state, c.f. Eqs. (2.29)–(2.31). In terms of the governing quantities this yields

$$
\begin{bmatrix}
p_{\text{trial}} \\
q_{\text{trial}} \\
(\cos 3\theta)_{\text{trial}}
\end{bmatrix}
=
\begin{bmatrix}
p(\Delta\varepsilon_v) \\[1ex]
\sqrt{\tfrac{3}{2}\boldsymbol{\xi}_{\text{trial}}^{\mathrm{T}}\boldsymbol{R}\boldsymbol{\xi}_{\text{trial}}} \\[2ex]
-\dfrac{27}{2}\dfrac{\{\xi_{xx}\xi_{yy}\xi_{zz} + \xi_{xy}\xi_{yz}\xi_{zx} - 2(\xi_{xx}\xi_{yz}^2 + \xi_{yy}\xi_{zx}^2 + \xi_{zz}\xi_{xy}^2)\}_{\text{trial}}}{q_{\text{trial}}^3}
\end{bmatrix}
\tag{2.50}
$$

in which $\boldsymbol{\xi}_{\text{trial}}$ is determined by

$$
\boldsymbol{\xi}_{\text{trial}} = [\boldsymbol{\xi}_n + 2\overline{G}_{\text{trial}}\boldsymbol{R}^{-1}\Delta\boldsymbol{\gamma}]; \quad \overline{G}_{\text{trial}} = \overline{G}(\Delta\varepsilon_v)
\tag{2.51}
$$

Furthermore,

$$
\Delta\kappa_{i,\text{trial}} = 0; \quad \Delta\lambda_{i,\text{trial}} = 0
\tag{2.52}
$$

A problem that remains is how to determine which flow system (cf. Eq. (2.48)) is active. In this chapter, a trial and error procedure has been adopted. It is assumed that the initial active flow systems are defined by the trial state ($f_{i,\text{trial}} \geq 0$). If, after completion, any $\Delta\lambda_i \geq 0$ or $f_i \leq 0$ is found, the number of active flow systems is adjusted accordingly and the return-mapping algorithm is restarted.

Remarks

- The algorithm presented can easily be adapted for explicit coupling effects (Feenstra, 1993; Lourenço, 1996). The set of residuals r_3 in Eq. (2.47) is then modified to

$$r_3 = \Delta\kappa_i - C_{ij}\Delta\kappa_j \left(\Delta\lambda_i \frac{\partial g_i}{\partial p}, \Delta\lambda_i m_i R\xi \right) \tag{2.53}$$

in which C_{ij} is a set of coupling terms, for example discussed in Lourenço (1996) regarding tensile and shear failure in masonry joints.

- The shear and bulk moduli indicated in Eqs. (2.37) and (2.38) may lead to a non-conservative elastic contribution, in which energy may be extracted from certain loading cycles (Zytinsky *et al.*, 1978). However, this fact may not be too important when monotonic loading is considered (Borja, 1991).

2.3.2   *Tangential linearization*

On basis of the integration algorithm, a consistent tangential operator is formulated. For this purpose, the stress update following from the solution of system of Eqs. (2.45)–(2.48) and the deviatoric stress update in Eq. (2.42) is written as

$$\sigma = \sigma_n + \xi(a(\varepsilon), \varepsilon) + \pi p(a(\varepsilon)) \tag{2.54}$$

which can be differentiated straightforwardly as

$$\frac{d\sigma}{d\varepsilon} = \frac{\partial\xi}{\partial\varepsilon} + \frac{\partial\xi}{\partial a}\frac{da}{d\varepsilon} + \pi \frac{\partial p}{\partial a}\frac{da}{d\varepsilon} = \frac{\partial\xi}{\partial\varepsilon} + \left[\frac{\partial\xi}{\partial a} + \pi\frac{\partial p}{\partial a} \right]\frac{da}{d\varepsilon} \tag{2.55}$$

Equation (2.55) contains only partial derivatives except for $da/d\varepsilon$. The latter quantities can be determined implicitly from the fact that the strain is constant during integration. Thus,

$$\frac{dr}{d\varepsilon} = \left[\frac{\partial r}{\partial a} + \frac{\partial r}{\partial\xi}\frac{\partial\xi}{\partial a} \right]\frac{da}{d\varepsilon} + \frac{\partial r}{\partial\xi}\frac{\partial\xi}{\partial\varepsilon} + \frac{\partial r}{\partial\varepsilon} = 0 \rightarrow$$

$$\frac{da}{d\varepsilon} = - \left[\frac{\partial r}{\partial a} + \frac{\partial r}{\partial\xi}\frac{\partial\xi}{\partial a} \right]^{-1} \left[\frac{\partial r}{\partial\xi}\frac{\partial\xi}{\partial\varepsilon} + \frac{\partial r}{\partial\varepsilon} \right] \tag{2.56}$$

in which the inverse term between brackets is available from the return-mapping algorithm, Eq. (2.49). Substitution of $da/d\varepsilon$ according to Eq. (2.56) into Eq. (2.55) gives the desired form for the consistent tangential operator

$$\frac{d\sigma}{d\varepsilon} = \frac{\partial\xi}{\partial\varepsilon} - \left[\frac{\partial\xi}{\partial a} + \pi\frac{\partial p}{\partial a} \right] \left[\frac{\partial r}{\partial a} + \frac{\partial r}{\partial\xi}\frac{\partial\xi}{\partial a} \right]^{-1} \left[\frac{\partial r}{\partial\xi}\frac{\partial\xi}{\partial\varepsilon} + \frac{\partial r}{\partial\varepsilon} \right] \tag{2.57}$$

2.4   Modeling of Clay

Cam-Clay models originate from the work of Roscoe and his co-workers at the University of Cambridge (Roscoe and Schofield, 1963; Schofield and Wroth, 1968). The original idea was further developed by Roscoe and Burland (1968) to the Modified Cam-Clay model; nowadays, the most widely used elasto-plastic model for the description of the mechanical behavior of clay.

The advantage of the Modified Cam-Clay model lies in its apparent simplicity and its capability to represent (at least qualitatively) the strength and deformation properties of clay realistically. Commonly observed properties such as an increasing stiffness as the material undergoes compression, hardening/softening and compaction/dilatancy behavior, and the tendency to eventually reach a state in which the strength and volume become constant are all captured by the Modified Cam-Clay model. Moreover, calibration of the model requires only a few conventional laboratory tests.

The material description starts most conveniently with the hydrostatic components. The increasing bulk stiffness for increasing compression is modeled by

$$\dot{p} = -\frac{p}{\lambda^*}\dot{\varepsilon}_v = K_t\dot{\varepsilon}_v \text{ loading} \tag{2.58}$$

$$\dot{p} = -\frac{p}{\kappa^*}\dot{\varepsilon}_v = K_t\dot{\varepsilon}_v \text{ unloading} \tag{2.59}$$

λ^* and κ^* will be referred to as the 'modified compression index' and the 'modified swelling index', respectively (Figure 2.2(a)). As unloading is assumed to be elastic Eq. (2.59) can be expressed using elastic components only

$$\dot{p} = -\frac{p}{\kappa^*}\dot{\varepsilon}_v^e = K_t\dot{\varepsilon}_v^e \tag{2.60}$$

For three-dimensional stress states also deviatoric components have to be specified. When Eq. (2.60) is taken as a point of departure the assumption of isotropy, together with a constant Poisson's ratio v, gives

$$G_t = \frac{3}{2}\frac{1-2v}{1+v}K_t \tag{2.61}$$

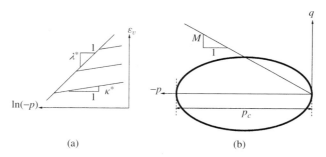

(a)   (b)

Figure 2.2   Material description for Modified Cam-Clay. (a) Compressive behavior. (b) Yield function and plastic potential

The yield function (f) is given by the expression

$$f = q^2 + M^2 p(p + p_c) \tag{2.62}$$

in which p_c is the preconsolidation pressure (Figure 2.2(b)). Associative plasticity is assumed so that Eq. (2.62) also defines the plastic potential. Consequently, the plastic volumetric strain rate $\dot{\varepsilon}_v^p$ is given by

$$\dot{\varepsilon}_v^p = \dot{\lambda}\frac{\partial f}{\partial p} = \dot{\lambda}M^2(2p + p_c) \tag{2.63}$$

The deviatoric plastic strain rate $\dot{\boldsymbol{\gamma}}^p$ is given by

$$\dot{\boldsymbol{\gamma}}^p = \dot{\lambda}\frac{\partial f}{\partial \boldsymbol{\xi}} = 3\dot{\lambda}R\boldsymbol{\xi} \tag{2.64}$$

At this point, it is convenient to define the isotropic Over Consolidation Ratio (OCR$_p$), which reflects the relationship between the initial preconsolidation pressure p_c and the current compressive pressure $-p$

$$\text{OCR}_p = -\frac{p_c}{p} \tag{2.65}$$

Combination of definition in Eq. (2.65) with the plastic flow direction in Eq. (2.63) leads to the observation that when OCR$_p < 2$ compaction is predicted while when OCR$_p > 2$ dilatancy is predicted by Eq. (2.63).

Hardening/softening is determined from combination of Eqs. (2.60) and (2.58) and the condition that during loading a stress point remains on the yield surface. This results in the following form for the evolution of p_c

$$\frac{\dot{p}_c}{p_c} = \frac{-\dot{\varepsilon}_v^p}{\lambda^* - \kappa^*} \tag{2.66}$$

Equation (2.66) shows that compaction leads to an increase of p_c (hardening) and dilatancy leads to a decrease of p_c (softening). This behavior is most conveniently illustrated with a conventional drained triaxial compression test. The stress path in the Rendulic plane, which is representative for this test, is shown in Figures 2.3 and 2.4. It is observed that when the initial value of the Over Consolidation Ratio is smaller than two ($p_{c,0}/p_0 < 2$) plastic compaction, and subsequently hardening behavior is observed (Figure 2.3). If, on the other hand, the initial Over Consolidation Ratio is significantly larger than two ($p_{c,0}/p_0 \gg 2$), plastic dilatancy and subsequently softening is predicted (Figure 2.4).

Figures 2.3 and 2.4 show that, for constant stress paths, eventually a state is reached where purely deviatoric flow is predicted. Subsequently, Eq. (2.66) predicts ideal plastic flow. Such a state is called the critical state and is identified by

$$\frac{q}{p} = -M \tag{2.67}$$

The Modified Cam-Clay model is most suitable for lightly overconsolidated clays, for which the compaction and hardening behavior are well predicted. A possible improvement which gives more accurate predictions for the coefficient of lateral earth pressure $K_0 =$

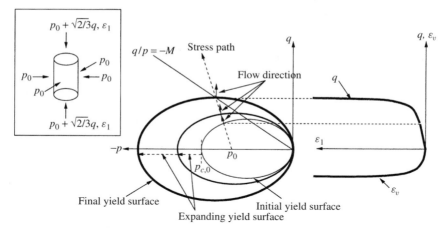

Figure 2.3 Drained triaxial compression $p_{c,0}/p_0 < 2$. The associative flow rule predicts compaction. Hence, Eq. (2.66) predicts hardening. Eventually, monotonic loading leads to a stress ratio $q/p = -M$, where purely deviatoric ideal plastic flow is predicted

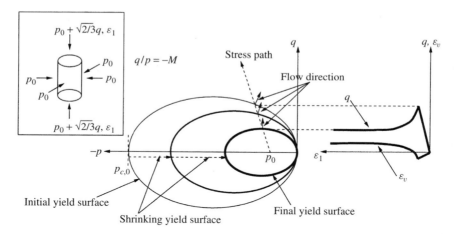

Figure 2.4 Drained triaxial compression $p_{c,0}/p_0 \gg 2$. The associative flow rule predicts dilatancy. Hence, Eq. (2.66) predicts softening. Eventually, monotonic loading leads to a stress ratio $q/p = -M$, where purely deviatoric ideal plastic flow is predicted

σ_h/σ_v can be found in Van Eekelen and Van den Berg (1994). For heavily overconsolidated clays, the prediction of the material behavior is less accurate (Wood, 1990).

2.4.1 Performance in laboratory tests

The deformation characteristics and the numerical performance are evaluated with four commonly known laboratory tests: Triaxial Compression (TC), Triaxial Extension (TE),

Pure Shear (PS) and Plane Strain Compression (PSC) (Figure 2.5). Instead of following the sign convention defined before, the sign convention for the stress components σ_1, σ_3 and for the strain component ε_1 is defined by Figure 2.5. All four tests have been simulated with a lightly overconsolidated material (initial isotropic Over Consolidation Ratio $OCR_{p,0} = 1.25$), and with a heavily over-consolidated material (initial isotropic Over Consolidation Ratio $OCR_{p,0} = 5$) with a cell pressure $\sigma_3 = 0.2$ MPa. The material parameters are adapted from Borja (1991) for Boston Blue Clay and are given by $\lambda^* = 0.032$, $\kappa^* = 0.013$, $M = 1.05$ and a Poisson's ratio $\nu = 0.2$.

The calculations have been performed with three different step sizes. The load-deformation and volume production-deformation curves are shown in Figure 2.6. As expected, all elementary tests predict hardening and compaction. The computed responses are nearly step size independent, which demonstrates the accuracy of the applied return-mapping algorithm. In Table 2.1, the average number of iterations required to reach an L_2-norm of 10^{-8} with respect to the initial L_2-norm of system of Eqs. (2.45)–(2.48) is displayed. Table 2.2 shows the convergence behavior in the equilibrium-finding iterative scheme for $OCR_{p,0} = 1.25$. It is observed that the convergence behavior is quadratic. Figure 2.7 shows the computed responses for $OCR_{p,0} = 5$. Except for the Plane Strain Compression test in Figure 2.7(d), which has a highly compressive stress path, all calculations show softening and dilatancy. As in the lightly overconsolidated case ($OCR_{p,0} = 1.25$) the computed responses are nearly step size independent. Table 2.3 shows the average number of iterations required to reach an L_2-norm of 10^{-8} with respect to the initial L_2-norm of system of Eqs. (2.45)–(2.48) for $OCR_{p,0} = 5$. Table 2.4 shows the convergence behavior in the equilibrium-finding iterative scheme for $OCR_{p,0} = 5$. Again, quadratic convergence is observed for the four laboratory tests.

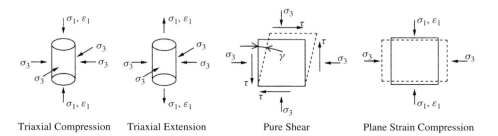

| Triaxial Compression | Triaxial Extension | Pure Shear | Plane Strain Compression |

Figure 2.5 Four common laboratory tests. In the pure shear test and the plane strain compression test the out-of-plane strains are zero

Table 2.1 Average number of iterations required to reach an L_2-norm 10^{-8} with respect to the initial L_2-norm of the system (2.45)–(2.48) per equilibrium iteration, $OCR_{p,0} = 1.25$

	CTC	TE	SS	PSC
10 steps	9	8.95	8.05	9
20 steps	7.91	7.87	7.02	7.93
200 steps	5.94	5.86	4.9	5.94

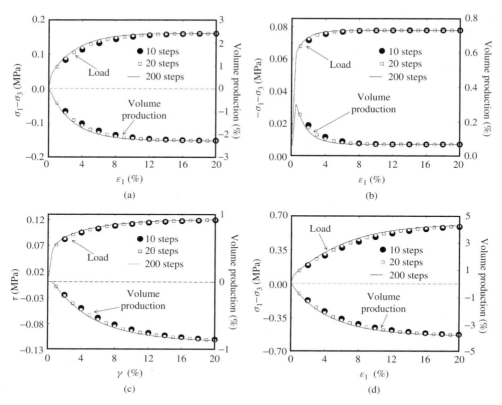

Figure 2.6   Simulation of laboratory tests with lightly overconsolidated material OCR$_{p,0}$ = 1.25. (a) Triaxial compression (b) Triaxial Extension. (c) Pure Shear. (d) Plane Strain Compression

Table 2.2   Force norm in the equilibrium equations for 20 steps, OCR$_{p,0}$ = 1.25

	CTC	TE	SS	PSC
Step 1	2.49×10^{-2}	2.93×10^{-1}	1.36×10^{-1}	2.79×10^{-2}
	6.16×10^{-5}	5.25×10^{-3}	1.47×10^{-3}	1.24×10^{-4}
	3.69×10^{-10}	1.77×10^{-6}	2.42×10^{-7}	2.63×10^{-9}
		2.02×10^{-13}		
Step 7	2.06×10^{-5}	4.34×10^{-6}	1.28×10^{-4}	1.70×10^{-4}
	2.13×10^{-10}	3.05×10^{-12}	2.48×10^{-9}	3.73×10^{-8}
Step 20	7.50×10^{-9}	2.10×10^{-12}	1.00×10^{-6}	7.85×10^{-7}

2.4.2   *Analysis of a guided pipe-jacking*

In The Netherlands, the Drinking Water Service Company of the province of South Holland has constructed a pipeline system of approximately 60 km using a guided pipe-jacking method. The resulting surface settlements have been measured (van der Broek

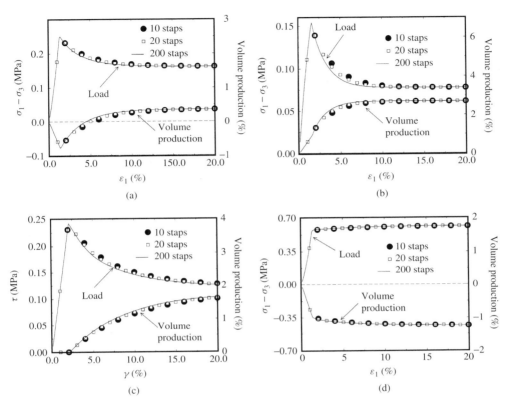

Figure 2.7   Simulation of laboratory tests with heavily overconsolidated material OCR$_{p,0}$ = 5. (a) Triaxial compression. (b) Triaxial Extension. (c) Pure Shear. (d) Plane Strain Compression

Table 2.3   Average number of iterations required to reach an L_2-norm 10^{-8} with respect to the initial L_2-norm of system (2.45)–(2.48) per equilibrium iteration, OCR$_{p,0}$ = 5

	CTC	TE	SS	PSC
10 steps	8.43	8.56	7.43	8.44
20 steps	7.62	7.35	6.63	7.57
200 steps	5.58	5.2	4.56	5.63

et al., 1996). A three-dimensional analysis is utilized for the prediction of the surface settlements.

Two top deposits have been modeled. The tunnel has a diameter of two meters, and is located in the lower deposit with its center at seven meters below the surface. The interface between the upper deposit and the lower deposit is located two meters below the surface.

Table 2.4   Force norm in the equilibrium equations for 20 steps, $OCR_{p,0} = 5$

	CTC	TE	SS	PSC
Step 3	1.30×10^{-2} 1.42×10^{-4} 1.64×10^{-8}	8.03×10^{-2} 2.96×10^{-3} 3.64×10^{-6} 5.54×10^{-12}	1.37×10^{-1} 1.13×10^{-2} 9.98×10^{-5} 7.97×10^{-9}	2.50×10^{-3} 1.15×10^{-5} 2.50×10^{-10}
Step 10	2.46×10^{-6} 3.72×10^{-12}	1.44×10^{-5} 3.42×10^{-11}	9.95×10^{-5} 3.91×10^{-9}	1.45×10^{-7}
Step 20	3.75×10^{-9}	6.78×10^{-10}	8.38×10^{-6} 1.93×10^{-11}	3.80×10^{-9}

For the upper deposit, a compression index $\lambda^* = 0.15$ and a swelling index $\kappa^* = 0.015$ have been assumed. For the lower deposit $\lambda^* = 0.15$ and $\kappa^* = 0.03$ are adopted. Furthermore, $M = 0.877$, a density of $15\,kN/m^3$, an initial isotropic Over Consolidation Ratio $OCR_{p,0} = 1.1$ and a Poisson's ratio $\nu = 0.2$ are adopted for both deposits. $\overline{B}$-elements (Hughes, 1980) have been applied.

Two types of analysis have been performed: (a) a drained analysis, in which it is assumed that the soil skeleton is completely responsible for the deformation behavior of the structure; (b) an undrained analysis, in which it is assumed that the deformations are determined by parallel action of soil and water content. Under undrained conditions the water content enforces a zero volume change upon the deformations, both in the elastic and in the plastic regime.

The calculation has been performed in three phases:

(1) **Drained, one step** The gravity load is applied. It has been assumed that the tunnel lining is already present but has no weight yet. The tunnel lining has been assumed as rigid. The weight of the soil is modeled by a point-load downwards. Moreover, it is assumed that the Tunnel Boring Machine is in-line with the tunnel lining. After the gravity load is applied, the horizontal stresses are calculated with a coefficient of lateral earth pressure $K_0 = \sigma_h/\sigma_v = 0.6$.

(2) **Drained/undrained, four steps** The weight of the tunnel lining is applied simultaneously with the removal of the weight of the soil in the tunnel in four steps. This has been simulated by application of a net load [WEIGHT TUNNEL LINING – WEIGHT SOIL] downwards. As the tunnel lining is lighter than the removed soil this results in an uplift of the tunnel lining and Tunnel Boring Machine (Figure 2.8(a)).

(3) **Drained/undrained, twenty-five steps** Pregrouting loss around the tunnel lining is simulated by contraction of the tunnel lining which has been assumed to be ten percent of the volume of the tunnel. Loss due to the boring process is simulated by a contraction of seven percent of the volume of the tunnel around the Tunnel Boring Machine (TBM) (Figure 2.8(b)). The tunnel boring front has been supported by an overburden pressure of 0.1 bar, typical for the guided pipe-jacking method.

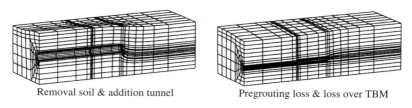

Removal soil & addition tunnel   Pregrouting loss & loss over TBM

Figure 2.8   Deformations at the different phases of the loading process

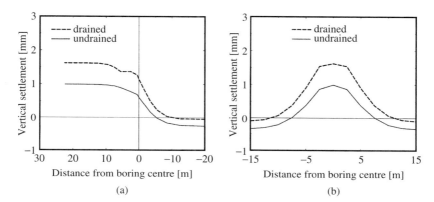

(a)   (b)

Figure 2.9   Surface settlements after the removal of soil and addition of the tunnel lining for the undrained analysis. (a) Settlements parallel to the tunnel. (b) Settlements 22.5 m behind the front of the Tunnel Boring Machine

Not all factors that influence the deformation have been taken into account; for instance, consolidation effects, the stiffness of tunnel lining and Tunnel Boring Machine and steering effects of the Tunnel Boring Machine have not been considered. However, the primary objectives of this simulation are: (a) to give at least a qualitative representation of the surface settlements; and (b) to assess the performance of the return-mapping algorithm presented before.

Figure 2.9 shows the surface settlements after phase two (removal of the soil and addition of the tunnel lining), and Figure 2.10 shows the surface settlements after phase 3 (contraction of the tunnel lining and Tunnel Boring Machine). It is observed from Figure 2.9 that the surface settlements under drained conditions are larger than under undrained conditions. This can be attributed to the fact that under undrained conditions the water content acts parallel to the soil skeleton which results in a stiffer behavior. From Figure 2.10, it is observed that the surface settlements under drained conditions are also larger than under undrained conditions. Since the loading in this phase is deformation controlled, this cannot be attributed to an increased stiffness due to the water content. Here, the smaller surface settlement under undrained conditions can be attributed to the fact that the water content enforces volume preserving deformations upon the soil skeleton. Since very lightly overconsolidated material has been adopted, drained conditions result in compaction and thus in larger surface settlements.

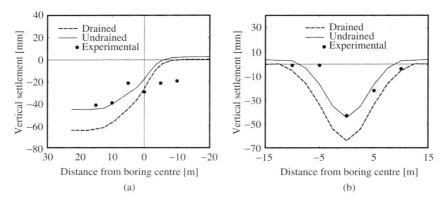

Figure 2.10   Surface settlements resulting from pregrouting loss and loss over the Tunnel Boring Machine. (a) Settlements parallel to the tunnel. (b) Settlements 22.5 m behind the front of the Tunnel Boring Machine

2.5   Modeling of Sand

While the Modified Cam-Clay model is able to capture a large number of phenomena that occur in clayey soils with only a few parameters, it cannot describe typical phenomena that occur in sands such as simultaneous hardening and dilatancy. A possible extension would be to assume that the plastic behavior in shear is independent from the behavior in compression. This approach was first suggested by Drucker *et al.* (1957), who suggested a spherical cap on the Drucker–Prager yield surface. Later, Dimaggio and Sandler (1971), Lade (1977), Molenkamp (1980) and Vermeer (1980) followed the same approach in which, in addition to shear failure, compressive deformations were modeled by a cap. More recent models integrate the shear and compressive behavior into a single surface (Desai, 1980, Kim and Lade, 1988, Lade and Kim, 1988a,b).

Here, the plastic behavior in shear and in compression will be modeled independently. This allows the formulation of a model which remains close to established concepts such as Rowe's stress-dilatancy theory (Rowe, 1962), and the formulation of the compressive behavior similar to the Modified Cam-Clay model.

Analogous to the Modified Cam-Clay model, an increasing bulk stiffness for increasing compression is modeled by

$$\dot{p} = -\frac{p}{\lambda^*}\dot{\varepsilon}_v = K_t\dot{\varepsilon}_v \text{ loading} \tag{2.68}$$

$$\dot{p} = -\frac{p}{\kappa^*}\dot{\varepsilon}_v = K_t\dot{\varepsilon}_v \text{ unloading} \tag{2.69}$$

with λ^* the 'modified compression index' and κ^* the 'modified swelling index' (Figure 2.11(a)). As elastic unloading is assumed Eq. (2.69) can be expressed in elastic components only

$$\dot{p} = -\frac{p}{\kappa^*}\dot{\varepsilon}_v^e = K_t\dot{\varepsilon}_v^e \tag{2.70}$$

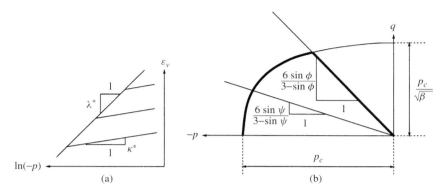

Figure 2.11 Material description for the double-hardening model. (a) Compressive behavior. (b) Yield function and plastic potential

Equation (2.70), together with the assumption of isotropy and a constant Poisson's ratio ν, gives the following expression for the tangent shear modulus G_t

$$G_t = \frac{3}{2}\frac{1-2\nu}{1+\nu}K_t \tag{2.71}$$

Shear failure is assumed to occur on a Drucker–Prager type failure surface

$$f_1 = \left(\frac{1-\alpha\cos 3\theta}{1-\alpha}\right)^{-n}q + \frac{6\sin\phi}{3-\sin\phi}p \tag{2.72}$$

in which ϕ is the friction angle, and α and n are parameters which include the effect of the Lode angle θ. A projection of the yield function in Eq. (2.72) on the deviatoric plane is shown in Figure 2.12. Optimal properties regarding convexity are obtained when $n = -0.229$ (van Eekelen, 1980). Convexity is then retained for $\alpha \le 0.7925$, which corresponds to a fit through all the corners of the Mohr–Coulomb criterion for a friction angle of 46.55°.

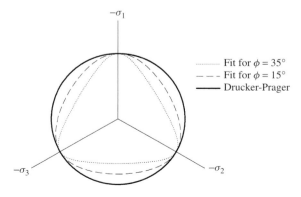

Figure 2.12 Yield function (2.72) in the deviatoric plane

Plastic flow on the shear failure surface given by Eq. (2.72) is assumed to be non-associative, and is determined by the Drucker–Prager plastic potential

$$g_1 = q + \frac{6\sin\psi}{3 - \sin\psi}p \tag{2.73}$$

in which ψ is the dilatancy angle. This assumption of a Drucker–Prager plastic potential is motivated by experimental results from Kim and Lade (1988) which indicate a plastic potential close to a Drucker–Prager contour in the deviatoric plane.

During compaction, plastic straining is described by the compression cap

$$f_2 = p^2 + \beta q^2 - p_c^2 \tag{2.74}$$

where associative plasticity is assumed. In Eq. (2.74), β is a shape parameter (van Langen, 1991) which is used to fit the prediction of the coefficient of lateral earth pressure $K_0 = \sigma_h/\sigma_v$. p_c is a measure for the current degree of over-consolidation and can be determined from an isotropic Over Consolidation Ratio (OCR_p) similar to the Modified Cam-Clay model

$$OCR_p = -\frac{p_c}{p} \tag{2.75}$$

or alternatively, from an Over Consolidation Ratio (OCR) which is based on the size of the cap in Eq. (2.74), and thus includes the effect of the effective deviatoric stress:

$$OCR = -\frac{p_c}{\sqrt{p^2 + \beta q^2}} \tag{2.76}$$

The yield functions given by Eqs. (2.72) and (2.74) and the plastic potential in Eq. (2.73) have been plotted in Figure 2.11(b).

The plastic strain rates are assumed to be an additive decomposition of plastic strain rates ε_1^p on the shear failure surface given by Eq. (2.72) and plastic rates ε_2^p on the compression cap in Eq. (2.74).

$$\dot{\varepsilon}^p = \dot{\varepsilon}_1^p + \dot{\varepsilon}_2^p \tag{2.77}$$

This leads to the following expression for the volumetric plastic strain rates

$$\dot{\varepsilon}_v^p = \dot{\lambda}_1\frac{\partial g_1}{\partial p} + \dot{\lambda}_2\frac{\partial f_2}{\partial p} = \dot{\lambda}_1\frac{6\sin\psi}{3 - \sin\psi} + 2p\dot{\lambda}_2 \tag{2.78}$$

and for the deviatoric plastic strain rates

$$\dot{\boldsymbol{\gamma}}^p = \dot{\lambda}_1\frac{\partial g_1}{\partial \boldsymbol{\xi}} + \dot{\lambda}_2\frac{\partial f_2}{\partial \boldsymbol{\xi}} = \dot{\lambda}_1\frac{3}{2}\frac{\boldsymbol{R}\boldsymbol{\xi}}{q} + \dot{\lambda}_2 3\boldsymbol{R}\boldsymbol{\xi} \tag{2.79}$$

Friction hardening is assumed according to

$$\sin\phi = \sin\phi(\gamma_{eff}^p) \tag{2.80}$$

in which γ_{eff}^p is defined as the effective deviatoric plastic strain on the failure surface (Vermeer, 1980), and is defined in a rate form as

$$\dot{\gamma}_{eff}^p = \sqrt{\tfrac{2}{3}\dot{\boldsymbol{\gamma}}_1^{pT}\boldsymbol{R}^{-1}\dot{\boldsymbol{\gamma}}_1^p} \tag{2.81}$$

Equation (2.81) leads to a convenient relationship between the rate of the plastic multiplier on the failure surface $\dot{\lambda}_1$ and $\dot{\gamma}^{\text{p}}_{\text{eff}}$

$$\dot{\gamma}^{\text{p}}_{\text{eff}} = \dot{\lambda}_1 \tag{2.82}$$

Dilatancy is assumed to obey Rowe's stress-dilatancy theory in triaxial compression

$$\sin \psi = \frac{\sin \phi - \sin \phi_{cv}}{1 - \sin \phi \sin \phi_{cv}} \tag{2.83}$$

in which ϕ_{cv} is identified as the so-called 'friction angle at constant volume' (Vermeer and de Borst, 1984). The evolution of hardening on the cap defined by Eq. (2.74) is derived from the loading function in Eqs. (2.68) and (2.70) and the condition that during compressive loading a stress point is located on the cap in Eq. (2.74). This results in the following form for the evolution of p_c:

$$\frac{\dot{p}_c}{p_c} = \frac{-\dot{\varepsilon}^{\text{p}}_{v,2}}{\lambda^* - \kappa^*} \tag{2.84}$$

in which $\dot{\varepsilon}^{\text{p}}_{v,2}$ is the plastic volumetric strain rate on the cap.

2.5.1 Performance in laboratory tests

The numerical performance of the proposed integration algorithm has been evaluated for four elementary tests, namely Conventional Triaxial Compression (CTC), Triaxial Extension (TE), Simple Shear (SS) and Plane Strain Compression (PSC) (Figure 2.13).

The adopted material parameters are $\kappa^* = 0.00573$, $\lambda^* = 0.00693$, $\nu = 0.18$, $\beta = 2/9$ and $\phi_{cv} = 30.66°$. Furthermore, the specimens are isotropically consolidated with a cell pressure $\sigma_3 = 0.2$ MPa and the shape factor $\alpha = 0.689$, which corresponds to a smooth fit through all the corners of the Mohr–Coulomb yield surface for a friction angle $\phi = 35°$. A multi-linear friction hardening diagram is used (Table 2.5).

The calculations have been performed with three different step sizes. The load-deformation curves are shown in Figure 2.14. It is observed that the solution is practically independent of the step size.

Table 2.6 shows the average number of iterations required to reach an L_2-norm of 10^{-8} with respect to the initial L_2-norm of system of Eqs. (2.45)–(2.48). Table 2.7 shows the

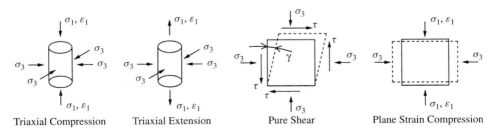

Triaxial Compression   Triaxial Extension   Pure Shear   Plane Strain Compression

Figure 2.13 Four common laboratory tests. In the pure shear test and the plane strain compression test the out-of-plane strains are zero

Table 2.5 Multi-linear friction hardening diagram

ϕ (degrees)	35.03	40.54	42.84	42.84
γ^p_{eff}	0.0	0.01	0.03	∞

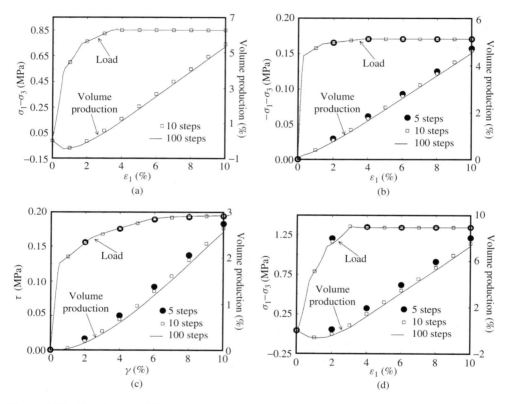

Figure 2.14 Simulation of laboratory tests. (a) Triaxial Compression. (b) Triaxial Extension. (c) Pure Shear. (d) Plane Strain Compression

Table 2.6 Average number of iterations required to reach an L_2-norm 10^{-8} with respect to the initial L_2-norm of the system (2.45)–(2.48) per equilibrium iteration

	CTC	TE	SS	PSC
5 steps	X	8.36	8.38	10.14
10 steps	7.8	5.75	8.88	8.5
100 steps	4.31	3.93	3.97	4

Table 2.7 Force norm in the equilibrium equations for ten steps

	CTC	TE	SS	PSC
	1.30×10^{-1}	5.61×10^{-1}	8.87×10^{-2}	4.94×10^{-2}
	1.67×10^{-2}	1.81×10^{-1}	3.52×10^{-3}	8.43×10^{-3}
Step 1	3.84×10^{-4}	8.80×10^{-3}	6.27×10^{-6}	3.25×10^{-4}
	2.16×10^{-7}	2.37×10^{-5}	1.07×10^{-11}	5.17×10^{-7}
		1.75×10^{-10}		
	3.82×10^{-3}	1.54×10^{-3}	3.19×10^{-2}	4.06×10^{-3}
Step 3	1.99×10^{-5}	6.07×10^{-7}	7.29×10^{-4}	7.32×10^{-5}
	5.49×10^{-10}		3.59×10^{-7}	1.13×10^{-7}
Step 10	3.60×10^{-7}	3.59×10^{-10}	8.31×10^{-6}	6.12×10^{-8}
			7.51×10^{-11}	

convergence in the force norm of the equilibrium equations. Except for the CTC test with the largest step size, for which no convergence could be obtained, excellent convergence behavior is observed both in the solution of the system of Eqs. (2.45)–(2.48) as well as in the solution of the equilibrium equations.

2.5.2 Analysis of the Leaning Tower of Pisa

The Leaning Tower of Pisa is one of the most challenging problems in geomechanics. The geotechnical profile and the essentially three-dimensional deformation behavior makes it an excellent test case to verify the applicability of the double-hardening model and the robustness and performance of the applied integration algorithm. A three-dimensional numerical simulation has been carried out in order to investigate the effect of the lead blocks that have been applied on the high-side of the tower in 1993 to reduce the inclination.

It is not the objective to give a quantitative representation of the deformations under the tower. The geotechnical conditions as well as the loading conditions on the foundation of the tower are simply too complicated. Moreover, the uncertainty in the constitutive parameters is too large to give any realistic prediction of the true deformations that occur in the subsoil under the tower. However, the three-dimensional character as well as the relatively clear classification of the constitutive parameters makes the Leaning Tower of Pisa an excellent benchmark for assessment of the performance of the adopted double-hardening model.

Calabresi *et al.* (1993) have performed triaxial tests on the soil deposits underneath the tower and have proposed Modified Cam-Clay parameters for the clayey deposits and Mohr–Coulomb frictional parameters for the sandy deposits. The material parameters proposed by Calabresi *et al.* (1993) are adapted to the double-hardening model (Table 2.8).

It is assumed that the respective friction angles do not change during the loading process. The parameter α is adjusted such that a fit through all the corners of the Mohr–Coulomb yield function is obtained for each deposit. Furthermore, a dilatancy angle $\psi = 0$ and a Poisson's ratio $\nu = 0.12$ are assumed for all deposits.

The foundation of the tower is assumed to have a density $\rho = 21.43$ kN/m^3, a Young's modulus of $E = 1.5 \times 10^6$ kPa, a Poisson's ratio $\nu = 0.12$ and a coefficient of lateral earth pressure $K_0 = 0.9231$.

Table 2.8   Material properties for the respective deposits under the tower. The material parameters proposed by Calabresi *et al.* (1993) are adapted to the double-hardening model

Depth [m]	E[kPa]	κ^*	λ^*	$\sin(\phi)$	OCR	K_0	ρ[kN/m^3]
0–1.4	–	0.00486	0.0481	0.555	4.2	0.9231	21.43
1.4–5.4	–	0.00495	0.049	0.554	3.0	0.7857	19.17
5.4–7.4	1.25×10^4	–	–	0.559	–	0.4493	15.40
7.4–17.8	–	0.014	0.107	0.446	1.35	0.6949	18.49
17.8–22	–	0.0151	0.0722	0.479	1.8	0.6667	19.64
22–24.4	1.8×10^4	–	–	0.564	–	0.4286	20.40
24.4–37	–	0.0106	0.0851	0.424	1.1	0.5625	21.66

Table 2.9   Material properties for the respective deposits under the tower. The compressive indices in Calabresi *et al.* (1993) are approximated by a linear elastic compressive stiffness based on the initial stress state (Grashuis, 1993)

Depth [m]	E[kPa]	$\sin(\phi)$	K_0	ρ[kN/m^3]
0–1.4	3×10^3	0.555	0.9231	21.43
1.4–5.4	3×10^3	0.554	0.7857	19.17
5.4–7.4	1.25×10^4	0.559	0.4493	15.40
7.4–17.8	8×10^3	0.446	0.6949	18.49
17.8–22	1.4×10^4	0.479	0.6667	19.64
22–24.4	1.8×10^4	0.564	0.4286	20.40
24.4–37	2.5×10^4	0.424	0.5625	21.66

Alternatively, a set of material properties has been used with a linear elastic approximation for the compressive behavior (Grashuis, 1993) (Table 2.9).

As for the guided pipe-jacking, the analysis has been performed with $\overline{B}$-elements (Hughes, 1980).

The vertical stress state under the tower is initialized with the self-weight of the soil skeleton. The horizontal stresses are then calculated with the coefficient of lateral earth pressure $K_0 = \sigma_h/\sigma_v$. Drained conditions have been assumed (i.e. no influence of the ground water on the deformation behavior). Hereafter, the loading history is simulated by three phases:

1.   The weight of the tower is applied in twenty-nine steps under drained conditions.

2.   The moment on the foundation due to the inclination of the tower is applied in twenty steps under drained conditions.

3.   The weight and the moment due to the lead-blocks are applied in eight steps. Both drained conditions and undrained conditions are simulated.

Figure 2.15 shows the incremental deformations after each loading phase for the material properties in Table 2.8. Figure 2.16 shows the applied moment versus the inclination of the tower. From Figure 2.16 it is observed that the material parameters listed in Table 2.8 give an inclination of approximately 1.2°, while the simulation with the material

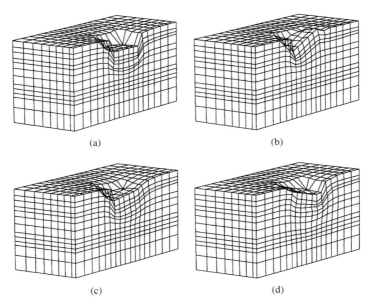

(a)   (b)

(c)   (d)

Figure 2.15   Incremental deformations for the material properties listed in Table 2.8 (double-hardening model for the clayey layers). (a) After application of the weight of the tower. (b) After application of the moment due to the inclination of the tower. (c) After application of the lead-blocks on the high side of the tower under drained conditions. (d) After application of the lead-blocks on the high side of the tower under undrained conditions

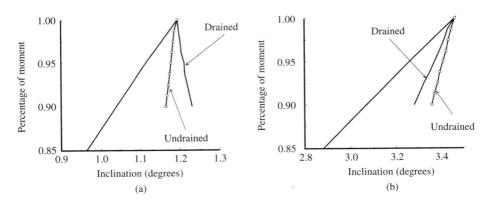

Figure 2.16   Moment versus inclination of the tower. (a) Utilizing the material parameters from Table 2.8. (b) Utilizing the material parameters from Table 2.9

parameters listed in Table 2.9 gives a much softer response (an inclination of approximately 3.5°). A salient feature which can be observed both from Figure 2.15(c), as well as from Figure 2.16(a) is that the material properties listed in Table 2.8 result in an increase of the inclination of the tower under drained conditions during application of the lead blocks.

2.6 Conclusions

A fully implicit Euler backward integration rule has been developed with the use of stress invariants and elastic secant moduli. The algorithm is conveniently implemented within a finite element framework and is applicable to a wide range of models that are representative for soils. Elementary tests under different monotonic loading conditions demonstrate the numerical performance and accuracy of the proposed integration algorithm. The robustness has been assessed by the analysis of realistic boundary value problems which indicate that indeed solutions can be obtained for relatively large loading steps.

Acknowledgements

This research has been carried out with the financial support of the Dutch Technology Foundation under grant DCT22.2930 and from the Center for Civil Engineering Research, Codes and Specifications (CUR-C94). We wish to express our sincere gratitude to Dr A.F. Pruijssers and Mr W.L.A.H. van den Broek, Dirk Verstoep BV, for supplying the measurements and the finite element model for the guided pipe-jacking and to Dr P. van den Berg, Delft Geotechnics for supplying the finite element model for the Leaning Tower of Pisa.

References

Borja, R. Cam-clay plasticity, Part II, Implicit integration of constitutive equations based on a nonlinear elastic stress predictor. *Comp. Meth. Appl. Mech. Engng.*, **88**, (1991) 225–240.

Britto, A.M and Gunn, M.J. *Critical State Soil Mechanics via Finite Elements*. Ellis Horwood, Chichester, (1987).

Calabresi, G., Rampello, S. and Callisto, L. The leaning tower of Pisa, geotechnical characterisation of the tower's subsoil within the framework of the critical state theory. *Studi e Recherche, Universita degli Roma 'La Sapienza'*, (1993).

de Borst, R and Feenstra, P.H. Studies in anisotropic plasticity with reference to the Hill criterion. *Int. J. Num. Meth. Engng.*, **29**, (1990) 315–336.

Desai, C.S. A general basis for yield, failure and potential functions in plasticity. *Int. J. Num. Anal. Meth. Geom.*, **4**, (1980) 361–375.

Desai, C.S., Sharma, K.G., Wathugala, G.W. and Rigby, D.B. Implementation of hierarchical single surface δ_0 and δ_1 models in finite element procedure. *Int. J. Num. Anal. Meth. Geom.*, **15**, (1991) 649–680.

DiMaggio, F.L and Sandler, I.V. Material model for granular soils. *J. Engng. Mech. Div. ASCE*, **97**, (1971) 935–950.

Drucker, D.C, Gibson, R.E. and Henkel, D.J. Soil mechanics and work hardening theories of plasticity. *Trans. ASCE*, **122**, (1957) 338–346.

Feenstra, P.H. *Computational aspects of biaxial stress in plane and reinforced concrete*. Dissertation, Delft University of Technology, (1993).

Grashuis, A.J. Pisa–Toren berekening en validatie Pluto-3D. *Report SE-57013*, Delft Geotechnics, (1993).

Hicks, M.A. MONICA-A computer algorithm for solving boundary value problems using the double hardening constitutive law MONOT: I. Algorithm development. *Int. J. Num. Anal. Meth. Geom.*, **19**, (1995) 1–27.

Hofstetter, G., Simo, J.C. and Taylor, R.L. A modified cap model: Closest point solution algorithms. *Comp. Struct.*, **46**(2), (1993) 203–214.

Hughes, T.J.R. Generalization of selective integration procedures to anisotropic and nonlinear media, *Int. J. Num. Meth. Eng.*, **15**, (1980) 1413–1418.

Kim, M.K. and Lade, P.V. Single hardening constitutive model for frictional materials I. Plastic potential function. *Computers and Geotechnics*, **5**, (1988) 307–324.

Koiter, W.T. Stress-strain relations, uniqueness and variational problems for elastic-plastic materials with singular yield surface. *Q. Appl. Math.*, **11**, (1953) 350–354.

Lade, P.V. Elasto-plastic stress-strain theory for cohesionless soil with curved yield surfaces. *Int. J. Sol. Struct.*, **13**, (1977) 1019–1035.

Lade, P.V. and Kim, M.K. Single hardening constitutive model for frictional materials II. Yield criterion and plastic work contours. *Computers and Geotechnics*, **6**, (1988a) 13–29.

Lade, P.V. and Kim, M.K. Single hardening constitutive model for frictional materials III. Comparisons with experimental data. *Computers and Geotechnics*, **6**, (1988b) 31–47.

Lourenço, P.B. *Computational strategies for masonry structures*. Dissertation, Delft University of Technology, (1996).

Matsuoka, H. and Nakai, T. Stress-deformation and strength characteristics of soil under three different principal stresses. *Proc. JSCE*, **232**, (1974) 59–70.

Molenkamp, F. Elasto-plastic double hardening model MONOT. *LGM Report CO-218595*, Delft Geotechnics, (1980).

Ortiz, M. and Popov, E.P. Accuracy and stability of integration algorithms for elastoplastic constitutive relations. *Int. J. Num. Meth. Engng.*, **28**, (1985) 461–474.

Roscoe, K.H. and Burland, J.B. On the generalized behaviour of 'wet' clay. *Engineering Plasticity*, **48**, (1968) 535–609.

Roscoe, K.H. and Schofield, A.N. Mechanical behaviour of an idealised 'wet' clay. *Proc. European Conf. on Soil Mechanics and Foundation Engineering*, Wiesbaden, **1**, (1963) 47–54.

Rowe, P.W. The stress-dilatancy relation for static equilibrium of an assembly of particles in contact. *Proc. Roy. Soc. London*, **A269**, (1962) 500–527.

Schellekens, J.C.J. and de Borst, R. The use of the Hoffmann yield criterion in finite element analysis of anisotropic composites. *Comp. Struct.*, **37**(6), (1990) 1087–1096.

Schofield, A.N. and Wroth, C.P. *Critical State Soil Mechanics*. McGraw Hill, London, (1968).

Simo, J.C. and Meschke, G. A new class of algorithms for classical plasticity extended to finite strains. Application to geomaterials. *Computational Mechanics*, **11**, (1993) 253–278.

Simo, J.C. and Taylor, R.L. A return mapping algorithm for plane stress elastoplasticity. *Int. J. Num. Meth. Engng.*, **22**, (1986) 649–670.

Simo, J.C. Kennedy, J.G. and Govindjee, S. Non-smooth multisurface plasticity and viscoplasticity. Loading/unloading conditions and numerical algorithms. *Int. J. Num. Meth. Engng.*, **26**, (1988) 2161–2185.

van den Broek, W.L.A.H., de Borst, R. and Groen, A.E. Two- and three-dimensional modelling of a guided pipe-jacking in soft soil. *Geotechnical Aspects of Underground Construction in Soft Ground*. Balkema, Rotterdam, (1996) 419–422.

van Eekelen, H.A.M. Isotropic yield surfaces in three dimensions for use in soil mechanics. *Int. J. Num. Anal. Meth. Geom.*, **4**, (1980) 89–101.

van Eekelen, S.J.M. and van den Berg, P. The Delft egg model, a constitutive model for clay. In: *DIANA Computational Mechanics '94*, Kusters, G.M.A. and Hendriks, M.A.N. (eds.), Kluwer Academic Dordrecht, (1994) 103–116.

van Langen, H. *Numerical analysis of soil-structure interaction*. Dissertation, Delft University of Technology, (1991).

Vermeer, P.A. *Formulation and analysis of sand deformation problems*. Dissertation, Delft University of Technology, (1980).

Vermeer, P.A. and de Borst, P. Non-associated plasticity for soils, concrete and rock. *Heron*, **29**(3), (1984) 1–64.

Wood, D.M. *Soil Behaviour and Critical State Soil Mechanics*. Cambridge University Press, Cambridge, (1990).

Zytinsky, M., Randolph, M.K., Nova, R. and Wroth, C.P. On modelling the unloading-reloading behaviour of soils. *Int. J. Num. Anal. Meth. Geom.*, **2**, (1978) 87–93.

3

An Asynchronous Parallel Algorithm for Consolidation Problems

X. Wang[1], R. Matteazzi[2], B. A. Schrefler[2] and **O. C. Zienkiewicz[3]**
[1] *Dalian University of Technology, Dalian, China*
[2] *Universita' di Padova, Italy*
[3] *University of Wales, Swansea, UK*

3.1   Introduction

Consolidation plays an important role in many soil mechanics problems, which is demonstrated by the vast amount of literature devoted to the solution of this problem since the pioneering work of Terzaghi (1951). There are two main areas where consolidation analysis is applied extensively. The first is connected with the physical loading of soil layers. This aspect comprises the transient analysis of footings, pile foundations, soil-structure interaction, embankments, large fill dams, etc. The second aspect of consolidation, connected with the change of hydraulic equilibrium in a system comprising aquifers and aquitards, has received attention only in recent times, when the effects of extensive groundwater withdrawal for industrial and agricultural purposes, and the effects of heavy oil and/or gas pumpage, became evident. Depending upon the geologic environment of the exploited aquifers and/or hydrocarbon reservoirs, heavy pumpage can result in large surface subsidence with possible property damage.

Subsidence analyses are computationally intensive because they involve problems of regional size and very long time spans. For example, in the case of the Groningen gas field, subsidence predictions for the year 2050 have been made from the year 1971. Further, there may be several reservoirs at different levels distributed over a large area, with possible interaction as far as surface subsidence is concerned. A typical case is the upper Adriatic region, where the depth of the location of pools ranges between 900 and

4000 m and the horizontal area involved is $19\,000\,\text{km}^2$. The different pools are not put in production at the same time, which further complicates the situation. Subsidence analysis, at least for the production phase (hence excluding rebound), is still carried out mainly with a linear elastic soil behavior. Corrections for capillary effects, in the case of the simultaneous presence of wetting and non-wetting fluids in the reservoir, can be made at a successive time.

The method presented here is meant for scenarios as described above. It is based on Biot's theory, where the pore pressure field and solid skeleton displacement field are coupled (Biot, 1941, 1962).

Instead of the usual staggered algorithms studied extensively by Zienkiewicz *et al.* (1988), Turska and Schrefler (1993), Felipa and Park (1980) and Saetta and Vitaliani (1991), which consider either a parallel or sequential solution, a new algorithm is presented here which can be considered as an asynchronous parallel algorithm. This algorithm is based on an efficient time integration strategy and a direct approach to solve pore pressure and soil skeleton deformation independently, while still preserving the coupled nature of the problem. The procedure is ideal for implementation on either a parallel computer or a cluster of workstations.

The procedure proposed has been implemented on such a cluster of workstations, and several numerical examples show the computational efficiency and the accuracy of the method that can be achieved.

3.2   Finite Element Formulation of the Problem

The governing equations for the consolidation problem are summarized below; further details of the derivation are given by Lewis and Schrefler (1987), Turska *et al.* (1993) and Zienkiewicz and Taylor (1985). Here the soil is assumed to be saturated with water. The equilibrium equation for the two-phase medium and the continuity equation for water in semi-discretized form are expressed in terms of solid displacement $\mathbf{u}$ and pressure $\mathbf{p}$ as

$$\mathbf{K}\dot{\mathbf{u}} - \mathbf{Q}\dot{\mathbf{p}} = \mathbf{F_u} \tag{3.1}$$

$$\mathbf{Q}^\mathrm{T}\dot{\mathbf{u}} + \mathbf{S}\dot{\mathbf{p}} + \mathbf{Hp} = \mathbf{F}_w \tag{3.2}$$

where $\mathbf{K}$ is the stiffness matrix for the soil skeleton, $\mathbf{S}$ and $\mathbf{H}$ are the compressibility and permeability matrices for the pore fluid, respectively. $\mathbf{Q}$ is the coupling matrix and the dot indicates differentiation with respect to time. The above matrices have the following form:

$$\mathbf{K} = \int_\Omega \mathbf{B}^\mathrm{T}\mathbf{D}_d\mathbf{B}\,d\Omega, \ \mathbf{Q} = \int_\Omega \mathbf{B}^\mathrm{T}\mathbf{m}\mathbf{N}_p\,d\Omega \tag{3.3a}$$

$$\mathbf{S} = \int_\Omega \mathbf{N}_p^\mathrm{T}\mathbf{Q}_b\mathbf{N}_p\,d\Omega, \ \mathbf{H} = \int_\Omega \nabla\mathbf{N}_p^\mathrm{T}k\nabla\mathbf{N}_p\,d\Omega \tag{3.3b}$$

where $\mathbf{N}_u$ and $\mathbf{N}_p$ are the shape functions for displacements and pressures, respectively, $\mathbf{B}$ is the strain matrix, $\mathbf{D}_d$ is the elastic matrix, $\mathbf{Q}_b$ is the fluid compressibility, and k is the permeability coefficient. The matrices $\mathbf{K}$, $\mathbf{S}$ and $\mathbf{H}$ are symmetric and positive definite.

Equation (3.1) is rewritten as

$$\dot{\mathbf{u}} = \mathbf{K}^{-1}(\mathbf{F}_u + \mathbf{Q}\dot{\mathbf{p}}) \tag{3.4}$$

Substituting Eq. (3.4) into Eq. (3.2) and then solving gives

$$\dot{\mathbf{p}} = \boldsymbol{B}\mathbf{p} + \overline{\boldsymbol{F}} \tag{3.5}$$

where

$$\boldsymbol{B} = -(\mathbf{S} + \mathbf{Q}^{\mathrm{T}}\mathbf{K}^{-1}\mathbf{Q})^{-1}\mathbf{H}$$
$$\overline{\boldsymbol{F}} = (\mathbf{S} + \mathbf{Q}^{\mathrm{T}}\mathbf{K}^{-1}\mathbf{Q})^{-1}(\mathbf{F}_w - \mathbf{Q}^{\mathrm{T}}\mathbf{K}^{-1}\mathbf{F}_u) \tag{3.6}$$

The discretization in the time domain for Eqs. (3.4) and (3.5) is carried out by the generalized trapezoidal method (generalized midpoint rule). This is an implicit method, where

$$\dot{\mathbf{p}}_{n+\theta} = (\mathbf{p}_{n+1} - \mathbf{p}_n)/\Delta t, \; \mathbf{p}_{n+\theta} = (1 - \theta)\mathbf{p}_n + \theta\mathbf{p}_{n+1} \tag{3.7}$$
$$\dot{\mathbf{u}}_{n+\theta} = (\mathbf{u}_{n+1} - \mathbf{u}_n)/\Delta t, \; \mathbf{u}_{n+\theta} = (1 - \theta)\mathbf{u}_n + \theta\mathbf{u}_{n+1} \tag{3.8}$$

with $0 \le \theta \le 1$. Equations (3.4) and (3.5) at time instant $t_{n+\theta}$ can be expressed as

$$\mathbf{p}_{n+1} = [\mathbf{I} + \Delta t(\mathbf{I} - \Delta t\theta\boldsymbol{B})^{-1}\boldsymbol{B}]\mathbf{p}_n + \Delta t(\mathbf{I} - \Delta t\theta\boldsymbol{B})^{-1}\overline{\boldsymbol{F}}_{n+\theta} \tag{3.9}$$
$$\mathbf{u}_{n+1} = \mathbf{u}_n + \mathbf{K}^{-1}\mathbf{Q}(\mathbf{p}_{n+1} - \mathbf{p}_n) + \Delta t\mathbf{K}^{-1}(\mathbf{F}_u)_{n+\theta} \tag{3.10}$$

The solution procedure adopted at every time-step is as follows: first, $\mathbf{p}_{n+1}$ is obtained by solving Eq. (3.9) with $\mathbf{p}_n$ given, then the new value for $\mathbf{p}_{n+1}$ together with $\mathbf{p}_n$ is substituted into Eq. (3.10) to obtain $\mathbf{u}_{n+1}$. This scheme corresponds to that proposed by Turska *et al.* (1993).

3.3 Parallel Computing Strategy

The indices used to evaluate parallel computing methods using p processors (or CPUs) are speedup S_p and efficiency η_c, where

$$S_p = \frac{\text{solution time taken by one processor}}{\text{solution time taken by } c \text{ processors}} \tag{3.11}$$

and

$$\eta_c = 100 \times \frac{S_p}{c} \tag{3.12}$$

The speedup is called linear if $S_p = c$, and the corresponding efficiency is 100%. Generally, this ideal target is difficult to reach, mainly because (1) most algorithms for solving problems were developed on the basis of sequential computing; they are not optimal for parallel computing; (2) it is difficult to balance the computational load distributed to the processors; and (3) data communication among processors is needed.

We now investigate a parallel implementation of the above algorithm (see Figure 3.1). The time interval of interest is $t \in (t_0, t_n)$; it may be similar to the total time history, or a significant part of it. This time interval has been divided into a number of subintervals (time steps) of the same length Δt. In general, it is assumed that $\mathbf{F}_u$ and $\mathbf{F}_w$ are slowly varying functions (even if this is not necessary, as shown by examples). As will be shown in the next section, the algorithm is consistent and unconditionally stable with parameter $\theta > \frac{1}{2}$ so that the time step Δt may be considered rather large for consolidation analysis. Furthermore, as can be seen from Eqs. (3.9) and (3.10), the response of $\mathbf{p}$ can be computed independently from $\mathbf{u}$, which is then evaluated, once $\mathbf{p}$ is known; it is to be noted that the number of pressure degree of freedom is usually much smaller than displacement degree of freedom.

For a conventional consolidation analysis, much time is spent on time integration, and the number n of time intervals Δt is much greater than the number of available processors c, i.e. $n \gg c$. Therefore, the number of intervals can be subdivided in the following way: let J be the integer part of n/c; then the global number of time steps is divided into c groups T_i ($i = 1, 2, \ldots, c$), each containing $S(T_i) = J + 1$ intervals if $i \leq n - cJ$ or $S(T_i) = J$ intervals if $i > n - cJ$. Then the time instants involved in the ith group are

$$T_i = [t_0, t_i, t_{i+c}, t_{i+2c}, \ldots t_{i+(S(T_i)-1)c}] \tag{3.13}$$

It should be noted that the first time step increment is $\Delta t_1 = i\Delta t$ (e.g. if $i = 1$ then $\Delta t_1 = \Delta t$; if $i = 2$ then $\Delta t_1 = 2\Delta t$; etc.), and that $\Delta t = c\Delta t$ for all subsequent time steps. So each group of intervals T_i has a different first time step length from the other groups, but for all subsequent integrations all groups use the same time increment Δt. Hence, considering this time-step integration, Eqs. (3.9) and (3.10) should be replaced as

$$\mathbf{p_{n+1}} = [\mathbf{I} + \Delta t_1(\mathbf{I} - \Delta t_1\theta\mathbf{B})^{-1}\mathbf{B}]\mathbf{p_n} + \Delta t_1(\mathbf{I} - \Delta t_1\theta\mathbf{B})^{-1}\overline{\mathbf{F}}_{n+\theta} \tag{3.14a}$$

$$\mathbf{u}_1 = \mathbf{u}_0 - \mathbf{K}^{-1}\mathbf{Q}(\mathbf{p}_1 - \mathbf{p}_0) + \Delta t_1\mathbf{K}^{-1}(\mathbf{F_u})_\theta \quad \text{for the first time-step} \tag{3.14b}$$

$$\mathbf{p_{n+1}} = [\mathbf{I} + c\Delta t(\mathbf{I} - c\Delta t\theta\mathbf{B})^{-1}\mathbf{B}]\mathbf{p_n} + c\Delta t(\mathbf{I} - c\Delta t\theta\mathbf{B})^{-1}\overline{\mathbf{F}}_{n+\theta} \tag{3.15a}$$

$$\mathbf{u}_{n+1} = \mathbf{u}_n - \mathbf{K}^{-1}\mathbf{Q}(\mathbf{p}_{n+1} - \mathbf{p}_n) + c\Delta t\mathbf{K}^{-1}(\mathbf{F}_u)_{n+\theta}$$

$$\text{for the } r\text{th } (r = 2, 3, \ldots S(T_i)) \text{ time-step} \tag{3.15b}$$

Compared to sequential computing with a single processor, the above process needs extra computing of $(\mathbf{I} - \Delta t_1\theta\mathbf{B})^{-1}$. This, however, involves a minor additional effort.

One advantage of the above procedure is the task balance among the c processors, particularly if n is a multiple of c. Except for data distribution at the beginning of the analysis, all processors work completely independently, which gives very high speedup and efficiency.

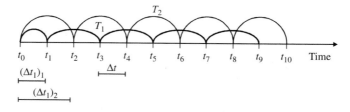

Figure 3.1   Scheme of parallel implementation

Clearly, the proposed strategy for consolidation analysis is a typical kind of asynchronous parallel method (Adeli, 1988; Lin *et al.*, 1997). It has been implemented on a cluster of workstations using the PVM (Parallel Virtual Machine) system (Sunderam, 1990; Geist and Sunderam, 1991; Geist *et al.*, 1994).

The PVM system is composed of a set of user-interface primitives and supporting software that together enable concurrent computing on loosely coupled networks of processing elements. PVM may be implemented on a hardware base consisting of a mixture of machine architectures, including single CPU system, vector machines and multiprocessors. These computing elements may be interconnected by one or more networks, which may themselves be different. Applications access these processing elements via a standard interface, namely a set of well-defined primitives that are embedded in procedural host languages. They are composed of one or more components that are sub-tasks at moderately large levels of granularity. During execution, multiple instances of each component may be initiated. The PVM user interface is strongly typed; support for operating in a heterogeneous environment is provided in the form of special constructs that selectively perform machine-dependent data conversions, where necessary. Inter-instance communication constructs include those for the exchange of data structures, as well as high-level primitives such as broadcast, barrier synchronization, mutual exclusion, global extreme, and rendezvous. Application programs view the PVM system as a flexible parallel computing resource that supports a message passing model of computation. PVM supports three general parallel programming models – tree computations, crowd computations and hybrid computation – in which crowd computing is the most common model for PVM applications. In this chapter, the master-slave (or host-node) model in crowd computing is adopted. The master program is only responsible for process spawning, collection and display of results. The slave programs perform the actual computations involved.

Programming a PVM like workstation networks is, for the most part, similar to programming a sequential computer, except that special attention has to be paid to the operations requiring communication between processors. Since most of the computations in the proposed algorithm for consolidation analysis can be performed independently, parallel implementation of the method comprises a mapping of time integration steps onto each processor of the workstation networks.

3.4   Consistency and Stability of the Method

If in Eq. (3.9) $\mathbf{p}_{n+1}$ and $\mathbf{p}_n$ are replaced by the corresponding exact values $\mathbf{p}_{n+1}^*$ and $\mathbf{p}_n^*$, respectively, we have the following expression:

$$\mathbf{p}_{n+1}^* - [\mathbf{I} + \Delta t(\mathbf{I} - \Delta t\theta\mathbf{B})^{-1}\mathbf{B}]\mathbf{p}_n^* - \Delta t(\mathbf{I} - \Delta t\theta\mathbf{B})^{-1}\overline{F}_{n+\theta} = \mathbf{r}_{n+1} \tag{3.16}$$

where $\mathbf{r}_{n+1}$ is the local traction error. If $||\mathbf{r}_{n+1}|| \to 0$ for $\Delta t \to 0$, then the algorithm for solving $\mathbf{p}$ defined by Eq. (3.9) is consistent.

Using a Taylor series expansion about $t_{n+\theta}$

$$\mathbf{p}_{n+1}^* = \mathbf{p}_{n+\theta} + (1 - \theta)\Delta t\dot{\mathbf{p}}_{n+\theta} + \tfrac{1}{2}(1 - \theta)^2\Delta t^2\ddot{\mathbf{p}}_{n+\theta} + O(\Delta t^3) \tag{3.17a}$$

$$\mathbf{p}_n^* = \mathbf{p}_{n+\theta} - \theta\Delta t\dot{\mathbf{p}}_{n+\theta} + \tfrac{1}{2}\theta^2\Delta t^2\ddot{\mathbf{p}}_{n+\theta} + O(\Delta t^3) \tag{3.17b}$$

Substituting Eqs. (3.17a) and (3.17b) into Eq. (3.16) gives

$$\mathbf{p}_{n+\vartheta} + (1-\theta)\Delta t \dot{\mathbf{p}}_{n+\theta} + \tfrac{1}{2}(1-\theta)^2 \Delta t^2 \ddot{\mathbf{p}}_{n+\theta} + O(\Delta t^3)$$

$$- [\mathbf{I} + \Delta t(\mathbf{I} - \Delta t\theta\mathbf{B})^{-1}\mathbf{B}][\mathbf{p}_{n+\theta} - \theta\Delta t\dot{\mathbf{p}}_{n+\theta} + \tfrac{1}{2}\theta^2 \Delta t^2 \ddot{\mathbf{p}}_{n+\theta} + O(\Delta t^3)]$$

$$- \Delta t(\mathbf{I} - \Delta t\theta\mathbf{B})^{-1}\overline{\mathbf{F}}_{n+\theta} = \mathbf{r}_{n+1} \tag{3.18}$$

Employing Eq. (3.5) results in

$$\tfrac{1}{2}\Delta t^2(1-2\theta)\ddot{\mathbf{p}}_{n+\theta} - \tfrac{1}{2}\theta^2 \Delta t^3(\mathbf{I} - \theta\Delta t\mathbf{B})^{-1}\mathbf{B}\ddot{\mathbf{p}}_{n+\theta} + O(\Delta t^4) = \mathbf{r}_{n+1} \tag{3.19}$$

as $\Delta t \to 0$, $||\mathbf{r}_{n+1}|| \to 0$. Similarly,

$$(\mathbf{r}_u)_{n+1} = \tfrac{1}{2}\Delta t(1-2\theta)(\ddot{\mathbf{u}}_{n+\theta} + \ddot{\mathbf{p}}_{n+\theta}) + O(\Delta t^3) \tag{3.20}$$

Again, as $\Delta t \to 0$, $||(\mathbf{r}_u)_{n+1}|| \to 0$.

To investigate the stability properties, we only need to show the stability of Eq. (3.9). Let $\lambda_1, \lambda_2, \ldots \lambda_n$ be distinct complex eigenvalues of matrix $\mathbf{B}$, where n is the rank of $\mathbf{B}$. Then the homogeneous form of Eq. (3.9) decomposes into n equations:

$$p_{n+1(j)} = [1 + \Delta t(1 - \theta\Delta t\lambda_j)^{-1}\lambda_j]p_{n(j)} \tag{3.21}$$

where $j = 1, 2, \ldots, n$, and $p_{n(j)}$ denote the scalar values which are obtained by premultiplying the vector $\mathbf{p}_n$ by the eigenvectors $\mathbf{\Psi}_j^T$ of $\mathbf{B}$:

$$p_{n(j)} = \mathbf{\Psi}_j^T \mathbf{p}_n \tag{3.22}$$

The stability condition can be written as

$$\left| \frac{1 + (1-\theta)\Delta t\lambda_j}{1 - \theta\Delta t\lambda_j} \right| < 1 \tag{3.23}$$

for $j = 1, 2, \ldots, n$. Denoting $\text{Re}(\lambda_j) = \lambda_r$ and $\text{Im}(\lambda_j) = \lambda_i$, inequality in Eq. (3.23) is equivalent to

$$2\lambda_r < (2\theta - 1)\Delta t(\lambda_r^2 + \lambda_i^2) \tag{3.24}$$

For $\theta > \tfrac{1}{2}$, if $\lambda_r < 0$, Eq. (3.24) is satisfied for all Δt and λ_i, i.e. the algorithm is unconditionally stable. If $\lambda_r > 0$, Eq. (3.23) gives a conditional stability with

$$\Delta t > \frac{2\lambda_r}{(2\theta - 1)(\lambda_r^2 + \lambda_i^2)} \tag{3.25}$$

For $\theta > \tfrac{1}{2}$, we have conditional stability when

$$\Delta t < \frac{-2\lambda_r}{(1 - 2\theta)(\lambda_r^2 + \lambda_i^2)} \tag{3.26}$$

Equation (3.26) shows that when $\theta < \tfrac{1}{2}$ and $\lambda_r > 0$, the algorithm is unstable.

3.5 Numerical Examples

Example 1: Single aquifer deformation due to withdrawal from a pumping well
A cylindrical region surrounding a well of radius ε is considered. The aquifer is confined at the top and bottom by impervious layers, and the well fully penetrates the aquifer. A numerical model based on the finite element mesh of Figure 3.2 was used to predict vertical and horizontal displacements and the change in excess pore pressure due to fluid withdrawal. The boundary at $r = \varepsilon$ is restrained from any lateral movement; however, it is free to move vertically. At $r = R(R = 1000\,\text{m})$, where R is taken sufficiently far from the well area so that there is limited response to pumping at that point, a constant head is assumed. At the top surface, the soil is free to move both vertically and horizontally. The bottom impervious layer is assumed fixed so that no vertical displacement can take place. The geometry, mesh used and boundary conditions of the example are shown in Figure 3.2, while the soil properties are represented in Table 3.1.

The outflow rate is uniform over the height of the well, hence resulting in a uniform pore pressure distribution along vertical lines.

First, we explain how the asynchronous method works and how to investigate its accuracy.

Let us consider a step history function for the outflow, as shown in Figure 3.3, which may be typical in subsidence problems (Lewis and Schrefler, 1987). A comparison of the resulting displacements of the top of the surface has been carried out for the proposed

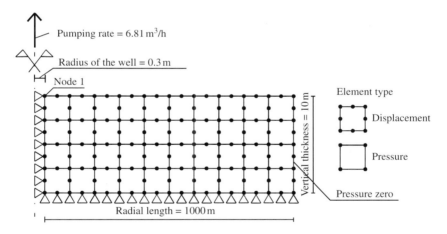

Figure 3.2 Example 1: geometry and finite element mesh used

Table 3.1 Soil Properties

Permeability	$K = 10^{-4}\,\text{m/s}$
Porosity	$\phi = 0.2$
Young's modulus	$E = 2.2 \times 10^4\,\text{KN/m}^2$
Poisson's ratio	$\nu = 0.1$
Specific weight of water	$\gamma = 9.81\,\text{KN/m}^3$
Water bulk modulus	$Kf = 2 \times 10^6\,\text{KN/m}^2$
Compressibility of water	$\beta = 5 \times 10^{-10}\,\text{N/m}^2$

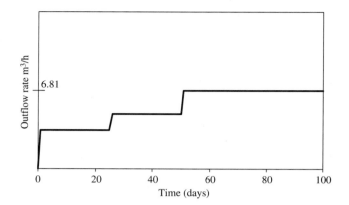

Figure 3.3   Example 1: step function of withdrawal rate

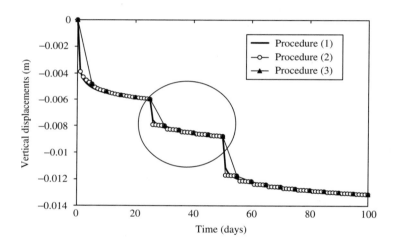

Figure 3.4   Step function: surface displacements of node 1 versus time

parallel asynchronous procedure and the usual sequential procedure. Figure 3.4 presents the results using three different types of solution methods:

(1)   A sequential procedure with $\Delta t = 1$ day and 100 time-steps.

(2)   A parallel procedure with the same temporal history, but using five processors with $\Delta t = 5$ days and 20 time-steps for each processor.

(3)   A sequential procedure with $\Delta t = 5$ days and 20 time-steps which, except for the overhead of communication, needs the same calculation time as in procedure (2).

Procedure (1), which is more accurate but five times more expensive than the other two, is taken as reference, and the accuracy of procedures (2) and (3) is compared with reference to procedure (1).

From Figure 3.4, it can be seen that the results of the proposed parallel procedure lie between the other two solutions. In particular, at least in this type of problem, the parallel procedure permits arriving at a solution with more information when compared to the 'equivalent' sequential procedure. Further, the results are closer to the reference solution of the more expensive sequential procedure. In Figure 3.5 it is possible to see the difference between parallel and sequential procedures more closely.

The method is also able to follow more complicated loading histories, as shown in the next loading case, the sinusoidal load function of Figure 3.6. Figure 3.7 shows that, as before, the parallel procedure provides more information than the 'equivalent' sequential procedure.

Furthermore, in this case it is possible to observe that as long as the forcing load varies, the solution of the proposed parallel procedure follows the reference solution very

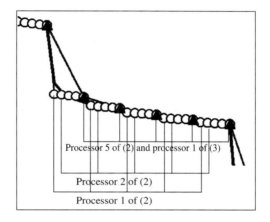

Figure 3.5   Detail of Figure 3.3

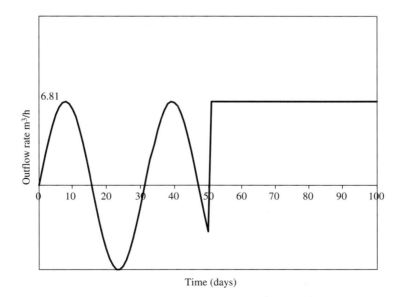

Figure 3.6   Example 1: sinusoidal function of withdrawal rate

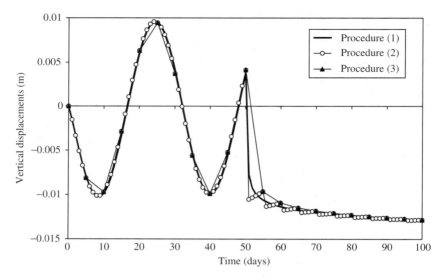

Figure 3.7   Sinusoidal function: surface displacements of node 1 versus time

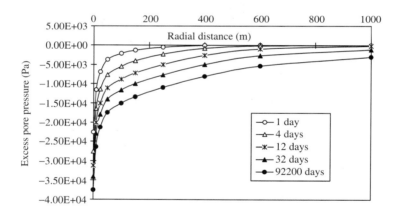

Figure 3.8   Constant function: excess pore pressure versus radial distance of the aquifer

well, while when the load function becomes constant the solutions of the parallel and the 'equivalent' sequential procedures become less accurate.

We now investigate, with a varying number of processors, a constant pumping rate outflow (as in Eq. (17)) to show the speedup and efficiency of the parallel implementation of the method. (Lewis *et al.* (1991))

Excess pore pressure versus radial distance is shown in Figure 3.8. The displacements for different time values are shown in Figure 3.9. Table 3.2 gives the computing time, speedup and efficiency. The speedup is close to linear, and the efficiencies are all above 95%. The speedup for this example is also demonstrated by Figure 3.10.

The method obviously works also for normal consolidation problems with physical loading as shown by the following two-dimensional consolidation.

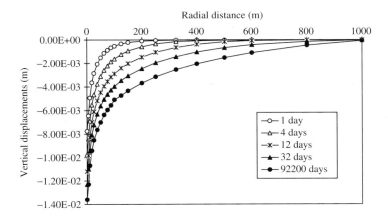

Figure 3.9 Constant function: vertical displacements versus radial distance at the top of the aquifer

Table 3.2 Example 1: 92 200 time-steps: speedup and efficiency

No. of processors	CPU time(sec)	Speedup	Efficiency(%)
1	9795.986	–	–
2	4919.596	1.99	99.56
3	3338.274	2.93	97.81
4	2518.052	3.89	97.26
5	2128.68	4.6	92.04

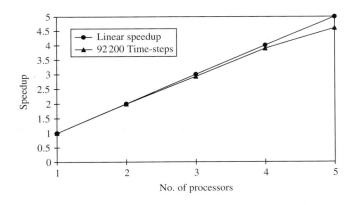

Figure 3.10 Example 1: speedup for parallel computing by using PVM

Example 2: Two-dimensional consolidation
This example problem is taken from Turska *et al.* (1993), and the finite element mesh is shown in Figure 3.11. In order to simulate a variable loading condition, which could influence the results, the load was varied linearly up to a value of time equal to 25 days, and then it was assumed constant. The problem was solved using the following parameters: Young modulus $E = 622.7$ KPa, Poisson's ratio $\nu = 0.4$ and permeability $\kappa_x = \kappa_y = 1.22 \times 10^{-5}$ m/day.

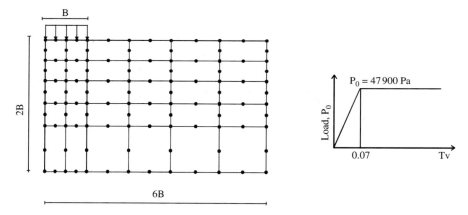

Figure 3.11 Finite element mesh for 2-D consolidation

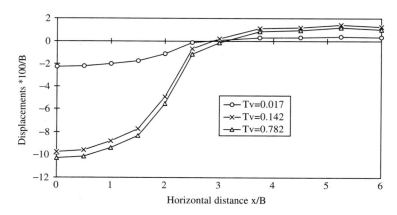

Figure 3.12 Example 2: surface settlements versus horizontal distance at three different times

The resulting settlements and pore pressure are shown in Figures 3.12 and 3.13, respectively. Table 3.3 gives the speedup and efficiency, showing again that this problem can be run efficiently on a cluster of workstations. The speedup for this example is also demonstrated in Figure 3.14.

3.6   Conclusion

An asynchronous parallel method for consolidation analysis has been presented in this chapter. The consistency and stability of this method are discussed. The algorithm is stable when $\theta > \frac{1}{2}$, but may require a lower limit for Δt in some conditions ($\lambda_r > 0$). The proposed strategy of time integration requires only little data communication, and makes the balance among processors easy. The method has been successfully implemented on a cluster of workstations. Numerical examples show that this method has high efficiency and speedup is close to linear. As distinct from synchronous parallel algorithms, there is not a marked loss of efficiency when increasing the number of processors.

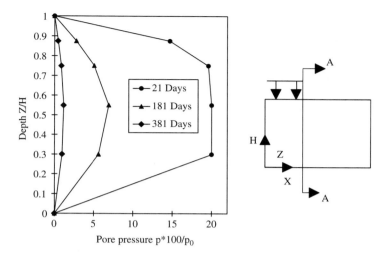

Figure 3.13   Example 2: pore pressure versus depth along section A-A at three different times

Table 3.3   Example 3: 1060 time-steps: speedup and efficiency

No. of processors	CPU time(sec)	Speedup	Efficiency(%)
1	66.457	–	–
2	34.235	1.9412	97.06
3	22.954	2.8953	96.51
4	18.034	3.6851	92.13
5	14.663	4.5324	90.65

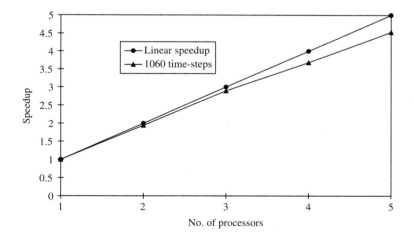

Figure 3.14   Example 2: speedup for parallel computing by using PVM

Although this algorithm is developed on the basis of workstations, it is also suited to other parallel computers.

Acknowledgements

The authors gratefully acknowledge the financing of this work from the International Scientific Co-operation Initiative 'Marie Curie' fellowship foundation of the European Commission and the National Natural Science Foundation of China. The computations were performed on a cluster of multiprocessor unix servers from Digital at the Department of Structural and Transportation Engineering of the University of Padua.

References

Adeli, H. (Ed). Parallel and distributed processing in structural engineering. *Proc. session sponsored by the structural division of the American Society of Civil Engineering in conjunction with the ASCE National convention*, Nashville, Tennessee, (May 1988).

Biot, M. A. General theory of three-dimensional consolidation. *J. Appl. Phys.*, **12**, (1941) 155–164.

Biot, M. A. Generalised theory of acoustic propagation in porous media. *J. Appl. Phys.*, **33**, (1962) 1482–1498.

Felippa, C. A. and Park, K. C. Staggered transient analysis procedures for coupled mechanical systems: formulation. *Comp. Methods Appl. Mech. Engrg.*, **24**, (1980) 61–111.

Geist, G. A. and Sunderam, V. *Network based concurrent computing on the PVM system.* Oak Ridge National Laboratory, ORNL/TM-11760, (1991).

Geist, G. A. *et al. PVM 3 User's guide and reference manual.* Oak Ridge National Laboratory, ORNL/TM-12187, (1994).

Lewis, R. W. and Schrefler, B. A. *The Finite Element Method in the Deformation and Consolidation of Porous Media.* Wiley, Chichester, (1987).

Lewis, R. W., Schrefler, B. A. and Simoni, L. Coupling versus uncoupling in soil consolidation. *Int. J. Num. and Analytical Methods in Geom.*, **15**, (1991) 533–548.

Lin, J., Wang, X. and Williams, F. W. Asynchronous parallel computing of structural non-stationary random seismic responses. *Int. J. Numer. Methods Engrg.*, **40**, (1997) 2133–2149.

Saetta, A., Schrefler, B. A. and Vitaliani, R. Solution strategies for coupled consolidation analysis in porous media. *I. J. Met. Num. Calculo Diseño Ingen.*, **7**, (1991) 55–66.

Sunderam, V. PVM: A framework for parallel distributed computing. *Concurrence: Practice and Experience*, (1990) 315–339.

Terzaghi, K. and Peck, R. B. *Solid Mechanics in Engineering Practice.* Wiley, New York, (1951).

Turska, E. and Schrefler, B. A. On convergence conditions of partitioned solution procedures for consolidation problems. *Comput. Methods Appl. Mech. Engrg.*, **106**, (1993) 51–64.

Turska, E., Wisniewski, K. and Schrefler, B. A. Error propagation of staggered solution procedures for transient problems. *Comput. Methods Appl. Mech. Engrg.*, **114**, (1993) 51–63.

Zienkiewicz, O. C. The coupled problems of soil-pore fluid-external fluid interaction, basis for a general geomechanics code. *Proc. Fifth Int. Conf. on Numerical Methods in Geomechanics*, Nagoya, Balkema Press, (1985) 1731–1740.

Zienkiewicz, O. C. and Taylor, R. L. Coupled problems, a simple stepping procedure. *Common. Appl. Numer. Methods*, **1**, (1985) 233–239.

Zienkiewicz, O. C., Paul, D. K. and Chan, A. H. C. Unconditionally stable staggered solution procedure for soil-pore fluid interaction problems. *Int. J. Numer. Methods Engrg.*, **26**, (1988) 1039–1055.

4

Estimation of Material Parameters with Discontinuities for Elliptic Models

Luther W. White
The University of Oklahoma, Oklahoma, USA

4.1   Introduction

The flow of a single phase incompressible fluid in a confined saturated porous medium may be modeled by means of a second order elliptic boundary value problem Eq. (4.2). The partial differential equation is based on introducing a flow potential function that includes pressure and gravitational effects into an expression of the conservation of mass encapsilated in Darcy's law (Chavent and Jaffre, 1986; Peaceman, 1977). In order to apply this model effectively in a particular context, it is necessary to determine certain material parameters expressing the geomechanical properties of the porous medium to transmit fluid. In the groundwater context this coefficient is known as the hydraulic conductivity whereas in the simulation of flow in oil reservoirs it is referred to as permeability (Chavent and Jaffre, 1986; Peaceman, 1977). These material parameters depend upon both rock and fluid properties. The coefficient is in fact a tensor that may not only depend upon the location within the porous medium, but may vary in a discontinuous manner from one location to another. Indeed, it is just such behavior that is of most interest in determining the geomechanical properties of the reservoir (Mitasova and Mitas, 1991). Typically the estimates of these material parameters are to be obtained using sparse measurements of the parameters themselves supplemented with measurements of a nonlinear function of the parameter, usually pressure.

The purpose of this paper is to describe an approach to estimate numerically material parameters with possible discontinuities in models in two and three spatial dimensions. The Regularized-Output-Least-Squares (the ROLS step) estimation procedure along with spline-based finite elements is used as a technique to detect sudden changes in behavior against a background. A method is then used that searches among a class of discontinuous functions to isolate the anomaly (the IA step). The two methods are used iteratively until no further anomalous behavior is detected. Essentially, the first procedure is used to detect a sudden change in behavior, and the second is used to isolate and refine a region containing it. For the first step we use a ROLS procedure in which the material parameter is approximated by linear combinations of certain spline functions. The regularization used is a H^2 seminorm (Gilbarg and Trudinger, 1983; Ladyzhenskaya and Ural'tseva, 1968; Maz'ja, 1985). This regularization gives sufficient compactness in admissible sets to obtain the existence of a solution to the associated minimization problem (White and Zhou, 1995). While regularization implies additional smoothing, it seems possible using ROLS to detect discontinuities and sudden changes so often exhibited by geological mappings.

Having detected an anomaly, we next attempt to isolate it by estimating its magnitude and a region containing it. This is the IA step. It is again formulated in terms of an output-least-squares minimization problem, but with an admissible set consisting of piece-wise constant functions specifically taylored to find discontinuous functions. The result of the procedure, when added to the ROLS estimate, determines a discontinuous coefficient with background. Intuitively, the discontinuous estimator is subtracted from the model coefficient thereby reducing the discontinuous behavior. Again the detection step is applied to test for further discontinuous behavior.

The regularized output least squares estimation procedure along with its differentiablility and resolution properties have been described previously (Banks and Kunisch, 1989; White and Zhou, 1995; White, 1997). The procedures we develop depend upon the ability of ROLS methods to detect certain behavior. The resolution of an estimation procedure refers to its ability to detect perturbations in estimated parameters. In the process of testing estimation problems, a standard procedure is to solve the forward problem to determine the state for a specified parameter. Using data obtained by making measurements on the state, one then applies the estimation procedure to attempt recover the parameter from the data. In this way a mapping, the so called recovery mapping, is determined from the parameter space into itself. The differentiability properties of the recovery mapping give us insight into the resolution properties of the estimation method. The study of the resolution properties of ROLS method is discussed in White (1997).

The focus here is on determining numerical estimates of parameters in models in two and three spatial dimensions. To illustrate our approach, we will present two examples. The first is a problem in two dimensions, involving the estimate of a single permeability function but with multiple regions of discontinuities. The second is a problem involving a model in three spatial dimensions with a diagonal permeability tensor. In both cases data available comes from measurements of permeability and pressure. The underlying mathematical model is an elliptic boundary value problem with no flow Neumann boundary conditions. Uniqueness of solutions is obtained by requiring pressure at certain locations be zero. This condition is incorporated into computational models by penalization. In Section 4.2 we describe salient features of elliptic Neumann boundary value problems in two and three dimensions, along with ROLS method. For ease in reading, proofs of the main results are consigned to an appendix. In Section 4.3 we describe the finite element formulation of the ROLS procedure. In Section 4.4 we set forth our procedure for delineating regions containing anomalous behavior that have been detected by the ROLS step. In Section 4.5 we present the results of numerical experiments for both two and three dimensional problems. Conclusions are stated in Section 4.6.

4.2   Formulation of the ROLS Estimation Problem

We consider problems in $\mathbf{R}^n$, $n = 2, 3$. However, for ease we pose problems in $\mathbf{R}^3$, since specialization to $\mathbf{R}^2$ is obvious. In this context, we view coordinates designated by x and y as denoting horizontal location, and z denoting vertical location. Thus, suppose that Ω is a domain in $\mathbf{R}^3$ with a Lipschitz boundary Γ, a unit cube, $\Omega = \mathcal{D}x(0, 1)$, where $\mathcal{D} = (0, 1) \times (0, 1)$, is sufficient for the purposes of our discussion. Moreover, suppose that there are well locations (x_i, y_i) for $i = 1, \ldots, N_w$, situated as indicated in Figure 4.1. Let the flow potential Φ be given by

$$\Phi(x, y, z) = p(x, y, z) + z\rho g \tag{4.1}$$

where p represents the pressure function, g is the gravitational acceleration, and ρ is fluid density. From Darcy's law (Chavent and Jaffre, 1986; Peaceman,1977) an incompressible fluid satisfies the following equation in Ω

$$-\nabla \cdot (K\nabla\Phi) = f \tag{4.2}$$

where the permeability K is a spatially dependent symmetric 3×3 matrix-valued function. For the purposes of this work, however, we take K to be a spatially dependent diagonal matrix of the form

$$K = \begin{bmatrix} k_1 & 0 & 0 \\ 0 & k_2 & 0 \\ 0 & 0 & k_3 \end{bmatrix} \tag{4.3}$$

and $k_i = k_i(x, y, z)$, such that there exists a positive constant μ such that

$$k_i \in L^\infty(\Omega) \text{ and } k_i \geq \mu \text{ a.e. in } \Omega \tag{4.4}$$

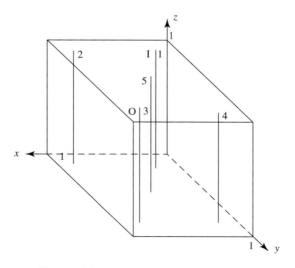

Figure 4.1   Reservoir and well placement

We assume the no-flow boundary condition

$$K\nabla\Phi \cdot n = 0 \text{ on } \Gamma \tag{4.5}$$

Remark 1   The forcing function $f \in L^2(\Omega)$ models the rate at which mass per unit volume of fluid enters into the domain Ω. In three dimensions, fluid is injected along a line segment (or for that matter a curve)

$$W_1 = \{(x_1, y_1, z): 0 < z < 1\} \tag{4.6}$$

within Ω located at (x_1, y_1). Mathematically, we represent f by

$$f(x, y, z) = \mathcal{F}(z)\delta(x, y) \tag{4.7}$$

where $\delta \in L^2(\mathcal{D})$ is a nonnegative function with support in a small ball centered at (x_1, y_1) with integral 1. The term $\mathcal{F}$ is a function of z only allows for injection of fluid at various depths. We may identify f with a continuous linear functional on the Sobolev space $H^1(\Omega)$ such that the action of f on an element $\phi \in H^1(\Omega)$, denoted by $\langle f, \phi \rangle$, is given by

$$\langle f, \phi \rangle = \int_0^1 \mathcal{F}(z) \left(\int_0^1 \int_0^1 \delta(x, y)\phi(x, y, z) \, dx \, dy \right) dz \tag{4.8}$$

Here the function δ may be thought of as an approximation to the identity (Neri, 1971). Note it is not necessary for the injection well to run the full depth of the domain Ω or for that matter to be a line segment.

At certain locations the pressure is held fixed. This is modeled by introducing Δ is a continuous linear functional on $H^1(\Omega)$ with the property that

$$\langle \Delta, \phi \rangle \neq 0 \text{ for any } \phi = \text{ nonzero constant} \tag{4.9}$$

Remark 2   For any positive number λ

$$\left[\int_\Omega |\nabla\phi|^2 dx + \lambda\langle\Delta, \phi\rangle^2 \right]^{1/2} \tag{4.10}$$

is a norm on $H^1(\Omega)$ equivalent to the $H^1(\Omega)$ norm, (cf Maz'ja, 1985).

The underlying boundary value problem seeks the solution Φ of Eqs. (4.2) and (4.5) given Eqs. (4.2), (4.3) and (4.7), subject to the condition

$$\langle \Delta, \Phi \rangle = 0 \tag{4.11}$$

The weak formulation of the problem is appropriate for our purposes. Hence, define the space

$$V_0 = \{\phi \in H^1(\Omega): \langle\Delta, \phi\rangle = 0\} \tag{4.12}$$

and note that V_0 is a Hilbert space with norm

$$||\phi|| V_0 = \left[\int_\Omega |\nabla\phi|^2 dx \right]^{1/2} \tag{4.13}$$

and associated inner product. Furthermore, from Remark 2 there is a positive number γ_0 such that for any $\phi \in V_0$

$$||\phi|| H^1(\Omega) \le \gamma_0 ||\phi|| V_0 \tag{4.14}$$

The weak formulation of the boundary value problem (cf. Gilbarg and Trudinger, 1983; Ladyzhenskaya and Ural'tseva, 1968) may be stated in terms of the following problem:

Find Φ in V_0 such that

$$\int_\Omega K\nabla\Phi \cdot \nabla\phi \, dx = \int_\Omega f\phi \, dx \tag{4.15}$$

for any ϕ in V_0. Of course, defining the functional

$$J(\phi) = \int_\Omega (K|\nabla\phi|^2 - 2f\phi) \, dx \tag{4.16}$$

an equivalent formulation of Eq. (4.15) maybe be given in terms of the minimization problem (Schultz, 1973):

Find $\Phi \in V_0$ such that

$$J(\Phi) = \text{infimum}\{J(\phi): \phi \in V_0\} \tag{4.17}$$

It is well known that there exists a unique solution $\Phi(K)$ to the problem (4.17). Denote by

$$Q_0 = \{K = (k_1, k_2, k_3) \text{ such that Eqs. (4.3) and (4.4) hold}\} \tag{4.18}$$

Observe that Q_0 is a nonempty closed convex set in $Q = (L^\infty(\Omega))^3$. The following result is key to existence, and its proof is given in the appendix.

Proposition 3   If $K \in Q_0$, then there exists a unique solution $\Phi(K) \in V_0$. Furthermore, the mapping $K \mapsto \Phi(K)$ is continuous from Q_0 with the norm topology into V_0 with the norm topology.

Remark 4   If the sequence $\{K^{(i)}\}_{i=1}^\infty$ is bounded in Q but converges only in $L^2(\Omega)^3$ then only weak convergence of the sequence $\{\Phi(K^{(i)})\}_{i=1}^\infty$ in V_0 holds (Gutman, 1990; White and Zhou, 1995).

To solve the elliptic problem numerically, it is convenient to give a penalty formulation to the problem to incorporate the constraint. To this end, define a family of functionals indexed by $\varepsilon \in (0, \infty)$

$$J_\varepsilon(\Phi) = \int_\Omega (K|\nabla\phi|^2 - 2f\phi) \, dx + \frac{1}{\varepsilon} \langle \Delta, \phi \rangle^2 \tag{4.19}$$

A family of minimization problems over $H^1(\Omega)$ is then posed as

Find $\Phi_\varepsilon \in H^1(\Omega)$ such that

$$J_\varepsilon(\Phi_\varepsilon) = \text{infimum}\{J_\varepsilon(\phi): \phi \in H^1(\Omega)\} \tag{4.20}$$

Referring to Remark 2, we see that for any $\varepsilon \in (0, \infty)$ there exists a positive real number $\lambda(\varepsilon)$ such that

$$\int_\Omega K|\nabla\phi|^2 dx + \frac{1}{\varepsilon}\langle\Delta, \phi\rangle^2 \geq \lambda(\varepsilon)\,\|\phi\|^2_{H^1(\Omega)} \tag{4.21}$$

Moreover, it is clear that for $1/(\mu)\varepsilon > 0$, there exists a positive number λ_0 independent of ε such that

$$\int_\Omega K|\nabla\phi|^2 dx + \frac{1}{\varepsilon}\langle\Delta, \phi\rangle^2 \geq \lambda_0\,\|\phi\|^2_{H^1(\Omega)} \tag{4.22}$$

The weak formulation of Eq. (4.20) is given as follows:
Find $\Phi \in H^1(\Omega)$ such that

$$\int_\Omega K\nabla\Phi \cdot \phi\,dx + \frac{1}{\varepsilon}\langle\Delta, \phi\rangle\langle\Delta, \phi\rangle = \int_\Omega f\phi\,dx \tag{4.23}$$

for any $\phi \in H^1(\Omega)$.

Proposition 5   For any $\varepsilon > 0$ there exists a unique solution Φ_ε to (4.17).
The limit behavior of the family of solutions $\{\Phi_\varepsilon: \varepsilon \in (0, \infty)\}$ as $\varepsilon \to 0$ is established in the following:

Proposition 6   Suppose that $\{\varepsilon_i\}_{i=1}^\infty$ is a sequence of positive real numbers such that $\varepsilon_i \to 0$ as $i \to \infty$. Then the sequence of solutions $\{\Phi_{\varepsilon_i}\}_{i=1}^\infty$ converges to Φ_0 in $H^1(\Omega)$. Moreover, $\langle\Delta, \Phi_\varepsilon\rangle = O(\varepsilon)$

We prove a slightly more general result. Define the functionals on $H^1(\Omega)$ by

$$J_\varepsilon(\phi, K) = \int_\Omega (K|\nabla\phi|^2 - 2f\phi)\,dx + \frac{1}{\varepsilon}\langle\Delta, \phi\rangle^2 \tag{4.24}$$

and

$$J_0(\phi, K) = \int_\Omega (K|\nabla\phi|^2 - 2f\phi)\,dx \tag{4.25}$$

so that

$$J_\varepsilon(\phi, K) = J_0(\Phi, K) + \frac{1}{\varepsilon}\langle\Delta, \phi\rangle^2. \tag{4.26}$$

Formulate the family of minimization problems for each $\varepsilon > 0$ by
Find $\Phi(\varepsilon, K) \in H^1(\Omega)$ such that

$$J_\varepsilon(\Phi(\varepsilon, K), K) = \text{infimum}\{J_\varepsilon(\phi, K): \phi \in H^1(\Omega)\} \tag{4.27}$$

The solution of (4.27)(ii) is denoted by $\Phi(0, K)$. The following convergence result is of importance for approximating the ROLS problem:

Proposition 7 Suppose that $K^{(i)} \to K_0$ in Q and $\varepsilon_i \to 0$ as $i \to \infty$. Set $\Phi_i = \Phi(\varepsilon_i, K_i)$ and $\Phi_0 = \Phi(0, K)$. Then $\Phi_i \to \Phi_0$ in $H^1(\Omega)$ and $\langle \Delta, \Phi_i \rangle = O(\varepsilon_i)$.

To specify the ROLS estimation problem, we wish to determine the coefficients k_1, k_2, and k_3 from within an admissible set

$$Q_{ad} = \{K = (k_1, k_2, k_3): k_i \in H^2(\Omega) \text{ and } k_i \geq \mu > 0\} \tag{4.28}$$

from measurements $k_{ij}^{(0)}$ of k_j at the locations (x_i, y_i, z_i) for $i = 1, \ldots, N_0$ with $j = 1, 2, 3$ and $(x_i, y_i, z_i) \in \Omega$. Recall that for Ω as specified, $H^2(\Omega)$ embeds compactly in $C^0(\Omega)$, (cf. Gilbarg and Trudinger, 1983; Ladyzhenskaya and Ural'tseva, 1968). We view $k_j^{(0)}$ as an N_0 column vector for each $j = 1, 2, 3$. In addition we also assume that measurements $\Phi_i^{(0)}$, for $i = 1, \ldots, N_0$ of Φ are available at these locations as well. These observations are modeled by means of continuous linear functionals defined on the spaces $H^1(\Omega)$ and $H^2(\Omega)$ (Parker, 1994). The action of these linear functionals are combined in such a way as to define operators from $H^1(\Omega)$ and $H^2(\Omega)$ into $\mathbf{R}^{N_0}$. Thus, suppose that there are given continuous linear functionals $\{\Delta_n\}_{n=1}^{N_0}$ on $H^1(\Omega)$ and $\{\Theta_n\}_{n=1}^{N_1}$ on $H^2(\Omega)$ to serve as observation functionals of Φ and k_i, $i = 1, 2, 3$ (Parker, 1994). From these functionals, we construct the operators $C_0: H^1(\Omega) \mapsto Z = \mathbf{R}^{N_0}$ and $C_1: H^2(\Omega) \mapsto Z = \mathbf{R}^{N_0}$ as

$$C_0 \Phi = \begin{bmatrix} \langle \Delta_1, \Phi \rangle \\ \cdot \\ \cdot \\ \cdot \\ \langle \Delta_{N_0}, \Phi \rangle \end{bmatrix} \tag{4.29}$$

and

$$C_1 \psi = \begin{bmatrix} \langle \Theta_1, \psi \rangle \\ \cdot \\ \cdot \\ \cdot \\ \langle \Theta_{N_1}, \psi \rangle \end{bmatrix} \tag{4.30}$$

respectively.

We define the fit-to-data functional $F: Q_{ad} \mapsto \mathbf{R}$ by

$$F(K) = |(C_0 \Phi(K) - \Phi^{(0)}|^2 + N(K) \tag{4.31}$$

where

$$N(K) = \sum_{j=1}^{3} \left\{ |C_1 k_j - k_j^{(0)}|^2 + \beta_1 \left\| \nabla k_j \right\|^2 + \beta_2 \left\| D^2 k_j \right\|^2 \right\} \tag{4.32}$$

The ROLS problem is stated in terms of the minimization problem:
Find $K_0 \in Q_{ad}$ such that

$$F(K_0) = \text{infimum}\{F(K): K \in Q_{ad}\} \tag{4.33}$$

The existence of a solution to (4.33) follows under the assumptions of our formulation and Proposition 3.

Proposition 8 The ROLS problem (4.33) has a solution.

Remark 9 By a similar argument, existence of ROLS estimators may be established for the penalized ε models as well. Denote a ROLS estimator for the ε problem by K_ε. It is easily demonstrated using Proposition 7 that if $\{\varepsilon_i\}_{i=1}^\infty$ is such that $\varepsilon_i \to 0$ as $i \to \infty$ and $K_i = K_{\varepsilon_i}$ are associated ROLS estimators, then $\{K_i\}_{i=1}^\infty$ is weakly sequentially compact in $H^2(\Omega)^3$ and any weak cluster point such that $K_i \to K_0$ is a ROLS estimator for the constrained problem (4.33).

4.3 Finite Dimensional ROLS

Having laid a basic theoretical framework for existence of solutions to the ROLS problem, we now present the finite dimensional formulation for the ROLS estimation procedure. Let ε be a fixed positive number and consider the variational equation

$$\int_\Omega K \nabla \Phi \cdot \nabla \phi \, dx + \frac{1}{\varepsilon} \langle \Delta, \Phi \rangle \langle \Delta, \phi \rangle = \int_\Omega f \phi \, dx \tag{4.34}$$

Suppose that $\{B_j\}_{j=1}^N$ and $\{b_j\}_{j=1}^M$ are linearly independent functions in $H^1(\Omega)$ and $H^2(\Omega)$, respectively. Express Φ and k_i for $i = 1, 2, 3$ as sums

$$\Phi = \sum_{j=1}^N c_j B_j \tag{4.35}$$

and

$$k_i = \sum_{j=1}^M a_j^{(i)} b_j. \tag{4.36}$$

Given the coefficient matrix K of the form expressed by Eq. (4.3), we seek Φ of the form expressed by Eq. (4.35) such that for a fixed ε

$$\int_\Omega K \nabla \Phi \cdot \nabla B_j \, dx + \frac{1}{\varepsilon} \langle \Delta, \Phi \rangle \langle \Delta, B_j \rangle = \int_\Omega f B_j \, dx \tag{4.37}$$

for $j = 1, \ldots, N$. Using the representations of Φ and $k_1, k_2,$ and k_3 above, Eq. (4.37) takes the form

$$\sum_{i=1}^N c_i \left\{ \sum_{n=1}^M \left[a_n^{(1)} \left(\int_\Omega b_n B_{ix} B_{jx} \, dx \right) + a_n^{(2)} \left(\int_\Omega b_n B_{iy} B_{jy} \, dx \right) \right. \right.$$
$$\left. \left. + a_n^{(3)} \left(\int_\Omega b_n B_{iz} B_{jz} \, dx \right) \right] + \frac{1}{\varepsilon} \langle \Delta, B_i \rangle \langle \Delta, B_j \rangle \right\} = \int_\Omega f B_j \, dx \tag{4.38}$$

Define the matrices for $i, j = 1, 2, \ldots, N$,

$$G_{ij}^{(1,n)} = \int_{\Omega} b_n B_{ix} B_{jx} \, dx \tag{4.39}$$

$$G_{ij}^{(2,n)} = \int_{\Omega} b_n B_{iy} B_{jy} \, dx \tag{4.40}$$

$$G_{ij}^{(3,n)} = \int_{\Omega} b_n B_{iz} B_{jz} \, dx \tag{4.41}$$

$$G_{ij}^{(0)} = \langle \Delta, B_i \rangle \langle \Delta, B_j \rangle \tag{4.42}$$

and the vector (again denoted by f) with components

$$f_j = \int_{\Omega} f B_j \, dx \, dy \, dz \tag{4.43}$$

The system Eq. (4.38) becomes

$$\left[\sum_{n=1}^{M} (a_n^{(1)} G^{(1,n)} + a_n^{(2)} G^{(2,n)} + a_n^{(3)} G^{(3,n)}) + \frac{1}{\varepsilon} G^{(0)} \right] c = f \tag{4.44}$$

The finite dimensional minimization problem is formulated starting from the functional $F(K)$ in Eqs. (4.31)–(4.32) with minimization problem Eq. (4.33).

The finite dimensional formulation of the fit-to-data functional is obtained by introducing the $N_0 \times N$ matrix Φ

$$\Phi_{ij} = \langle \Delta_i, B_j \rangle \tag{4.45}$$

for $i = 1, \ldots, N_0$ and $j = 1, \ldots, N$, the $M \times M$ matrix

$$H_{ij} = \int_{\Omega} \nabla b_i \cdot \nabla b_j \, dx \tag{4.46}$$

for $i, j = 1, \ldots, M$, the $N_1 \times M$ matrix

$$\Psi_{ij} = \langle \Theta_i, b_j \rangle \tag{4.47}$$

for $i = 1, \ldots, N_1$ and $j = 1, \ldots, M$, the column N_0 pressure measurement vector

$$\Phi^{(0)} = \begin{bmatrix} \Phi_1^{(0)} \\ \cdot \\ \cdot \\ \cdot \\ \Phi_{N_0}^{(0)} \end{bmatrix} \tag{4.48}$$

the column N_1 permeability measurement vectors

$$\mathbf{K}^{(i)} = \begin{bmatrix} k_{i1}^{(0)} \\ \cdot \\ \cdot \\ \cdot \\ k_{iN_1}^{(0)} \end{bmatrix} \tag{4.49}$$

for $i = 1, 2, 3$, and the M column vector

$$\mathbf{a}^{(i)} = \begin{bmatrix} a_1^{(i)} \\ \cdot \\ \cdot \\ \cdot \\ a_M^{(i)} \end{bmatrix} \tag{4.50}$$

Define the $M \times 3$ matrix

$$A = [\mathbf{a}^{(1)}, \mathbf{a}^{(2)}, \mathbf{a}^{(3)}] \tag{4.51}$$

Let

$$\Phi_2 = \Phi^* \Phi \text{ and } \Psi_2 = \Psi^* \Psi \tag{4.52}$$

where * denotes transposition. We also set

$$G(\mathbf{a}^{(1)}, \mathbf{a}^{(2)}, \mathbf{a}^{(3)}) = \sum_{n=1}^{M} (a_n^{(1)} G^{(1,n)} + a_n^{(2)} G^{(2,n)} + a_n^{(3)} G^{(3,n)}) \tag{4.53}$$

The functional $F(\cdot)$ may thus be viewed as being defined on $(\mathbf{R}^M)^3$, and is expressed as

$$F(A) = c^* \Phi_2 c - 2\Phi^{(0)*} \Phi c + |\Phi^{(0)}|^2 + \mathbf{a}^* (H + \Psi_2) \mathbf{a} - 2\mathbf{K}^* \Psi \mathbf{a} + |\mathbf{K}|^2 \tag{4.54}$$

where N-vector c satisfies the equation

$$\left(G(A) + \frac{1}{\varepsilon} G^{(0)} \right) c = f \tag{4.55}$$

We seek that coefficient matrix A minimizing the functional $F(A)$ of Eq. (4.54) subject to the condition Eq. (4.55). Of course, there are further constraints on the matrix A reflecting the constraints in the definition of the admissible set Q_{ad}. However, in test problems we treat the minimization as an unconstrained problem (White, 1997). In fact, a simple steepest descent method may be used for minimization. Towards this end, we differentiate the functional $F(A)$ to obtain the partial Frechet derivatives for $j = 1, 2, 3$

$$D_j F(A) \delta^{(j)} = [\Phi_2 c - \Phi^* \Phi^{(0)}]^* D_j c(A) \delta^{(j)} + [(\Psi_2 + H) \mathbf{a}^{(j)} - \Psi^* K^{(j)}] \delta^{(j)} \tag{4.56}$$

where $D_j c(A) \delta^{(j)}$ is the solution of the equation

$$\left[G(A) + \frac{1}{\varepsilon} G^{(0)} \right] [D_j c(A) \delta^{(j)}] = -\sum_{n=1}^{M} \delta_n^{(j)} G^{(j,n)} c(A). \tag{4.57}$$

Introduce the equation

$$\left[G(A) + \frac{1}{\varepsilon} G^{(0)} \right] \pi(A) = \Phi_2 c(A) - \Phi^* \Phi^{(0)} \tag{4.58}$$

and the column M-vector

$$\Xi^{(j)}(A) = \begin{bmatrix} \pi(A)^* G^{(j,1)} c(A) \\ \cdot \\ \cdot \\ \pi(A) G^{(j,M)} c(A) \end{bmatrix} \tag{4.59}$$

Premultiplying Eq. (4.57) by $\pi(A)^*$ and using (4.58), we may express the partial derivative of the functional F with respect to $\mathbf{a}^{(j)}$ for $j = 1, 2, 3$ as follows:

$$D_j F(A) = (\Psi_2 + H)\mathbf{a}^{(j)} - \Psi^* \mathbf{K}^{(j)} - \Xi^{(j)}(A) \tag{4.60}$$

We note that the vector $\pi(A)$ need only be calculated once, as it is independent of j. The derivatives of F are then obtained by means of matrix multiplications.

4.4 Estimation of the Discontinuity

Having detected a discontinuity by means of the regularized output least squares method (actually by any procedure), our next step is to isolate and obtain some estimate of it. We proceed by considering an example in which the admissible permeability functions k_i, $i = 1, 2, 3$ have a discontinuity determined by two regions within Ω parameterized by two real numbers a and b. We denote these two regions by $\Omega(a, b)$ and $\Omega \backslash \Omega(a, b)$. Actually, the region $\Omega(a, b)$ also involves a third parameter c measuring the extent of the three dimensional subdomain in the z-direction. However, for convenience we will assume that c can be determined directly from the data. An admissible permeability function k_i, for $i = 1, 2, 3$, is also parameterized constants k_{i1} and k_{i2} modeling the magnitude of the parameter in the two regions. Hence, k_i takes the form

$$k_i(x, y, z) = k_{i0}(x, y, z) + k_{ie}(x, y, z) \tag{4.61}$$

where

$$k_{ie}(x, y) = k_{i1} \text{ if } (x, y) \in \Omega(a, b), \quad \text{and } k_{i2} \text{ otherwise} \tag{4.62}$$

It is assumed that the function k_{i0} is known. The function k_{i0} represents the ROLS estimate capturing background behavior. To fix ideas, let us suppose that $\Omega(a, b)$ is a rectangular solid of the form $(x_0 - a, x_0 + a) \times (y_0 - b, y_0 + b) \times (z_0 - c, z_0 + c)$. However, we assume that in fact c is known from core data. Introduce the characteristic function $(x, y, z) \mapsto \xi(x, y, z; a, b)$ of the set $\Omega(a, b)$, where

$$\xi(x, y, z; a, b) = 1 \text{ if } (x, y, z) \in \Omega(a, b), \text{ and } 0 \text{ otherwise} \tag{4.63}$$

We now express

$$k_{ie}(x, y, z) = (k_{i1} - k_{i2})\xi(x, y, z; a, b) + k_{i2} \tag{4.64}$$

The stiffness matrices are given as

$$(G_l)_{ij} = \int_\Omega k_{l2} \nabla B_i \cdot \nabla B_j \, dx + (k_{l1} - k_{l2}) \int_{z_0-c}^{z_0+c} \int_{x_0-a}^{x_0+a} \int_{y_0-b}^{y_0+b} \nabla B_i \cdot \nabla B_j \, dx \tag{4.65}$$

for $l = 1, 2, 3$. Setting

$$(G_0)_{ij} = \int_\Omega \nabla B_i \cdot \nabla B_j \, dx \tag{4.66}$$

$$(G_{l1})_{ij} = \int_\Omega k_{l0} \nabla B_i \cdot \nabla B_j \, dx \tag{4.67}$$

and

$$(G_{12})_{ij}(a, b) = \int_{z_0-c}^{z_0+c} \int_{x_0-a}^{x_0+a} \int_{y_0-b}^{y_0+b} \nabla B_i \cdot \nabla B_j \, dx \tag{4.68}$$

we obtain an equation

$$\left\{ \sum_{l=1}^{3} [G_{l1} + (k_{l1} - k_{l2})G_{12}(a, b) + k_{l2}G_0] + \frac{1}{\varepsilon} G^{(0)} \right\} c = f \tag{4.69}$$

with the approximating solution Φ expressed as $\Phi = \sum_{i=1}^{N} c_i B_i$. Note that $k_{l1}, k_{l2}, a,$ and b are unknowns to be determined from the data.

Define the following functionals:

$$j_0(a, b, K) = \sum_{j=1}^{N_0} (\langle \Delta_j, \Phi \rangle - \Phi_j^{(0)})^2 \tag{4.70}$$

and

$$\mathcal{N}(a, b, K) = \sum_{i=1}^{3} \sum_{j=1}^{N_1} (\langle \Theta_j, k_i \rangle - k_{ij}^{(0)})^2 \tag{4.71}$$

Also, define the column N_1 vector functions Ψ_1 and Ψ_2 of a and b by

$$\Psi_1(a, b)_j = \langle \Theta_j, \xi(\cdot; a, b) \rangle \tag{4.72}$$

and

$$\Psi_2(a, b)_j = \langle \Theta_j, 1 - \xi(\cdot; a, b) \rangle \tag{4.73}$$

and the vector κ_i for $i = 1, 2, 3$ by

$$\kappa_{ij} = \langle \Theta_j, k_{i0} \rangle - k_{ij}^{(0)} \tag{4.74}$$

The functional $\mathcal{N}(a, b, K)$ may be written as

$$\mathcal{N}(a, b, K) = \sum_{i=1}^{3} (k_{i1} \Psi_1(a, b) + k_{i2} \Psi_2(a, b) + \kappa_i)^* (k_{i1} \Psi_1(a, b) + k_{i2} \Psi_2(a, b) + \kappa_i) \tag{4.75}$$

Since the basis functions used to approximate the state are the same as in the previous section, the functional j_0 takes the form

$$j_0(a, b, K) = c^* \Phi_2 c - 2\Phi^{(0)*} \Phi_c + |\Phi^{(0)}|^2 \tag{4.76}$$

With a and b fixed, $\mathcal{N}$ is differentiable with respect to the remaining variables. Define the criterion function

$$j(a, b, K) = j_0(a, b, K) + \mathcal{N}(a, b, K) \tag{4.77}$$

We look for a and b in a specified range that is obtained from information determined from the ROLS step, say $a \in [a_0, a_1]$ and $b \in [b_0, b_1]$. In fact, we allow a and b to take on values

$$a = a_0 + (i - 1)\frac{(a_1 - a_0)}{L_1} \tag{4.78}$$

for $i = 1, \ldots, L_1 + 1$ and

$$b = b_0 + (i - 1)\frac{(b_1 - b_0)}{L_2} \tag{4.79}$$

for $i = 1, \ldots, L_2 + 1$. Fixing the values of a and b, we then minimize the functional $j(a, b, K)$ with respect to K. This step may be accomplished by means of a steepest descent method. The solution is taken to be the a, b, and K giving the smallest value of the criterion. The derivatives may be calculated in a straightforward manner similar to those in the previous section.

4.5   A Numerical Example

We consider problem in which we specify a coefficient matrix $K(x, y)$ and generate pressure data from the associated solution by solving the problem Eqs. (4.2)–(4.4) for $\Phi = p$ by finite elements for a specific forcing function f. Using this data we then attempt to recover K. We first consider a problem in two spatial dimensions. Towards this end, let $\mathcal{D} = (0, 1) \times (0, 1)$, and suppose that measurements of pressure and permeability can be made at locations (0.175, 0.175), (0.835, 0.175), (0.5, 0.5), (0.175, 0.835), and (0.835, 0.835). For a test permeability function we use the following:

$$K_{\text{test}}(x, y) = \begin{cases} 8 + 3.5\cos(x + y), & \text{for } (x, y) \in (0, 0.3) \times (0, 0.3) \\ 4 + 2.5\cos(x + y - 2), & \text{for } (x, y) \in (0.75, 1) \times (0.75, 1) \\ 2 + \cos(x + y), & \text{otherwise} \end{cases} \tag{4.80}$$

shown in Figure 4.2. This function is motivated by permeability mappings arising in connection with fluvial domains in which regions with higher permeability are separated by a region of lower permeability. The high permeability regions result from the formation of sandbars while the lower permeability areas result from the silting of the river. Further, we suppose that $p = 0$ at the point (0.835, 0.835) and that fluid is injected at the point (0.175, 0.175). The resulting pressure function obtained by means of a finite element

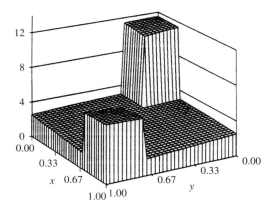

Figure 4.2   Test permeability

solution is portrayed in Figure 4.3. For the approximations to the pressure, we use tensor products of cubic B-splines (Schultz, 1973) defined on a uniform mesh determined by subdividing (0,1) into seven subintervals. Since imposing Neumann boundary conditions improves accuracy, we use 64 basis functions for approximating pressure adjusted to incorporate the Neumann boundary condition. For approximating the parameter, we again use tensor products of cubic B-splines but defined on a mesh determined by subdividing (0,1) into five equal subintervals. Imposing no boundary conditions, we then use 64 basis functions to approximate the parameter.

Using data at the observation points, we apply the (ROLS) method as a detection procedure. The regularizing parameters β_1 and β_2 are set at 10^{-4}. These parameters are determined to be small, but sufficiently large that the Hessian be positive definite. The results are in Figure 4.4.

We now search with the (IA) method for a coefficient of the form

$$K_1(x, y) = \begin{cases} k_1 \text{ if } (x, y) \in (0, a) \times (0, b) \\ k \text{ otherwise} \end{cases} \tag{4.81}$$

using the technique discussed in the previous section. The result is illustrated in Figure 4.5.

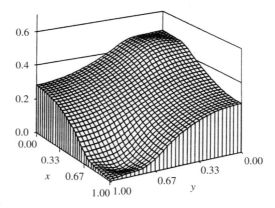

Figure 4.3   Test pressure

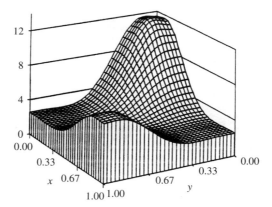

Figure 4.4   ROLS estimate of permeability

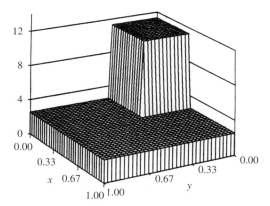

Figure 4.5   A first estimate of permeability

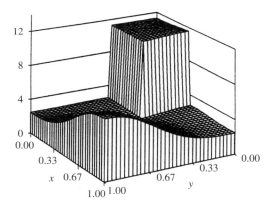

Figure 4.6   ROLS update of the permeability

Apply further detection by again using the (ROLS) method to estimate the coefficient K_D where the permeability has the form

$$K(x, y) = K_1(x, y) + K_D(x, y) \tag{4.82}$$

The result is portrayed in Figure 4.6.

Again, use the (IA) procedure this time by searching for discontinuities based on the rectangle $(b, 1) \times (b, 1)$. This yields the Figure 4.7.

Use the (ROLS) method to look for further discontinuities. This step results detects no substantial variations. Hence, the (ROLS) solution determines the background. The outcome of this step is depicted in Figure 4.8.

Finally, the resulting pressure from the estimated coefficient is portrayed in Figure 4.9.

As a three dimensional example, we attempt to recover a diagonal matrix-valued permeability function K from permeability and pressure measurements. Let $\Omega = \mathcal{D} \times (0, 1)$ and

$$\Omega(a, b) = (0.25, 0.75) \times (0.25, 0.75) \times (0.25, 0.75)$$

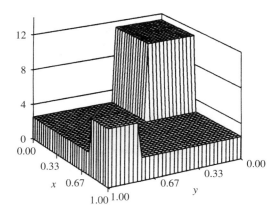

Figure 4.7 IA update of the permeability

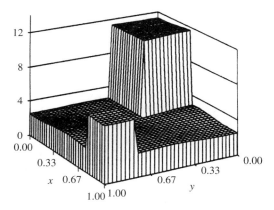

Figure 4.8 ROLS update of the permeability (final estimate)

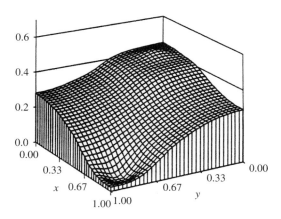

Figure 4.9 Estimated pressure

Suppose that measurements of pressure and permeability are made at locations

$(0.175, 0.175, z), (0.835, 0.175, z), (0.5, 0.5, z), (0.175, 0.835, z)$ and $(0.835, 0.835, z)$

For a test permeability matrix function we use the following:

$$k_{1\,\text{test}}(x, y, z) = \begin{cases} 5 + \sin \pi(x + y + z) \text{ for } (x, y, z) \in \Omega/\Omega(a, b) \\ 10 + 2 \sin \pi(x + y + z), \text{ for } (x, y, z) \in \Omega(a, b) \end{cases} \qquad (4.83)$$

$$k_{2\,\text{test}}(x, y, z) = \begin{cases} 6 + \cos \pi(x + y + z) \text{ for } (x, y, z) \in \Omega/\Omega(a, b) \\ 11 + 2 \cos \pi(x + y + z), \text{ for } (x, y, z) \in \Omega(a, b) \end{cases} \qquad (4.84)$$

$$k_{3\,\text{test}}(x, y, z) = \begin{cases} 7 + \sin \pi(x + y + z) \text{ for } (x, y, z) \in \Omega/\Omega(a, b) \\ 12 + 2 \sin \pi(x + y + z), \text{ for } (x, y, z) \in \Omega(a, b) \end{cases} \qquad (4.85)$$

shown in Figures 4.10–4.12. In the plots we have graphed all functions as functions of x and y on $\mathcal{D}$ with $z = 1/2$. Further, we suppose that $p = 0$, that is, $\Phi = z$, at the point

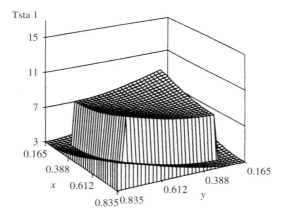

Figure 4.10 $k_{1\,\text{test}}(X, Y, Z), Z = 0.5$

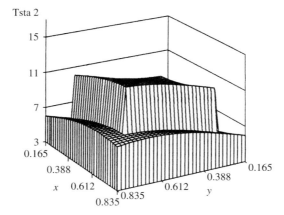

Figure 4.11 $k_{2\,\text{test}}(X, Y, Z), Z = 0.5$

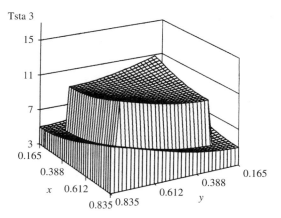

Figure 4.12 $k_{3\,\text{test}}(X, Y, Z), Z = 0.5$

(0.835, 0.835). It is also supposed that fluid is injected along the segment $(0.175, 0.175, z)$ for $z \in (0, 1)$. For the approximations to the pressure, we use tensor products of cubic B-splines (Schultz, 1973) defined on a uniform mesh determined by subdividing $(0,1)$ into 4 subintervals. Since imposing Neumann boundary conditions improves accuracy, we use $N = 343$ basis functions $\{B_j\}_{j=1}^{N}$ for approximating pressure adjusted to incorporate the Neumann boundary condition. For approximating the parameter, we again use tensor products of cubic B-splines, but defined on a mesh determined by subdividing $(0,1)$ into three equal subintervals. We use $M = 216$ basis functions $\{b_j\}_{j=1}^{M}$ to approximate each the three parameter functions. Data is obtained by making five measurements at each of the five wells of each of the components of the permeability matrix and the pressure at the observation wells. Thus, 100 measurements (i.e. 75 measurements of the parameters and 25 of the pressure) are to be used to determine 648 variables in the parameter functions.

Using the regularized output least squares method as a detection procedure to obtain $k_{i0}(x, y, z)$ results in Figures 4.13–4.15. Based on this result, we search for a coefficient

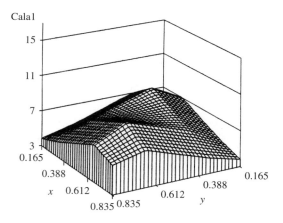

Figure 4.13   ROLS estimate of $k_1, Z = 0.5$

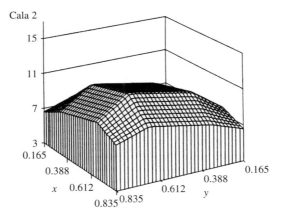

Figure 4.14 ROLS estimate of k_2, $Z = 0.5$

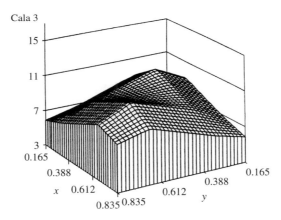

Figure 4.15 ROLS estimate of k_3, $Z = 0.5$

of the form

$$k_i(x, y, z) = k_{i0}(x, y, z) + k_{ie}(x, y, z) \tag{4.86}$$

for $i = 1, 2, 3$ using the technique discussed in Section 4.4 to estimate k_{ie}. Based on the measurements we take $c = 0.25$ to be known. Thus, in essence, the cross section of the set $\Omega(a, b)$ is to be determined from the pressure data. The results are depicted in Figures 4.16–4.18.

To aid in comparison in this case, we introduce the relative error. The relative error in an estimated quantity K by

$$\text{rel error } (K) = \frac{||K_{\text{test}} - K_{\text{test}}||}{||K_{\text{test}}||} \tag{4.87}$$

we find that the relative error after the ROLS step is

$$\text{rel error } (K) = 0.17 \tag{4.88}$$

$$\text{rel error } (\Phi) = 0.0758 \tag{4.89}$$

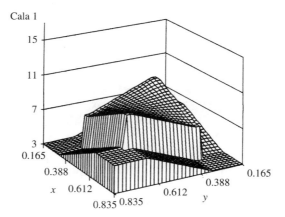

Figure 4.16   IA update of k_1, $Z = 0.5$

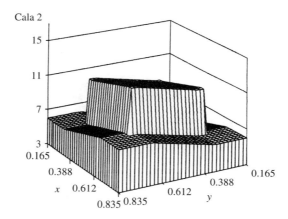

Figure 4.17   IA update of k_2, $Z = 0.5$

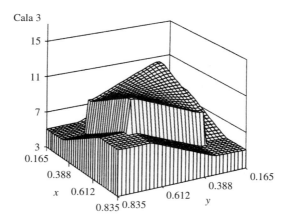

Figure 4.18   IA update of k_3, $Z = 0.5$

On the other hand, the relative error after the (IA) step is found to be

$$\text{rel error } (K) = 0.083 \tag{4.90}$$

$$\text{rel error } (\boldsymbol{\Phi}) = 0.055 \tag{4.91}$$

4.6 Conclusions

We have applied the output least square estimation technique as a detection tool for the estimation of a discontinuous permeability tensor using permeability measurements and pressure data. In addition, we introduced a method to estimate the location and magnitude of a jump discontinuity. We presented a numerical example for the location of discontinuities in a permeability function in the presence of a background. By alternating detection and discontinuity estimation procedures if necessary, it seems to be possible to construct coefficients with discontinuities in the presence of a background function.

Appendix: Proof of the Main Results

For convenience we collect the proofs of the main results.

Proposition 3 If $K \in Q_0$, then there exists a unique solution $\boldsymbol{\Phi}(K) \in V_0$. Furthermore, the mapping $K \mapsto \boldsymbol{\Phi}(K)$ is continuous from Q_0 with the norm topology into V_0 with the norm topology.

Proof We begin by observing that for any $K \in Q_0$ the existence of a unique solution is well known. Moreover, the estimate

$$||\boldsymbol{\Phi}(K)||\, V_0 \le \frac{\Gamma_0}{\mu}\, ||f|| \tag{4.92}$$

holds for any such $K \in Q_0$. Suppose that $\{K^{(i)}\}_{i=1}^{\infty} \subset Q_0$ has the property that $K^{(i)} \to K$ in Q as $i \to \infty$. It follows that $K \in Q_0$. Hence, there is a unique solution $\boldsymbol{\Phi}(K) \in V_0$ associated with K. To show that $\boldsymbol{\Phi}(K^{(i)}) \to \boldsymbol{\Phi}(K)$ weakly in V_0 as $i \to \infty$, we note from the above estimate there is a subsequence again $\boldsymbol{\Phi}_i = \boldsymbol{\Phi}(K^{(i)})$ such that $\boldsymbol{\Phi}(K^{(i)}) \to \boldsymbol{\Phi}$ weakly in V_0 and $\boldsymbol{\Phi} \in V_0$. Suppose that $\boldsymbol{\Phi} \in V_0$ then

$$\int_{\Omega} K^{(i)} \nabla \boldsymbol{\Phi}_i \cdot \nabla \phi \, dx \to \int_{\Omega} K \nabla \boldsymbol{\Phi} \cdot \nabla \phi \, dx \tag{4.93}$$

it follows that

$$\int_{\Omega} K \nabla \boldsymbol{\Phi} \cdot \nabla \phi \, dx = \int_{\Omega} f \boldsymbol{\Phi} \, dx \tag{4.94}$$

for any $\phi \in V_0$. Thus, from uniqueness $\boldsymbol{\Phi} = \boldsymbol{\Phi}(K)$, and furthermore, the sequence entire sequence converges to $\boldsymbol{\Phi}$.

Norm convergence in V_0 may be seen as follows. Note that for any $\phi \in V_0$,

$$\int_{\Omega} (K^{(i)} \nabla \boldsymbol{\Phi}_i - K \nabla \boldsymbol{\Phi}) \cdot \nabla \phi \, dx = 0 \tag{4.95}$$

for any i. Rewriting and setting $\phi = \Phi_i - \Phi$, we obtain

$$\int_\Omega K^{(i)} |\nabla(\Phi_i - \Phi)|^2 dx = \int (K - K^{(i)}) \nabla\Phi \cdot \nabla(\Phi_i - \Phi) \, dx \qquad (4.96)$$

The estimate

$$||\Phi_i - \Phi|| \, V_0 \leq \frac{\gamma_0 \, ||f||}{\mu^2} ||K^{(i)} - K||_Q \qquad (4.97)$$

implies the convergence in V_0.

Proposition 7   Suppose that $K^{(i)} \to K_0$ in Q and $\varepsilon_i \to 0$ as $i \to \infty$. Set $\Phi_i = \Phi(\varepsilon_i, K_i)$ and $\Phi_0 = \Phi(0, K)$. Then $\Phi_i \to \Phi_0$ in $H^1(\Omega)$ and $\langle \Delta, \Phi_i \rangle = O(\varepsilon_i)$.

Proof   Note that for $1/(\mu) \geq \varepsilon$ and $K \in Q_0$

$$J_\varepsilon(\phi, K) \geq \lambda_0 \, ||\phi||^2_{H^1(\Omega)} - 2 \, ||f|| \, ||\phi||_{H^1(\Omega)} \qquad (4.98)$$

Let $\Phi_0 \in V_0 \subset H^1(\Omega)$ be such that

$$J(\Phi_0, K_0) = \text{ infimum}\{J_0(\phi, K_0): \phi \in V_0\} \qquad (4.99)$$

Then

$$J_{\varepsilon_i}(\Phi_0, K^{(i)}) = J_0(\Phi_0, K^{(i)}) \qquad (4.100)$$

$$\geq J_{\varepsilon_i}(\Phi_i, K^{(i)}) \geq \lambda \, ||\Phi_i||^2_{H^1(\Omega)} - 2 \, ||f|| \, ||\Phi_i||_{H^1(\Omega)} \qquad (4.101)$$

Since

$$J_0(\Phi_0, K^{(i)}) = \int_\Omega (K^{(i)} |\nabla\Phi_0|^2 - 2f\Phi_0) \, dx \to \int_\Omega (K_0 |\nabla\Phi_0|^2 - 2f\Phi_0) \, dx \qquad (4.102)$$

it follows that there exists a positive constant M_0 such that for any i

$$M_0 \geq ||\Phi_i||_{H^1(\Omega)} \qquad (4.103)$$

Hence, there is a subsequence again denoted by $\{\Phi_i\}^\infty_{i=1}$ such that $\Phi_i \to \Phi$ weakly in $H^1(\Omega)$. Moreover, since $K^{(i)} \to K_0$ in Q, it follows from the convergence of

$$\varepsilon_i \int_\Omega (K^{(i)} \nabla\Phi_i \cdot \nabla\phi - f\phi) \, dx + \langle \Delta, \Phi_i \rangle \langle \Delta, \Phi \rangle = 0 \qquad (4.104)$$

for any $\phi \in H^1(\Omega)$ that

$$\langle \Delta, \Phi_i \rangle = O(\varepsilon_i) \qquad (4.105)$$

and the limit Φ satisfies

$$\Phi \in V_0 \qquad (4.106)$$

On the other hand, if $\Phi \in V_0$, taking the limit of

$$\int_\Omega (K_0 \nabla\Phi_i \cdot \nabla\phi - f\phi) \, dx = 0 \qquad (4.107)$$

where $\phi \in V_0$ implies

$$\Phi = \Phi_0 \tag{4.108}$$

by uniqueness. Accordingly, it follows that the sequence $\{\Phi_i\}_{i=1}^{\infty}$ converges and

$$\Phi_i \to \Phi_0 \text{ weakly in } H^1(\Omega) \tag{4.109}$$

To obtain convergence in $H^1(\Omega)$, set $\phi = \Phi_i - \Phi_0$ in Eq. (4.23) and regroup to obtain

$$\int_{\Omega} [K^{(i)}\nabla\Phi_i \cdot \nabla\Phi_i - f\Phi_i] \, dx = \frac{1}{\varepsilon_i} \langle \Delta, \Phi_i \rangle^2 + \int_{\Omega} (K^{(i)}\nabla\Phi_i \cdot \nabla\Phi_0 - f\Phi_0) \, dx \tag{4.110}$$

Taking the limit, we see that

$$\int_{\Omega} K^{(i)}|\nabla\Phi_i|^2 dx \to \int_{\Omega} K_0|\nabla\Phi_0|^2 dx \tag{4.111}$$

Thus, from

$$\int_{\Omega} K^{(i)}|\nabla(\Phi_i - \Phi_0)|^2 dx + \frac{1}{\varepsilon_i} \langle \Delta, \Phi_i \rangle^2 \geq \lambda \, ||\Phi_i - \Phi_0||^2_{H^1(\Omega)} \tag{4.112}$$

and

$$\int_{\Omega} K^{(i)}|\nabla(\Phi_i - \Phi_0)|^2 dx = \int_{\Omega} K^{(i)}|\nabla\Phi_i|^2 dx - 2\int_{\Omega} K^{(i)}\nabla\Phi_i \cdot \nabla\Phi_0 \, dx$$
$$+ \int_{\Omega} K^{(i)}|\nabla\Phi_0|^2 dx \tag{4.113}$$

we obtain the convergence of the sequence $\{\Phi_i\}_{i=1}^{\infty}$ to Φ_0 in $H^1(\Omega)$.

Proposition 8 The ROLS problem Eq. (4.33) has a solution.

Proof Existence is a consequence of the observation that $K \mapsto F(K)$ is lower semi-continuous with respect to the weak topology on $H^2(\Omega)^3$, and the observation that the set

$$Q_{ad} \cap \{K : d + 1 \geq F(K)\} \tag{4.114}$$

where $d = \text{infimum}\{F(K) : K \in Q_{ad}\}$ is weakly compact in $H^2(\Omega)^3$.

References

Banks, H.T. and Kunisch, K. *Estimation Techniques for Distributed Parameter Systems.* Birkhauser, Boston, (1989).

Chavent, G. and Jaffre, J. *Mathematical Models and Finite Elements for Reservoir Simulation.* North Holland, New York, (1986).

Gilbarg, D. and Trudinger, N. *Elliptic Partial Differential Equations of Second Order, 2nd ed.* Springer-Verlag, New York, (1983).

Gutman, S. Identification of discontinuous parameters in flow equations. *SIAM J. Cont. and Opt.,* **28**, (1990) 1049–1060.

Ladyzhenskaya, O.A. and Ural'tseva, N. *Linear and Quasilinear Elliptic Equations.* Academic Press, New York, (1968).

Maz'ja, V. *Sobolev Spaces.* Springer-Verlag, New York, (1985).

Mitasova, H. and Mitas, L. Interpolation by regularized splines with tension: I. Theory and implementation. *Math. Geology,* **25**, (1991) 641–655.

Neri, U. *Singular Integrals.* Springer-Verlag. New York, (1971).

Parker, R.L. *Geophysical Inverse Theory.* Princeton University Press, Princeton, NJ, (1994).

Peaceman, D. *Fundamentals of Numerical Reservoir Simulation.* Elsevier, New York, (1977).

Schultz, M. *Spline Analysis.* Prentice Hall, Englewood Cliffs, NJ, (1973).

White, L. and Zhou, J. Continuity and uniqueness of regularized output least squares optimal estimations. *J. Math Anal. Applic.,* **196**, (1995) 53–83.

White, L. Resolution of regularized output least squares estimation procedures. *J. Appl. Math. Comp.,* **81**, (1997) 139–172.

PART 2

Constitutive Modeling and Testing

5

Constitutive and Numerical Modeling of Liquefaction

M. Pastor[1], **A. H. C. Chan**[2] and **O. C. Zienkiewicz**[3]

[1] *CETA/CEDEX Madrid, Spain*
[2] *University of Birmingham, UK*
[3] *University of Wales, Swansea, UK*

5.1   Introduction

Saturated soils and other two phase media have been the subject of much investigation, both experimentally and numerically, over a number of years. Soils in particular have received considerable attention because of the vital role they play in foundations, dams, dykes and pavement construction. The interaction of soil and pore fluid can be strong enough, due to the build up of the pore pressure, to lead to a catastrophic material softening, a phenomenon known as 'liquefaction'. Liquefaction occurs frequently in saturated loose granular materials under earthquake and other dynamic loading such as blasting. The potential consequences of liquefaction can be illustrated by the near collapse of the Lower San Fernando dam near Los Angeles during the 1971 earthquake (Seed *et al.*, 1975, 1988). This failure fortunately did not involve any loss of life as the level to which the dam 'slumped' still contained the reservoir. Had this been but a few feet lower, the over-topping of the dam would indeed have caused a major catastrophe, with the flood hitting a densely populated area of Los Angeles.

It is evident from the literature (e.g. Seed *et al.*, 1975, 1988) that the example quoted above involved the interaction of pore water pressure and the soil skeleton. Perhaps one particular feature of this interaction that is often overlooked relates to the 'weakening' of the soil-fluid composite during the periodic motion such as that involved in an earthquake. It is this feature, rather than the overall acceleration forces, that caused the slumping of the Lower San Fernando dam. What appears to have happened is that, during the motion, the interstitial pore pressure increased, thus reducing the inter-particle forces in solid phase of the soil and causing a loss of strength.

Granular particles such as sand are liable to compact during shaking. However, this reduction in volume is often prevented by the lack of drainage during the period of shaking. A near undrained condition resulted and the pore pressure was forced to rise to counter such contractive behavior. This results in the reduction of the effective stress and eventually, for loose sand, may lead to a phenomena called 'initial liquefaction', which is defined as zero effective stress state in the soil. On the other hand, for a denser material, the state of zero effective stress state may never occur, and cycles of alternative contraction and dilation may result. This is termed 'cyclic mobility'.

A qualitative, and, if possible, a quantitative prediction of the phenomena leading to permanent deformation or unacceptably high build up of pore pressures, is therefore essential to guarantee the safe behavior of such structures. In the analysis of such dynamic behavior, the usual decoupled and factor of safety approach may not be most appropriate.

For very slow phenomena with adequate drainage, drained static behavior can be assumed. The behavior of the two phases, i.e. soil skeleton (the deformable porous solid) and water (the incompressible pore fluid), decouples and solutions can be found separately for the soil skeleton and pore fluid via usual mechanics and effective stress principles even for nonlinear problems. On the other hand, if the loading is applied very rapidly and drainage is prevented, an assumption of undrained conditions can be made, and the pore pressure can be calculated via the bulk modulus of the fluid, and again a single set of field equations need to be solved. However, under transient consolidation and dynamic conditions, such decoupling does not occur.

Returning to the example of the Lower San Fernando dam, according to the strict sense of factor of safety on force equilibrium, the dam essentially failed because a substantial part of the upstream side of the dam collapsed and slid into the reservoir. This would be the result if a Newmark type sliding block analysis were performed implying that the action due to the earthquake loading exceeded the shear resisting capacity of the soil. However, because the motion was arrested, a complete failure did not occur.

The designer can demand that the factor of safety on force equilibrium should never be allowed to fall below unity. However, this could lead to extremely conservative and costly design. On the other hand, the soil-structure system may be so stiff that a brittle failure may occur at high levels of excitation. By contrast, a more flexible ductile system with adequate energy absorption mechanism which is allowed to move and 'fail' may survive a similar level of excitation with minor damages to the structure and its human inhabitants.

To model liquefaction behavior in a saturated soil/structure system during dynamic excitation, there are three main components:

(a) The establishment of an adequate mathematical framework to describe the phenomenon.

(b) The establishment of a numerical (discrete) approximation procedure.

(c) The establishment of an adequate constitutive relationship for the material behavior.

Each of these items is a major topic on its own, and involves a certain degrees of approximation. There are also different levels of agreement amongst researchers on approaches to be taken in each of these topics. In this chapter, the mathematical framework, the process of numerical approximation and the definition of an adequate constitutive relationship are first presented. The discussion then turns to the validation of the numerical procedure using analytical solutions and physical experiments. Then, the basic equations used in the numerical modeling are introduced, together with the basic information for the numerical analyses. Comparisons with two centrifuge experimental results are given, before the final discussions and conclusions.

5.2   Mathematical Framework for the Dynamic Behavior of Saturated Soil

In this area, there is almost total agreement on the approach to be taken. Taking, for instance, the VELACS project (Arulanandan and Scott, 1993), which stands for the Verification of Liquefaction Analysis by Centrifuge Studies (the project is introduced later in this chapter). Most of the numerical predictions for the centrifuge tests, except the test involving modeling a level ground with single layer of sand, used formulations (Smith, 1994; Chan *et al.*, 1994), which can be traced back to the original Biot's dynamic formulation (Biot, 1956). The Biot formulation provides a mathematical description of physical behavior which can adequately describe the transient behavior of saturated soil for most geomechanics applications, except notably fast pile driving and explosive events.

The basic equations for the Biot formulation are:

(a)   The equilibrium equation of the soil-pore fluid mixture.

(b)   The equilibrium equation for the pore fluid which is a generalization of the Darcy's equation to include the acceleration of the soil skeleton.

(c)   The conservation of mass for the pore fluid.

(d)   The concept of effective stress.

(e)   The constitutive equation.

The formulation emphasized the two phase nature of saturated soil. The effective stress equation emphasized the disparity between the natures of the two media, so the simple distribution of stress between the two media according to the ratio of volume would not be correct. Lastly, the use of two equilibrium equations emphasized that the two phases are acting separately, and a fully drained or an undrained analysis would not be appropriate in general (Zienkiewicz *et al.*, 1980).

In dynamic analysis, a fully drained calculation assumes no change in pore water pressure occurred, usually on the basis that permeability is high. However, a high permeability may lead to a high fluid velocity which, in turn, would be accompanied with a high fluid acceleration. The high level of fluid velocity and acceleration would introduce substantial changes in pore water pressure and invalidate the original assumption of the calculation.

On the other hand, a fully undrained calculation during earthquake shaking could be justified for soils with relatively low permeability such as silt and clay, but not in the case of sand, where drainage can be substantial during the short duration of seismic activity. Furthermore, this assumption of no drainage breaks down during the consolidation stage. To obtain a correct spatial distribution of permanent settlement, the build-up, the redistribution and drainage of the pore water pressure has to be as accurate as possible.

These simplified fully drained and fully undrained analyses were employed, in most cases, due to the limited availability of inexpensive computing power. However, the situation is now changing and, for most of the examples quoted in this chapter, the results could be obtained on a 486DX2 IBM compatible personal computer with less than an hour of calculation.

Only a brief outline of the two-dimensional formulation is given in this section for brevity. Further details can be found in Zienkiewicz *et al.* (1990, 1998) and Chan (1988). The formulation is based on the fully implicit u-p approximation of the Biot (1956) formulation for a saturated porous medium, possibly interacting with a solid component (e.g. a structure).

Using the continuity, momentum and generalized Darcy equations, the complete Biot equation governing deformable porous media can be expressed in two-dimensional Cartesian co-ordinates as

$$\frac{\partial \sigma_{xx}}{\partial x} + \frac{\partial \tau_{xy}}{\partial y} + \rho g_x - \rho \ddot{u}_x - \rho_f \ddot{w}_x = 0 \tag{5.1a}$$

$$\frac{\partial \tau_{yx}}{\partial x} + \frac{\partial \sigma_{yy}}{\partial y} + \rho g_y - \rho \ddot{u}_y - \rho_f \ddot{w}_y = 0 \tag{5.1b}$$

where σ_{xx}, σ_{yy} are the normal total stresses (tensile positive), which are further separated into the effective stresses σ'_{xx}, σ'_{yy} acting on the soil and pore pressure p (compression positive) for the fluid in the pores, g_x and g_y are the body acceleration, ρ and ρ_f are the average density of the soil-pore fluid mixture and the density of the pore fluid, respectively, u_x and u_y are the displacement of the soil, and $\ddot{w}_x$, $\ddot{w}_y$ are the averaged relative fluid acceleration. By including porosity n in the definition of the averaged relative fluid velocity

$$\dot{w}_x = n(\dot{U}_x - \dot{u}_x) \tag{5.2a}$$

$$\dot{w}_y = n(\dot{U}_y - \dot{u}_y) \tag{5.2b}$$

their value conforms with the one used in the original Darcy equation (see e.g. Craig, 1992), with U_x and U_y being the displacement of the fluid. The generalized Darcy equation can be expressed as

$$\begin{pmatrix} \dot{w}_x \\ \dot{w}_y \end{pmatrix} = \frac{v_k}{v_f \rho_k g_k} \begin{pmatrix} k_{xx} & k_{xy} \\ k_{yx} & k_{yy} \end{pmatrix} \begin{pmatrix} -\dfrac{\partial p}{\partial x} + \rho_f g_x - \rho_f \ddot{u}_x - \dfrac{\rho_f \ddot{w}_x}{n} \\[2mm] -\dfrac{\partial p}{\partial y} + \rho_f g_y - \rho_f \ddot{u}_y - \dfrac{\rho_f \ddot{w}_y}{n} \end{pmatrix} \tag{5.3}$$

where the permeability tensor consists of k_{xx}, k_{xy}, k_{yx} and k_{yy}. For material with isotropic permeability k (unit $=$ length/time):

$$k_{xx} = k_{yy} = k$$

$$k_{xy} = k_{yx} = 0 \tag{5.4}$$

Furthermore, v_k, ρ_k and g_k are the viscosity of the fluid, density of fluid and gravitational acceleration at which the permeability is measured, and v_f is the viscosity of the fluid actually being used. For example, for a centrifuge experiment conducted at 78 times normal gravity level using 80cs silicone oil, g_x is 0, g_y is $78g_k$, where $g_k = g = 9.81\,\text{ms}^{-2}$ and v_k is 1 centistoke (for water) and v_f is 80 centistoke (for 80cs silicone oil). Lastly, the continuity equation for the fluid phase can be expressed as

$$\frac{\dot{p}}{Q} + \frac{\partial \dot{u}}{\partial x} + \frac{\partial \dot{u}}{\partial y} + \frac{\partial \dot{w}}{\partial x} + \frac{\partial \dot{w}}{\partial y} = 0 \tag{5.5}$$

where Q is the averaged bulk modulus defined, with K_s being the bulk modulus of the soil grains and K_f the bulk modulus of the pore fluid as:

$$\frac{1}{Q} = \frac{1-n}{K_s} + \frac{n}{K_f} \tag{5.6}$$

5.3 Numerical (Discrete) Approximation Procedure

The primary variables in the equation set (5.1), (5.3) and (5.5) can be identified as u, w and p. Alternatively, by the use of Eq. (5.2), the primary variables can be changed to u, U and p. There is little difference to choose between these two sets. On the other hand, the set of basic unknowns can be reduced further. By eliminating pore pressure p between Eqs. (5.3) and (5.5), the so-called u-w formulation (because u and w are the primary variables) results. Again, with the use of Eq. (5.2), the formulation can be transformed into a u-U formulation, where U is the displacement of the pore fluid and w is the averaged relative fluid displacement defined in accordance with the Darcy's law.

Nevertheless, the number of independent variables can be reduced further. If the averaged relative fluid acceleration is neglected from Eqs. (5.1) and (5.3), and $\dot{w}_x$, $\dot{w}_y$ are eliminated between Eqs. (5.3) and (5.5), the u-p formulation is recovered. By neglecting further the solid acceleration in Eqs. (5.1) and (5.3), the standard consolidation equation is obtained. Lastly, if both the solid and fluid velocities are neglected, the resulting equations are now decoupled, and they are the static solid equation and the steady state seepage equation.

Although the u-p formulation neglects the averaged relative fluid acceleration, it has been found to be satisfactory for static, consolidating and slow dynamic loading including most seismic calculations, and it has been widely adopted. The most popular implementation is the fully Implicit procedure (Chan, 1988; Zienkiewicz *et al.*, 1990, 1998; Parra-Colmenares, 1996). Other variants included the Explicit u/Implicit p formulation (Zienkiewicz *et al.*, 1982; Leung, 1984) and Implicit-Implicit staggered (Zienkiewicz *et al.*, 1987, 1988; Paul, 1982).

For faster dynamic conditions, the u-U or the u-w formulation would be more appropriate with the retention of the averaged relative fluid acceleration. Again, a number of variants exist for this strategy namely Implicit u-U (Shiomi, 1983; Zienkiewicz and Shiomi, 1984; Lacy, 1986; Prévost, 1982, 1987), the fully Implicit u-w (Ghaboussi and Momen, 1982) and the fully Explicit u-w (Chan *et al.*, 1991).

The u-U-p formulation, although being formulated by Zienkiewicz and Shiomi (1984), is a relatively new development (Anandarajah, 1990; Sandhu *et al.*, 1990; Gajo *et al.*, 1994). With three sets of independent variables, the procedure imposed a higher computational requirement, but as it contains no extra terms in addition to the u-U or the u-w

formulation, they should share similar converged results. However, as the formulation has a mixed character, it could have a better convergency characteristic, but this has yet to be shown.

In this chapter, we shall focus on the Implicit u-p formulation for its simplicity. By using the finite element method (Zienkiewicz and Taylor, 1989, 1991) for spatial discretization, the discretized u-p formulation is as follows:

The dynamic form:

$$\mathbf{M\ddot{u}} + \mathbf{P(u)} - \mathbf{Qp} = \mathbf{f_u} \tag{5.7a}$$

$$\mathbf{G\ddot{u}} + \mathbf{Q^T\dot{u}} + \mathbf{S\dot{p}} + \mathbf{Hp} = \mathbf{f_p}$$

The consolidation form:

$$\mathbf{P(u)} - \mathbf{Qp} = \mathbf{f_u} \tag{5.7b}$$

$$\mathbf{Q^T\dot{u}} + \mathbf{S\dot{p}} + \mathbf{Hp} = \mathbf{f_p}$$

The static and steady state form:

$$\mathbf{P(u)} - \mathbf{Qp} = \mathbf{f_u} \tag{5.7c}$$

$$\mathbf{Hp} = \mathbf{f_p}$$

where $\mathbf{M}$ is the mass matrix, $\mathbf{G}$ is the dynamic coupling matrix, $\mathbf{S}$ is the compressibility matrix, $\mathbf{H}$ is the permeability matrix (sometimes a viscous damping term $\mathbf{C\dot{u}}$ is added to the first equation in Eq. (6.7a), $\mathbf{u}$ and $\mathbf{p}$ are the vectors for the nodal value of u and p, respectively, $\mathbf{Q}$ is the coupling matrix, $\mathbf{P(u)}$ is the nonlinear internal force vector given by:

$$\mathbf{P(u)} = \int_{\omega} \mathbf{B^T}\underline{\sigma}'\, d\Omega \tag{5.8}$$

where $\mathbf{B}$ is the usual displacement-strain transformation matrix for the finite element method, $\underline{\sigma}'$ is a vector of all effective stress components at the integration (Gauss) points and Ω is the domain concerned. Equations (5.5)–(5.8), together with the undrained version of Eqs. (5.7a) and (5.7c), formed the basis of the implicit finite element program DIANA-SWANDYNE II (Chan, 1995a). The time discretization is performed using the generalized Newmark method presented by Katona and Zienkiewicz (1985).

Due to the highly nonlinear behavior of soil, the determination of the effective stress for each integration points presented the greatest challenge to the numerical modeling of liquefaction. There is a great diversity in approaches, and even language used, in the formulation of the constitutive relationship for soil. In the next section, we shall summarize the work that we have performed over the past decade.

5.4   An Adequate Constitutive Relationship for the Material Behavior

5.4.1   Introduction

During the last few decades, a great effort has been devoted to (i) improving the knowledge of how soil behaves under complex loading paths, and (ii) developing suitable constitutive models able to reproduce most salient features of soil behavior found in experiments.

During the early 1970s, two basic facts were well established. First, residual conditions took place at failure surfaces of the Mohr–Coulomb type; and second, yield surfaces exhibited hardening dependent on density.

Models based on the Critical State concept were able to reproduce with accuracy the behavior of cohesive soils under monotonic loading conditions. However, they failed to model the behavior of granular materials, and the response of soils (cohesive and granular) under cyclic loading.

At the same time, industry demanded simulation tools for problems encountered in areas such as offshore or earthquake engineering, where catastrophic failure such as that of the Lower San Fernando dam in 1971 or the collapse of buildings at Niigata during the 1964 earthquake (Seed and Idriss, 1967) could take place. Laboratory tests brought to light the phenomena of liquefaction (Castro, 1969) and cyclic mobility (Ishihara *et al.*, 1975) under cyclic loading.

Concerning constitutive modeling, two lines of research were followed. The first approach focused on extending Critical State models to sand, which was done by Nova (1977) and Wilde (1977), while the second approach concentrated on how to reproduce plastic strain and densification caused by cyclic loading. Here the problem was that classical plasticity models were unable to reproduce plastic strain within the yield surface, as the predicted behavior is elastic after a first cycle of loading if amplitude is not increased.

A simple yet efficient solution consisted of introducing the densification induced by cyclic shearing as an 'autogenous volumetric strain', and suitable densification laws were proposed in Cuéllar *et al.* (1977) and Zienkiewicz *et al.* (1982).

The second approach consisted in extending the theory of plasticity beyond the limits imposed in the classical formulation. The first successful theory was the multi-surface kinematic-hardening model proposed by Mroz in (1967), where a set of 'loading surfaces' within an outer 'boundary' surface was postulated. Since then, further development and improvements have taken place (see, for instance, Mroz *et al.*, 1978; Prévost, 1977; and di Prisco *et al.*, 1993b).

The number of surfaces allows tracking of loading events such as the maximum stress level reached, or points at which stress has reversed. Large intensity loading events erase lower intensity events. An elastic domain may also be postulated, corresponding to the volume enclosed by the inner surface. As the stress is increased from an initial state, the surfaces reached by the stress path translate until a new loading surface is attained. This movement must comply with a rule which ensures that surfaces never intersect each other.

Other elastoplastic, kinematic or anisotropic hardening models have shown to perform well in modeling liquefaction and other cyclic loading phenomena (Ghaboussi and Momen, 1982; Hirai, 1987; Aubry *et al.*, 1982). However, the price to pay in numerical computations is high, and simplified versions were sought.

If the number of surfaces is reduced to two (i.e. the outer or consolidation and the inner or yield), a field of hardening modulii can still be described by prescribing the variation between both surfaces. This model was independently proposed by Krieg (1975) and Dafalias and Popov (1975), and evolved to what is known today as 'Bounding Surface Theory'. A similar approach, the 'subloading surface model', was proposed by Hashiguchi and co-workers (1977).

Subsequently, the Generalized Plasticity Theory proposed by Zienkiewicz and Mroz (1984) proved to be an adequate framework in which simple models could be produced.

The purpose of this section is to present the Generalized Plasticity Theory as a framework within which constitutive models able to reproduce most salient features of soil behavior can be developed such as the model proposed by Zienkiewicz *et al.* (1985) and Pastor *et al.* (1985, 1990).

5.4.2   Generalized Plasticity

If material response does not depend upon the velocity at which stresses vary, the relation between the increments of stress and strain can be written as

$$d\varepsilon = C : d\sigma \tag{5.9}$$

where C is a fourth order tensor, homogeneous of degree zero in $d\sigma$. Before continuing, some basic properties of C will be described.

We consider a uniaxial loading-unloading-reloading test shown schematically in Figure 5.1, where the constitutive tensor C is a scalar, the inverse of the slope at the point considered.

As can be seen, the slope depends upon the stress level, being smaller at higher stresses. However, if one compare the slopes at points A_1, A_2 and A_3, one would find that they are not the same, and C is dependent on past history (stresses, strains, modification of material microstructure, etc.).

Taking a closer look at point C, it can be seen that for a given point, different slopes are obtained in loading and unloading, which implies a dependence on the direction of stress increment. This dependence is only on the direction, as C is a homogeneous function of degree zero on $d\sigma$. Therefore, for this simple one-dimensional case, it is possible to write

$$d\varepsilon_L = C_L : d\sigma \tag{5.10a}$$

and

$$d\varepsilon_U = C_U : d\sigma \tag{5.10b}$$

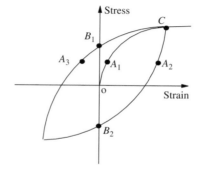

Figure 5.1   General stress:strain behavior

Moreover, if we consider two infinitesimal stresses and $-d\sigma$, we will obtain

$$d\varepsilon_L = C_L : d\sigma \qquad (5.10\text{c})$$

$$d\varepsilon_U = C_U : d\sigma \qquad (5.10\text{d})$$

with the global result of

$$d\varepsilon = d\varepsilon_L + d\varepsilon_U \neq 0 \qquad (5.11)$$

This kind of constitutive law has been defined by Darve (1990) as incrementally nonlinear.

There are different alternatives to introduce the dependence on the direction of the stress increment, among which it is worth mentioning the multilinear laws proposed by Darve and co-workers in Grenoble (1982), or the hypoplastic laws of Dafalias (1986) or Kolymbas (1991). However, the simplest algorithm consists of defining in the stress space a normalized direction n for any given state of stress σ, such that all possible increments of stress are separated into two classes, loading and unloading:

$$d\varepsilon_L = C_L : d\sigma \quad n : d\sigma > 0 \qquad (5.12\text{a})$$

$$d\varepsilon_U = C_U : d\sigma \quad n : d\sigma < 0 \qquad (5.12\text{b})$$

Neutral loading corresponds to the limit case, for which

$$n : d\sigma = 0 \qquad (5.12\text{c})$$

This is the starting point of the Generalized Theory of Plasticity, introduced by Zienkiewicz and Mroz (1984) and later extended by Pastor and Zienkiewicz (Zienkiewicz *et al.*, 1985; Pastor *et al.*, 1985, 1986, 1990).

Introduction of this loading-unloading discriminating direction defines a set of surfaces which is equivalent to those used in Classical Plasticity, as will be shown later, but these surfaces need never be explicitly defined.

Continuity between loading and unloading states requires that constitutive tensors for loading and unloading be of the form

$$C_L = C^e + \frac{1}{H_L} n_{gL} \otimes n \qquad (5.13\text{a})$$

and

$$C_U = C^e + \frac{1}{H_U} n_{gU} \otimes n \qquad (5.13\text{b})$$

where n_{gL} and n_{gU} are arbitrary tensors preferably of unit norm, and H_L and H_U are two scalar functions defined as loading and unloading plastic moduli. It can be very easily verified that both laws predict the same strain increment under neutral loading where both expressions are valid, and hence non-uniqueness is avoided. As for such loading, the increments of strain using the expressions for loading and unloading are

$$d\varepsilon_L = C_L : d\sigma = C^e : d\sigma \qquad (5.14\text{a})$$

and

$$d\varepsilon_U = C_U : d\sigma = C^e : d\sigma \qquad (5.14\text{b})$$

Material behavior under neutral loading is reversible, and it can be therefore regarded as elastic. Indeed, tensor C^e characterizes elastic material behavior, and it can be very easily checked that any infinitesimal cycle of stress $(d\sigma, -d\sigma)$, where $d\sigma$ corresponds to neutral loading conditions, and results in zero accumulated strain.

This suggests that the strain increment can be decomposed into two parts, as follows:

$$d\varepsilon = d\varepsilon^e + d\varepsilon^p \tag{5.15}$$

where

$$d\varepsilon^e = C^e : d\sigma \tag{5.16}$$

and

$$d\varepsilon^p = \frac{1}{H_{L/U}} (n_{gL/U} \otimes n) : d\sigma \tag{5.17}$$

We note that irreversible-plastic deformations have been introduced without the need for specifying any yield or plastic potential surfaces, nor hardening rules. All that is necessary to specify are two scalar functions H_L and H_U and the directions, n_{gL} and n_{gU}, and n.

To account for the softening behavior of material (i.e. when H_L is negative), definitions of loading and unloading have to be modified as follows:

$$d\varepsilon_L = C_L : d\sigma \quad n : d\sigma^e > 0 \tag{5.18a}$$

$$d\varepsilon_U = C_U : d\sigma \quad n : d\sigma^e < 0 \tag{5.18b}$$

where $d\sigma^e$ is given by

$$d\sigma^e = C^{e^{-1}} : d\varepsilon \tag{5.19}$$

5.4.3   *A simple model for sand liquefaction (Pastor and Zienkiewicz, 1986)*

Following experimental results reported by Frossard (1983) in drained triaxial tests, dilatancy can be approximated by a linear function of the stress ratio η

$$d_g = (1 + \alpha)(M_g - \eta) \tag{5.20}$$

It can be seen that dilatancy is zero at the line

$$\eta = M_g \tag{5.21}$$

which coincides with the projection of the Critical State line on the p'-q plane.

This line has also been referred to as the 'characteristic state line' (Habib and Luong, 1978) or the 'line of phase transformation' (Ishihara *et al.*, 1975) and plays an important role in modeling sand behavior as will be shown later. It has to be noted that this line is not the Critical State line, which will be reached at residual conditions. Whether Critical State Line existed or not, has been a matter of discussion during past years, due to the difficulty in obtaining homogeneous specimens at failure after shear bands have developed. However, recent experiments carried out at Grenoble by Desrues (1997) have shown that inside the shear band a critical void ratio is reached.

During a test, this line can be crossed a first time, with the specimen is still far from the residual state. If shearing continues, the stress path will finally approach the Critical State line.

Therefore, the condition $\eta = M_g$ represents two different states at which dilatancy is zero, the 'Characteristic State' and the Critical State.

Direction of plastic flow n_{gL} can be determined in the triaxial space by similar procedures used in cohesive soils, giving

$$n_g = \begin{pmatrix} n_{gv} \\ n_{gs} \end{pmatrix} \tag{5.22}$$

with

$$n_{gv} = \frac{d_g}{\sqrt{1 + d_g^2}} \tag{5.23a}$$

$$n_{gs} = \frac{1}{\sqrt{1 + d_g^2}} \tag{5.23b}$$

It is important to notice that use of non-associative flow rules is necessary for the modeling of unstable behavior within the hardening region, and therefore direction n should be specified as different from n_{gL}.

We chose to do so by writing

$$n = \begin{pmatrix} n_v \\ n_s \end{pmatrix} \tag{5.24}$$

with

$$n_v = \frac{d_f}{\sqrt{1 + d_f^2}} \tag{5.25a}$$

$$n_s = \frac{1}{\sqrt{1 + d_f^2}} \tag{5.25b}$$

where

$$d_f = (1 + \alpha)(M_f - \eta) \tag{5.26}$$

Again, while M_g depends upon Lode's angle in the manner suggested by Zienkiewicz and Pande (1977), M_f maintains a constant ratio with M_g.

It should be noted that both directions have been defined without reference to any yield or plastic potential surfaces, though, of course, these can be established *a posteriori*. In fact, it is possible to integrate both expressions to obtain both plastic potential and yield surfaces (note that the contribution of Lode angle variation has not been included in the definitions of the n and n_g tensors).

$$f = \left\{ q - M_f \, p' \left(1 + \frac{1}{\alpha} \right) \left[1 - \left(\frac{p'}{p_c'} \right)^{\alpha} \right] \right\} \tag{5.27a}$$

$$g = \left\{ q - M_g p' \left(1 + \frac{1}{\alpha} \right) \left[1 - \left(\frac{p'}{p'_g} \right)^{\alpha} \right] \right\} \tag{5.27b}$$

where the size of both surfaces is characterized by the integration constants p'_c and p'_g. Similar yield surfaces were proposed by Nova (1982).

To derive a suitable expression for the plastic modulus H_L it is necessary to take into account well-established physical facts:

(i)   Residual conditions take place at the Critical State line

$$\left(\frac{q}{p'} \right)_{res} = M_g \tag{5.28}$$

However, failure does not necessarily occur when this line is first crossed.

(ii)   The frictional nature of material response requires the establishment of a boundary separating impossible states from those that are permissible.

A convenient law was introduced by Pastor and Zienkiewicz (1986) in the form

$$H_L = H_0 p' H_f \{ H_v + H_s \} \tag{5.29}$$

where

$$H_f = \left(1 + \frac{\eta}{\eta_f} \right)^4 \tag{5.30}$$

together with

$$\eta_f = \left(1 + \frac{1}{\alpha} \right) M_f \tag{5.31}$$

limit the possible states, and where

$$H_v = \left(1 - \frac{\eta}{M_g} \right) \tag{5.32a}$$

$$H_s = \beta_0 \beta_1 e^{-\beta_0 \xi} \tag{5.32b}$$

The model developed so far needs to be completed by including plasticity during unloading and a memory function to describe past events. Concerning the former, the response is usually characterized as isotropic and elastic in most Classical Plasticity Models, which is not always very accurate.

In fact, it can be observed from experiments that higher pore pressures than those corresponding to elastic unloading appear. Figure 5.2 depicts the results obtained by Ishihara and Okada (1982) on undrained shearing of loose sands under reversal of stress. Isotropic elastic unloading is characterized by zero volumetric plastic strain, and as under undrained conditions volume is constant, volumetric elastic strain should also be zero and, therefore, p' should not change (a variation in p' causes a change in volumetric elastic strain). Instead of unloading along a vertical line, the stress path turns towards the origin, which indicates higher pore pressures than isotropic elastic. This phenomenon

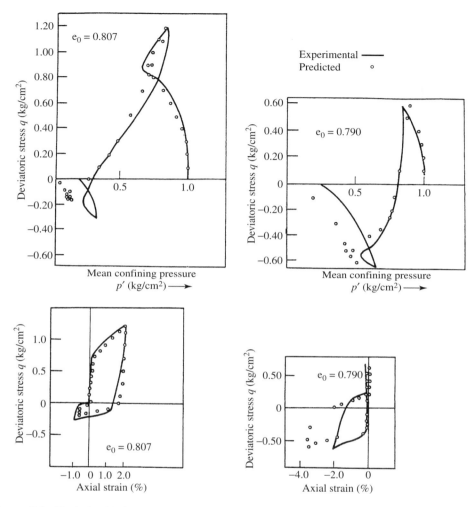

Figure 5.2   Undrained behavior of loose sand under reversal of stress (experiments after Ishihara and Okada (1982))

depends on the stress ratio η_u from which unloading takes place, its importance increasing with it.

Two possible explanations are possible:

(i)   Either the material structure has changed after having crossed the Characteristic State line, and the new distribution of contacts makes the specimen anisotropic (Bahda, 1997).

(ii)   Plastic deformations develop during unloading.

If we assume that plastic strains appear upon unloading, and that they are of contractive nature, a simple expression for the plastic modulus fulfilling these requirements was proposed by Pastor *et al.* (1990):

$$H_u = H_{u0} \left(\frac{M_g}{\eta_u} \right)^{\gamma_u} \quad \text{for} \quad \left| \frac{M_g}{\eta_u} \right| > 1 \tag{5.33a}$$

$$H_u = H_{u0} \quad \text{for} \quad \left| \frac{M_g}{\eta_u} \right| \le 1 \tag{5.33b}$$

and extends the range of the model so far proposed hierarchically.

To determine the direction of plastic flow produced upon unloading, we note that irreversible strains are contractive (densifying) in nature. Direction n_{gU} can thus be provided by

$$n_{gU} = \begin{pmatrix} n_{guv} \\ n_{gus} \end{pmatrix} \tag{5.34}$$

where

$$\eta_{guv} = - \left| n_{gs} \right| \tag{5.35a}$$

and

$$\eta_{gus} = +n_{gs} \tag{5.35b}$$

Concerning reloading, it is necessary to take into account history of past events. Here, we will modify the plastic modulus introducing a discrete memory factor H_{DM} as

$$H_{DM} = \left(\frac{\zeta_{\max}}{\zeta} \right)^{\gamma} \tag{5.36}$$

where ξ was defined above as

$$\zeta = p' \cdot \left\{ 1 - \left(\frac{1+\alpha}{\alpha} \right) \frac{\eta}{M} \right\}^{1/\alpha} \tag{5.37}$$

and γ is a new material constant.

Finally, plastic modulus is given by

$$H_L = H_0 . p' . H_f (H_v + H_s) H_{DM} \tag{5.38}$$

Alternatively, the model can be further elaborated, as shown in Pastor *et al.* (1987, 1993), by improving the way in which history of past events is taken into account. To this end, two components are introduced:

(i)   A surface defining the maximum level of stress reached.

(ii) The point at which last reversal took place.

Directions n and n_g, and the plastic modulus H_L, depend upon the relative position of the stress state C with respect to the point at which the load was reversed, i.e. stress state B, and an image point D defined on the same mobilized stress surface as B.

To obtain the values of H_L, n and n_g, suitable interpolation rules are used. In particular, n is interpolated from $-n$ to n using a linear law. The direction of plastic flow is again obtained by defining a suitable dilatancy at stress state C, d_{gC}, which is interpolated from an initial value d_{g0} to

$$d_{g0} = (1 + \alpha)(M_g - \eta_D) \tag{5.39}$$

The initial value of the dilatancy at the reversal point d_{g0} is given by

$$d_{g0} = (1 + \alpha)(M_g - C_g \eta_B) \tag{5.40}$$

where the constant $C_g (0 < C_g < 1)$ varies with the density, C_g being close to zero for medium-loose sands.

The plastic modulus is interpolated between an initial value H_{U0} and its final value at the image point on the mobilized stress surface H_D. The initial value can be assumed to be infinite to decrease a possible accumulation of plastic strain under very low amplitude cycles:

$$H = H_{U0} + f.(H_D - H_{U0}) \tag{5.41}$$

where f is an interpolation function depending on the relative position of points B, C and D, and which is 1 when C and D coincide.

Concerning the rule to obtain the image stress point D, there are several alternative possibilities. For instance, it can be obtained as the intersection of the straight line joining the reversal and the stress point with the mobilized stress surface, as depicted in Figure 5.3.

This interpolation law provides a smooth transition between unloading to reloading. In fact, unloading may be considered as a new loading process. It is important to note that direction of plastic flow and unit vector n will not be functions of the stress state only, but of the past history as well.

Finally, the influence of sand densification under cyclic loading can be taken into account by introducing in the plastic modulus a factor H_d:

$$H_d = e^{-\gamma_d \varepsilon_v^p} \tag{5.42}$$

It should be mentioned that since the simple models we have described here were proposed, several improvements and modifications have been introduced, particularly at

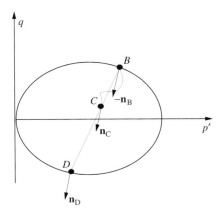

Figure 5.3   Interpolation rule

CERMES (Paris) (Saitta, 1994), where research recently completed has succeeded in including state parameters describing in a consistent way the behavior of sand under different conditions of confining pressure and relative density (Bahda, 1997; Bahda *et al.*, 1997).

5.5   Assessment of Model Performance

To illustrate the predictive capability of the proposed model, we next consider several sets of experiments reported in the literature (Castro, 1969; Saada and Bianchini, 1989), and which cover the basic features of granular soils behavior under monotonic and cyclic loading.

At one end of the density range, we find that very loose sands exhibit liquefaction under undrained shearing. Liquefaction of loose sand under undrained monotonic loading is characterized by a peak shear stress followed by a dramatic decrease of strength. During the process, shown schematically in Figure 5.4, the pore pressure is continuously increasing as the sand tends to contract. The asymptotic value towards which it tends is the confining pressure, and therefore, the effective mean confining pressure p' is close to zero at the end of the test. The soil then behaves as a viscous fluid in which foundations can sink.

If several cycles of loading with an amplitude smaller than the peak are applied, the tendency to densify causes a pore pressure increase and a shift of the stress path towards the origin, until the moment arrives at which liquefaction takes place in the same manner that was described for monotonic loading. Denser sands do not liquefy, but exhibit a phenomenon known as 'cyclic mobility'. Failure is progressive, since the stress path approaches the critical state.

Liquefaction of loose sands has attracted the attention of both experimental and constitutive researchers since the pioneering work of Castro in 1969. During recent years, the effort has focused on topics like state parameters to characterize a sand, effects of initial conditions, final equilibrium state, etc., both from the constitutive and experimental points

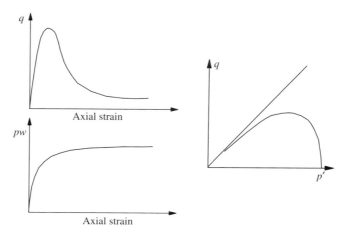

Figure 5.4   Liquefaction of very loose sand (schematic diagram)

of view. Here we could mention the work of Nova and co-workers (di Prisco *et al.*, 1993a; Doanh *et al.*, 1997; Canou *et al.*, 1994; Bahda, 1997).

This seems to contradict the fact that a peak exists, and the material can be thought of as being softened. However, in a frictional material, strength has to be analysed in terms of mobilized stress ratios, rather than deviatoric stress, and no peak is presented by this parameter.

This behavior can be considered unstable in the sense of Drucker (1959):

$$d\sigma^T.d\varepsilon^P < 0 \qquad (5.43)$$

thus having

$$d\sigma^T \left(\frac{1}{H} n_g \cdot n^T \right) d\sigma < 0 \qquad (5.44)$$

If such a feature is to be modeled with a positive plastic modulus, the associated plasticity theory has to be abandoned, choosing

$$n_g \neq n \qquad (5.45)$$

Figures 5.5–5.7 show, respectively, the stress paths, deviatoric stress versus axial strain and pore pressures obtained by Castro (1969), together with the model predictions, which agree well with the experimental data.

At the other end of the density range, peaks exist in deviatoric stress during *drained shear of very dense sands*, this effect developing progressively as density is increased.

The factor H_s is introduced in the expression giving plastic modulus to account for:

- Crossing of the Characteristic State line ($\eta = M_g$) without immediately producing failure.

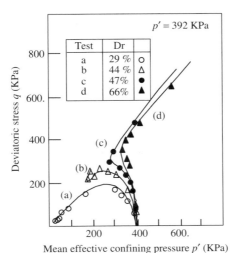

Figure 5.5   Undrained behavior of Banding sand (Castro, 1969). Computed results shown in solid lines (Deviatoric Stress versus Mean Effective Stress)

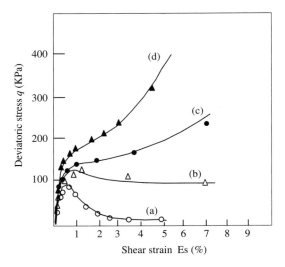

Figure 5.6   Undrained behavior of Banding sand (Castro, 1969) Computed results shown in solid
lines (Deviatoric Stress versus Shear Strain)

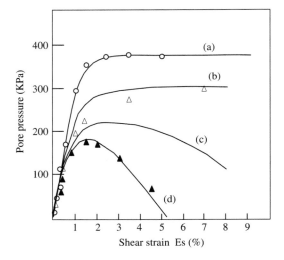

Figure 5.7   Undrained behavior of Banding sand (Castro, 1969). Computed results shown in solid
lines (Pore Pressure versus Shear Strain)

- Reproduction of softening
- Residual conditions take place at the Critical State line.

To illustrate the role of plastic modulus in the transition from softening to hardening
regimes, let us consider a drained triaxial test, such as that illustrated in Figure 5.8. During
the first part of the path, both H_v and H_s are positive and decrease in a monotonous way.
At $\eta = M_g$ (i.e. when crossing the characteristic state line), H_v becomes zero, while H_s

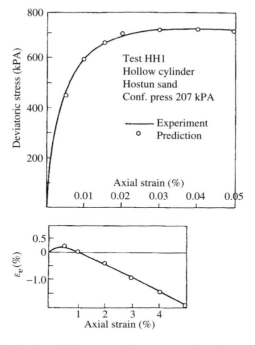

Figure 5.8 Drained behavior of Hostun sand in compression (Saada and Bianchini, 1989)

is still positive. If the process continues, a moment arrives at η_p where

$$H_v + H_s = 0 \tag{5.46}$$

with

$$\eta_p > M_g \tag{5.47}$$

If the test is run under the displacement control mode, the deviatoric stress does not change for an infinitesimal variation of the strain

$$dp' = dq = 0 \tag{5.48a}$$

$$d\varepsilon_s \neq 0 \tag{5.48b}$$

$$d\varepsilon_v \neq 0 \tag{5.48c}$$

Meanwhile, H_s has decreased, and consequently the plastic modulus becomes negative. The soil has entered the softening regime, and from this moment the deviatoric stress will present a descending branch.

The deviatoric strain hardening function H_s will vanish as deformation progresses, reaching a final asymptotic value of zero at $\eta = M_g$, this time at the Critical State line. During the softening process the following inequalities will hold:

$$d\sigma^T . d\varepsilon^p < 0 \tag{5.49}$$

and

$$d\sigma_T \left(\frac{1}{H} n_g . n^T \right) d\sigma < 0 \tag{5.50}$$

It can be seen that there is no need on this occasion for non-associativeness to ensure the existence of peaks as H is negative, and in fact, very dense sands may exhibit the limiting associative behavior with

$$M_f = M_g \tag{5.51}$$

The ratio M_f / M_g seems to be dependent on relative density, and in Pastor *et al.* (1985) a suitable relation was proposed:

$$\frac{M_f}{M_g} = D_r \tag{5.52}$$

where D_r is the relative density. Figures 5.8 and 5.9 show model predictions for dense and loose sand response in drained conditions (Saada and Bianchini, 1989).

Care should be taken when analyzing the results of tests in general, and in the case of dense sands in particular, as failure localizes along narrow zones referred to as shear bands. From the moment of their inception, the specimen is no longer homogeneous, and the experimental results correspond to a boundary value problem rather than a homogeneous body. However, the following facts should be stressed:

- Even if the specimen is not homogeneous, softening must exist for the sample to exhibit a peak.

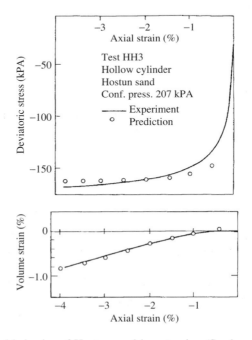

Figure 5.9   Drained behavior of Hostun sand in extension (Saada and Bianchini, 1989)

- The overall response is governed by the ratio between the width of the shear band and the length of specimen. This effect is similar to what can be observed in numerical computations, and which has been referred to as *mesh-dependence*.

- Experimental evidence seems to indicate the existence of a residual Critical State. To obtain it, it is simpler to use loose than very dense specimens.

Undrained shearing of medium-loose to dense sands For this intermediate characteristic, once the Characteristic State Line is reached, an upturn in the stress path is produced as the soil changes from contractive to dilative behavior. If the material is isotropic, determination of CSL position can easily be performed from a point at which the undrained stress path has a vertical tangent in p'-q space; then the following relationships hold:

$$d\varepsilon_v^p = 0 \tag{5.53}$$

and

$$d\varepsilon_v^e = 0 \tag{5.54}$$

as

$$dp' = 0 \tag{5.55}$$

Figures 5.5–5.7 show how relative density influences the undrained behavior of sand, together with predictions of the proposed model.

At this point, a model has been produced such that:

(i)   it reproduces the most salient features of sand under monotonic shearing;

(ii) it is very simple, as no surface is involved and consistency conditions do not have to be fulfilled;

(iii) it is computationally efficient in finite element codes, as the stress point does not have to be brought back to the yield surface and tangent moduli are easily established.

It is possible now to model *cyclic phenomena as liquefaction and cyclic mobility* which appear in loose and medium sands under cyclic loading, and which are responsible for catastrophic failure of structures subjected to earthquakes.

Both phenomena are largely caused by the overall tendency of medium and loose sands to densify when subjected to drained cyclic shearing. If the load is applied fast enough or the permeability is relatively small, this mechanism causes progressive pore pressure build up leading to failure.

In the case of very loose sands, liquefaction takes place following a series of cycles in which the stress path migrates towards lower confining pressures. Figure 5.10 shows the results obtained by Castro (1969) in his pioneering work.

Denser sands do not exhibit liquefaction but *cyclic mobility*. Failure here is progressive, since the stress path approaches the Characteristic State Line by its shift caused by pore pressure build up. Deformations during unloading cause the stress path to turn towards the origin, and strains produced during the next loading branch are of higher amplitude.

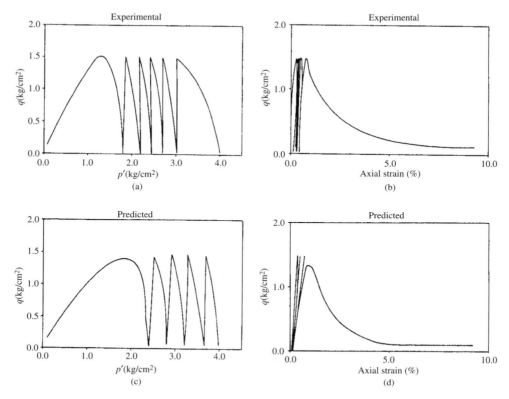

Figure 5.10 Liquefaction of loose banding sand under cyclic loading (top). Experimental data from Castro (1969) (bottom) Prediction

Figure 5.11 shows both the experimental results obtained by Tatsuoka (1972) on Fuji River sand.

5.6   Parametric Identification and Analysis Procedure

Having established the numerical formulation and constitutive relationship, in this section, issues are outlined that are of practical importance in the application of these procedures, namely a general analysis procedure which ensure static equilibrium is established before the dynamic analysis and parametric identification for the constitutive model used. Since the examples given in this chapter both related to the VELACS exercise, the procedure outlined in this section also applies well to the modeling of centrifuge tests, but the extension to prototype analysis is obvious.

5.6.1   *General analysis procedure*

This analysis procedure is applicable to all the predictions performed by the authors:

(1)   The information about the model or the prototype is analysed and the key data noted.

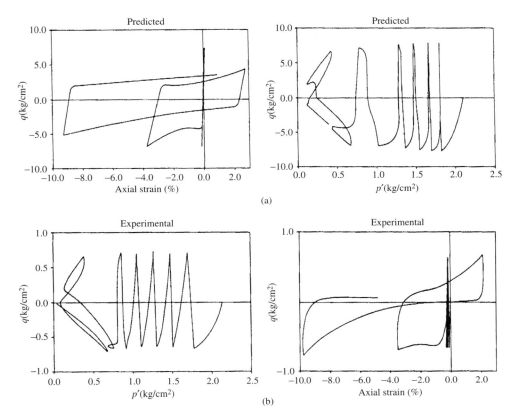

Figure 5.11 Cyclic mobility of loose Niigata sand (top). Prediction (bottom). Experimental data from Tatsuoka (1972)

(2) A finite element mesh is generated using a pre-processor. Time required for the subsequent dynamic and consolidation analysis are also taken into consideration so that a reasonable mesh is chosen.

(3) The appropriate boundary conditions are then applied at the boundaries. For centrifuge models, special attention is given to the modeling of the laminar box behavior and tied nodes for both solid displacement and fluid pressure are used.

(4) The hydrostatic pressures on impermeable solid–fluid interface are assumed to be constant throughout the analysis. They are prescribed at the fluid nodes concerned, and the pressures on the solid phase are also applied as total stress.

(5) The appropriate permeability and gravitational acceleration are then included for the prototype or the centrifuge model. The models are represented at the model scale so the appropriate acceleration level is the centrifugal acceleration imposed.

(6) A static analysis is performed to determined the initial stress state of the model. A K_o value of 0.4 is assumed for the cohesionless material. To avoid tensile stress

and high stress ratio, a Mohr–Coulomb elasto-perfectly plastic model is used for the initial analysis with a reduced friction angle of say 25° when the critical state friction angle was determined to be 30°.

(7)   The outputs of the static analysis are considered carefully to check if the initial pore pressure distribution is reasonable, and also if the stress state is acceptable.

(8)   A no-earthquake dynamic run is then performed to check if the initial stress state is in correct equilibrium condition. If it is not, a new static initial analysis is performed with modified parameters to obtain equilibrium.

(9)   When the initial stress state is acceptable, a linear elastic analysis is performed to note the basic behavior of the finite element mesh.

(10) Then a nonlinear analysis is performed for the earthquake stage with the supplied horizontal and vertical earthquake with proper scaling. The dynamic analyses were performed using a Generalized Newmark (Katona and Zienkiewicz, 1985) scheme with nonlinear iterations using initial linear elastic tangential global matrix. The constitutive model used is Pastor–Zienkiewicz (1986) mark-III model. The parameters used are described in Section 5.6.3. The time step used is usually equal to a simple multiple of the earthquake spacing. The choice of the time step depends on the number of the stations in the earthquake input and the frequency of the input. Void ratio (i.e. permeability) and other geometric properties were kept constant during the analysis. Rayleigh damping of (minimum) 5% is usually applied at the dominant frequency in the earthquake or earthquake-like motion input to enhance the energy dissipation characteristic of the constitutive model.

(11) The earthquake phase of the analysis is then plotted to check for anomaly.

(12) The consolidation then follows the dynamic analysis. Usually a larger time step is used for the consolidation analysis; a gradual change in time step is used to avoid numerical instability. The full dynamic equation is used for the consolidation stage of the analysis with the appropriate mass matrix.

(13) The results are first plotted using a simple post-processing program to check its validity. If the results do not seem reasonable, the dynamic analysis is repeated with another set of numerical parameters, iteration schemes, etc., until reasonable and numerically stable results are obtained.

(14) Various plots are then performed for the final report. Since total quantities (e.g. pore pressure and displacement) are used in the program, post-processing is required to obtain the excess pore water pressure and relative displacement required by the specification.

(15) Other post-processing parameters (e.g. excess pore water pressure ratio, spectral analysis and response spectrum) are calculated for the reporting purpose.

5.6.2   *Description of the precise method of determining each coefficient in the numerical model*

The determination of each coefficient of the Pastor–Zienkiewicz mark III model follows the procedure outlined in Section 5.5 of Chan (1988), and is being reproduced in this section. As drained monotonic, undrained monotonic and undrained cyclic tests are the most widely available tests in common engineering applications, they are chosen in the parametric determination process. During earthquake and other rapid loading, the undrained test is more relevant. The tests should be performed with samples having relative density around the intended relative density.

In this section, the procedure to identify each of the parameters required by the model is illustrated, and each parameter is discussed in turn:

(1) Mg (dimensionless): can be estimated from the graph plotting stress ratio versus the shear strain or axial strain. Mg is approximately equal to the maximum value of stress ratio the test reaches. It can also be estimated from the q versus p' plot, with a tangent drawn from the origin to the residual stress path in an undrained triaxial test. Mg corresponds to the maximum slope obtained by this method. Mg can also be obtained from drained test using the intercept of dilatancy versus stress ratio plot. In the VELACS exercise, the stress ratio plot was used.

(2) M_f (dimensionless): can be determined by matching the shape of the stress path in the q versus p' plane in an undrained triaxial test. Alternatively, it can be obtained by matching the critical stress ratio that the soil changes from contractive to dilative behavior in the case of dense sand. As suggested in Eq. (5.52), the value of D_R. Mg can serve as a good starting point for the evaluation of its value. For the VELACS analysis, the critical stress ratio is used.

(3) α_g (dimensionless): can be obtained from the slope of graph between dilatancy and stress ratio over Mg graph. However, this value is usually taken as 0.45, and it is also used in this exercise.

(4) α_f (dimensionless): is usually taken to be the same as α_g so that the loading locus and plastic potential are having the same shape.

(5) K_{ev0c} (dimension of stress): represents the value of bulk modulus at the mean effective stress p'_0. It can be obtained by matching the initial slope of the mean effective stress p' or pore pressure versus axial strain plot in an undrained test. Its value can be adjusted so that a better match of the curve of pore pressure versus axial strain can be obtained. In the VELACS exercise, this is done so that the end point in the predicted curves stayed close to the experimental data.

(6) K_{es0c} (dimension of stress): represents the value of three times of Shear modulus at the mean effective stress p'_0. It can be obtained by matching the initial slope of deviatoric stress q versus axial strain plot in an undrained test. Its value can be adjusted so that a better match of the curve of q versus axial strain can be obtained. In the VELACS exercise, this is done so that the end point in the predicted curves stayed close to the experimental data.

(7) β_0 (dimensionless): is usually taken as 4.2 and this value is used here.

(8) β_1 (dimensionless): is usually taken as 0.2 and this value is used here.

(9) Ho (dimensionless): is determined by fitting the curves in p' or q versus axial strain plot. It can be found by matching the shape of the q versus p' plot for undrained tests also.

(10) Huo (dimensionless): is determined by matching the initial slope of the first unloading curve.

(11) γ_u (dimensionless): is determined by matching the rate of change of slope of the first unloading curve or by matching the number of cycles in a series of loading and unloading tests. The second method is used in this exercise.

(12) γ_{DM} (dimensionless): is determined by matching the slope of the first reloading curve, or by matching the number of cycles in a series of loading and unloading. The second method is used in this exercise.

(13) p'_0 (dimension of stress): is the initial mean effective stress of the undrained triaxial test.

5.6.3 *Parameters identification for Pastor–Zienkiewicz Mark III model*

There are quite a number of parameters in the Pastor–Zienkiewicz mark III model which require definition. Two CIUC (Isotropically Consolidated followed by Undrained Compression test) experimental results using the Nevada sand starting from 40 kPa were chosen to identify the parameters for the VELACS exercise. The experimental results were taken from Arulmoli *et al.* (1992), which provided the standard soil model test results for the numerical predictors. The 40 kPa ones were chosen because they are close to the mean effective stress value at the middle of the centrifuge model. One undrained monotonic and cyclic test is taken from each of the loose sand ($Dr = 40\%$) and dense sand ($Dr = 60\%$) experimental data sets, respectively.

The comparison of the constitutive model and the physical undrained triaxial tests has been given in Chan *et al.* (1992a,b). These results are produced using soil model subroutine for DIANA-SWANDYNE II (Chan, 1995a) interfaced with a soil model testing program SM2D (Chan, 1995b):

(1) Loose sand: ($Dr = 40\%$). Experiment CIUC4051 was used. The parameters obtained are as follows. $Mg = 1.15$, $M_f = 1.03$, $\alpha_f = \alpha_g = 0.45$, $K_{ev0c} = 770$ kPa, and $K_{es0c} = 1155$ kPa. Elastic modulus is considered proportional to the mean effective stress, $\beta_0 = 4.2$, $\beta_1 = 0.2$, $p'_0 = 4$ kPa, Ho $= 600$, Huo $= 4000$ kPa, $\gamma_u = 2$, and $\gamma_{DM} = 0$.

(2) Dense sand: ($Dr = 60\%$). Experiment CIUC6012 was used. The parameters obtained are as follow: $Mg = 1.32$, $M_f = 1.30$, $\alpha_f = \alpha_g = 0.45$, $K_{ev0c} = 2000$ kPa, and $K_{es0c} = 2600$ kPa. Elastic modulus is considered proportional to the mean effective stress, $\beta_0 = 4.2$, $\beta_1 = 0.2 p'_0 = 4$ kPa, Ho $= 750$, Huo $= 40\,000$ kPa, $\gamma_u = 2$, and $\gamma_{DM} = 4$.

5.7 Validation of the Numerical Procedure

Having established the numerical procedure, it is imperative to establish the extent of validity of the assumptions made during the three stages of numerical formulation. This can be performed individually for the three stages:

(1) The constitutive model can be compared with laboratory experiments such as the conventional triaxial apparatus, true triaxial apparatus, hollow cylinder and shear

box tests. The constitutive model can be tested against other constitutive models. A parametric sensitivity analysis can also be performed.

(2) The numerical approximation can be tested against analytical solutions obtained using linear elastic material such as the Terzaghi consolidation and solutions for dynamic (cyclic) steady state (Zienkiewicz *et al.*, 1980; Simon *et al.*, 1984). However, for such analytical solutions to be used correctly, the basic assumption of the formulation must be respected.

(3) However, to test the validity of the assumptions made in the initial mathematical framework and the behavior of the constitutive model in the boundary value problems, one has to resort to the comparison with physical results from prototype or model experiments. It is generally accepted that centrifuge experiments are able to provide a good representation of the prototype behavior with proper scaling of stress level within the model (Schofield, 1980). In this chapter, the displacement evolution with time of numerical calculations using the implicit u-p formulation are compared with centrifuge experiments performed on the Cambridge Geotechnical Centrifuge and the centrifuge in California Institute of Technology, USA. Various comparisons have been made:

(i) A simple level sand bed with loose and dense on each side contained in a shear stack to model the far-field shear beam condition. The test was performed by Scott *et al.* (1993) and formed the model No. 3 of the VELACS project. It was duplicated at the University of California, Davis, USA (Farrell and Kutter, 1993) and Rensselaer Polytechnic Institute, USA (Taboada and Dobry, 1993).

(ii) A submerged quay wall retaining loose soil contained in a rigid box. The test was performed by Zeng (1993), and formed the model No. 11 of the VELACS project.

VELACS is the acronym for Verification of Liquefaction Analysis by Centrifuge Studies funded by National Science Foundation, USA. The main objective of the project is to verify numerical prediction procedures using centrifuge tests. A total of nine centrifuge tests were planned. The numerical predictors, invited from all over the world, were given the model configuration and planned input and density of the soil used in the tests; they were then asked to submit their predictions before the tests were begun. The Class A prediction results were kept by a third party, so the centrifuge experiments were performed without the knowledge of the submitted numerical predictions. In October 1993, a symposium was held at the University of California, Davis, and the numerical predictions were compared with the centrifuge results, and the centrifuge tests were also compared with their duplicate tests (Arulanandan and Scott, 1993).

5.8 Experimental Set-up and Results

5.8.1 VELACS Model No. 3

The set-up of the experiment, taken from Scott *et al.* (1993), are given in Figures 5.12 and 5.13. The experimental results and numerical predictions of excess pore pressure trace

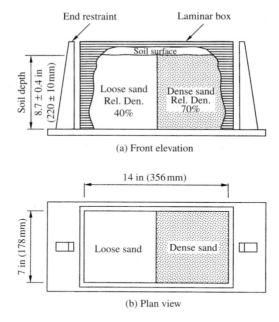

(a) Front elevation

(b) Plan view

Figure 5.12   Geometry of test specimen

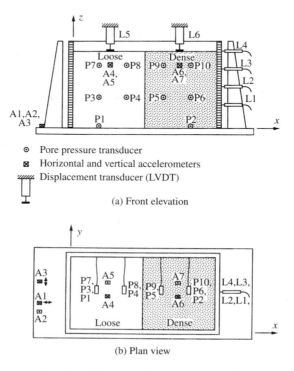

⊙   Pore pressure transducer

⊠   Horizontal and vertical accelerometers

▥   Displacement transducer (LVDT)

(a) Front elevation

(b) Plan view

Figure 5.13   Test specimen instrumentation

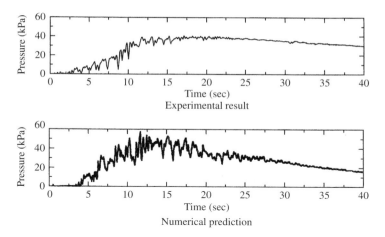

Figure 5.14 Excess pore water pressure for location P2

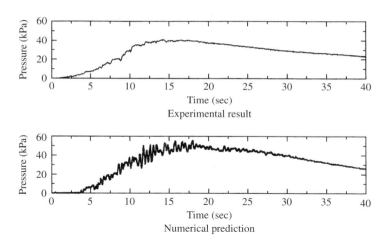

Figure 5.15 Excess pore water pressure for location P3

for transducers P2, P3 and P9 are given in Figures 5.14–5.16, respectively. Overall, the comparison is excellent for all levels of the model.

The experimental results and numerical predictions of horizontal LVDT L1 to L4 are given in Figures 5.17 and 5.18, respectively. In this case, the results were given in prototype scale. Other numerical results can be found in Chan *et al.* (1993b). The predictions show good comparison with experimental results, except that the amplitude of the horizontal displacements seemed to indicate an opposite direction to the experimental results. This could be due to the positive direction of the LVDT being placed in the negative *x*-direction. The long term values for horizontal displacement (L1–L4) were slightly higher than the experimental values. In general, the numerical predictions for this test are very good.

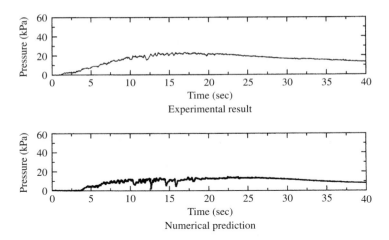

Figure 5.16   Excess pore water pressure for location P9

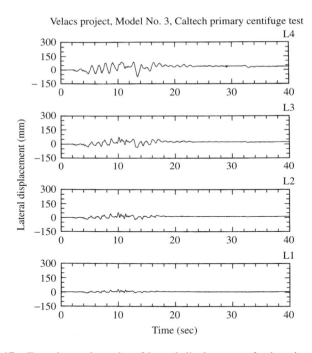

Figure 5.17   Experimental results of lateral displacement for locations L1 to L4

5.8.2   *VELACS Model No. 11*

The experimental set-up, taken from Zeng (1993), is shown in Figure 5.19. Reasonable results have been obtained in the Class A prediction (1993a), and even better agreement was obtained by Madabhushi and Zeng (1994) in their Class C prediction of two similar tests, also using DIANA-SWANDYNE II, when a layer of slip element was used behind

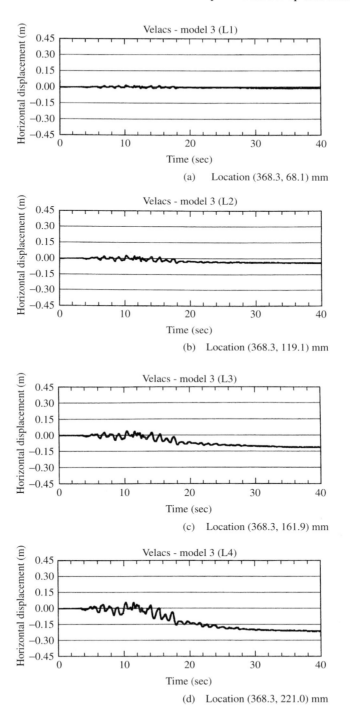

(a)   Location (368.3, 68.1) mm

(b)   Location (368.3, 119.1) mm

(c)   Location (368.3, 161.9) mm

(d)   Location (368.3, 221.0) mm

Figure 5.18   Numerical predictions of lateral displacement for locations L1 to L4

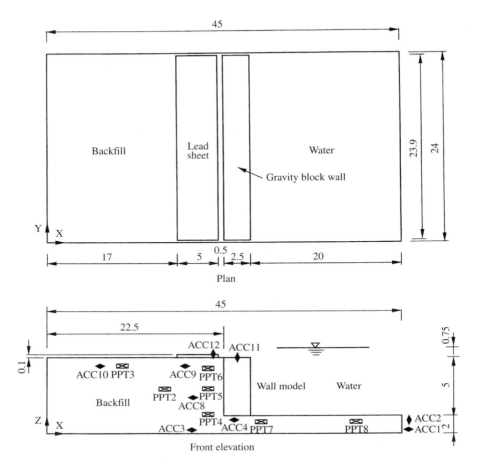

Figure 5.19   Centifugal model

the submerged quay wall. The displacement and rotation of the quay for the experimental and numerical results are given in Figure 5.20 for dry and Figure 5.21 for saturated condition and very good agreement has been observed. This type of quay wall failure shares a similar mechanism to the ones which failed on the Port Island of Kobe during the Hyogoken-Nanbu earthquake on 17th January 1995.

5.9   Discussion and Conclusion

A complete framework for the numerical modeling of liquefaction behavior is introduced in this chapter. The mathematical framework is based on the *u-p* approximation of the generalized Biot formulation for dynamic saturated soil and pore fluid interaction. The Pastor–Zienkiewicz mark III model is introduced in detail, together with the method of parametric identification. Possible extensions to the model are also discussed in order to enhance its predictive capabilities. Numerical examples are compared with centrifuge scaled model tests, and excellent results are obtained.

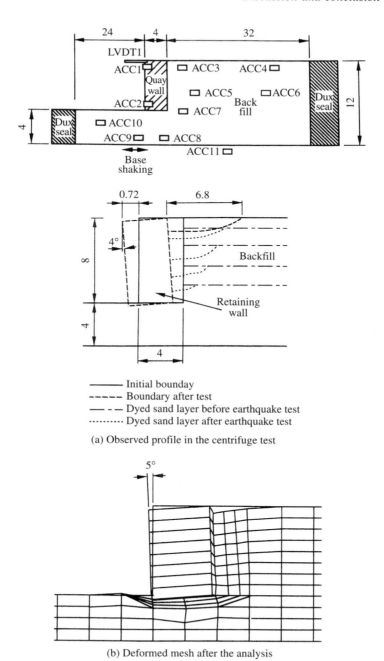

---- Initial bounday
----- Boundary after test
— - — Dyed sand layer before earthquake test
··········· Dyed sand layer after earthquake test

(a) Observed profile in the centrifuge test

(b) Deformed mesh after the analysis

Figure 5.20 (a) Schematic section of the centrifuge model XZ7; (b) post-test profile of the centrifuge model XZ7

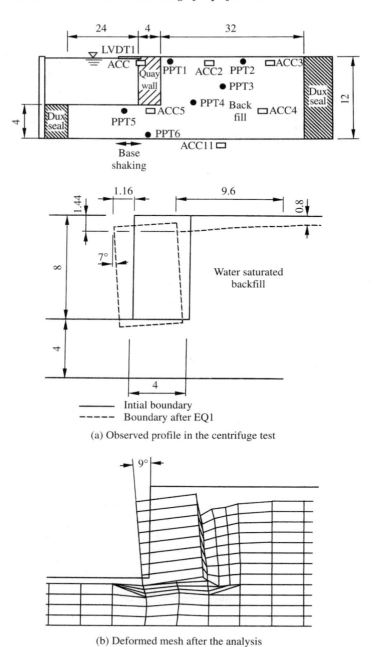

(a) Observed profile in the centrifuge test

(b) Deformed mesh after the analysis

Figure 5.21   (a) Schematic section of the centrifuge model XZ9; (b) post-test profile of the centrifuge model XZ9

References

Anandarajah, A. *HOPDYNE – A Finite Element Computer Program for the Analysis of Static, Dynamic and Earthquake Soil and Soil-Structure Systems*. Internal Report, Johns Hopkins University, Baltimore, (1990).

Arulanandan, K. and Scott, R. F. (eds.) *Proc. VELACS Symposium*, Vol. 1, A. A. Balkema, Rotterdam, (1993).

Arulmoli, K., Muraleetharan, M. M. H. and Fruth, L. S. *VELACS Laboratory testing program – Soil data report*. Earth Technology Corporation, Project No. 90-0562, Irvine, California, (1992).

Aubry, D., Hujeux, J. C., Lassoudi, F. and Meimon, Y. A double memory model with multiple mechanisms for cyclic soil behaviour. *Int. Symp. Num. Models in Geomechanics*, Dungar, R., Pande, G. N. and Studer J. A. (eds.), Balkema, Rotterdam, (1982).

Bahda, F. *Etude du comportement du sable au triaxial: experience et modélisation*. PhD, Thesis, CERMES, ENPC Paris, (1997).

Bahda, F., Pastor, M. and Saitta, A. A double hardening model based on generalized plasticity and state parameters for cyclic loading of sands. In: *NUMOG VI*, Pande, G. N. and Pietruszczak, S. (eds.), (1997).

Biot, M. A. Theory of propagation of elastic waves in a fluid-saturated porous solid, part I – low-frequency range. *J. Acoust. Soc. Am.*, **28**(2), (1956) 168–178.

Canou, J., Bahda, F., Saitta, F. and Dupla, J. C. Initiation of sand liquefaction under monotonic and cyclic loading. *XIII Int. Conf. Soil Mech. Found. Eng.*, New Delhi, India, (1994) 1297–1300.

Castro, G. *Liquefaction of sands*. PhD Thesis, Havard University, Cambridge, MA, (1969).

Chan, A. H. C. *A Unified Finite Element Solution to Static and Dynamic Geomechanics Problems*. PhD Thesis, University College of Swansea, Wales, (1988).

Chan, A. H. C. User Manual for DIANA SWANDYNE-II, School of Civil Engineering, University of Birmingham, (December 1995a).

Chan, A. H. C. *User manual for SM2D – Soil Model Tester for 2-dimensional application*. School of Civil Engineering, University of Birmingham, (December 1995b).

Chan, A. H. C., Famiyesin, O. O. and Muir Wood, D.A Fully Explicit u-w Schemes for Dynamic Soil and Pore Fluid Interaction. *APCOCM*, **1**, (December 1991) 881–887.

Chan, A. H. C., Famiyesin, O. O. R. and Muir Wood, D. Report No. CE-GE92-23-0: *Numerical Simulation Report for the VELACS Project – General Description*, Department of Civil Engineering, Glasgow University, Glasgow, (1992a).

Chan, A. H. C., Famiyesin, O. O. R. and Muir Wood, D. Report No. CE-GE92-23-1: *Numerical Simulation Report for the VELACS Project – Class A prediction of RPI model 1*, Department of Civil Engineering, Glasgow University, Glasgow, (1992b).

Chan, A. H. C., Famiyesin, O. O. and Muir Wood, D. Numerical Prediction for Model No. 11. In: *Verification of numerical procedures for the analysis of soil liquefaction problems, UC Davis, 17-20 Oct, Vol. 1*, Arulanandan, K. and Scott, R. F. (eds.), A. A. Balkema, Rotterdam, (1993a) 909–931.

Chan, A. H. C., Famiyesin, O. O. and Muir Wood, D. Numerical Prediction for Model No. 3. In: *Verification of numerical procedures for the analysis of soil liquefaction problems, UC Davis, 17-20 Oct, Vol. 1*, Arulanandan, K. and Scott, R. F. (eds.), A. A. Balkema, Rotterdam, (1993b) 489–510.

Chan, A. H. C., Siddharthan, R. and Ito, K. Overview of the Numerical Predictions for VELACS Model No. 3. In: *Verification of numerical procedures for the analysis of soil liquefaction problems, UC Davis, 17-20 Oct*, Arulanandan, K. and Scott, R. F. (eds.), A. A. Balkema, Rotterdam, (1994) 1443–1454.

Craig, R. F. *Soil Mechanics* (5th edn). Chapman & Hall, London, (1992).

Cuéllar, V., Bazant, Z. P., Krizek, R. J. and Silver, M. L. Densification and hysteresis of sand under cyclic shear. *J. Geotech. Engrg.*, **103**, (1977) 918.

Dafalias, Y. F. Bounding surface plasticity. I: Mathematical foundation and hypoplasticity. *J. Engrg. Mech. Div.*, **112**, (1986) 966–987.

Dafalias, Y. F. and Popov, E. P. A model for nonlinearly hardening materials for complex loading. *Acta Mechanica*, **21**, (1975) 173–192.

Darve, F. (ed.) *Geomaterials. Constitutive Equations and Modelling.* Elsevier Applied Science, Amsterdam, (1990).

Darve, F. and Labanieh, S. Incremental constitutive law for sands and clays: simulation of monotonic and cyclic tests. *Int. J. Num. Anal. Geomech.*, **6**, (1982) 243–275.

Desrues, J. Private Communication (1997).

Di Prisco, C., Matiotti, R. and Nova, R. Theoretical investigation of the undrained stability of shallow submerged slopes. *Géotechnique*, **45**(3), (1993a) 479–496.

Di Prisco, C., Nova, R. and Lanier, J. A mixed isotropic kinematic hardening constitutive law for sand. In: *Modern Approaches to Plasticity*, Kolymbas, D. (ed.), Balkema, Rotterdam, (1993b) 83–124.

Doanh, T., Ibraim, E. and Matiotti, R. Undrained instability of very loose Hostun sand in triaxial compression and extension. Part I: Experimental observations. *J. Cohesive-Frictional Materials*, (1997).

Drucker, D. C. A definite of stable inelastic material. *J. Appl. Mech.*, **26**, (1959) 101–106.

Farrell, T. M. and Kutter, B. L. Experimental results of Model No. 3. In: *Verification of numerical procedures for the analysis of soil liquefaction problems, UC Davis, 17-20 Oct*, Arulanandan, K. and Scott, R. F. (eds.), A. A. Balkema, Rotterdam, (1993) 463–469.

Frossard, E. Une équation d'ecoulement simple pour less materiaux granulaires. *Géotechnique*, **33**(1), (1983) 21–29.

Gajo, A., Saetta, A. and Vitaliani, R. Evaluation of three- and two-field finite element methods for the dynamic response of saturated soil. *Int. J. Num. Meth. Engrg.*, **37**, (1994) 1231–1247.

Ghaboussi, J. and Momen, H. Modeling and analysis of cyclic behavior of sands. *Soil Mechanics – Transient and Cyclic Loads* (1982) 313–346.

Habib, P. and Luong, M. P. Sols pulvurulents sous chargement cyclique. In: *Materiaux and Structures Sous Chargement Cyclique, Ass. Amicale des Ingenieurs Anciens Eléves de l'Ecole Nationale des Ponts et Chausses*, Palaiseau, 28–29 (September 1978) 49–79.

Hashiguchi, K. and Ueno, M. Elastoplastic Constitutive laws of granular materials. In: *Constitutive Equations of Soils, 9th. Int. Congr. Soil Mech. Found. Engng*, Murayama, S. and Schofield, A. N. (eds.), JSSMFE, Tokyo, (1977) 73–82.

Hirai, H. An elastoplastic constitutive model for cyclic behaviour of sands. *Int. J. Num. Anal. Geomech.*, **11**, (1987) 503–520.

Ishihara, K. and Okada, S. Effects of large preshearing on cyclic behaviour of sand. *Soils and Foundations*, **22**(3), (1982) 109–125.

Ishihara, K., Tatsuoka, F. and Yasuda, S. Undrained deformation and liquefaction of sand under cyclic stresses reversing direction. *Soils and Foundations*, **15**(1), (1975) 29–44.

Katona, M. G. and Zienkiewicz, O. C. A unified set of single step algorithms Part 3: The Beta-m method, a generalisation of the Newmark scheme. *Int. J. Num. Meth. Engrg.*, **21**, (1985) 1345–1359.

Kolymbas, D. An outline of hypoplasticity. *Archive of Applied Mechanics – Ingenieur Archiv*, **61**, (1991) 143–151.

Krieg, R. D. A practical two-surface plasticity theory. *J. Appl. Mech.*, **42**, (1975) 641–646.

Lacy, S. J. *Numerical Procedures for Nonlinear Transient Analysis of Two-phase Soil Systems.* PhD Dissertation, Department of Civil Engineering, Princeton University, Princeton, NJ, (1986).

Leung, K. H. Earthquake response of saturated soils and liquefaction. PhD Dissertation, University College of Swansea, Wales, (1984).

Madabhushi, S. P. G. and Zeng, X. An analysis of the seismic behaviour of quay walls. In: *Verification of Numerical Procedures for the Analysis of Soil Liquefaction Problems, UC Davis, 17-20 Oct, Vol. 2*, Arulanandan, K. and Scott, R. F. (eds.), A. A. Balkema, Rotterdam, (1994) 1593–1606.

Mroz, Z. On the description of anisotropic work-hardening. *J. Mech. Phys. Solids*, **15**, (1967) 163–175.

Mroz, Z., Norris, V. A. and Zienkiewicz, O. C. An anisotropic hardening model for soils and its application to cyclic loading. *Int. J. Num. Anal. Geomech.*, **2**, (1978) 203–221.

Nova, R. On the hardening of soils. *Arch. Mech.*, **29**(3), (1977) 445–458.

Nova, R. A constitutive model for soil under monotonic and cyclic loading. In: *Soil Mechanics – Transient and Cyclic Loads*, Pande, G. N. and Zienkiewicz, O. C. (eds.), Wiley, Chichester, (1982) 343–375.

Parra-Colmenares, E. J. *Numerical Modeling of Liquefaction and Lateral Ground Deformation Including Cyclic Mobility and Dilation Response in Soil Systems*. PhD Thesis, Rensselaer Polytechnic Institute, Troy, New York, (1996).

Pastor, M. and Zienkiewicz, O. C. A generalised plasticity hierarchical model for sand under monotonic and cyclic loading. *NUMOG II*, Ghent, April (1986) 131–150.

Pastor, M., Zienkiewicz, O. C. and Leung, K. H. Simple model for transient soil loading in earthquake analysis, Part II. Non-associative models for sands. *Int. J. Num. Anal. Geomech.*, **9**, (1985) 477–489.

Pastor, M., Zienkiewicz, O. C. and Chan, A. H. C. A generalized plasticity continuous loading model for Geomaterials. In: *Numerical Methods in Engineering: Theory and Approximation*, Zienkiewicz, O. C., Pande, G. N. and Middleton, J. (eds.), Martinus Nijhoff, (1987).

Pastor, M., Zienkiewicz, O. C. and Chan, A. H. C. Generalised plasticity and the modelling of soil behaviour. *Int. J. Num. Anal. Geomech.*, **14**, (1990) 151–190.

Pastor, M., Zienkiewicz, O. C., Xu, Guang-D. and Peraire, J. Modelling of sand behaviour: cyclic loading, anisotropy and localization. In: *Modern Approaches to Plasticity*, Kolymbas, D. (ed.), Elsevier, Holland, (1993) 469–492.

Paul, D. K. *Efficient Dynamic Solutions for Single and Coupled Multipled Field Problems*. PhD Dissertation, University College of Swansea, Wales, (1982).

Prevost, J. H. Mathematical modeling of montonic and cyclic undrained clay behavior. *Int. J. Num. Anal. Geomech.*, **1**(2), (1977) 195–216.

Prévost, J. H. Nonlinear transient phenomena in saturated porous media. *Comp. Meth. Appl. Mech. Engrg.*, **30**, (1982) 3–18.

Prévost, J. H. *DYNAFLOW: A Nonlinear Transient Finite Element Analysis Program*. Report 81-SM-1, Department of Civil Engineering, Princeton University, Princeton, NJ, (1987).

Saada, A. and Bianchini, G. (eds.) Constitutive equations for granular non-cohesive soils. *Proc. Int. Workshop*, Cleveland, 22–24 July 1988, A. A. Balkema, Rotterdam, (1989).

Saitta, A. *Modélisation élastoplastique du comportement mécanique des sols. Application á la liquefaction des sables et á la sollicitation déxpansion de cavité*. PhD Thesis, CERMES, ENPC Paris, (1994).

Sandhu, R. S., Shaw, H. L. and Hong, S. J. A three-field finite element procedure for analysis of elastic wave propagation through fluid-saturated soils. *Soil Dyn. Earth. Engrg.*, **9**, (1990) 58–65.

Schofield, A. N. Cambridge Geotechnical Centrifuge Operations – 20th Rankine Lecture. *Géotechnique*, **30**(3), (1980) 227–268.

Scott, R. F., Hushmand, B. and Rashidi, H. Model No. 3 primary test description and test results. In: *Verification of numerical procedures for the analysis of soil liquefaction problems, UC Davis, 17-20 Oct*, Arulanandan, K. and Scott, R. F. (eds.), A. A. Balkema, Rotterdam, (1993) 435.

Seed, H. B. and Idriss, I. M. Analysis of Soil Liquefaction: Niigata Earthquake. *ASCE SM*, **93**(SM3), (1967).

Seed, H. B., Lee, K. L., Idriss I. M. and Makdisi, F. I. Analysis of slides of the San Fernando dams during the earthquake of February 9, 1971. *J. Geotech. Engrg.*, **101**(GT7), (1975) 651–688.

Seed, H. B., Seed, R. B., Harder, L. F. and Jong, H. L. *Re-evaluation of the slide in the Lower San Fernando Dam in the Earthquake of February 9, 1971*. Report No. UCB/EERC-88/04, April, University of California, Berkeley, (1988).

Shiomi, T. *Nonlinear Behaviour of Soils in Earthquake – C/Ph/73/83*. PhD Dissertation, University Collage of Swansea, Wales, (1983).

Simon, B. R., Zienkiewicz, O. C. and Paul, D. K. An analytical solution for the transient-response of saturated porous elastic solids. *Int. J. Num. Anal. Geomech.*, **8**(4), (1984) 381–398.

Smith, I. M. An overview of numerical procedures used in the VELACS project. *Proc. VELACS Symp.*, UC Davis, 17–20 October 1994, Vol. 2, 1321–1338.

Taboada V. M. and Dobry R., Experimental results of attempted duplication of Model No. 3 at RPI in Verification of numerical procedures for the analysis of soil liquefaction problems, UC Davis, 17–20 Oct, 1993, A.A. Balkema, Rotterdam.

Tatsuoka, F. *Shear Test on a Triaxial Apparatus – a Fundamental Study*. PhD Thesis, Tokyo University, (1972).

Wilde, P. Two-invariants dependent model of granular media. *Arch. Mech.*, **29**, (1977) 799–809.

Zeng, X. Experimental results of Model No. 11. In: *Verification of Numerical Procedures for the Analysis of Soil Liquefaction Problems, UC Davis, 17-20 Oct, Vol. 1*, Arulanandan, K. and Scott, R. F. (eds.), A. A. Balkema, Rotterdam, (1993) 895–908.

Zienkiewicz, O. C. and Mroz, Z. Generalized plasticity formulation and applications to geomechanics. In: *Mechanics of Engineering Materials*, Desai, C. S. and Gallagher, R. H. (eds.), Wiley, New York, (1984) 655–679.

Zienkiewicz, O. C. and Pande, G. N. Some useful forms of isotropic yield surfaces for soil and rock mechanics. In: *Finite Elements in Geomechanics*, Gudehus, G. (ed.), Wiley, Chichester, (1977) 179–190.

Zienkiewicz, O. C. and Shiomi, T. Dynamic behaviour of saturated porous media: The generalized Biot formulation and its numerical solution. *Int. J. Num. Anal. Geomech.*, **8**, (1984) 71–96.

Zienkiewicz, O. C. and Taylor, R. L. *The Finite Element Method – Volume 1: Basic Formulation and Linear Problems*, 4th edn., McGraw-Hill, London, (1989).

Zienkiewicz, O. C. and Taylor, R. L. *The Finite Element Method – Volume 2: Solid and Fluid Mechanics, Dynamics and Non-linearity*, 4th edn. McGraw-Hill, London, (1991).

Zienkiewicz, O. C., Chang, C. T. and Bettess, P. Drained, undrained, consolidating and dynamic behaviour assumptions in soils. *Géotechnique*, **30**(4), (1980) 385–395.

Zienkiewicz, O. C., Leung, K. H., Hinton, E. and Chang, C. T. Liquefaction and permanent deformation under dynamic conditions – Numerical solution and Constitutive relations. In: *Soil Mechanics – Transient and Cyclic Loads*, Pande, G. N. and Zienkiewicz, O. C. (eds.), Wiley, Chichester, (1982).

Zienkiewicz, O. C., Leung, K. H. and Pastor, M. Simple model for transient soil loading in earthquake analysis. Part I. Basic model and its application. *Int. J. Num. Anal. Geomech.*, **9**, (1985) 453–476.

Zienkiewicz, O. C., Paul, D. K. and Chan, A. H. C. Numerical solution for the total response of saturated porous media leading to liquefaction and subsequent consolidation. *NUMETA 87 Conference*, Swansea, July 6–10 (1987).

Zienkiewicz, O. C., Paul, D. K. and Chan, A. H. C. Unconditionally stable staggered solution procedure for soil-pore fluid interaction problems. *Int. J. Num. Meth. Engrg.*, **26**, (1988) 1039–1055.

Zienkiewicz, O. C., Chan, A. H. C., Pastor, M., Paul, D. K. and Shiomi, T. Static and Dynamic Behaviour of Geomaterials – A rational approach to quantitative solutions, Part I – Fully Saturated Problems. *Proc. Roy. Soc. Lond.*, **A429**, (1990) 285–309.

Zienkiewicz, O. C., Chan, A. H. C., Pastor, M., Schrefler, B. A. and Shiomi, T. *Computational Geomechanics with Special Reference to Earthquake Engineering*. Wiley, Chichester, (1998).

6

Viscoplasticity of Geomaterials

N. D. Cristescu and **O. Cazacu**
University of Florida, Gainesville, FL, USA

6.1 Introduction

The slow deformation in metals, mainly at high temperatures, has been studied both from theoretical and experimental standpoints since the beginning of this century. The theory of viscoplasticity of metals, rooted in the works of Ludwik and Prandtl (for a history of the main concepts see Nadai, 1950, 1963, and Bell, 1973) has been extensively developed, mainly in the second half of the century. What concerns geomaterials, such as rocks and soils, their slow deformation and motion were observed since the beginning of mankind. However, a scientific approach and description of these phenomena are of relatively recent date. While, for metals, inelasticity can be explained in principle by means of the mechanics and physics of dislocation nucleation and propagation, for most rocks it is mainly the mechanisms of closure and/or opening of microcracks (and sometimes of pores) and their multiplication which explain the inelastic properties: compressibility and/or dilatancy, damage, creep, failure, etc. Dislocation mechanisms also have a role in individual crystal deformation. Irreversible volumetric deformation during creep is responsible for an increased complexity of the constitutive laws for geomaterials.

The first to report that sandstone is dilatant in uniaxial compression tests seems to be Bauschinger in 1879 (see Bell, 1973). Afterwards, Bridgman (1949) has found that in uniaxial compression tests, soapstone, marble and diabase are dilatant at high applied stresses. He was the first to mention that dilatancy is produced by 'rapid creep'. Also, he suggested that irreversible compressibility is due to closing of pores, while dilatancy is due to the opening of pores. Further pioneering experimental work concerning compressibility and/or dilatancy of rocks is due to Brace *et al.* (1966) and Bieniawski (1967), among others. For early papers concerning compressibility and/or dilatancy of rocks, see, for example, Cristescu (1989), while for soils see Schofield and Wroth (1968) and Wood (1990).

The following review is devoted to time effects on the mechanical properties of geomaterials. First, experimental data needed for the formulation of a constitutive equation are presented, then the procedure required for explicit formulation of the constitutive equation is outlined, and finally, an example of application of this constitutive equation to underground excavation is shown.

6.2   Experimental Foundation

Let us first review several types of diagnostic tests which are revealing the fundamental properties exhibited by rocks. We will present shortly the traditional tests usually
performed on rocks, such as: uniaxial tests, triaxial tests (both Kármán tests and true
triaxial tests), and creep tests. From these tests, conclusions regarding the overall mechanical behavior of the considered rock can be drawn, as well as suggestions concerning the
type of constitutive equation that would describe the observed behavior.

In the first place uniaxial compression tests are performed. As an example, Figure 6.1
shows uniaxial stress-strain curves for schist corresponding to three different loading rates,
and the results of an uniaxial creep test. During the creep test, the axial stress was held
constant at various levels for 3 days, 4 days, 6 days, 10 days and 12 days, respectively. Let
us note that creep was observed starting from low values of the applied stress. Also, it can
be seen that the whole uniaxial stress-strain curve raises with the increase of the loading
rate. The failure is dependent on the loading history (in Figure 6.1, the failure points are
represented by 'x'). In the creep test, failure was reached at much lower stress levels
than in the uniaxial standard tests. Similar results have been obtained by Hansen *et al.*
(1984) and Hunsche and Schultze (1994), among others. Since the stress-strain curves
are loading-rate dependent from the smallest values of the applied stress, and creep is
recorded from the smallest applied stresses, it can be concluded that for most geomaterials
the yield stress can be considered to be zero (see also Allemandou and Dusseault, 1996).

Since most rocks subjected to a constant stress deform by creep, this phenomenon
is of great significance for mining and civil engineering applications. Some peculiar
characteristics of creep behavior are illustrated in Figure 6.2. The uniaxial creep test on
marclay (Cristescu, 1993b) was performed with stepwise increase of the stress. For low
values of the applied stress the volume decreases (compressibility), while for higher stress
values the volume increases (dilatancy).

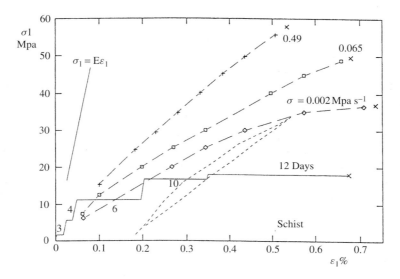

Figure 6.1   Uniaxial stress-strain curves for schist for various loading rates (Cristescu, 1986)

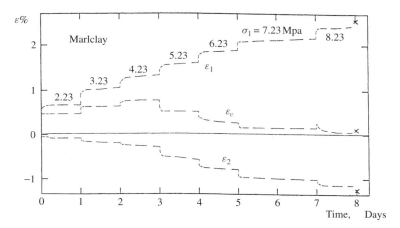

Figure 6.2   Creep curves for marclay in uniaxial stress tests showing compressibility at low stresses and dilatancy at higher stresses

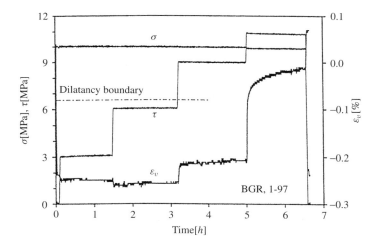

Figure 6.3   Results of a true triaxial creep test on a cubic rock salt sample

The results of a true triaxial creep test on rock salt, obtained by Hunsche (see Cristescu and Hunsche, 1998), are presented in Figure 6.3. After the hydrostatic loading, the volumetric creep deformation was 0.23% (compressibility). Then the mean stress σ was kept constant, and the octhaedral stress τ was increased in steps. For $\tau = 3$ MPa, and $\tau = 6$ MPa, the rock is compressible, while for $\tau = 9$ MPa and $\tau = 11$ MPa, it is dilatant. Similar results were reported by Maranini (1997) for limestone in axisymmetric triaxial creep tests, at low confining pressures.

Relaxation tests are important for the characterization of the rock behavior, since stress reduction (and subsequent relaxation) also occurs in the vicinity of underground openings. Relaxation tests are uniaxial or triaxial tests where, after a certain amount of deformation, the relative movement of the endfaces of the specimen is stopped ($\dot{\varepsilon}_1 = 0$), and kept so for a long time interval. During this time interval, the decrease (relaxation) of the axial stress is recorded. Figure 6.4 shows the results of a triaxial relaxation test on rock salt, consisting of four subsequent loading steps (each at $\dot{\varepsilon} = $ constant) with four succeeding relaxation phases. A significant stress relaxation was observed at each phase. Since the creep behavior after a relaxation test is governed mainly by transient creep, the results of such tests can also be used for the validation of transient creep laws.

The measurement of the elastic properties of most rocks is rather difficult because the test results can be strongly influenced by the rheological behavior of the material. This problem can be overcome by performing dynamic tests where viscous effects have no chance to interfere. Since dynamic tests are difficult to interpret correctly, in most cases the elastic parameters are measured from quasi-static experiments with unloading-reloading cycles. However, during unloading-reloading cycles significant hysteresis loops are usually observed. The determination of the elastic moduli from such tests yields much smaller

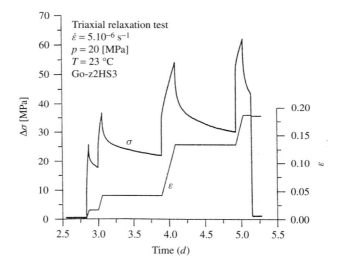

Figure 6.4   Triaxial relaxation test on rock salt

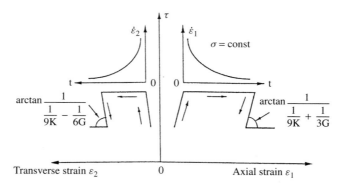

Figure 6.5   Static procedure for the determination of the elastic parameters from unloading processes following short term creep periods

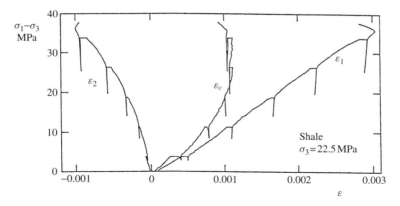

Figure 6.6 Stress-strain curves obtained in a triaxial test on shale; the unloadings followed a period of creep of several minutes

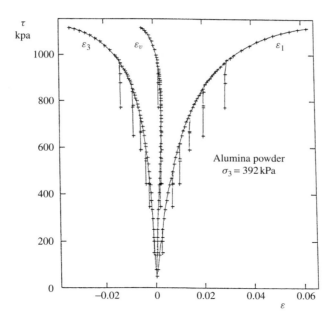

Figure 6.7 Stress-strain curves for alumina powder; the partial unloading followed a stress relaxation of 15 minutes

values than in the dynamic tests. An experimental technique that allows a reasonable good separation of viscous effects from unloading was proposed by Cristescu (1989). Thus, before passing from loading to unloading one has to keep the stress constant for a certain time interval, in order to permit the material to reach by creep a quasi-stable state. In this way, the hysteresis loops can be practically eliminated, and from the unloading slopes the elastic parameters can be evaluated with more accuracy (see Figure 6.5). An example is shown in Figure 6.6 for a shale (Nawrocki *et al.*, 1997). At different stress levels, the rock was allowed to creep until the rate of deformation components become nearly zero, and then small unloading-reloading cycles were performed. The reason for performing only partial unloading is the thickness of the specimen. During a complete unloading additional phenomena would have interfered including kinematic hardening in the opposite direction. Similarly, in a strain controlled test, a short period of relaxation, before a partial unloading followed by reloading permit a more precise determination of the elastic moduli (see Figure 6.7). The partial unloading and reloading follow the same path.

6.3   Compressibility/Dilatancy Boundary

Depending on the stress state the volume of a rock specimen increases (dilatancy) or decreases (compressibility) due to the mechanisms of opening/closing of microcracks. Figure 6.8 shows the variation of the volumetric strain with the octahedral stress for rock salt in true triaxial tests. In these tests, the load was first applied hydrostatically (hydrostatic phase). When the desired mean stress level σ has been reached, the three principal stresses were changed linearly in time so that σ is held constant and the octahedral stress τ is increased linearly (deviatoric phase) until failure. It is clearly seen that for all tests there exists a minimum of the volume. These stress states can be used to define the compressibility/dilatancy boundary (Cristescu, 1985a,b).

The shape of the compressibility/dilatancy boundary for rock salt as determined from the above described true triaxial tests is shown in Figure 6.9. The dash line means: measurable dilatancy occurs only above this horizontal line (this horizontal line has been determined from tests not shown in the diagram). Let us note that in the neighborhood of the C/D boundary (along which $\dot{\varepsilon}_v^I = 0$) the volumetric behavior is practically elastic, this domain of incompressibility (no irreversible volumetric strain variation) enlarges for high values of σ. Thus, in this range of σ, the shape of the C/D boundary cannot be precisely determined, and it is more like a band (see Figure 6.10). However, to reduce the complexity of the constitutive model, it can be considered that the C/D boundary is

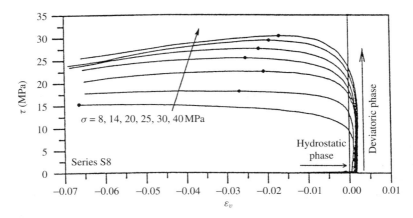

Figure 6.8   Volume change during the hydrostatic phase and part of the deviatoric phase in true triaxial compression tests, σ is the mean stress at minimum volume

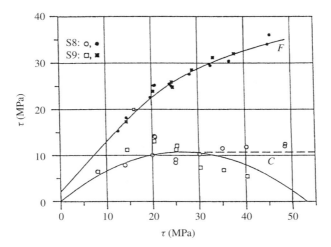

Figure 6.9 Failure surface F and dilatancy boundary C for two types of rock salt (Cristescu and Hunsche, 1997)

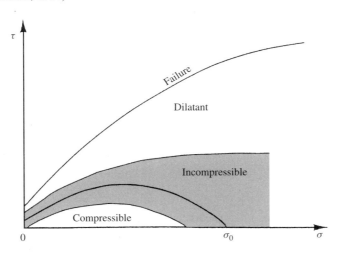

Figure 6.10 Domains of compressibility and dilatancy, separated by a domain of irreversible incompressibility

a curve in the (σ, τ) plane, represented by

$$X(\sigma, \tau) = 0 \qquad (6.1)$$

The C/D boundary is an important long term safety boundary, because creep failure and increasing permeability are inevitable under stress conditions above this boundary, whereas this is not the case below.

Some rocks are only slightly compressible, thus the compressibility is disregarded and only the incompressible and the dilatant domains are considered in the formulation of constitutive laws (see Chan *et al.*, 1992; Aubertin *et al.*, 1996). Rocks of high initial porosity are generally compressible up to failure. Under hydrostatic loading the pore collapse takes place (see for example, Andersen *et al.* (1992), Mowar *et al.* (1994) and Maranini (1997)). In the (σ, τ) plane, an 'initiation of pore collapse' boundary, may be defined and the rock behavior modeled accordingly.

6.4   Failure Surface

As already mentioned, failure of geomaterials is loading-history dependent. If the final loading level is reached in a great number of small successive steps of constant stress, failure occurs generally for higher strains and at lower stresses than in the case of fewer loading steps (see Figure 6.1). However, in the case of fewer loading steps, the time to failure is generally shorter than in the tests with a larger number of loading steps or in those in which a small constant loading rate is used. Besides time, many other factors such as specimen shape and size, platen friction, humidity, temperature may influence the measurement and determination of intact rock strength (Paterson, 1978). However, short term failure criteria formulated solely in terms of stresses may be very useful for predicting three-dimensional strength properties in the field. The term *failure* is used to describe either the stress state at which macrofracture occurs, or the peak stress attained during ductile deformation. For isotropic rock, a failure criterion is given by

$$F(\sigma, \tau, \Delta) = 0 \tag{6.2}$$

where σ is the mean stress, τ is the octahedral stress and Δ is the third stress invariant. For instance, for rock salt such a condition can be expressed as (Cristescu, 1994, using the data by Hunsche)

$$Y(\sigma, \tau) = -r\frac{\tau}{\sigma_*} - s\left(\frac{\tau}{\sigma_*}\right)^6 + \tau_0 + \frac{\sigma}{\sigma_*} = 0 \tag{6.3}$$

where r, s and τ_0 are material constants, while $\sigma_* = 1\,\text{MPa}$ (the unit stress in the tests used for the identification of the parameters). A comprehensive review of the experimental evidence, together with strength criteria for isotropic rocks, can be found in Lade (1993).

The study of the influence of anisotropy on the strength characteristics is a topic which has attracted a great deal of interest. Since the symmetry most frequently encountered is transverse isotropy, experimental tests are generally carried out on cylindrical specimens corred in several directions with respect to the strata planes and subjected to axisymmetric compressive stresses. Experimental test results have been reported by a great number of authors (Donath, 1961, 1964, 1972; Hoek, 1964; McLamore and Gray, 1967; Akai *et al.*, 1970; Allirot and Boehler, 1979; Homand *et al.*, 1993, Niandou *et al.*, 1997, etc.). From the strength data obtained, it follows that the compressive strength is a function of both the confining pressure and the orientation angle θ (θ being the angle between the normal to the strata planes and the direction of the maximum compressive stress). Theories for describing the continous variation of strength with orientation have been proposed by Jaeger (1960), McLamore and Gray (1967) and Ramamurthy (1993). However, these theories require a large amount of curve fitting, and cannot be applicable to truly three-dimensional stress states. Anisotropic failure criteria, making use of strength tensors, were proposed by Pariseau (1972), Nova (1980) and Theocaris (1989a,b, 1991), among others. A general theory of the flow and fracture of anisotropic solids was developed by Boehler and Sawczuk (1970, 1977) and Boehler (1987) in the framework of the theory of invariance. Following an approach similar to Boehler (1975), a generalization of the Mises–Schleicher criterion to transversely isotropic conditions was proposed by Cazacu (1995) (see also Cazacu et al., 1998, and Cristescu and Cazacu, 1995). The invariant form of the anisotropic Mises–Schleicher criterion (AMS) is given by

$$\frac{3}{2} \operatorname{tr} \left(\mathbf{\Sigma}'\right)^2 - \frac{m}{3} \operatorname{tr} \mathbf{\Sigma} - 1 = 0, \tag{6.4}$$

where m is a material constant, and $\mathbf{\Sigma}'$ is the deviator of the second order tensor $\mathbf{\Sigma}$, defined by

$$\Sigma_{ij} = B_{ijkl}\sigma_{kl}. \tag{6.5}$$

The anisotropic form of the criterion is based on a generalization of the second invariant of the deviatoric stress, and of the mean stress obtained through the introduction of the fourth order tensor $\mathbf{B}$. In contrast to Boehler (1975), all five components of this fourth order tensor are considered as independent strength parameters. To ensure that the AMS failure surface has the same shape for any orientation of the principal stresses system (X_1, X_2, X_3) with respect to the structural system (S_1, S_2, S_3), the following relationship between the engineering strengths of the material must be fulfilled:

$$\frac{1}{Q^2} = \frac{4}{Y_T Y_C} - \frac{1}{X_T X_C} \tag{6.6}$$

where Q is the shear strength in the symmetry plane (S_2, S_3); X_C, X_T are the uniaxial compressive and tensile strengths along S_1, while Y_C, Y_T are the uniaxial compressive and tensile strengths along S_2. If Eq. (6.6) holds, the failure surface is an elliptic paraboloid in the three-dimensional space of the principal stresses. For any orientation θ, the intersection of the failure surface with the triaxial plane $(\sigma_3, \sqrt{2}\sigma_1 = \sqrt{2}\sigma_2)$ is a parabola. Thus, the failure surface is 'open' on the compression side, showing that the axial stress may be increased without limit, if the confining pressure is increased proportionally. The failure curve is 'closed' on the tensile side, the hydrostatic tensile strength being

$$p = \frac{1}{2(1/Y_C - 1/Y_T) + (1/X_C - 1/X_T)} \tag{6.7}$$

In a plane stress test, on the other hand, the failure curve is 'closed' and the strength can only achieve a finite magnitude. As an example, in Figure 6.11 is shown the intersection of the failure surface with the plane $\sigma_1 = \sigma_2$, for $\theta = 0°$ and $\theta = 90°$, for a diatomite (data after Allirot and Boehler, 1979). The intersections of the parabola corresponding to $\theta = 0°$ with the σ_3 axis are at the uniaxial compressive strength (Y_C), and the uniaxial tensile strength $(-Y_T)$, respectively. The intersections with the $(\sigma_1 = \sigma_2, \sigma_3 = 0)$ axis represent the biaxial failure strengths in compression and tension, respectively. Similarly, for $\theta = 90°$, the intersection with the σ_3 axis are at the uniaxial compressive strength (X_C), and the uniaxial tensile strength $(-X_T)$, respectively. From Eq. (6.17) it follows that if the condition $2X_T < Y_C$ is fulfilled, as it is the case for most rocks, then $|p| < Y_T$. Similarly, if condition $2X_T < X_C$ is satisfied then $|p| < X_T$. Thus, the absolute value of the hydrostatic tensile strength p is lower than Y_T and X_T. The AMS criterion models satisfactorily the failure characteristics under tensile stresses, thus no cutoffs are necessary. For the determination of the five independent parameters of the criterion only two

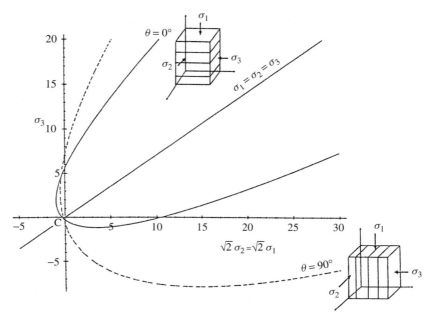

Figure 6.11   Intersection of the AMS criterion with the triaxial plane $(\sigma_3, \sqrt{2}\sigma_1 = \sqrt{2}\sigma_2)$ for $\theta = 0°$ and $\theta = 90°$ (data after Allirot and Boehler, 1979)

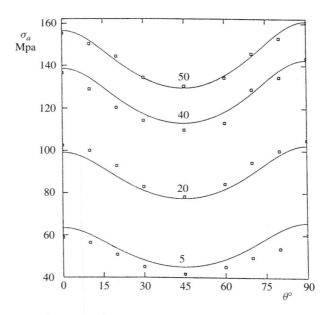

Figure 6.12   Comparison between the experimental results on Tournemire shale and the predictions of the AMS criterion (data after Niandou *et al.*, 1997)

types of tests need to be performed: (i) uniaxial compression and tensile tests in the S_1 and S_2 direction, respectively; and (ii) shear test in the (S_1, S_2) plane. Since for rocks shear, tests are difficult to perform, the parameter of the criterion related to the shear strength in the (S_1, S_2) plane can be estimated by least square fit, using the compression strength values at a given confining pressure for the orientations $\theta = 0°$, $\theta = 90°$, and at least another intermediate orientation. As an example, in Figure 6.12 is shown the variation of the peak axial stress σ_a with the orientation θ at several confining pressures, for Tournemire shale (data after Niandou *et al.*, 1997). The solid lines correspond to theoretical predictions, while symbols correspond to the data. The influence of the confining pressure on the strength characteristics is well described although only the test results for 50 MPa confining pressure were used for the fitting.

6.5 Formulation of a Viscoplastic Constitutive Equation

The constitutive equation for geomaterials must be formulated with respect to a configuration that is not stress-free. Indeed, if $\sigma^P(X)$ is the 'primary' stress state existing *in situ* at the location X, the 'secondary' stress state after the excavation, $\sigma^S(X,t)$, may vary in time. If, during and after the excavation, the deformation of the walls is not prevented, due to the stress difference $\sigma^R(X,t) = \sigma^S(X,t) - \sigma^P(X)$, the strain field around the opening will also vary. This 'relative strain', denoted by $\varepsilon^R(X,t)$, is the deformation of the rock with respect to the configuration existing *in situ* before excavation. $\varepsilon^R(X,t)$, and the corresponding displacement $u^R(X,t)$ may be measured in field, and should be obtained from σ^R with the help of the constitutive equation. Therefore, the constitutive equation has to be formulated with respect to the actual, *in situ*, reference configuration.

Based on the experimental evidence, several constitutive hypotheses may be assumed:

(1) The displacements and material rotations are small. Thus, the strain rate is the sum of the elastic (reversible) strain rate $\dot{\varepsilon}^E$, and the irreversible strain rate $\dot{\varepsilon}^I$:

$$\dot{\varepsilon} = \dot{\varepsilon}^E + \dot{\varepsilon}^I \tag{6.8}$$

(2) Since in most geomaterials, the two extended body seismic waves can propagate, the instantaneous response is elastic and described by

$$\dot{\varepsilon}^E = \frac{\dot{\sigma}}{2G} + \left(\frac{1}{3K} - \frac{1}{2G} \right) \dot{\sigma} \mathbf{1}, \tag{6.9}$$

where the shear and bulk moduli G and K may depend upon stress and/or strain invariants, or on some damage parameter (history of damage evolution); σ is the Cauchy stress, σ is the mean stress and $\mathbf{1}$ is the second order unit tensor.

(3) The irreversible part of the rate of deformation, due to transient creep may be described by

$$\dot{\varepsilon}^I_T = k_T \left\langle 1 - \frac{W(t)}{H(\sigma)} \right\rangle \frac{\partial F}{\partial \sigma} \tag{6.10}$$

where $H(\sigma)$ is the yield function, with

$$H(\sigma) = W(t) \tag{6.11}$$

the equation of the stabilization boundary (which is the locus of the stress states at the end of transient creep when stabilization takes place, i.e. when $\dot{\varepsilon}^I = 0$, $\dot{\sigma} = 0$) with:

$$W(T) = \int_0^T \boldsymbol{\sigma}(t) \cdot \dot{\boldsymbol{\varepsilon}}^I(t)\,dt = \int_0^T \sigma(t)\dot{\varepsilon}^I_v(t)\,dt + \int_0^T \boldsymbol{\sigma}'(t) \cdot \dot{\boldsymbol{\varepsilon}}^{I\prime}(t)\,dt$$

$$= W_v(T) + W_D(T) \tag{6.12}$$

the irreversible stress work per unit volume at time T, used as internal state variable. k_T is a viscosity parameter, while the symbol $\langle \; \rangle$, known as Macauley bracket, is used to denote the positive part of a function (i.e. $\langle A \rangle = 1/2(A + |A|)$). $F(\boldsymbol{\sigma})$ is a viscoplastic potential that controls the orientation of $\dot{\varepsilon}^I_T$. If F coincides with H we say that the constitutive equation is 'associated' to the yield function H. Otherwise, the constitutive equation is said to be 'non-associated', as often happens for most geomaterials. Alternatively, the orientation of the viscoplastic strain rate, can be given by a second order valuated function $N(\boldsymbol{\sigma})$ (see Cazacu, 1995 and Cazacu et al., 1997), and the flow rule has the form:

$$\dot{\boldsymbol{\varepsilon}}^I_T = k_T \left\langle 1 - \frac{W(t)}{H(\sigma)} \right\rangle N(\sigma) \tag{6.13}$$

A generalization for finite strains of (6.10) is due to Cleja-Tigoiu (1991).

To describe *steadystate creep*, one can adapt accordingly either the function H or can add to Eq. (6.10) an additional term, such as

$$\dot{\boldsymbol{\varepsilon}}^I_S = k_S \frac{\partial S}{\partial \sigma} \tag{6.14}$$

where $S(\boldsymbol{\sigma})$ is a viscoplastic potential for steady state creep and k_S is a viscosity coefficient for steady state creep, which may possibly depend upon stress invariants and possibly on damage. Since the Macaulay bracket $\langle \; \rangle$ is absent in Eq. (6.14) the creep described by such a term will last as long as stress is applied. If the geomaterial is also compressible, the function $S(\boldsymbol{\sigma})$ must satisfy some restrictive conditions, ensuring that the volumetric creep during compressibility is of transient nature only. Remember that steady state creep means that creep takes place at constant $\dot{\varepsilon}$ under nearly constant stress state. Transient and steady state creep are quite often difficult to distinguish. One can describe them either by a single term in the constitutive equation (as mentioned above), or by two additive terms. However, if the rock exhibits both compressibility and dilatancy, then the transient creep term must describe compressibility in the compressibility domain and dilatancy in the dilatancy domain. However, the steady state term must describe either dilatancy or incompressibility. Thus, it is quite difficult to model both transient and steady state creep using a single potential.

Let us examine what types of properties can describe such a constitutive equation. If the irreversible deformation is due solely to transient creep, from Eq. (6.10) we obtain for the irreversible volumetric rate of deformation:

$$(\dot{\varepsilon}^I_v)_T = k_T \left\langle 1 - \frac{W(t)}{H(\sigma)} \right\rangle \frac{\partial F}{\partial \sigma} \cdot \mathbf{1} \tag{6.15}$$

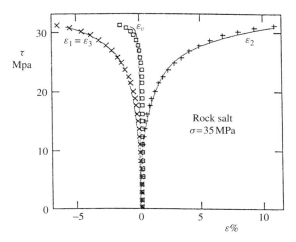

Figure 6.13 Triaxial stress-strain curves for rock salt; model prediction (solid line) in comparison with data

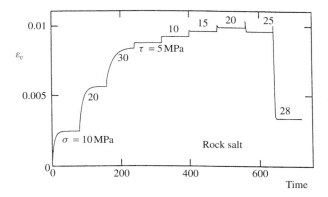

Figure 6.14 Simulation of the volumetric behavior in a true triaxial test (Cristescu, 1994a)

Therefore, the volumetric behavior is governed by the orientation of the normal to the surface $F(\boldsymbol{\sigma}) = $ constant at the point representing the actual stress state. If the projection of the normal $(\partial F/\partial \boldsymbol{\sigma}) \cdot \mathbf{1}$ on the σ-axis is pointing towards the positive orientation of this axis, the stress state produces irreversible compressibility, otherwise it produces dilatancy. If the normal is orthogonal to the σ-axis, the volumetric behavior is elastic. The procedure to determine from triaxial compression data the constitutive functions involved in Eq. (6.10) (or Eq. (6.13)) and Eq. (6.15) can be found in Cristescu (1989, 1991, 1993a, 1994) and in Cristescu and Hunsche (1998). Let us only point out here that in the specific mathematical expressions of S, and of the viscoplastic potential F (or in the expression of N) for a given rock, the short-term failure condition as well as the expression of the C/D boundary are incorporated. In Figure 6.13 is shown a comparison between the model prediction and the data in a deviatoric part of a true triaxial test on rock salt. Figure 6.14 shows the predicted volumetric strain versus time in a numerical true triaxial test. The model describes the volume compressibility during the hydrostatic part of the test, the nearly elastic behavior for small values of τ, and volume dilatancy at higher values of τ.

To characterize the viscoplastic behavior of geomaterials, a generalized yield function was used with Perzyna's theory by Desai and co-workers (see, for example, Desai and Zhang, 1987; Desai and Varadarajan, 1987; Chie and Desai, 1994; Desai *et al.*, 1995; Samtani *et al.*, 1996; etc.). This yield function permits hierarchical development to incorporate progressive complexities such as associative and non-associative responses, anisotropic hardening, strain softening and fluid pressure. It depends upon all stress invariants and involves a relatively small number of material constants.

To describe the mechanical behavior of rock salt, a viscoplastic model with mixed hardening that also describes the directional damage development during inelastic straining was proposed by Aubertin and co-workers (for the latest version of the model, see Aubertin *et al.*, 1996).

6.6   Evolutive Damage in Rocks

This section is devoted to the study of long term damage and failure of compressible/dilatant rocks (i.e. rocks exhibiting both compressibility and dilatancy). For such rocks, the progressive damage that ultimately leads to failure is closely related to the same mechanisms that produce dilatancy.

The irreversible volumetric change as well as failure is strongly time-dependent (see Figure 6.1). Slower rates produce more compressibility and more dilatancy. Also, the dilatancy at failure diminishes with increasing strain rate (see Figure 6.15). For slower loading rates, the rock can sustain much more damage before ultimate failure. The time to failure is significantly increasing with the decrease of the applied stress (see Kranz

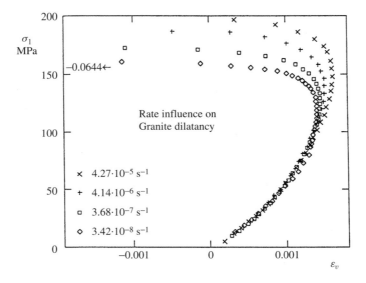

Figure 6.15   Rate influence on dilatancy and failure (last point on each curve) for granite

et al., 1982). This time increase seems to be asymptotic, i.e. increasing very much when the stress approaches a certain level (see also Lajtai and Dzik, 1996; Tharp, 1996). A correlation between dilatancy and the magnitude of the elastic wave velocities was also observed. For granite tested under uniaxial stress, Terada and Yanagidani (1986) have shown that the *P*-wave velocity increases until dilatancy sets up; after dilatancy begins, the velocity increase due to closing of microcracks compensates the velocity decrease due to growing of newly initiated microcracks; at higher stress levels the velocity decrease is greatly accelerated, immediately before faulting localization occurs. Similar results have been obtained by Scott *et al.* (1993) for sandstone.

The irreversible volumetric stress work, W_v (see Eq. (6.12)) is related to the energy of microcracking. Let us follow the variation of W_v during a true triaxial test (see Figure 6.16(a)). In the hydrostatic stage W_v increases with increasing σ. However, if the test is carried out up to very high pressures σ, W_v may reach a constant maximum value $W_{v(\text{Max})}$ that corresponds to the hydrostatic pressure σ_0, at which all pores and microcracks are closed (see Figure 6.16(a),(b)). In the deviatoric stage, W_v continues to increase until the C/D boundary is reached, then it decreases (see Figure 6.16(a),(c)). The total decrease of the irreversible stress work W_v, starting from its maximal value $W_{v(\text{Max})}$,

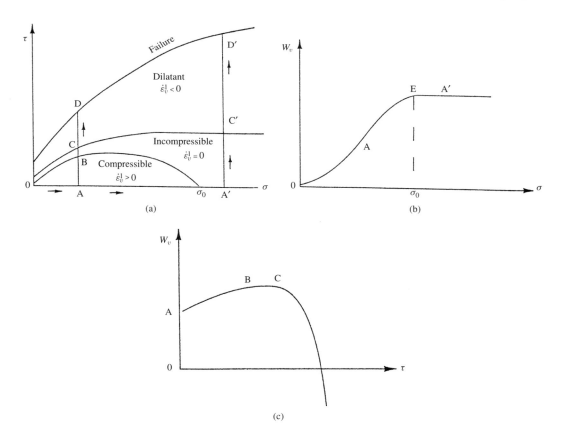

Figure 6.16 Schematic diagram of the variation of W_v in a true triaxial test. (a) Loading path shown in the constitutive plane; (b) hydrostatic stage of the test; (c) deviatoric stage

is a damage indicator (see Cristescu, 1986). Thus, *this damage parameter is a measure of the energy release due to microcracking during dilatancy.*

Let us introduce the notation

$$d(t) = W_{v(\max)} - W_v(t) \tag{6.16}$$

for the *damage parameter of the rock* at time t. By using the constitutive equation (see Eqs. (6.12) and (6.14)), it follows that the damage evolution law is given by

$$\dot{d}(t) = -\dot{W}_v(t) = -\sigma(t)\dot{\varepsilon}_v^I(t) = -k_T \left\langle 1 - \frac{W(t)}{H(\sigma)} \right\rangle \frac{\partial F}{\partial \sigma} \sigma - k_S \frac{\partial S}{\partial \sigma} \sigma$$

$$\text{if } \frac{\partial F}{\partial \sigma} < 0 \text{ and } \frac{\partial S}{\partial \sigma} < 0 \tag{6.17}$$

Thus, $\dot{d}$ depends upon the stress state and on the loading history. *The energetic parameter which characterizes the failure threshold* is defined as *the total energy release due to microcracking during the whole dilatancy process*, i.e. $d_f = W_{v(\mathrm{Max})} - W_{v(\mathrm{Failure})}$. For instance, for rock salt, the estimates of d_f obtained from true triaxial tests on several types of rock salt (Hunsche, 1996) are shown in Figure 6.17. From these results, it follows that d_f does not depend upon the Lode parameter, and apparently neither on the loading rate. d_f ranges between 0.5 and 1 MPa (an average value of 0.71 MPa was estimated by Cristescu (1994)). For granite, Cristescu (1986) reported a value of 0.167 MPa (data after Sano *et al.*, 1981). Values for d_f for other rocks are given by Cristescu (1989). Generally, d_f can be considered to be a material constant.

The above concepts can be used to predict creep failure. Indeed, by making use of the constitutive equation, we get

$$W_v(t) = \sigma \varepsilon_v^I(t) = \frac{\left\langle 1 - \dfrac{W_T(t)}{H(\sigma)} \right\rangle \dfrac{\partial F}{\partial \sigma} \sigma}{\dfrac{1}{H} \dfrac{\partial F}{\partial \sigma} \cdot \sigma} \left\{ 1 - \exp\left[\frac{k_T}{H} \frac{\partial F}{\partial \sigma} \cdot \sigma(t_0 - t) \right] \right\}$$

$$+ k_s \frac{\partial S}{\partial \sigma} \sigma(t - t_o) + W_v^P \tag{6.18}$$

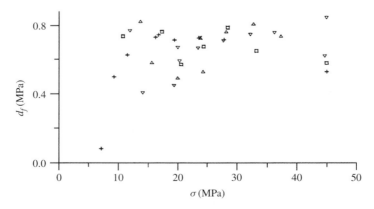

Figure 6.17   Values of d_f for rock salt as obtained from a series of triaxial tests on rock salt

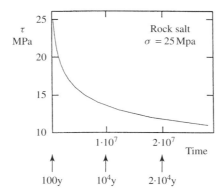

Figure 6.18   Time to creep failure for rock salt (Cristescu, 1993b, 1996d)

where t_o is the time of the beginning of the test (just after loading) and W_v^P represents the initial 'primary' value of W_v at time t_o. As an example, the predicted time to failure for rock salt in true triaxial creep tests for $\sigma = 25$ MPa and various values of τ is shown in Figure 6.18. If the value of τ is close to the short-term failure surface (which in this case is at $\tau = 26$ MPa), then the time to failure is close to zero. For lower values of τ the time to failure increases, and tends towards infinity for stress states in the vicinity of the C/D boundary. The time shown in Figure 6.18 is dimensionless, because of the incertitude in the determination of k_S (the value of k_T is not important in long-term creep tests). For $10^{-5}s^{-1} < k_s < 10^{-4}s^{-1}$, at τ around 21 MPa creep failure is expected after 100 years, at $\tau = 13$ MPa after 10^4 years, and when τ approaches 11 MPa the time to failure tends towards infinity. Therefore, for each σ, the time to failure depends upon $\tau - \tau_{C/D}$, the octahedral shear overstresses (above the C/D boundary). Thus, failure is possible anywhere in the dilatancy domain if stationary creep produces microcracking leading ultimately to a loss of cohesion (see also Rokahr and Staudtmeister, 1983). Since in stationary creep laws power functions τ^m are generally involved, stress states involving a high value of τ will produce a fast creep failure. Thus, for any practical purposes, one has to check when and where (i.e. at what location around an excavation, for example) creep failure will take place (this is expected at high values of τ). Also, progressive failure may slowly spread into the rock mass. Examples are given in the next section.

6.7   Rectangular-Like Galleries

The elastic stress distribution around a rectangular-like gallery is more difficult to be found as compared to the case of circular tunnel cross sections. Two approaches are generally used: the complex variable method (see Savin, 1961; Jaeger and Cook, 1979; and Gerçek, 1993), and the method suggested by Greenspan (1944) for plane stress case. The later approach was changed and adapted to the plane strain case by Massier (1995a,b). Thus, with Massier's method, the stress, strain, and displacement distribution around a rectangular like gallery excavated in an elastic rock can be found. It is question of rectangular in shape cross section gallery, with more or less rounded corners. The 'corners' of the cross sections, or an elongated shape of the cross section, as well as the ratio σ_h/σ_v of the far field stresses, introduce a stress concentration. Here σ_h is the in plane horizontal far field stress. In those locations, around the excavation where, due to stress concentration, we have very high values of τ, a very fast convergence by creep, as well as a fast evolutive damage, are to be expected. An analysis of the damage initiation zones as a function of the geometry of the cross section and of the far field stresses can be found in Cristescu and Paraschiv (1995a,b, 1996), Cristescu (1996), and Cristescu and Hunsche (1998, Ch. 10). We present here a few examples.

To study where around the gallery the rock becomes dilatant, where it is compressible and where failure is to be expected (as well as a possible estimation of the volume and shape of the zone of failure) the procedure described in the previous sections is followed. We intend to analyse the state just after excavation, which will be considered to be 'initial data' for creep, creep convergence, and creep failure. For the analysis the following 'ingredients' are needed:

(a) The *compressibility/dilatancy boundary:*

$$X(\sigma, \tau) = 0 \qquad (6.19)$$

For $X(\sigma, \tau) > 0$ the rock is compressible, whereas for $X(\sigma, \tau) < 0$ it is dilatant (see the dash-dot line in Figure 6.19

(b) The *short-term failure surface* (solid line in Figure 6.19):

$$Y(\sigma, \tau) = 0 \qquad (6.20)$$

(c) The *initial yield surface* (shown as interrupted line in Figure 6.19) is obtained assuming that the initial primary stress state σ^P (shown as a rhombus in Figure 6.19) is a point located on the initial stabilization boundary $H(\sigma^P) = W^P$ for transient creep. However, the constitutive equation describes also steady-state creep, which may be due either to a stress variation produced by an excavation, or to a slow tectonic motion produced by the primary stress state. For this reason steady-state

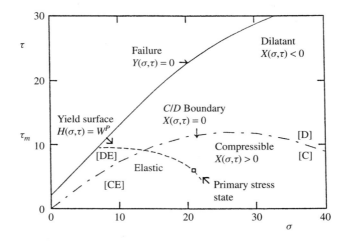

Figure 6.19 Possible stress-states around a cavern. The interrupted line represents the initial yield surface and the rhombus a possible primary stress belonging to this surface. The dash-dot line is the compressibility/dilatancy boundary

creep in Figure 6.19 produces volumetric dilatancy for stress states corresponding to points belonging to the domain labeled [DE], while from the point of view of the transient creep an 'unloading' is taking place in the same domain. Similarly, [CE] is an 'elastic' domain for transient creep, but an 'incompressible' domain for steady-state creep. If no excavation is performed, a primary stress state represented by a point located in a domain [C] or [CE] will produce a steady-state creep (a slow tectonic motion, or land slides, for example) with no irreversible volumetric changes, while a primary stress state located in the domains [D] or [DE] will result in a steady-state creep producing irreversible dilatancy besides a change in shape.

We consider that the response of the rock to any sudden loading or unloading is *elastic*. As already mentioned, 'loading' means a stress change $\sigma(t_o) \rightarrow \sigma(t)$, with $\sigma(t) \neq \sigma(t_o)$ for which $H(\sigma(t)) > H(\sigma(t_o))$, while 'unloading' means $H(\sigma(t)) < H(\sigma(t_o))$. The 'neutral' loading (i.e. $H(\sigma(t)) = H(\sigma(t_o))$) does not play an important role. If the initial primary stress state is on the stabilization boundary W^P is obtained from $H(\sigma^P) = W^P$. Let us choose a contour of the cross section of the gallery by assigning specific values to the constants P (which governs the horizontal dimensions), Q (governing the vertical dimensions), and R (governing the rounding off of the corners) (see Figures 6.20 and 6.21). Then, for a given depth h, and a certain ratio σ_h/σ_v of the far field stresses, one can plot a 'map' in the $\sigma\tau$-plane, showing the boundaries and regions defined by the concepts (a)–(c) described above. If we assume that the excavation is performed quite fast, the stress distribution which results just after excavation is computed using Massier's elastic solution. Thus, the stress distribution along the gallery contour and in the rock neighboring the gallery are obtained (see the map in Figure 6.20). Further, to show the cavern contour as well as the boundaries of various significant regions, a second 'map' in the xy-plane is plotted. For instance, Figure 6.20 shows where around the gallery the rock becomes dilatant, where compressible, and where an unloading is taking place, as well as the amount of rock involved in short-term failure. In [CE] transient creep unloading takes place, while steady state creep is not changing the volumetric strain; however, in [DE]

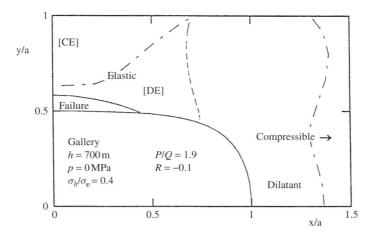

Figure 6.20 Domains of compressibility, dilatancy and failure around a gallery for $\sigma_h/\sigma_v = 0.4$ and pressure $p = 0\,\mathrm{MPa}$

steady state creep is producing dilatancy and transient creep–unloading (unloading regions are defined by $H(\sigma(t)) < H(\sigma^P)$). How much the dilatancy domain is spreading into the rock mass can be determined quite accurately. Also, the location of stress concentration (maximal values of τ) due to the geometry of the cross section and to the ratio of the far field stresses can be determined. For instance, Figure 6.21 shows the stress distribution along the contour for several internal pressures. For $p = 0$, the stress concentration on the contour is around $\beta = 40°$. There, a short-term failure will be followed by a creep failure progressing in time and spreading into the rock mass. Even for $p = 5\,\text{MPa}$, the stress concentration at $\beta = 45°$ involves quite high values of τ, and thus creep failure is to be expected quite soon.

As already mentioned, the stress concentration depends mainly upon three factors: concentration induced by the elongated shape of the cross-section, concentration induced by the ratio of the in-plane far field stresses, and concentration induced by the possibly existing 'corners' of the cross-section of the contour. If these three factors produce around the contour a combined stress concentration in the very same location, then there is a big danger of a very early creep failure. To avoid this superposition of effects we can either change the shape of the cross-section, or adapt the shape of the cavity to the ratio σ_h/σ_v (rounding more or less the corners, or choosing less elongated cross sections), if that is possible, or try to accommodate to a more favorable ratio of the in-plane far field stresses. The latter can be envisaged in those locations where the horizontal far field stresses are distinct in various horizontal directions. Then, if the design of the cavern project allows it, one can orient the excavation of the gallery (used for storage, for example) in the most favorable direction with respect to the stress concentration produced by the ratio of the in-plane far field stresses. Thus, we can envisage the choosing of an appropriate shape of the cross-section and an appropriate orientation of the gallery axis with respect to the existing horizontal far field stresses, in order to reduce the accumulation in the same

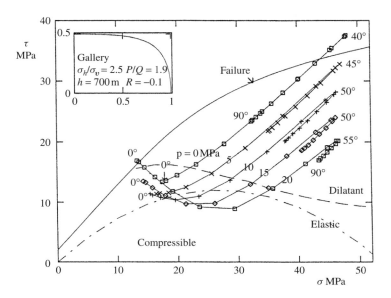

Figure 6.21   Stress distribution along the contour for several internal pressures

location (or neighboring locations) of the effects of stress concentration due to several factors involved. Let us now discuss the *influence of the ratio σ_h/σ_v on the stability* of a horizontal, elongated in shape, cavern or gallery.

Figure 6.22 shows the case of such a cavern, excavated in rock salt, with an elongated cross-section in the horizontal direction, at depth $h = 200$ m. The shape of the cavern is shown in Figure 6.22(b). For $p = 0$ the location of creep failure initiation is shown in Figure 6.22(a), for various possible ratios σ_h/σ_v. Thus, if $\sigma_h \le \sigma_v$ the stress concentration is always at the tip $x = a$ of the contour, i.e. the stress concentration induced by the ratio σ_h/σ_v and by the elongated shape of the cross sections combine and are located at the very same place of the contour. If the ratio σ_h/σ_v takes larger values, this location shifts counterclock-wise along the contour as shown in Figure 6.22(a). Figure 6.22(c) shows by $+++$ the time up to the initiation of the creep failure, as function of location of creep failure initiation β. Thus for $\beta = 0°$ and $\sigma_h/\sigma_v \le 1$ this time is shorter. When the location

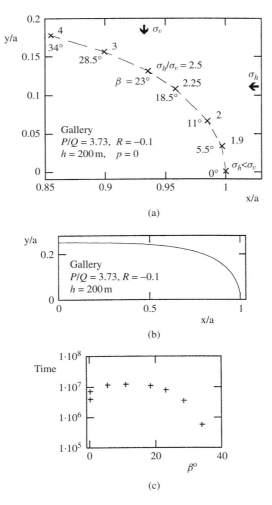

Figure 6.22   Location of the initiation of creep failure (see text)

for induced stress concentration due to elongated cross-section shape, and far field stresses slightly separate, the time to creep failure initiation increases. For very high values of σ_h/σ_v (i.e. for $\sigma_h/\sigma_v > 3$) this time starts decreasing again, since the stress concentration induced by far field stresses starts being dominant. Thus, in those locations where in various directions σ_h takes values between $\sigma_{h\,min}$ and $\sigma_{h\,max}$, one can orient the gallery so as to increase its stability by simply estimating the optimal σ_h with $\sigma_{h\,min} \leq \sigma_h \leq \sigma_{h\,max}$, also taking into account the geometry and depth.

6.8   Conclusions

This chapter summarizes the efforts directed in the last two decades to formulate general constitutive equations for rocks, which account for time effects (creep and relaxation, loading rate effects) and both compressibility and dilatancy. First, we presented the diagnostic tests to be performed to identify the main features of rock behavior under compression. Those tests suggested which type of constitutive equations is most appropriate, and provide the information needed for determination of the material coefficients involved in the constitutive equations. A brief description of a general procedure for determination of non-associated elastic/viscoplastic constitutive model for rocks was outlined; additional details and examples on real rocks (are given by Cristescu (1989) and Cristescu and Hunsche (1998)). Special attention was given to the characterization of the elastic response, a method for accurate determination of the elastic parameters. It was also shown that the constitutive model accounts for progressive damage, which is related to the energy of microcracking during dilatancy.

   Examples of application of these elastic/viscoplastic models to describe creep, evolutive damage, closure, and failure of a variety of underground openings such as boreholes, tunnels, caverns of various shapes, excavated at various depths and for various possible primary stress fields, were given. Finally, analyses of the stability of these openings for different cross-sectional geometry and various distributions of the far field stresses were also considered.

Acknowledgement

The authors would like to acknowledge the financial support of the Engineering Research Center (ERC) for Particle Science and Technology at the University of Florida, The National Science Foundation (NSF) grant #EEC-94-02989, and the Industrial Partners of the ERC.

References

Akai, K., Yamamoto, K. and Arioka, M. Experimentelle Forschung über anisotropische Eigenschaften von kristallinen Schiefern. *Proc. 2nd. Int. Congr. on Rock Mechanics*, Belgrade, vol. **II**, (1970) 181.

Allemandou, X. and Dusseault, M. B. Procedures for cyclic creep testing of salt rock, results and discussion. In: *The Mechanical Behavior of Salt: Proc. of the Third Conference*, Ghoreychi, M. Berest, P., Hardy, Jr. H. R., and Langer. M. (Eds.). Trans Tech Publ, (1996) 207–220.

Allirot, D. and Boehler, J. P. Évolution des propriétés mécaniques d'une roche stratifiée sous pression de confinement. *Proc. 4th Int. Soc. Rock Mech.*, **1**, Balkema, Rotterdam, (1979) 15–22.

Andersen, M. A., Foged, N. and Pedersen, H. F. The rate-type compaction of a weak North Sea chalk. *Rock Mechanics. Proc. 33rd U.S. Symposium.* Tillerson, J. R and Wawersik, W. R. (Eds.), Balkema, (1992) 253–261.

Aubertin, M., Sgaoula, J., Servant, S., Julien, M. R., Gill, D. E. and Ladanyi, B. An up-to-date version of SUVIC-D for modeling the behavior of salt. In: *The Mechanical Behavior of Salt. Proc. of the Fourth Conference.* Aubertin. M and Hardy Jr, M. R. (Eds.), Trans. Tech. Publ., Clausthal-Zellerfeld, (1996) 205–220.

Bell, J. F. *The Experimental Foundations of Solid Mechanics.* Handbuch der Physik, vol. VI a/1, Springer-Verlag, Berlin, (1973).

Bieniawski, Z. T. Mechanics of brittle fracture of rock. Part I – Theory of the fracture process; Part II – Experimental studies; Part III – Fracture in tension and under long-term loading. *Int. J. Rock Mech. Min. Sci.*, **4**, (1967) 365–430.

Boehler, J. P., and Sawczuk, A. (1970) Équilibre limite des sols anisotropes. *J. de Mécanique*, **9**, (1970) 5–33.

Boehler, J.-P. *Contributions théoriques et expérimentales à l'étude des milieux plastiques anisotropes.* Thése de Doctorat des Sciences, Grenoble, France, (1975).

Boehler, J. P. and Sawczuk, A. On yielding of oriented solids. *Acta Mechanica*, **27**, (1977) 185–206.

Boehler, J.-P. Yielding and failure of transversely isotropic solids. In: *Applications of Tensor Functions in Solid Mechanics.* Boehler, J.-P. (Ed.) Courses and Lectures No. 292, CISM Udine, Springer-Verlag, (1987) 3–140.

Bridgman, P. W. Volume changes in the plastic stages of simple compression. *J. Appl. Phys.* **20**, (1949) 1241–1251.

Brace, W. F., Paulding, B. W. Jr. and Scholz, C. Dilatancy in the fracture of crystalline rocks. *J. Geophys. Res.* **71**(16), (1966) 3939–3953.

Cazacu, O. *Contribution à la modelisation elasto-viscoplastique d'une roche anisotrope.* PhD Disertation. University of Science and Technology Lille, France, (1995).

Cazacu, O., Jin, J. and Cristescu, N. D. A new constitutive model for alumina powder compaction. *KONA, Powder and Particle.* **15**, (1997) 103–112.

Cazacu, O., Cristescu, N. D., Shao, J. F. and Henry, J. P. A new anisotropic failure criterion for transversely isotropic solids. *Mechanics of Cohesive-Frictional Materials*, **3**, (1998) 89–103.

Chan, K. S., Bodner, S. R., Fossum, A. F. and Munson, D. E. A constitutive model for inelastic flow and damage evolution in solids under triaxial compression. *Mechanics of Materials*, **14**, (1992) 1–14.

Chia, J. H. and Desai, C. S. Constitutive modelling of thermoviscoplastic response of rock salt. In: *Computer Methods and Advances in Geomechanics: Proc. 8th Conf. IACMAG*, Siriwardane, H. J. and Zaman, M. M. (Eds.), Balkema, Rotterdam, (1994) 555–560.

Cleja-Tigoiu, S. Elasto-viscoplastic constitutive equations for rock-type materials (finite deformation). *Int. J. Engng. Sci.*, **29**, (1991) 1531–1544.

Cristescu, N. Rock plasticity. In: *Plasticity Today; Modelling, Methods and Applications.* Sawczuk, A. and Bianchi, G. (Eds.), Elsevier, (1985a) 643–655.

Cristescu, N. Viscoplastic creep of rocks around horizontal tunnels. *Int. J. Rock Mech. Min. Sci.& Geomech. Abstr.* **22**(6), (1985b) 453–459.

Cristescu, N. Damage and failure of viscoplastic rock-like materials. *Int. J. Plasticity*, **2**(2), (1986) 189–204.

Cristescu, N. *Rock Rheology*, Kluwer Academic, Dordrecht, (1989).

Cristescu, N. Nonassociated elastic/viscoplastic constitutive equations for sand. *Int. J. Plasticity*, **7**, (1991) 41–64.

Cristescu, N. A general constitutive equation for transient and stationary creep of rock salt. *Int. J. Rock Mech. Min. Sci.& Geomech. Abstr.* **30**(2), (1993a) 125–140.

Cristescu, N. Rock rheology. In: *Comprehensive Rock Engineering.* Hudson, J. A. (Ed.), Pergamon Press, Vol. 1, (1993b) 523–544.

Cristescu, N. Viscoplasticity of geomaterials, In: *Visco-Plastic Behavior of Geomaterials*, Cristescu, N. D. and Gioda, G. (Eds.), Springer-Verlag, New York, (1994) 103–207.

Cristescu, N. and Cazacu, O. Viscoplasticity of anisotropic rock. *Fifth Int. Symposium on Plasticity and its Current Applications*, Gordon and Breach, (1995) 499–502.

Cristescu, N. and Paraschiv, I. The optimal shape of rectangular-like caverns and galleries. *Int. J. Rock Mech. Min. Sci.*, **32**, (1995a) 285–300.

Cristescu, N. and Paraschiv, I. Optimum design of large cavern. *Proc. 8th Int. Congr. on Rock Mechanics*, Balkema, Rotterdam, (1995b) 923–928.

Cristescu, N. Evolutive damage in rock salt. In: *Mechanical Behavior of Salt: Proc. of the Fourth Conf.*, Aubertin, M. and Hardy Jr., H. R (Eds.), Trans. Tech. Publ., Clausthal-Zellerfeld, (1996) 131–141.

Cristescu, N. and Paraschiv, I. Creep and creep damage around large rectangular-like caverns. *Mechamics of Cohesive-Frictional Materials*, **1**, (1996) 165–197.

Cristescu, N. D. and Hunsche, U. *Time Effects in Rock Mechanics*. Wiley, Chichester, (1998).

Desai, C. S. and Varadarajan, A. A constitutive model for quasi-static behavior of rock salt. *J. Geophys. Res.*, **92**(B11), (1987) 11445–11456.

Desai, C. S. and Zhang, D. Viscoplastic model for geologic materials with generalized flow rule. *Int. J. Numer. Analit. Meth. in Geomech.* **11**, (1987) 603–620.

Desai, C. S., Samtani, N. C. and Vulliet, L. Constitutive modeling and analysis of creeping slopes. *J. Geotechnical Eng.*, **121**, (1995) 43–56.

Donath, F. A. Experimental study of shear failure in anisotropic rocks. *Geol. Soc. Am. Bull.* **72**, (1961) 985.

Donath, F. A. Strength variation and deformational behavior in anisotropic rock. In: *State of Stress in the Earth's Crust*. Judd, W. R. (Ed.), Elsevier, Amsterdam, (1964) 281–297.

Donath, F. A. Effects of cohesion and granularity on deformational behavior of anisotropic rock. In: *Studies in Mineralogy and Precambian Geology*. Doc, B. R. and Smith, D. K. (Eds.), Vol. 135, Geological Society of America, Boulder, (1972) 95–128.

Gerçek, H. Qualitative prediction of failures around non-circular openings. In: *Assessment and Prevention of Failure in Rock Engineering*. Paşamehmetoğlu, A. G., Kawamoto, T. Whittaker, B. N. and Aydan, Ō (Eds.), Balkema, Rotterdam, (1993) 727–732.

Greenspan, M. Effect of a small hole on the stresses in a uniformly loaded plate. *Quart. J. Appl. Math.* **2**, (1944) 60–71.

Hansen, F. D., Mellegard, K. D. and Senseny, P. E. Elasticity and strength of ten natural rock salt. In: *The Mechanical Behavior of Salt: Proc. First Conf.* Hardy, Jr., H. R and Langer, M. (Eds.), Trans. Tech. Publ., Clausthal-Zellerfeld, (1984) 71–83.

Hoek, E. Fracture of anisotropic rock., *J. S. Afr. Inst. Min. Metall.*, **64**(10), (1964) 501–518.

Homand, F., Morel, E., Henry, J. P., Cuxac, P. and Hammade. Characterization of the moduli of elasticity of an anisotropic rock using dynamic and static methods. *Int. J. Rock Mech. Min. Sci. & Geomech. Abstr.*, **30**, (1993) 527–535.

Hunsche, U. and Schulze, O. Das Kriechverhalten von Steinsalz. (Creep of rocksalt). *Kali und Steinsalz*, **11**, (1994) 238–255 (in German).

Jaeger, J. C. Shear failure of anisotropic rocks. *Geol. Mag.*, **97**, (1960) 65–72.

Jaeger, J. C. and Cook, N. G. W. *Fundamentals of Rock Mechanics*. Chapman & Hall, London, (1979).

Kranz, R. L., Harris, W. J. and Carter, N. L. Static fatigue of granite at 200°C. *Geophys. Res. Lett.*, **9**(1), (1982) 1–4.

Lade, P. V. Rock strength criteria: The theories and the evidence. In: *Comprehensive Rock Engineering*. Hudson, J. A. (Ed.), Pergamon Press, Vol. 1, Fundamentals, (1993) 255–284.

Lajtai, E. Z. and Dzik, E. J. Searching for the damage threshold in intact rock. In: *Rock Mechanics. Tools and Techniques: Proc. 2nd NARMS*. Aubertin, M., Hassani, F. and Mitri, H. (Eds.), Balkema, Rotterdam, (1996) 701–708.

Maranini, E. *Comportamento Viscoplastico dei Materiali Rocciosi: Studio Sperimentale, Leggi Costitutive, Applicazioni*. Tesi di Dottorato, Universita degli Studi di Ferrara, Dottorato di Ricerca in Geologia Applicata, Italia, (1997).

Massier, D. *In-situ* creep of a linear viscoelastic rock around some noncircular tunnels. Part I. Theoretical solution. *Rev. Roum. Sci. Techn., Ser. Mec. Appl.*, **40**(2–3), (1995a) 413–423.

Massier, D. *In-situ* creep of a linear viscoelastic rock around some noncircular tunnels. Part II. Boundary stress and displacement concentrations. *Rev. Roum. Sci. Techn., Ser. Mec. Appl.*, **40**(4–6), (1995b) 485–502.

McLamore, R. and Gray, K. E. The mechanical behavior of anisotropic sedimentary rocks. *J. of Engrg. Industry,* **89**, (1967) 62–73.

Mowar, S., Zaman, M., Stearns, D. W. and Roegiers, J.-C. Pore collapse mechanisms in Cordoba Cream limestone. In: *Rock Mechanics Models and Measurements. Challenges from Industry: Proc. 1st NARMS.* Nelson, P. P and Laubach, S. E. (Eds.), Balkema, Rotterdam, (1994) 767–773.

Nadai, A. *Theory of Flow and Fracture of Solids.* Vol. 1, McGraw-Hill, New York, (1950).

Nadai, A. *Theory of Flow and Fracture of Solids.* Vol. 2, McGraw-Hill, New York, (1963).

Nawrocki, P. A., Cristescu, N. D., Dusseault, M. B. and Bratli, R. K. Experimental methods for determining constitutive parameters for non-linear rock modelling. *Proc. Nineth Int. Conf. on Computer Methods and Advances in Geomechanics,* Balkema, Rotterdam, (1997) 831–836.

Niandou, H., Shao, J. F., Henry, J. P. and Fourmaintraux, D. Laboratory Investigation of the Mechanical Behavior of Tournemire Shale. *Int. J. Rock Mech. Min. Sci. & Geomech. Abstr.,* **34**, (1997) 3–16.

Nova, R. The failure of transversely anisotropic rocks in triaxial compression. *Int. J. Rock Mech. Min. Sci. & Geomech. Abstr.,* **17**, (1980) 325–332.

Pariseau, W. G. Plasticity theory for anisotropic rocks and soils. *Proc. 10th Symposium on Rock Mechanics,* AIME, (1972) 267–295.

Paterson, M. S. *Experimental Rock Deformation – The Brittle Field.* Springer-Verlag, Berlin, (1978).

Ramamurthy, T. Strength and modulus responses of anisotropic rocks. In: *Comprehensive Rock Engineering.,* Hudson, J. A. (Ed.), Pergamon Press, vol. 1, Fundamentals, (1993) 313–329.

Rokahr, R. B. and Staudtmeister, K. Creep rupture criteria for rock salt. *Sixth Int. Salt Institute, Symposium on Salt,* vol. 1, (1983) 455–462.

Samtani, N. C., Desai, C. S. and Vulliet, L. An interface model to describe viscoplastic behavior. *Int. J. Numerical and Analyt. Methods in Geomech.,* **20**, (1996) 231–252.

Sano, O., Itô, I. and Terada, M. Influence of strain rate on dilatancy and strength of Oshima granite under uniaxial compression. *J. Geophys. Res.,* **86**(B10), (1981) 9299–9311.

Savin, G. N. *Stress Concentration Around Holes.* McGraw Hill, New York, (1961).

Schofield, A. and Wroth, P. *Critical State Soil Mechanics.* McGraw-Hill, New York, (1968).

Scott Jr., T. E., Ma, Q. and Roegiers, J.-C. Acoustic velocity changes during shear enhanced compaction of sandstone. In: *Rock Mechanics in the 1990s, The 34th U.S. Symp. on Rock Mechanics.* Haimson, B. C. (Ed.), *Int. J. Rock Mech. Min. Sci. & Geomech. Abstr.,* **30**, (1993) 763–769.

Terada, M. and Yanagidani, T. Application of Ultrasonic Computer Tomography to Rock Mechanics. *Ultrasonic Spectroscopy and its Applications to Materials Science,* Report of Special Project Research, Kyoto. (1986) 205–210.

Tharp, T. M. A fracture mechanics analysis of stand-up time for mine roof beams. In: *Rock Mechanics: Proc. 2nd NARMS.* Aubertin, M., Hassani, F. and Mitri. H. (Eds.), Balkema, Rotterdam, (1996) 1177–1184.

Theocaris, P. S., The elliptic paraboloid failure surface for transversely isotropic materials off-axis loaded. *Rheologica Acta,* **28**, (1989a) 154–165.

Theocaris, P. S. The paraboloid failure surface for the general orthotropic material. *Acta Mechanica,* **79**, (1989b) 53–79.

Theocaris, P. S. The elliptic paraboloid failure criterion for cellular solids and brittle foams. *Acta Mechanica,* **89**, (1991) 93–121.

Wood, D. M. *Soil Behaviour and Critical State Soil Mechanics.* Cambridge University Press, Cambridge, (1990).

7

Gradient and Cosserat Models: Theory and Computation

Hans-B Mühlhaus[1], Rene de Borst[2] and Bert Sluys[2]

[1]*CSIRO, Division of Exploration & Mining, Nedlands WA, Australia*
[2]*Delft University of Technology, The Netherlands*

7.1 Introduction

It has long been recognized that internal degrees-of-freedom play an important role in the continuum mechanical modeling of material behavior. However, it has been less widely appreciated that these same internal degrees-of-freedom, and in particular the granular rotation or spin, may need to be included explicitly in the kinematical description of flow and deformation processes. The inclusion of non-standard degrees-of-freedom leads to a so-called generalized continuum theory. In the case of independent rotational degrees-of-freedom (besides the spin of a material element of the continuum), one speaks of a Cosserat continuum.

From a microscopic point of view, coupling between the average granular velocity **v** and the spin field **w** is a manifestation of noncentral intergranular forces. A hint to a macroscopic description of the same effect emerges from the observation that a line element of flowing granular material will rotate with the vorticity 1/2 curl **v**, but in doing so will meet the resistance or stress due to the internal spin if vorticity and spin are not synchronized. Such resistance is a function of 1/2 curl **v** − **w**. This observation was first made in a gas dynamics context by Born (1920), who suggested antisymmetric stresses as the appropriate resistance. Born seemed unaware of the work of the brothers Cosserat (1909).

The recent renaissance of generalized continuum theories was mainly triggered by difficulties of conventional continuum models in dealing with strain non-uniformities and strain softening. Indeed, an adequate choice of the additional degrees-of-freedom and/or higher order gradient terms restores the well-posedness of the differential problem, and makes it possible to describe spatial features of the deformation or flow behavior (Walgraef and Aifantis, 1984; Triantafillydes and Aifantis, 1987; Mühlhaus and Vardoulakis, 1987; de Borst and Mühlhaus, 1993). Desai *et al.* (1996) and Desai and Zhang (1997) used a nonlocal variant of Desai's Disturbed State Concept (DSC) model (see Desai, 1995, for an overview) in the numerical analysis of an indentation problem. The nonlocality was introduced by replacing the local strain tensor of the original model by an averaged strain tensor. The authors demonstrated that the resulting finite-element solution is insensitive with respect to mesh refinement. As indicated before, when used in connection with problems involving strain softening, standard models typically exhibit a pathological mesh dependency. This dependency is totally unrelated to the usual reduction of the discretization error upon mesh refinement.

Mühlhaus and Vardoulakis (1987) assumed that slow granular flow can be described within the framework of an incremental Cosserat plasticity theory. In a linear instability analysis, they predicted shear band thicknesses of the order of 10–20 average grain diameters, a result which is in good agreement with the range of shear band thicknesses observed in experiments. However, this does not prove that granular materials are indeed Cosserat media. It merely shows that higher order gradients, and in this case rotation gradients, become significant once the characteristic wavelength of the macroscopic deformation pattern is of the order of 10–20 grain diameters. Indeed, Jenkins (1993) and Mühlhaus and Oka (1996) have derived continuum theories for slow deformations of granular materials where the higher order stress depends upon gradients of the relative deformation gradient, rather than the gradient of the granular spin alone, as it would in Cosserat-type theories.

7.2   Gradient and Cosserat Continuum Models

In this section we use the special case of a granular medium as a starting point and motivation for the derivation of a specific polar continuum model. Consider two spherical granules 1 and 2 having translational velocities c_1 and c_2 and angular velocities s_1, and s_2, respectively. The total relative velocity at the point of contact is

$$g_{21} = c_{21} + \frac{D}{2}(k \times s_{21}) \tag{7.1}$$

where $c_{21} = c_2 - c_1$, $s_{21} = s_{12} = s_1 + s_2$, D is the diameter of the granules and k is the unit vector along the center line from granule 1 to granule 2. We decompose the granule velocities c_α and spins s_α, $\alpha = (1, 2)$ into average values v_α and ω_α and fluctuations (deviations from the average values).

In view of the desired continuum theory, we assume that

$$v_{21} = v_2 - v_1 = D[k \cdot \nabla]v + \frac{D^2}{2}[k \cdot \nabla][k \cdot \nabla]v + \cdots \tag{7.2}$$

$$\omega_{21} = \omega_2 - \omega_1 = D[k \cdot \nabla]\omega + \frac{D^2}{2}[k \cdot \nabla][k \cdot \nabla]\omega + \cdots \tag{7.3}$$

With Eqs. (7.2) and (7.3), the mean field part of g_{21} is given by:

$$\bar{g}_{21} = D\left(1 + \frac{D}{2}[k \cdot \nabla]\right)\gamma k \tag{7.4}$$

where the matrix elements of γ are

$$\gamma_{ij} = [\nabla v - w^c]_{ij} = v_{i,j} - W^c_{ij} \tag{7.5}$$

and W^c is the spin tensor corresponding to ω (viz $\omega \times k = W^c k$).

We base the derivation of our continuum theory on the expression for the average specific power:

$$\dot{w} = \frac{n}{2}\langle \mathbf{F}_{21} \cdot \bar{\mathbf{g}}_{21} \rangle \tag{7.6}$$

where $\rho = mn$ is the density, m is the mass of a granule, n is the number of granules per unit volume, the factor $1/2$ considers the fact that one contact is shared by two granules and $\langle 1 \rangle = 1$ is a suitable averaging operator.

Next we insert Eq. (7.4) into Eq. (7.6), and write the results as follows:

$$\dot{w} = \sigma_{ij}\gamma_{ij} + m_{ijk}\gamma_{ij,k} \tag{7.7}$$

where

$$\sigma_{ij} = \frac{n}{2}D\langle F_{21_i}k_j \rangle \tag{7.8}$$

$$m_{ijk} = \frac{n}{4}D^2\langle F_{21_i}k_jk_k \rangle \tag{7.9}$$

If higher order velocity gradients are neglected Eq. (7.7) can be written as

$$\dot{w} = \sigma_{ij}\gamma_{ij} - m_{ijk}W^c_{ij,k} \tag{7.10}$$

Equation (7.10) can be represented alternatively in terms of second rank axial tensor fields as

$$\dot{w} = \sigma_{ij}\gamma_{ij} + \mu_{nk}\kappa_{nk} \tag{7.11}$$

where

$$W^c_{ij} = -e_{ijm}\omega_m, \quad \kappa_{mk} = \omega_{m,k} \tag{7.12}$$

$$m_{ijk} - m_{jik} = e_{ijn}\mu_{nk} \tag{7.13}$$

and e_{ijn} ($e_{123} = 1$) designates the permutation symbol. Insertion of Eq. (7.13) into Eq. (7.9) yields

$$\mu_{lk} = \frac{n}{4}D^2 e_{ijl}\langle F_{21_i}k_jk_k \rangle \tag{7.14}$$

The averages in Eqs. (7.6)–(7.9) and (7.14) are defined as

$$\langle (\cdot) \rangle = \int_{4\pi} d\mathbf{k}A(\mathbf{k}, \mathbf{r}, t)(.), \quad A(\mathbf{k}) = A(-\mathbf{k}) \tag{7.15}$$

where $\mathbf{r}$ is the current position vector of a material point. The function $A(\cdot)$ accounts for the orientation distribution of the contacts between adjacent granules, so that $A\,d\mathbf{k}$ is the probable number of contacts in the element $d\mathbf{k}$ centered at $\mathbf{k}$ (e.g. Oda *et al.*, 1982). If the distribution of collisions is isotropic and independent of time and position, then $A = k/(4\pi)$, where k is the average number of contacts per granule, the so-called *coordination number*. We now turn to the derivation of the field equations. By integration and application of the divergence theorem, the right-hand side of Eq. (7.7) is converted into

$$-\int_v \tau_{ij,j}v_i\,dV + \int_v e_{ijk}\tau_{ij}\omega_k\,dV + \int_A \tau_{ij}n_jv_i\,dA + \int_A m_{ijk}n_k\gamma_{ij}\,dA \tag{7.16}$$

where we have introduced the notation

$$\tau_{ij} = \sigma_{ij} - m_{ijk,k} \tag{7.17}$$

The stress τ_{ij} has a concrete, physical significance to which we return in the next section. Equation (7.16) suggests the form

$$\dot{W}_{\text{ext}} = -\int_v \rho(\dot{v}_i v_i + (D^2/10)\dot{\omega}_i\omega_i)\,dV + \int_A t_i v_i\,dA + \int_A t_{ij}\gamma_{ij}\,dA$$

$$+ \int_v \rho g_i v_i\,dV + \int_v \rho \mu_i \omega_i\,dV \tag{7.18}$$

for the power of the external forces; t_i and t_{ij} are the surface tractions of the stress and higher order stress tensors (cf. Eq. (7.21)). In the definition of $\dot{W}_{\text{ext}}$ we have included the inertial terms; ρg_i and $\rho \mu_i$ are volume forces and couples, respectively. From the invariance properties of $\int_v \dot{w}\,dV - \dot{W}_{\text{ext}} = 0$, it follows that

$$\tau_{ij,j} + \rho g_i = \rho \dot{v}_i, \quad -e_{ijk}\tau_{ij} + \rho \mu_k = \rho(D^2/10)\dot{\omega}_k \tag{7.19}$$

We note that $\tau_{ij} = \tau_{ji}$ if $\mu_k = 0$ and $(D^2/10)\dot{\omega}_k$. Structurally the present theory is obviously very different from a Cosserat theory. A Cosserat theory is obtained if higher order velocity gradients are neglected. In the latter case, the field equations are obtained from Eq. (7.11) as

$$\sigma_{ij,j} + \rho g_i = \rho \dot{v}_i, \quad \mu_{ik,k} - e_{ilm}\sigma_{lm} + \rho \mu_i = \rho(D^2/10)\dot{\omega}_i \tag{7.20}$$

and in Eq. (7.18) the term $\int_A t_{ij}\gamma_{ij}\,dA$ is replaced by $\int_A m_i\omega_i\,dA$.

It remains to specify the boundary conditions. We first consider the general case. From the surface integrals appearing in Eq. (7.18), we can deduce the boundary conditions necessary for a complete specification of the mechanical problem. The first of these terms already involves **v** directly, and hence will cause no special difficulty. The last term can be written as

$$\int_A t_{ij}\gamma_{ij}\,dA = \int_A m_{ijk}n_k\gamma_{ij}\,dA = \int_A m_{ijk}n_k v_{i,j}\,dA + \int_A m_{ijk}n_k e_{ijn}\omega_n\,dA \tag{7.21}$$

and the second term on the RHS of Eq. (7.21), which involves ω directly, also causes no special difficulties. However, the remaining term is considerably less accommodating, due to the fact that tangent derivatives of **v** are determined by specifying **v** on the boundary, while the normal derivatives must be specified separately. Thus, the first term on the RHS of Eq. (7.21) must be decomposed into two parts, and the part determined by **v** must be written so that the integrand contains **v** explicitly. For brevity, define $\phi_{ij} = m_{ijk}n_k$. Then the integrand of the term of interest is

$$\phi_{ij}v_{i,j} = \phi_{ij}(v_{i,j} - v_{i,m}n_m n_j) + \phi_{ij}v_{i,m}n_m n_j$$

$$= [\partial_j - n_j n_m \partial_m]\phi_{ij}v_i - (\phi_{ij,j} - \phi_{ij,m}n_m n_j)v_i + \phi_{ij}v_{i,m}n_m n_j \tag{7.22}$$

The surface divergence theorem

$$\int_S [\partial_j - n_j n_m \partial_m]a_j\,dA = \int_S a_m n_m[\partial_j - n_j n_m \partial_m]n_j\,dA \tag{7.23}$$

follows from decomposing the vector **a** into tangent and normal parts, and noting that the integral of the surface divergence of the tangent part vanishes by virtue of the closedness of **S**. Applying the chain rule to the remaining normal part shows that one of the terms in the chain rule vanishes identically, leading to Eq. (7.23). Putting all this together leads to the three essential and corresponding natural boundary condition pairs

$$\tau_{ij}n_j + m_{ijk,n}n_j n_k n_n +$$

$$[(n_{n,n} - n_{n,m}n_n n_m)m_{ijk} + m_{ijm}n_{m,k}]n_j n_k$$

$$- m_{ijk,j}n_k - m_{ijk}n_{k,j} \qquad \text{or} \quad v_i \qquad (7.24)$$

$$m_{ijk}n_j n_k \qquad \text{or} \quad v_{i,j}n_j$$

$$m_{ijk}n_k e_{ijn} \qquad \text{or} \quad \omega_n$$

In a Cosserat continuum no gradient terms appear in the expression for the power of the external force, so that the deduction of the boundary conditions is straightforward. The two essential and natural boundary condition pairs read

$$\sigma_{ij}n_j \quad \text{or} \quad v_i \qquad (7.25)$$

$$m_{ij}n_j \quad \text{or} \quad \omega_i \qquad (7.26)$$

7.3 Interpretation

Here we outline a heuristic interpretation of how stress tensors of the form

$$\tau_{ij} = \sigma_{ij} - m_{ijk,k} \qquad (7.27)$$

as in Eq. (7.17) can come about. A similar interpretation was presented by Mühlhaus (1995) within the context of laminated materials. We consider a characteristic volume $V^c = (2a, 2b, 2c)$ which, for convenience, we assume as rectangular with sides parallel to the coordinate axes (x_1, x_2, x_3). In the absence of volume forces, we have the identity

$$\int_{V^c} \sigma_{ij}^m \, dV = \int_{A^c} \sigma_{ik}^m n_k x_j \, dA = \int_{A^c} \sigma_{ik}^m x_j e_{mnk} \, dx_m \, dx_n \qquad (7.28)$$

where $\sigma_{ij}^m(\mathbf{r})$ designates the microstress distribution within V^c. In particular,

$$\int_{V^c} \sigma_{12}^m \, dV = b \int (\sigma_{12}^m(x_2 = b) + \sigma_{12}^m(x_2 = -b)) \, dx_1 \, dx_3$$

$$+ \int ((\sigma_{11}^m(x_1 = a) - \sigma_{11}^m(x_1 = -a))x_2) \, dx_2 \, dx_3$$

$$+ \int ((\sigma_{13}^m(x_3 = c) - \sigma_{13}^m(x_3 = -c))x_3) \, dx_1 \, dx_2 \qquad (7.29)$$

This expression can be written as

$$\tau_{12} = \sigma_{12} - m_{12k,k} \qquad (7.30)$$

where

$$\tau_{12} = \frac{1}{8abc} \int \sigma_{12}^m \, dV \tag{7.31}$$

$$\sigma_{12} = \frac{1}{4ac} \int \sigma_{12}^m (x_2 = 0) \, dx_1 \, dx_3 \tag{7.32}$$

$$m_{122,2} = \frac{-b^2}{8ac} \int \sigma_{12,22}^m (x_2 = 0) \, dx_1 \, dx_3 \tag{7.33}$$

$$m_{121,1} = \frac{-1}{8abc} \int ((\sigma_{11}^m (x_1 = a) - \sigma_{11}^m (x_1 = -a))x_2) \, dx_2 \, dx_3$$

$$= \frac{-1}{4bc} \int \sigma_{11,12}^m (x_1 = 0, x_2 = 0)x_2^2 \, dx_2 \, dx_3 \tag{7.34}$$

$$m_{123,3} = \frac{-1}{8abc} \int ((\sigma_{13}^m (x_3 = c) - \sigma_{13}^m (x_3 = -c))x_2) \, dx_1 \, dx_2$$

$$= \frac{-1}{4ab} \int \sigma_{13,32}^m (x_2 = 0, x_3 = 0)x_2^2 \, dx_1 \, dx_2 \tag{7.35}$$

Analogous expressions are obtained for the other components. The result suggests the interpretation of τ_{ij} as the volume average of the microstress σ_{ij}^m and σ_{ij} as the area average of σ_{ij}^m. The volume average is symmetric if σ_{ij}^m is symmetric, however σ_{ij} is nonsymmetric in general if σ_{ij}^m is inhomogeneous within V^c. In a Cosserat Continuum, the only non-vanishing component of the higher order stress tensor is $m_{121} - m_{211} = \mu_{31}$. For the situation illustrated in Figure 7.1, it follows from $\tau_{21} = \tau_{12}$ that

$$\mu_{31,1} + \sigma_{21} - \sigma_{12} = 0 \tag{7.36}$$

That is, in this particular case the symmetry condition $\tau_{21} = \tau_{12}$ is equivalent to the moment equilibrium condition of the Cosserat continuum.

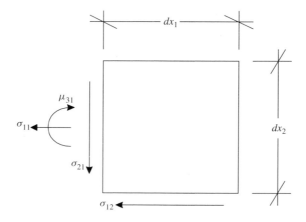

Figure 7.1   Stresses and couple stresses at the element (dx_1, dx_2)

7.4   Cosserat and Gradient Elasticity

To facilitate the treatment of nonlinearities, we formulate the constitutive relations in rate form. Differentiation of Eq. (7.8) gives

$$\dot{\sigma}_{ij} = \frac{n}{2}D\langle \dot{F}_{21_i}k_j\rangle + \frac{n}{2}D\langle F_{21_i}\dot{k}_j\rangle + S_{ij} \tag{7.37}$$

where

$$S_{ij} = \int_{4\pi} dk \frac{d}{dt}\left(\frac{n}{2}DA\right)F_{21_i}k_j \tag{7.38}$$

The unit vector $\mathbf{k}$ is parallel to the connecting line between the mass centers of granules 1 and 2. By differentiation of $k_i = (x_i^2 - x_i^1)/|(x_i^2 - x_i^1)|$, we obtain

$$\dot{k}_i = \tilde{\omega}_{ij}k_j \tag{7.39}$$

where

$$\tilde{\omega}_{ij} = v_{i,j} - \delta_{ij}(v_{k,l}k_k, k_l) + O(D^2) \tag{7.39a}$$

The above relationship suggests the following symmetric definition of the objective stress rate:

$$\overset{\nabla}{\sigma}_{ij} = \frac{n}{2}D\langle \overset{\nabla}{F}_{21_i}k_j\rangle + S_{ij} \tag{7.40}$$

where

$$\overset{\nabla}{\sigma}_{ij} = \dot{\sigma}_{ij} - \frac{n}{2}D\langle \tilde{\omega}_{jk}F_{21_i}k_k\rangle - \frac{n}{2}D\langle \tilde{\omega}_{ik}F_{21k}k_j\rangle \tag{7.40a}$$

and

$$\overset{\nabla}{F}_{21_i} = \dot{F}_{21_i} - \tilde{\omega}_{ik}F_{21k} \tag{7.40b}$$

A more tractable definition is obtained by replacing $\tilde{\omega}_{ij}$ by its average over the solid angle, viz

$$\tilde{\omega}_{ij} \rightarrow \frac{1}{4\pi}\int_{4\pi} d\mathbf{k}\tilde{\omega}_{ij} = v_{i,j} - \frac{1}{3}v_{k,k}\delta_{ij} \tag{7.41}$$

The definition Eq. (7.40) then becomes

$$\overset{\nabla}{\sigma}_{ij} = \dot{\sigma}_{ij} - \tilde{\omega}_{jk}\sigma_{ik} - \tilde{\omega}_{ik}\sigma_{kj} \tag{7.42}$$

Assuming Eq. (7.41), the rate of m_{ijk} (Eq. (7.9)) is obtained as

$$\overset{\nabla}{m}_{ijk} = \dot{m}_{ijk} - \tilde{\omega}_{il}m_{ljk} - \tilde{\omega}_{jm}m_{imk} - \tilde{\omega}_{kn}m_{ijn} \tag{7.43}$$

and

$$\overset{\nabla}{m}_{ijk} = \frac{n}{4}D^2\langle \overset{\nabla}{F}_{21_i}k_jk_k\rangle + S_{ijk} \tag{7.44}$$

The tensor S_{ijk} is defined in analogy to Eq. (7.38). The above definitions reduce to the corresponding Jaumann rates if we equate $\tilde{\omega}_{ij}$ to the spin W_{ij} of an infinitesimal element of the continuum.

For illustration we consider the simple model

$$\overset{\nabla}{F}_{21} = K_t(\overline{\mathbf{g}}_{21} - (\overline{\mathbf{g}}_{21} \cdot \mathbf{k})\mathbf{k}) + K_n(\overline{\mathbf{g}}_{21} \cdot \mathbf{k})\mathbf{k} \tag{7.45}$$

K_t and K_n are incremental normal and tangential contact stiffnesses and $\bar{\mathbf{g}}_{21}$ is the mean field part of the relative velocity vector. Next we insert Eq. (7.45) into Eqs. (7.40) and (7.44). The result is:

$$\overset{\triangledown}{\sigma}_{ij} = \frac{nD^2}{2}[(K_n - K_t)\langle k_i k_j k_n k_m \rangle + K_t \delta_{in} \delta_{rm} \langle k_r k_j \rangle]\gamma_{nm} + S_{ij} \tag{7.46}$$

$$\overset{\triangledown}{m}_{ijk} = \frac{nD^4}{8}[(K_n - K_t)\langle k_i k_j k_n k_m k_k k_s \rangle + K_t \delta_{in} \delta_{rm} \langle k_j k_k k_r k_s \rangle]\gamma_{nm,s} + S_{ijk} \tag{7.47}$$

In the derivation of Eqs. (7.46) and (7.47) we have assumed for simplicity that the contact stiffnesses are independent of $\mathbf{k}$. If such a dependency exists, then K_n and K_t have to be included into the argument of the averaging operator. The Jaumann rate of the couple stress tensor in Eq. (7.14) is obtained as

$$\overset{\triangledown}{\mu}_{ij} = \frac{nD^4}{8}K_t[\langle k_j k_k \rangle \kappa_{ik} - \langle k_i k_j k_k k_l \rangle \kappa_{lk}] + M_{ij} \tag{7.48}$$

where M_{ij} is the couple stress analogue of Eq. (7.38).

If the contact distribution is isotropic then $A = k/(4\pi)$ and (e.g., Mühlhaus and Oka, 1996)

$$\frac{1}{k}\langle k_i k_j \rangle = \frac{1}{3}\delta_{ij}, \quad \frac{1}{k}\langle k_i k_j k_k k_l \rangle = \frac{1}{15}(\delta_{ij}\delta_{kl} + \delta_{ik}\delta_{jl} + \delta_{il}\delta_{jk}) \tag{7.49}$$

and

$$\frac{1}{k}\langle k_i k_j k_k k_l k_m k_n \rangle = \frac{1}{7}(\delta_{in}\delta_{jklm} + \delta_{jn}\delta_{klmi} + \delta_{kn}\delta_{lmij} + \delta_{ln}\delta_{mijk} + \delta_{mn}\delta_{ijkl}) \tag{7.50}$$

where

$$\delta_{ijkl} = \frac{1}{k}\langle k_i k_j k_k k_l \rangle \tag{7.51}$$

The relative importance of the higher order stresses m_{ijk} and μ_{ij} is obviously determined by the ratio of the characteristic length of the microstructure ($= D$ in the present example) to the length scale of the macroscopic deformation pattern.

7.5   Laminated Materials

The Cosserat Continuum Theory (CCT) (Cosserat, 1909) has been applied successfully in the analysis of materials composed of elastic layers with alternating elastic coefficients (e.g., Biot, 1967, used a CCT with constraint rotations; in Mühlhaus, 1985, 1993, the rotations are free). However, difficulties are encountered with the proper representation of certain limit cases. In the following, a brief outline of a model is given which is derived within the framework of the relative gradient theory outlined in the preceding section (Eqs. (7.7), (7.17)–(7.19); see also Mühlhaus (1995) for further details). The theory is a generalization of the CCT and related theories, and at the same time remedies some of their shortcomings.

The basic ideas are explained by considering the example in Figure 7.2(a),(b). In Figure 7.2(a), it is assumed that the layers are perfectly smooth, i.e. the shear stiffness of the interface vanishes. We identify the Cosserat rotation ω with the cross-sectional rotation

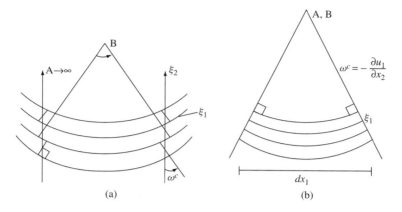

Figure 7.2   Bending of a layered material

of the layers. In this case, there is no choice for the couple stress-curvature relation other than

$$m = \frac{Eh^2}{12(1 - v^2)} \kappa, \quad \kappa = \frac{\partial \omega^c}{\partial x_1} \tag{7.52}$$

where h is the layer thickness. In Figure 7.2(b), we consider continuous pure bending of the layered system. Since, for pure bending, the local shear stress vanishes, the structural bending stiffness is independent of the interface properties. Continuity of the deformation requires that

$$\omega = -\frac{\partial u_1}{\partial x_2} \tag{7.53}$$

and couple stresses should not be present in this case. This, however, contradicts Eq. (7.52), according to which $m \neq 0$. We propose a modified definition for the curvature which reconciles the two cases. We assume similar to Eq. (7.52), that

$$m = \frac{Eh^2}{12(1 - v^2)} \kappa^{\text{rel}}, \quad \text{but } \kappa^{\text{rel}} = \frac{\partial}{\partial x_1} \left(\frac{\partial u_1}{\partial x_2} + \omega \right) \tag{7.54}$$

When the derivative of u_1 with respect to x_2 vanishes as in Figure 7.2(a), then $\kappa^{\text{rel}} = \kappa$. In this case, the definitions in Eqs. (7.52) and (7.54) coincide. In pure bending of the layered system, as in Figure 7.2(b), Eq. (7.53) holds and, accordingly, $\kappa^{\text{rel}} = 0$ and $m = 0$. Motivated by this simple model, we shall derive a consistent continuum theory for layered materials. It is understood that the assumptions in Eq. (7.54) are purely *ad hoc* and, therefore, constitutive by nature.

The relations in Eq. (7.55) may be written in the terms of the stress and deformation measures of the previous section as

$$m_{121} = \frac{Eh^2}{12(1 - v^2)} \kappa_{121}, \quad \kappa_{121} = \gamma_{12,1} \tag{7.55}$$

i.e. the layered medium considered here can be represented as an orthotropic relative gradient continuum (Mühlhaus, 1996).

We close this section with a brief discussion of a micro-mechanical counterpart to our card-deck model (Figure 7.2) in connection with dislocation induced lattice curvature. However, in contrast to a physical deck of cards, here the couple stresses do not vanish upon pure bending. In this case, a CCT-type theory is applicable.

Kröner (1963) considered a crystal of infinite dimensions filled with straight edge dislocations of macroscopically constant density (Figure 7.3). The glide planes are parallel with the spacing $2d$. The stress field within the crystal is $2d$ – periodic. The corresponding couple stress-lattice curvature relation reads

$$m = \frac{1}{12} \left(\frac{\pi}{2}\right)^2 \frac{E(2d)^2}{(1-v^2)} \kappa^{\text{lattice}} \tag{7.56}$$

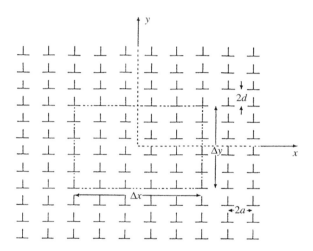

Figure 7.3 Scheme of a macroscopically constant density of edge dislocations (after Kröner, 1963). Demarcated: a volume element $\Delta V = \Delta x \Delta y \Delta z$. The dislocation arrangement in ΔV is stabilized by the outside dislocations

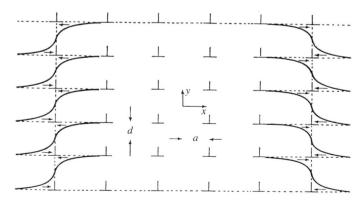

Figure 7.4 Distribution of microscopical stress σ_{xx} in ΔV. The quantity σ_{xx} does not change the sign in the x-direction (Kröner, 1963)

The lattice curvature is related to the non-vanishing (z, x) component α of Nye's dislocation density tensor as

$$\alpha = \frac{b}{4ad} = -\kappa^{\text{lattice}} \tag{7.57}$$

where b is the length of the Burgers vector and $2a$ is the average distance between the dislocations along the glide plane. The differences between the bending stiffnesses in Eqs. (7.52) and (7.55) are explained by the nonlinearity of the stress distribution between the glide planes. For more details, the reader is referred to Kröner's paper (see also Fleck *et al.*, 1994).

7.6   Cosserat Plasticity

The main carrier of inelastic deformations in dense granular assemblies are relative movements along and across well defined glide planes. The situation is not unlike the one in continuous solids such as metals. During inelastic flow, the particles move over each other, which may lead to an increase (dilatancy) or decrease (contractancy) of the overall volume of the assembly, depending upon whether the initial density was above or below a certain critical value. For a detailed discussion of the micromechanical mechanisms of dense granular flow, we refer to the review article by Adams and Briscoe (1993). The flow resistance is proportional to the resolved normal stress of the active glide planes, and the coefficient of proportionality is commonly designated as pressure sensitivity or mobilized friction coefficient. It should be mentioned that the pressure sensitivity correlates only weakly with the interparticle friction coefficient (e.g. Thornton, 1993) and is non-zero even in the hypothetical case of ideally smooth particles.

The thermodynamic driving force of the glide process depends upon the average resolved shear stress of the glide plane and, in the case of rough granules, on the average eccentricity thereof with respect to the mass centers of the granules. Within the framework of the Cosserat Theory assumed here, the influence of the load eccentricity is considered by the couple stress tensor.

We envisage a single yield criterion model. The question then is, how to define an expression for the thermodynamic driving force involving both the stress tensor and the couple stress tensor in a physical meaningful way? In Mühlhaus and Vardoulakis (1987), the driving force is defined as the work conjugate to a measure for the average relative velocity between neighboring granules. Here we wish to follow a different route: we base our yield model Hencky's principle of maximum work of distortion, going back to the early days of the theory of plasticity (e.g. Hill, 1950, p. 3). In this theory, it is assumed that during yielding the deviatoric part of the elastic strain energy is equal to the elastic strain energy at yield in uniaxial tension. For elastic-ideal plastic materials, the resulting yield criterion is obviously identical to the von Mises yield criterion. One way of generalizing this concept to granular materials is to assume that during yielding the ratio of the deviatoric part of the elastic strain energy to the volumetric part of the strain energy is constant (constant pressure sensitivity), or is a function of the equivalent plastic strain. For a standard continuum the resulting criterion is identical to the so-called generalized von Mises criterion, which is also known as the Drucker–Prager yield criterion.

For simplicity, we restrict ourselves to assemblies of identical spherical grains. We assume that the elastic part of the deformations is infinitesimal, and that the contact stiffnesses are constant. The integrated forms of the constitutive relationships in Eqs. (7.12) and (7.15) are obtained as

$$\sigma_{ij} = 2\mu\varepsilon_{ij}^e + \lambda\varepsilon_{kk}^e\delta_{ij} + 2(\mu - \lambda)(\Omega_{ij}^e - \Omega_{ij}^{ec}) \tag{7.58}$$

$$m_{ij} = (\mu - \lambda)\frac{d^2}{10}(2k_{[ij]}^e + 3k_{ij}^{'e}) \tag{7.59}$$

where

$$(\cdot)_{ij}' = (\cdot)_{ij} - \tfrac{1}{3}(\cdot)_{kk}\delta_{ij}, \quad (\cdot)_{ij} = \tfrac{1}{2}((\cdot)_{ij} + (\cdot)_{ji}) \tag{7.59a}$$

and

$$(\cdot)_{[ij]} = \tfrac{1}{2}((\cdot)_{ij} - (\cdot)_{ji}) \tag{7.59b}$$

ε_{ij}^e, Ω_{ij}^e, Ω_{ij}^{ce} and k_{ij}^e designate the elastic parts of the strain, continuum rotation, micro-rotation and micro-curvature tensors, and λ, μ are Lamé's constants. We write the corresponding deviatoric and volumetric strain energies, $\rho\psi^D$ and $\rho\psi^V$, respectively, expressed in terms of stress and couple stress, as follows:

$$\rho\psi^D = \frac{1}{2}\frac{\tau^2}{\mu}, \quad \rho\psi^V = \frac{1}{2}\frac{p^2}{K} \tag{7.60}$$

where K designates the bulk modulus, and

$$\tau = \left(\frac{1}{2}\sigma_{(ij)}'\sigma_{(ij)}' + \frac{1}{2}a\sigma_{[ij]}\sigma_{[ij]} + \frac{a}{D^2}\left(\frac{10}{3}m_{(ij)}'m_{(ij)}' + \frac{10}{5}m_{[ij]}m_{[ij]}\right)\right)^{1/2}, \quad p = \frac{1}{3}\sigma_{kk}, \tag{7.61}$$

where $a = \mu/(\mu - \lambda)$. For $m_{ij} = 0$ and $\sigma_{ij} = \sigma_{ji}$, the equivalent shear stress τ becomes $\tau = (\frac{1}{2}\sigma_{ij}'\sigma_{ij}')^{1/2}$, the second deviatoric stress invariant of the standard continuum. According to the preceding discussion, the yield criterion is assumed as

$$f = \tau - \bar{\sigma} + \alpha p \tag{7.62}$$

where α is the pressure sensitivity, $\bar{\sigma}$ represents some cohesion in the material and $f \leq 0$ for plastically admissible stress states. To complete the description of inelastic granular flow, we require a flow rule. In conformity with tradition, we assume the existence of a plastic potential g, which, in analogy to Rudnicki and Rice's (1975) model, we write as

$$g = \tau + \beta p \tag{7.63}$$

where β is the so-called *dilatancy function*.

7.7   An Algorithm for Cosserat Plasticity

Having laid out the basic structure of a Cosserat elasto-plastic continuum model, we now set out to describe an algorithm for computations within the framework of the finite element method. For simplicity, we restrict the treatment to the two-dimensional case.

However, the extension to three dimensions is straightforward (Groen *et al.*, 1994). For plane-strain condition, expression (7.61) for τ reduces to

$$\tau = \left[\tfrac{1}{2}(\sigma_{xx}'^2 + \sigma_{yy}'^2 + \sigma_{zz}'^2) + \tfrac{1}{4}(1+a)(\sigma_{xy}^2 + \sigma_{yx}^2) + \tfrac{1}{2}(1-a)\sigma_{xy}\sigma_{yx} \right.$$
$$\left. + \frac{8a}{3}\left(\left(\frac{m_{zx}}{D}\right)^2 + \left(\frac{m_{zy}}{D}\right)^2 \right) \right]^{1/2} \tag{7.64}$$

In fact, this can be conceived as a special case of the more general expression for τ derived by Mühlhaus and Vardoulakis (1987)

$$\tau = (a_1\sigma_{ij}'\sigma_{ij}' + a_2\sigma_{ij}'\sigma_{ji}' + a_3 m_{ij}m_{ij}/D^2)^{1/2} \tag{7.65}$$

with a_1, a_2, a_3 coefficients which can, for instance, be determined as has been done previously. For plane-strain conditions, the Mühlhaus/Vardoulakis expression specializes as

$$\tau = \left[\frac{1}{2}(\sigma_{xx}'^2 + \sigma_{yy}'^2 + \sigma_{zz}'^2) + a_1(\sigma_{xy}^2 + \sigma_{yx}^2) + 2a_2\sigma_{xy}\sigma_{yx} + a_3\left(\left(\frac{m_{zx}}{D}\right)^2 + \left(\frac{m_{zy}}{D}\right)^2 \right) \right]^{1/2} \tag{7.66}$$

and upon comparison with Eq. (7.64), we observe that the particular Cosserat plasticity model developed above is retrieved when setting

$$a_1 = \frac{1}{4}(1+a), \quad a_2 = \frac{1}{4}(1-a), \quad a_3 = \frac{8a}{3}$$

For algorithmic convenience, we cast τ into a matrix-vector format. Specifically, we introduce the projection vector

$$\pi^T = \left[\tfrac{1}{3}, \tfrac{1}{3}, \tfrac{1}{3}, 0, 0, 0, 0\right] \tag{7.67}$$

the pseudo stress and strain vectors

$$\sigma^T = \left[\sigma_{xx}, \sigma_{yy}, \sigma_{zz}, \sigma_{xy}, \sigma_{yx}, \frac{m_{zx}}{D}, \frac{m_{zy}}{D}\right] \tag{7.68}$$

$$\varepsilon^T = [\varepsilon_{xx}, \varepsilon_{yy}, \varepsilon_{zz}, \varepsilon_{xy}, \varepsilon_{yx}, \kappa_{zx}D, \kappa_{zy}D] \tag{7.69}$$

and the projection matrix

$$\mathbf{P} = \begin{bmatrix} \frac{2}{3} & -\frac{1}{3} & -\frac{1}{3} & 0 & 0 & 0 & 0 \\ -\frac{1}{3} & \frac{2}{3} & -\frac{1}{3} & 0 & 0 & 0 & 0 \\ -\frac{1}{3} & -\frac{1}{3} & \frac{2}{3} & 0 & 0 & 0 & 0 \\ 0 & 0 & 0 & 2a_1 & 2a_2 & 0 & 0 \\ 0 & 0 & 0 & 2a_2 & 2a_1 & 0 & 0 \\ 0 & 0 & 0 & 0 & 0 & 2a_3 & 0 \\ 0 & 0 & 0 & 0 & 0 & 0 & 2a_3 \end{bmatrix}. \tag{7.70}$$

With Eqs. (7.67)–(7.70) the yield formation f and the plastic potential attain an appealingly compact form:

$$f = \left(\tfrac{1}{2}\sigma^T\mathbf{P}\sigma\right)^{1/2} + \alpha\sigma^T\pi - \bar{\sigma} \tag{7.71}$$

$$g = \left(\tfrac{1}{2}\sigma^T\mathbf{P}\sigma\right)^{1/2} + \beta\sigma^T\pi \tag{7.72}$$

Defining the flow rate in a standard fashion

$$\dot{\varepsilon}^p = \dot{\lambda}\frac{\partial g}{\partial \sigma} \tag{7.73}$$

with $\dot{\lambda}$ a plastic multiplier determined from the consistency condition $\dot{f} = 0$, the plastic strain rates are then elaborated as

$$\dot{\varepsilon}^p = \dot{\lambda}\left(\frac{\mathbf{P}\sigma}{2\left(\frac{1}{2}\sigma^T\mathbf{P}\sigma\right)^{1/2}} + \beta\pi\right) \tag{7.74}$$

It now remains to identify the plastic strain measure γ^p (the hardening parameter) for a plasticity theory in a Cosserat medium. For this purpose, we first recall the conventional strain-hardening hypothesis:

$$\dot{\gamma}^p = \left[\frac{2}{3}\dot{e}^p_{ij}\dot{e}^p_{ij}\right]^{1/2} \tag{7.75}$$

with $\dot{e}^p_{ij}$ the plastic deviatoric strain-rate tensor. For uniaxial stressing, $\dot{\gamma}^p$ reduces to the uniaxial plastic strain rate, $\dot{\gamma}^p = \dot{\varepsilon}^p_{xx}$. Since there are no couple-stress effects in uniaxial loading, we require that any modification to Eq. (7.75) for Cosserat media does not affect the result for pure uniaxial loading. Considering this prerequisite, a possible generalization, analogous to Eq. (7.65), is to postulate that (Mühlhaus and Vardoulakis, 1987; de Borst, 1993)

$$\dot{\gamma}^p = [b_1\dot{e}^p_{ij}\dot{e}^p_{ij} + b_2\dot{e}^p_{ij}\dot{e}^p_{ji} + b_3\dot{\kappa}^p_{ij}\dot{\kappa}^p_{ij}D^2]^{1/2} \tag{7.76}$$

with $b_1 + b_2 = \frac{2}{3}$ so that definition in Eq. (7.75) for the strain-hardening hypothesis in a non-polar solid can be retrieved.

For the case of planar deformation, $\dot{\gamma}^p$ can be elaborated as

$$\dot{\gamma}^p = \left[\frac{2}{3}[(\dot{e}^p_{xx})^2 + (\dot{e}^p_{yy})^2 + (\dot{e}^p_{zz})^2] + b_1(\dot{e}^p_{xy})^2 + 2b_2\dot{\varepsilon}^p_{xy}\dot{\varepsilon}^p_{yx}\right.$$

$$\left. + b_1(\dot{\varepsilon}^p_{yx})^2 + b_3[(\dot{\kappa}^p_{xz}D)^2 + (\dot{\kappa}^p_{yz}D)^2]\right]^{1/2} \tag{7.77}$$

Introduction of the matrix

$$Q = \begin{bmatrix} \frac{2}{3} & -\frac{1}{3} & -\frac{1}{3} & 0 & 0 & 0 & 0 \\ -\frac{1}{3} & \frac{2}{3} & -\frac{1}{3} & 0 & 0 & 0 & 0 \\ -\frac{1}{3} & -\frac{1}{3} & \frac{2}{3} & 0 & 0 & 0 & 0 \\ 0 & 0 & 0 & \frac{3}{2}b_1 & \frac{3}{2}b_2 & 0 & 0 \\ 0 & 0 & 0 & \frac{3}{2}b_2 & \frac{3}{2}b_1 & 0 & 0 \\ 0 & 0 & 0 & 0 & 0 & \frac{3}{2}b_3 & 0 \\ 0 & 0 & 0 & 0 & 0 & 0 & \frac{3}{2}b_3 \end{bmatrix} \tag{7.78}$$

allows the rate of the hardening parameter $\dot{\gamma}^p$ to be written in a similar format as the yield function:

$$\dot{\gamma}^p = \left[\frac{2}{3}(\dot{\varepsilon}^p)^T\mathbf{Q}\dot{\varepsilon}^p\right]^{1/2} \tag{7.79}$$

with the vector $\dot{\varepsilon}^p$ assembling the plastic strain-rate components.

We next introduce the flow rule given by Eq. (7.74) into expression (7.79) for the rate of the hardening parameter. Since $\dot{\lambda}$ and $\bar{\sigma}$ are always non-negative, and as $\mathbf{Q}\pi = \mathbf{0}$, the result is given by

$$\dot{\gamma}^p = \dot{\lambda} \left(\frac{\sigma^{\mathrm{T}} \mathbf{P} \mathbf{Q} \mathbf{P} \sigma}{\sigma^{\mathrm{T}} \mathbf{P} \sigma} \right)^{1/2} \tag{7.80}$$

If the parameters a_1, a_2, a_3 and b_1, b_2, b_3 are chosen such that

$$\mathbf{P} \mathbf{Q} \mathbf{P} = \mathbf{P} \tag{7.81}$$

Equation (7.80) reduces to exactly the same format as obtained in standard plasticity theory

$$\dot{\gamma}^p = \dot{\lambda} \tag{7.82}$$

We shall now integrate the governing rate equations with a Euler Backward algorithm. This algorithm is relatively simple, and has been proven to be accurate and stable (de Borst, 1993). For this purpose, we first define an 'elastic' trial stress state

$$\sigma_t = \sigma_0 + \mathbf{D}^e \Delta \varepsilon \tag{7.83}$$

σ_0 being the stress at the beginning of the finite load increment, $\mathbf{D}^e$ being the Hookean elastic stiffness matrix, which can for instance be derived differentiating Eq. (7.60) twice and inverting, and $\Delta \varepsilon$ the finite strain increment at the given point. This trial stress is computed assuming that plastic effects do not take place. In reality, they do take place and the total strain increment is partitioned into an elastic part $\Delta \varepsilon^e$ and a plastic part $\Delta \varepsilon^p$:

$$\Delta \varepsilon = \Delta \varepsilon^e + \Delta \varepsilon^p \tag{7.84}$$

Between the stress increment $\Delta \sigma$ and the elastic strain increment $\Delta \varepsilon^e$, we have a bijective relationship

$$\Delta \sigma = D^e \Delta \varepsilon^e \tag{7.85}$$

Furthermore, the expression for the plastic strain rate in Eq. (7.74) is integrated using a single-point Euler backward rule. This results in

$$\Delta \varepsilon^p = \Delta \lambda \left[\frac{\mathbf{P} \sigma_n}{2 \sqrt{\frac{1}{2} \sigma_n^{\mathrm{T}} \mathbf{P} \sigma_n}} + \beta \pi \right] \tag{7.86}$$

where the subscript n refers to the value at the end of the loading step. By definition, σ_n is given by

$$\sigma_n = \sigma_0 + \Delta \sigma \tag{7.87}$$

so that combination of Eqs. (7.83)–(7.87) results in

$$\sigma_n = \sigma_t - \Delta \lambda \left[\frac{\mathbf{D}^e \mathbf{P} \sigma_n}{2 \sqrt{\frac{1}{2} \sigma_n^{\mathrm{T}} \mathbf{P} \sigma_n}} + \beta \mathbf{D}^e \pi \right] \tag{7.88}$$

We next use the condition that at the end of the loading step, the yield condition must be satisfied: $f(\sigma_n, \gamma_n) = 0$. Then, Eq. (7.88) transforms into

$$\sigma_n = \sigma_t - \Delta\lambda \left[\frac{\mathbf{D}^e \mathbf{P}\sigma_n}{2[\bar{\sigma}(\gamma_n^p) - \alpha\pi^T\sigma_n]} + \beta\mathbf{D}^e\pi \right] \tag{7.89}$$

A complication now arises, since we wish to express σ_n as a function of σ_t and $\Delta\lambda$, while in Eq. (7.89), σ_n also occurs in the denominator of the second term on the right-hand side. Therefore, we express $\pi^T\sigma_n$ as a function of $\pi^T\sigma_t$ and $\Delta\lambda$ by premultiplying Eq. (7.89) by the projection vector π. This gives

$$\pi^T\sigma_n = \pi^T\sigma_t - \Delta\lambda\beta K \tag{7.90}$$

with $K = \pi^T\mathbf{D}^e\pi$. In the case of isotropic elasticity, K is the bulk modulus. Substitution of this identity into Eq. (7.89) results in the desired formulation:

$$\sigma_n = \mathbf{A}^{-1}[\sigma_t - \Delta\lambda\beta\mathbf{D}^e\pi] \tag{7.91}$$

where

$$\mathbf{A} = 1 + \frac{\Delta\lambda\mathbf{D}^e\mathbf{P}}{2[\bar{\sigma}(\gamma_n^p) + \Delta\lambda\alpha\beta K - \alpha\pi^T\sigma_t]} \tag{7.92}$$

Substitution in the yield condition $f(\sigma_n, \gamma_n) = 0$ then results in a nonlinear equation in $\Delta\lambda$, which can be solved using a local iterative procedure.

For the derivation of a properly linearized set of tangential moduli, we will restrict ourselves to the format of a pressure-dependent J_2 micro-polar plasticity theory that obeys the constraint in Eq. (7.82). Differentiating Eq. (7.88) then yields

$$\dot{\sigma} = \mathbf{H}\left[\dot{\varepsilon} - \dot{\lambda}\frac{\partial g}{\partial\sigma}\right] \tag{7.93}$$

where

$$\frac{\partial g}{\partial\sigma} = \sqrt{\frac{1}{2}}\frac{\mathbf{P}\sigma}{\sqrt{\sigma^T\mathbf{P}\sigma}} + \beta\pi \tag{7.94}$$

and

$$\mathbf{H}^{-1} = [\mathbf{D}^e]^{-1} + \Delta\lambda\sqrt{\frac{1}{2}}\frac{\sigma^T\mathbf{P}\sigma\mathbf{P} - \mathbf{P}\sigma\sigma^T\mathbf{P}}{\sqrt{\sigma^T\mathbf{P}\sigma}} \tag{7.95}$$

Since $f = f(\sigma, \gamma)$, the consistency condition $\dot{f} = 0$ can be elaborated as

$$\left(\frac{\partial f}{\partial\sigma}\right)^T\dot{\sigma} + \frac{\partial f}{\partial\gamma^p}\dot{\gamma}^p = 0 \tag{7.96}$$

Introducing the hardening modulus

$$h(\gamma^p) = \frac{\partial\bar{\sigma}}{\partial\gamma^p} \tag{7.97}$$

and using the yield condition in Eq. (7.62), we obtain

$$\left(\frac{\partial f}{\partial\sigma}\right)^T\dot{\sigma} + h\dot{\lambda} = 0 \tag{7.98}$$

Equation (7.98) can be combined with Eq. (7.93) to give the explicit consistent tangential stiffness relation

$$
\dot{\sigma} = \left[\mathbf{H} - \frac{\mathbf{H}\dfrac{\partial g}{\partial \sigma}\left(\dfrac{\partial f}{\partial \sigma}\right)^{\mathrm{T}}\mathbf{H}}{h + \left(\dfrac{\partial f}{\partial \sigma}\right)^{\mathrm{T}}\mathbf{H}\dfrac{\partial g}{\partial \sigma}} \right] \dot{\varepsilon}
\tag{7.99}
$$

7.8　Examples: Shear Layer and Biaxial Test

The model discussed in the previous section has been implemented in a six-noded triangular plane strain element. This element has 18 degrees-of-freedom due to the fact that each node has three degrees-of-freedom, two translations and one rotation.

To illustrate the effectiveness of the Cosserat model in predicting physically realistic solutions for mode-II failure problems, the shear layer of Figure 7.5 has been analysed. It is assumed that the shear layer is infinitely long in both the negative and the positive x-direction. The discretization of the shear layer, which has a height $H = 100\,\mathrm{mm}$, is shown in Figure 7.5 for the case of 20 elements.

Basically, the problem is one-dimensional, and could also have been analysed using line elements. Use of two-dimensional elements requires that linear constraint equations be added to the set of algebraic equations which result after discretization of the continuum. They have to be applied to the displacements in the x-direction, as well as to the rotational degrees-of-freedom, since all displacements in the y-direction are prevented (isochoric motion). The bottom of the shear layer is fixed ($u_x = 0$, $u_y = 0$), and the upper boundary is subjected to a shear force σ_{xy} (per unit area), which has been controlled using a standard arc-length technique. The additional boundary condition $\omega_2 = 0$ has been enforced at the lower and upper boundaries.

The standard elastic moduli have been chosen as shear modulus $\mu = 4000\,\mathrm{MPa}$ and Poisson's ratio $\nu = 0.25$. The initial yield strength has been taken equal to $\bar{\sigma} = 100\,\mathrm{MPa}$, with no friction or dilatancy ($\alpha = \beta = 0$), while a linear softening diagram has been used with a hardening (softening) modulus $h = -0.125\,\mu$. For the Cosserat continuum the additional material constant $D = 12\,\mathrm{mm}$ has been inserted. First, the standard values for a_1, a_2, a_3 and b_1, b_2, b_3 have been utilized.

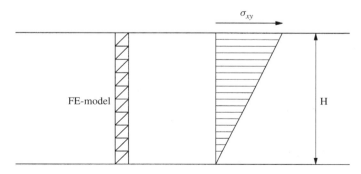

Figure 7.5　Infinitely long shear layer; applied loading and finite element layout

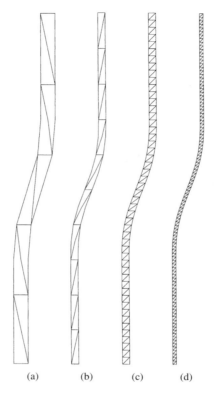

Figure 7.6   Incremental displacement patterns at a residual load level of $\sigma_{xy}/\bar{\sigma} = 28$. (a) 10 elements; (b) 20 elements; (c) 100 elements; (d) 200 elements

In contrast to a standard continuum, a homogeneous strain state is not obtained under pure shear loading for a Cosserat continuum, at least not with the essential boundary conditions listed above. Already in the elastic regime, a boundary layer with a height that is proportional to D develops at the upper and lower parts of the shear layer. As a consequence, the localization develops smoothly and gradually in the middle of the shear layer, without the need for introducing imperfections or for adding a part of the eigenvector to the homogeneous solution.

When the discretization is refined to such an extent that more than one set of two triangular elements is placed over the localization zone, the width of the numerically predicted localization zone becomes constant (Figure 7.6). Figure 7.7(a) shows also that the load-displacement curve converges to a physically realistic solution.

A major question that needs to be addressed when introducing higher-order continuum models is the determination of the additional material parameters. The role of the internal length scale D is obvious, since it governs the brittleness and width of the localization zone. A smaller value for D implies a smaller width of the localization zone, and a steeper post-peak response as is shown in Figure 7.7(b). The influence of the ratio $a_1 : a_2$ is shown in Figure 7.7(c). Apart from the 'standard' set ($a_1 = a_2 = 1/4$), the so-called kinematic model of Mühlhaus and Vardoulakis ($a_1 = 3/8$, $a_2 = 1/8$) and the static model of Mühlhaus and Vardoulakis ($a_1 = 3/4$, $a_2 = -1/4$) have been used. It appears that the

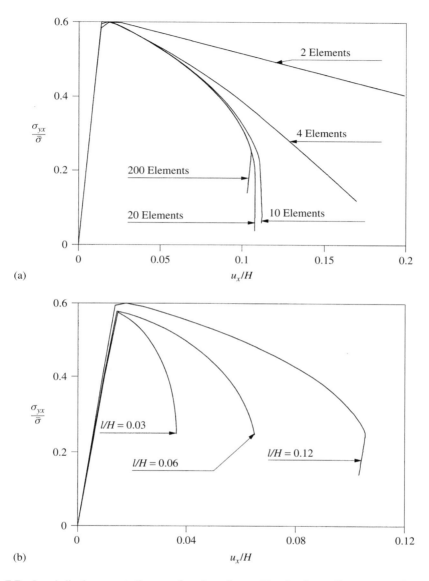

Figure 7.7   Load-displacement diagram for shear layer. Results for a Cosserat continuum with strain softening. (a) The effect of mesh refinement; (b) the role of the characteristic length D

specific choice of these parameters has little influence on the results in the first part of the post-peak regime, but that beyond some critical point in the post-peak regime, the gradual evolution of the localization zone as observed for the standard model breaks down for the kinematic model, and at an even earlier stage for the static model. Beyond these points, the kinematic and static model react in a very brittle manner. Finally, Figure 7.7(d) shows the influence of the softening modulus on the structural behavior.

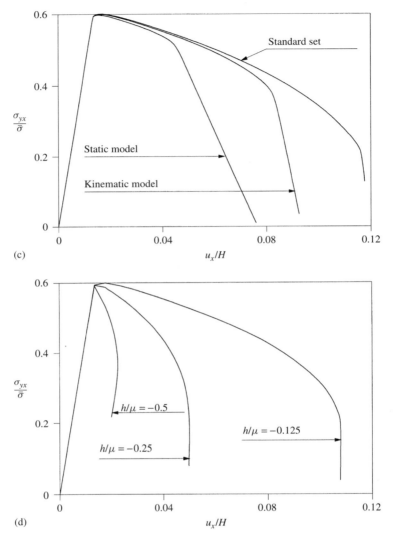

Figure 7.7 (*continued*) (c) influence of the parameter set (a_1, a_2, a_3); (d) influence of the softening modulus h

To demonstrate the effectiveness of the Cosserat continuum in two-dimensional boundary value problems, where frictional sliding is the prevailing failure mechanism, a plane-strain biaxial test has been simulated. The specimen that has been considered has a width $B = 60\,\text{mm}$ and a height $H = 180\,\text{mm}$. Smooth boundary conditions ($u_y = 0$) have been assumed at the upper and lower boundaries, and natural boundary conditions have been assumed at all sides for the rotations. A Drucker–Prager yield condition with a non-associated flow rule was employed. The material data were as follows: shear modulus $\mu = 1000\,\text{MPa}$, Poisson's ratio $\nu = 0.2$, $D = 6\,\text{mm}$, $\alpha = 1.2/\sqrt{3}$, $\beta = 0$ and

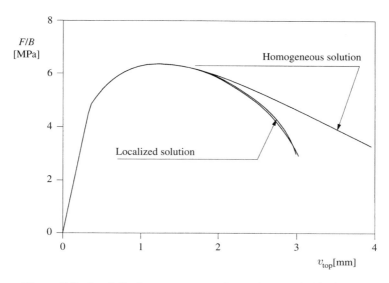

Figure 7.8   Load-displacement curves for a plane-strain biaxial test

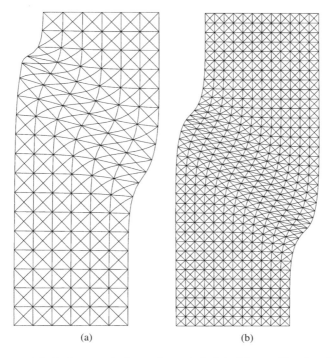

Figure 7.9   Incremental displacements for a plane-strain biaxial test (Drucker–Prager plasticity) (a) Medium mesh with 432 elements; (b) fine mesh with 1728 elements

$\bar{\sigma} = 1.2(1 - 25\gamma^p)$. For the coefficients a_1, a_2 and a_3, the 'standard' values have been analysed.

Three different meshes have been adopted, with 108, 432 and 1728 six-noded triangular elements, respectively. The load-displacement curves for all three discretizations are indistinguishable, as is shown in Figure 7.8. This figure also shows the solution that is obtained under the assumption of homogeneous deformations (no localization).

Because in pure compression rotational degrees-of-freedom are not activated, an imperfect element (5% reduction in the initial yield strength) has been inserted in the model at the left boundary near the horizontal center line. From this point, two shear bands initially propagate, but later only one band persists. The incremental displacement at this stage are shown in Figure 7.9 for the medium (432 elements) and the fine mesh (1728 elements), respectively.

7.9   Conclusion

Starting from the virtual power expression for dense assemblies of spherical granules, we derive expressions for the average stress tensor, momentum and angular momentum balances and corresponding variational boundary conditions. We show that the stress tensor can be decomposed into two parts with distinct physical significance. The first part can be interpreted as an area averaged stress tensor, which is nonsymmetric in general. The second part is obtained as the divergence of a third order tensor which can be interpreted as the difference between the area and the volume averages of the stress tensor. The underlying field theory is general although its derivation is motivated by the situation in granular materials. Applications of the theory within the context of layered materials are also discussed.

Turning to computational aspects, we derive a symbolic solution of a general mixed boundary value problem in the form of a finite element formulation and corresponding solution algorithm. The problem of mesh sensitivity of the regularized boundary value problem is discussed. The performance of the numerical method is illustrated by means of solutions for static strain localization problems.

References

Adams, M. J. and Briscoe, B. J. Deterministic micromechanical modelling of failure or flow in discrete planes of densely packed particles assemblies: Introductory principles. In: *Granular Matter*, Mehta, A. (Ed.), *Springer-Verlag*, (1994) 259–291.

Aifantis, E. C. The physics of plastic deformation. *Int. J. Plasticity*, **3**, (1987) 211–247.

Biot, M. Rheological stability with couple stresses and its application to geological folding. *Proc. Royal Society. London*, **A2298**, (1967) 402–423.

Born, M. The mobility of electrolytic ions. *Zeitschrift der Physik*, **1**, (1921) 221–249.

de Borst, R. A generalisation of J_2-flow theory for polar continua. *Computer Meth. Appl. Mech Engrg.*, **103**, (1993) 347–362.

de Borst, R. and Mühlhaus, H.-B. Gradient dependent plasticity: formulation and algorithmic aspects. *J. Num. Meth. Engng.*, **35**, (1992) 521–539.

Chapman, S. and Cowling, T. G. *The Mathematical Theory of Non-uniform Gases*, 3rd edition. Cambridge University Press, Cambridge, (1970).

Cosserat, E. and Cosserat, F. (1909) *Theorie des corps deformables.* A Herrmann et fils, Paris, (1909).

Desai, C. S. and Zhang, W. Computational aspects of disturbed state constitutive models. *Int. J. Comput. Meth. Appl. Math. Engrg.*, (1997).

Desai, C. S., Basaran, C. and Zhang, W. Numerical algorithms and mesh dependence in the disturbed state concept. *Int. J. Num. Meth. Engng.*, (1997).

Desai, C. S. Constitutive modelling using the disturbed state as microstructure self-adjustment concept. In: *Continuum Models for Materials with Microstructure*, Mühlhaus, H. B (Ed.), Wiley, Chichester, (1995) 239–296.

Fleck, N. A., Muller, G. M., Ashby, M. F. and Hutchinson, J. W. Strain gradient plasticity: theory and experiment. *Acta. Metall. Mater.*, **42**, (1994) 474–487.

Goldstein, H. *Classical Mechanics*, 2nd edition. Addison-Wesley, Reading, MA, (1980).

Groen, A. E., Schellekens, J. C. J. and de Borst, R. Three-dimensional finite element studies of failure in soil bodies using a Cosserat continuum. In: *Computer Methods and Advances in Geomechanics*, Siriwardane, H. J. and Zaman, M. M., (eds.), Balkema, Rotterdam, (1994) 581–586.

Haff, P. K. Grain flow as a fluid-mechanical phenomenon. *J. Fluid Mech.*, **134**, (1983) 401–430.

Hill, R. *The Mathematical Theory of Plasticity.* Clarendon Press, Oxford, (1950).

Jenkins, J. T. Anisotropic elasticity for random arrays of identical spheres. In: *Modern Theory of Anisotropic Elasticity and Applications*, Wu, Y (ed.). SIAM, Philadelphia, (1991).

Jenkins, J. T. and Richman, M. W. Kinetic theory for plane flows of a dense gas of identical, rough. Inelastic, circular disks. *Phys. Fluids*, **28**, (1985) 3485–3494.

Jenkins, J. T. and Savage, S. B. A theory for the rapid flow of identical, smooth, nearly elastic particles. *J. Fluid Mech.*, **130**, (1983) 187–202.

Kröner, E. On the physical reality of torque stresses in continuum mechanics. *Int. J. Engng. Sci.*, **1**, (1963) 261–278.

Kröner, E. Mechanics of generalised continua. *Proc. IUTAM Symp. Generalised Cosserat Continuum and the Continuous Theory of Dislocations with Applications.* Springer-Verlag, Berlin, (1968).

Lun, C. K. K. Kinetic theory for granular flow of dense, slightly inelastic, slightly rough spheres. *J. Fluid Mech.*, **233**, (1991) 539–559.

Lun, C. K. K. and Bent, A. A. Numerical simulation of inelastic frictional spheres in simple shear flow. **258**, (1994) 335–353.

Lun, C. K. K. and Savage, S. B. A simple kinetic theory for granular flow of rough, inelastic, spherical particles. *Trans. ASME E: J. Appl. Mech.*, **54**, (1987) 47–53.

McCoy, B. J., Sandler, S. I. and Dahler, J. S. Transport properties of polyatomic fluids. iv. The kinetic theory of a dense gas of perfectly rough spheres, *J Chem. Phys.*, **45**, (1966) 3485–3512.

Mindlin, R. D. Microstructure in linear elasticity. *Arch. Rat. Mech. Anal.*, **16**, (1964) 15–78.

Mühlhaus, H.-B. A relative gradient model for laminated materials. In: *Continuum Models for Materials with Microstructure*, Mühlhaus, H.-B. (ed.). Wiley, Chichester, (1995).

Mühlhaus, H.-B. and Aifantis, E. C. A variational principle for gradient plasticity. *Int. J. Solids and Structures*, **28**, (1991) 217–231.

Mühlhaus, H.-B. and Oka, F. Dispersion and wave propagation in discrete and continuous models for granular materials. *Int. J. Solids and Structures*, **33**, (1996) 2841–2858.

Mühlhaus, H.-B. and Vardoulakis, I. The thickness of shear bands in granular materials. *Geotechnique*, **37**, (1987) 217–283.

Mühlhaus, H.-B, Continuum models for layered and blocky rock. In: *Comprehensive Rock Engineering, Vol. II: Analysis and Design Methods.* Pergamon Press, Oxford, (1993) 209–230.

Rudnicki and Rice. Conditions for the localisation of deformation in pressure sensitive dilatant materials. *J. Mech. Phys. Solids*, **23**, (1975) 371–394.

Savage, S. B. Numerical simulation of couette flow of granular materials: spatio-temporal coherence of 1/f noise. In: *Physics of Granular Media*, Bideau, J and Dodds, J. (eds.). Nova Science, New York, (1992) 343–362.

Steinmann, P. A micropolar theory of finite deformation and finite rotation multiplicative elasto-plasticity. *Int. J. Solids Structures*, **31**, (1994) 1063–1084.

Thornton, C. and Sun, G. Axisymmetric compression of 3D polydisdysperse systems of spheres. In: *Powders and Grains 93*, Thornton, C. (ed.). Balkema, Rotterdam, (1993) 129–134.

Triantafyllidis, N. and Aifantis, E. C. A gradient approach to the localisation of deformation-I. Hyperelastic materials. *J. Elasticity*, **16**, (1986) 225–238.

Vardoulakis, I. and Sulem, J. *Bifurication Analysis in Geomechanics*. Chapman & Hall, London, (1995) 334–423.

Walgraef, D. and Aifantis, E. C. On the formation and stability of dislocation patterns – I: One dimensional considerations; II: Two-dimensional considerations; III: Three dimensional considerations. *Int. J. Engng Sci.*, **23**, (1985) 1351–1372.

8

Micromechanics of Granular Materials as Structured Media

Ching S. Chang
University of Massachusetts, Amherst, MA, USA

8.1   Introduction

In the past, the stress-strain behavior of granular materials such as soils, powders or ceramic materials, has been derived from concepts of continuum mechanics that treats the material as a continuum and homogeneous. Recently, a different direction has been pursued in which the stress-strain relationships are derived from a microstructural mechanics approach that treats the granular material as a collection of particles. Along this line, some earlier attempts were made by Duffy and Mindlin (1957), Duffy (1959) and Hendron (1963). Lately, the approaches can be classified into two categories, namely, direct computer simulation (Cundall and Strack, 1979; Kishino, 1988; Ting *et al.*, 1989; Ng, 1989; Chang and Misra, 1989), and microstructural mechanics (Walton, 1987; Jenkins, 1987; Emeriault and Cambou, 1996; Bathurst and Rothenburg, 1988; Chang, 1987). The direct simulation method is solely a discrete analysis, which aims to simulate the movement of every discrete particle of an assembly. The method of microstructural mechanics bridges discrete and continuum analysis. It treats the discrete material as an equivalent continuum, and it aims for a stress-strain model explicitly in terms of the properties and the packing structure of discrete particles.

In the microstructural mechanics approach, the material behavior is investigated at two different scales: the micro-scale behavior is referred to the local interaction between two particles; and the macro-scale behavior is referred to the stress-strain relationship of a representative volume of particle assembly. The size of a representative volume must be sufficient to contain a large number of particles such that its behavior is representative of the assembly. However, it is noted that the notion of a representative volume is equivalent to that of an 'infinitesimal element (or material point)' in classic mechanics. Therefore, it is useful now for the reader to keep a clear perspective that the macro-scale denoted here is only in a relative sense to the size of a granular particle; the magnitude of the macro-scale is merely the size of a 'material point'.

In the microstructural mechanics approach, the objective is to derive the stress-strain behavior of a particle assembly based on the local behavior of particle interaction. Thus, in this approach, it is necessary to have the following:

(1)   A contact law that describes the local behavior of particle interaction.

(2)   A micro-macro relationship that links the micro-scale variables and the macro-scale variables.

In what follows, we describe different types of micro-macro relationships, the form of contact law, and methods of deriving stress-strain relationships. Then a brief review is given for different types of stress-strain models based on this approach, including elastic models, visco-elastic models, elasto-plastic models, and higher-order models.

8.2   Links Between Micro and Macro Variables

Generally, we use discrete variables such as contact forces and particle displacement to describe the micro-scale behavior, and continuum variables such as stress and strain to describe the macro-scale behavior. Links between micro and macro variables can be expressed in two ways, namely, micro-to-macro transformation and macro-to-micro hypothesis. In a micro-to-macro transformation, the macro-scale variable (e.g. stress) is determined from the micro-scale variables (e.g. contact forces). In a macro-to-micro hypothesis, the micro-scale variables (e.g. contact forces) are determined from the macro-scale variable (e.g. stress). It is noted that the two operations are in opposite directions, as shown in Figure 8.1.

It is obvious that the total number of micro-variables is much more than the total number of macro-variables. Therefore, the essential difference between the two relationships is that, in the micro-to-macro transformation, we determine a small number of macro-scale variables based on a large number of micro-scale variables. On the other hand, in the macro-to-micro hypothesis, we determine a large number of micro-scale variables based on a small number of macro-scale variables. Therefore, the micro-to-macro transformation can be regarded as an averaging process. Once the manner of averaging is defined, the expression is unique. However, its reciprocal relationship (i.e. the macro-to-micro relationship) cannot theoretically exist, because a small number of known macro-variables are not enough to predict the large number of unknown micro-variables. It is clear that there are many sets of micro-scale variables which can be averaged to the same macro-scale variable. The macro-to-micro relationship can only exist in an approximate fashion. Therefore, we termed the relationships as *macro-to-micro hypotheses*.

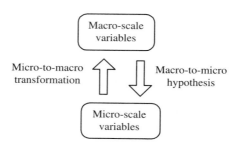

Figure 8.1   Schematic illustration of two types of relationship between micro- and macro-variables

8.2.1   Micro-to-macro transformation

There are two types of micro-to-macro transformation: kinematic transformation links the variables of strain and displacement; and static transformation links the variables of stress and forces (Chang and Gao, 1996a).

Static transformation

The stress for a representative volume is defined as the volume average of particle stresses. Based on the principle of force equilibrium and Gauss theorem, the integral of stress over a representative volume V can be transformed into a surface integral. The surface integral can then be written in a summation form (Chang and Liao, 1990). Thus, the volume average of stress is

$$\Delta\sigma_{ij} = \frac{1}{V}\sum_c \Delta f_j^c l_i^c \tag{8.1}$$

where f_j is the force between two particles in contact, and l_i is the branch vector joining centroids of the two particles. The superscript c represents the contact point. The summation is over all contact points in the representative volume. It is noted that Eq. (8.1) is derived directly from force equilibrium. Therefore, it is not limited to linear displacement field, even though the same expression has also been derived from an assumption of linear displacement field by Christofferson *et al.* (1981).

Kinematic transformation

The strain of a representative volume is defined as the gradient of displacement field. The displacement field is the continuum field established from the displacement of each particle centroid. Since the displacement field is generally nonlinear, the strain is defined based on a linear least square approximation of the displacement field. The following relationship can be derived using a linear least square fit to the displacement field (Chang and Liao, 1994; Liao *et al.*, 1997):

$$\Delta\varepsilon_{ij} = \frac{1}{V}\sum_c A_{jk} l_k^c \Delta\delta_i^c \tag{8.2}$$

where δ_i is contact displacement, and the tensor A_{jk} is a fabric tensor defined by

$$A_{jk} = \left[\frac{1}{V}\sum_c l_j^c l_k^c\right]^{-1} \tag{8.3}$$

8.2.2   Macro-to-micro hypothesis

There are two types of macro-to-micro hypotheses, namely kinematic hypothesis and static hypothesis (Chang and Gao, 1996a). The macro-to-micro hypotheses can be derived from the micro-to-macro transformation using the principle of virtual work. The stress and strain are defined in such a way that the principle of virtual potential work holds, i.e.

$$W = \sigma_{ij}\Delta\varepsilon_{ij} = \frac{1}{V}\sum_c f_j^c \Delta\delta_j^c \tag{8.4}$$

or the principle of virtual complementary work holds, i.e.

$$W = \Delta\sigma_{ij}\varepsilon_{ij} = \frac{1}{V}\sum_c \Delta f_j^c \delta_j^c \tag{8.5}$$

Kinematic hypothesis

Considering the principle of virtual potential work, if the stress in Eq. (8.4) is substituted by the volume average quantity of Eq. (8.1), then

$$W = \frac{1}{V}\sum_c f_j^c l_i^c \Delta\varepsilon_{ij} = \frac{1}{V}\sum_c f_j^c \Delta\delta_j^c \tag{8.6}$$

To satisfy Eq. (8.6), the following relationship is hypothesized

$$\Delta\delta_j^c = \Delta\varepsilon_{ij} l_j \tag{8.7}$$

The kinematic hypothesis can be used to determine the relative movement of two particles from a macro-strain. This hypothesis corresponds to the underlying assumption of linear displacement field of particle centroids. The hypothesis has been used by many investigators for micro-mechanics modeling (Bathurst and Rothenburg, 1988; Chang, 1987; Jenkins, 1987; Walton, 1987).

Static hypothesis

Considering the principle of virtual complementary work, if the strain in Eq. (8.5) is substituted by the volume average quantity of Eq. (8.2), then

$$W = \frac{1}{V}\sum_c \Delta\sigma_{ij} A_{jk} l_k^c \delta_i^c = \frac{1}{V}\sum_c \Delta f_j^c \delta_j^c \tag{8.8}$$

To satisfy Eq. (8.8), the following relationship is hypothesized:

$$\Delta f_i = \Delta\sigma_{ij} A_{jk} l_k \tag{8.9}$$

The static hypothesis can be used to determine the contact force between two particles from a macro-stress. This linear form of static hypothesis is identical to that proposed by Chang and Liao (1994) and Chang and Gao (1996a) derived from variational principle. Another form of static hypothesis can be found in the work by Emeriault and Cambou (1996), in which the static hypothesis is in a series expansion form based on a representation theory (Spencer, 1987). As mentioned earlier, the hypotheses only give approximate relationships. Solutions based on the kinematic and static hypotheses represent the upper and lower estimates of the true solution.

8.3   Contact Law for Two Particle Interaction

A contact law describes the interaction behavior of two particles in contact, including the modes of compression, sliding, rolling and twisting (Chang and Liao, 1990). For simplicity, we neglect the effect of particle rotation in this section. The contact law, as schematically shown in Figure 8.2, describes the relationship between the relative displacement $\Delta \delta_i^{nm}$, and the contact force Δf_i^{nm}, between the nth and mth particles.

It can be expressed in a general incremental relationship as

$$\Delta f_i^{;nm} = K_{ij}^{nm} \Delta \delta_j^{nm}; \quad \Delta \delta_i^{nm} = H_{ij}^{nm} \Delta f_j^{nm} \tag{8.10}$$

in which the contact stiffness tensor K_{ij}^{nm} and flexibility tensor H_{ij}^{nm} can be expressed by two scalar quantities: the shear stiffness k_s^{nm} representing the resistance to sliding; and the normal stiffness k_n^{nm} representing the resistance to compression of two particles. The inverse of the contact stiffness (i.e. flexibility) is denoted by h_n^{nm} and h_s^{nm}. We have the following expressions:

$$K_{ij}^{nm} = k_n^{nm} n_i^{nm} n_j^{nm} + k_s^{nm} (s_i^{nm} s_j^{nm} + t_i^{nm} t_j^{nm}) \tag{8.11}$$

$$H_{ij}^{nm} = h_n^{nm} n_i^{nm} n_j^{nm} + h_s^{nm} (s_i^{nm} s_j^{nm} + t_i^{nm} t_j^{nm}) \tag{8.12}$$

where n, s and t are the basic unit vectors of the local coordinate system constructed at each contact. The vector n is the outward normal to the contact plane. The other two orthogonal vectors s and t are on the contact plane. In a spherical coordinate system, a vector can be defined by two angles γ and β as shown in Figure 8.3, e.g.

$$\vec{n} = \sin \gamma \cos \beta \vec{i} + \sin \gamma \sin \beta \vec{j} + \cos \gamma \vec{k}$$

$$\vec{s} = \cos \gamma \cos \beta \vec{i} + \cos \gamma \sin \beta \vec{j} - \sin \gamma \vec{k}$$

$$\vec{t} = -\sin \beta \vec{i} + \cos \beta \vec{j} \tag{8.13}$$

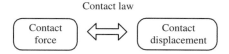

Figure 8.2   Schematic of the contact law

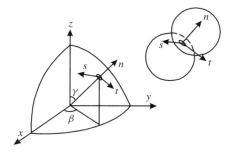

Figure 8.3   Local coordinate system

8.4   Stress-Strain Relationship of the Particle Assembly

The stress-strain relationship for a particle assembly can be derived explicitly in terms of contact stiffness (Chang *et al.*, 1992a). The method of establishing the stress-strain relationship can be based either on kinematic hypothesis or on static hypothesis, as schematically shown in Figures 8.4 and 8.5, respectively. These two methods are described below.

Based on kinematic hypothesis

The stress-strain relationship for a particle assembly is expressed with the strain as an independent variable, given by

$$\Delta\sigma_{ij} = C_{ijkl}\,\Delta\varepsilon_{kl} \tag{8.14}$$

where the constitutive tensor C_{ijkl} can be derived from the following three relationships:

(1)   Static micro-to-macro transformation (Eq. (8.1)).
(2)   Contact law (Eq. (8.10)).
(3)   Kinematic macro-to-micro hypothesis (Eq. (8.7)).

Using these three equations, it leads to the derived stiffness tensor C_{ijkl}, given by

$$C_{ijkl} = \frac{1}{2V}\sum_{n}\sum_{m} l_i^{nm} K_{jk}^{nm} l_l^{nm} \tag{8.15}$$

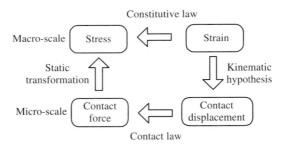

Figure 8.4   Approach for constitutive law based on kinematic hypothesis

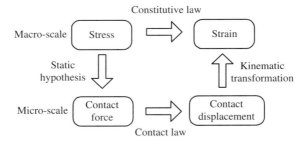

Figure 8.5   Approach for constitutive law based on static hypothesis

where l_i is the branch vector as defined in Eq. (8.1) that joins centroids of the two particles. K_{jk} is the contact stiffness defined in Eq. (8.10). Note that the stiffness tensor is derived from a kinematic hypothesis, which assumes a linear displacement field. Under the linear displacement field, the local compatibility is satisfied for the whole representative volume. Therefore, the constitutive tensor is corresponding to the Voigt upper bound.

Based on static hypothesis

We now express the stress-strain relationship for a particle assembly in a form opposite to that given in Eq. (8.14), with stress as the independent variable. Thus

$$\Delta\varepsilon_{ij} = S_{ijkl}\Delta\sigma_{kl} \tag{8.16}$$

The constitutive tensor can be derived from the following three relationships:

(1) Kinematic micro-to-macro transformation (Eq. (8.2)).
(2) Contact law (Eq. (8.10)).
(3) Static macro-to-micro hypothesis (Eq. (8.9)).

Using these three equations, it leads to the derived stiffness tensor S_{ijkl}, expressed as

$$S_{ijkl} = A_{im}A_{kn}\frac{1}{V}\sum_{c} l_m^c S_{jl}^c l_n^c \tag{8.17}$$

Note that the result is derived from a static hypothesis, which assumes a field of contact forces. Although the contact force field is consistent with the applied stress, it is not guaranteed to satisfy the local equilibrium for all particles within the representative volume. Therefore, the constitutive tensor is not necessarily a lower bound. It can only be regarded as a lower estimate.

8.5   Elastic Model

In this section, the framework of microstructural mechanics approach described in previous sections is illustrated by examples of deriving stress-strain relationships for a linear elastic model and a stress-dependent elastic model.

8.5.1   *Linear elastic model*

In general, the overall stress-strain behavior of the assembly is augmented from the behavior of two-particle interaction. In this section, the two particle interaction is considered to be linear elastic, which leads to a linear elastic stress-strain relationship for the assembly.

For a suitably large representative volume with a large number of particles, any summation over all inter-particle contacts can be expressed in an integral form using a directional distribution density function $\xi(\gamma, \beta)$ for inter-particle contacts. Let F be a quantity dependent on the orientation of contact; the summation of such a function over all contacts can be written as

$$\frac{1}{N} \sum_{c=1}^{N} F^c = \int_0^{2\pi} \int_0^{\pi} F(\gamma, \beta) \xi(\gamma, \beta) \sin \gamma \, d\gamma \, d\beta \tag{8.18}$$

where N is the total number of inter-particle contacts, and angles γ and β are defined in a spherical coordinate as shown in Figure 8.3. Expression of the directional distribution density function $\xi(\gamma, \beta)$ depends upon the packing structure. It satisfies

$$1 = \int_0^{2\pi} \int_0^{\pi} \xi(\gamma, \beta) \sin \gamma \, d\gamma \, d\beta \tag{8.19}$$

For an isotropic packing structure, the density function is $1/4\pi$. For an anisotropic condition, the function $\xi(\gamma, \beta)$ can be efficiently expressed using a spherical harmonics expansion (Chang and Misra, 1996b). Using Eq. (8.18), the summation form in Eq. (8.15) can be written in an integral form as follows:

$$C_{ijkl} = \frac{N}{V} \int_0^{2\pi} \int_0^{\pi} l_i(\gamma, \beta) K_{jl}(\gamma, \beta) l_k(\gamma, \beta) \xi(\gamma, \beta) \sin \gamma \, d\gamma \, d\beta \tag{8.20}$$

For an isotropic packing structure with equal size granules, the integral in Eq. (8.20) becomes

$$C_{ijkl} = \frac{Nr^2}{\pi V} \int_0^{2\pi} \int_0^{\pi} n_i(\gamma, \beta) K_{jl}(\gamma, \beta) n_k(\gamma, \beta) \sin \gamma \, d\gamma \, d\beta \tag{8.21}$$

The functions of $n_i(\gamma, \beta)$ and $K_{jl}(\gamma, \beta)$ are given in Eqs. (8.11) and (8.13). After the integration process of Eq. (8.21), the constitutive constants for an isotropic packing of equal sized particles can thus be derived explicitly in terms of contact stiffness (Chang et al., 1992a). Similarly, the stiffness tensor S_{ijkl} based on static hypothesis in Eq. (8.17) can also be derived (Chang and Gao, 1996a). The derived Young's modulus, Poisson's ratio and shear modulus are listed as follows.

Based on the kinematic hypothesis

$$E_{eff} = \frac{2k_n}{r\psi} \left(\frac{2 + 3\alpha}{4 + \alpha} \right); \quad G_{eff} = \frac{k_n}{5r\psi}(2 + 3\alpha) \tag{8.22}$$

$$v_{eff} = \frac{1 - \alpha}{4 + \alpha} \tag{8.23}$$

where $\alpha = k_s/k_n$, the dimensionless packing parameter $\psi = 3V/2Nr^3$, N is the total number of inter-particle contacts in the representative volume V, and r is the radius of the particles. The value of V/N of a packing can be obtained from the void ratio, the coordination number and the particle size. For a granular assembly, let the void ratio e be the ratio of void volume to solid volume, and the coordination number n be the average number of contact points per particle. The packing parameter ψ, can be expressed as follows:

$$\psi = \frac{4\pi(1 + e)}{n} \tag{8.24}$$

The packing parameter is 1.4 for a very dense packing and 5 for a very loose packing.

For the possible range of stiffness ratio α from 0 to ∞, the Poisson's ratio ranges from 1/4 to -1. The range of Young's modulus is $(1 \sim 6)\, k_n/r\psi$. The range of shear modulus is $(2/5 \sim \infty)\, k_n/r\psi$.

Based on the static hypothesis

$$E_{eff} = \frac{10k_n}{r\psi}\left(\frac{\alpha}{2+3\alpha}\right); \quad G_{eff} = \frac{5k_n}{r\psi}\left(\frac{\alpha}{3+2\alpha}\right) \tag{8.25}$$

$$v_{eff} = \frac{1-\alpha}{2+3\alpha} \tag{8.26}$$

For the possible range of stiffness ratio α from 0 to ∞, the Poisson's ratio ranges from 1/2 to $-1/3$. The range of Young's modulus is $(0 \sim 10/3)\, k_n/r\psi$. The range of Shear modulus is $(0 \sim 5/2)\, k_n/r\psi$. Given the same stiffness ratio α, the moduli based on the static hypothesis are smaller than those based on the kinematic hypothesis.

8.5.2 Stress-dependent elastic model

For bonded particulates such as concrete or cemented sand, the modulus is relatively independent of confining stress. The material may behave as a linear elastic model. However, for unbounded sand, the elastic modulus has been found to be stress sensitive, especially in the range of low confining stress. A good ability of predicting the stress-dependent elastic modulus is important in seismic analysis and in geotechnical engineering design. The microstructural approach discussed here is a rational way to facilitate such a prediction model. Using the microstructural approach, one can model the pressure sensitivity behavior of granular material at the level of inter-particle contact, in which the stiffness between two particles varies due to a change of contact area. This section will briefly illustrate the derivation of a stress-dependency model.

Non-linear contact law

The contact area between two non-conforming elastic bodies depends upon the magnitude of the inter-particle force. A normalized expression for the normal stiffness of a contact between two spherical particles may be written as a power law, given by (Chang *et al.*, 1992; Chang and Ma, 1990)

$$\frac{k_n}{P_a} = C_n \left(\frac{f_n}{P_a r^2}\right)^{\eta} \tag{8.27}$$

where r is the radius of particles, P_a is the atmospheric pressure, C_n and η are dimensionless parameters depending upon the mineral type and surface properties of the particles, and f_n is the magnitude of the inter-particle normal force at the contact plane. Based on the classic Hertz theory, $\eta = 1/3$ and $C_n = (3M^2)^{1/3}$, where

$$\frac{1}{M} = \frac{1-v_1}{2G_1} + \frac{1-v_2}{2G_2} \tag{8.28}$$

G_1 and G_2 are the shear moduli, v_1 and v_2 are the Poisson's ratio of the contiguous spheres.

The tangential force at the contact results in deformation caused by the development of slip over a part of the contact surface. Clearly, when the tangential force exceeds the frictional strength at the contact, sliding takes place. Considering this, a general expression for the tangential stiffness of two dissimilar non-linear inelastic rough particles can be written as a function of the contact force and the particle properties as follows (Mindlin and Deresiewicz, 1953):

$$K_r = C_r K_n \left(1 - \frac{f_r}{f_n \tan \Phi_u} \right)^\eta \tag{8.29}$$

where C_r and η are material constants, ϕ_u is particle-to-particle friction angle, and f_r is the resultant tangential force at the contact. According to Hertz–Mindlin theory,

$$C_r = \frac{2(1-v)}{2-v} \tag{8.30}$$

in which the value of C_r for smooth elastic spheres ranges between 0.67 to 1.

However, in natural sands the particles are not perfectly rounded, and have rough contact surfaces. Compared with the case of smooth surfaces, these asperities at contact tend to undergo plastic deformation, resulting in a plastic yield at a much lower value of contact pressure. The yielding also causes the pressure distribution at the contact to be more uniform rather than being parabolic. The pressure distribution in addition to non-linear inelastic material behavior result in exponent η to be higher and constants C_n and C_r to be lower than that obtained from Hertz theory for smooth elastic bodies. Experimental data on sands suggests that the exponent η range from 1/3 to 1/2.

Stress-strain relationship

Considering an isotropic packing with equal sized particle is under an application of isotropic confining stress σ_{3c}, then the contact forces are equal for all contacts, and can be estimated as

$$f_n = \frac{3V\sigma_{3c}}{2rN} = \psi r^2 \sigma_{3c} \tag{8.31}$$

According to Eqs. (8.27) and (8.31), the contact stiffness can thus be expressed in terms of the applied confining pressure σ_{3c}. With the non-linear contact law, the corresponding shear modulus and Poisson's ratio are listed as follows:

Based on the kinematic hypothesis

$$\frac{G_{eff}}{P_a} = \left(\frac{2}{5} + \frac{3}{5} C_r \right) C_n \psi^{\eta-1} \left(\frac{\sigma_{3c}}{P_a} \right)^\eta \tag{8.32}$$

$$v_{eff} = \frac{1 - C_r}{4 + C_r} \tag{8.33}$$

Based on the static hypothesis

$$\frac{G_{eff}}{P_a} = \left(\frac{5C_r}{3 + 2C_r}\right) C_n \psi^{\eta - 1} \left(\frac{\sigma_{3c}}{P_a}\right)^\eta \tag{8.34}$$

$$v_{eff} = \frac{1 - C_r}{2 + 3C_r} \tag{8.35}$$

It is noted that the effective shear modulus derived via the two hypotheses yields a similar expression. The value is influenced significantly by the value of C_r. The shear modulus obtained via the static hypothesis is smaller than that obtained via the kinematic hypothesis. The two estimates provide a range that bounds the true shear modulus.

This can be demonstrated by depicting Eqs. (8.32) and (8.34) together with the shear modulus measurements (Figure 8.6) from Ottawa sand (Hardin and Richart Jr, 1963). These experimental results were obtained under a confining pressure of 41.62 psi. (0.287 MPa). The prediction in Figure 8.6 utilizes Hertz–Mindlin theory for C_n, C_r and η, with the correspondent values $G = 20963$ Mpa, and $v = 0.15$, which are realistic values for sand particles. As Figure 8.6 shows, Eqs. (8.32) and (8.34) reasonably capture the scatter as well as the trend of the measurements.

Another example of interest is to compare the derived equations with the empirical equations established by Hardin and Black (1966) for shear modulus of sands. The empirical equation is based on experimental results for samples made of round grained sand and tested under low amplitude of vibration in a resonant column device, given by

$$\left(\frac{G_{eff}}{P_a}\right) = \frac{686(2.17 - e)^2}{1 + e} \left(\frac{\sigma_{3c}}{P_a}\right)^{1/2} \tag{8.36}$$

This empirical equation resembles the derived shear modulus in Eqs. (8.32) and (8.34). By selecting appropriate values of parameters, C_n, C_r and η, Eqs. (8.32) and (8.34) can be made identical to the Hardin and Black's equation (Chang *et al.*, 1992a).

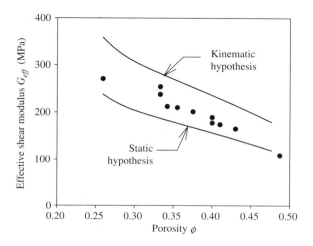

Figure 8.6   Predicted range of shear modulus compared with experiments

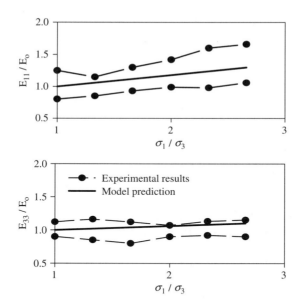

Figure 8.7 Prediction of moduli due to stress-induced anisotropy

Under anisotropic loading conditions, the stress-dependent modulus also becomes orientation-dependent, and thus exhibits stress-induced anisotropy. The microstructural approach mentioned here can easily be extended for this case. Under an anisotropic stress, the contact force is not the same in all orientations. Since contact stiffness depends upon inter-particle forces, the stress-induced anisotropy is thus modeled. An example is shown in Figure 8.7.

A simple closed form equation for modulus under a general anisotropic condition cannot be obtained. The solid curves in Figure 8.7 are numerically computed moduli compared with experimentally measured results for a cubic sample. The stress of the cubic sample is initially under an isotropic stress condition. Then the stress is increased in the major principal direction, while the stresses in the minor directions are kept to be constant. The moduli E_{11} and E_{33} are along the orientations correspondent to the principal stresses σ_1 and σ_3. The moduli in principal directions of the cubical sample were measured from the p-wave velocity (Kopperman *et al.*, 1982). In Figure 8.7, the moduli are normalized to the reference modulus E_0 at isotropic stress condition. Under the change of principal stress, the modulus in the principal direction increases with the stress ratio, while the modulus remains almost same in the minor principal direction. For details in calculation procedure, the reader is refered elsewhere (Emeriault and Chang, 1997; Chang *et al.*, 1989, 1992a).

8.6 Visco-Elastic and Elasto-Plastic Models

We now illustrate the microstructural approach for inter-particle contact behavior other than elastic, such as time-dependent deformation due to visco-elastic contact or plastic deformation due to sliding between particles. In the case of a visco-elastic behavior, a contact law has been developed for compression and tangential compliance (Zhu *et al.*, 1996a,b) of two elastic particles connected by a visco-elastic binder. The contact law has been applied to the derivation of rheological constitutive relationship for a granular assembly viewed as a discrete system of particles connected with equivalent springs and dashpots (Chang and Gao, 1996b).

The mechanism of sliding between particles is of general interest because it is responsible for some intriguing behavior of sand, such as the inter-locking, plastic flow and the volume dilation during shear. The role of particle sliding can be illustrated from the behavior of a Body Centered Cubic (BCC) packing of spheres. For the configuration of body centered cubic packing, each particle has eight neighbors and the void ratio is 0.47.

Because of its simple packing structure, the behavior of BCC packing can be derived in explicit expressions. It is interesting to note that, under a one-dimensional compression, the applied vertical force is arched into horizontal forces through the inclined inter-particle contact. The magnitude of horizontal force depends on the packing structure and the shear forces mobilized on the contact. The ratio of stresses, corresponding to the coefficient of earth pressure at rest, is given by

$$K_o = \frac{\sigma_x}{\sigma_z} = \frac{1}{2}\tan(\alpha - \phi_\mu)\tan\alpha \tag{8.40}$$

where ϕ_μ is the inter-particle frictional angle and the packing angle α is 54.7°. According to Eq. (8.40), $K_o = 1$ for frictionless particles ($\phi_\mu = 0$). For particles with average friction ($\phi_\mu = 20°$), $K_o = 0.49$ and for particles with very rough surface ($\phi_\mu = 30°$), $K_o = 0.32$.

During one-dimensional compression, particles are generally undergo shear sliding, thus the deformation is elasto-plastic. For a subsequent unloading condition, inter-particle response immediately become elastic. The particle sliding is locked-in, which produces a permanent settlement and generates a higher value of earth pressure coefficient K_o. The locked-in stress plays an important role in deformation behavior. We denote K_o^{OC} and K_o^{NC}, respectively, for the unloading and loading conditions. It can be seen that (Chang 1994b)

$$K_o^{OC} = K_o^{NC}(OCR(1 - b) + b)) \tag{8.41}$$

$$b = \frac{\tan(\alpha - \kappa)}{\tan(\alpha - \phi_\mu)}; \quad \tan\kappa = C_r\tan\alpha \tag{8.42}$$

where OCR is the over-consolidation ratio defined as the ratio of the stress before unloading over the current stress. This equation is applicable when the inter-particle contacts are in elastic range. As unloading continues, particles overcome the locked-in shear forces and start to slide in the opposite directions. The value of K_o continues to increase until passive sliding occurs. When passive sliding occurs, the value of K_o remains to be a constant denoted as $(K_o)_{max}$, given by

$$(K_o)_{max} = \tfrac{1}{2}\tan(\alpha + \phi_\mu)\tan\alpha \tag{8.43}$$

A widely used empirical expression for the effect of OCR on the earth pressure coefficient is given as follows (Alpan, 1967; Schmid, 1966):

$$K_o^{OC} = K_O^{NC}OCR^m \tag{8.44}$$

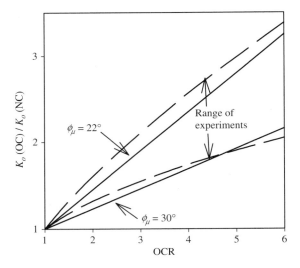

Figure 8.8   Prediction of K_o for different OCR ratios

The value of m varies from 0.4 to 0.68 (Alpan, 1967; Schmertmann, 1975; Sherif *et al.*, 1974). Using these values, the empirical equation is calculated and plotted in dash lines in Figure 8.8. The calculated results of Eq. (8.41) using inter-particle friction angle equal to 22° and 30°, give approximately the same range.

Although the BCC packing is not a realistic microstructure for soil, the behavior of this packing show reasonable agreement with the behavior observed from experiments for sands. This simple example illustrates the important roles of particle sliding on the magnitude of strain, the lateral stress due to arching, and the OCR effect due to locked-in stress. This example also provides insights to the effects of inter-particle stiffness and inter-particle friction angle on the stress-strain response of the assembly.

A realistic packing structure for soil should be of a random type. Elasto-plastic deformation of the randomly packed granules needs to account for the frictional sliding between contacts. The derivation of stress-strain relationship is difficult because, in the case of particle sliding, the displacement field of particle centroids is far from linear, even within a representative volume (i.e. the infinitesimal element). Thus, the overall behavior is no longer a straight volume average of the local behavior (Chang, 1993; Chang and Misra, 1990a; Duffy and Mindlin, 1957). To account for the non-linear displacement field in a representative volume, several attempts have been considered. One attempt is to separate the deformation into elastic and plastic parts. Displacement field for the elastic part is assumed to be linear. The plastic part of deformation, resulting from the sliding between particles, is not linear and is assumed to occur only along few orientations based on the contact forces and contact frictional resistance. The plastic deformation is then added to the elastic part to yield the total displacement field (Liao *et al.*, 1995).

Another approach is to divide the representative volume into a set of cells (i.e. micro-elements). Each cell represents a particle group and has its distinctive configuration. A statistical set of cells of different configurations is used to characterize the packing structure of the granular material. Upon deformation, the displacement field in each cell is linear. However, their gradients are allowed to be different depending on the stiffness of each cell, as long as the strains for all cells add up to the macro-strain of the assembly. Thus, the constraint of linear displacement field is relaxed in the cell model. The key issue now of this model is to have a reliable method that correctly determines the strain distribution for cells under the application of a given macro-strain. Two methods have been used in the analysis of strain distribution of cells, namely the self-consistent method and the statistical method. Being conceptually viewed as a cell structure, the granular material in certain respects is analogous to the polycrystalline material. The self-consistent method used in polycrystalline was applied to the granular material to relate the strain of each cell and the macro-scale strain of the representative element. Consequently, the stress-strain relationship can be derived (Chang *et al.*, 1992b). A typical predicted stress-strain curve for a biaxial stress condition is shown in Figure 8.9 (Chang *et al.*, 1992b).

Although the predicted curve resembles the behavior of sand, the self-consistent method is cumbersome because it requires us to know a Green's function. Unfortunately, the three-dimensional Green's function for a general anisotropic media is not available. Therefore, the method is only limited to a two-dimensional situation. For this reason, a canonical ensemble average technique used in statistical mechanics was adopted for developing the stress-strain relationship. A partition function was used to relate the strain of each cell and the macro-scale strain of the representative element. On this basis, an elasto-plastic constitutive theory was derived considering the strain fluctuation and particle sliding (Chang, 1993, 1996a; Chang *et al.*, 1992a). The method was applied to the analysis of an idealized three-dimensional random packing. Figure 8.10 shows the prediction of lateral

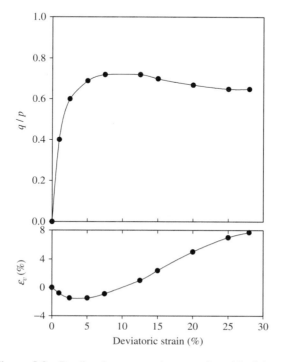

Figure 8.9 Predicted stress-strain curve for a biaxial test

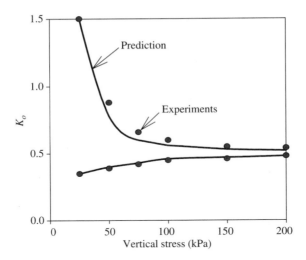

Figure 8.10   Prediction of K_o under one-dimensional compression for different initial stress conditions

stress change in a one-dimensional loading condition compared with experiments for Toyoura sand (Okochi and Tatsuoka, 1984). The model was also used to predict the plastic yielding surface and the non-associated plastic flow behavior for this packing under a different stress path (Chang *et al.*, 1992a). The method has the predictive potential that captures salient features of the behavior for an assembly of particles due to particle sliding.

8.7 Higher-Order Models

Another interesting and fundamental issue in granular mechanics is on the method of representing a discrete system by an equivalent continuum. The classic continuum theory has chosen the first-order derivative of displacement field as the descriptor of deformation. However, for a medium with finite size of particles, it requires continuum fields not only for the displacement, but also the rotation of discrete particles. Furthermore, the continuum field need not be linear within the representative volume (infinitesimal element). The variables may include higher-order derivatives of displacement and rotation (Chang and Gao, 1995a,b; Chang and Liao, 1990). Therefore, by selecting suitable continuum variables, one can model a discrete granular system by a generalized continuum of various type, for example, the micro-polar continuum, the high-gradient continuum, the non-local continuum, etc. Different types of models capture different physical mechanisms of granular materials. For example, a micro-polar continuum captures the rotation mode of particles, and thus models the consequence of asymmetric stress in granular material (Chang and Ma, 1990). A high-gradient continuum, on the other hand, models the effects of heterogeneous strain field (Chang and Gao, 1995a,b). Different from the classic continua in solid mechanics, the higher-order continua display the effects of internal characteristic length. When the particle size is relatively large compared to the sample size, the notion of classic continua is no longer adequate to represent the granular system.

8.8   Summary

In this brief review, a framework of microstructural mechanics approach is illustrated. In this approach, it is necessary to have a relationship between the macro-scale variables and the micro-scale variables. A review of two types of relationship is given: the micro-to-macro transformation and the macro-to-micro hypothesis. These relationships are important elements in micromechanics modeling. The stress-strain laws derived from different hypotheses yield to different results. The stress-strain models discussed here include elastic models, visco-elastic models, elasto-plastic models, and higher-order models. The application of the microstructural mechanics approach has been shown in the derived stress-strain relationships for granular materials.

References

Alpan, I. The empirical evaluation of the coefficient K_o and K_{or}. *Soils and Found*, **7**(1), (1967) 31–40.

Bathurst, R. J. and Rothenburg, L. Micromechanical aspects of isotropic granular assemblies with linear contact interactions. *J. Appl. Mech.*, **55**(1), (1988) 17–23.

Chang, C. S. Micromechanical modeling of constitutive relations for granular materials. *Micromechanics of Granular Materials*, Satake M. and Jenkins, J. T. (eds.). Elsevier, Amsterdam, (1987) 271–279.

Chang, C. S. Micromechanics modeling for deformation and failure of granulates with frictional contacts. *Mechanics of Material*, **16**(1–2), (1993) 13–24.

Chang, C. S. Inter-particle properties and elastic moduli for sand. *Mechanics of Granular Material*, International Society of Soil Mechanics and Foundation Engineering, (1994a) 7–14.

Chang, C. S. Settlement and compressibility for sand under one-dimensional loading condition. In: *Vertical and Horizontal Deformations of Foundations and Embankments*, Yeung A. T. and Felio, G. Y. (eds.), Vol. 2. ASCE, (1994b) 1298–1311.

Chang, C. S., Chang, Y. and Kabir, M. Micromechanics modelling for the stress-strain behavior of granular soil – I: theory. *J. Geotechnical Engineering*, **118**(12), (1992) 1959–1974.

Chang, C. S. and Gao, J. Second-gradient constitutive theory for granular material with-random packing structure. *Int. J. Solids and Structures*, **32**(16), (1995a) 2279–2293.

Chang, C. S. and Gao, J. Non-linear dispersion of plane wave in granular media. *Int. J. Non-linear Mechanics*, **30**(2), (1995b) 111–128.

Chang, C. S. and Gao, J. Kinematic and static hypotheses for constitutive modelling of granulates considering particle rotation. *Acta Mechanica*, **115**(1–4), (1996a) 213–229.

Chang, C. S. and Gao, J. Rheological modeling of randomly packed granules with visco-elastic binders of Maxwell type. *Computer and Geotechnique* (1996b).

Chang, C. S., Kabir, M. and Chang, Y. Micromechanics modelling for the stress-strain behavior of granular Soil – II: Evaluation. *J. Geotechnical Engineering*, **118**(12), (1992a) 1975–1994.

Chang, C. S. and Liao, C. Constitutive relations for particulate medium with the effect of particle rotation. *Int. J. Solids and Structures*, **26**(4), (1990) 437–453.

Chang, C. S. and Liao, C. L. Estimates of elastic modulus for media of randomly packed granules. *Appl. Mech. Rev.*, **47**(1), (1994) 197–206.

Chang, C. S. and Ma, L. Modelling of discrete granulates as micropolar continuum. *J. Engrg Mech.*, **116**(12), (1990) 2703–2721.

Chang, C. S. and Misra, A. Computer simulation and modelling of mechanical properties of particulates. *J. Computer and Geotechniques*, **7**(4), 269–287.

Chang, C. S. and Misra, A. Application of uniform strain theory to heterogeneous granular solids. *J. Engrg Mech.*, **116**(10), (1990a) 2310–2328.

Chang, C. S. and Misra, A. Packing structure and mechanical properties of granulates. *J. Engrg mech.*, **116**(5), (1990b) 1077–1093.

Chang, C. S., Misra, A. and Acheampong, K. Elastoplastic deformation of granulates with frictional contacts. *J. Engrg Mech.*, **118**(8), (1992) 1692–1708.

Chang, C. S., Sundaram, S. S. and Misra, A. Initial moduli of particulated mass with frictional contacts. *Int. J. Numerical and Analytical Methods in Geomechanics*, **13**(6), (1989) 626–641.

Christoffersen, J., Mehrabadi, M. M. and Nemat-Nasser, S. A micromechanical description of granular material behavior *J. Appl. Mech.*, **48**(2), (1981) 339–344.

Cundall, P. A. and Strack, O. D. L. A discrete numerical model for granular assemblies. *Geotechnique*, **29**(1), (1979) 47–65.

Duffy, J. A differential stress-strain relation for the hexagonal close packed array. *J. Appl. Mech. Trans ASME* (1959) 88–94.

Duffy, J. and Mindlin, R. D. Stress-strain relations and vibrations of granular media. *J. Appl. Mech.*, **24**(4), (1957) 585–593.

Emeriault, F. and Cambou, B. Mechanical modeling of anisotropic non-linear elasticity of granular medium. *Int. J. Solids and Structures*, **33**(18), (1996) 2591–2607.

Emeriault, F. and Chang, C. S. Anisotropic elastic moduli of granular assemblies from micromechanical approach. *J. Engrg Mech.*, **123**(12), (1997) 1289–1293.

Hardin, B. O. and Black, W. L. Sand stiffness under various triaxial stresses. *J. Soil Mech. and Foundation Engrg.*, **92**, (1966) 27–42.

Hardin, B. O. and Richart, Jr. F. E. Elastic wave velocities in granular soils. *J. Soil Mech. Found. Div.*, **89**, (1963) 33–65.

Hendron, A. J. *The behavior of sand under one-dimensional compression*. PhD thesis, Department of Civil Engineering, University of Illinois, Urbana, (1963).

Jenkins, J. D. Volume change in small strain axisymmetric deformations of a granular material. In: *Micromech. Granular Materials*, Satake. M. and Jenkins, J.T. (eds.). Elsevier, Amsterdam, (1987) 245–252.

Johnson, K. L. *Contact Mechanics*. Cambridge University Press, England, (1985).

Kishino, Y. Disc model analysis of Granular media. In: *Micromechanics of Granular Materials*, Satake, M. and Jenkins, J. T. (eds). Elsevier, Amsterdam, (1988) 143–152.

Kopperman, S. E., Stokoe, K. H. and Knox, D. P. Effect of state of stress on velocity of low amplitude compression waves propagating along principal stress directions in dry sand. *Geotechnical. Engrg. Report* GR82-22, University of Texas, Austin, (1982).

Liao, C. L., Chang, T. and Young, D. Elastic-plastic constitutive modelling of granular assembly *Int. J. Solids and Structures*, **32**(8), (1995) 1121–1133.

Liao, C. L., Chang, T. P., Young, D. and Chang C. S. Stress-strain relationship for granular materials based on hypothesis of best fit. *Int. J. Solids and Structures*, **34**(31–32), (1997) 4087–4100.

Mindlin, R. D. and Deresiewicz, H. H. Elastic spheres in contact under varying oblique forces *J. Appl. Mech.*, **20**(3), (1953) 327–344.

Misra, A. and Chang, C. S. Effective elastic moduli of heterogeneous granular solids. *Int. J. Solids and Structures*, **30**(18), (1993) 2547–2566.

Ng, T. T. *Numerical simulation of granular soil under monotonic and cyclic loading: A particulate mechanics approach*. PhD Dissertation, Renesselaer Polytechnic Institute, Troy, (1989).

Okochi, Y. and Tatsuoka, F. Some factors affecting K-values of sand measured in triaxial cell. *Soils and Foundations*, **24**(3), (1984) 52–68.

Schmid, B. Discussion of *Earth Pressure at Rest Related to Stress History*, by E. W. Brook and H. O. Ireland. *Can. Geot. J.*, **3**(4), (1966) 239–242.

Schmertmann, J. H. Measurement of in situ shear strength. *Specialty Conference In Situ Measurement of Soil Properties*, Raleigh, Vol. 2, (1975) 57–138.

Sherif, M. A., Ishibashi, I. and Ryden, D. E. Coefficient of lateral earth pressure at rest in cohesionless soils. *Soil Engg. Research Report No. 10*, University of Washington, Seattle, Washington, (1974).

Spencer, A. J. M. Isotropic polynomial invariants and tensor functions. In: *Application of Tensor Functions in Solid Mechanics*, Boehlel, J. P. (ed.), Springer-Verlag, Berlin, (1987).

Ting, J. M., Corkum, B. T., Kauffman, C. R. and Greco, C. Discrete numerical model for soil mechanics *J. Geotech. Engrg.* **115**(3), (1989) 379–398.

Walton, K. The effective elastic moduli of a random packing of spheres. *J. Mech. Physics of Solids*, **35**(2), (1987) 213–226.

Zhu, H., Chang, C. S. and Rish, J. W. Tangential and normal compliance for conforming binder contact. I: Elastic binder *Int. J. Solids and Structures*, **33**(29), (1996a) 4337–4349.

Zhu, H., Chang, C. S. and Rish, J. W. Tangential and normal compliance for conforming binder contact. II: Visco-elastic binder *Int. J. Solids and Structures*, **33**(29), (1996b) 4351–4363.

9

Constitutive Modeling for Fluid-Saturated Porous Solid by the Asymptotic Homogenization Method

Kenjiro Terada[1], Takashi Ito[2] and **Noboru Kikuchi[3]**

[1]*The University of Tokyo, Japan*
[2]*Toyota College of Technology, Aichi, Japan*
[3]*The University of Michigan, Ann Arbor, MI, USA*

9.1 Introduction

When focusing our attention to the mechanical behavior of fluid-saturated porous geomaterials in linear quasi-static situations, the pioneering work of Terzaghi (1926) and the three-dimensional extensions by Biot (1941, 1956, 1962) provide basic concepts for modeling the fluid-solid coupling phenomena encountered in geomechanics, geophysics, etc., and have invoked many extensions and more appropriate numerical techniques in this context (see, for example, Desai and Christian (1977), Desai and Gallagher (1984) and Lewis and Schrefler (1987)). The effect of the pore fluid is usually expressed in terms of the so-called effective stress principle, stating that the pore pressure reduces the effective value of the mean normal stress of the solid skeleton. Although a large number of efforts have been made to include more complicated features in the constitutive relations of the materials (e.g. see Bazant (1985) or Desai *et al.* (1987)), there are few alternative approaches that offer major advantages over the classical or mechanistic ones.

The essential features in mechanics observed in the microscopic scale seems to have been ignored in their development. In most of the mechanics theories for fluid-solid mixtures, the mechanical behavior of microstructures is described only via the volumetric proportions of the respective phases, and they are therefore sometimes referred to as the 'volume fraction' theories, (see e.g. de Boer and Ehlers, 1988). However, what characterizes the solid-fluid coupling phenomenon should be the interaction between the fluid and solid phases at a microscopic scale. More specifically, the time-dependent overall mechanical behavior of the mixture is the consequence of the microscopic flow of viscous fluids, and the microscopic stress distribution in the solid phase must be much more complicated, affected by the geometrical configuration of the microstructure as well as the pore pressure. Therefore, it is necessary for macroscopic constitutive modeling to have more proper descriptions involving the phase interaction in a microscopic scale.

In this context, it seems appropriate to use the notion of Representative Volume Element (RVE), which has been extensively utilized in the mechanics of composites, including micromechanics (see, for example, Hashin, 1983; Christensen, 1979; and Mura, 1982). However, little attention has been paid to the characterization of fluid-saturated porous solids by using the RVE approach. As for geomaterials, Darcy's law (which characterizes the flow through the porous media) can barely be derived by the RVE analysis assuming a Stokes flow in a microscopic scale (see Bear, 1967). The reason may be that the geometry of microstructures is, in general, unknown for the mixture, and that averaging over the RVE must be performed irrespective of governing equations that hold for field variables defined at a microscopic scale.

On the other hand, the mathematical theory of asymptotic homogenization enables the derivation of the macroscopic governing equations from the microscopically assumed field equations, based on the RVE concepts. Keller (1980) derived the Darcy's law by applying the method of asymptotic expansion to the Stokes flow problem defined in a microscopic pore region, and Sanchez-Palencia (1980) followed this idea and characterized many homogenization results in mechanics as the asymptotic behavior of composites by introducing periodic microstructures, namely unit cells, as RVEs. The homogenization method of the fluid saturated elastic solid or fluid with rigid suspensions is the direct extension, and extensive studies can be found in, for example, Burridge and Keller (1981), Lévy (1979, 1987), Sanchez-Hubert (1980), and Arbogast, Douglas and Hornung (1990). However, there have not been any numerical simulations, and even engineering applications have not been discussed probably because of the complexity of the formulations.

The main purpose of this chapter is to re-examine the global-local nature of the quasi-static deformation of a fluid-saturated solid structure by using the homogenization method. We first derive the homogenization formulae for fluid-saturated porous elastic solids from the microscopically defined system, leading to the fact that the homogenized governing equations are identical to the classical Biot's equations for consolidation of settlements. Several numerical examples are presented to illustrate the homogenization and localization capability of the method, and the applicability of the method to problems in geotechnical engineering is discussed. In the last section, a newly developed geometry modeling technique for microstructures is introduced to take into account the geometrical features of fairly complicated microstructures. The novel modeling method is expected to provide some insight into the mechanics of geomaterials, keeping the connection between the overall mechanical behavior and the micromechanical behavior. This work is the first attempt to take into account the microstructures of fluid-saturated porous in the constitutive modeling for consolidation problems in geomechanics.

9.2   Formulation of the Homogenization

9.2.1   *Problem setting*

Let us consider a homogeneous, linearly elastic porous solid saturated with a single Newtonian fluid occupying domain Ω^ε (see Figure 9.1(a)). Here the symbol ε indicates the size of the microscopic heterogeneity, and is introduced as a nondimensional parameter

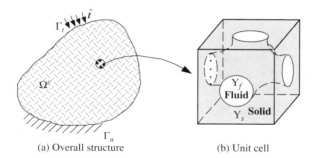

(a) Overall structure   (b) Unit cell

Figure 9.1 Fluid saturated porous elastic solid structure and its periodic microstructure

such that

$$\varepsilon = \frac{l}{L} \tag{9.1}$$

where L is the characteristic length of the overall structure and l is that of its representative microstructure in which the constitutive relations are introduced. Assuming that the whole domain is the assemblage of periodically aligned microstructures, we identify a unit cell shown in Figure 9.1(b) with a Representative Volume Element (RVE), in which both the solid and the fluid phases are assumed to be open and connected so that the fluid can move through the RVE boundaries.

In a quasi-static situation, the equilibrium of such a composite body with some mechanical stimuli may be described by

$$\frac{\partial \sigma_{ij}^{\varepsilon}}{\partial x_j} + \rho^{\varepsilon} f_i = 0 \quad \text{in } \Omega^{\varepsilon} \tag{9.2}$$

where σ^{ε} is the stress tensor and $\rho^{\varepsilon} f$ the body force. Here, every index takes from one to three in the current three-dimensional setting, and the summation convention is adopted for the dummy indices, as usual.

Although the fluid and solid are mixed in a microscopic scale, the boundary value problem can be defined once the displacement and/or the force boundary conditions are given and the appropriate constitutive equations are chosen for each constituent material (fluid and solid) in the unit cell. Along with the Stokes condition for the bulk viscosity, the stress within the Newtonian fluid phase, $\sigma^{f\varepsilon}$, is related to the deformation via the following constitutive relation:

$$\sigma_{ij}^{f\varepsilon} = -p^{\varepsilon}\delta_{ij} + 2\varepsilon^2\mu^{\varepsilon}\left(\dot{\varepsilon}_{ij} - \frac{1}{3}\dot{\varepsilon}_{pp}\delta_{ij}\right) = -p^{\varepsilon}\delta_{ij} + \varepsilon^2\mu^{\varepsilon}D\frac{\partial v_i^{\varepsilon}}{\partial x_j} \quad \text{in } \Omega_f^{\varepsilon} \tag{9.3}$$

where δ_{ij} is the Kronecker delta symbol and ε_{kh} is defined as a function which transforms a vector field into the symmetric part of its gradient with respect to x, i.e. for a velocity field v^{ε},

$$\varepsilon_{kh}(v^{\varepsilon}) = \frac{1}{2}\left(\frac{\partial v_k^{\varepsilon}}{\partial x_h} + \frac{\partial v_h^{\varepsilon}}{\partial x_k}\right) \tag{9.4}$$

provides the strain rate tensor that is equivalent to $\dot{\varepsilon}(u^{\varepsilon})$ for the displacement field, u^{ε}. Here, we also have defined the operator D, that produces the shear components of the strain rate tensor:

$$D : \dot{\varepsilon}_{ij} \mapsto \dot{\varepsilon}_{ij} - \tfrac{1}{3}\dot{\varepsilon}_{pp}\delta_{ij} \tag{9.5}$$

Note that the actual viscosity μ^ε of the fluid is replaced by $\varepsilon^2\mu^\varepsilon$, due to the dimensional consideration under the current quasi-static situation (see, for example, Levy, 1979, 1987).

On the other hand, the stress in the solid phase $\sigma^{s\varepsilon}$ may be related to the infinitesimal strain, such that

$$\sigma_{ij}^{s\varepsilon} = E_{ijlm}^\varepsilon \varepsilon_{lm}(u^\varepsilon) = E_{ijlm}^\varepsilon \frac{\partial u_l^\varepsilon}{\partial x_m} \quad \text{in } \Omega_s^\varepsilon \tag{9.6}$$

where E_{ijlm}^ε is the elasticity tensor that satisfies both the symmetry and positiveness conditions so that the problem is well-posed (see Oleinik., 1992).

The velocity of fluid $\hat{v}^f$ (or the displacement $\hat{u}^s$) may be applied to some parts of the boundary Γ_u as

$$v_i^\varepsilon = \hat{v}_i^f \quad \text{on } \Gamma_u^f \text{ (or } u_i^\varepsilon = \hat{u}_i^\varepsilon \quad \text{on } \Gamma_u^s) \tag{9.7}$$

and the surface force (or traction) $\hat{t}^f$ (or $\hat{t}^s$) must be prescribed on the remaining traction boundary Γ_t as

$$\sigma_{ij}^{f\varepsilon} n_j^f = \hat{t}_i^f \quad \text{on } \Gamma_t^f \text{ (or } \sigma_{ij}^{s\varepsilon} n_j = \hat{t}_i \quad \text{on } \Gamma_t^s) \tag{9.8}$$

Here, the sub- or superscript 's' on a quantity stands for solid, and 'f' stands for fluid, as they do in the following sections.

On the interface Σ between the solid and fluid, both phases interact with each other via the following condition:

$$v_i^\varepsilon = \frac{\partial u_i^\varepsilon}{\partial t} \quad \text{on } \Sigma \tag{9.9}$$

$$\sigma_{ij}^{f\varepsilon} n_j^f = -\sigma_{ij}^{s\varepsilon} n_j^s \quad \text{on } \Sigma \tag{9.10}$$

where n^f (or n^s) is the outward unit normal of the interface.

9.2.2 *Asymptotic expansions with micro- and macroscopic scales*

Under the above setting, let us characterize the macro- and micromechanical behaviors of the solid-fluid mixture by the asymptotic homogenization method. To measure the scales represented by L and l, we introduce two different coordinate systems, x and y, which are related through ε to each other by

$$y = \frac{x}{\varepsilon} \tag{9.11}$$

Here, the macroscopic scale denoted by x measures the quantities in the domain Ω, at which the heterogeneities are invisible and the microscopic scale y is a magnified scale to measure microscopic heterogeneities. Following the conventional procedure of the homogenization method, we consider the case where the material properties are periodic with respect to y, say Y-periodic (see Benssousan, Lions and Papanicoulau, 1978). More precisely, if we denote by Y a unit parallelepiped cell; the periodic structure is supposed to have periods εY.

The variational form for Stoke's flow problem for the fluid phase becomes

$$-\int_{\Omega_f^\varepsilon} p^\varepsilon \frac{\partial w_i^\varepsilon}{\partial x_j} dx + \varepsilon^2 \int_{\Omega_f^\varepsilon} \left(\mu D \frac{\partial v_i^\varepsilon}{\partial x_j}\right) \frac{\partial w_i^\varepsilon}{\partial x_j} dx = \int_{\Omega_f^\varepsilon} \rho_f f_i w_i^\varepsilon dx + \int_{\Gamma_f^\varepsilon} \hat{t}_i w_i^\varepsilon dx$$

$$+ \int_\Sigma \sigma_{ij}^f n_j^f w_i^\varepsilon ds \ \forall w^\varepsilon \qquad (9.12)$$

where w^ε is the admissible function satisfying the homogeneous boundary conditions. Similarly, the variational form for the solid phase can be written as

$$\int_{\Omega_s^\varepsilon} E_{ijlm} \frac{\partial u_l^\varepsilon}{\partial x_m} \frac{\partial w_i^\varepsilon}{\partial x_j} dx = \int_{\Omega_f^\varepsilon} \rho_f f_i w_i^\varepsilon dx + \int_{\Gamma_i^s} \hat{t}_i w_i^\varepsilon dx + \int_\Sigma \sigma_{ij}^s n_j^s w_i^\varepsilon ds \ \forall w^\varepsilon \qquad (9.13)$$

Note that, due to the interface condition in Eq. (9.10), the combination of Eqs. (9.12) and (9.13) is equivalent to the coupled problem defined by Eq. (9.2) in the whole domain, i.e. the mechanical coupling occurs though the interface between the fluid and the solid phases.

Now, let us introduce the asymptotic expansions as solutions of Eqs. (9.12) and (9.13) as

$$\begin{cases} u^\varepsilon(x,t) = u(x,y,t) = u^0(x,t) + \varepsilon u^1(x,y,t) + \varepsilon^2 u^2(x,y,t) & (9.14a) \\ v^\varepsilon(x,t) = v(x,y,t) = v^0(x,y,t) + \varepsilon v^1(x,y,t) + \varepsilon^2 v^2(x,y,t) & (9.14b) \\ p^\varepsilon(x,t) = p(x,y,t) = p^0(x,y,t) + \varepsilon p(x,y,t)^1 + \varepsilon^2 p^2(x,y,t) & (9.14c) \end{cases}$$

as well as their variations

$$w^\varepsilon(x,t) = w(x,y,t) = w^0(x,y,t) + \varepsilon w^1(x,y,t) + \varepsilon^2 w^2(x,y,t) \qquad (9.15)$$

where the superscript introduced in each term indicates the order of dependency on both x and/or y, but not the powers of ε. Here, note that the 0th order term of displacement field u^0, of the solid phase is a function of x only since we already know that the term represents the average displacement independent of y, namely the macroscopic deformation, of the homogenized structure, see for example, Sanchez-Palencia (1980). Due to the dependence on both x and y, the derivative with respect to the scale for the original structure must be evaluated by

$$\left.\frac{\partial}{\partial x_j}\right|_{\Omega^\varepsilon} = \left.\left(\frac{\partial}{\partial x_j} + \frac{1}{\varepsilon}\frac{\partial}{\partial y_j}\right)\right|_{\Omega\times Y} \qquad (9.16)$$

9.2.3 Equations for the fluid phase

After substituting the expanded form of the velocity and the pressure fields into Eq. (9.12), we can recognize that the following equation must be satisfied in the microscopic scale for the fluid phase:

$$\left(\rho_f f_i - \frac{\partial p^0}{\partial x_i}\right) \int_{Y_f} W_i \, dy + \mu \int_{Y_f} D\left(\frac{\partial v_i^{\text{rel}}}{\partial y_j}\right) \frac{\partial W_i}{\partial y_j} dy = 0 \ \forall W(y) \qquad (9.17)$$

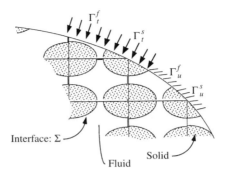

Figure 9.2   Definitions of the boundaries and the interfaces

where v^{rel} is the divergence-free velocity field relative to the solid deformation rate, $\dot{u}^0$. Here, Y_f is the domain of the fluid phase within the unit cell (see Figure 9.2). Noting that there exists a unique solution κ^k, for each k, which is Y-periodic and divergence free in Y_f such that

$$\mu \int_{Y_f} D \frac{\partial \kappa_i^k}{\partial y_j} \frac{\partial W_i}{\partial y_j}\, dy = \int_{Y_f} W_k\, dy \; \forall W(y) \tag{9.18}$$

we recognize the solutions of Eq. (9.17) must be of the following form:

$$v_i^{\text{rel}}(y,x,t) = \left(\rho_f f_k - \frac{\partial p^0(x,t)}{\partial x_k} \right) \kappa_i^k(y) \tag{9.19}$$

from which the velocity field of the fluid phase can be expressed by

$$v_i^0(x,y,t) = \frac{\partial u_i^0(x,t)}{\partial t} + \left(\rho_f f_k - \frac{\partial p^0}{\partial x_k}(x,t) \right) \kappa_i^k(y) \tag{9.20}$$

The macroscopic velocity field of the fluid phase in the mixture is obtained by averaging the velocity field within the cell:

$$V_i = \langle v_i^0(x,y,t)\rangle = \frac{\partial u_i^0(x,t)}{\partial t} + \left(\rho^f f_k - \frac{\partial p^0}{\partial x_k}(x,t) \right) \langle \kappa_i^k(y)\rangle \tag{9.21}$$

where $\langle \cdot \rangle$ indicates the volume average over the unit cell. If we denote the averaged (apparent) flow velocity and the stationary permeability coefficients by, respectively,

$$V_i = \langle v_i^{\text{rel}}\rangle \quad \text{and} \quad K_{ik} = \langle \kappa_i^k\rangle. \tag{9.22}$$

Equation (9.21) can be recognized as the generalized Darcy's law:

$$V_i = \frac{\partial u_i^0(x,t)}{\partial t} - K_{ik}\frac{\partial P}{\partial x_k} \tag{9.23}$$

where P defines the following generalized pressure gradient:

$$\frac{\partial P}{\partial x_k} := -\rho^f f_k + \frac{\partial p^0}{\partial x_k}(\boldsymbol{x},t).$$
(9.24)

9.2.4   Equations for the solid phase

Substitution of the asymptotic expansion of the displacement field and its variation into Eq. (9.13) also yields the governing equation for the solid phase:

$$\int_\Omega \left[\frac{1}{|Y|} \int_{Y_s} E_{ijlm} \left(\frac{\partial u_l^0}{\partial x_m} + \frac{\partial u_l^1}{\partial y_m} \right) \left(\frac{\partial w_i^0}{\partial x_j} + \frac{\partial w_i^1}{\partial y_j} \right) dy \right] dx$$

$$= \frac{|Y_s|}{|Y|} \int_\Omega \rho^s f_i w_i^0 \, dx + \int_{\Gamma_t^s} \hat{t}_i w_i^0 \, ds - \int_\Omega p^0 \left[\frac{1}{|Y|} \int_{Y_s} \frac{\partial w_k^1}{\partial y_k} dy \right] ds \; \forall \boldsymbol{w}^0(\boldsymbol{x}), \, \boldsymbol{w}^1(\boldsymbol{y})$$
(9.25)

where the quantities denoted by $|.|$ represents the volume of the domain. It should be noted that the divergence theorem has been used for the last term associated with the interaction with the fluid phase, and only this term is different from linear elasticity formulation. Therefore, the macroscopic and microscopic problems can be derived in the same way as the linear elasticity formulation (see, for example, Guedes and Kikuchi, 1991).

The microscopic problem is given by the following variational form:

$$\int_\Omega \left[\frac{1}{|Y|} \int_{Y_s} E_{ijlm} \left(\frac{\partial u_l^0}{\partial x_m} + \frac{\partial u_l^1}{\partial y_m} \right) \frac{\partial w_i^1}{\partial y_j} dy \right] dx = -\int_\Omega p^0 \left[\frac{1}{|Y|} \int_{Y_s} \frac{\partial w_j^1}{\partial y_j} dy \right] dx \; \forall \boldsymbol{w}^1(\boldsymbol{y})$$
(9.26)

Due to the linearity, the microscopic displacement is found to be of the form

$$u_l^1(\boldsymbol{x}, \boldsymbol{y}, t) = -\chi_l^{kh}(\boldsymbol{y}) \frac{\partial u_k^0}{\partial x_h}(\boldsymbol{x}, t) - p^0(\boldsymbol{x}, t) \phi_l(\boldsymbol{y}) + c_l^1(\boldsymbol{x})$$
(9.27)

knowing that the characteristic deformations, χ_l^{kh} and ϕ_l, are the solutions of the following problems, respectively:

$$\int_{Y_s} E_{ijlm} \frac{\partial \chi_l^{kh}(\boldsymbol{y})}{\partial y_m} \frac{\partial w_i^1}{\partial y_j} dy = \int_{Y_s} E_{ijkh} \frac{\partial w_i^1}{\partial y_j} dy \forall \boldsymbol{w}^1(\boldsymbol{y})$$
(9.28)

$$\int_{Y_s} E_{ijlm} \frac{\partial \phi_l(\boldsymbol{y})}{\partial y_m} \frac{\partial w_i^1}{\partial y_j} dy = \int_{Y_s} \frac{\partial w_k^1}{\partial y_k} dy \forall \boldsymbol{w}^1(\boldsymbol{y})$$
(9.29)

where Y^s is the domain of the solid phase within a unit cell. Here, $c_l^1(\boldsymbol{x})$ is regarded as a constant within a unit cell.

The macroscopic problem can be separated from Eq. (9.25) as follows:

$$\int_\Omega \left[\frac{1}{|Y|} \int_{Y_s} E_{ijlm} \left(\frac{\partial u_l^0}{\partial x_m} + \frac{\partial u_l^1}{\partial y_m} \right) dy \right] \frac{\partial w_i^0}{\partial x_j} dx = \frac{|Y_s|}{|Y|} \int_\Omega \rho_s f_i w_i^0 \, dx$$

$$+ \int_{\Gamma_t^s} \hat{t}_i w_i^0 \, ds \; \forall \boldsymbol{w}^0(\boldsymbol{x})$$
(9.30)

Substitution of Eq. (9.27) into this yields

$$\int_{\Omega} E^H_{ijkh} \frac{\partial u^0_k(x)}{\partial x_h} \frac{\partial w^0_i}{\partial x_j} dx - \int_{\Omega} p^0 Q^H_{ij} \frac{\partial w^0_i}{\partial x_j} dx = \int_{\Omega} f^{sH}_i w^0_i dx + \int_{\Gamma^s_t} \hat{t}_i w^0_i ds \ \forall w^0(x)$$
(9.31)

where the homogenized elasticity tensor, the coefficient associated with the fluid pressure and the average body force are respectively defined as

$$E^H_{ijkh} = \frac{1}{|Y|} \int_{Y_s} \left(E_{ijkh} - E_{ijlm} \frac{\partial \chi^{kh}_l}{\partial y_m} \right) dy, Q^H_{ij} = \frac{1}{|Y|} \int_{Y_s} E_{ijlm} \frac{\partial \phi_l}{\partial y_m} dy, \text{ and}$$

$$f^{sH}_i = \frac{|Y_s|}{|Y|} \rho_s f_i$$
(9.32)

9.2.5   Continuity equation–mass conservation of the fluid

The pressure distribution within the fluid phase is induced from the volume change of the overall structure. When the flux of pressure, i.e. the pressure gradient, produces the flow of fluid from one microstructure to another, its mass must be conserved. Assuming the incompressibility, the weak form of the continuity equation is given by

$$\int_{\Omega^\varepsilon_f} \frac{\partial v^\varepsilon_i}{\partial x_i} \theta \, dx = 0 \ \forall \theta(x)$$
(9.33)

where $\theta(x)$ is the variation of a scalar function. Substituting the asymptotic expansion of the velocity field and using the Darcy's law in Eq. (9.23) derived above, we reach a system of differential equations for poroelasticity as

$$\int_{\Omega} \frac{\partial}{\partial x_i} \left(\frac{\partial u^0_i}{\partial t} \right) \theta \, dx + \int_{\Omega} K_{ik} \frac{\partial P}{\partial x_k} \frac{\partial \theta}{\partial x_i} dx + \int_{\Gamma_u} \hat{q} \theta \, ds = 0 \ \forall \theta(x)$$
(9.34)

along with Eq. (9.31), in which the primal variables to be obtained are the average displacement $u^0(x)$, and pressure $p^0(x)$. Here $\hat{q}$ is the flow flux, which must be specified on Γ_u.

9.2.6   Micro- and macroscopic homogenized constitutive law

Using the two characteristic deformations χ^{kh} and ϕ, the macroscopic strain and the pressure are localized in the microscopic stress, such that

$$\sigma^0_{ij} = \sigma^{s0}_{ij} + \sigma^{f0}_{ij}$$
(9.35)

where

$$\sigma^{s0}_{ij} = \left(E_{ijkh} - E_{ijlm} \frac{\partial \chi^{kh}_l}{\partial y_m} \right) \frac{\partial u^0_k}{\partial x_h} - p^0 E^s_{ijlm} \frac{\partial \phi_l}{\partial y_m} \text{ in } Y_s \text{ and } \sigma^{f0}_{ij} = -p^0 \delta_{ij} \text{ in } Y_f \quad (9.36)$$

The macroscopic stress for the mixture is given by the volume average of the microscopic stress over the whole unit cell domain, such that

$$\langle \sigma_{ij}^0 \rangle = \langle \sigma_{ij}^{s0} \rangle + \langle \sigma_{ij}^{f0} \rangle = E_{ijkh}^H \frac{\partial u_k^0}{\partial x_h} - p^0(\delta_{ij}n + Q_{ij}^H) \tag{9.37}$$

where

$$\langle \sigma_{ij}^{s0} \rangle = E_{ijkh}^H \frac{\partial u_k^0}{\partial x_h} - p^0 Q_{ij}^H, \; \langle \sigma_{ij}^{f0} \rangle = -n\,p^0 \delta_{ij} \text{ and } n = \frac{|Y^f|}{|Y|} \tag{9.38}$$

It is easy to see that the effective stress, celebrated in soil mechanics, is the same as the macroscopic stress $\langle \sigma_{ij}^{s0} \rangle$.

9.3 Numerical Examples: Micro- and Macromechanical Analyses

In this section, we present the whole procedure of the local and global analyses using the above formulation. The FE model used in the microscopic problems, defined by Eqs. (9.18), (9.28) and (9.29) is shown in Figure 9.3, and is the same through this numerical analyses. Young's modulus of the solid phase is assumed as 30 MPa and Poisson's ratio as 0.25, and the viscosity 10^{-3} Pa.s is used for the fluid phase.

The homogenized material constants obtained from Eqs. (9.22) and (9.32) are calculated as

$$\mathbf{K} = \begin{bmatrix} 0.84 & 0 & 0 \\ & 0.84 & 0 \\ \text{sym.} & & 0.84 \end{bmatrix} (\text{m}^4/\text{N.s}) \tag{9.39}$$

$$\mathbf{E}^H = \begin{bmatrix} 6.04 & 0.735 & 0.735 & 0 & 0 & 0 \\ & 6.04 & 0.735 & 0 & 0 & 0 \\ & & 6.04 & 0 & 0 & 0 \\ & & & 0.760 & 0 & 0 \\ & \text{sym.} & & & 0.760 & 0 \\ & & & & & 0.760 \end{bmatrix} (\text{MPa}) \tag{9.40}$$

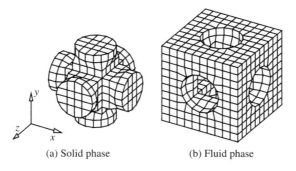

(a) Solid phase   (b) Fluid phase

Figure 9.3   FE unit cell model

$$
Q = \begin{bmatrix} 0.251 & 0 & 0 \\ & 0.251 & 0 \\ \text{sym.} & & 0.251 \end{bmatrix} \tag{9.41}
$$

The evaluation indicates that the until cell used in this example provides an isotropic homogenized material when viewing the overall structure in the macroscopic scale. Since the structure is replaced by the homogenized structure, it is no longer necessary to consider the microstructure in solving the macroscopic problems governed by Eqs. (9.31) and (9.34).

Because there is no difference between the homogenized problem and Biot's, the conventional implementation methods developed for the consolidation problem can be utilized for finite element discretization (e.g. Sandhu, 1981; Siriwardane and Desai, 1986). In the example below, the C^1 interpolation for the displacement field and C^0 for pressure is adopted for the purposes of numerical stability, (see Sandhu *et al.*, 1977). The FE model for the macroscopic problem is given in Figure 9.4(a) together with the boundary conditions and the deformed geometry. The distributed load is given on one portion of the top surface, and then the deformation and Darcean flow occur as shown in Figure 9.4(b).

Figure 9.5 shows the stress and pressure changes in time, which shows the same tendency as that of usual consolidation problems. That is, at the initial stage, the homogenized structure has pressure corresponding to the initial deformation, and the solid skeleton sustains less loading. As the fluid flows out from the permeable boundary, the pressure value is reduced and, on the other hand, the solid phase sustains more and more stress. At the end of this process, all the loading is taken by the solid skeleton, and the pore pressure tends to drop to zero.

After solving the macroscopic problem, we can compute both the stress and velocity distributions within the unit cell by combining the characteristic functions and the actual macroscopic responses. This process is referred to as localization, which is only allowed for the global-local approach with the asymptotic homogenization method. The localization involves an evaluation of the microscopic deformation from the macroscopic one, and that of the microscopic flow field from the pressure gradient at any *point* in the overall

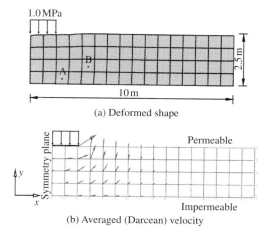

Figure 9.4   FE analysis model for the macroscopic problem, and the results

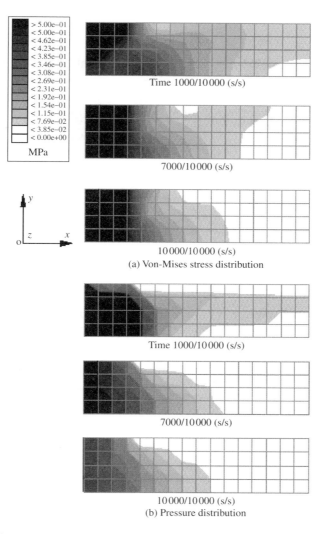

Figure 9.5   Variation of stress and pressure distributions in time

structure. Typical examples of the micromechanical responses are shown in Figures 9.6 and 9.7, in which Points A and B correspond to the points indicated in Figure 9.4(a). In Figure 9.6, the overall strain values at the middle stage of the macroscopic (transient) analysis are given to the unit cell, and Figure 9.7 shows the corresponding velocity vectors.

Note that, although these micromechanical behaviors are time-dependent, we can evaluate both quantities at any point in the overall structure, and at any time, using the characteristic deformations that are unique through the analysis. It should also be noted that the maximum value of the microscopic stress is larger than the average value over a unit cell structure. This fact implies that the excessive stress values, which cannot appear in the macroscopic analyses, might induce the local defects of the solid phase and, eventually, may have the overall geotechnical structure collapsed under drained or undrained

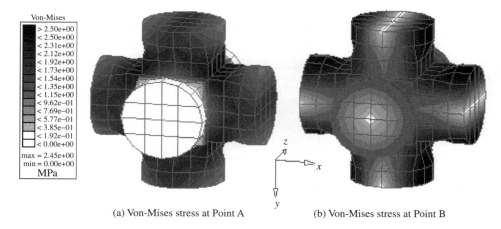

(a) Von-Mises stress at Point A   (b) Von-Mises stress at Point B

Figure 9.6   Microscopic stress distributions at different macroscopic locations

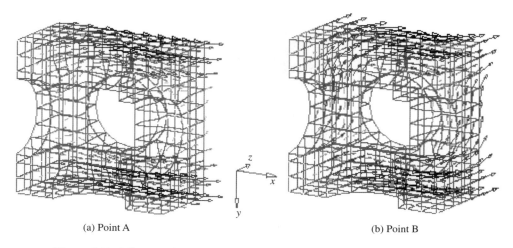

(a) Point A   (b) Point B

Figure 9.7   Microscopic velocity distribution of different macroscopic locations

conditions. Thus, this type of analysis could provide the possibility of making predictive simulations in geotechnical engineering problems with possible loading conditions.

This example justifies that the homogenization method can be used in the global-local analyses of geomaterials saturated with fluids, in which the micromechanical behavior is crucial. However, the microstructures of geomaterials are seldom formed in a regular geometrical configuration, as in Figure 9.3, and the nonuniformity may affect their behavior. Therefore, it is necessary to obtain more realistic microstructural models in order to evaluate the actual responses of a unit cell, and to shed light on the failure mechanism of an overall geotechnical structure. The modeling technique presented in the next section may be helpful in this context.

As a final remark, the two-dimensional description is inadequate for the current formulation from the viewpoint of RVE analyses, because the solid and fluid phases must be open and connected. This also implies that, in the numerical analysis for fluid-saturated porous solids, the constitutive modeling with microstructures makes sense only in the three-dimensional case. However, the three-dimensional identification of the microscopic geometrical configuration is quite difficult, and even two-dimensional modeling has not been done in the homogenization analysis.

9.4 Combination with Digital Image-Based Geometry Modeling Technique

Actual geomaterials have far more complicated geometrical configurations than those we have used in the previous examples. It is, therefore, preferable to include the effects of this complexity of the microstructure in the numerical analysis of the FEM so that the localization results are more realistic. In this section, by using the unit cell model of a geomaterial-like microstructure made artificially, we shall simulate the homogenization and localization.

To construct a realistic FE geometry model, we use the Digital Image-Based (DIB) FE modeling technique (see Hollister and Riemer (1993) or Terada, Miura and Kikuchi (1996)). Figure 9.8 shows the FE model that was generated by this modeling technique, and represents the virtual geometrical configuration of a unit cell in terms of voxels of a 3-D image, in which the volume fraction of the fluid phase was measured as 11%. Taking into account the specific effects of geometry, the numerical example presented below will show the localization capability.

After solving Eqs. (9.18), (9.28) and (9.29), the homogenized constants are calculated using Eqs. (9.22) and (9.32) as follows:

$$
\mathbf{E}^{H} =
\begin{bmatrix}
304 & 106 & 122 & 1.88 & -3.55 & 5.43 \\
 & 282 & 114 & 3.40 & -1.10 & 5.71 \\
 & & 394 & 5.65 & -5.03 & 2.79 \\
 & & & 106 & 2.29 & -1.132 \\
 & \text{sym.} & & & 110 & 1.60 \\
 & & & & & 99.4
\end{bmatrix}
\text{MPa}
\tag{9.42}
$$

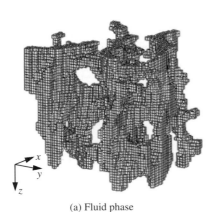 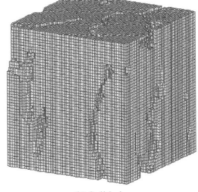

(a) Fluid phase   (b) Solid phase

Figure 9.8 Digitized FE unit cell model (data is taken by scanning an actual bone microstructure and is given by Professor Hollister at the University of Michigan)

$$\mathbf{Q}^H = \begin{bmatrix} -0.3499 & 0.0139 & -0.009678 \\ & -0.3786 & 0.01094 \\ \text{sym.} & & -0.2512 \end{bmatrix} \qquad (9.43)$$

$$\mathbf{K}^H = \begin{bmatrix} 0 & 0 & 0 \\ & 1.861 & 0.1533 \\ \text{sym.} & & 1.903 \end{bmatrix} \text{(m}^4\text{/N.s)}. \qquad (9.44)$$

Assuming that the macroscopic analysis has been done, let us localize the macroscopic deformation into the microscopic ones. The following two cases are considered in the computation of localization process:

$$\varepsilon = \begin{bmatrix} 0 & 0 & 0 \\ & 0 & 0 \\ \text{sym.} & & 0 \end{bmatrix}; \quad p = 100\,\text{MPa}; \quad \frac{\partial p}{\partial y} = 1.0 \times 10^{-4},$$

$$\frac{\partial p}{\partial x} = \frac{\partial p}{\partial z} = 0.0\,\text{N/m}^3$$

$$\varepsilon = \begin{bmatrix} 0 & 0 & 0 \\ & 5.0 & 0.5 \\ \text{sym.} & & 5.0 \end{bmatrix} \times 10^{-4}; \quad p = 0\,\text{MPa}; \quad \frac{\partial p}{\partial y} = \frac{\partial p}{\partial z} = 0.5 \times 10^{-4},$$

$$\frac{\partial p}{\partial x} = 0.0\,\text{N/m}^3$$

The stress and velocity distributions within a unit cell are given in Figures 9.9 and 9.10, respectively. As can be seen from these figures, both distributions are completely irregular. Thus, taking into account the specific effects of geometry, we may predict the portion that could be damaged by either the excess stress or the viscous fluid flow during the consolidation process.

However, as can be seen from the velocity distribution in Figure 9.10, there are some regions that do not have fluid flow even if the fluid region seems connected and open to neighboring unit cells. This unrealistic result is probably due to the fact that the model does not satisfy the geometrical consistency on the cell boundaries, in which the periodicity is postulated in the actual computations. As a result, it is possible that the fluid region in one microstructure is not connected with that of a neighboring one. In this situation, our digitized microstructure cannot be a unit cell any more. This is the reason why the *x*-component in the permeability matrix in Eq. (9.44) reveals zero permeability.

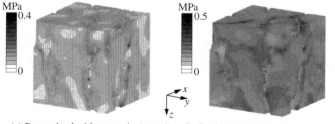

(a) Pressurized without strain (case 1)   (b) Strained without pressure (case 2)

Figure 9.9   Von-Mises stress distribution by the localization

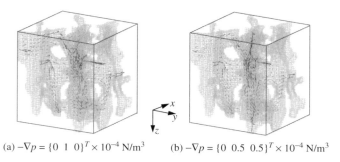

(a) $-\nabla p = \{0 \ \ 1 \ \ 0\}^T \times 10^{-4}$ N/m^3 (b) $-\nabla p = \{0 \ \ 0.5 \ \ 0.5\}^T \times 10^{-4}$ N/m^3

Figure 9.10   Velocity distribution by the localization (when the macroscopic pressure gradients are applied)

Nonetheless, the inconsistency may not be essential once the RVE, namely the unit cell region, is taken as being large enough. Thus, fairly complicated microstructures of geomaterials can be used as a unit cell by taking a large enough region of the RVE. This issue is discussed theoretically in the literature regarding the homogenization for random media (e.g. see Sab, 1992).

We have presented the numerical example of the homogenization for fluid-solid mixtures by using the digitized mesh of the microstructure whose morphology is very complicated. When the constitutive models derived from microstructures by the homogenization methods are used in poroelasticity analyses, the stress field in the solid phase and the flow field in the fluid phase could be accurate in the sense that the FE geometry model can possess the geometrical feature of actual geomaterials.

9.5   Conclusions

The constitutive modeling for fluid-saturated porous elastic solids have been studied for quasi-static situations by the homogenization method, and several numerical examples were presented to illustrate the global-local computation for characterizing the mechanical behaviors. Note that any conventional approaches for industrial composites cannot achieve this result, and that any volume fraction theory cannot provide the micromechanical response from the macroscopic deformation or flow. As far as the authors know, neither the application of this kind of constitutive modeling to geotechnical engineering, nor the numerical simulations, can be found in the literature.

The formulation was based upon the assumption that the viscosity in the microscopic scale is of the order ε^2, so that the fluid can flow though the passage made of an elastic solid skeleton, but large enough to stick to the solid phase at the interface. With the two-scale asymptotic expansions of the variables, the homogenization method converted the field equation defined in the fluid (i.e. Stoke's flow) into the field equation of a flow through porous media, called Darcy's law. On the other hand, the solid phase behaved like a porous elastic medium constrained by both the volumetric deformation of the solid and the seepage flow. Therefore, these results justified the Biot's classical theory of soil consolidation by introducing a periodic microstructure. All the material parameters used in the macroscopic equations were obtained from the corresponding microscopic problems and, therefore, reflected well the microstructural configurations.

Furthermore, the homogenization method provided additional information about the micromechanical responses to the macroscopic excitations, which could not be obtained by conventional formulations. That is, the localization capability enabled the evaluation of microscopic stress in the solid phase and the velocity distribution within the fluid phase, both of which helped us to study the interaction between the solid and fluid phases. It was found that both the macromechanical strain and the hydrostatic pressure caused a nonuniform stress distribution within the microstructure, which could explain the failure mechanism of the skeleton itself. At the same time, the effects of flowing viscous fluids on the micromechanical responses can be estimated from the velocity profile within the unit cell, which cannot be observed from the macroscopic analyses.

Of course, the direct applications of the current formulation to geotechnical engineering problems is not straightforward, since it is difficult to identify or define the microstructural configuration of geomaterials, the material properties and the constitutive relations of the constituents. In addition, geomaterials are not necessarily continua even in a microscopic scale, and the strict description of physical phenomena is not an easy task in general. Nonetheless, the asymptotic homogenization method presented in this chapter proposes a new approach to the constitutive modeling for fluid-saturated porous geomaterials. The multiple scale constitutive modeling would be a great help in attacking formidable problems in geomechanics. It is expected to take a positive approach to various unsolved problems arising from the micromechanical behavior of geomaterials by extending the asymptotic homogenization method to dynamic or/and nonlinear problems, etc.

Acknowledgements

The authors were supported, during this research, by NSF MSS-93-01807, US Army TACOM: DAAE07-93-C-R125, US Navy ONR: N00014-94-1-0022, AFOSR-URI program: DoD-G-F49620-93-0289 and the Japanese Ministry of Education, Science and Culture.

References

Arbogast, T., Douglas, Jr., J and Hornung, U. Derivation of the double porosity model of single phase flow via homogenization theory. *SIAM J. Math. Anal*, **21**(4), (1990) 823–836.

Bazant, Z. (ed.) *Mechanics of Geomaterials*. Wiley, New York, (1985).

Bear, J. *Dynamics of Fluids in Porous Media*. Dover, New York, (1967).

Benssousan, A., Lions, J.-L. and Papanicoulau, G. *Asymptotic Analysis for Periodic Structures*. North-Holland, Amsterdam, (1978).

Biot, M. A. General theory of three-dimensional consolidation. *J. Appl. Physics*, **12**, (1941) 154–164.

Biot, M. A. Theory of deformation of a porous viscoelastic anisotropic solid. *J. Appl. Physics*, **27**, (1956) 459–467.

Biot, M. A. Mechanics of deformation and acoustic propagation porous media. *J. Appl. Physics*, **27**, (1962) 1482–1498.

Burridge, R. and Keller, J. B. Poroelasticity equations derived from microstructure. *J. Acoustical Soc. Am.*, **70**(4), (1981) 1140–1146.

Christensen, R. M. *Mechanics of Composite Materials*. Wiley-Interscience, New York, (1979).

Desai, C. S. and Christian, J. T. (eds.) *Numerical Methods in Geotechnical Engineering*. McGraw-Hill, New York, (1977).

Desai, C. S. and Gallagher, R. G. (eds.) *Mechanics of Engineering Materials*. Wiley, New York, (1984).

Desai, C. S., Krempl, E., Kiousis, P. D. and Kundu, T. (eds.) *Constitutive Laws for Engineering Materials Theory and Applications, Vol. 1.* Elsevier, New York, (1987).

de Boer, R. and Ehlers, W. A historical review of the formulation of porous media theories. *Acta Mechanica*, **74**, (1988) 1–8.

Guedes, J. M. and Kikuchi, N. Preprocessing and postprocessing for materials based on the homogenization method with adaptive finite element methods. *Comput. Methods in Appl. Mech. Engrg.*, **83**, (1991) 143–198.

Hashin, Z. Analysis of composite materials – a survey. *J. Appl. Mech.*, **50**, (1983) 481–505.

Hollister, S. J. and Riemer, B. A. Digital image based finite element analysis for bone microstructure using conjugate gradient and Gaussian filter technique. *Math. Methods in Medical Imaging II*, **2035**, (1993) 95–106.

Keller, J. B. Darcy's law for flow in porous media and the two phase method. In: *Nonlinear Partial Differential Equations in Engineering and Applied Science*, Sternberg, R. L., Kalinowski, A. J. and Papadakis, J. S. (eds.) Office of Naval Research, (1980) 429–443.

Lévy, T. Propagation of waves in a fluid-saturated porous elastic solid. *Int. J. Engrg. Sci.*, **17**, (1979) 1005–1014.

Lévy, T. Fluids in porous media and suspension. In: *Homogenization Techniques for Composite Media: Lecture Note on Physics*, **272**, Sanchez-Palencia, E. and Zaoui, A. (eds.). Springer-Verlag, Berlin, (1987) 63–119.

Lewis, R. W. and Schrefler, B. A. (eds.) *The Finite Element Method in the Deformation and Consolidation of Porous Media.* Wiley, Chichester, (1987).

Mura, T. *Micromechanics of Defects in Solids.* Martinus Nijhoff, Netherlands, (1982).

Oleinik, O. A. *Mathematical Problems in Elasticity and Homogenization.* North-Holland, Amsterdam, (1992).

Sandhu, R. S., Liu, H. and Singh, K. J. Numerical performance of some finite element schemes for analysis of seepage in porous elastic media. *Int. J. Numer. Anal. Meth. Geomech.*, **1**, (1977) 177–194.

Sanchez-Palencia, E. *Non-homogeneous Media and Vibration Theory: Lecture Notes in Physics*, **127**, Springer-Verlag, Berlin, (1980).

Sanchez-Hubert, J. Asymptotic study of the macroscopic behaviour of a solid-liquid mixture. *Math. Methods in the Appl. Sci.*, **2**, (1980) 1–11.

Sandhu, R. S. Finite element analysis of coupled deformation and fluid flow in porous media. In: *Numerical Methods in Geomechanics*, Martins, J. B. (ed.). *Proc. NATO Advanced Study Institute*, Reidel, New York, (1981) 203–227.

Sab, K. On the homogenization and the simulation of random materials. *Euro. J. Mech., A/Solids II*, **5**, (1992) 585–607.

Siriwardane, H. J. and Desai, C. S. Two numerical schemes for nonlinear consolidation. *Int. J. Numer. Meth. Engrg.*, **17**, (1986) 405–426.

Terzaghi, K. *Principles in Soil Mechanics.* McGraw-Hill, New York, (1926).

Terada, K. Miura, T. and Kikuchi, N. Digital image-based modeling applied to the homogenization analysis of composite materials. *Computational Mechanics* (submitted 1996).

10

Unified Disturbed State Concept and HiSS Plasticity Models

G. Wije Wathugala
Louisiana State University, Baton Rouge, Louisiana, USA

10.1 Introduction

Many problems in engineering involve predicting the response of a material due to external excitations. The general solution to these problems often involves the definition of a constitutive model to describe the behavior of the material. The most common problem of this kind is predicting the behavior of materials under mechanical and other environmental loadings. In this chapter, we discuss a novel approach called the Disturbed State Concept (DSC), to develop unified constitutive models for many engineering materials including clay, sand, rocks, concrete, metal, ceramics and interfaces. In this chapter, we concentrate on applications of DSC to geomaterials. Applications to other materials are described in a later chapter, while a detailed treatment of the DSC is given by Desai (2000).

In general, a constitutive model for an engineering material describes an idealized material which behaves according to a given mathematical model. If the behavior of this ideal material under the loading condition of interest is the same or very close to that of the real material, then we can expect good predictions from the constitutive model. In general, the real materials are discontinues and have flaws. For example, in the case of sand, it consists of randomly packed grains of different sizes and shapes. However, the ideal material in general is continuous and does not have flaws. Therefore, at the micro scale, the behavior of the ideal material and the real material may not be similar. However, in general, they can be similar at the macro scale. Most of the constitutive models for engineering materials come under the above category. For many problems, good predictions at the macro scale can be obtained from traditional models. However, for some materials under certain situations such as microcracking, strain softening and shear band formations, if we do not account for the discontinuous nature of the material, reasonable predictions in any scale may not be possible. The DSC allows a weighted representation of the noncontinuous processes (micro cracking, microstructural changes, rearrangement of grain packing structure in a granular material, etc.) observed in these materials. This ability of DSC to capture certain noncontinuous processes facilitates the development of unified and general constitutive models for classes of materials.

In DSC, any given material is assumed to be composed of two phases, Relative Intact (RI) and Fully Adjusted (FA) state. Initially, at the unstrained stage, material has only the RI phase. During loading, microstructural changes occur in the material, which results in transforming some parts of RI to FA. The response that we see during experiments is the average response from these two phases. They can have different stresses and strains. Both phases behave differently, i.e. they are represented using different constitutive models. Here, even very simple constitutive models for the two phases can produce complex average responses. Due to a stress differential between the fully adjusted part and the intact part of the material, a material moment exists. Furthermore, due to there being different strains in the intact and the fully adjusted parts, there is a relative strain in the material. The relative motion within the material accounts for additional energy dissipation from the material. DSC also facilitates accounting for the energy dissipated due to intact part deformations, as in traditional continuum mechanics.

10.2 Review

The Disturbed State Concept (DSC) is related to the concept of *disturbance* or *correction* that was proposed by Desai (1974) to characterize the behavior of over-consolidated clays. Desai postulated that the response of an over-consolidated soil can be expressed in terms of its response in its normally consolidated state (curve 1, Figure 10.1) as the reference state, with the influence of over-consolidation (curve 2, Figure 10.1) being treated as disturbance.

An analogous concept, called the Residual Flow Procedure (RFP), was proposed by Desai (1976) in the context of free surface flow in porous materials. Here, the residual (or disturbance) is related to the constitutive pressure-permeability relation in which the permeability for the fully saturated material is treated as the reference state. Then the difference between the permeability for the fully saturated material and the permeability for the unsaturated material is treated as the residual. Westbrook (1985) and Bruch (1991) have shown that the algorithms resulting from the RFP are similar to those from the

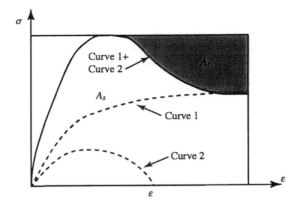

Figure 10.1 Observed stress-strain behavior of over-consolidated clay as composed of behaviors of normally consolidated soil and of part causing over-consolidation, (after Desai, 1974)

variational inequality procedures subsequently developed by Alt (1977), Baiocchi *et al.* (1975) and Oden and Kikuchi (1980).

Frantziskonis and Desai (1987) used the concept of separating complex material behavior into two separate simple behaviors, as proposed by Desai (1974), and the concept of damage used by many others (Kachanov, 1958; Krajcinovic and Fonseka, 1981) to develop a plasticity-based constitutive model for strain softening of concrete. Desai (1987) proposed a generalized hypothesis that lead to the DSC, and its subsequent applications, including idealizations for the RI and FA parts and definitions for disturbance to represent degradation or softening, and heading or stiffening (Desai, 2000).

Wathugala and Desai (1987) developed a formulation for using *critical state* to define the FA phase. Here, the RI phase was modeled using a generalized version of the nonlinear elastic hyperbolic model (Duncan and Chang, 1970; Duncan *et al.*, 1980). They used the model to predict the stress-strain response of Sacramento river sand at various initial densities and confining pressures (Wathugala and Desai, 1989). The DSC has been developed and used for a number of materials and interfaces/joints. For example, Desai and Ma (1992) applied the DSC for rock joints and interfaces, while Katti and Desai (1995) applied it to saturated clays. Armaleh and Desai (1994) applied the DSC to characterize the behavior of Leighton Buzzard (LB). Navayogarajah *et al.* (1992) added a strain softening capability to an interface model using DSC. Desai and Woo (1993) implemented a variation of the DSC for nonlinear dynamic analysis. Desai and Toth (1996) applied DSC to the modeling of ceramics composites and cemented sands, and presented the mathematical properties and comparison of DSC with other models. DSC was exploited for modeling microelectronics packaging materials (solders) and interfaces by Chia and Desai (1994), Desai *et al.* (1997) and Basaran *et al.* (1998). Recently, Pal (1997) and Pal and Wathugala (1997, 1999) used the DSC concept to develop an interface model for geosynthetics–sand interfaces, and predicted the behavior of pullout tests. Pal (1997) also used the DSC interface model in predicting the load deformation behavior of an instrumented geosynthetics retaining wall (Bathurst *et al.*, 1993).

10.3   Disturbed State Concept

In the DSC, it is assumed that external loading (mechanical, thermal and other environmental) cause disturbances or changes in the microstructure of the materials. Initially, the material is in the RI state. As the magnitude of the external disturbances increases, a part of it approaches the Fully Adjusted (FA) state at randomly distributed locations in the material through a process of *self-adjustment* of its microstructure, which is similar to the idea of self-organization (Bak and Tang, 1989; Aifantis, 1989). The growth of the FA phase of the material is modeled by the *disturbance function D*, and may depend upon many factors, such as the trajectory of plastic strains, plastic work and entropy. Henceforth, at any given time the material is composed of randomly distributed clusters of material at RI and FA states. Consequently, the observed response of the material is defined by a weighted average of the response of the RI and FA parts (Figure 10.2). The response of the RI part and the response of the FA (critical) part are the reference responses of the material.

Desai (1992) and Desai and Ma (1992) explained the main idea behind the DSC by the following analogy: 'Consider a cube of ice which melts under a given temperature includes a mixture of ice and water. The fully adjusted state is analogous to water. The

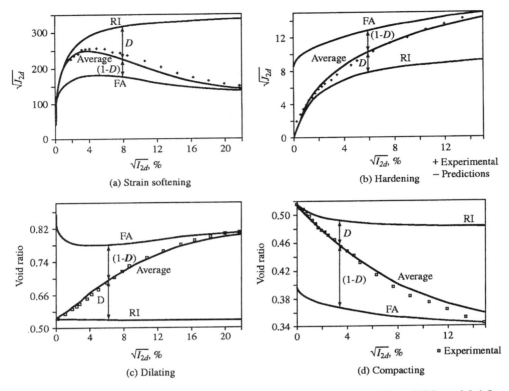

Figure 10.2 Predicted and observed behavior of Sacramento river sand from DSC model (after Wathugala and Desai, 1989)

behavior of the mixture can be expressed in terms of its two reference responses for ice and for water.' The deviation of the observed response from the reference Relative Intact (RI) response constitutes the disturbance that is caused due to applied forces, and is affected by factors such as friction, anisotropy, microcracking, damage, fracture and creep. For geomaterials, the average response is a weighted average of the response in the FA (critical state) and RI (basic model) phases, as shown in Figure 10.2. It also illustrates how the same principle can be used to capture strain softening, strain hardening, shear dilation and compaction of geomaterials under different densities and confining pressures.

10.3.1 Relative Intact (RI) state of the material

This material state is that which excludes the influence of the factors that are considered as causing the disturbance. 'For example, if microcracking and subsequent softening is considered as the disturbance, the response of the material without microcracking can be treated as the intact state' (Desai, 1992, 1994, 1995, 2000). The intact state of the material may vary with respect to the initial hydrostatic pressure, the initial density, and some other intrinsic factors. Therefore, the intact state is a relative definition. The intact state response can be determined experimentally or by approximation. If the intact response is obtained by experiments, the material needs to be tested under different initial conditions, such as densities and hydrostatic pressures. Hence, the intact response may be expressed as a function of the mean pressure and the material density. For the highest possible density of the material at a very high hydrostatic pressure, the intact response will approach an asymptotic value.

To approximate an intact response of the material, one of the following models may be used: (i) linear elastic, (ii) nonlinear elastic, (iii) continuously hardening elasto-plastic, and other suitable models. It is assumed that the deviation of the response from the elastic or continuously hardening elasto-plastic response is caused by disturbances such as microcracking, leading to damage and softening. The DSC can be also used to define healing or stiffening (Desai, 2000).

Another approximate procedure for determination of the intact response of the material is the elimination of characteristics which cause the material to degrade. For example, a set of parameters in the Hierarchical Single Surface (HiSS) plasticity model δ_1 defines the nonassociative (frictional) behavior of granular material. The response of the material without the parameter that defines the frictional part of the behavior, the associative HiSS model δ_o, can describe the reference intact state behavior. When the associative plasticity model δ_o is used to represent the intact response, the introduction of the disturbance function (D) into the constitutive equation leads to an effect similar to the use of non-associative plasticity for frictional granular materials. Details of the HiSS models, which can be used to model the RI state, are given in Section 10.4. The general incremental stress-strain relationship in the RI phase may be given by

$$d\sigma^i_{mn} = C^i_{mnop} d\varepsilon^i_{op} \qquad (10.1)$$

where $d\sigma^i_{mn}$ and $d\varepsilon^i_{op}$ are incremental stress and strain tensors for the RI phase, respectively, C^i_{mnop} and is the constitutive tensor for the RI phase.

10.3.2 Fully Adjusted (FA) state of the material

The *Fully Adjusted* (FA) state of the material is an asymptotic state in which the material may no longer be further disturbed. At this state the disturbance (or disorder) is at its maximum, and the material cannot transform into another state. In the fully adjusted state the material may be assumed to carry one of the following loads: (a) cannot carry any load and it has zero strength like a void in the classical continuum damage model (Kachanov, 1986); (b) can carry hydrostatic stress but no shear stress at all (like a constrained liquid), i.e.

$$\sigma^c_{ij} = \tfrac{1}{3}\sigma^c_{kk}\delta_{ij} \neq 0_{ij}, \quad S^c_{ij} = 0_{ij}, \quad d\,S^c_{ij} = 0_{ij} \qquad (10.2)$$

where S_{ij} is the deviatoric stress tensor, σ_{kk} is the trace of the stress tensor, δ_{ij} is the Kronecker delta function, $\delta_{ij} = 1$ when $i = j$ and $\delta_{ij} = 0$ when $i \neq j$; and (c) can carry hydrostatic stress and shear stresses reached up to that point, but cannot carry any additional shear stress. The material will deform in shear with zero volume change, similar to the classical *critical state* (Casagrande, 1936; Roscoe *et al.*, 1958). In this state, void ratio e confining stress and shear stresses have a unique relationship. These relationships for soils may be expressed as

$$\sqrt{J_{2D}^c} = \overline{m}J_1^c \quad \text{or} \quad \sqrt{\frac{1}{2}S_{ij}^c S_{ij}^c} = \overline{m}\sigma_{kk}^c \tag{10.3}$$

and

$$e^c = e_0^c - \lambda \ln \left(\frac{J_1^c}{3p_a} \right) \tag{10.4}$$

where $\overline{m}$, λ and e_0^c are material parameters and p_a is the atmospheric pressure. The material at this state cannot carry any additional shear stress if there is no increase in J_1. At this stage, material is free to deform in shear without any increase in the shear stress; i.e., incremental deviatoric strain tensor, $dE_{ij}^c \neq 0$ when $dS_{ij}^c = 0$.

The general incremental stress-strain relationship in the FA phase may be given by

$$d\sigma_{mn}^c = C_{mnop}^c d\varepsilon_{op}^c \tag{10.5}$$

where $d\sigma_{mn}^c$ and $d\varepsilon_{op}^c$ are the incremental stress tensor and strain tensor for the FA phase, respectively, and C_{mnop}^c is the constitutive tensor for the FA phase. It can be computed from the constitutive model for the FA state.

10.3.3 Disturbance function

The disturbance in the material is defined by the ratio of the area (or volume) of the fully adjusted material to the total area (or volume) of the material. In general, the disturbance will be a tensor. In traditional damage mechanics, damage effect tensors of a higher order have been used. Onat and Leckie (1988) have suggested that only even order tensors be used for damage tensors for metals. For example, Voyiadjis and Thiagarajan (1996) used a fourth-order tensor for metal matrix composites. Here we define the disturbance as a zeroth-order tensor or a scalar for simplicity as

$$D = \frac{V^c}{V} \tag{10.6}$$

where V^c is the fully adjusted volume and V is the total volume in an element of material. Wathugala and Desai (1987, 1989), Armaleh and Desai (1994) and Katti and Desai (1995) defined the disturbance for granular materials as

$$D = \frac{M^c}{M} = \frac{V_s^c}{V_s} \tag{10.7}$$

where M is the mass of soil solids and V_s is the volume of soil solids. The disturbance in the material can be expressed in terms of internal variables such as the trajectory of the plastic strains or plastic work, density, shear wave velocity, interface roughness, temperature, number of cycles experienced or entropy.

The progress of the disturbance in the material may be represented by one of the following functions, and some material internal variables for these functions: (i) a Weibull type exponential function from statistical theory of strength (Kachanov, 1986); (ii) the energy dissipated in the system (Ang and Kwok, 1987; Chia and Desai, 1994); (iii) a criterion based on thermodynamics (Muhlhaus, 1994). Frantziskonis and Desai (1987), and

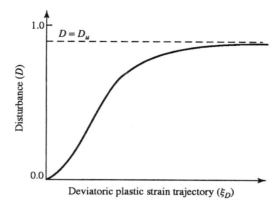

Figure 10.3 Schematic of disturbance function vs. trajectory of plastic deviatoric strain

many others, have used a Weibull type function for disturbance (D), shown in Figure 10.3, and given by

$$D = D_u(1 - e^{-A\xi_D^Z})$$

(10.8)

where D_u, A are Z material constants, and ξ_D is the trajectory of plastic deviatoric strains, given by

$$\xi_D = \int \sqrt{dE_{ij}^p \, dE_{ij}^p}$$

(10.9)

where dE_{ij}^p is the increment of deviatoric plastic strain tensor. Wathugala and Desai (1989) have used a similar function with one less material constant, as given by

$$D = D_u(1 - \text{sech}(A\xi_D))$$

(10.10)

Desai and Toth (1996) used the following function to define the growth of disturbance in a cemented sand and a ceramic composite:

$$D = D_u\left[1 - \left\{1 + \left(\frac{\xi_D}{h}\right)^w\right\}^{-s}\right]$$

(10.11)

where h, w and s are material parameters.

10.3.4 Strains

Classical damage mechanics formulations and the early DSC model by Frantziskonis and Desai (1987) assumed the same strain tensor for all the phases of the material. However, in the DSC, strains in the RI and FA parts can be different. A formulation based on this assumption was developed by Wathugala and Desai (1989), Armaleh and Desai (1994), and Katti and Desai (1995):

$$\varepsilon_{ij}^a = (1 - D)\varepsilon_{ij}^i + D\varepsilon_{ij}^c$$

(10.12)

A differential form of Eq. (10.12) is given by

$$d\varepsilon_{ij}^a = (1 - D)d\varepsilon_{ij}^i + Dd\varepsilon_{ij}^c + dD(\varepsilon_{ij}^c - \varepsilon_{ij}^i) \tag{10.13}$$

The change in disturbance dD can be expressed as a function of $d\varepsilon_{ij}^i$ (see the appendix)

$$dD = R_{st}d\varepsilon_{st}^i \tag{10.14}$$

where R_{st} can be found from the constitutive model for the RI phase and the disturbance function. R_{st} for HiSS models is given by Eq. (A.10.15).

Void ratios (e) for different phases in the granular material may be given by

$$e^a = \frac{V_v}{V_s}; \quad e^i = \frac{V_v^i}{V_s^i} \quad \text{and} \quad e^c = \frac{V_v^c}{V_v^c} \tag{10.15}$$

$$e^a = \frac{V_v^i + V_v^c}{V_s} = \left(\frac{V_v^i}{V_s^i}\right)\left(\frac{V_s^i}{V_s}\right) + \left(\frac{V_v^c}{V_s^c}\right)\left(\frac{V_s^c}{V_s}\right)$$

$$= \left(\frac{V_v^i}{V_s^i}\right)\left(\frac{V_s - V_s^c}{V_s}\right) + \left(\frac{V_v^c}{V_s^c}\right)\left(\frac{V_s^c}{V_s}\right) \tag{10.16}$$

where V_v and V_s are the volumes of voids and solids of a soil element, respectively. Substituting Eqs. (10.15) and (10.7) into Eq. (10.16) yields

$$e^a = (1 - D)e^i + De^c \tag{10.17}$$

The differential form of Eq. (10.17) is given by

$$de^a = (1 - D)de^i + Dde^c + dD(e^c - e^i) \tag{10.18}$$

The incremental volumetric strain is given by

$$d\varepsilon_v^a = \frac{de^a}{1 + e_0^a} \tag{10.19}$$

where e_0 is the initial void ratio. Substitution of Eq. (10.19) into (10.18) leads to

$$d\varepsilon_v^a = (1 - D)d\varepsilon_v^i + Dd\varepsilon_v^c + dD(\varepsilon_v^c - \varepsilon_v^i) \tag{10.20}$$

At the critical state, granular materials do not resist deviatoric strains, and therefore it is assumed that they are fully controlled by the deviatoric strains of the intact (RI) phase. It was assumed that

$$dE_{ij}^a = dE_{ij}^i = dE_{ij}^c = dE_{ij} \tag{10.21}$$

where dE_{ij} is the incremental deviatoric strain tensor. Total average strain is given by

$$d\varepsilon_{ij}^a = \frac{\delta_{ij}}{3}d\varepsilon_v^a + dE_{ij}^a \tag{10.22}$$

Substituting Eqs. (10.20) and (10.21) into (10.22), the incremental average strain tensor may be given by

$$d\varepsilon_{ij}^a = (1 - D)d\varepsilon_{ij}^i + Dd\varepsilon_{ij}^c + \frac{\delta_{ij}}{3}dD(\varepsilon_v^c - \varepsilon_v^i) \tag{10.23}$$

10.3.5 Stresses

From theory of means, the average stress tensor may be given by

$$\sigma_{ij}^a = (1 - D)\sigma_{ij}^i + D\sigma_{ij}^c \tag{10.24}$$

Similarly, deviatoric stress tensor in the different phases is related by

$$S_{ij}^a = (1 - D)S_{ij}^i + DS_{ij}^c \tag{10.25}$$

The incremental form of Eq. (10.24) is given by

$$d\sigma_{ij}^a = (1 - D)d\sigma_{ij}^i + Dd\sigma_{ij}^c + dD(\sigma_{ij}^c - \sigma_{ij}^i) \tag{10.26}$$

The following assumptions were made, as an approximation:

$$J_1^a = J_1^c = J_1^i = J_1 \tag{10.27}$$

and the deviatoric stress tensor in all the phases proportional to each other, i.e.

$$S_{ij}^c = kS_{ij}^i \tag{10.28}$$

where k is a scalar variable. The second invariant of the deviatoric stress tensors at FA and RI can be related using Eq. (10.28) as follows:

$$J_{2D}^c = \tfrac{1}{2}S_{ij}^c S_{ij}^c = k^2 \left(\tfrac{1}{2}S_{ij}^i S_{ij}^i\right) = k^2 J_{2D}^i \tag{10.29}$$

From Eq. (10.29), k can be expressed as

$$k = \sqrt{\frac{J_{2D}^c}{J_{2D}^i}} \tag{10.30}$$

Difference form of Eq. (10.28) is given by

$$dS_{ij}^c = k\, dS_{ij}^i + dkS_{ij}^i \tag{10.31}$$

Decomposition of the incremental stress tensor in Eq. (10.26) yields

$$d\sigma_{ij}^a = DdS_{ij}^c + (1 - D)dS_{ij}^i + dD(S_{ij}^c - S_{ij}^i) + [D + (1 - D)]\delta_{ij}\frac{dJ_1}{3} + \frac{dD\delta_{ij}(J_1^c - J_1^i)}{3}$$

$$= DdS_{ij}^c + (1 - D)dS_{ij}^i + dD(S_{ij}^c - S_{ij}^i) + \delta_{ij}\frac{dJ_1}{3} \tag{10.32}$$

Substitution of Eqs. (10.28) and (10.31) into (10.32) yields

$$d\sigma_{ij}^a = [D(k - 1) + 1]dS_{ij}^i + \frac{dJ_1\delta_{ij}}{3} + S_{ij}^i[dD(k - 1) + dkD]$$

Further rearranging yields

$$d\sigma_{ij}^a = [D(k - 1) + 1]d\sigma_{ij}^i - D(k - 1)\frac{dJ_1\delta_{ij}}{3} + S_{ij}^i[dD(k - 1) + dkD] \tag{10.33}$$

10.3.6   Stress-strain relationship

It is possible to express dk in terms of incremental intact strain tensor (see the appendix) as

$$dk = T_{st} d\varepsilon_{st}^i \tag{10.34}$$

where T_{st}, when FA is governed by Eq. (10.3), as in the critical state, is given by

$$T_{st} = \overline{m} \left\{ \frac{2J_{2D}^i \delta_{ij} - J_1^i S_{ij}^i}{2J_{2D}^{i\,3/2}} \right\} C_{ijst}^i \tag{10.35}$$

where C_{ijst}^i is the elasto-plastic constitutive stiffness tensor for intact materials. Substituting Eqs. (10.14) and (10.34) into (10.33) yields

$$d\sigma_{ij}^a = \left\{ [D(k-1)+1]C_{ijkl}^i - D(K-1)\frac{C_{ttkl}^i \delta_{ij}}{3} + S_{ij}^i [R_{kl}(k-1) + T_{kl}D] \right\} d\varepsilon_{kl}^i \tag{10.36}$$

Eq. (10.36) can be written in the succinct form as

$$d\sigma_{ij}^a = C_{ijkl}^{DSC} d\varepsilon_{kl}^i \tag{10.37}$$

where the effective constitutive tensor for the DSC is given by

$$C_{ijkl}^{DSC} = L_{ijkl} + M_{ijkl} \tag{10.38}$$

where

$$L_{ijkl} = \left\{ [D(k-1)+1]C_{ijkl}^i - D(k-l)\frac{C_{ttkl}^i \delta_{ij}}{3} \right\} \quad \text{and} \tag{10.39}$$

$$M_{ijkl} = S_{ij}^i [R_{kl}(k-1) + T_{kl}D] \tag{10.40}$$

10.3.7   Finite element formulation

The general DSC formulations can be implemented in finite element procedures to solve boundary value problems. It can be done in a general manner so that the program can be used with or without disturbance. Using the Newton–Raphson iterative procedure, the incremental form of the equilibrium equation of the displacement-based finite element method can be written as (Basaran and Desai, 1994)

$$\sum_m \int_V [B]^T \{d^{r-1}\sigma_{n+1}^a\} dV = \{Q_{n+1}\} - \sum_m \int_V [B]^T \{^{r-1}\sigma_n^a\} dV \tag{10.41}$$

where m is the number of elements, V is the element volume, $[B]$ is the displacement-strain transformation matrix, and $\{d^r \sigma_n^a\}$ is the increment of average (observed) stress at load step n and iteration r. $\{Q_{n+1}\}$ is the global nodal load vector, and $\{^r \sigma_n^a\}$ is the total stress vector at load step n and the r iteration. If it is assumed that the displacement field of the system is controlled by the intact displacement field, $\{q\}$, then substituting Eq. (10.38) into Eq. (10.41) yields

$$\sum_m \int_V [B]^T [L + M][B]\{dq^r_{i(n+1)}\}\, dV = \{Q_{n+1}\} - \sum_m \int_V [B]^T \{^{r-1}\sigma^a_n\}\, dV \qquad (10.42)$$

Desai and Toth (1996) found that $[L + M]$ becomes nonpositive definite in the softening region. However, they also found that $[L]$ is always positive definite. Desai and Woo (1993) found that stable solutions near the softening region can be found by moving $[M]$ to the right-hand side (RHS) of Eq. (10.42), as below:

$$\sum_m \int_V [B]^T [L][B]\{dq^r_{i(n+1)}\}\, dV = \{Q_{n+1}\} - \sum_m \int_V [B]^T \{^{r-1}\sigma^a_n\}\, dV$$

$$- \sum_m \int_V [B]^T [M][B]\{dq^r\}\, dV \qquad (10.43)$$

In Eq. (10.43), $\{dq^r\}$ appears on both sides of the equation. Therefore, it necessary to perform iterations with a trial value of $\{dq^r\}$ in the solution procedure.

An alternative but simplified algorithm is also developed in which the RI response is first evaluated, and then the observed response is calculated using an iterative procedure in which disturbance (based on the observed response) is modified progressively (Desai, 2000). This procedure avoids negative definite constitutive and stiffness matrices, and provides consistent solutions.

10.4　Hierarchical Single Surface (HiSS) Models

As indicated in Section 10.3.1, Hierarchical Single Surface (HiSS) models may be used to describe the intact (RI) behavior of the material. Furthermore, the HiSS models have been found to be highly successful for the characterization of elasto-plastic hardening response of materials. Hence, HiSS models are described below.

10.4.1　Introduction

The hierarchical single surface modeling approach (Desai, 1980; Desai et al., 1986) allows for progressive development of models of higher grades corresponding to different levels of complexities. Here, the model for initially isotropic material, hardening isotropically with associative plasticity, is treated as the basic, δ_0 model that involves zero deviation from normality of the increment of plastic strain to the yield surface F. Models of higher grades, isotropic hardening with nonassociative response due to friction (δ_1), and nonassociative response due to factors such as friction and induced anisotropy (δ_2), are obtained by superimposing modifications or corrections to the basic δ_0 model. The δ^* series of the HiSS models is developed to capture the behavior of cohesive soils in addition to all the previous capabilities of HiSS models (Wathugala, 1990; Wathugala and

Desai, 1993). Visco-plastic behavior is simulated (Desai and Zhang, 1987; Desai, Samtani and Vulliet, 1995) by using the approach proposed by Perzyna (1966). Chia and Desai (1994), Basaran and Desai (1994) and Desai *et al.* (1997) have incorporated the capability to capture termal effects for metal alloys (solders). The above hierarchical approach is applied to characterize behavior of both 'solids' and 'discontinuities'; the latter is based on the specialization of the model for solids (Desai and Fishman, 1991). These models have been developed and verified for many engineering materials, including clay, sand, soft rocks, concrete and Pb/Sn solders. Basic information about the models and their principal references are given in Table 10.1.

DSC models and applications for modeling of interfaces and joints are given by Desai (1995, 2000), Desai and Fishman (1991), Desai and Ma (1992), Navayogarajah *et al.* (1992) and Pal and Wathugala (1999).

10.4.2   General incremental stress-strain relationship

A general incremental stress-strain relationship for any type of loading may be expressed as

$$d\sigma_{ij} = C^*_{ijkl} d\varepsilon_{kl}$$

or

$$d\varepsilon_{ij} = D^*_{ijkl} d\sigma_{kl} \tag{10.44}$$

Here the superscript * can be *VL*, *RL* or *UL*, depending on virgin loading, reloading or unloading, respectively. C_{ijkl} and D_{ijkl} are elasto-plastic constitutive stiffness and compliance tensors, respectively. The general form of these tensors are given by Wathugala (1990) as

$$D^*_{ijkl} = D^e_{ijkl} + \frac{n^Q_{ij} n^R_{kl}}{H^*} \tag{10.45}$$

and

$$C^*_{ijkl} = C^e_{ijkl} - \frac{C^e_{ijnm} n^Q_{nm} n^R_{op} C^e_{opkl}}{H^* + n^R_{rs} C^e_{rstu} n^Q_{tu}} \tag{10.46}$$

where C_{ijkl} and D^e_{ijkl} are elastic constitutive stiffness and compliance tensors, respectively. The task of the constitutive model is to define the terms n^R_{ij}, n^Q_{ij} and H^* for all the loading situations. Traditionally, n^R_{ij} is used to define the loading criteria. Loading, unloading and neutral loading due to a stress increment $d\sigma_{ij}$ is defined as: loading $n^R_{ij} d\sigma_{ij} > 0$; unloading $n^R_{ij} d\sigma_{ij} < 0$; and neutral loading $n^R_{ij} d\sigma_{ij} = 0$. n^Q_{ij} is in the direction of incremental plastic strain tensor; H^* is the plastic modulus. In traditional plasticity models, n^R_{ij} and n^Q_{ij} during virgin loading are defined as unit normal tensors to the yield surface and potential surface, respectively. The plasticity modulus H^{VL} is obtained using the hardening function and the consistency condition (see the appendix). In the δ^* series models, these quantities are also defined inside the yield surface so that they can capture plastic strains observed inside the yield surface for many geologic materials. In the following sections, different functions used to define these quantities in different HiSS models are described.

Table 10.1   Various HiSS models

	Name	Potential function, Q	Hardening	Non-virgin loading	Basic references	Materials
Rate Independent	δ_0	Associative, $Q \equiv F$	Eq. (10.53)[1]	elastic	Desai *et al.* (1986)	dense sand, soft rocks, concrete
			Eq. (10.54)		Desai and Hashmi (1989)	sand
	δ_0^*	Associative, $Q \equiv F$ for virgin loading	Eq. (10.56)	inelastic reloading, elastic unloading	Wathugala (1990), Wathugala and Desai (1993)	clay
		$Q \equiv R$ for reloading	Eq. (10.55)		Wathugala (1990)	sand, clay
	$\delta_{0\theta}{}^2$	$Q \equiv F$	Eq. (10.53)	elastic	Chia and Desai (1994), Desai *et al.* (1997)	Pb/Sn solders
	δ_1	Eq. (10.59)	Eq. (10.53)	elastic	Desai *et al.* (1986), Frantziskonis *et al.* (1986)	sand
	δ_2	A moving surface	Eq. (10.53)	inelastic with a moving surface	Desai *et al.* (1986), Somasundaram and Desai (1988)	sand
	δ_2^*	A moving surface	Eq. (10.56)	inelastic reloading, elastic unloading	Wathugala (1990), Wathugala and Desai (1991), Titi (1996), Titi and Wathugala (1996)	clay
Dynamic	δ_v	Associative, $Q \equiv F$	Eq. (10.53)	elastic	Desai and Zhang (1987), Desai and Varadarajan (1987)	rocksalt
	δ_v	Associative, $Q \equiv F$	(10.56)	elastic	Desai, Samtani and Vulliet (1995)	clay
	$\delta_{v\theta}$	Associative, $Q \equiv F$	(10.53)	elastic	Chia and Desai (1994), Desai *et al.* (1997)	Pb/Sn solders
	δ_{od}	$Q \equiv F$	Eqs. (10.53), (10.54), (10.56)	Nonlinear elastic based on disturbance	Katti and Desai (1995), Park and Desai (1997), Shao and Desai (1998), Desai *et al.* (1997)	clay, sand

[1] These equations are given later.
[2] θ denotes temperature.

10.4.3   Yield function

The same single surface yield function is used for all the HiSS models. In this section, we present the history of its development and the common form used now. Desai (1980) proposed a general polynomial function of direct invariants of stress tensor (J_1, $\sqrt{J_2}$ and $\sqrt[3]{J_3}$), and showed that truncated forms of the function can represent most of the yield and potential functions in use for geological materials. Baker and Desai (1984) showed that, by introducing joint invariants $K_i (i = 1, 2$ and 3: $K_1 = \sigma_{ij}\varepsilon_{ij}^p$, $K_2 = \sigma_{ij}\varepsilon_{jk}^p\varepsilon_{ki}^p$ and $K_3 = \sigma_{ij}\sigma_{jk}\varepsilon_{ki}^p$) into the yield and potential functions, induced anisotropy can be modeled. It was also shown that translations, rotations and distortions of the yield and potential surfaces are equivalent to introducing joint invariants into these functions. Desai and Faruque (1984) studied the following yield function, which is a truncated form of the general yield function proposed earlier (Desai, 1980):

$$F \equiv J_{2D} + \alpha_1 J_1^2 - \beta_1 J_1 J_3^{1/3} - \gamma_1 J_1 - k_1^2 \tag{10.47}$$

where J_{2D} is the second invariant of the deviatoric stress tensor; J_1, J_2 and J_3 are first, second and third invariants of the stress tensor; and α_1, β_1, γ_1 and k_1 are material response functions or constants. This model has been used successfully to model laboratory behavior of silty sand (Desai and Faruque, 1984). It was also used to predict a series of full scale field tests on grouted anchors in sand (Desai, Muqtadir and Scheele, 1986).

Desai, Somasundaram and Frantziskonis (1986) proposed a compact form for the yield function in Desai (1980) from which Eq. (10.47) can be obtained as a special case. The nondimensional form of this function, proposed by Desai and Salami (1987) and Desai and Wathugala (1987) is given below:

$$F \equiv \frac{J_{2D}}{p_a^2} - F_b F_s \tag{10.48}$$

where p_a is the atmospheric pressure with stress units, F_b is the basic function which describes the shape of F in the $J_1 vs \sqrt{J_{2D}}$ space, and is given by (Figure 10.4)

$$F_b \equiv \alpha_{ps} \left(\frac{J_1 + J_{1s}}{p_a} \right)^n + \gamma \left(\frac{J_1 + J_{1s}}{p_a} \right)^2 \tag{10.49}$$

F_s is the shape function which describes the shape of F in octahedral planes, and given by

$$F_s = (1 - \beta S_r)^m \tag{10.50}$$

where S_r is a stress ratio, here

$$S_r = -\sin(3\theta) = \frac{\sqrt{27}}{2} J_{2D} J_{3D}^{-(3/2)} \tag{10.51}$$

where θ is the Lode angle (Desai and Siriwardane, 1984). β, γ and m are material response functions (assumed to be material constants for most of the versions) associated with the ultimate behavior. It was found that $m = -0.5$ gives satisfactory results for most of the geologic materials, and therefore it is adopted here. $J_{1s}/3$ is the bonding (tensile)

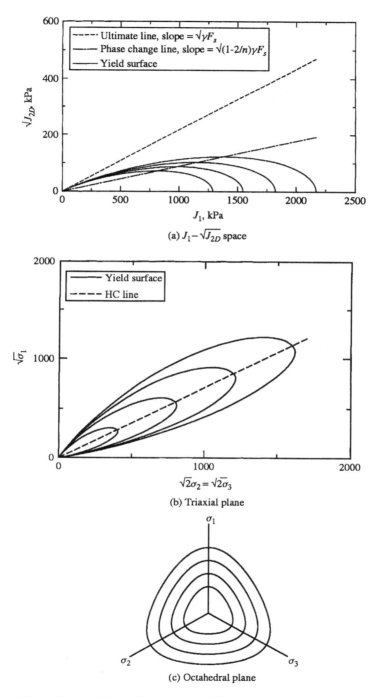

Figure 10.4 Shape of the yield surfaces in different stress spaces (here $J_{1s} = 0$)

strength. For cohesionless materials, $J_{1s} = 0$. For materials that possess tensile strength, such as concrete, metals and rocks, positive J_{1s} values are found from laboratory tests. n is the phase change or transition (from contraction to dilation) parameter α_{ps} is the growth or hardening function. Use of the symbol α is more common for the hardening function in HiSS models with elastic non-virgin loadings. In HiSS models which can simulate inelastic strains inside the yield surface (δ_2 and δ^* series models), α_{ps} is used to distinguish it from the variable α_r. Since, we are presenting all the models in a single modular formulation, α_{ps} is used here.

The shape of the yield surface in different stress spaces are shown in Figure 10.4. Yield surface becomes a straight line with a slope of $\sqrt{\gamma F_s}$ in the $J_1 - \sqrt{J_{2D}}$ space, when $\alpha_{ps} \rightarrow 0$ (Figure 10.4(a)). This line is called the ultimate line. The line that passes through the crest of all the yield surfaces is also a straight line with the slope of $\sqrt{(1 - 2/n)\gamma F_s}$ (Figure 10.4(a)). For associative plasticity models, depending on whether the stress point is below or above this line, the model predicts compressive or dilative plastic volumetric strains. Therefore, this line is also called the phase change line. In the δ^* series model for clays, this line also becomes the critical state line. If the material of interest shows a curved ultimate line or critical state line, it is possible to modify the yield function to accommodate this deviation (Desai and Salami, 1987; Desai and Varadarajan, 1987).

10.4.4 Hardening function

The hardening function α_{ps} can be a function of various internal variables related to the plastic deformations. In general, it is defined as

$$\alpha_{ps} \equiv \alpha_{ps}(\xi, \xi_D, \xi_V) \tag{10.52}$$

where ξ, ξ_D and ξ_V are trajectories of total, deviatoric and volumetric plastic strains, respectively. Several hardening functions have been used with HiSS models. The most common one for dilative materials is that of Desai, Somasundaram and Frantziskonis (1986):

$$\alpha_{ps} = \frac{a_1}{\xi^{\eta_1}} \tag{10.53}$$

where a_1 and η_1 are hardening parameters. Desai and Hashmi (1989) proposed a hardening function in terms of ξ and ξ_D as

$$\alpha_{ps} = b_1 \exp\left[-b_2\xi\left(1 - \frac{\xi_D}{b_3 + b_4\xi_D}\right)\right] \tag{10.54}$$

where b_1, b_2, b_3 and b_4 are hardening parameters. This function modeled proportional stress paths better than Eq. (10.53). Wathugala and Desai (1993) have proposed a general hardening function for sand and clays:

$$\alpha_{ps} = \frac{h_1}{\left\{\xi_V + h_3\xi_D^{h_4}\right\}^{h_2}} \tag{10.55}$$

where h_1, h_2, h_3 and h_4 are hardening parameters. They also proposed a simplified form of the above hardening function for clays by adopting $h_3 = 0$. This yields

$$\alpha_{ps} = \frac{h_1}{\xi_V^{h_2}} \tag{10.56}$$

To avoid shrinking yield surfaces and expansion of yield (or pre-stress) surface during nonvirgin loadings, Wathugala and Desai (1993) defined $d\xi_V$ as

$$d\xi_v^P = \frac{1}{\sqrt{3}}\langle d\varepsilon_v^P \rangle \tag{10.57}$$

where $\langle \ \rangle$ are the McAuley's brackets; i.e. $\langle d\varepsilon_v^P \rangle = d\varepsilon_v^P$ for $d\varepsilon_v^P > 0$, and $\langle d\varepsilon_v^P \rangle = 0$ otherwise. Wathugala and Desai (1993) only used $d\varepsilon_v^P$ due to due to virgin loadings in Eq. (10.57).

All these functions (Eqs. (10.53), (10.54), (10.55) and (10.56)) conform to the general function given in Eq. (10.52). The plastic modulus for virgin loading, H^{VL}, can be found from the consistency condition ($dF = 0$) in plasticity. H^{VL} for the general hardening function given in Eq. (10.52) is given by (Wathugala, 1990; appendix Eq. (A10.8))

$$H^{VL} = \frac{\dfrac{\partial F}{\partial \alpha_{ps}}}{\left(\dfrac{\partial F \partial F}{\partial \sigma_{ij} \partial \sigma_{ij}}\right)^{0.5}} \left(\frac{\partial \alpha_{ps}}{\partial \xi} + \frac{\partial \alpha_{ps}}{\partial \xi_D}(n_{Dij}^Q n_{Dij}^Q)^{1/2} + \frac{\partial \alpha_{ps}}{\partial \xi_v}\frac{\langle n_{kk}^Q \rangle}{\sqrt{3}}\right) \tag{10.58}$$

where $\langle \ \rangle$ are the McAuley's brackets.

10.4.5 Potential function

Traditionally, n_{ij}^Q is defined as the unit normal to a potential surface at the current stress state. The simplest approach is to use the loading surface as the potential surface (i.e. $n_{ij}^R = n_{ij}^Q$). This is called the associative plasticity, and results in a symmetric constitutive tensor. In the δ_0 model, we use $F \equiv Q$. In the δ_0^* model we use $R \equiv Q$.

Non-associative model, δ_1

Shear dilation predicted from the δ_0 model is higher than that observed in laboratory tests on dense sand. Therefore, Frantziskonis *et al.* (1986) developed δ_1 the model, where Q was defined as a modified version of the basic yield function as suggested by Desai and Siriwardene (1980). Here, Q can be obtained by swapping α_{ps} in Eq. (10.49) by α_Q. Here, α_Q is defined as

$$\alpha_Q = \alpha_{ps} + \kappa(\alpha_0 - \alpha_{ps})(1 - r_v) \tag{10.59}$$

where $r_v = \xi_v/\xi$ and α_0 is the α_{ps} at the beginning of shear loading, and is κ a material parameter associated with the dilation angle.

10.4.6   Nonvirgin loadings

In traditional plasticity-based models, nonvirgin loadings are considered elastic. This is the case for δ_0 and δ_1 models. Inelastic strains are observed in nonvirgin loading for many geomaterials. They are smaller than that observed during virgin loading, but it is important to account for them, especially during cyclic loadings. If the model does not predict inelastic strains during nonvirgin loadings, it may not be able to predict shear induced pore pressures during undrained nonvirgin loading. As indicated earlier, nonvirgin loading can be separated into reloading and unloading using the loading criteria given earlier.

Reference surface, R

The convex reference surface (R), which passes through the current stress point in the stress space, as shown in Figure 10.5 is used to distinguish unloading from reloading. The shape of R is similar to that of the F surface, and is given by

$$R \equiv \frac{J_{2D}}{p_a^2} - F_{br} F_s \tag{10.60}$$

where

$$F_{br} \equiv \alpha_r \left(\frac{J_1}{p_a} \right)^n + \gamma \left(\frac{J_1}{p_a} \right)^2 \tag{10.61}$$

The value of α_r is obtained by equating $R = 0$ and substituting current stresses into Eq. (10.60) All other symbols have the same meaning as that for the yield surface.

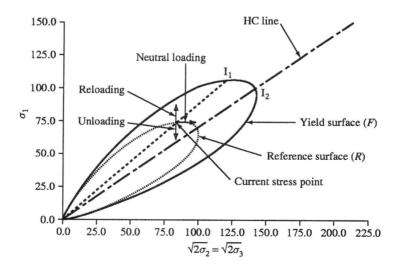

Figure 10.5   Yield surface (F) and reference surface (R) in triaxial plane ($\sigma_1 - \sqrt{2}\sigma_3$)

Interpolation function for reloading

The plastic modulus for nonvirgin loadings is usually obtained through separate interpolation procedures. Wathugala and Desai (1993) used the following simple interpolation function to evaluate the reloading plastic modulus, H^{RL}:

$$H^{RL} = H_{I_1}^{VL} + H_{I_2}^{VL} r_1 \left(1 - \frac{\alpha_{ps}}{\alpha_r}\right)^{r_2} \tag{10.62}$$

where r_1 and r_2 are material parameters, and $H_{I_1}^{VL}$ and $H_{I_2}^{VL}$ are virgin plastic moduli at points I_1 and I_2 (Figure 10.5) on the prestress surface. The image point I_1 is located at the intersection between the radial line passing through the current stress and the prestress surface. The point I_2 is located at the intersection of the hydrostatic compression line and the prestress surface. The effect of the distance from the current stress point to the prestress surface is included in Eq. (10.62) through the ratio (α_{ps}/α_r). When stress point reaches the prestress surface, $\alpha_r \to \alpha_{ps}$ and $H^{RL} \to H_{I_1}^{VL}$. This ensures smooth transition from reloading to virgin loading. Introduction of $H_{I_2}^{VL}$ into the Eq. (10.62) ensures that H^{RL} is always greater than zero. This is necessary, since a zero plastic modulus represents perfect plasticity, and should not occur inside the yield surface.

10.5 Determination of Material Parameters

Detailed procedures for finding the DSC parameters are given by Desai (2000). In this part, the procedures to obtain the material constants for the DSC models will be discussed. The elastic material constants Young's modulus, E and Poisson's ratio v, are common for all material models, hence their determination is not discussed here.

10.5.1 The ultimate state constants, γ and β

At the ultimate failure surface, the hardening parameter α is zero. Therefore, F becomes

$$F = J_{2D} - (\gamma J_1^2)(1 - \beta S_r)^{-0.5} = 0 \tag{10.63}$$

This can be arranged as,

$$\left[\frac{J_{2D}}{J_1^2}\right]_{ult}^{-2} \gamma^2 - \beta S_r = 1 \tag{10.64}$$

The final equation has two unknowns, hence the ultimate stresses for at least two stress paths are necessary to calculate γ and β. If more than two stress paths are available, a least square procedure is used to calculate γ and β. In the δ^* series models for clay, normally consolidated clays do not reach the ultimate line, but fail at the phase change line (or in this case, critical state line). For those materials, a different procedure is used to find ultimate parameters from the slope of the critical state line (Wathugala, 1990).

From conventional friction angles

The ultimate constants are related to conventional friction angles in the τ–σ stress space (Mohr–Coulomb) (Desai and Wathugala, 1987; Desai 1989), by

$$\beta = \frac{1 - \left(\dfrac{\tan\theta_C}{\tan\theta_E}\right)^{-4}}{1 + \left(\dfrac{\tan\theta_C}{\tan\theta_E}\right)^{-4}} \qquad (10.65a)$$

$$\sqrt{\gamma} = \frac{\tan\theta_C}{(1-\beta)^{-0.25}} = \frac{\tan\theta_E}{(1+\beta)^{-0.25}} \qquad (10.65b)$$

where

$$\tan\theta_c = \frac{2}{\sqrt{3}}\left(\frac{\sin\phi_C}{3-\sin\phi_C}\right) \quad \text{and} \quad \tan\theta_E = \frac{2}{\sqrt{3}}\left(\frac{\sin\phi_E}{3+\sin\phi_E}\right) \qquad (10.65c)$$

θ_C and θ_E are the slopes of the ultimate envelopes $\sqrt{J_{2D}} - J_1$ space for compression and extension tests, respectively. ϕ_C and ϕ_E are the slopes of the ultimate envelopes in $\tau - \sigma$ space, for compression and extension tests, respectively.

Normally consolidated clays fail when stress reaches the critical state line. Therefore, for these materials the above method is not directly applicable. However, Wathugala (1990) has found that β can be obtained from Eq. (10.65a), and γ can be computed from

$$\sqrt{\gamma} = \left(\frac{n}{n-2}\right)^{1/2} \frac{\tan\theta_C}{(1-\beta)^{-0.25}} = \left(\frac{n}{n-2}\right)^{1/2} \frac{\tan\theta_E}{(1+\beta)^{-0.25}} \qquad (10.66)$$

10.5.2   *Phase change parameter, n*

The slope of the line connecting the crest of all the yield surfaces, or *the phase change line*, is given by Wathugala (1990) as

$$S_{pc} = \frac{\sqrt{J_{2D}}}{J_1} = \left[\left(1-\frac{n}{2}\right)\gamma F_s\right]^{1/2} = \left(1-\frac{n}{2}\right)^{1/2} S_{ult} \qquad (10.67)$$

where S_{pc} and S_{ult} are slopes of phase change line and ultimate line in the $\sqrt{J_{2D}} - J_1$ space, respectively. For associative plasticity models, the phase change line can be located from the stresses at the state where material changes from positive to negative (dilative) plastic volumetric strains. For the non-associative model, δ_1, the above method is not theoretically correct. However, it was found that the error in using the same approach is very small. When the slope of the phase change line is found, n can be computed from Eq. (10.67).

10.5.3   *Hardening parameters, a_1 and η_1*

When the stress point is on the yield surface (i.e. yielding), $F = 0$. For this case, Eq. (10.48) can be rearranged as

$$\alpha = \left[\gamma - \left(\frac{J_{2D}}{J_1^2} \right) \frac{1}{F_s} \right] \left(\frac{p_a}{J_1} \right)^{n-2} \tag{10.68}$$

The trajectories of plastic strains (ξ, ξ_v and ξ_D) can be calculated directly from the laboratory stress-strain data. Now it is possible to develop simple least square procedures to determine the hardening parameters for any hardening function. Details on how to find hardening parameters for different hardening functions defined in Section 10.4.4 are as follows: Eq. (10.54) is given in Desai and Hashmi (1989) and Desai and Wathugala (1987); Eq. (10.55) in Wathugala (1990); Eq. (10.56) in Wathugala and Desai (1993). Details on how to find hardening parameters for the basic hardening function, Eq. (10.53), are given below.

Taking the logarithm of both sides of the hardening function, $\alpha = a_1/\xi^{\eta_1}$, yields

$$[1 - \ln \xi] \begin{Bmatrix} \ln a_1 \\ \eta_1 \end{Bmatrix} = \ln \alpha \tag{10.69}$$

α and ξ can be calculated for each point on the observed stress-strain curve using the above two equations. When all the points are considered, this results in a set of simultaneous equations, in the form

$$[A]\{X\} = \{b\} \tag{10.70}$$

where

$$[A] = \begin{bmatrix} 1 & \ln \xi_1 \\ 1 & \ln \xi_2 \\ \cdot & \cdot \\ \cdot & \cdot \\ 1 & \ln \xi_n \end{bmatrix} \quad \{X\} = \begin{Bmatrix} \ln a_1 \\ \eta_1 \end{Bmatrix} \quad \{b\} = \begin{bmatrix} \ln \alpha_1 \\ \ln \alpha_2 \\ \cdot \\ \cdot \\ \ln \alpha_n \end{bmatrix} \tag{10.71}$$

This set of equations is solved for $\{X\}$ using a least square procedure.

Calculation of the history parameters, such as α_{ps}, α, ξ, ξ_v, ξ_D, at the first load step in the finite element method requires special consideration (Desai *et al.*, 1991). In general, they are computed from the analytical solution for proportional stress paths (Wathugala, 1990). The initial value of α or α_{ps} can be calculated from Eq. (10.68). For the δ_0 model with the basic hardening function, the trajectory of plastic strains, ξ, can be computed from

$$\xi = \left(\frac{a_1}{\alpha} \right)^{1/(\eta_1)} \tag{10.72}$$

10.5.4 Disturbance function parameters, D_u, A and Z

From Eq. (10.25) J_{2D}^a may be expressed as

$$J_{2D}^a = \tfrac{1}{2} S_{ij}^a S_{ij}^a = \tfrac{1}{2}[(1 - D)S_{ij}^i + D S_{ij}^c][(1 - D)S_{ij}^i + D S_{ij}^c]$$
$$= \tfrac{1}{2}(1 - D)^2 S_{ij}^i S_{ij}^i + \tfrac{1}{2} D^2 S_{ij}^c S_{ij}^c + \tfrac{1}{2}(1 - D)D S_{ij}^c S_{ij}^i \tag{10.73}$$

For materials which satisfy Eq. (10.28) (i.e. granular materials and any other material with zero deviatoric stresses for the FA phase), Eq. (10.73) can be further simplified to

$$J_{2D}^a = (1 - D)^2 J_{2D}^i + D^2 J_{2D}^c + (1 - D)D\sqrt{J_{2D}^i J_{2D}^c}$$

$$\sqrt{J_{2D}^a} = (1 - D)\sqrt{J_{2D}^i} + D\sqrt{J_{2D}^c} \qquad (10.74)$$

From Eq. (10.74), disturbance D can be expressed as

$$D = \frac{\sqrt{J_{2D}^i} - \sqrt{J_{2D}^a}}{\sqrt{J_{2D}^i} - \sqrt{J_{2D}^c}} \qquad (10.75)$$

From the stresses at the residual, D_u may be found from

$$D_u = \frac{\left(\sqrt{J_{2D}^i} - \sqrt{J_{2D}^a}\right)_{ult}}{\left(\sqrt{J_{2D}^i} - \sqrt{J_{2D}^c}\right)_{ult}} \qquad (10.76)$$

Eq. (10.8) may be rearranged as

$$\left(1 - \frac{D}{D_u}\right) = e^{-A\xi_D^Z} \qquad (10.77)$$

Taking the logarithm twice, Eq. (10.77) yields (Desai and Ma, 1992; Desai, 2000)

$$\ln\left(-\ln\left(1 - \frac{D}{D_u}\right)\right) = \ln A + Z\ln(\xi_D) \qquad (10.78)$$

For all the points in the stress path, D can be calculated from Eq. (10.75). The trajectory of deviatoric plastic strains can be computed from Eq. (10.9). Substituting these values into Eq. (10.78) for each stress point yields a set of simultaneous equations. They can be solved for A and Z using a least square procedure. Alternatively, D may be computed from intact, average and critical void ratios (Wathugala and Desai, 1987)

$$D = \frac{e^a - e^i}{e^c - e^i} \qquad (10.79)$$

10.5.5 *Critical state parameters for FA*

The first step is to estimate the stresses and void ratios for the FA phase. Traditionally, this is computed at the ultimate where most of the material is at FA. If we know the value of D_u we can compute the stresses and void ratios from the average stresses and void ratios. Then, the parameter $\bar{m}$ can be found from Eq. (10.3). The parameters e_0^c and λ can be computed from Eq. (10.4).

10.6 Applications

DSC models have been used for many geomaterials and other engineering materials. Typical applications for geomaterials are listed in Table 10.2. Details of other materials are given by Chia and Desai (1994).

Table 10.2 Typical applications of DSC models for geomaterials

Material	RI model	FA model	References	Figures
Sacramento River sand (Lee and Seed, 1967)	Hyperbolic model	Critical state	Wathugala and Desai (1989)	Figure 10.2
Leighton Buzzard sand	HiSS δ_0 model	Critical state	Armaleh and Desai (1994)	Figure 10.6
Sabine clay	HiSS δ_o^* model	Critical state	Katti and Desai (1995)	Figure 10.7
Rock joints (various)	HiSS δ_{0i} model	Critical state	Desai and Ma (1992)	Figure 10.8
RMC sand (Bathurst et al., 1993)	HiSS δ_{oi}^* model	Critical state	Pal (1997), Pal and Wathugala (1997)	Figure 10.9

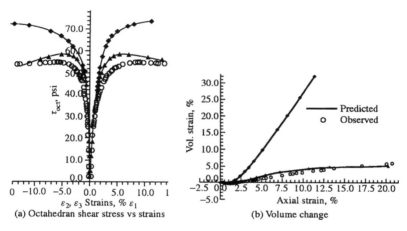

Figure 10.6 Comparison of triaxial test data on Leighton Buzzard sand vs. predictions (after Armaleh and Desai, 1994)

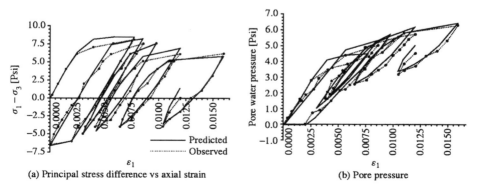

Figure 10.7 Comparison of test data with predictions for cyclic triaxial tests on Sabine Clay (after Katti, 1991)

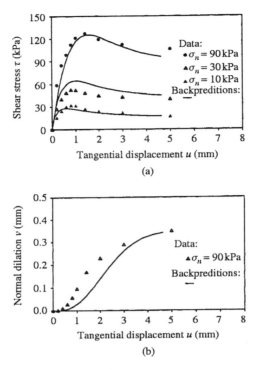

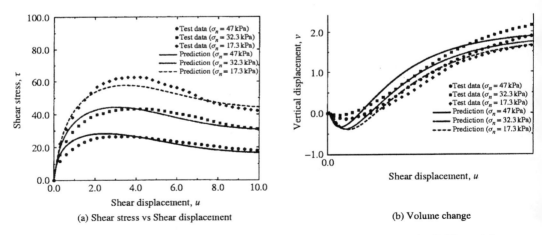

Figure 10.8   Comparison of predictions and observations, data from Bandis *et al.* (1981) (after Desai and Ma, 1992)

(a) Shear stress vs Shear displacement

(b) Volume change

Figure 10.9   Comparison of direct shear test data on RMC sand and predictions by DSC model (after Pal, 1997)

Acknowledgements

Financial support from Louisiana Board of Regents under the grant LEQSF (1994–97)-RD-A-08 is gratefully acknowledged. Doctoral student Mr Prasad Samarajiva assisted with the preparation of this chapter.

Appendix A1: Derivation of H^{VL}

From the flow rule in plasticity

$$d\varepsilon_{ij}^P = \lambda n_{ij}^Q \qquad (A.10.1)$$

where λ is a positive constant of proportionality and n_{ij}^Q is a unit tensor in the direction of incremental plastic strain tensor. Substituting Eq. (A.10.1) into the definition of the incremental plastic strain trajectories, $d\xi$, $d\xi_D$ and $d\xi_V$ yields

$$d\xi = (d\varepsilon_{ij}^P d\varepsilon_{ij}^P)^{0.5} = \lambda(n_{ij}^Q n_{ij}^Q)^{0.5} = \lambda \qquad (A.10.2)$$

$$d\xi_D = (dE_{ij}^P dE_{ij}^P)^{0.5} = \lambda(n_{Dij}^Q n_{Dij}^Q)^{0.5} \qquad (A.10.3)$$

$$d\xi_V = \frac{1}{\sqrt{3}}\langle\varepsilon_{kk}\rangle = \frac{\lambda}{\sqrt{3}}\langle n_{kk}^Q\rangle \qquad (A.10.4)$$

where $\langle\varepsilon_{kk}\rangle$ is the McAuley's brackets, and n_{Dij}^Q is the deviatoric part of n_{ij}^Q.

Applying the consistency condition, $dF = 0$ to the yield function, yields

$$dF = \frac{\partial F}{\partial\sigma_{ij}}d\sigma_{ij} + \frac{\partial F}{\partial\alpha_{ps}}\left[\frac{\partial\alpha_{ps}}{\partial\xi}d\xi + \frac{\partial\alpha_{ps}}{\partial\xi_D}d\xi_D + \frac{\partial\alpha_{ps}}{\partial\xi_V}d\xi_V\right] = 0 \qquad (A.10.5)$$

Substituting Eqs. (A.10.2), (A.10.3) and (A.10.4) into (A.10.5) and rearranging yields

$$\lambda = \frac{\dfrac{\partial F}{\partial\sigma_{mn}}d\sigma_{mn}}{\dfrac{\partial F}{\partial\alpha_{ps}}\left[\dfrac{\partial\alpha_{ps}}{\partial\xi} + \dfrac{\partial\alpha_{ps}}{\partial\xi_D}(n_{Dij}^Q n_{Dij}^Q)^{0.5} + \dfrac{\partial\alpha_{ps}}{\partial\xi_V}\langle n_{kk}^Q\rangle\right]} \qquad (A.10.6)$$

Plastic modulus H^{VL} is related to λ by

$$H^{VL} = \frac{dn_{ij}^F d\sigma_{ij}}{\lambda} \qquad (A.10.7)$$

Substituting Eq. (A.10.6) into (A.10.7) yields

$$H^{VL} = \frac{\dfrac{\partial F}{\partial\alpha_{ps}}\left(\dfrac{\partial\alpha_{ps}}{\partial\xi} + \dfrac{\partial\alpha_{ps}}{\partial\xi_D}\left(n_{Dij}^Q n_{Dij}^Q\right)^{0.5} + \dfrac{\partial\alpha_{ps}}{\partial\xi_V}\langle n_{kk}^Q\rangle\right)}{\left(\dfrac{\partial F\partial F}{\partial\sigma_{mn}\partial\sigma_{mn}}\right)^{0.5}} \qquad (A.10.8)$$

Appendix A2: Derivation of R_{st}

To develop the equivalent constitutive tensor for the DSC formulation, it is useful to express the term dD in terms of $d\varepsilon_{kl}^i$, the incremental total strain tensor for the intact part. The disturbance is a function of the trajectory of the deviatoric plastic strain of the RI part, hence

$$dD = \frac{\partial D}{\partial \xi_D^i} d\xi_D^i \qquad (A.10.9)$$

In this section, all the quantities refer to the RI phase, and therefore the superscript i is omitted for clarity. From Hooke's law and the linear decomposition of total strain to elastic and plastic strains

$$d\sigma_{ij} = c_{ijkl}^e d\varepsilon_{kl}^e = c_{ijkl}^e (d\varepsilon_{kl} - d\varepsilon_{kl}^p) \qquad (A.10.10)$$

Substituting Eqs. (A.10.2), (A.10.3), (A.10.4) and (A.10.10) into (A.10.5) yields

$$dF = \frac{\partial F}{\partial \sigma_{ij}} c_{ijkl}^e (d\varepsilon_{kl} - d\varepsilon_{kl}^p) + \lambda \frac{\partial F}{\partial \alpha_{ps}} \left(\frac{\partial \alpha_{ps}}{\partial \xi} + \frac{\partial \alpha_{ps}}{\partial \xi_D} \left(n_{Dij}^Q n_{Dij}^Q \right)^{0.5} + \frac{\partial \alpha_{ps}}{\partial \xi_V} \langle n_{kk}^Q \rangle \right) = 0 \qquad (A.10.11)$$

Substituting Eqs. (A.10.1) and (A.10.8) into (A.10.11) and rearranging yields

$$\lambda = \frac{n_{ij}^F c_{ijst}^e d\varepsilon_{st}}{n_{mn}^F c_{mnop}^e n_{op}^Q - H^{VL}} \qquad (A.10.12)$$

Substituting Eqs. (A.10.3) and (A.10.12) into (A.9.9) yields

$$dD = \frac{dD}{d\xi_D} \frac{(n_{Dtu}^Q n_{Dtu}^Q)^{0.5} (n_{ij}^F c_{ijst}^e d\varepsilon_{st})}{n_{mn}^F c_{mnop}^e n_{op}^Q - H^{VL}} \qquad (A.10.13)$$

Equation (A.10.13) can be written in the following form:

$$dD = R_{st} d\varepsilon_{st}^i \qquad (A.10.14)$$

where R_{st} is given by the following equation:

$$R_{st} = \frac{dD}{d\xi_D} \frac{(n_{Dtu}^Q n_{Dtu}^Q)^{0.5} n_{ij}^F c_{ijst}^e}{n_{mn}^F c_{mnop}^e n_{op}^Q - H^{VL}} \qquad (A.10.15)$$

Appendix A3: Derivation of T_{kl}

From Eq. (10.30),

$$k = \sqrt{\frac{J_{2D}^c}{J_{2D}^i}} \qquad (A.10.16)$$

Substituting Eq. (10.3) in Eq. (A.10.16) gives

$$k = \frac{mJ_1^{(i)}}{\sqrt{J_{2D}^{(i)}}} \tag{A.10.17}$$

$$dk = \frac{m[2J_{2D}^{(i)}dJ_1^{(i)} - J_1^{(i)}dJ_{2d}^{(i)}]}{2(J_{2D}^{(i)})^{3/2}} \tag{A.10.18}$$

For intact material, T_{kl} is derived from the equations

$$d\sigma_{ij}^{(i)} = C_{ijkl}^{(i)}d\varepsilon_{kl}^{(i)}$$

$$dJ_1^{(i)} = d\sigma_{ii}^{(i)}$$

$$= C_{iikl}^{(i)}d\varepsilon_{kl}^{(i)} \tag{A.10.19}$$

$$dJ_{2D}^{(i)} = S_{ij}^{(i)}\left[C_{ijkl}^{(i)} - \frac{\delta_{ij}}{3}C_{ttkl}^{(i)}\right]d\varepsilon_{kl}^{(i)} \tag{A.10.20}$$

$$dk = \frac{m[2J_{2D}^{(i)}C_{iikl}^{(i)} - J_1^{(i)}S_{ij}^{(i)}C_{ijkl}^{(i)}]}{2J_{2D}^{(i)\,3/2}}d\varepsilon_{kl}^{(i)} \tag{A.10.21}$$

$$dk = T_{kl}d\varepsilon_{kl}^{(i)} \tag{A.10.22}$$

where

$$T_{kl} = \frac{m[2J_{2D}^{(i)}C_{iikl}^{(i)} - J_1^{(i)}S_{ij}^{(i)}C_{ijkl}^{(i)}]}{2J_{2D}^{(i)\,3/2}} \tag{A.10.23}$$

References

Aifantis, E.C. *Plasticity and Self Organization.* Advances in Constitutive Laws in Engineering Materials, Ed Jinghang & Murakami, (1980).

Alt, H.W. A free boundary problem associated with the flow of ground watermark. *Rat. Mech. Analysis,* **64**, (1977) 111–126.

Ang, A. H. and Kwok, Y. H. A seismic damage model for masonary structures. In: *Stochastic Structural Mechanics,* Lin, Y. K and Schueller, G. I. (eds.). US-Austria Joint Seminar, Boca Raton, Florida, (1987).

Armaleh, S. H. and Desai, C. S. Modeling and testing of cohesionless material using the disturbed state concept. *J. Mech. Behavior of Materials,* **5**(3), (1994) 279–295.

Baiocchi, C., Comincioli, and Maione, U. Unconfined flow through porous media. *Meccanica J. (Italian Association of Theoretical and Applied Mechanics),* **10**, (1975) 151–155.

Bak, P. and Tang, C. Earthquakes as a self-organized phenomenon. *J. Geophys. Res.,* **94**(B11), (1989) 15635–15637.

Baker, R. and Desai, C.S. Induced anisotropy during plastic straining. *Int. J. Num. Anal. Methods in Geomech.,* **8**(2), (1984).

Basaran, C. and Desai, C.S. *Finite Element Thermomechanical Analysis of Electronic Packaging Problems Using Disturbed State Constitutive Models.* Report to NSF, The University of Arizona, (1994).

Basaran, C., Desai, C.S. and Kundu, T. Thermomechanical finite element analysis of electronics packaging problems using disturbed state constitutive models, Part I & II. *Trans. ASME J. Electr. Packaging,* **120**(1), (1998) 41–53.

Bathurst *et al*. A database of results from an incrementally constructed geogrid-reinforced soil wall test. *Proceedings of Soil Reinforcement: Full Scale Experiments of the 80*'s, ISSMFE/ENPC Paris, France, (1993).

Bandis, A.C., Lumsden, and Barton, N.R. Experimental studies of scale effects on the shear behavior of rock joints. *Int. J. Rock Mechanics and Min. Sci. Gemech. Abstracts*, **18**, (1981) 1–21.

Bruch, J. C. Fixed domain methods for free and moving boundary flow in porous media. *J. Transport in Porous Media*, **6**, (1991) 627–649.

Casagrande, A. Characteristics of cohesionless soils affecting the stability of slopes and earth fills. *J. Boston Soc. Civil Eng*. Reprinted in Contributions to Soil Mechanics 1925–1940, BSCE, 257–276.

Chia, J. and Desai, C. S. *Constitutive Modeling of Thermomechanical Response of Materials in Semiconductor Devices with Emphasis on Interface Behavior*. Report to NSF, Department of Civil Engineering and Engineering Mechanics, The University of Arizona, (1994).

Desai, C. S. A consistent finite element technique for work-softening behavior. In: *Proc. Int. Conf. on Comp. Meth. in Nonlinear Mech.*, Oden, J. T. *et al*. (eds.). University of Texas at Austin, TX, (1974).

Desai, C.S. Finite element residual schemes for unconfined flow. *J. Num. Meth. Engrg.*, **10**, (1976) 1415.

Desai, C. S. A general basis for yield, failure and potential functions in plasticity. *Int. J. Num. and Anal. Meth. in Geomech.*, **4**, (1980) 361–375.

Desai, C.S. *Further on Unified Hierarchical Models Based on Alternative Correction or 'Damage' Approach*. Internal Research Report, Department of Civil Engineering and Engineering Mechanics, University of Arizona, Tucson, A2, (1987).

Desai, C.S. Modeling and testing: implementation of numerical models and their application in practice. In: *Numerical Methods and Constitutive Modeling in Geomechanics*, Desai, C. and Gioda, G. (eds.). Springer-Verlag, New York, (1989).

Desai, C.S. *The Disturbed State as a Phase Transformation Through Self-Adjustment Concept for Modeling of Mechanical Response of Materials and Interfaces*. Report, Department of Civil Engineering and Engineering Mechanics, The University of Arizona, Tucson, AZ, (1992).

Desai, C.S. Hierarchical single surface and the disturbed state constitutive models with emphasis on geotechnical applications. In: *Geotechnical Engineering: Emerging Trends in Design and Practice*, Saxena, K.R. (ed.). Oxford & IBH Publishing Co. Pvt. Ltd., New Delhi, (1994).

Desai, C.S. Constitutive modeling using the disturbed state as microstructure self adjustment concept. In: *Continuum Models for Materials with Microstructure*, Muhlhaus, H.B. (ed.). Wiley, Chichester, (1995).

Desai, C.S. *Mechanics of Materials and Interfaces: The Disturbed State Concept*. CRC Press, Boca Raton, Florida, Under publication.

Desai, C.S., Chia, J., Kundu, T. and Prince, J. Thermomechanical response of materials and interfaces in electronic packaging, Part I & II. *J. Electr. Packaging*, **119**(4), (1997) 294–309.

Desai, C.S. and Faruque, M.O. Constitutive model for geologic materials. *J. Geotech. Engrg.*, **110**(9), (September 1984) 1391–1408.

Desai, C.S. and Fishman K.L. Plasticity-based constitutive model with associated testing for joints. *Int. J. Rock Mechanics and Mining Sci.*, **28**(1), (1991) 12–26.

Desai, C.S., Galagoda, H.M. and Wathugala, G.W. Hierarchical modeling for geologic materials and discontinuities-joints and interfaces. In: *Proc. of Second International Conference on Constitutive Laws for Engineering Materials: Theory and Applications*, Desai *et al*. (eds.). Elsevier, New York, Vol. 1 (1987) 81–95.

Desai C.S. and Hashmi Q.S.E. Analysis, evaluation, and implementation for a nonassociative model for geologic materials. *Int. J. Plasticity*, **5**(4), (1989) 397–420.

Desai, C. S. and Ma, Y. Modeling of joints and interfaces using the disturbed state concept. *Int. J. Num. and Anal. Meth. Geomech.*, **16**, (1992) 623–653.

Desai, C.S., Muqtadir, A. and Scheele, F. Interaction analysis of anchor soil systems. *J. Geotech. Eng.*, **112**(GT5), (May 1986).

Desai, C.S., Park, I.J. and Shao, C. Fundamental yet simplified model for liquefaction instability. *Int. J. Num. Analyt. Methods in Geomech.*, **22**, (1998) 721–748.

Desai, C.S. and Salami, M.R. Constitutive model for rocks. *J. Geotech. Engrg.*, **113**, (May 1987) 407–423.

Desai, C.S., Samtani, N.C. and Vulliet, L. Constitutive modeling and analysis of creeping slopes. *J. Geotech. Engrg.*, **121**, (1995) 43–56.

Desai, C.S, Sharma, K.G., Wathugala, G.W. and Rigby, D.B. Implementation of Hierarchical Single Surface δ_0 and δ_1 models in finite element procedures. *Int. J. Num. Analy. Methods in Geomech.*, **15**, (1991) 649–680.

Desai, C.S. and Siriwardane, H.J. A concept of correction functions to account for nonassociative characteristics of geologic media. *Int. J. Num. Anal. Methods in Geomech.*, **4**, (1980) 377–387.

Desai, C. S. and Siriwardane, H. J. *Constitutive Laws for Engineering Materials: With Emphasis on Geological Materials*. Prentice-Hall, Englewood Cliffs, NJ, (1984).

Desai, C.S., Somasundaram, S. and Frantziskonis, G. A hierarchical approach for constitutive modeling of geologic materials. *Int. J. Num. Anal. Meth. in Geomech.*, **10**, (1986) 225–257.

Desai, C.S. and Toth, J. Disturbed state constitutive modeling based on stress-strain and nondestructive behavior. *Int. J. Solids Structures*, **33**(11), (1996) 1619–1650.

Desai, C.S. and Varadarajan, A. A constitutive model for quasi-static behavior of rock salt. *J. Geophys. Res.*, **92**(11), (1987) 11,445–11,456.

Desai, C.S. and Wathugala, G.W. Hierarchical and unified models for solids and discontinuities (joints/interfaces). *Short Course Notes, 2nd Int. Conf. on Constitutive Laws for Engineering Materials: Theory and Applications*, Tucson, AZ, (1987).

Desai, C. S. and Woo, L. Damage model and implementation in nonlinear dynamic problems. *J. Computational Mech.*, **11**(3), (1993) 189–199.

Desai, C. S. and Zhang, D. Viscoplastic model for geologic materials with generalized flow rule. *Int. J. Num. Anal. Meth. in Geomech.*, **11**, (1987) 603–620.

Duncan, J.M. and Chang, C.Y. Nonlinear analysis of stress and strain in soils. *J. Soil Mechanics of Foundations*, **96**(SM5), (1970).

Duncan, J.M., Byrne, P., Wong, K.S. and Mabry, P. *Strength, Stress-Strain and Bulk Modulus Parameters for Finite Element Analyses of Stresses and Movements in Soil Masses*. Report No. UCB/GT/80-01 Department of Civil Engineering, University of California, Berkeley, CA, (1980).

Eringen, A. C. Theory of micropolar continuum. *Proc. 9th Midwestern Conf.*, University of Wisconsin, Madison, (1965) 23–40.

Frantziskonis, G., Desai, C.S. and Somasundaram, S. Constitutive behavior for nonassociative behavior. *J. Eng., Mech.*, **9**, (1986) 932–946.

Frantziskonis, G. and Desai, C.S. Constitutive model with strain softening. *Int. J. Solids and Structures*, **23**(6), (1987) 733–750.

Kachanov, L.M. The theory of creep. English translation edited by A.J. Kennedy. National Lending Library, Boston, (1958).

Kachanov, L.M. *Introduction of Continuum Damage Mechanics*. Martinus Nijhoff, Dordrecht, The Netherlands, (1986).

Katti D. R. *Modeling including associated testing of cohesive soils using the disturbed state concept*. PhD dissertation, The University of Arizona, Tucson, AZ, (1991).

Katti D. R. and Desai C. S. Modeling and testing of cohesive soils under disturbed-state concept. *J. Engrg. Mech.*, **121**(5), (1995) 648–658.

Krajcinovic, D. and Fonseka, G.U. The continuous damage theory for brittle materials, Part I and II. *J. Appl. Mech.*, **48**, (1981) 809–824.

Lee, K.L. and Seed, H.B. Drained Strength Characteristics of Sands. *J. Soil Mech. & Fdn. Engrg.: Proc. ASCE.* (November 1967) 117–141.

Muhlhaus, H. B. A thermodynamic criteria for damage. *Proc. 8th Int. Conf. of Int. Assoc. for Computer Methods and Advances in Geomechanics*, West Virginia, (May 1994) 22–23.

Navayogarajah, N., Desai, C.S. and Kiousis, P.D. Hierarchical Single Surface model for static and cyclic behavior of interfaces. *J. Engrg. Mech.*, **118**(5), (1992) 990-1-1011.

Onat, E.T. and Leckie F.A. Representation of mechanical behavior in the presence of changing internal structure. *J. Appl. Mech.*, **55**, (1988) 1–10.

Oden J. and Kikuchi. Theory of variational inequalities with applications to problems of flow through porous media. *Int. J. Eng. Sci.*, **18**(10), (1980) 1173–1284.

Pal, S. *Numerical Simulation of Geosynthetic Reinforced Earth Structure using Finite Element Methods*. PhD Dissertation, Louisiana State University, Baton Rouge, LA, (1997).

Pal, S. and Wathugala, G.W. *Analysis of Geosynthetic Reinforced Soil Structures Through Numerical Simulations*. Report to Louisiana Transportation Research Center (LTRC), (October 1997).

Pal, S. and Wathugala, G.W. DSC constitutive model for sand-geosynthetic interfaces. In: *Computer Methods and Advances in Geomechanics: Proceedings of the 9th Int. Conf. of the International Association for Computer Methods and Advances in Geomechanics*, Yuan, J. (ed.). Wuhan, China, Vol. 2. A.A. Balkema, Rotterdam, (1997) 875–880.

Pal, S. and Wathugala, G.W. Disturbed state model for sand-geosynthetic interfaces and application to pull-out tests. *Int. J. Num. Anal. Meth. in Geomech.* **23**, (1999) 1873–1892.

Park, I.J. and Desai, C.S. *Analysis of Liquefaction in Pile Foundations and Shake Table Tests Using the Disturbed State Concept*. Report to NSF, Department of Civil Engineering and Engineering Mechanics, University of Arizona, Tucson, AZ, (1997).

Perzyna, P. Fundamental problems in visco-plasticity. In: *Recent Advances in Applied Mechanics*. Academic Press, New York, 1966.

Roscoe, K. H., Schofield, A. and Wroth, C. P. On the yielding of soils. *Geotechnique*, **8**, (1958) 22–53.

Shao, C. and Desai, C.S. *Implementation of DSC Model for Dynamic Analysis of Soil-Structure Interaction Problems*. Report to NSF, Department of Civil Engineering and Engineering Mechanics, University of Arizona, Tucson, AZ, (1998).

Somasundaram, S. and Desai, C. S. Modeling and testing for anisotropic behavior of soils. *J. Engrg. Mech.*, **114**, (1988) 1177–1194.

Titi, H.H. *The Increase in Shaft Capacity with Time for Friction Piles Driven into Saturated Clay*. PhD Dissertation, Louisiana State University, Baton Rouge, LA, (1996).

Titi, H.H. and Wathugala, G.W. *Theoretical and Laboratory Study of Increase in Bearing Capacity of Driven Piles in Saturated Clay with Time (Pile Setup), Phase I: Theoretical and Numerical Study*. Report submitted to the National Science Foundation, November, (1996).

Voyiadjis, G.Z. and Thiyagarajan, G. A damage cyclic model for metal matrix composites. In: *Damage and Interface Debonding in Composites*, Voyiadjis G.Z. and Allen D.H. (eds.). Elsevier, Amsterdam, (1996) 107–133.

Wathugala, G.W. *Finite Element Dynamic Analysis of Nonlinear Porous Media with Applications to Piles in Saturated Clays*. PhD Dissertation, University of Arizona, Tucson, AZ, (May 1990).

Wathugala, G.W. and Desai, C.S. *Constitutive Model for Soils with Strain Softening and Shear Dilation*. Unpublished Internal Research Report, Department of Civil Engineering and Engineering Mechanics, University of Arizona, Tucson, AZ, (1987).

Wathugala, G.W. and Desai, C.S. 'Damage' based constitutive model for soils. *Proc. 12th Canadian Congress of Applied Mechanics (CANCAM'89)*, Ottawa, May 28–June 1 (1989) 530–531.

Wathugula, W. and Desai, C. S. Hierarchical Single Surface model for anisotropic hardening cohesive soils. In: *Computer Methods and Advances in Geomechanics*, Beer, Brooker and Carter (eds.). Balkema, Rotterdam, (1991) 1245–1249.

Wathugula, W. and Desai, C. S. Constitutive model for cyclic behavior of clays. I: Theory. *J. Geotech. Engrg.*, **119**(4), (1997) 714–729.

Westbrook, D. R. Analysis of Inequality and Residual Flow Procedure and an Iterative Scheme for Free Surface seepage. *Int. J. Num. Meth. in Engrg.*, **21**, (1985) 1791.

11

Laboratory Testing for the Validation of Constitutive Models

Adel S. Saada

Case Western Reserve University, Cleveland, OH, USA

11.1   General Review

The goal of laboratory testing has been, and still is, to study the behavior of a given soil under conditions similar to those encountered in the field, and to obtain the parameters needed to describe its behavior in a set of constitutive equations. Those equations are used to solve problems of practical importance such as bearing capacity, slope stability, and excavation. Plasticity of the Coulomb type was, and is still, successfully used in design, and the two coefficients, ϕ' and c, are the two parameters that are most sought by foundation engineers. The fact that the expression of the vertical stress due to a vertical load applied on the surface of a semi-infinite linear elastic medium does not contain any elasticity coefficient, has gone a long way in marrying the profession to the use of a linear elastic perfectly plastic model.

Coulomb's coefficients (c and ϕ') were obtained over the years using a variety of shearing devices. Failure was often the stage at which the extracted information was considered significant. Load-deformation characteristics were used for comparison purposes among various soils; and not to develop constitutive relations.

A brief history of soil strength testing was given by Saada and Townsend (1981) in a first state-of-the-art paper published by ASTM. This brief history was followed by a description of the various direct shear tests and devices, simple shear tests and devices and finally three-dimensional tests and devices. Table 11.1 lists the devices described in this paper.

At this stage it is important to remember that a soil specimen being tested represents a point where a given state of stress is present. This can only be true if the stresses are uniformly distributed throughout the specimen and on its boundaries. Thus, the manner in

Table 11.1   Summary of laboratory shear equipment

Equipment	Consolidation	Comments
Translatory direct shear	K_0	The normal and shear stresses and
Torsional direct shear	K_0	strains are obtained as averages
Solid specimens		from measured or applied values
Hollow specimens		of axial and tangential forces and
Simple shear	K_0	displacements. Other components are
		calculated based on various assumptions.
Triaxial on solid cylinders	isotropic, K_0, anisotropic	The axial stresses and strains are computed as averages from axial forces and displacements. Lateral stresses are applied and lateral deformations measured or computed.
Triaxial on prismatic specimens	isotropic, K_0, anisotropic	The surface stresses and deformations are either applied or computed as average from surface forces and displacements.
Triaxial on hollow specimens with the same inner and outer pressures	isotropic, K_0, anisotropic	The axial and torsional shear stresses and strains are computed as averages from axial forces and torques and from axial displacements and rotations. Other components of strain are either measured or calculated based on various assumptions.

which the loads are applied becomes of paramount importance. End effects and geometry play a major role in any attempt at simulating field conditions, or just at interpreting recorded results.

11.1.1   Direct shear tests and devices

The translatory direct shear devices are either rectangular or circular in shape (Figure 11.1). They have been used for years to determine values of c and ϕ' with the assumption that failure occurred freely on a horizontal plane subjected to a specific normal stress and that stresses were uniformly distributed there. Hvorslev (1937) was the first to dramatically show the shortcomings of these devices in which the stress distribution is far from being uniform. Attempts at obtaining the residual strength were not particularly successful because of the small range of deformations that can be reached in those tests. Skempton (1964) used multiple reversal to accumulate large deformations.

   The torsional direct shear tests on solid or hollow specimens allow for large shearing deformation but still suffer from stress non-uniformities due both to the end effects and to the geometry of the specimen. Residual strength is probably the only quantity that would be worth getting with such devices. They demand too many assumptions that are hard to justify.

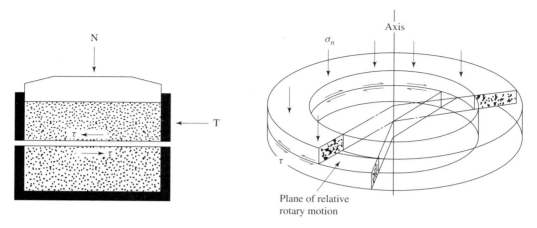

Figure 11.1   Direct shear tests (Saada and Townsend, 1981)

11.1.2   Simple shear tests and devices

Such devices were introduced to induce a condition of simple shear in a soil specimen (Figure 11.2). An earlier apparatus designed by Kjellman (1951) uses a cylindrical specimen confined by a rubber membrane and a series of thin and evenly spaced rings. In 1966 the Norwegian Geotechnical Institute (NGI) refined Kjellman's device by replacing the rings with thin wires embedded in the rubber membrane. Roscoe (1953) used a square specimen in a box with hinged ends imposing on the specimen a condition of simple shear strain; leaving the to rigid faces the task of providing the necessary boundary stresses.

A variety of studies some experimental and some computational have shown these devices to be less than satisfactory in providing stress or strain uniformity in the specimens

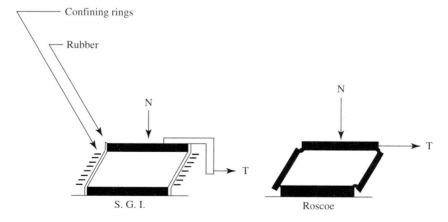

Figure 11.2   Simple shear tests (Saada and Townsend, 1981)

being tested. Those studies have been reviewed by Saada and Townsend (1981) with the conclusion that, at best, these tools can be used to compare descriptively different soils. It is suggested that their proper place is in the hands of designers who have already conducted a large number of such tests and built structures which behaved satisfactorily.

11.1.3   Triaxial tests on solid circular cylinders

In their classical treatise Bishop and Henkel (1962) cover in great detail all the aspects of this test (Figure 11.3). Its main limitation is that two out of the three principal stresses are always equal. An axial load is generally applied to the specimen through a solid piston, and together with a spherical state of stress one can cover a wide variety of stress paths. Measurements of deformations, applied stresses, loads and pore water pressures are presently totally automated. Controls, electro-pneumatic or electro-hydraulic result in a faithful adherence to a predetermined testing program. A variety of ingenious bushings and feedback systems ensure that the loads or deformations applied are faithfully transmitted to the specimen.

The uniformity of the deformations as well as of the stresses have been the subject of many a discussion (Saada and Townsend, 1981). The friction that develops at the end platens and the restriction it imposes on the free expansion or contraction of the ends of the test specimen results in unknown and nonuniform stresses within the soil. Under compression, barreling takes place and under extension necking. The use of shorter specimens with greased near frictionless platens seems to alleviate this situation. The complexities of this problem are also exacerbated by the fact that the kind of soil used has a major influence on those nonuniformities. Saturated clays whose behavior approaches that of a continuum behave differently from sands. Using short bulky specimens of sands, even with greased ends does not insure a uniform stress distribution even under hydrostatic stresses. Granular

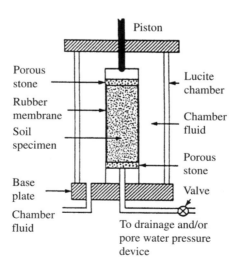

Figure 11.3   Triaxial cell (Bishop and Henkel, 1962)

frictional materials will arch causing a nonhomogeneous stress distribution as one moves from the loaded boundaries toward the center of the specimen. Membrane corrections are often made when one is testing medium to coarse sands.

In spite of its many drawbacks, this test is the most popular today in soil mechanics. It is conceptually very simple and, with or without corrections, it allows one to conduct comparative studies among various soils and obtain design parameters that are used in practice.

11.1.4 Devices that can apply three different principal stresses to prismatic specimens

Among the earliest prismatic specimens are the plane strain ones where one face of a cube or a parallelepiped is prevented from moving by means of rigid greased plates. Table 11.2 lists most of the plane strain devices built by various institutions. Figure 11.4 shows schematically the system of stresses acting on a prism subjected to a plane strain condition. Notice that the various faces can be subjected to controlled stress through the use of a membrane or to controlled displacement through the use of a rigid plate. The device by Hambly (1969) uses rigid plates and is of the controlled deformation type. The device of Al-Husseini (1971) controls the deformation vertically and the stress horizontally through the use of a membrane.

To rotate the principal stresses under plane strain conditions Arthur *et al.* (1977) designed the directional shear cell in which a cube is subjected to normal and shearing stresses on four of its faces while two others are not allowed to deform (Figure 11.5). This apparatus can apply only small values of shearing stresses even though some enhancements have been made at the University of Colorado. Some photoelastic studies conducted at the author's laboratory have shown that, even under the best of conditions it is extremely difficult, if not impossible, to obtain uniform shearing stress distributions on the faces of the test specimens.

Table 11.2 Summary of plane strain shear devices

Designer	Location
Lorenz *et al.*	Technische Universitat, Berlin
Bjerrum and Kummencji	NGI
Christensen	DGI
Leussink and Wittke	Technical University of Karlsruhe
Marsal *et al.*	Comision Federal de Electricedad, Mexico
Duncan and Seed	University of California
Wood	Imperial College
Dickey *et al.*	MIT
Al-Husseini and Wade	Georgia Institute of Technology
Al-Husseini	WES
Hambly	Cambridge University
Campanella and Vaid	University of British Columbia
Ichihara and Matsuzawa	Nagoya University

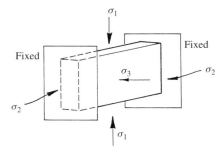

Figure 11.4 Plane strain condition (Al-Husseini, 1971)

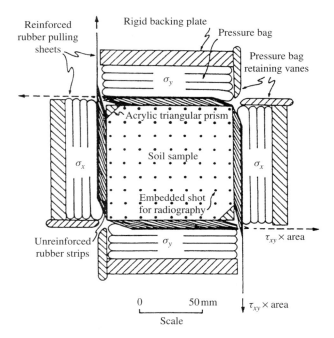

Figure 11.5 Directional shear cell (Arthur, 1988)

It is appropriate to mention here the plane strain device of Vardoulakis (1990) in which the bottom of the specimen can move horizontally on ball bearings. Such a device is primarily used to study localization and bifurcation in soils.

The name true triaxial device seems to be used in conjunction with cubical soil specimens subjected to three principal stresses, three principal displacements or a mixture of both. Saada and Townsend (1981) describes most of the ones used by a number of research institutions and individuals with their advantages and disadvantages. When the stresses are applied through membranes the displacement of the faces is usually not uniform. When the displacements are applied to the faces the stresses are generally not uniform. In this category the author prefers the true triaxial apparatuses with rigid platens over the ones with flexible pressure membranes (Figures 11.6, 11.7). This preference is due to observations

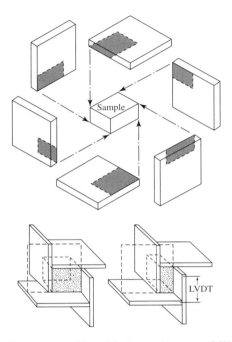

Figure 11.6 True triaxial apparatus with rigid platens (Avery and Wood, 1988; Lanier, 1988)

recently made on one of the latter devices in which the sides were manufactured of heavy transparent plexiglass. Video photographs of the deformation of the faces of the cube indicate that, even before one reaches the peak the non-uniformities become unacceptable to the point of placing in question the validity of the recorded deformations. Table 11.3 compares the various types of apparatuses with their advantages and disadvantages.

The proceedings of the Symposium on Advanced Triaxial Testing of Soil and Rock, which was presented in June 1986 by ASTM, contain a wealth of information on devices that use prismatic specimens. The description of the Cambridge True Triaxial Apparatus by Avery and Wood (1988) and that of the Grenoble Tri-Tri by Lanier (1988) are to be noted. More will be said later about the results of tests conducted with such devices.

11.1.5 Devices that can apply three different principal stresses to hollow circular cylinders

Thin long hollow circular cylinders subjected to combinations of hydrostatic, axial and torsional stresses have been used extensively since the mid-1960s to study the behavior of soils under combined stresses and in particular the influence of the direction of the principal stresses with respect to the axis of symmetry of cross anisotropic soils. Two state-of-the-art papers by Saada and Townsend (1981) and Saada (1988) give extensive details about hollow cylinders torsional devices, their advantages and their limitations.

True triaxial apparatuses with solid platens and thin long hollow cylinder torsional devices will be examined in more detail in the following sections. Several points that deserve attention need to be mentioned at this stage:

Table 11.3   A comparison of three types of boundary conditions in a multiaxial cubical test apparatus

Strain Controlled: All Rigid Boundaries	Stress Controlled: All Flexible Boundaries	Mixed (Rigid and Flexible) Boundary Condition
	Advantages	
(a) Strains can be measured accurately	(a) normal, principal stresses assured on the loading faces	(a) boundary interference is usually avoided by assigning the rigid boundary as the compressive deviator direction and stress-controlled flexible boundary as extension deviator direction
(b) Uniform strains are possible	(b) uniform stress distribution over all faces is possible	
(c) Large uniaxial strains can be achieved	(c) large strain can be achieved in three dimensions without significant boundary interference	
(d) Complicated and predetermined strain paths can be readily modeled	(d) complicated and predetermined stress paths can be readily followed	(b) pore water pressure and other facilities are easily accommodated
(e) Pressure cells and pore water pressure instrumentation can easily be accommodated in the loading platens	(e) shear distortions are possible and measurable	(c) stress or strain paths can be easily followed if a predetermined selection of specimens orientation with respect to apparatus axes is allowed
		(d) plane strain tests can be modeled
	Disadvantages	
(a) The uniformity of stresses induced is difficult to verify	(a) interference at boundaries can occur if proper precautions are not taken	(a) complicated predetermined stress or strain paths are difficult or impossible to follow
(b) Apparatus will not accommodate or allow shear distortions	(b) the uniformity of large strains can be difficult to maintain if proper lubrication is not performed	(b) uniformity of stress and strain fields in directions normal to rigid and flexible boundaries difficult to ascertain
(c) Loading platen interference occurs at large multiaxial strain states	(c) difficult to follow predetermined strain paths	(c) heterogeneous stress and strain fields occur near boundaries
(d) Difficult to follow predetermined stress paths	(d) plane strain experiments can only be achieved through stepwise corrections of stress state normal to plane	(d) apparatus is usually large and unwieldy
(e) Apparatus is usually large and unwieldy		(e) operation is usually extremely complicated
(f) Operation is usually complicated	(e) pore water pressure facilities are not easily accommodated	(f) apparatus will not allow shear distortions

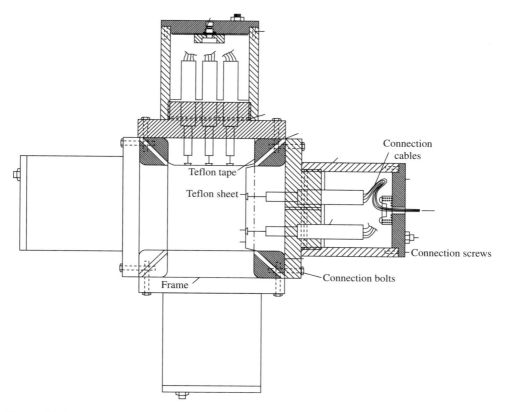

Figure 11.7   Multiaxial cubical apparatus with flexible boundaries (Saada and Townsend, 1981)

1. Prior to 1963, the mechanical properties of cross anisotropic soils (K_o consolidated) were studied by testing specimens cut at various inclinations to the direction of consolidation. As shown by Saada (1970), such tests when conducted in standard triaxial cells resulted in unknown bendings and shears being applied to the specimen (Figure 11.8). Those end effects can be avoided by using thin hollow cylinders subjected to fixed ratios of axial and torsional stresses. To each ratio there corresponds a specific inclination β of the major principal stress to the direction of consolidation. Thus, rather than inclining the specimen, one would incline the stresses.

2. If one keeps the same pressure inside and outside the cylinder one is assured of uniform normal stresses across the thickness of the specimen; provided the specimen is thin. On the other hand, the shearing stresses due to torsion cannot be uniform, except possibly at the full plastic stage, if the material does respond as an elasto- or visco-plastic material. With the proper proportions one can reduce these non-uniformities, and thus the errors to a minimum. The magnitudes of the errors can be bracketed and taken into account in the calculations.

3. Much has been made of the parameter $b = (\sigma_2 - \sigma_3)/(\sigma_1 - \sigma_3) = \sin^2 \beta$, where the σs are the principal stresses and β is the inclination of the major principal stress on the axis of symmetry of the material in a thin hollow cylinder subjected to combinations

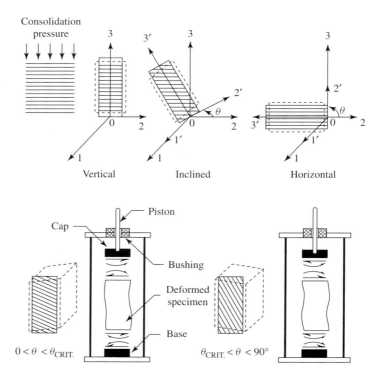

Figure 11.8 Bending moments and shearing forces introduced by testing inclined anisotropic specimens (Saada, 1970)

of axial and torsional stresses. b is a parameter that helps locate the position and study the influence of the intermediate principal stress with respect to the major and the minor ones.

If one keeps the same pressure inside and outside the hollow cylinder, the variation of b can only be obtained at the expense of a variation of the inclination β. This is a drawback should one wish to study the response of cross anisotropic materials. Figure 11.9 illustrates the various possibilities offered by the different apparatuses used in creating combined states of stress. Notice that if one uses different inside and outside pressures, the uniformity assumption becomes totally untenable. Rather than having a specimen with a uniform and known state of stress and strain, one has to solve a boundary value problem just to find them at the various points. Under such conditions, which constitutive equation should one use? Let us not forget that we are conducting a test to help develop this same mathematical model!

In the opinion of the author, hollow cylinder tests with different inner and outer pressures are of no value in developing constitutive relations. At best, they can be used to validate a given set of relations when those relations are applied to solve the problem of the hollow cylinder subjected to specific boundary conditions, and in which a number of responses are being measured. If one wishes to study the influence of b without rotating the principal stresses, then a cubic configuration could be used.

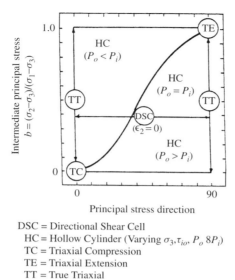

DSC = Directional Shear Cell
HC = Hollow Cylinder (Varying $\sigma_3, \tau_{io}, P_o \, 8P_i$)
TC = Triaxial Compression
TE = Triaxial Extension
TT = True Triaxial

Figure 11.9 Combination of band β achieved in various devices (Saada, 1988)

11.1.6 Summary and conclusions

Each of the devices and tests mentioned in the previous sections has had, and is still having, its uses. They have served us well, and many continue to do so in the hands of experienced practitioners. It is important to emphasize once more that to generate and validate constitutive relations through strength tests, the specimen must be looked at as a point. The closest that this situation has been arrived at is within the thin long hollow cylinder and the true triaxial apparatus with solid faces of the type used at the University of Grenoble by Lanier and co-workers (1988), or at Cambridge (Avery and Wood, 1988). The reasons for this opinion are as follows.

During an international workshop on constitutive equations for granular noncohesive soils held at Case Western Reserve University (Saada and Bianchini, 1987), tests were conducted on the same materials both in the hollow cylinder device and in the cubical device with solid faces. Those tests were used to develop as well as to validate some thirty different constitutive relations by modelers from all over the world. Predictions were made on similar and different stress paths in both apparatuses. Stresses and strains were compared by authors and organizers (Lanier *et al.*, 1989), as well as the validity of the various models for complex stress paths (Bianchini *et al.*, 1991). The results of over 200 tests were compiled in a database (Saada and Bianchini, 1993), which was distributed to the international community. The results were found to be extremely satisfactory in spite of the difference in the boundary conditions, the application of load and deformation and the fact that the tests were conducted at two different institutions, namely Case Western Reserve University and the University of Grenoble, and by two different teams. The author is not aware of any other effort of this kind and of this magnitude and whose results have been made public. The two devices complement each other and what cannot be done with one can be done with the other.

11.2   The Hollow Cylinder Test

11.2.1   *Triaxial cells, loading systems and tranducers*

Modern cells allow for a variety of heights and diameters. It is advisable to have large cells so that samples obtained with standard Shelby tubes can be accommodated. Pistons capable of applying axial and torsional loads require special bushings to minimize friction, and transducers placed inside and outside the cell measure loads, stresses and deformations. Figure 11.10 shows in a schematic way the cell used in the author's laboratory. Tatsuaka (1988) and Hight *et al.* (1993) describe recent devices and testing systems that are used in conjunction with the hollow cylinder.

Loading devices presently use full or partial feedback controls and computer technology allows for applying loads, deformation and stresses at any ratio so as to obtain nearly every conceivable stress or strain path. As previously mentioned, a change in *b* can only be obtained at the expense of a rotation of the principal stresses. In the very small ranges, electromagnets are used to induce resonance with relatively small amplitudes in resonant column type of tests. In this case, accelerometers are used to monitor modulii and damping.

11.2.2   *Sample configuration and preparation*

Clay specimens can be prepared in the laboratory from a powder mixed with de-aired, distilled water under vacuum. The consolidation of the slurry yields blocks from which solid cylinders are cut and cored (Saada, 1988). The two configurations shown in

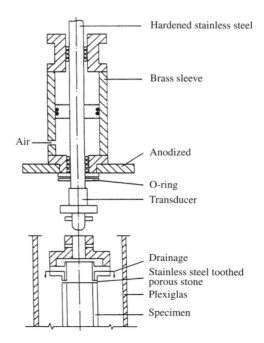

Figure 11.10   The case cell for axial and torsional stresses on hollow cylinders

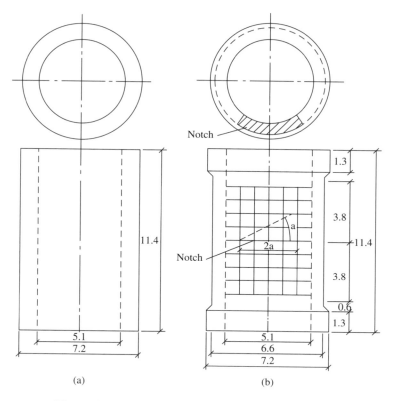

Figure 11.11   Specimen geometry (dimensions in cm)

Figure 11.11 can and have been used. The dogbone shape helps minimize end effects. Sometimes a square grid is stamped on the specimen to help study bifurcation and shear banding. Similar specimens can be made with clay obtained undisturbed with a Shelby tube.

Sand specimens are prepared by pluviation or tanping. Pluviation is followed usually by vibration. In either case a mechanical anisotropy may result. Both configurations shown in Figure 11.11 are possible here too (Saada, 1994; Liang *et al.*, 1997).

Sometimes, in special studies, notches on discontinuities are induced in the walls of the cylinder to initiate cracks and study their propagation (Saada, 1994).

The geometry, namely the height and the radii, affects the uniformity of the stress distribution. Radial frictional forces are developed at the ends of the specimen if it has a tendency to expand or contract. This tendency is always present when the volume and the length change. The writer reached the conclusion (Saada and Townsend, 1981) that to have in a specimen a central zone free from end effects the following criteria should be satisfied:

$$H \geq 5.44\sqrt{R_o - R_i} \quad \text{and} \quad n = \frac{R_i}{R_o} \leq 0.65$$

where H is the height and R_o and R_i are the outer and inner radii. Other researchers have used different ratios (Saada, 1988).

11.2.3   Stresses and strains

In the thin hollow cylinder, the combination of axial and torsional loads leads to major and minor principal stresses that are inclined to the axis of symmetry of the material (Figure 11.12). Test results are interpreted in terms of averages. Such averages have generally been calculated with the assumption that the work done by the applied forces and torques is equal to the sum of the work done by the stresses and strains involved. They are as follows (Figure 11.12):

$$\text{Average axial stress } \sigma_z = \frac{F_{ax}}{\pi(R_o^2 - R_i^2)} + \frac{P_o R_o^2 - P_i R_i^2}{R_o^2 - R_i^2} \tag{11.1}$$

$$\left.\text{Average } \sigma_r = \frac{P_o R_o + P_i R_i}{R_o + R_i}\right\} = P \text{ for } P_i = P_o \tag{11.2}$$

$$\text{Average } \sigma_\theta = \frac{P_o R_o - P_i R_i}{R_o - R_i} = P \text{ for } P_i = P_o \tag{11.3}$$

$$\text{Average } \tau_{\theta z} = \frac{3M_t}{2\pi(R_o^3 - R_i^3)} \tag{11.4}$$

$$\text{Average } \varepsilon_z = \frac{\Delta H}{H} \tag{11.5}$$

$$\text{Average } \varepsilon_r = \frac{u_o - u_i}{R_o - R_i} \tag{11.6}$$

$$\text{Average } \varepsilon_\theta = \frac{u_o + u_i}{R_o + R_i} \tag{11.7}$$

$$\text{Average } \gamma_{\theta z} = \frac{2\theta(R_o^3 - R_i^3)}{3H(R_o^2 - R_i^2)} \tag{11.8}$$

The average values of σ_z and σ_θ are based on equilibrium equations only, and average values of ε_z and $\gamma_{\theta z}$ are based on strain compatibility only; all are therefore independent of the constitutive laws of the material being tested. The expression of the average value of σ_r is based on a linear elastic stress distribution; the average σ_z is based on a uniform stress distribution; the average ε_r and ε_θ are based on a linear variation of radial displacement across the wall. The equation of equilibrium in the radial direction yields

$$\frac{d\sigma_r}{dr} + \frac{\sigma_r - \sigma_\theta}{r} = 0 \tag{11.9}$$

and if $P_o = P_i = P$, it is quite proper to assume that there is no variation in the value of the normal radial stress σ_r across the thickness. Thus $(d\sigma_r)/(dr) = 0$, and the equation of equilibrium above yields $\sigma_r = \sigma_\theta = P$. The compatibility equation in the $[r, \theta]$ plane yields

$$\frac{d\varepsilon_\theta}{dr} + \frac{1}{r}(\varepsilon_\theta - \varepsilon_r) = 0 \tag{11.10}$$

and if ε_θ is assumed to be uniform across the thickness, then $\varepsilon_r = \varepsilon_\theta$. This leads to a linear variation of the radial displacement across the wall of the cylinder.

The assumption that $\varepsilon_r = \varepsilon_\theta$ places restrictions on the shape of the potential function, if one chooses to use the classical theory of plasticity in expressing the stress-strain relation of the soil. This issue is discussed further by Saada (1988) and Lade *et al.* (1996) additional to the fact that there are other theories besides classical plasticity, the author believes that this assumption is far more acceptable than many of the commonly made ones; especially when the accuracy of the needed measurements to differentiate between ε_r and ε_θ is extremely questionable in view of the magnitude of the displacements involved.

11.2.4   Stress paths

Complicated stress paths are needed not only to simulate real conditions in the field, but also to test the validity of models calibrated through the conduct of a few simple experiments. Paths that often occur in the field involve vertical normal and horizontal shearing stresses; fixed or variable. Such conditions are easily duplicated in the hollow cylinder device. In dynamic behavior, the normal and shearing stresses can be in or out of phase or duplicate the spectrum that occurs during an earthquake (Liang, 1995).

In the laboratory, one often starts from a state of spherical stress where $\sigma_r = \sigma_\theta = \sigma_z$. The addition of axial and torsional stresses causes a rotation of the major and minor principal stresses (Figure 11.12). The intermediate principal stress is radial and equal to the pressure in the cell. If the ratio $\Delta\tau_{\theta z}/\Delta\sigma_z$ remains constant, the inclination β remains constant and one has a case of proportional stress; if not, σ_1 and σ_3 rotate. Tests can be conducted also keeping the mean stress constant, in addition to keeping a fixed

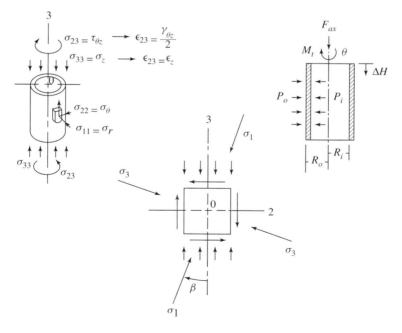

Figure 11.12   Notation

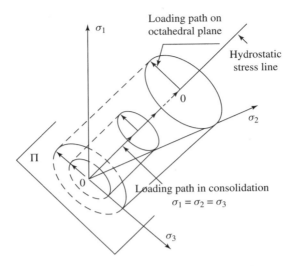

Figure 11.13   Stress paths and projections on the Π plane

direction of the principal stresses. This is done by adjusting the cell pressure (i.e. σ_2) while the axial load is being increased (or decreased) so that $(\sigma_z + \sigma_r + \sigma_\theta)$ remains fixed while the torque is being applied. If the cell pressure remains constant the test has been referred to as a direct test and the stress path moves from octahedral plane to octahedral plane. On the other hand, if the mean stress remains constant the test is referred to as a *generalized* test and the stress path remains on one octahedral plane. One can thus have the same inclination of the principal stress, but at a different mean stress, thus separating the effects of confinement and anisotropy. In both cases b is the same but σ_2 is different (Figure 11.13).

A clear advantage of the use of hollow cylinders is that they easily lend themselves to dynamic testing, a thing hard to realize with the true triaxials of the prismatic type. Indeed, the most significant resonant column tests have been conducted on hollow cylinders. Often the modulii and the damping under small strain are obtained first, then the test continued as a quasi static test. Controls have been built in to handle both types of data and provide synchronized loading (Saada, 1988; Alarcon *et al.*, 1986). One can apply cycles of extension and compression at any inclination of the principal stresses; or one can generate a mohr circle of fixed diameter moving along the normal stress axis by applying sinusoidal axial and torsional stresses with a ninety degrees phase difference.

Saada (1988) gives examples of stress paths both static and dynamic. Phenomena such as necking occur when extension is the dominant mode of deformation. There, the formation of shear bands also affects the results of the test (Lade *et al.*, 1996). Recently, the hollow cylinder has been used to study localization and bifurcation in both clay and sand soils (Saada *et al.*, 1994; Liang *et al.*, 1997). Grids printed on the soil itself were used to follow the deformation of hollow cylinders, the initiation and propagation of shear bands, as well as to measure the strains in the vicinity of and within the shear bands. The use of video-imaging techniques has gone a long way in providing non-invasive techniques for detailed observation and deformation measurements for both clay and sand specimens.

11.3   The True Triaxial Test with Six Rigid Boundaries

Arthur (1988), in his state-of-the-art paper, provides an excellent description of cubical devices and their versatility. Lanier (1988) provides a description of the apparatus, and some of the tests that have been conducted with the true triaxial of the University of Grenoble. At the end of Section 9.1, Lanier expressed his preference for this type of true triaxial apparatus because of his experience with it. This kind of device is not as common as the hollow cylinder but can create some paths that it can claim as typically its own. It will be referred to in the following as TTA.

11.3.1   Stress and strains

In this TTA the deformations are controled, the stresses measured and a feedback system allows for either a programmed strain path or a programmed stress path. One has to assume, however, that both stresses and strains are homogeneous within the specimen and that the principal axes of stress and strain coincide. Homogeneity is lost when shear bands are formed in the material. Induced anisotropy will prevent the principal stresses and strain to coincide in direction. In spite of this, the TTA is the only apparatus that allows control of the three principal stresses and strains.

   The specimen used by Saada and Bianchini (1993) and at the University of Grenoble (Lanier, 1988) is about 10×10 cm, and is enclosed in a rubber membrane that slides against the solid faces. This friction is much reduced or nearly eliminated through the use of proper lubricants. Figure 11.6 shows the fashion in which the faces of the cube move. Figure 11.14 shows rotating stress paths and the resulting strain paths which have become the trademark of this device. The database previously mentioned (Saada and Bianchini, 1993) contains the results of a very large number of tests conducted with this device.

   For large strains shear banding occurs under plane strain conditions. Such an instability which usually occurs near the peak results in unreliable data at this stage of the experiment.

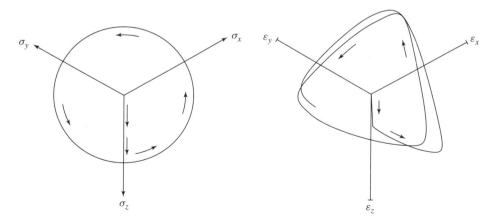

Figure 11.14   Circular stress path and corresponding strain path (Lanier, 1988)

11.4 Conclusions

There exist today a large number of constitutive equations, each with advantages and disadvantages; and more seem to spring up every year. They are validated with a limited number of experiments and, more often than not, are used in practice by their inventor on practically an exclusive basis. Some models are more general than others, and are looked at as supermodels whereby each of the existing ones becomes one of their particular cases (Katti and Desai, 1995). In all cases, whether the user of a model is looking at elastic modulii, plastic potential functions, flow coefficients or failure, he will have to test its validity and know its limitations under specific circumstances dictated by the problem at hand. The thin long hollow cylinder apparatus complemented by the TTA with rigid boundaries, driven by the proper controls, can simulate most of the conditions met in the field or the laboratory. It is today the instrument of choice in covering in as accurate a way as presently possible most of the stress and strain paths needed by the majority of researchers.

References

Alarcon, A., Chameau, J. L. and Leonards, G. A. A new apparatus for investigating the stress-strain characteristics of sands. *Geotechnical Testing J.*, **9**(4), (1986) 204–212.

Al-Husseini, M. M. *Investigation of Plane Strain Shear Testing, Report No. 1*, Technical Report 5-71-2 USAE, Waterways Experiment Station, Vicksburg, MI, (March 1971).

Arthur, J. R. F. *State of the Art: Cubical Devices: Versatility and Constraints*, ASTM STP 977, (1988) 743–765.

Arthur, J. R. F., Chua, S. K. and Dunstan, T. Induced anisotropy in sands. *Geotechnique*, **27**(1), (1977) 1–30.

Avery, D. W. and Wood, D. M. *The Cambridge True Triaxial Apparatus, Advanced Triaxial Testing of Soil and Rock*, ASTM STP 977, (1988) 796–805.

Bianchini, G., Saada, A., Puccini, P., Lanier, J. and Zitouni, Z. Complex stress paths and validation of constitutive models. *Geotechnical Testing J.*, **14**(1), (1991) 13–25.

Bishop, A. W. and Henkel, D. J. *The Measurement of Soil Properties in the Triaxial Test*. Arnold, London, (1962).

Bjerrum, L. and Landva, A. Direct simple shear tests on Norwegian quick clay. *Geotechnique*, **16**(1), (1966) 1–20.

Drescher, A., Vardoulakis, I. and Han, C. A biaxial apparatus for testing soils. *Geotechnical Testing J.*, **13**(3), (1990) 226–234.

Hambly, E. C. and Roscoe, K. H. Observations and prediction of stresses and strains during plane strain of wet clay. *Proc. 7th ICSMEF*, Mexico, Vol. 2, (1969) 173–181.

Hight, D. W., Gens, A. and Symes, M. J. The development of a new hollow cylinder apparatus for investigating the effects of principal stress rotation in soils. *Geotechnique*, **33**(4), (1983) 355–383.

Hvorslev, M. J. *On the Strength Properties of Remolded Cohesive Soils*. PhD Thesis, (1937).

Katti, D. R. and Desai, C. S. Modelling and testing of cohesive soil using disturbed state concept. *J. Engrg. Mech.*, **121**(5), (1995) 648–658.

Kjellman, W. Testing the shear strength of clay in Sweden. *Geotechnique*, **2**(3), (1951) 225–235.

Lade, P., Yamamuro, J. A. and Skyers, B. D. Effects of shear band formation in triaxial extension tests. *Geotechnical Testing J.*, **19**(4), (1996) 398–340.

Lanier, J. *Special Stress Paths Along the Limit Surface of a Sand Specimen with the Use of a True Triaxial Apparatus*, ASTM STP 977, (1988) 859–869.

Lanier, J., Zitouni, Z., Saada, A., Puccini, P. and Bianchini, G. Comportement Tridimensionnel des Sables: Comparaison d' Essais Veritablement Triaxiaux et d' Essais sur Cylindre Creux. *Revue Française de Géotechnique*, **49**, (1989) 67–76.

Liang, L. *Development of an energy method for evaluating the liquefaction potential of a soil deposit.* Thesis, Submitted in Partial Fulfillment of the Requirements for the Degree of Doctor of Philosophy, Case Western Reserve University, (1995).

Liang, L., Saada, A. S., Figueroa, J. L. and Cope, C. T. The use of digital image processing in monitoring shear band development. *Geotechnical Testing J.*, **20**(3), (1997).

Roscoe, K. H. An apparatus for the application of simple shear to soil samples. *Proc. 3rd ICSMFE*, Zurich, Vol. 1, (1953) 186.

Saada, A. S. *State of the Art Paper: Hollow Cylinder Torsional Devices: Their Advantages and Limitations*, ASTM STP 977, (1988) 766–795.

Saada, A. S. Testing of anisotropic clay soils. *J. Soil Mechanics and Foundation Division*, **96**(SM5), (1970) 1847–1852.

Saada, A. S. and Bianchini, G. F. (Eds.) *Proceedings of the International Workshop of Constitutive Equations for Granular Non-Cohesive Soils*, Cleveland, OH, (1987).

Saada, A. S. and Bianchini, G. F. *The Case Data Base for Geotechnical Laboratory Testing.* Department of Civil Engineering, Case Western Reserve University, (1993).

Saada, A. S., Bianchini, G. F. and Liang, L. Cracks, bifurcation and shear bands propagation in saturated clays. *Geotechnique*, **44**(1), (1994) 35–64.

Saada, A. S. and Townsend, F. C. *State of the Art: Laboratory Strength Testing of Soils, Laboratory Shear Strength of Soils*, ASTM STP 740, (1981) 7–77.

Skempton, A. W. Long term stability of clay slopes. *Geotechnique*, **14**(2), (June 1964) 77–102.

Tatsuoka, F. *State of the Art: Some Recent Developments in Triaxial Testing Systems for Cohesionless Soils*, ASTM STP 977, (1988) 7–67.

12

Nondestructive Testing Techniques for Material Characterization

T. Kundu

University of Arizona, Tucson, AZ, USA

12.1   Introduction

Nondestructive testing of materials has been practiced for many years. When one visually inspects an object for irregularities, one is carrying out a nondestructive testing of the object. Since the visual inspection technique is many centuries old, we can say that the nondestructive testing (NDT) technique is older than destructive testing techniques. However, one should bear in mind that although the visual inspection technique is an old method, most of the sophisticated NDT techniques practiced today are relatively new, and some of these have been developed in the latter part of this century.

There are many NDT techniques available today for characterizing different types of materials used in engineering, medical and bioscience applications. A partial list of these techniques includes visual inspection, liquid penetration, acoustic emission, conventional ultrasonics, acoustic microscopy, atomic force microscopy, optical/electronic/tunnel microscopy, photoelasticity, optical holography, acoustical holography, moiré interferometry, laser speckle interferometry, thermal shearography, X-radiography, gamma radiography, neutron radiography, tomography, nuclear magnetic resonance, magnetic field perturbation, electrical and eddy current method, chemical methods, microwave technique, and the Lamb wave technique. However, even this list does not cover all NDT techniques. Naturally, no investigator can be an expert of all these NDT techniques, and it is impossible to describe all these methods in one chapter. Here some of the traditional NDT methods are briefly described, and a few relatively new methods are described in more detail.

12.2   Review of Some Traditional NDT Techniques

A brief review of some NDT techniques is given in this section. Interested readers can read more about these techniques from the references provided.

12.2.1   Visual inspection method

Visual inspection is the oldest nondestructive testing method, and in the opinion of many industries, 80% of defects found are located by visual inspection (Bray and McBride, 1992). Visual inspections are often used to detect potential abnormalities in an object. Then, more sophisticated techniques are used to obtain the size and exact location of the defect. However, there is always a large degree of uncertainty associated with the visual inspection technique. The reliability of this technique depends upon the minimum allowable defect size among other factors. Yonemura (1976) measured the uncertainty for a single observer, between observers, and between groups of observers, and concluded "No matter how consistent an inspector may be repeating his performance day after day, unless his performance meets some minimum specified performance level, the performance is unacceptable".

Visual inspection is simple, fast and economic; that is why an object is generally visually inspected before other more sophisticated NDT methods are used. The primary disadvantage of visual inspection is that the small defects and internal defects in an opaque object cannot be detected by this technique.

12.2.2   Liquid penetrant testing

Liquid penetrant testing is a nondestructive technique for detecting surface discontinuities in solid materials. Cracks and porosity open to the surface can be found by this technique. The penetrant is drawn into very small openings of cracks and pores by capillary action when it is applied to the surface of the object. After an adequate dwell time, all of the excess liquid is removed from the surface. Then a powder is applied on the surface, which absorbs the residual penetrant from the surface cracks. Thus, it gives a clear visual indications of surface discontinuities. A major advantage of this technique is that shape, size or other geometric factors of test objects are not too important. However, it is only a surface inspection method and integrity inside the material cannot be assessed by this technique. Detailed descriptions on this technique can be found in McMaster (1982), Bray and Stanley (1989) and Borucki and Jordan (1989).

12.2.3   Radiography

Radiography is an effective nondestructive testing method for detecting internal flaws of a material. A radiographic image is a two-dimensional distribution of the intensity of Gamma rays, X-rays, or neutrons that have passed through the material. The object attenuates the radiation according to the mass, type and size of internal defects. Thus, a radiographic image shows the distribution of internal flaws of a material. One advantage of radiography over other NDT techniques is that a wide range of material thickness levels can be assessed, extending from thin metal foils and slices of bonding materials to perhaps a foot of steel and several feet of solid rocket propellant material (Bray and McBride, 1992). Complex shapes can be radiographed if both sides of the material are accessible; on one side the source and on the other side the recording plate are placed. Different kinds of metallic and nonmetallic materials can be radiographed. However, the primary disadvantages of this technique are that radiation is a health hazard, it is expensive, mainly due to the use of a silver halide film for recording the image, some processing time is needed before the image can be seen and analysed, delaminations and cracks perpendicular to the propagating beam are almost always undetectable. Based on what kind of beam is passed through the material, the radiography techniques can be classified into three groups, X-radiography, Gamma-radiography, and neutron-radiography.

X-rays are a form of electromagnetic radiations that, due to their very short wavelengths, can penetrate materials that are opaque to light. X-rays are capable of ionizing photographic films and forming latent image, which are later developed. However, X-rays and Gamma rays are capable of ionizing human tissues also and thus are considered health hazards. X-rays are generated by high-velocity electrons striking a metallic target, which, in the flaw detection X-ray system, is usually tungsten embedded in copper. Gamma rays differ from X-rays in their source. While X-rays are electromagnetic waves generated electrically, Gamma rays are emitted while artificial and naturally occurring isotopes decay. Isotopes which are capable of generating Gamma rays are Cobalt 60, Cesium 137, Iridium 192 and Thulium 170. For half-life and penetration capabilities of these isotopes, and a more elaborate description of the X-radiography and Gamma-radiography the readers are referred to Halmshaw (1982, 1987) and Bray and McBride (1992).

In neutron-radiography, or N-radiography, neutrons are used instead of X-rays or Gamma rays for penetrating the materials. Neutrons can be obtained from nuclear reactors, from particle accelerators, from artificially produced radioactive sources, from spontaneous fission of isotopes, and by subcritical neutron multiplications. Neutrons interact with the atomic nucleus, and their attenuation is proportional to material density and neutron absorption of the cross section while X-rays interact with atomic orbiting electrons, and X-ray attenuation is proportional to material density and atomic number. Consequently, one method cannot replace the other, they complement each other. A detailed description of neutron or N-ray detection techniques can be found in McIntire and Bryant (1985).

12.2.4   *Magnetic resonance testing*

Magnetic resonance is found in systems that possess magnetic moments and angular momenta (Slichter, 1980). Nuclear Magnetic Resonance (NMR) and Electron Paramagnetic Resonance (EPR) spectroscopies are the two main types of magnetic resonance testing techniques which have been used in the past by chemists and molecular physicists to study intra- and intermolecular structures and reaction mechanisms. NMR was discovered in 1945; since then it has become a major tool for the investigation of the microscopic structure and properties of matter. NMR is useful for determining molecular structure, and for monitoring chemical reaction mechanisms and rates without affecting the structure, mechanism, or rates of the material or reaction being investigated. NMR has been used in medical imaging in a manner similar to X-ray computed axial tomography (House and Lauterburg, 1979). However, NMR imaging uses no ionizing radiation, and thus it is nonhazardous. Outside the medical field, NMR has been used for detecting explosives in packages (King *et al.*, 1979), and for determining moisture content of composite materials (Matzkanin, 1981). EPR and NMR techniques are very similar; the main difference is that EPR is microwave induced and NMR is radio-frequency (rf) induced transitions of magnetic energy levels of atomic nuclei. EPR is particularly useful for determining molecular structures and monitoring chemical reactions without affecting the samples being examined. Out of these two techniques, so far only NMR has been used as an imaging tool. For a more detailed description of NMR principles and applications, readers are referred to Bennett and Swartzendruber (1986).

12.2.5   Eddy current technique

When an alternating electric current (ac) flows in a coil of many turns of fine wire, it generates an electric field with very little current flow. When such an energized coil is placed near the surface of a conducting material, eddy currents are induced in the material. These currents are proportional to the electrical conductivity of the material. The induced current in the material flows in a direction opposite to the current in the coil. When current in the coil reverses in direction, the induced current reverses. Material properties as well as defects such as cracks and voids, affect the magnitude and phase of the induced current. Analysing the eddy currents the material conditions can be assessed. The primary advantage of this technique is that mechanical contact is not required between the eddy current transducers and the test object. Eddy current penetration depth can be controlled by adjusting the frequency of energizing current. The penetration depth of the eddy current is related to the material resistivity, relative permeability and frequency in the following manner (Bray and McBride, 1992):

$$\alpha = 50.3(\rho/\mu f)^{1/2} \tag{12.1}$$

where α = depth of current penetration in millimeters, ρ = material resistivity (micro-ohm-cm), μ = dimensionless relative permeability and f = frequency in Hz. The method has a high sensitivity to small discontinuities at the surface and near the surface. The instrumentation for eddy current testing is relatively low cost for most applications. However, the main limitation of this technique is that its application is limited to electrically conductive materials only. Interested readers are referred to Fisher (1989), and special issues of *Material Evaluation* volumes 47 and 49.

12.2.6   Thermal technique

In thermal techniques the specimen is heated by a heat source and then the heat source is removed. The heat source can heat the specimen by radiation (heat lamp, lasers, etc.), convection (blowers, ovens, etc.), or conduction (contact hot plates, etc.). After some time, a sensor scans the specimen and records its temperature. The temperature sensors can be contact or noncontact type. Noncontact sensors detect Infrared Radiation (IR) emanating from the test object. Two of the first publications in NDT using IR scanning and detection of flaws were by Beller (1965) and Florant (1964). If a homogeneous and uniform specimen surface is uniformly heated, then the surface temperature at a given time during the cooling process should also be uniform. However, in presence of an internal flaw – inclusion, void or crack – the surface temperature becomes different near the flaw, and thus the flaw can be identified. Interested readers can read Hardy and Bolen (1989) and Lawson and Sabey (1970), where thermal inspection techniques have been discussed in greater detail.

12.2.7   *Optical holography*

Holography is an exciting technique developed to obtain dramatic realism in photography. It also provides the capability for accurately measuring irregularities on the surface of a component. These irregularities often are caused by flaws beneath the surface, and thus this technique sometimes can indirectly detect internal flaws. However, the technique does not penetrate the surface of opaque objects, since the testing medium consists of light rays. The holographic process involves the recording of a complete wave front so that all of the information about a surface is retained and made available for subsequent retrieval. It is a two-step method: the first step is called formation or recording; the second step is called reconstruction, which consists of placing the photographic film into a coherent beam of light and producing an image of the original object.

Optical holographic interferometry detect flaws in a material by recording measurements in unstressed and stressed conditions. Comparison of holographic fringes created on a hologram reveals flaws or discontinuities.

Various techniques are used in the construction of holograms. Three major hologram types are Fresnel holograms, Fraunhofer holograms, and Fourier holograms. Description of these three holography techniques can be found in Caulfield (1979). Modern holographic applications are not limited to optics; the method of holography applies to all waves – electron waves, acoustic waves, X-rays and microwaves.

12.2.8   *Ultrasonic technique*

Ultrasonic waves are vibrational waves having frequencies higher than hearing range of the human ear, which is typically 20 kHz. The upper range of these waves can be as high as 15 or 30 GHz (1 GHz = 10^9 Hz). However, for most flaw detection applications, the frequency generally varies from 200 kHz to 20 MHz. Ultrasonic techniques are useful for the detection of internal flaws, as well as for the determination of material properties. More information about the applications of ultrasonic techniques for material characterization can be found in Bar-Cohen and Mal (1989) for the inspection of metals and composites, and Desai, Jagannath and Kundu (1995) for the inspection of geologic materials. Using ultrasonic techniques, flaws in metals and nonmetals can be detected, flaw distance from the surface can be measured, flaws can be detected in thick or thin materials, flaw imaging is possible and material properties can be determined. When both surfaces of the object are not accessible, then ultrasonic techniques can also be used. However, complex geometries are difficult to inspect using the conventional ultrasonic technique, and small and tight or closed cracks are also difficult to detect.

Basic concepts

The speed of the wave propagation (C), frequency (f) and wave length (λ) are related in the following manner:

$$\lambda = \frac{C}{f} \tag{12.2}$$

There are two main types of elastic waves – body or bulk waves and surface waves – that propagate in a material. Body waves are of two types – longitudinal waves and shear waves. Longitudinal waves are also known as compressional waves, dilatational waves, acoustic waves, pressure waves, primary waves or P-waves. Shear waves are called distortional waves, secondary waves or S-waves. Based on the polarization direction (vertical or horizontal), the S-wave can be classified as an SV-wave or SH-wave.

Longitudinal wave speed (α) in a bulk material is given by

$$\alpha = \sqrt{\frac{E(1 - \nu)}{\rho(1 + \nu)(1 - 2\nu)}} = \sqrt{\frac{\lambda + 2\mu}{\rho}} \tag{12.3}$$

where E = Young's modulus, ν = Poisson's ratio, ρ = density, λ = Lame's first constant, and μ = Lame's second constant = shear modulus.

The shear wave speed β is given by

$$\beta = \sqrt{\frac{\mu}{\rho}} \tag{12.4}$$

where μ and ρ are identical to those in Eq. (12.3).

The surface wave, which is also known as the Rayleigh wave, propagates along the surface of a material and decays very quickly with depth. It has a speed slightly less than the shear wave speed. The ratio C_R/β (C_R is the Rayleigh wave speed, and β is the shear wave speed) varies from 0.87 to 0.95 as the Poission's ratio varies from 0 to 0.5, as shown in Table 12.1.

Elastic waves which have plane wave fronts are called plane waves; similarly, cylindrical and spherical waves have cylindrical and spherical wave fronts, respectively. In an infinite isotropic medium, a point source generates spherical waves and a line source generates cylindrical waves. It can be shown from the Fourier analysis that both spherical and cylindrical waves can be mathematically reconstructed by superimposing a large number of plane waves propagating in different directions.

When a plane P- or S-wave strikes a plane interface between two linear elastic solids, then some of the elastic energy is reflected back and the rest is transmitted into the second material. For P- or S-wave incidence, reflected and transmitted waves in general contain both P- and S-waves. Thus, in the presence of a plane interface or boundary, longitudinal waves can generate shear waves, and *vice versa*. This phenomenon is called *mode conversion*. However, when a P-wave strikes a plane interface vertically (normal

Table 12.1   Variation of C_R/β with the Poisson's ratio ν

ν	0.0	0.05	0.15	0.25	0.30	0.40	0.50
C_R/β	0.87	0.88	0.90	0.92	0.93	0.94	0.95

incidence), then reflected and transmitted energies contain only P-waves. Reflected and transmitted powers in this case are given by

$$\frac{P_r}{P_i} = \left(\frac{Z_1 - Z_2}{Z_1 + Z_2}\right)^2 \qquad (12.5a)$$

$$\frac{P_t}{P_i} = \frac{4Z_1 Z_2}{(Z_1 + Z_2)^2} \qquad (12.5b)$$

where P_r, P_t and P_i are reflected, transmitted and incident powers, respectively. Z_1 and Z_2 are acoustic impedances of materials 1 and 2, respectively. Acoustic impedance Z is the product of density (ρ) and P-wave speed (α), $Z = \rho \alpha$.

If a plane P-wave, propagating in material 1, strikes the interface of materials 1 and 2 at an angle θ_1, measured from the vertical line through the interface, then the reflected S-wave makes an angle γ_1 and transmitted P and S waves in material 2 make angles θ_2 and γ_2, respectively, as shown in Figure 12.1(a). These angles are related to the P- and S-wave speeds of materials 1 and 2 in the following manner:

$$\frac{\sin \theta_1}{\alpha_1} = \frac{\sin \theta_2}{\alpha_2} = \frac{\sin \gamma_1}{\beta_1} = \frac{\sin \gamma_2}{\beta_2} \qquad (12.6)$$

where subscripts 1 and 2 are used to indicate wave speeds in materials 1 and 2, respectively. Equation (12.6) is known as the Snell's law.

Experiment

Ultrasonic waves are generated by mechanical vibrations of piezoelectric crystals, excited by the electric current flow. Piezoelectric crystals are mounted in a casing with some

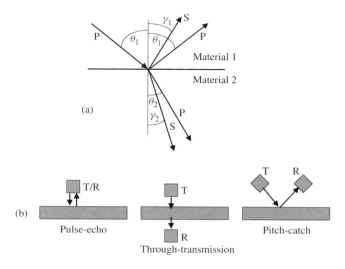

Figure 12.1 (a) Reflected and transmitted P and S waves generated by an incident plane P-wave at the plane interface of two solids; (b) three modes of ultrasonic testing

damping material and protective coatings to build ultrasonic transducers. Detailed transducer design is beyond the scope of this chapter, but interested readers can refer to Krautkramer and Krautkramer (1983), Bray and Stanley (1989), and Silk (1984) for more information.

Transducers generate uniform ultrasonic waves beyond the near field. Within the near field, the waves are not uniform. This phenomenon can be explained by Huygen's principle. For disk-shaped transducer elements with $\lambda \ll D$, the near field length N is approximately given by

$$N = \frac{D^2}{4\lambda} \tag{12.7}$$

where D is the diameter of the transducer and λ is the ultrasonic wave length. For full confidence in an ultrasonic inspection, the specimen should be placed beyond the near field region.

The width of an ultrasonic beam increases as it propagates. This is a function of λ/D. The divergence angle (φ) of the beam can be calculated from

$$\varphi = 2\sin^{-1}\left(1.2\frac{\lambda}{D}\right) \tag{12.8}$$

To inspect a material, ultrasonic transducers are used in one of the three common modes of operation: pulse-echo mode, through-transmission mode and pitch-catch mode (see Figure 12.1(b)). In the pulse-echo mode, one transducer is used both as a transmitter and a receiver. In the through-transmission mode, two transducers are used: one transmits the signal and the second receives the signal – the object under inspection is placed in between these two transducers. In the pitch-catch arrangement, two transducers, one transmitter and one receiver, are used on the same side of the specimen. In the presence of an internal defect the reflected signal changes. Hence, analysing the reflected signal the material can be assessed. For inspecting a specimen, placed horizontally in the xy plane, the transducers are moved horizontally. At every position the transmitter sends the ultrasonic energy into the specimen, and the receiver receives the reflected or transmitted energy depending on the transducer arrangement. When the transducers scan the specimen, if the entire reflected/transmitted wave form is recorded at every position of the transducers and displayed on the oscilloscope screen, then the recording system is called an A-scan recording. However, in most applications only the peak value of the complete reflected/transmitted wave form is recorded and plotted against the x, y coordinates of the receiver. In the presence of a defect, the peak value changes and thus the defective zone becomes visible in the peak value generated image. The ultrasonic image generated in this manner is called the C-scan image. Ultrasonic images of different horizontal cross-sections at various depths of the specimen can be generated by recording the reflected signal at different time intervals or time gates. The vertical cross-sections of the horizontally placed specimen can be imaged by the B-scan recording. In this case, the reflected signal strength is plotted against the time of arrival.

12.2.9 *Acoustic Emission (AE) technique*

Here sensors (transducers) are used to listen to objects under inspection. If cracks have formed inside a material then acoustic energy is released because of the relaxation of the stress at the point of crack formation. The released acoustic energy propagates through the material to its surface, where it can be detected by the transducers mounted there. Hence, formation or propagation of cracks inside the material can be identified. Besides the crack initiation and growth, other mechanisms which can produce acoustic emission are dislocation movements, twinning, phase changes, the fracture of brittle inclusions or surface films, fiber breakage and delaminations in composite materials, chemical activity and cavitation. Emission energies due to these disturbances can range from the motion of a few dislocations in metals to that required to cause catastrophic cracking of structures, such as during an earthquake. It is a passive technique; it simply listens to the sound generated inside the material because of some change occurring there. Hence, it cannot detect a stable crack that is not growing, and the crack size cannot be determined by this technique. The great sensitivity of the AE monitoring system sometimes presents a problem if electrical interference and ambient noise are not properly filtered out of the emission signals. The advantage of this technique is that since it is a passive system, no equipment is required to excite a pulse, and hence the instrumentation is relatively simple and inexpensive. Since the sensors can be located remotely, these can be used when the object is in operation in hostile environments. A relatively large volume of material can be inspected at a reasonable cost; it is also suitable for long-term in-service monitoring. It can locate the crack initiation and growth points, and these regions can then be inspected more carefully by other techniques, such as ultrasonics and radiography, to assess the material in those locations.

AE signals can be detected by accelerometers, piezoelectric transducers, air-gap capacitive transducers, optical/laser sensors, and magnetostriction transducers. This technique has been used in the past to give adequate warning before the catastrophic failure of 2014-T6 aluminum (Hartbower *et al.*, 1971) and fiber-reinforced plastic composites (Mitchell, 1981). In most materials, the number and intensity of the released acoustic signal increases significantly near the failure point, hence it can serve as a good warning tool before the failure. The AE technique has been found to be effective in detecting leaks in above-ground storage tanks (Miller, 1990; Nordstrom, 1990) and in in-flight monitoring of airplanes (Martin, 1980).

12.2.10   *Acousto-ultrasonics*

This term was first used by Vary (Vary and Bowels, 1979; Vary, 1988). Here, two piezo-electric sensors are attached to the surface of the specimen. One of the transducers that acts as a source is excited by an ultrasonic pulser. The other transducer acts as a sensor in a fashion typical of that used for acoustic emission monitoring. Thus, this technique combines the ultrasonic excitation and AE monitoring. Although piezoelectric transducers are the most common type of excitation sources used in acousto-ultrasonic experiments, several other type excitations are also in use. Examples of such sources would include pencil lead breaks, breaking of glass capillaries, pulsed-laser excitation, the tapping of a hammer, and others. These excitations result in acoustic wave propagation within the object. The wave propagation within the object depends upon the material properties and structure of the object. After these waves are detected by the acoustic receivers mounted on the surface of the object, the data is analysed to interpret the condition of the object. The main advantage of the acousto-ultrasonic technique over the acoustic emission technique is that stable cracks which are not growing can also be detected by this technique.

12.2.11   *Moiré interferometry*

The moiré interferometry technique is used to measure displacement, rotation, curvature and strain of a body. In this technique, with the aid of some special patterns known as the moiré patterns, effects of the distortions of the body are magnified to provide a visual picture. The term 'moiré' is a French word that means the patterns observed in watered silks. In moiré interferometry, the visual pattarns are produced by the superposition of two regular motifs that geometrically interfere. Regular motifs can be a set of equispaced parallel lines, rectangular arrays of dots, concentric circles or equispaced radial lines. One needs to understand the basic theory of geometric moiré to measure displacements and strains from the moiré patterns formed by the moiré interferometry. The theory of geometric moiré is described in a simple manner in the following section. For a more detailed description of the geometric moiré and the moiré interferometry, readers are referred to Parks (1993) and Post (1993).

Theory of in-plane moiré

The most common use of moiré patterns is to determine the displacements and strains of a plane parallel to the plane of analysis. In-plane moiré patterns can be formed using very simple gratings, such as equispaced, parallel lines. If one plate remains fixed (reference motif or grating) and the other plate moves with the specimen (specimen motif or grating), then the interference pattern, formed by superimposing the two gratings, changes with the movement of the specimen grating. When the specimen grating moves, rotates or deforms, then alternately dark and bright fringes are formed, as shown in Figure 12.2. These fringe patterns are called the moiré patterns. In Figure 12.2, the line spacing in one grating (reference grating) has been kept unchanged in all four pictures, but the other grating (specimen grating) has been subjected to a 3° rotation (top-left), 20° rotation (top-right), 10% stretching (bottom-left) and 5% stretching after 2° rotation (bottom-right). If the lines in individual patterns are equally spaced and the specimen movement is uniform, then the fringes in the moiré patterns are also equally spaced, as seen in Figure 12.2. The movement of the specimen can be determined from the angle of inclination of the fringes and the distance between the two dark or light fringes (pitch of the moiré pattern).

Basic equations

Let us consider two gratings of pitch (center to center distance between two successive lines) g and g'. If one grating rotates by an angle θ with respect to the other, then it can be shown (Chiang, 1982; Parks, 1993) that the moiré patterns have a spacing of d between the two successive fringe lines and a slope of ϕ, measured from the reference grating of pitch g . d and ϕ are related to g, g' and θ in the following manner:

$$d = \frac{gg'}{\sqrt{g^2 \sin^2 \theta + (g' - g \cos \theta)^2}} \qquad (12.9a)$$

$$\tan \phi = -\frac{g \sin \theta}{(g' - g \cos \theta)} \qquad (12.9b)$$

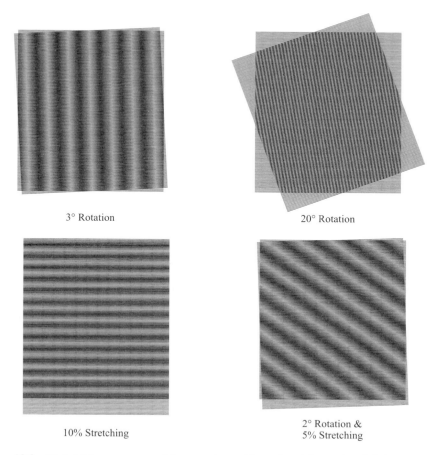

3° Rotation

20° Rotation

10% Stretching

2° Rotation &
5% Stretching

Figure 12.2   Moiré fringes generated by superimposition of undeformed and deformed gratings

Inverting the above relations, one obtains

$$g' = \frac{d}{\sqrt{1 + (d/g)^2 + 2(d/g)\cos\varphi}} \qquad (12.10a)$$

$$\theta = \arctan\frac{\sin\phi}{d/g + \cos\phi} \qquad (12.10b)$$

In the above equations, g is known from the reference grating, and d and ϕ can be measured (Durelli, 1970; Morse, Durelli and Sciammarella, 1961) from the moiré patterns. Hence, one can obtain g' and θ from Eqs. (12.10a) and (12.10b). θ is a measure of the rotation of the specimen and g' is a measure of the stretching of the specimen. After knowing g', the strain is obtained directly as

$$\varepsilon = \frac{g' - g}{g} \qquad (12.11)$$

From Eq. (12.9), one can see that for pure rotation (then $g = g'$),

$$d = \frac{g}{2\sin(\theta/2)} \qquad (12.12)$$

Clearly, for small rotations the denominator of the right-hand side of Eq. (12.12) is small, then d is large. Hence, small rotations can be measured more accurately by moiré patterns. Similarly for pure stretching ($\theta = 0$), Eqs. (12.9a) and (12.9b) can be simplified as

$$d = \frac{gg'}{g' - g} \qquad (12.13)$$

Again, for small stretching the denominator of the right-hand side of the above equation is small, and thus d is large. Hence, small strains can be measured more accurately than large strains from the moiré patterns. For large θ or g' the spacing d becomes small and difficult to measure. For example, when θ is 60° the spacing d is equal to g, for g' equal to $2g$ (100% strain) $d = 2g$. Clearly, if d is as small as g or $2g$, one cannot see it with the naked eyes or measure it accurately. In fact, for rotations which are greater than 30°, and strains greater than 30%, the moiré fringe patterns disappear.

The moiré pattern formed by horizontal gratings is a measure of veritcal displacements, and the moiré pattern formed by vertical gratings is a measure of horizontal displacements. If two moiré patterns, both corresponding to horizontal displacements, are overlapped after giving one pattern a small uniform displacement in the horizontal direction, then a new pattern is formed because of the interference between the two moiré patterns. The new pattern is known as the *super-moiré*, and is a measure of the derivatives of displacements, or strains. If the original grating is in the x-direction, and one moiré pattern is given a shift in the x-direction, then the super-moiré shows the variation of the strain ε_{xx}; if both these directions are y, then the super-moiré shows the variation of ε_{yy}, and if one direction (grating or shift) is x and the other is y, then the super-moiré is a measure of ε_{xy}.

Experimental set-up

To measure material deformation by the moiré interferometry technique, the specimen grating film is attached to the specimen. The reference grating is generated by interfering two laser beams infront of the specimen, as shown in Figure 12.3. In the overlapping

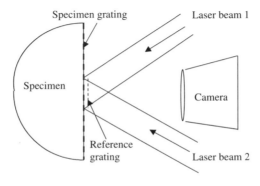

Figure 12.3   Schematic of moiré laser interferometry experimental set up

region of the two laser beams, alternating dark and bright fringes are formed from the varying phase difference of the two arriving beams. A camera is placed in front of the specimen to photograph the combined effect of the two gratings. Due to the material deformation as the specimen grating moves with respect to the reference grating, the moiré pattern seen by the camera changes. From these moiré patterns, photographed by the camera, the material deformation is obtained. A more rigorous description of the moiré interferometry technique can be found in Post (1985, 1993), Sciammarella (1982) and Theocaris (1969).

12.3   Recent Developments

In this section, nondestructive testing techniques which have been recently developed are briefly described.

12.3.1   *Lamb wave technique*

Certain types of internal defects in multilayered solids which cannot be seen very clearly by conventional ultrasonic C-scan imaging techniques can be detected easily when the plate is scanned by leaky Lamb waves (Kundu *et al.*, 1996; Kundu and Maslov, 1997; Maslov and Kundu, 1997). A leaky Lamb wave generated image is denoted as the L-scan image, where L stands for the leaky Lamb wave, to distinguish it from the conventional C-scan image. Chimenti and Martin (1991) first attempted to generate the L-scan image with some success. The first step here is to place two transducers, one transmitter and one receiver, with an angle of inclination, in a pitch-catch arrangement over the specimen, as shown in Figure 12.4, to generate leaky Lamb waves in the specimen. Chimenti and Martin (1991) placed the receiver in the null zone, and then they scanned the specimen. The null zone position changes in presence of an internal defect, so when a defect is encountered the receiver voltage amplitude is altered and the image of the defect is generated. The major problem with this arrangement is that the null zone position is very sensitive to the plate thickness. Hence, a small change in the plate thickness alters the receiver voltage amplitude significantly. To avoid this problem, one needs to filter the L-scan generated data through a special filter, called the MFq filter (see Chimenti and Martin (1991) for a detailed description). This signal processing helps to minimize the effect of the plate thickness variation on the null zone, but apparently retains the sensitivity to defects of interest. Karpur *et al.* (1995) and Kundu *et al.* (1996) avoided the problem imposed by the variation in plate thickness by placing the receiver beyond the null zone and the specularly reflected zone. Thus, only propagating leaky Lamb waves are received by the receiver. With this arrangement the investigators successfully detected certain types of internal defects in multilayered composite plates. Some of these defects which could be seen very clearly in L-scan images were not very clear in C-scan images.

   For efficiently generating the L-scan image, one should first decide which leaky Lamb mode must be used. In the presence of defects such as broken fibers and delamination, the stresses in the defective region are significantly reduced. Hence, if these defects are located in a region where the stress level is high for certain Lamb modes, then the presence of the defects affects these modes significantly by releasing the high stress. These modes

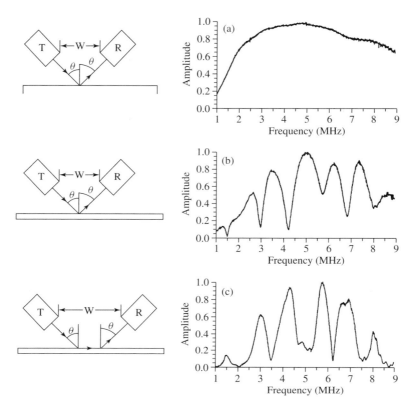

Figure 12.4   Reflected wave spectra (right column) for different transmitter-receiver-specimen orientations shown in the left column. For a thick aluminum block reflector no Lamb waves are generated (a); for a plate in the focus position (b), dips correspond to the leaky Lamb waves, for a plate in the defocus position (c) peaks correspond to the leaky Lamb waves (After Kundu *et al.*, 1995)

would then be most effective in producing the image of the defect. Hence, stress levels inside the plate for various Lamb modes are to be theoretically computed first.

Theory

For computing internal stresses in the plate, one needs to study the mechanics of elastic wave propagation in a multilayered solid. A number of investigators over the last few decades have studied this problem. Thomson (1950), Haskell (1953), Dunkin (1965), Schwab and Knopoff (1970), Kundu and Mal (1985), Mal (1988), Mal, Yin and Bar-Cohen (1991), and Lévesque and Piché (1992), among others, studied various aspects of this problem. A good review of some of these investigations can be found in Brekhovskikh and Godin (1990) and Chimenti (1996). A brief theory of wave propagation in a multilayered solid is given below.

Let us consider the two-dimensional elastic wave propagation problem in the xz plane. The P- and S-wave potentials, denoted by ϕ and ψ, respectively, can be expressed in the forms:

$$\phi = [ae^{ivz} + be^{-ivz}]e^{ikx-i\omega t} \tag{12.14a}$$

$$\psi = [ce^{i\eta z} + de^{-i\eta z}]e^{ikx-i\omega t} \tag{12.14b}$$

where the wave number $k = (\omega/\alpha_0) \sin(\theta_0) = (\omega/\alpha_s) \sin(\theta_1) = (\omega/\beta_s) \sin(\theta_s)$, $v = (\omega/\alpha_s)\cos(\theta_1)$, $\eta = (\omega/\beta_s)\cos(\theta_s)$, ω is the signal frequency in rad/sec, α_0, α_s and β_s are longitudinal wave speed in water, longitudinal wave speed in solid and shear wave speed in solid, respectively. For brevity, in the subsequent expressions the common factor $e^{ikx-i\omega t}$ is omitted.

The particle displacement can be obtained from the wave potentials

$$u_x = ik\phi - \frac{\partial\psi}{\partial z} \tag{12.15a}$$

$$u_z = \frac{\partial\phi}{\partial z} + ik\psi \tag{12.15b}$$

and the stress components are given by

$$\sigma_{xz} = -2\rho k\beta_s^2 \left(\gamma\psi - i\frac{\partial\phi}{\partial z}\right) \tag{12.16a}$$

$$\sigma_{zz} = 2\rho k\beta_s^2 \left(\gamma\phi + i\frac{\partial\psi}{\partial z}\right) \tag{12.16b}$$

$$\sigma_{xx} = 2(\sigma_{zz} + 2ik\rho\beta_s^2 u_x)\left(1 - \frac{\beta_s^2}{\alpha_s^2}\right) - \sigma_{zz} \tag{12.16c}$$

where $\gamma = k - \omega^2/(2k\beta_s^2)$.

For a multilayered plate immersed in water, one can assign a superscript m to displacement and stress components of the mth layer. From the continuity of normal and shear stresses and displacements across the interface between two adjacent elastic layers, one can relate the displacement-stress vectors of any two layers in the following form:

$$\begin{pmatrix} u_x^m \\ u_z^m \\ \sigma_{zz}^m \\ \sigma_{xz}^m \end{pmatrix} = A_m.A_{m+1}.A_{m+2}\ldots A_{n-1} \begin{pmatrix} u_x^n \\ u_z^n \\ \sigma_{zz}^n \\ \sigma_{xz}^n \end{pmatrix} \tag{12.17}$$

where A_m is a 4×4 layer matrix or propagator matrix of the mth layer. The superscripts m and n of the elements of the displacement-stress vectors of Eq. (12.17) correspond to the mth and nth layers, respectively. Expressions for the individual elements of the layer matrix can be found in Thomson (1950), Haskell (1953) or Brekhovskikh and Gobin (1990).

If fluid half-spaces are present above and below the multilayered plate, then wave potentials for the upper and lower half-spaces can be written as

$$\phi_U = e^{-iv_U z} + Re^{iv_U z} \tag{12.18a}$$

$$\phi_L = Te^{-iv_L z} \tag{12.18b}$$

where R and T are plane wave reflection and transmission coefficients of the multilayered plate, immersed in the fluid; subscripts U and L of Y indicate upper and lower fluid half-spaces, respectively.

To compute the stress and displacement components inside the plate, their values at the top and bottom liquid-solid interfaces are to be obtained first. It should be noted here that $\sigma_{xz} = 0$ at both top and bottom liquid-solid interfaces, and u_z, σ_{zz} are continuous across both these interfaces, so that their values can be computed in terms of R and T from the fluid potentials given in Eqs. (12.18a) and (12.18b) and displacement/stress-potential relations given in Eqs. (12.15) and (12.16). Since horizontal displacements on the fluid and solid sides of the interface are not necessarily the same, because of possible slippage at the interface, two horizontal components of displacement at the top and bottom surfaces of the plate are unknown along with R and T. These four unknowns are solved from the four scalar equations given in Eq. (12.17). Thus, one gets

$$R = \frac{M_{32} - i\omega Z_L M_{33} + i\omega Z_u(M_{22} - i\omega Z_L M_{23})}{M_{32} - i\omega Z_L M_{33} - i\omega Z_U(M_{22} - i\omega Z_L M_{23})} \tag{12.19a}$$

$$T = \frac{-2i\omega Z_L \rho_U/\rho_L}{M_{32} - i\omega Z_L M_{33} - i\omega Z_U(M_{22} - i\omega Z_L M_{23})} \tag{12.19b}$$

where $M_{ij} = J_{ij} - J_{i1} J_{4j}/J_{41}$, J is a 4×4 matrix obtained by multiplying layer matrices of all layers. $Z_L = \rho_L \alpha_L/\cos\theta_L$ and $Z_U = \rho_U \alpha_U/\cos\theta_U$ are the acoustic impedances of the lower and upper fluid half spaces, respectively. After computing the stress and displacement components at the top and bottom surfaces of the plate, their values at any other interface inside the plate can be obtained from Eq. (12.17). To get these values at a point within a layer, an artificial boundary can be introduced that goes through that point, and then Eq. (12.17) can be used.

It is well known that the above formulation works well for low frequencies or thin plates, however, they suffer from a severe numerical loss-of-precision problem as the frequency or the plate thickness increases. To avoid the precision problem at high frequencies the delta matrix modification (Dunkin, 1965; Kundu and Mal, 1985; Lévesque and Piché, 1992; Castings and Hosten, 1994) must be adopted. In the delta-matrix technique, the 4×4 matrices of Eq. (12.17) are replaced by their 6×6 delta matrices whose elements are formed by the determinants of the 2×2 submatrices within the original 4×4 matrices.

Experiment

The first step of generating an L-scan image is to produce propagating leaky Lamb waves in the specimen. For this purpose two transducers are placed in a pitch-catch arrangement over the plate specimen, as shown in Figure 12.4. The transmitter is excited by the tone burst excitation. The excitation frequency is then varied continuously from a minimum value to a maximum value within the bandwidth of the transducers. The reflected signal is received by the receiver; the signal amplitude is then displayed on an oscilloscope screen as a function of the frequency. When the transmitters and the reflector are positioned such that the reflected energy is maximum or very close to maximum, then the reflected amplitude spectrum has shapes, as shown in Figures 12.4(a) or 12.4(b). If no Lamb waves are generated, then the spectrum looks like the plot shown in Figure 12.4(a), which has been generated by a couple of broadband 5 MHz transducers taking an aluminum block as the reflector, thus no Lamb waves are excited in the specimen. However, if leaky Lamb waves are generated at some frequencies, then at those frequencies dips are observed, as shown in Figure 12.4(b), which is produced by the same set of transducers at the same orientation, but the aluminum block is replaced by a thin composite plate, where leaky Lamb waves are excited at some frequencies. If the transducers are moved further down (i.e. the specimen-transducer distance is reduced without altering the distance between the two transducers), then because of defocusing, the reflected amplitude spectrum changes its shape and magnitude. In this defocused position peaks are observed at the frequencies producing the Lamb waves, as shown in Figure 12.4(c). This is because when the defocusing is high, the specularly reflected beam cannot reach the receiver, but the leaky waves can. That is why we observe peaks in Figure 12.4(c) at the frequency values for which we observed dips in Figure 12.4(b). Hence, the frequencies corresponding to the leaky Lamb wave modes can be obtained from the dips (Figure 12.4(b)) or peaks (Figure 12.4(c)) of the reflected signal spectra. Then the leaky Lamb wave speed or the phase velocity can be obtained from Snell's Law:

$$C_L = \frac{\alpha_U}{\sin \theta} \tag{12.20}$$

where C_L is the Lamb wave phase velocity, α_U is the longitudinal wave speed in the coupling fluid (for water it is equal to 1.49 km/sec), and θ is the angle of inclination of the transducers, i.e. the angle between the vertical axis and the transducer axis.

If the transducer angle is changed, the corresponding leaky Lamb wave speed is changed, hence the dip positions along the frequency axis vary. Thus, the Lamb wave dispersion curves can be experimentally generated by monitoring the transducer angles and dips of the reflected signal spectra. After selecting a specific leaky Lamb mode for scanning, that mode is generated by first setting the transducer angle (θ), then proper defocusing is done by vertically moving the transducers and observing the leaky Lamb wave peaks, and finally the frequency is set at a value corresponding to a leaky Lamb wave peak of interest. The specimen is then scanned with this transmitter-receiver arrangement.

The specimen which is scanned by leaky Lamb waves is a five-layer composite plate is fabricated with internal defects. This five-layer metal matrix composite plate has a dimension of $80 \times 33 \times 1.97 \text{ mm}^3$. Five layers or plies of SCS-6 fibers in Ti-6Al-4V matrix are oriented in $0°$ and $90°$ directions in alternate layers. SCS fibers are produced by Textron Inc. This fiber has a carbon core of about 25 µm in diameter; two concentric layers of silicon carbide surround the carbon core; and finally, two very thin (couple of microns or so) layers of carbon coating are placed on the outside. The overall fiber diameter is about 152 µm. The fibers in the top, bottom and middle layers are oriented in the $0°$ direction or along the length of the plate. The other two plies are in the $90°$ direction, or along the width of the plate. The composite is made by foil-fiber-foil technique. The internal flaws, shown in Figure 12.5, are intentionally introduced in the plate during the fabrication process. The first and the fifth layers of fibers do not have any flaw. The left part of the second layer fibers ($90°$) is coated with boron nitride to impede the formation of good bonding between the fibers and the matrix, as schematically shown in Figure 12.5. The fibers in the third layer ($0°$) are intentionally broken near the middle. The fourth layer ($90°$) has two areas of missing fibers. On the left side, five fibers and on the right side ten fibers are removed. The missing fiber zones have approximate widths of 1.2 mm and 2.4 mm on the left- and right-hand sides, respectively.

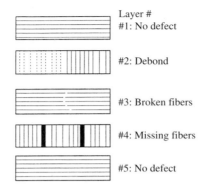

Figure 12.5   Schematic of defects in different layers in the five-layer plate specimen (after Kundu *et al.*, 1995)

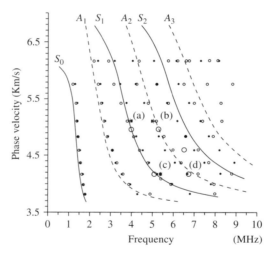

Figure 12.6   Theoretical dispersion curves for a 1.97 mm thick plate whose $\alpha_s = 6.85$ km/sec, $\beta_s = 3.62$ km/sec, and $\rho = 4.1$ gm/cc are shown by solid lines (symmetric modes) and dotted lines (antisymmetric modes). Circles ($\circ$) and stars ($*$) are experimentally obtained leaky Lamb wave dispersion curves when the wave propagates across (90°) and along (0°) the top layer fiber direction respectively. Four bigger circles marked as (a), (b), (c) and (d) show the frequency-phase velocity combinations used for generating L-scan images (after Maslov and Kundu, 1997)

Numerical and experimental results

The average longitudinal wave speed (α_s), shear wave speed (β_s) and density (ρ) of the plate are found to be $\alpha_s = 6.85$ km/sec, $\beta_s = 3.62$ km/sec and $\rho = 4.1$ gm/cc. Figure 12.6 shows the Lamb wave dispersion curves for this plate. The solid and dotted lines correspond to the symmetric and antisymmetric modes obtained theoretically with the measured values of α_s, β_s and ρ. The circles ($\circ$) and stars ($*$) in Figure 12.6 are experimentally obtained data points that correspond to the propagation of different leaky Lamb wave modes along 90° (perpendicular to the fiber direction of the top layer) and 0° (parallel to the fiber direction of the top layer) directions, respectively. It should be noted in this figure that at lower frequencies (less than 6 MHz), experimental points match very well with the theoretical curves.

To study which mode is most efficient for detecting defects at a specific depth, the stress components inside the plate are computed for two different Lamb modes, S_1 and A_2. Figures 12.7 and 12.8 show σ_{zz} (a) and σ_{xz} (b) for S_1 and A_2 modes, respectively. It should be noted here that stress and velocity components are plotted against depth and phase velocity. Since frequency changes with the phase velocity for every Lamb mode (see dispersion curves of Figure 12.6), the frequency corresponding to different phase velocities are different, and can be obtained from the dispersion curves. It should be noted that for the antisymmetric mode (Figure 12.8), σ_{zz} is small near the middle of the plate, but σ_{xz} is large in that region. The situation is reverse for the symmetric mode (Figure 12.7). It should also be noted that for a phase velocity slightly larger than 5 km/sec, the σ_{xz} component is almost zero for the entire plate for both modes.

In Figure 12.9 four images are shown. These images are generated by propagating leaky Lamb waves in the 90° direction (across the fiber direction of the top layer). The top two images correspond to the S_1 and bottom two correspond to the A_2 modes. Corresponding points (frequency and phase velocity) for these four images on the dispersion curves are shown in Figure 12.6 by four circles marked by (a), (b), (c) and (d). For (a) and (b) the incident angle is 17.5° (corresponding phase velocity = 4.96 km/sec), and for (c) and (d) the incident angle is 21° (corresponding phase velocity is 4.16 km/sec). What is interesting to note here is that, for the same symmetric mode (S_1), image (a) does not show the internal defects very clearly, except for the broken fibers, the dark band zone, marked by an arrow, near the middle of the plate, but image (c) shows missing fibers,

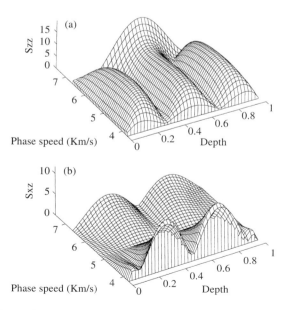

Figure 12.7   Variations of σ_{zz} (a) and σ_{xz} (b) for the first symmetric (S_1) mode of Lamb wave propagation in the plate (after Maslov and Kundu, 1997)

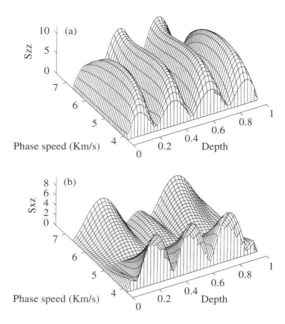

Figure 12.8   Variations of σ_{zz} (a) and σ_{xz} (b) for the second anti-symmetric (A_2) mode of Lamb wave propagation in the plate (after Maslov and Kundu, 1997)

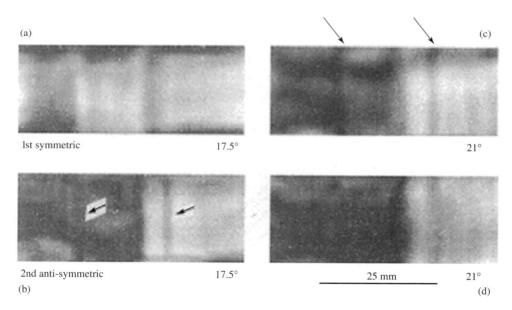

Figure 12.9   L-scan images generated by S_1 (top) and A_2 (bottom) modes of Lamb wave propagation. Incident angles are 17.5° (left column) and 21° (right column) (after Maslov and Kundu, 1997)

marked by two arrows, as well as the debond zone, the dark region on the left-hand side. The A_2 mode, on the other hand, shows the debond zone and missing fibers marked by two arrows in image (b) clearly. However, missing fiber zones are not so clear in image (d) although the Lamb mode is still A_2. These observations can be easily explained by the internal stress patterns shown in Figures 12.7 and 12.8.

From Figure 12.7(b), one can see that for the S_1 mode, near the phase velocity of 5 km/sec σ_{xz} is almost zero. Figure 12.7(a) shows that in second (depth 0.2–0.4) and fourth (depth 0.6–0.8) layers, normal stress σ_{zz} is also very small, close to zero. Hence, defects in the second and fourth layers (such as missing fibers) cannot be seen in image (a). However, σ_{zz} is comparatively large in the middle layer (depth 0.4–0.6), and the broken fiber defect in this layer can be seen by the vertical dark band near the middle of the plate. Image (c) corresponds to the same Lamb mode as (a), but for this image the phase velocity is 4.16 km/sec. This image shows missing fibers and debond zone more clearly. This is because σ_{xz} is large in second (depth 0.2–0.4) and fourth (depth 0.6–0.8) layers for the phase velocity equal to 4.16 km/sec (Figure 12.7(b)). Debond and missing fibers release this stress, and hence affect the reflected signals significantly. That is why missing fibers and the debond zone can be clearly seen in this image. Missing fibers and the debond zone can be seen in image (b) also. This is because in Figure 12.8(a) one can see that for the A_2 mode near the phase velocity of 5 km/sec, σ_{zz} is large at depth 0.4 (the debond position), it is also large in the fourth layer (depth 0.6–0.8) that contains the missing fibers. Thus, although for the phase velocity 4.96 km/sec the shear stress component σ_{xz} is small (Figure 12.8(b)), the defects could be seen due to the release of σ_{zz} component of stress in presence of the defects. Finally, in image (d) one can see the debond zone, but missing fibers are not very clearly seen. In this case, in the fourth layer, the layer containing the missing fibers, σ_{xz} varies strongly from a minimum value near the top of the layer to a large value at the bottom of the layer (see Figure 12.8(b)). In this case, possibly the combined effect of σ_{zz} and σ_{xz} blurred the image of the missing fibers. However, here the image of the debond zone can be clearly seen because at the debond position, at depth 0.4, σ_{zz} is large (Figure 12.8(a)) and σ_{xz} is small (Figure 12.8(b)). Debond releases σ_{zz}, producing a clear image of the debond zone.

This experiment shows that to generate clear images of defects at a specific depth of a multilayered solid, one not only needs to specify the appropriate leaky Lamb mode, but also the correct frequency (or phase velocity). Completely different images are generated when the signal frequency is changed without changing the leaky Lamb wave propagation mode. The main advantage of this technique is that it is very effective in detecting defects in multilayered plate type structures, such as pavements. However, the experimental setup here is more complicated than conventional ultrasonic C-scan type setup.

12.3.2 *Acoustic microscopy*

Since the acoustic microscope was first introduced by Lemons and Quate (1973), it has been used widely over the last two decades for different purposes. Material scientists use this apparatus for characterization of both isotropic (Atalar, 1979; Briggs *et al.*, 1982) and anisotropic materials (Kushibiki *et al.*, 1982, 1991; Hildebrand and Lam, 1983; Kim *et al.*, 1992). Development of line focus acoustic lenses (Kushibiki *et al.*, 1982, 1991) helped scientists in studying material anisotropy by acoustic microscopes. Using this instrument, scientists also measure the coating thickness and its properties (Weglein and Mal,

1991; Kundu, 1992; Yu and Boseck, 1994), characterize different types of bonds such as adhesive, diffusion and kissing bonds (Sinclair, 1980; Derby *et al.*, 1983; Cognard, 1990; Nagy, 1991), detect surface and internal defects in metals and ceramic materials including semiconductors, ceramic capacitors, electronic packaging, electronic circuit boards and Integrated Circuit (IC) chips (Weglein, 1983; Miller, 1985; Lawrence *et al.*, 1990; Moore, 1992; De Liso *et al.*, 1993), obtain residual stress patterns (Meeks *et al.*, 1989), surface roughness (Gopalan *et al.*, 1989), and plastic deformations (Ishikawa *et al.*, 1989; Weaver and Briggs, 1985) in materials. Different types of materials characterized by this instrument are ceramics (Clarke *et al.*, 1985), polymers (Tucker and Wilson, 1980; Maev, 1988; Fagan *et al.*, 1989), composites (Karim and Kundu, 1989; Lawrence *et al.*, 1993), High Temperature Super Conducting (HTSC) materials (Bukhny *et al.*, 1990), fiber optics (Jen *et al.*, 1989; Gadomski and Boseck, 1990), rocks (Rodriquez and Briggs, 1990), and biological tissues and cells (Ueno and Itoth, 1981; Pavin *et al.*, 1990; Kundu *et al.*, 1991, 1992; Scherba *et al.*, 1994).

In addition to characterizing the solid and viscous materials, acoustic microscopes are also useful for characterizing the liquid, which is used as the coupling fluid at the lens tip. It can measure sound velocity in a liquid very accurately (10 ppm) (Guillon, 1989). Low temperature application of this instrument is also being investigated. Yamanaka *et al.* (1990) showed that its sensitivity at $-30°C$ is much better than that at the ambient temperature. They designed a Variable Low Temperature Scanning Acoustic Microscope (VLTSAM) using a methanol coupler which is capable of operating between $+30°C$ and $-94°C$. Today a maximum resolution of 15 nm can be achieved by an acoustic microscope using a 15.3 GHz acoustic lens in the liquid helium coupling fluid near $0°K$ temperature (Rugar *et al.*, 1982; Hadimioglu and Foster, 1984; Moulthrop *et al.*, 1992). With this microscope one can image objects which show little or no contrast on a scanning electron microscope. Conventionally acoustic microscopes are used to characterize surfaces and near surface regions, because they penetrate into the material a distance of the order of Rayleigh wavelength (Atalar, 1985). To improve the penetration property of the acoustic microscope, one can reduce the acoustic signal frequency (Gilmore *et al.*, 1986; Kundu, 1988; Awal *et al.*, 1992), low frequency (10–100 MHz) acoustic microscopes have been popular for many industrial applications.

Because of the wide application of this instrument, it has been extensively used in both materials engineering and medical science, and more than a thousand research papers have been published on this subject. All these works cannot be reviewed here, but a comprehensive list may be obtained from several review articles and books (Khuri-Yakub, 1993; Briggs, 1992).

Working principle

A schematic diagram of the acoustic microscope is shown in Figure 12.10(a). The transducer T is mounted on the top of a lens rod, that is typically made of sapphire for high frequency (>100 MHz) or quartz for low frequency (<100 MHz) microscopes. Longitudinal ultrasonic waves are generated by the transducer T. The waves propagate through the lens rod and are then focused to a point (by a concave spherical lens located at the end of the lens rod) or to a line (by a concave cylindrical lens). When a specimen is placed between the lens and the focal point of the converging beams, then the beams are reflected by the specimen and go back to the transducer after traveling through the

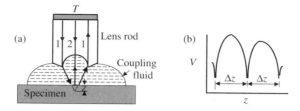

Figure 12.10   (a) Schematic of an acoustic microscope, (b) a typical $V(z)$ plot

coupling fluid and passing through the lens-fluid interface twice. The lens rod geometry is such that only two types of ultrasonic beams can return to the transducer. The beam which propagates along the central axis of the lens (marked by number 2 in Figure 12.10(a)) do not deviate from its path, and is reflected back to the transducer following the same path, as shown in the figure. The beam 1 that strikes the specimen at the Rayleigh critical angle (Weglein, 1979; Bertoni, 1984; Kundu *et al.*, 1985; Kundu and Mal, 1986; Kundu, 1988, 1992) generates leaky Rayleigh waves along the coupling fluid-specimen interface, and then goes back to the transducer. The transducer now works as a receiver, which receives two reflected beams having different phases. The phase difference between these two beams changes with the defocus distance, the distance between the lens focal point and the reflecting surface. Depending upon the phase difference between these two beams, constructive or destructive interference may result. Hence, the receiver voltage (V) versus the defocus distance (z) shows an oscillating pattern as shown in Figure 12.10(b). This plot is commonly known as the $V(z)$ curve. Peaks and dips in the $V(z)$ curve correspond to the constructive and destructive interference positions, respectively. The Rayleigh wave speed C_R in the specimen can be related to the spacing distance Δz in the following manner (Weglein, 1979; Kundu and Mal, 1985):

$$C_R = \alpha_f \left[\frac{\alpha_f}{f\,\Delta z} - \left(\frac{\alpha_f}{2f\,\Delta z} \right)^2 \right]^{-1/2} \tag{12.21}$$

where f is the signal frequency and α_f is the longitudinal wave speed in the coupling fluid.

Thus, one elastic property of the material can be easily obtained from the above equation. To obtain other properties such as the P-wave speed, Young's modulus, ultrasonic attenuation coefficient, or film thickness, one needs to invert the complete $V(z)$ curve by the simplex algorithm (Kundu, 1992) or similar optimization techniques.

Figures 12.11 and 12.12 show an acoustic image of a biological cell and its property variation along a scan line. It has been obtained at room temperature using a high frequency (1 Ghz) Leica acoustic microscope (ELSAM).

12.3.3   *Ultrasonics for geologic materials*

Although the field of ultrasonics has been well developed, its applications have been limited mostly to the characterization of metals, ceramics, fiber reinforced polymers, and ceramic/metal matrix composites. Its use to assess geologic materials such as soil or rock has been meager. This is because high frequency (MHz) ultrasonic waves commonly used

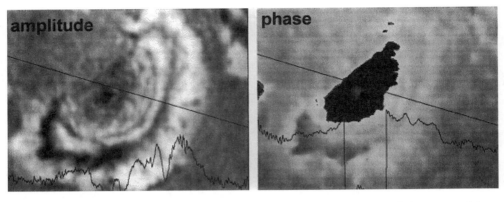

Figure 12.11   Acoustic image of a biological cell on a silicone substrate. It is generated by a
1 Ghz Leica acoustic microscope. The left-hand figure is the amplitude image, and
the right-hand figure is the phase image. The black line in the amplitude image
shows the scan line along which the cell profile and acoustic properties of the cell
are computed and shown in Figure 12.12 (after Kundu *et al.* 2000)

in ultrasonics testings attenuate very quickly inside soil and damaged rocks. In the past,
geologists have used low frequency (a few Hz) elastic waves to detect big faults and
natural resources such as oil and coal inside the earth. Use of ultrasonic waves to detect
internal macro-cracks in a rock sample is also common practice, because these waves
can propagate a long distance through intact rocks and then is reflected by macro-cracks.
However, when a large number of micro-cracks appear inside a rock/soil sample, one
cannot distinguish among individual cracks. To assess such samples, it is important to
know the crack density instead of the size and location of individual cracks. P-wave speed
(α_0) in the intact material and (α_d) in the damaged material can be related to the crack
density (C_d) of the damaged material in the following manner (Hudson, 1981; Sayers
et al., 1990):

$$C_d = \frac{\dfrac{15(1 - v_0)}{2(1 - 2v_0)}\left(1 - \dfrac{\alpha_d^2}{\alpha_0^2}\right)}{\left(\dfrac{32(1 - v_0)}{3(2 - v_0)} + \dfrac{8(1 - v_0)(3 - 2v_0 + 7v_0^2)}{3(1 - 2v_0)^2}\right)} \tag{12.22}$$

where v_0 is the Poisson's ratio of the intact material. Thus, the crack density in a damaged
material can be obtained from the Poisson's ratio of the intact state and ultrasonic wave
speeds in the damaged and intact states. Often, stress-induced damage is not isotropic,
and hence the P-wave speed in x, y and z directions are different. In that case, α_d is the
rms value of the wave speeds in three mutually perpendicular directions,

$$\alpha_d = \sqrt{\frac{\alpha_x^2 + \alpha_y^2 + \alpha_z^2}{3}} \tag{12.23}$$

Desai *et al.* (1995) measured the crack density as well as the stress-induced anisotropies
(A_α and A_a) in a simulated rock sample from the velocity and attenuation of P-waves in
three mutually perpendicular directions by defining the material anisotropies A_α and A_a

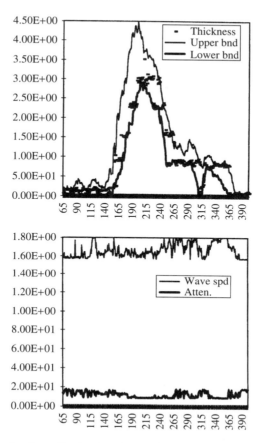

Figure 12.12   Computed cell profile (top) and acoustic properties (bottom) of the biological cell
along the scan line shown in Figure 12.11. Thin and thick curves of the top figure
are upper and lower limits of cell thickness, and dots are converged values from
the computation. Cell thickness is shown in micron. Ultrasonic wave speed (thinner
curve of the bottom figure) is in km/s and attenuation (thicker curve of the bottom
figure) is in $(\text{Mhz}^{-2}\text{m}^{-1})$. Note that both average wave speed and attenuation are
relatively smaller in the thicker portion (nucleus) of the cell

in the following manner:

$$A_\alpha = \sqrt{\sum_{i=1}^{3}\left[1 - \left(\frac{\alpha_i}{\alpha_a}\right)\right]^2}$$

$$A_a = \sqrt{\sum_{i=1}^{3}\left[1 - \left(\frac{\alpha_i}{\alpha_\alpha}\right)\right]^2} \tag{12.24}$$

where α_i and a_i are wave speeds and attenuations in three directions (x, y and z) in the
damaged material, and α_a and a_a are average wave speed and attenuation, respectively.

These two definitions of material anisotropy show similar variations of the anisotropy as a function of the applied load, although their numerical values differ (Desai *et al.*, 1995). It gives a quick and efficient way of measuring stress induced anisotropy in a material.

Damage and other types of disturbance (D_v) inside a material, subjected to external loadings, have been defined by Desai and Toth (1996) in terms of ultrasonic wave speeds through the material in the following manner:

$$D_v = \frac{V_i - V_a}{V_i - V_c} \tag{12.25}$$

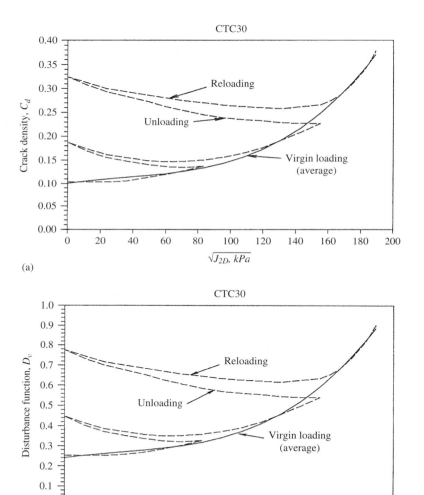

(a)

(b)

Figure 12.13   Variations of C_d and D_v for CTC 30–cemented sand. (a) $\sqrt{J_{2D}}$ vs C_d; (b) $\sqrt{J_{2D}}$ vs D_v. (after Desai and Toth, 1996)

where V_i is the wave speed in the intact material, V_c is the wave speed in the fully adjusted or completely disturbed state of the material, and V_a is the wave speed in the current state of the material. The disturbance function (D_v), defined by Eq. (12.25), when plotted against applied load, shows similar variation as the crack density. C_d and D_v have been plotted in Figure 12.13 for a conventional triaxial compression test. Similarities between the two plots are obvious, and it proves that Eq. (12.25) can correctly measure the disturbance/damage of the material. It is interesting to see in Figure 12.13 that the crack density and disturbance function increases continuously with the virgin loading. During unloading it increases further, partly due to opening of the cracks as the stress is reduced. Upon reloading, the disturbance function decreases, and then on reaching the point of unloading and virgin loading, it increases. Similar trends have been observed by Sayers *et al.* (1990) and Desai *et al.* (1995). Thus, ultrasonic wave speeds and attenuation can provide an indirect measure of damage and induced anisotropy in a material.

12.4   Concluding Remarks

Nondestructive testing is an important step of material characterization. In spite of the development of many advanced NDT techniques for characterizing metals and composites for so-called high-tech (electrical and aeronautical) applications, use of such advanced techniques for assessing geologic and other materials used for civil engineering applications have been meager. This is because most rocks and soils are comparatively less expensive, and did not require sophisticated characterization in the past. However, demands on these materials are increasing, and their characterization beyond simple Hooke's law and Von-Mises' criterion is necessary for efficient use of these materials. Some new techniques, such as the Lamb wave technique, developed for characterizing multilayered composites can, in principle, be extended to assess multilayered pavements. However, extending NDT techniques which are being used for characterizing mostly homogeneous metals and composites to very inhomogeneous and highly irregular soils and rocks is not an easy task. Hence, there is a good potential for research and development in this area. Investigators are currently applying nondestructive testing techniques beyond its traditional application of looking inside materials for detecting internal material defects. Now, many investigators are interested in material property determination, quantitative characterization of internal structures, and remaining life prediction of a structure by modern NDT techniques.

Acknowledgment

This work was partially supported by a grant from the National Science Foundation, CMS-9523349 and NATO Linkage grant, HTECH.LG931353. The author would like to thank Drs P. Karpur, T. E. Matikas and P.D. Nicolaou for providing the composite specimen. The acoustic image of the cell was obtained in Dr Bereiter-Hahn's laboratory in Germany.

References

Atalar, A. Material characterization in acoustic microscopy theory. *J. Appl. Phys.*, **50**, (1979) 8237–8239.

Atalar, A. Penetration depth of the scanning acoustic microscope. *IEEE Trans. Sonics and Ultrasonics*, **32**, (1985) 164–167.

Awal, M. A., Mahalanobis, A. and Kundu, T. Low frequency acoustic microscopy and pattern recognition for studying damaged and anisotropic composites and material defects. *J. Nondestr. Eval.*, **11**(1), (1992).

Bar-Cohen, Y. and Mal, A. K. Ultrasonic inspection In: *Nondestructive Evaluation and Quality Control, Metals Handbook*, 17, 9th ed. ASM International, Matrials Park, OH, (1989) 231–277.

Beller, W. *Missiles-Rockets*, **16**, (1965) 22–27.

Bennett, S. and Swartzendruber, L. Nuclear magnetic resonance. In: *Materials Characterization, Metals Handbook*, Vol. 10, 9th ed. ASM International, Materials Park, OH, (1986) 277–286.

Bertoni, H. L. Ray-optical evaluation of $V(z)$ in the reflection acoustic microscope. *IEEE Trans. Sonics and Ultrasonics*, **31**, (1984) 105–116.

Borucki, J. S. and Jordan, G. Liquid penetrant inspection. In: *Nondestructive Evaluation and Quality Control, Metals Handbook*, Vol.17, 9th ed. ASM International, Materials Park, OH, (1989) 71–88.

Bray, D. E. and McBride, D. *Nondestructive Testing Techniques*. Wiley, New York, (1992).

Bray, D. E. and Stanley, R. K. *Nondestructive Evaluation*. McGraw-Hill, New York, (1989).

Briggs, G. A. D. *Acoustic Microscopy*. Oxford University Press, (1992).

Briggs, G. A. D., Illett, C. and Somekh, M. G. Acoustic microscopy for material studies. In: *Acoustical Imaging*, Ash, E. A. and Hill, C. R. (eds.). Plenum Press, London, (1982) 89–99.

Brekhovskikh, L. M. and Gobin, O. A. *Acoustics of Layered Media, Plane and Quasi Plane Waves*. Springer-Verlag, Berlin, (1990) 87–103.

Bukhny, A., Chernosatonskii, L. A. and Maev, R. G. Acoustic imaging of high temperature superconducting materials. *J. Microscopy*, **160**, (1990) 299–313.

Castings, M. and Hosten, B. Delta Operator technique to improve the Thomson–Haskell method stability for propagation in multilayered anisotropic absorbing plates. *J. Acoust. Soc. Am.*, **95**, (1994) 1931–1941.

Caulfield, H. J. *Handbook of Optical Holography*. Academic Press, New York, (1979).

Chiang, F. P., In: *Manual of Engineering Stress Analysis*, Kobayashi, A. S. Prentice-Hall, Englewood Cliffs, NJ, (1982).

Chimenti, D. E. Guided waves in plates and their use in materials characterization. *Appl. Mech. Rev.*, in press (1996).

Chimenti, D. E. and Martin, R. W. Nondestructive evaluation of composite laminates by leaky Lamb waves. *Ultrasonics*, **29**, (1991) 13–21.

Clarke, L. R., Chou, C.-H., Khuri-Yakub, B. T. and Marshall, D. B. Acoustic evaluation of grinding damage in ceramic materials. *IEEE Ultrason. Symp*, New York, (1985) 979–982.

Cognard, J., Sathish, S., Kulik, A. and Gremaud, G. Scanning acoustic microscopy of the cellular structure of the interphase in a metal adhesive bond. *J. Adhes.*, **32**, (1990) 45–49.

De Liso, G., Muschitello, M. and Stucchi, M. Failure analysis of encapsulated electronic devices by means of scanning ultrasonic microscopy technique. *Scanning*, **15**, (1993) 236–242.

Derby, B., Briggs, G. A. D. and Wallach, E. E. Non-destructive testing and acoustic microscopy of diffusion bonds. *J. Mater. Sci.*, **18**, (1983) 2345–2353.

Desai, C. S., Jagannath, S. V. and Kundu, T. Mechanical and ultrasonic anisotropic response of soil. *ASCE J. Engrg. Mech.*, **121**(6), (1995) 744–752.

Desai, C. S. and Toth, J. Disturbed state constitutive modeling based on stress-strain and nondestructive behavior. *Int. J. Solids Structures*, **33**(11), (1996) 1619–1650.

Dunkin, J. W. Computation of modal solutions in layered elastic media at high frequencies. *Bull. Seismological Soc. of Am.*, **55**, (1965) 335–358.

Durelli, A. J. and Parks, V. J. *Moiré Analysis of Strain*. Prentice-Hall, Englewood Cliffs, NJ, (1970).

Fagan, A. F., Bell, J. M. and Briggs, G. A. D. Acoustic microscopy of polymers and polymer composites. In: *Fractography and Failure Mechanisms of Polimers and Composites*, Roulin-Moloney, A. C. (ed.). Elsevier, Oxford, (1989) 213–230.

Fisher, J. L. Remote field eddy current In: *Metals Handbook, Nondestructive Evaluation and Quality Control*, Vol. 17. American Society for Metals, Metals Park, OH, (1989) 195–201.

Florant, L. *Instrument Society of America*, **11**, (1964) 61–66.

Gadomski, A. and Boseck, C. Scanning acoustic microscopy. A possibility of application in investigation of optical glasses and fibers. *Optica Applic.*, **20**, (1990) 303–320.

Gilmore, R. S., Tam, K. C., Young, J. D. and Howard, D. R. Acoustic microscopy from 10 to 100 MHz for industrial applications. *Phil. Trans. Royal Soc.*, **A320**, (1986) 215–235.

Gopalan, K., Roberts, R. A., Markovich, J. G. and Vaitekunas, J. J. High-speed digital peak detector and averager for acoustic microscopy. *IEEE Instrumentation and Measurement Technology Conf.*, IEEE, Piscataway, NJ, (1989) 54–56.

Guillon, F. Measurement of sound velocity in liquid using the acoustic material signature. *Ultrasonics*, **27**(1), (1989) 26–30.

Hadimioglu, B. and Foster, J. S. Advances in superfluid helium acoustic microscopy. *J. Appl. Phys.*, **56**, (1984) 1976–1980.

Halmshaw, R. *Industrial Radiography*. Applied Science, London, (1982).

Halmshaw, R. *Nondestructive Testing*. Arnold, London, (1987).

Hardy, G. and Bolen, J. Therman Inspection. In: *Metals Handbook: Nondestructive Testing and Quality Control*, Vol.17, 9th ed. ASM International, Materials Park, OH, (1989).

Hartbower, C. E., Reuter, W. G., Morais, C. F. and Crimmins, P. P. *Correlation of Stress Wave Emission Characteristics with Fracture in Aluminum Alloys*. Aerojet Solid Propulsion Company, (1971).

Haskell, N. A. The dispersion of surface waves on multilayered media. *Bull. Seismological Soc. Am.*, **43**, (1953) 17–34.

Hildebrand, J. A. and Lam, L. K. Directional acoustic microscopy for observation of elastic anisotropy. *Appl. Phys. Lett.*, **42**, (1983) 413–415.

House, W. V. and Lauterbur, P. C. *Nuclear Magnetic Resonance Zeugmatography Imaging*, 356/SPIE, Vol.173, *Application of Optical Instrumentation in Medicine VII*, (1979).

Hudson, J. A. Wave-speed and attenuation of elastic waves in material containing cracks. *Geophys. J. Roy. Astron. Soc.*, **64**, (1981) 133–150.

Ishikawa, I., Semba, T., Kanda, H., Katakura, K., Tani, Y. and Sato, H. Experimental observation of plastic deformation areas, using an acoustic microscope. *IEEE Trans. on Ultrason. Ferroelectric and Freq. Controls*, **36**, (1989) 274–279.

Jen, C. K., Neron, C., Bussiere, J. F., Li, L., Lowe, R. and Kushibiki, J. Characterization of cladded glass fibers using acoustic microscopy. *Appl. Phys. Lett.*, **55**, (1989) 2485–2487.

Karim, M. R. and Kundu, T. Acoustic material signature of fiber reinforced composite solids. *Composite Material Technology, Twelfth Annual Energy Sources Technology Conf. & Exhibition*, Houston, TX, January 22–25 (1989) 131–134.

Karpur, P., Benson, D. M., Matikas, T. E., Kundu. T. and Nicolaou, P. D. An approach to determine the experimental transmitter-receiver geometry for the reception of leaky lamb waves. *Materials Evaluation*, **53**, (1995) 1348–1352.

Khuri-Yakub, B. T. Scanning acoustic microscopy. *Ultrasonics*, **31**, (1993) 361–372.

Kim, J. O., Achenbach, J. D., Mirkarimi, P. B., Shinn, M. and Barnett, S. A. Elastic constants of single-crystal transition-metal nitride films by line-focused acoustic microscopy. *J. Appl. Phys.*, **72**, (1992) 1805–1811.

King, J. D., Rollwitz, W. L. and Matzkanin, Magnetic resonance methods for NDE. In: *Proc. 12th Symp. on NDE*, Kundu, T. (ed.). *Acoustic Microscopy at Low Frequency: ASME J. Appl. Mech.*, **55**, (1988) 545.

Krautkramer, J. and Krautkramer, H. *Ultrasonic Testing of Materials, 3rd ed.* Springer-Verlag, Berlin, (1983).

Kundu, T. A theoretical analysis of acoustic microscopy with converging acoustic beams. *J. Appl. Phys. B.: Photophysics and Laser Chemistry*, **46**, (1988) 325–331.

Kundu, T. A complete acoustic microscopical analysis of multilayered specimens. *ASME J. Appl. Mech.*, **59**, (1992) 54–60.

Kundu, T. Inversion of the acoustic material signature of multilayered solids. *J. Acoust. Soc. Am.*, **91**, (1992) 591–600.

Kundu, T., Bereiter-Hahn, J. and Hillmann, K. Measuring elastic properties of cells by evaluation of scanning acoustic microscopy $V(z)$ values using simplex algorithm. *Biophys. J.*, **59**, (1991) 1194.

Kundu, T., Bereiter-Hahn, J. and Hillmann, K. Calculating acoustical properties of cells: influence of surface topography and liquid layer between cell and substrate. *J. Acoust. Soc. Am.*, **91**(5), (1992) 3008–3017.

Kundu, T., Bereiter-Hahn, J. and Karl, I., Cell Property Determination from the Acoustic Microscope Generated Voltage Versus Frequency Curves, *Biophysical Journal*, in press, (2000).

Kundu, T. and Mal, A. K. Elastic waves in a multilayered solid due to a dislocation source. *Wave Motion*, **7**, (1985) 459–471.

Kundu, T. and Mal, A. K. Acoustic material signature of a layered plate. *Int. J. Engrg. Sci.*, **24**, (1986) 1819–1829.

Kundu, T., Mal, A. K. and Weglein, R. D. Calculation of the acoustic material signature of a layered solid. *J. Acoust. Soc. Am.*, **77**, (1985) 353–361.

Kundu, T. and Maslov, K. Material interface inspection by Lamb waves. *Int. J. Solids and Structures*, **34**, (1997) 3885–3901.

Kundu, T., Maslov, K., Karpur, P., Matikas, T. E. and Nicolaou, P. D. A Lamb wave scanning approach for mapping of defects in [0/90] titanium matrix composites. *Ultrasonics*, **34**, (1996) 43–49.

Kushibiki, J., Ohkubo, A. and Chubachi, N. Material characterization by acoustic line-focus beam. *Acoustical Imaging*, **12**, (1982) 101.

Kushibiki, J., Takahashi, H., Kobayashi, T. and Chubachi, N. Characterization of $LiNbO_3$ crystals by line-focus-beam acoustic microscope. *Appl. Phys. Lett.*, **58**, (1991) 2622–2624.

Lawrence, C. W., Scruby, C. B. and Briggs, G. A. D. Acoustic microscopy of ceramic-fiber composites. *I.*, *II&III* : Glass-matrix, glass-ceramic matrix, & metal matrix composites. *J. Mater. Sci.*, **28**, (1993) 3635–3660.

Lawrence, C. W., Scruby, C. B., Briggs, G. A. D. and Dunhill, A. Crack detection in silicon nitride by acoustic microscopy. *NDT International*, **23**, 3–10.

Lawson, W. D. and Sabey, J. W. Infrared techniques. In: *Research Techniques in Nondestructive Testing, Vol.1*. Academic Press, London, (1970) 443–479.

Lemons, R. A. and Quate, C. F. A scanning acoustic microscope. *Proc. IEEE Symp.*, (1973) 18–20.

Lévesque, D. and Piché, L. A robust transfer matrix formulation for the ultrasonic response of multilayered absorbing media. *J. Acoust. Soc. Am.*, **92**, (1992) 452–467.

Maev, R. G. Scanning acoustic microscopy of polymeric materials and biological substances– Review. *Archives of Acoustics*, **12**, (1998) 13–43.

Mal, A. K. Wave propagation in layered composite laminates under periodic surface loading. *Wave Motion*, **10**, (1988) 257–266.

Mal, A. K., Yin, C.-C. and Bar-Cohen, Y. Ultrasonic nondestructive evaluation of cracked composite laminates. *Composite Engrg.*, **1**, (1991) 85–101.

Martin, G. C. In-flight acoustic emission monitoring. *Proc. Conf. Mechanics of Nondestructive Testing*, Blacksburg, VA, (1980).

Maslov, K. and Kundu. T. Selection of Lamb modes for detecting internal defects in composite laminates, *Ultrasonics*, **35**, (1997) 141–150.

Mater. Eval. Special Issue on 'Remote-Field Eddy Current Testing', **47**(1), (1989).

Mater. Eval. Special Issue on 'Eddy Current Testing', **49**(1), (1991).

Matzkanin, G. A. *Investigation of the Effect of Moisture on the Mechanical Properties of Organic Matrix Composite Materials Using Nuclear Magnetic Resonance.* AVRAD-COM Report No. TR-81-F-5, Contract No. DLA 900-79-C-1266, (1981).

McIntire, P. and Bryant, L. *Nondestructive Testing Handbook, Vol.3: Radiography and Radiation Testing,* 2nd ed. American Society for Nondestructive Testing, Columbus, OH, (1985).

McMaster, R. C. *Nondestructive Testing Handbook, Vol.2, 2nd ed.,* American Society for Nondestructive Testing, Columbus, OH, (1982).

Meeks, S. W., Peter, D., Horne, D., Young, K. and Novotny, V. Microscopic imaging of residual stress using a scanning phase-measuring acoustic microscope. *Appl. Phys. Lett.,* **55**, (1989) 1835–1837.

Miller, A. J. Scanning acoustic microscopy in electronics research. *IEEE Trans. Sonics and Ultrasonics.,* **32**, (1985) 313.

Miller, R. K. *Mater. Evaluation,* **48**(6), (1990) 822–824, 826–827, 829.

Mitchell, J. R. *FRP Storage Tank Testing with Acoustic Emission.* Plastics Seminar, Technical Report TR-107-69, 55, Physical Acoustics Corp., Princeton, NJ, (1981).

Moore, T. M. C-mode acoustic microscopy applied to integrated circuit package inspection. *Solid State Phys.,* **35**, (1992) 411–421.

Morse, S., Durelli, A. J. and Sciammarella, C. A. Geometry of Moiré fringes in strain analysis. *Trans. ASCE,* **126**(1), (1961).

Moulthrop, A. A., Muha, M. S., Hadimioglu, B., Silva, C. P. and Kozlowski, G. C. Acoustic microscopy and nonlinear effects in pressurized superfluid helium. *IEEE Trans. Ultrason. Ferroelect. & Freq. Control,* **39**(2), (1992) 204–211.

Nagy, P. B. Ultrasonic detection of kissing bonds at adhesive interfaces. *J. Adhesion Sci. & Tech.,* **5**(8), (1991) 619–630.

Nordstrom, R. *Mater. Evaluation,* **48**(2), (1990) 251–254.

Parks, V. J. Geometric Moiré. In: *Handbook on Experimental Mechanics,* 2nd ed. Kobayashi, A. S. (ed.). VCH Publishers, New York, (1993) 267–296.

Pavlin, C. J., Sherar, M. D. and Foster, S. Subsurface ultrasound microscopic imaging of the intact eye. *Ophthamology,* **97**, (1990) 244–250.

Post, D. Moiré interferometry for deformation and strain studies. *Opt. Engrg.,* **24**(4), (1985) 663–667.

Post, D. Moiré interferometry. In: *Handbook on Experimental Mechanics,* 2nd ed. Kobayashi, A. S. (ed.). VCH Publishers, New York, (1993) 297–364.

Rodriquez, A., Briggs, G. A. D. and Montoto, M. Acoustic microscopy of rocks. *J. Microscopy,* **160**, (1990) 21–29.

Rugar, D., Foster, J. S. and Heiserman, J. Acoustic microscopy at temperatures less than 0.2 K. In: *Acoustical Imaging,* Ash, E. and Hill, C. R. (eds.). Plenum Press, New York, (1982) 13–25.

Sayers, C. M., van Munster, J. G. and King, M. S. Stress-induced ultrasonic anisotropy in Berea sandstone. *Int. J. Rock Mechanics and Min. Sci.,* **27**, (1990) 429–436.

Scherba, G., Hoagland, P. A. and O'Brian, D. O. Acoustic microscopy: A study of contrast in fresh tissue. *IEEE Trans. Ultrason. Ferroelectric and Freq. Controls,* **41**, (1994) 451–457.

Schwab, F. and Knopoff, L. Surface wave dispersion computation. *Bull. Seismological Soc. Am.,* **60**, (1970) 321–344.

Scimmarella, C. A. The Moiré method, a review. *Exper. Mech.,* **22**(11), (1982) 418–433.

Silk, M. G. *Ultrasonic Transducers for Nondestructive Testing.* Adam Hilger, Bristol, (1984).

Sinclair, D. A. and Ash, E. A. Bond integrity evaluation using transmission scanning AM-1. *Electron. Lett.,* **16**, (1980) 880–882.

Slichter, C. P. *Principles of Magnetic Resonance.* Springer-Verlag, New York, (1980).

Theocaris, P. S. *Moiré Fringes in Strain Analysis.* Pergamon Press, Elmsford, NY, (1969).

Thomson, W. T. Transmission of elastic waves through a stratified solid medium. *J. Appl. Phys.,* **21**, (1950) 89–93.

Tucker, P. A. and Wilson, R. G. Acoustic microscopy of polymers. *J. Polym. Sci.,* **18**, 97–103.

Vary, A. The acousto-ultrasonic approach. In: *Acousto-Ultrasonic: Theory and Application*, Duke, J. C. (ed.). Plenum Press, New York, (1988) 1–21.

Vary, A. and Bowels, K. J. An ultrasonic-acoustic technique for nondestructive evaluation of fiber composite quality. *Polym. Eng. Sci.*, **19**(5), (1979) 373–376.

Ueno, E. and Itoh, K. Scanning acoustic microscope and ultrasound tomography of breast tumors. *JSUM Proc.*, **38**, (1981) 237–238.

Weaver, J. M. R. and Briggs, G. A. D. Acoustic microscopy techniques for observing dislocation damping. *J. Physique*, **12**(C10), (1985) 743–750.

Weglein, R. D. A model for predicting acoustic material signature. *Appl. Phys. Lett.*, **34**, (1979) 179–181.

Weglein, R. D. Integrated circuit inspection via acoustic microscopy. *IEEE Trans. Sonics and Ultra-sonics*, **30**, (1983) 40–42.

Weglein, R. D. and Mal, A. K. Elastic characterization of diamond films by acoustic microscopy. *Surface and Coatings Technology*, **47**, (1991) 677–686.

Yonemura, G. T. *Considerations and Standards for Visual Inspection Techniques*. NBSIR76-1142, National Bureau of Standards, Washington, DC, (1976).

Yu, Z. and Boseck, S. Some considerations in layer-thickness measurement for semiconductor technology by acoustic microscopy. *Optik*, **96**, (1994) 83–92.

PART 3

Modeling and Simulation

13

On the Mathematical Modeling of Certain Fundamental Elastostatic Contact Problems in Geomechanics

A. P. S. Selvadurai
McGill University, Montreal, QC., Canada

13.1   Introduction

The classical problem of the axisymmetric indentation of an elastic halfspace by a rigid indentor with a smooth flat base is a problem which has attracted the attention of eminent mathematicians and mechanicists alike. This celebrated problem was first posed by Boussinesq (1886), who reduced it to a problem in potential theory. Boussinesq's approach to the formulation of the contact problem was very much guided by his observations concerning the elasticity problem related to the normal loading of the surface of a halfspace by a concentrated force. Boussinesq reduced this problem to that of finding a single harmonic function with the characteristics of a simple layer in electrostatics, distributed over the plane region and possessing an intensity proportional to the applied normal loading. This ability to pose elasticity problems as their counterparts in potential theory became one of the remarkable contributions by Boussinesq to the developments in the mathematical theory of elasticity. Although Love (1927) and others (see for example the references cited by, Korenev (1960), Galin (1961), Ufliand (1965), Hetenyi (1966), de Pater and Kalker (1975), Selvadurai (1979a), Gladwell (1980) and Johnson (1985)) made seminal contributions to studies related to contact problems associated with elastic media, a completely novel approach to the formulation and solution of this classical contact problem was proposed by Harding and Sneddon (1945). Through the application of integral transform techniques, Harding and Sneddon were able to reduce their analysis of this classical contact problem to the solution of a single integral equation of the Abel type. The significant achievement of this work is that the force-indentation relationship, and the contact stress distribution at the base of the rigid circular indentor, can be evaluated in exact closed forms, and in forms identical to those obtained by Boussinesq.

Since these seminal developments, the study of this contact problem has been extended to cover numerous modifications, including transverse isotropy of the elastic medium, material non-homogeneity with special reference to axial variation in the shear modulus (e.g. see Selvadurai (1996)), interaction of the indentor and loads applied to the elastic medium, adhesion and frictional effects at the indentor-elastic medium interface, flexibility of the indentor, influence of crack development in the halfspace region at the boundary of the indentor, poroelastic behavior of the halfspace region (e.g. see Agbezuge and Deresiewicz (1975), Chiarella and Booker (1975) and Selvadurai and Yue (1994) Yue and Selvadurai (1995a,b)), effects of reinforcement located within the halfspace region, effects of a cracked region beneath the indentor, and the influence of a rigid boundary located at a finite depth. The subject matter related to this classical contact problem is too extensive to be adequately reviewed within the context of this chapter. Excellent expositions, including references to the literature, are given in the texts and articles cited previously.

The reasons for the seminal nature of this contact problem within the scope of geomechanics, both applied and computational, are twofold. First, the solution to this classical problem has significant application potential in geotechnical engineering. It provides a convenient formula for the estimation of elastic settlements of foundations, with a circular planform, within the working load range. Although rigid foundations with a circular planform are a rarity, the solution developed for the load-settlement response can be conveniently adopted to estimate the settlement of square rigid foundations. The application potential has been recognized by geomechanicians, geotechnical engineers and mathematicians alike (e.g. see Krynine (1941), Terzaghi (1943), Sneddon (1951), Tschebotarioff (1973), Gibson (1974), Poulos and Davis (1975) and Davis and Selvadurai (1996)). Secondly, the solution to this classical problem in contact geomechanics has served as a useful benchmark problem for the calibration of solutions derived from finite element, boundary element (or boundary integral equations) and coupled finite element-boundary element methods. The accuracy of these numerical schemes are tested in their ability to model both regions of semi-infinite extent and singular stress states at the boundary of the rigid indentor (e.g. see Zienkiewicz and Taylor (1989) and Desai and Christian (1977)). The objective of this contribution is to outline some salient features pertaining to this classical contact problem in geomechanics, and to document recent developments which extend the range of applicability of the classical model.

13.2   The Classical Contact Problem

We shall examine the axisymmetric problem of the indentation of the surface of a halfspace ($r\varepsilon(0, \infty)$; $z\varepsilon(0, \infty)$) by a rigid circular indentor with a smooth base (Figure 13.1). The indentation is due to an axial load P, which causes the circular indentor to settle by an amount Δ. To formulate the resulting boundary value problem, we need to assign boundary conditions on $z = 0$ and certain regularity conditions as $r, z \to \infty$. The boundary conditions governing the problem are

$$u_z(r, 0) = \Delta; \quad 0 \leq r < a \tag{13.1}$$

$$\sigma_{zz}(r, 0) = 0; \quad a < r < \infty \tag{13.2}$$

$$\sigma_{rz}(r, 0) = 0; \quad 0 < r < \infty \tag{13.3}$$

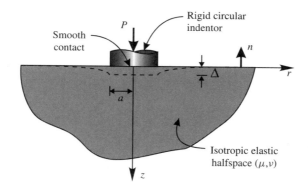

Figure 13.1   The classical contact problem in geomechanics

where

$$\mathbf{u} = (u_r, 0, u_z) \tag{13.4}$$

and

$$\boldsymbol{\sigma} = \begin{bmatrix} \sigma_{rr} & 0 & \sigma_{rz} \\ 0 & \sigma_{\theta\theta} & 0 \\ \sigma_{rz} & 0 & \sigma_{zz} \end{bmatrix} \tag{13.5}$$

are, respectively, the displacement vector and the Cauchy stress tensor referred to the cylindrical polar coordinate system (r, θ, z). The zero elements in Eqs. (13.4) and (13.5) are a direct consequence of the assumed state of axial symmetry. The boundary conditions in Eqs. (13.1) to (13.3) are the only 'known' conditions associated with the indentation problem. The condition in Eq. (13.3) stems from the frictionless nature of the contact and conditions in Eqs. (13.1) and (13.2) correspond to 'mixed' boundary conditions, where either displacements or tractions are prescribed on complementary subsets of the plane boundary $z = 0$. In addition to these conditions, certain regularity conditions need to be prescribed on the displacements and/or stresses as $r, z \to \infty$. These regularity conditions represent constraints that are based on the physical requirements of the problem, and are necessary and sufficient to make the boundary value problem well posed. The requirement that the stress field vanishes as $r, z \to \infty$ is a requirement which is consistent with physical observations. Furthermore, the stresses themselves may vanish as $(r^2 + z^2)^{1/2}$ becomes large (say in comparison to 'a', the radius of the indentor), but the resultant of tractions on a closed surface which includes but does not intersect or touch the region $r < a$ must be equivalent to a force vector with null components in the r and θ directions, and a component of magnitude P in the z-direction.

   To proceed with the solution of the problem, we need to resort to either representation of the displacement field or the stress field in terms of arbitrary functions. With three-dimensional problems, including problems which exhibit axial symmetry, the procedures utilized largely involve representations of the displacement field in terms of specific functions. The development of these representations are fully discussed by Truesdell (1959), Sternberg (1960) and Gurtin (1972). Two of the most commonly used representations are attributed to Love (1927) and Papkovich and Neuber (e.g. see Westergaard, 1952). Simplified versions of these are also given by Green and Zerna (1968). Particular advantage can be taken of the fact that one of the boundary conditions relate to vanishing shear tractions at all points on the plane $z = 0$. Green and Zerna (1968) have shown that when shear tractions vanish on the plane $z = 0$, the resulting boundary value problem in classical isotropic elasticity can be formulated in terms of a single harmonic function $\phi(r, z)$ which satisfies

$$\nabla^2 \phi = 0 \tag{13.6}$$

where

$$\nabla^2 = \frac{\partial^2}{\partial r^2} + \frac{1}{r}\frac{\partial}{\partial r} + \frac{\partial^2}{\partial z^2} \tag{13.7}$$

is Laplace's operator for the axisymmetric case. The displacement and stress components relevant to the formulation of the boundary conditions in Eqs. (13.1) and (13.2) in terms of $\phi(r, z)$ take the forms

$$2\mu u_z = z\frac{\partial^2 \phi}{\partial z^2} - 2(1 - \nu)\frac{\partial \phi}{\partial z} \tag{13.8}$$

$$\sigma_{zz} = z\frac{\partial^3 \phi}{\partial z^3} - \frac{\partial^2 \phi}{\partial z^2} \tag{13.9}$$

where μ is the linear elastic shear modulus and ν is Poisson's ratio. Using Eqs. (13.8) and (13.9), the boundary conditions in Eqs. (13.1) and (13.2) give

$$u_z(r, 0) = -\frac{(1 - \nu)}{\mu}\frac{\partial \phi}{\partial z} = \Delta; \quad 0 \leq r < a \tag{13.10}$$

$$\sigma_{zz}(r, 0) = -\frac{\partial^2 \phi}{\partial z^2} = 0; \quad a < r < \infty \tag{13.11}$$

with the implicit condition that for any choice of ϕ the components of the shear traction σ_{rz} are given by

$$\sigma_{xz} = z\frac{\partial^3 \phi}{\partial x \partial z^2}; \quad \sigma_{yz} = z\frac{\partial^3 \phi}{\partial y \partial z^2} \tag{13.12}$$

and these vanish on $z = 0$.

13.2.1 Solution in terms of a complex-valued function

The elasticity problem is now reduced to the determination of a potential function $\partial\phi/\partial z$ with prescribed values of $\partial\phi/\partial z$ and $\partial^2\phi/\partial z^2$ on subsets of $z = 0$. We can further simplify the mixed boundary value problem by introducing a function $\Omega(r, z)$ which is related to $\phi(r, z)$ through

$$\Omega(r, z) = -\frac{(1 - \nu)}{\mu}\frac{\partial \phi}{\partial z} \tag{13.13}$$

such that

$$\nabla^2 \Omega = 0 \tag{13.14}$$

The mixed boundary conditions given by Eqs. (13.10) and (13.11) can now be written as

$$\Omega = \Delta; \quad 0 \leq r \leq a \tag{13.15}$$

$$\frac{\partial \Omega}{\partial z} = 0; \quad a < r < \infty \tag{13.16}$$

which represent the classical problem in potential theory solved by Boussinesq. The problem is analogous to finding the potential due to a perfectly conducting disk occupying the region $0 \leq r \leq a$; $z = 0$ which is maintained at a prescribed potential distributed about its center. The solution of this mixed boundary value problem can be approached in a variety of ways, including methods based on potential theory, the complex function approach and Hankel transform techniques. In the complex function approach, the solution can be represented in the form

$$\Omega(r, z) = \frac{1}{2} \int_{-a}^{a} \frac{g(t) \, dt}{[r^2 + (z + it)^2]^{1/2}} \tag{13.17}$$

where $g(t)$ is a real continuous even function of t and for $0 \leq t \leq a$,

$$[r^2 + (z + it)^2]^{1/2} = \xi e^{(1/2)i\eta}; \quad [r^2 + (z - it)^2]^{1/2} = \xi e^{-(1/2)i\eta} (\xi \geq 0)$$
$$\xi^2 \cos \eta = r^2 + z^2 - t^2; \quad \xi^2 \sin \eta = 2zt (0 \leq \eta \leq \pi) \tag{13.18}$$

The function $\Omega(r, z)$ given by Eq. (13.17) is real, and the Laplacian of the function can be obtained by carrying out differentiations within the integral sign provided the resulting integrands are continuous functions of (r, z, t). This is satisfied provided $t \neq \pm iz \pm r$ $(0 \leq t \leq a)$, and these conditions are satisfied everywhere except in the indentor region $r \leq a$; $z = 0$. Hence, $\Omega(r, z)$ is harmonic everywhere expect on the indentor region $r \leq a$; $z = 0$, since the Laplacian of the integrand is zero for each value of t. It can also be shown that (e.g. see Green and Zerna, 1968)

$$|\Omega(r, z)| \leq \frac{1}{[r^2 + (z - a)^2]^{1/2}} \int_0^a |g(t)| \, dt \tag{13.19}$$

which implies that

$$\Omega(r, z) = 0((r^2 + z^2)^{-(1/2)}) \tag{13.20}$$

which satisfies the regularity condition as $(r, z) \to \infty$ and that the boundary condition represented by Eq. (13.16) is identically satisfied by Eq. (13.17) and the boundary condition in Eq. (13.15) gives

$$\Delta = \int_0^r \frac{g(t) \, dt}{\sqrt{r^2 - t^2}}; \quad 0 \leq r \leq a \tag{13.21}$$

The integral equation (Eq. (13.21)) can be solved for $g(t)$

$$\Delta \int_0^u \frac{r \, dr}{\sqrt{u^2 - r^2}} = \int_0^u \frac{r \, dr}{\sqrt{u^2 - r^2}} \int_0^r \frac{g(t) \, dt}{\sqrt{t^2 - r^2}}$$

$$= \int_0^u g(t) \, dt \int_t^u \frac{r \, dr}{\sqrt{(u^2 - r^2)(r^2 - t^2)}} = \frac{\pi}{2} \int_0^u g(t) \, dt \tag{13.22}$$

or

$$g(t) = \frac{2\Delta}{\pi} \frac{d}{dt} \int_0^t \frac{r \, dr}{\sqrt{t^2 - r^2}} = \frac{2\Delta}{\pi} \tag{13.23}$$

In the general case when the indentor profile has an arbitrary axisymmetric shape $w(r)$, the equivalent form for Eq. (13.23) is given by

$$g(t) = \frac{2}{\pi} \frac{d}{dt} \int_0^t \frac{rw(r)\,dr}{\sqrt{t^2 - r^2}} = \frac{2w(0)}{\pi} + \frac{2t}{\pi} \int_0^t \frac{w'(r)\,dr}{\sqrt{t^2 - r^2}} \qquad (13.24)$$

where $w'(r)$ is the derivative. This completes the formal analysis of the problem; once $g(t)$ is known, $\Omega(r, z)$ is determined from Eq. (13.17), and the stresses are determined from Eq. (13.16). The normal stress beneath the indentor is given by

$$\sigma_{zz}(r, 0) = \frac{\mu}{(1 - \nu)} \left\{ \frac{\partial \Omega}{\partial z} \right\}_{z=0} \qquad (13.25)$$

Omitting details, it can be shown that

$$\sigma_{zz}(r, 0) = \frac{2\Delta\mu}{\pi(1 - \nu)} \frac{1}{r} \frac{\partial}{\partial r} \int_r^a \frac{t\,dt}{\sqrt{t^2 - r^2}} = -\frac{2\Delta\mu}{\pi(1 - \nu)\sqrt{a^2 - r^2}}; \quad 0 < r < a \qquad (13.26)$$

The load-displacement relationship for the rigid circular indentor can be obtained by making use of the equilibrium equation

$$P + 2\pi \int_0^a \sigma_{zz}(r, 0)r\,dr = 0 \qquad (13.27)$$

which gives

$$P = \frac{4\mu a\Delta}{(1 - \nu)} \qquad (13.28)$$

13.2.2 Solution in terms of Boussinesq's fundamental solutions for a halfspace region

The contact problem can also be formulated by using Boussinesq's fundamental solution to the problem of the normal loading of a halfspace region by a concentrated force P_0 (Figure 13.2). The boundary condition associated with this problem can be presented in

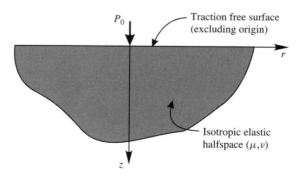

Figure 13.2   Boussinesq's problem for the concentrated normal loading of the halfspace

the modified form as follows:

$$\sigma_{zz}(r, 0) = \frac{P_0}{\pi a^2}; \quad 0 < r < a \tag{13.29}$$

$$\sigma_{zz}(r, 0) = 0; \quad a < r < \infty \tag{13.30}$$

$$\sigma_{rz}(r, 0) = 0; \quad 0 < r < \infty \tag{13.31}$$

where P_0 is the magnitude of the total force acting within the region $0 < r < a; z = 0$. The solution for the concentrated force can be recovered as a special case when $a \to 0$. In addition to these boundary conditions, the displacement and stress fields should reduce to zero as $(r^2 + z^2)^{1/2} \to \infty$. Here again, in the case of a concentrated force P_0 which acts at the origin, the resultant of forces acting on any closed surface which encompasses the origin or touches it should be equivalent to a force of magnitude P_0 which acts in the axial direction. The solution to this problem can be obtained via several techniques which were cited previously and given in standard texts on elasticity and integral transforms (e.g. see Sneddon (1951), Westergaard (1952), Fung (1965), Timoshenko and Goodier (1970), Little (1973), Davis and Selvadurai (1996) and Selvadurai (2000)).

The results of importance here are expressions for the displacement component u_z and the stress components σ_{zz} and σ_{rz}. For the concentrated force problem, these are given by

$$u_z(r, z) = \frac{P_0}{2\pi R} \left[rz - \frac{(1 - 2\nu)r}{(R + z)} \right] \tag{13.32}$$

$$\sigma_{zz}(r, z) = -\frac{3P_0 z^3}{2\pi R^5} \tag{13.33}$$

$$\sigma_{rz}(r, z) = -\frac{3P_0 rz^2}{2\pi R^5} \tag{13.34}$$

where $R^2 = r^2 + z^2$.

It is evident that the displacement and stress components satisfy the boundary and regularity conditions applicable to concentrated force problem, and that the stress state is singular at the origin. Also, we can show that, for example

$$P_0 + 2\pi \int_0^\infty \sigma_{zz}(r, \ell) r \, dr = 0 \tag{13.35}$$

where ℓ is any depth at which the stresses are computed.

In the formulation of the contact problem which makes use of Boussinesq's fundamental solution, we begin with a result which satisfies the traction boundary condition in Eq. (13.3) and the regularity conditions. We formulate the indentation problem by considering the contact stress distribution $q(\rho, \varphi)$ which is an unknown function. Referring to Figure 13.3, the elemental load acting on the elemental area $dA(= \rho \, d\rho \, d\varphi)$ is given by $q(\rho, \varphi) \rho \, d\rho \, d\varphi$. If we denote the displacement at the surface of the halfspace due to this elemental load by $dw(r, \theta)$, then from Boussinesq's solution for the axial displacement of the halfspace region we have

$$dw(r, \theta) = \frac{(1 - \nu)}{2\pi\mu} \frac{q(\rho, \varphi)\rho \, d\rho \, d\varphi}{[r^2 + \rho^2 - 2r\rho \cos(\theta - \varphi)]^{1/2}} \tag{13.36}$$

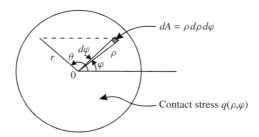

Figure 13.3   Platform for the contact region with non-uniform contact stress $q(\rho, \varphi)$

Integrating this result over the contact region, we have

$$w(r, \theta) = \frac{(1 - v)}{2\pi\mu} \int_0^{2\pi} \int_0^a \frac{q(\rho, \varphi)\rho \, d\rho \, d\varphi}{[r^2 + \rho^2 - 2r\rho\cos(\theta - \varphi)]^{1/2}} = \Delta = \text{const.} \qquad (13.37)$$

This represents the single integral equation for the determination of the unknown contact stress distribution $q(\rho, \varphi)$. The contact stresses should also satisfy the equation of equilibrium

$$\int_0^{2\pi} \int_0^a q(r, \theta)r \, dr \, d\theta + P = 0 \qquad (13.38)$$

where P is the external load applied to the indentor. Owing to the imposed axial symmetry of the contact problem $q(\rho, \varphi) = q(\rho)$, and Eq. (13.37) can be written as

$$\int_0^a \rho q(\rho) \, d\rho \int_0^{2\pi} \frac{d\varphi}{[r^2 + \rho^2 - 2r\rho\cos(\theta - \varphi)]^{1/2}} = \frac{2\pi\mu\Delta}{(1 - v)} \qquad (13.39)$$

A result of Copson (1947) can be used to express the second integral in Eq. (13.39) in the form

$$\int_0^{2\pi} \frac{d\varphi}{[r^2 + \rho^2 - 2r\rho\cos(\theta - \varphi)]^{1/2}} = 4 \int_0^{\min(r, \rho)} \frac{dt}{[(\rho^2 - t^2)(r^2 - t^2)]^{1/2}} \qquad (13.40)$$

where $\min(r, \rho)$ refers to the minimum values of r and ρ. The integral equation (Eq. (13.39)) can now be written as

$$\int_0^r \rho q(\rho) \, d\rho \int_0^\rho \frac{dt}{[(\rho^2 - t^2)(r^2 - t^2)]^{1/2}} + \int_r^a \rho q(\rho) \, d\rho \int_0^\rho \frac{dt}{[(\rho^2 - t^2)(r^2 - t^2)]^{1/2}}$$

$$= \frac{\pi\mu\Delta}{2(1 - v)}; \quad 0 \le r < 1 \qquad (13.41)$$

Inverting the order of integration, we have

$$\int_0^r \frac{dt}{[r^2 - t^2]^{1/2}} \int_t^a \frac{\rho q(\rho) \, d\rho}{[\rho^2 - t^2]^{1/2}} = \frac{\pi\mu\Delta}{2(1 - v)} \qquad (13.42)$$

To solve this integral equation, we set

$$S(r) = \int_r^a \frac{uq(u) \, du}{[u^2 - r^2]^{1/2}} \qquad (13.43)$$

The integral equation (Eq. (13.42)) can now be written as

$$\int_0^r \frac{S(t)\,dt}{[r^2 - t^2]^{1/2}} = \frac{\pi\mu\Delta}{2(1-v)}; \quad 0 \le r < 1 \tag{13.44}$$

Here again, we can make use of the procedures discussed in Section 13.2.1. We have

$$S(t) = \frac{2}{\pi}\left(\frac{\pi\mu\Delta}{2(1-v)}\right)\frac{d}{dt}\int_0^t \frac{r\,dr}{[t^2 - r^2]^{1/2}} = \frac{\mu\Delta}{(1-v)} \tag{13.45}$$

The stress distribution beneath the indentor is given by

$$\sigma_{zz}(r, 0) = \frac{2}{\pi r}\frac{d}{dr}\int_r^a \frac{tS(t)\,dt}{[t^2 - r^2]^{1/2}} = -\frac{2\mu\Delta}{\pi(1-v)[a^2 - r^2]^{1/2}} \tag{13.46}$$

The results for the load-displacement behavior of the rigid circular indentor can be obtained by considering the equation of equilibrium for the indentor.

13.2.3 Solution based on the Hankel transform approach

An entirely novel approach for the solution of the classical indentation problem was proposed by Harding and Sneddon (1945). This approach is based on the use of Hankel transforms for the formulation of the problem. While it is possible to formulate the mixed boundary value problem posed by Eqs. (13.1) to (13.3) by making use of a Hankel transform development of the partial differential Eq. (13.6), we shall consider here an alternative formulation based on Love's strain function $\Phi(r, z)$ (Love (1927)), which is governed by the biharmonic equation

$$\nabla^2\nabla^2\Phi(r, z) = 0 \tag{13.47}$$

The displacements and stresses in the elastic medium can be expressed as derivatives of $\Phi(r, z)$ in the forms

$$2\mu\, u_r = -\frac{\partial^2\Phi}{\partial r\partial z}; \quad 2\mu\, u_z = 2(1-v)\nabla^2\Phi - \frac{\partial^2\Phi}{\partial z^2} \tag{13.48}$$

and

$$\sigma_{zz} = \frac{\partial}{\partial z}\left[(2-v)\nabla^2\Phi - \frac{\partial^2\Phi}{\partial z^2}\right] \tag{13.49}$$

$$\sigma_{rz} = \frac{\partial}{\partial r}\left[(1-v)\nabla^2\Phi - \frac{\partial^2\Phi}{\partial z^2}\right] \tag{13.50}$$

The nth-order Hankel transform of $\Phi(r, z)$ is

$$\overline{\Phi}^n(\xi, z) = \mathcal{H}_n\{\Phi(r, z); \xi\} = \int_0^\infty r\Phi(r, z)J_n(\xi r)\,dr \tag{13.51}$$

and the corresponding inversion theorem is

$$\Phi(r, z) = \mathcal{H}_n^{-1}\{\overline{\Phi}^n(\xi, z); r\} = \int_0^\infty \xi \overline{\Phi}^n(\xi, z) J_n(\xi r) \, d\xi \tag{13.52}$$

Application of the Hankel transform of order zero, to the governing partial differential equation (Eq. (13.47)) reduces it to the following ordinary differential equation for $\overline{\Phi}^0(\xi, z)$:

$$\left\{ \frac{d^2}{dz^2} - \xi^2 \right\}^2 \overline{\Phi}^0(\xi, z) = 0 \tag{13.53}$$

The general solution of Eq. (13.53) is

$$\overline{\Phi}^0(\xi, z) = [A(\xi) + B(\xi)z]e^{-\xi z} + [C(\xi) + D(\xi)z]e^{\xi z} \tag{13.54}$$

where A, B, C and D are either arbitrary functions of ξ or arbitrary constants. These are determined by considering the boundary conditions and regularity conditions applicable to the indentation problem. Considering the regularity conditions applicable to the contact problem referred to a halfspace region, it is evident that if the displacements and stresses are to remain bounded as $z \to \infty$, we require

$$C(\xi) = D(\xi) = 0 \tag{13.55}$$

The expression for σ_{rz} can be written as

$$\sigma_{rz} = \int_0^\infty \xi^2 \left[v \frac{d^2\overline{\Phi}^0}{dz^2} + \xi^2(1 - v)\overline{\Phi}^0 \right] J_1(\xi r) \, d\xi \tag{13.56}$$

where $\overline{\Phi}^0$ now corresponds to the result Eq. (13.54), where $C(\xi)$ and $D(\xi)$ are set to zero. To satisfy the boundary condition in Eq. (13.3) pertaining to the zero shear stress boundary condition at $z = 0$, we require

$$A(\xi) = \frac{2vB(\xi)}{\xi} \tag{13.57}$$

The mixed boundary conditions given by Eqs. (13.1) and (13.2) can now be reduced to the system of dual integral equations

$$\int_0^\infty M(\beta) J_0(\rho\beta) \, d\beta = w_0^*; \quad 0 \le \rho \le 1 \tag{13.58}$$

$$\int_0^\infty \beta M(\beta) J_0(\rho\beta) \, d\beta = 0; \quad 1 < \rho < \infty \tag{13.59}$$

where

$$\beta = \xi a; \quad \rho = \frac{r}{a}; \quad M(\beta) = \beta^2 B(\xi)$$

$$w_0^* = -\frac{\mu a^3 \Delta}{(1 - v)} \tag{13.60}$$

For the solution of the system of dual integral equations defined by Eqs. (13.58) and (13.59) we introduce the finite Fourier cosine transform of $M(\beta)$ defined by

$$M(\beta) = \int_0^1 F(t)\cos(\beta t)\,dt \tag{13.61}$$

where $F(t)$ is an arbitrary function, such that Eq. (13.59) is identically satisfied for any choice of $F(t)$. If we consider the transform

$$\mathcal{H}_0\{\xi^{-1}\mathcal{F}_c[g(t);\xi];r\} = \sqrt{\frac{2}{\pi}}\int_0^\infty g(t)\,dt \int_0^\infty J_0(\xi r)\cos(\xi t)\,d\xi \tag{13.62}$$

we note that

$$\int_0^\infty J_0(\xi r)\cos(\xi t)\,d\xi = \frac{H(r-t)}{[r^2-t^2]^{1/2}} \tag{13.63}$$

where

$$H(r-t) = \begin{cases} 1 & r > t \\ 0 & r < t \end{cases} \tag{13.64}$$

Hence,

$$\frac{2}{\pi}\int_0^\infty J_0(\xi r)\,d\xi \int_0^\infty g(t)\cos(\xi t)\,dt = \sqrt{\frac{2}{\pi}}\int_0^\infty \frac{g(t)H(r-t)\,dt}{[r^2-t^2]^{1/2}}$$

$$= \sqrt{\frac{2}{\pi}}\int_0^r \frac{g(t)\,dt}{[r^2-t^2]^{1/2}} \tag{13.65}$$

Similarly, we can show that

$$\mathcal{H}_0\{\xi^{-1}\mathcal{F}_s[g(t);\xi];r\} = \sqrt{\frac{2}{\pi}}\int_0^r \frac{g(t)\,dt}{[t^2-r^2]^{1/2}} \tag{13.66}$$

and

$$\{\mathcal{F}_s[g'(t);\xi];r\} = -\xi\mathcal{F}_c\{g(t);\xi\} \tag{13.67}$$

$$\{\mathcal{F}_c[g'(t);\xi];r\} = \xi\mathcal{F}_s\{g(t);\xi\} \tag{13.68}$$

and Eq. (13.68) is valid provided $g(0) = 0$. Using the above results, we can show that

$$\mathcal{H}_0\{\mathcal{F}_c[g(t);\xi];r\} = -\sqrt{\frac{2}{\pi}}\int_r^\infty \frac{g(t)\,dt}{[t^2-r^2]^{1/2}} \tag{13.69}$$

$$\mathcal{H}_0\{\mathcal{F}_s[g(t);\xi];r\} = \sqrt{\frac{2}{\pi}}\int_0^r \frac{g(t)\,dt}{[r^2-t^2]^{1/2}} \tag{13.70}$$

with $g(0) = 0$. Now if we identify

$$g(t) = F(t)H(1-t) \tag{13.71}$$

then we can reduce the dual system to

$$\mathcal{H}_0\left\{\beta^{-1}\int_0^1 F(t)\cos(\beta t)d(t); \rho\right\} = \int_0^\rho \frac{F(t)\,dt}{[r^2 - t^2]^{1/2}}; \quad 0 \le \rho \le 1 \quad (13.72)$$

$$\mathcal{H}_0\left\{\int_0^1 F(t)\cos(\beta t)d(t); \rho\right\} = 0; \quad 1 < \rho < \infty \quad (13.73)$$

and the finite Fourier cosine transform in Eq. (13.61) identically satisfies the integral equation (Eq. (13.59)) and reduces Eq. (13.58) to a single integral equation

$$\int_0^\rho \frac{F(t)\,dt}{[\rho^2 - t^2]^{1/2}} = w_0^* \quad (13.74)$$

The solution of Eq. (13.74) is a classical problem in mechanics and mathematics, and the details of the method of solution can be found in standard texts on integral equations. Furthermore, the similarity between Eqs. (13.21), (13.44) and (13.74) is self-evident. We can obtain the results for the load (P) vs. displacement relationship for the rigid circular indentor in exactly the same way. We can summarize the results of engineering interest in the following forms: the displacements of the surface of the halfspace are given by

$$u_z(r, 0) = \begin{cases} \Delta; & 0 \le r \le a \\ \dfrac{2}{\pi}\Delta\sin^{-1}\left(\dfrac{a}{r}\right); & a \le r < \infty \end{cases} \quad (13.75)$$

$$\sigma_{zz}(r, 0) = \begin{cases} \dfrac{P}{2\pi a[a^2 - r^2]^{1/2}}; & 0 < r < a \\ 0 & a < r < \infty \end{cases} \quad (13.76)$$

Also, by considering the equilibrium of the rigid circular indentor we have

$$P = \frac{4\mu a\Delta}{(1 - \nu)} = \frac{2Ea\Delta}{(1 - \nu^2)} \quad (13.77)$$

where E is Young's modulus of the elastic medium.

13.3   Extensions to the Classical Contact Problem

The versatility of the classical solution for the contact problem developed in the previous section has been demonstrated through applications to a variety of problems, including the analyses of the interaction between a rigid indentor and a Mindlin force (Selvadurai, 1978) and interaction between a uniformly loaded elastic plate and an elastic halfspace (Selvadurai, 1980a). Prior to the presentation of the essentials of these applications, we note here the generalized results which have been summarized by Sneddon (1965). Consider the case where the axisymmetric profile of the indentor is given by

$$u_z(r, 0) = w(r); \quad 0 \le r \le a \quad (13.78)$$

with the additional boundary conditions defined by Eqs. (13.2) and (13.3). We further assume that the form of $w(r)$ is such that, during indentation, complete contact is established over the region $0 \le r \le a$. The contact stress distribution over the contact region is given by

$$\sigma_{zz}(r, 0) = \frac{\mu}{(1 - v)r} \frac{\partial}{\partial r} \int_r^a \frac{tg(t)\,dt}{[t^2 - r^2]^{1/2}}; \quad 0 < r < a \tag{13.79}$$

where

$$g(t) = \frac{2}{\pi} \frac{d}{dt} \int_0^t \frac{rw(r)\,dr}{[t^2 - r^2]^{1/2}} \tag{13.80}$$

Also, by considering the equilibrium of the indentor under the action of the external force P and the contact stresses $\sigma_{zz}(r, 0)$, we can show that

$$P = \frac{2\pi\mu}{(1 - v)} \int_0^a g(t)\,dt \tag{13.81}$$

13.3.1   Interaction between a rigid indentor and a Mindlin force

In this section, we consider the axisymmetric problem of the smooth complete indentation of a halfspace by a rigid circular indentor in the presence of a Mindlin force of magnitude P_M which acts at a distance c below the surface of the halfspace (Figure 13.4). The rigid circular indentor is subjected to an external load P which maintains complete contact over the region $0 \le r \le a$ without separation. The displacement of the rigid circular punch under the combined action of the external force P and the internal Mindlin force P_M is denoted by Δ_0^*. Considering Mindlin's problem, for the internal loading of a halfspace region under the action of purely the Mindlin force P_M with traction free surface, $z = 0$ is given by (Mindlin, 1936)

$$u_z^M(r, 0) = \frac{k_1 c^2}{(r^2 + c^2)^{1/2}} + \frac{k_2 c^4}{(r^2 + c^2)^{3/2}} \tag{13.82}$$

where

$$k_1 = \frac{P_M(1 - v)}{2\pi\mu c^2}; \quad k_2 = \frac{P_M}{4\pi\mu c^2} \tag{13.83}$$

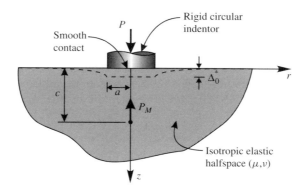

Figure 13.4   Interaction between a Mindlin force and a rigid circular indentor with a smooth flat base

The general expression for $w(r)$ corresponding to Eq. (13.78) is given by

$$w(r) = \Delta_0^* + u_z^M(r, 0) \tag{13.84}$$

Also by considering Eq. (13.84) and the result in Eq. (13.80) we have

$$g(t) = \frac{2}{\pi} \left[\Delta_0^* + \frac{(k_1 + k_2)c^3}{(t^2 + c^2)} - \frac{2k_2 c^2 t^2}{(t^2 + c^2)^2} \right] \tag{13.85}$$

and by considering the equilibrium constraint given by Eq. (13.81) in terms of $g(t)$, we obtain

$$\Delta_0^* = \frac{P(1 - v)}{2a\mu} \left[1 - \frac{P_M}{P} \left\{ \frac{2}{\pi} \tan^{-1}\left(\frac{a}{c}\right) + \frac{ac}{\pi(1 - v)(a^2 + c^2)} \right\} \right] \tag{13.86}$$

The contact stresses are obtained from Eq. (13.79), i.e.

$$
\begin{aligned}
\sigma_{zz}(r, 0) = -\frac{P}{2\pi a\sqrt{a^2 - r^2}} &\left[1 + \frac{P_M}{P} \left\{ -\frac{2}{\pi} \tan^{-1}\left(\frac{a}{c}\right) + \frac{ac}{\pi(1 - v)(a^2 + c^2)} \right. \right. \\
&+ \frac{ac}{\pi(1 - v)(r^2 + c^2)^2} \left[(1 - 2v)(r^2 + c^2) + c^2 \right. \\
&+ ([1 - 2v](r^2 + c^2) + 3c^2) \left[\frac{a^2 - r^2}{r^2 + c^2} \right]^{1/2} \tan^{-1} \left[\frac{a^2 - r^2}{r^2 + c^2} \right]^{1/2} \\
&\left. \left. \left. + \frac{c^2\{2a^2 + c^2 - r^2\}}{(a^2 + c^2)} \right] \right\} \right]; \quad 0 < r < a
\end{aligned}
\tag{13.87}
$$

The first observation (Selvadurai, 1978) is that the results given by Eqs. (13.86) and (13.87) are *exact closed form solutions* for the net displacement Δ_0^* of the indentor in the presence of the combined action of P and P_M, and for the contact stresses in the region $0 < r < a$. In the limiting case when either $c \to \infty$ or $P_M \to 0$, the results obtained from Eqs. (13.86) and (13.87) reduce to the classical results developed previously for the axisymmetric indentation purely due to the external force P. In the limit when $P_M = P$ and $c \to 0$, the rigid circular punch is subjected to a doublet of concentrated forces, which gives $\Delta_0^* \equiv 0$. The contact stresses given by Eq. (13.87) also tend to zero as $c \to 0$ and $P_M = P$.

This solution represents the first formal mathematical analyses of the 'Cable Jacking Test' discussed by Stagg and Zienkiewicz (1967) and Jaeger and Cook (1976). Selvadurai (1979b) also extended this particular contact problem to include effects of a non-uniform line load of Mindlin forces which acts along the axis of symmetry.

13.3.2 Interaction between a rigid circular indentor and an external surface load

We consider the problem of the indentation of an isotropic elastic halfspace by a rigid circular indentor with a smooth base in the presence of a concentrated force Q^* which

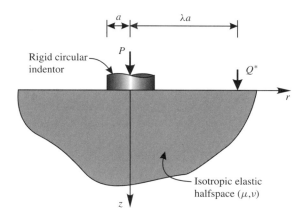

Figure 13.5   Interaction between the centrally loaded rigid circular indentor and an external surface
load

acts normal to the surface of the halfspace at a distance λa from the center of the rigid
circular indentor (Figure 13.5).

This problem was first examined by Selvadurai (1980b), who presented an exact closed
form solution for the resulting settlement of the rigid circular indentor. In the presence of
the externally placed force Q^*, the contact problem is, of course, non-axisymmetric. For
the solution of the problem, we first consider the contact problem where the rigid circular
indentor is subjected to an eccentric load Q which acts at a distance ζa from its center.
The action of the eccentric loading will induce a central displacement

$$u_z(r, 0) = w_0; \quad 0 \le r \le a \tag{13.88}$$

and a rotation ϑ_0 which corresponds to an axial displacement

$$u_z(r, 0) = \vartheta_0 r \cos \theta; \quad 0 \le r \le a \tag{13.89}$$

The remaining boundary conditions associated with the two sub-problems defined by
Eqs. (13.88) and (13.89) correspond to Eqs. (13.2) and (13.3). A Hankel transform
approach yields the following sets of dual integral equations for two unknown functions
$A_i(\xi)$; i.e.

$$\mathcal{H}_0\{\xi^{-1}A_1(\xi); r\} = w_0; \quad 0 \le r \le a \tag{13.90}$$

$$\mathcal{H}_0\{A_1(\xi); r\} = 0; \quad a < r < \infty \tag{13.91}$$

and

$$\mathcal{H}_1\{\xi^{-1}A_2(\xi); r\} = \vartheta_0 r; \quad 0 \le r \le a \tag{13.92}$$

$$\mathcal{H}_1\{A_2(\xi); r\} = 0; \quad a < r < \infty \tag{13.93}$$

The solution of the system of dual integral equations defined by Eqs. (13.90) and (13.91)
have been presented in Section 13.3. The dual system defined by Eqs. (13.92) and
(13.93) has been examined by Bycroft (1956) and Florence (1961) in connection with

the eccentrically loaded rigid indentation problem. The method of solution of the latter problem follows the general approach presented in Section 13.2.3. The result of importance to the subsequent analyses deals with relationship between the eccentric load Q and the displacements within and exterior to the indentor region. The exact closed form result for axial displacement is given by

$$u_z(r, \theta, 0) = \frac{Q(1 - v^2)}{2Ea} \left[1 + \frac{3\zeta_0 r}{2a} \cos \theta \right]; \quad 0 \le r \le a \tag{13.94}$$

$$u_z(r, \theta, 0) = \frac{Q(1 - v^2)}{2Ea} \left[\frac{2}{\pi} \sin^{-1} \left(\frac{a}{r} \right) + \frac{3\zeta_0 r}{2a} \left\{ 1 - \frac{2}{\pi} \tan^{-1} \left\{ \frac{\sqrt{r^2 - a^2}}{a} \right\} \right. \right.$$

$$\left. \left. - \frac{2a}{\pi} \frac{\sqrt{r^2 - a^2}}{r} \right\} \right]; \quad a \le r \le \infty \tag{13.95}$$

We now consider the problem of the rigid circular indentor resting in smooth contact with an isotropic elastic halfspace and subjected to an external concentrated force Q^* at the location $(\lambda a, 0, 0)$. It is further assumed that the indentor is subjected to a sufficiently large axisymmetric force P to maintain the contact stress at the smooth interface compressive for all choices of Q^* and λa. We shall disregard the effect of P, since this represents an additive solution and restrict attention between the indentor and the perturbing force Q^*. To examine this problem, we superimpose on the rigid indentor a central force $\overline{Q}$ and a moment $\overline{M}$ about the y-axis such that the indentor experiences zero displacement in the contact region (Figure 13.6). The associated contact problem is defined by the mixed boundary value problem

$$u_z(r, \theta, 0) = 0; \qquad 0 \le \theta \le 2\pi; \quad 0 \le r \le a \tag{13.96}$$

$$\sigma_{zz}(r, \theta, 0) = -p(r, \theta); \quad 0 \le \theta \le 2\pi; \quad a < r < \infty \tag{13.97}$$

$$\sigma_{rz}(r, \theta, 0) = 0; \qquad 0 \le \theta \le 2\pi; \quad 0 < r < \infty \tag{13.98}$$

where $p(r, \theta)$ is an even function of θ. Following the Hankel transform developments of the equations of equilibrium presented by Muki (1960), it can be shown that when the

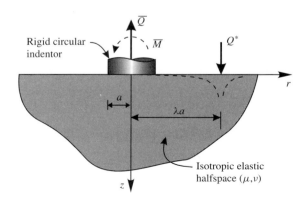

Figure 13.6   Corrective force resultants on the rigid indentor

condition (13.98) pertaining to shear tractions on $z = 0$ are satisfied, the expressions for u_z and σ_{zz} on $z = 0$ can be expressed in the following forms:

$$u_z(r, \theta, 0) = 2(1 - \nu) \sum_{m=0}^{\infty} \mathcal{H}_m[\xi^{-2} \Psi_m(\xi); r] \cos(m\theta) \tag{13.99}$$

$$\sigma_{zz}(r, \theta, 0) = -2\mu \sum_{m=0}^{\infty} \mathcal{H}_m[\xi^{-1} \Psi_m(\xi); r] \cos(m\theta) \tag{13.100}$$

where $\Psi_m(\xi)$ are unknown functions to be determined by satisfying the mixed boundary conditions (13.96) and (13.97), respectively. Assuming that $p(r, \theta)$ can be expressed in the form

$$p(r, \theta) = 2\mu \sum_{m=0}^{\infty} g_m(r) \cos(m\theta) \tag{13.101}$$

the mixed boundary conditions (13.96) and (13.97) yield the set of dual integral equations

$$\mathcal{H}_m[\xi^{-2} \Psi_m(\xi); r] = 0; \qquad 0 \leq r \leq a \tag{13.102}$$

$$\mathcal{H}_m[\xi^{-1} \Psi_m(\xi); r] = g_m(r); \qquad a < r < \infty \tag{13.103}$$

The solution of the dual system of Eqs. (13.102) and (13.103) is given by several authors, including Noble (1958) and Sneddon and Lowengrub (1969), and the details will not be pursued here. It is sufficient to note that for the concentrated external loading of the halfspace

$$[g_0(r); g_m(r)] = \frac{Q^*(1 + \nu)a\delta(r - \lambda a)}{4\pi^2 E r}[1{:}2] \tag{13.104}$$

where $\delta(r - \lambda a)$ is the Dirac delta function, and the contact stress distribution at the interface region $0 \leq r \leq a$ is given by

$$\sigma_{zz}(r, \theta, 0) = \frac{2Q^*}{\pi^2} \sum_{m=0}^{\infty} \left(\frac{r}{\lambda a}\right)^m \cos(m\theta) \int_a^{\infty} \frac{tH(\lambda a - t)\,dt}{\sqrt{\lambda^2 a^2 - t^2}(t^2 - r^2)^{3/2}} \tag{13.105}$$

and $H(\lambda a - t)$ is the Heaviside step function. The explicit expression for σ_{zz} in the contact region of the indentor is given by

$$\sigma_{zz}(r, \theta, 0) = \frac{Q^*\sqrt{\lambda^2 - 1}}{a^2\pi^2\sqrt{\rho^2 - 1}\{\lambda^2 + \rho^2 - 2\lambda\rho \cos\theta\}} \tag{13.106}$$

where $\rho = r/a$. Also, it can be verified that the displacement and stress fields associated with this solution reduce to zero as $r, z \to \infty$. It may be noted that the distribution of normal stress beneath the 'restrained indentor' due to the external force Q^* (as defined by Eq. (13.106)) has a form identical (except for a multiplicative constant) to that of the distribution of charge density on a thin conducting disk of radius a due to a point charge located at a distance λa (e.g. see Jeans, 1925). The force resultants $\overline{Q}$ required to maintain zero axial displacement (u_z) in the contact region $0 \leq r \leq a$ is given by

$$\overline{Q} = 2\pi \int_0^a \sigma_{zz}(r, \theta, 0)r\,dr = \frac{2Q^*}{\pi} \sin^{-1}\left(\frac{1}{\lambda}\right) \tag{13.107}$$

Similarly, the moment $\overline{M}$ required to maintain zero displacement in the contact region is given by

$$\overline{M} = 2 \int_0^a \int_0^\pi \sigma_{zz}(r, \theta, 0) r^2 \cos \theta \, dr \, d\theta$$

$$= Q^* \lambda a \left[1 - \frac{2}{\pi} \tan^{-1} \sqrt{\lambda^2 - 1} - \frac{2}{\pi} \frac{\sqrt{\lambda^2 - 1}}{\lambda^2} \right] \qquad (13.108)$$

By subjecting the rigid circular indentor to force and moment resultants in the opposite sense (to $\overline{Q}$ and $\overline{M}$), we can render the rigid indentor free from external force resultants (except of course for the central force P (Figure 13.5)). The displacement field due to these counteracting force and moment resultants can be obtained from Eqs. (13.94), (13.95), (13.107) and (13.108). We have

$$u_z(r, \theta, 0) = \frac{Q^*(1 - \nu^2)}{2aE} \left[\frac{2}{\pi} \sin^{-1} \left(\frac{1}{\lambda} \right) + \frac{3}{2} \lambda \rho \cos \theta \left\{ 1 - \frac{2}{\pi} \tan^{-1} \sqrt{\lambda^2 - 1} \right. \right.$$

$$\left. \left. - \frac{2}{\pi} \frac{\sqrt{\lambda^2 - 1}}{\lambda^2} \right\} \right]; \quad 0 \le r \le a \qquad (13.109)$$

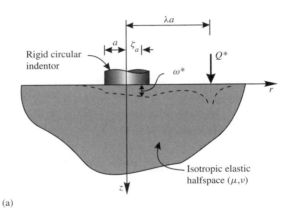

(a)

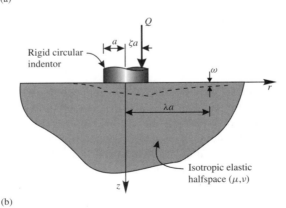

(b)

Figure 13.7   Reciprocal states for indentation problems

Consider the displacement w^* of the rigid circular indentor at the location $(\zeta a, 0, 0)$ due to the external force Q^* applied at $(\lambda a, 0, 0)$ (Figure 13.7(a)). From the results presented above, we have

$$w^* = \frac{Q^*(1 - \nu^2)}{2aE} \left[\frac{2}{\pi} \sin^{-1}\left(\frac{1}{\lambda}\right) + \frac{3\lambda\zeta}{2} \left\{ 1 - \frac{2}{\pi} \tan^{-1}\sqrt{\lambda^2 - 1} - \frac{2}{\pi} \frac{\sqrt{\lambda^2 - 1}}{\lambda^2} \right\} \right]$$

(13.110)

Similarly, the surface displacement of the elastic halfspace at the exterior location $(\lambda a, 0, 0)$ due to the loading of the rigid indentor by a concentrated force Q which is applied at $(\zeta a, 0, 0)$ is denoted by w (Figure 13.7(b)). By comparing the result for w, which is given by Eq. (13.95) with Eq. (13.110), it is evident that the two states satisfy Betti's classical reciprocal relationship (see also Selvadurai (1981) and Davis and Selvadurai (1996))

$$Q^* w = Q w^*$$

(13.111)

The solution to the interaction between the rigid circular foundation and the externally placed concentrated force serves as the Green's function, which can be used to generate solutions for cases involving arbitrary loaded areas. Examples of such applications are given by Selvadurai (1982, 1983).

13.4   The Embedded Indentor

The class of problem which deals with the indentation of an elastic medium by either a fully or partially embedded indentor has several useful applications, especially with reference to estimating the load-settlement behavior of disk shaped anchors and end bearing deep foundations in the working load range. While the geometry of such foundations can be quite complex, the idealization of the foundation by a disk shaped region, which is either rigid or possesses finite flexibility, makes the problem mathematically tractable, and lends itself to generation of results of practical value. In this section, we shall consider two such axisymmetric problems referred to an infinite space region and a halfspace region.

13.4.1   The axial indentation of an infinite space

We consider the problem of an isotropic elastic infinite region which is bounded internally by a rigid disk shaped anchor region of radius a which is located in the plane $z = 0$ (Figure 13.8). The disk anchor is subjected to an axial force P at its center which induces a displacement $\widetilde{\Delta}_0$ in the axial direction. The rigid disk shaped region is fully bonded to the surrounding elastic medium which restricts the radial displacement to zero within the indentation region. Since the medium is of infinite extent, the state of deformation in the medium induced by the axial movement of the disk indentor exhibits a state of asymmetry about the plane $z = 0$. We can therefore pose the embedded indentor problem for the infinite space region as a mixed boundary value problem for a halfspace region (say $z \geq 0$), where the plane $z = 0$ is subjected to the following mixed boundary

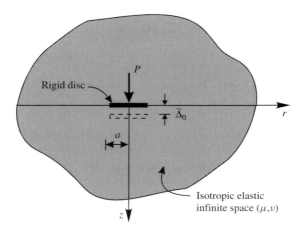

Figure 13.8   The axial indentation of an infinite space

conditions:

$$u_z(r, 0^+) = \tilde{\Delta}_0; \quad 0 \le r \le a \tag{13.112}$$

$$\sigma_{zz}(r, 0^+) = 0; \quad a < r < \infty \tag{13.113}$$

$$u_r(r, 0^+) = 0; \quad 0 \le r < \infty \tag{13.114}$$

where $z = 0^+$ refers to the region on the plane $z = 0$, which corresponds to the halfspace $z \ge 0$. A comparison of Eqs. (13.112) to (13.114) with Eqs. (13.1) to (13.3) indicates that the mixed boundary conditions applicable to the plane $z = 0^+$ are identical, and that the zero shear traction boundary condition given by Eq. (13.3) is now replaced by the zero radial displacement constraint. The analyses of the resulting mixed boundary value problem referred to the infinite space region can be approached via the same techniques that were outlined in 13.3 in connection with the solution of the indentation problem referred to the elastic region. We shall briefly outline the salient features of the analyses which employs a Hankel transform approach. Considering the solution scheme in terms of Love's strain function, the solution for $\overline{\Phi}^0(\xi, z)$ applicable to the domain $z \ge 0$ is obtained by setting $C(\xi) = D(\xi) = 0$ in Eq. (13.54). The resulting expression can be utilized to obtain the following integral expression for the radial displacement u_r;

$$2\mu u_r(r, z) = \int_0^\infty \xi^2[-\xi A(\xi) + B(\xi)(1 - \xi z)]e^{-\xi z} J_1(\xi r)\, d\xi \tag{13.115}$$

For the radial displacement to satisfy the boundary condition given by Eq. (13.114) for all r, we require

$$B(\xi) = \xi A(\xi) \tag{13.116}$$

The resulting expressions for the displacement $u_z(r, z)$ and the axial stress component $\sigma_{zz}(r, z)$ can be used to express the mixed boundary conditions given by Eqs. (13.112) and (13.113) in the following forms:

$$\int_0^\infty \xi^3 A(\xi) J_0(\xi r) \, d\xi = -\frac{2\mu \widetilde{\Delta}_0}{(3 - 4\nu)}; \quad 0 \le r \le a \tag{13.117}$$

$$\int_0^\infty \xi^4 A(\xi) J_0(\xi r) \, d\xi = 0; \quad a < r < \infty \tag{13.118}$$

The similarity between the system of dual integral equations, namely Eqs. (13.117) and (13.118) for the embedded indentor and Eqs. (13.58) and (13.59) for the indentation problem for a halfspace region, is clear. The procedure for the solution of the system of dual integral equations defined by Eqs. (13.117) and (13.118) is identical to that described in 13.3.3. A result of some interest to applications in geotechnical engineering relates to the load-displacement behavior of the embedded indentor. This can be evaluated by considering the equation of equilibrium for the embedded indentor under action of $\sigma_{zz}(r, 0^+)$ and $\sigma_{zz}(r, 0^-)$ with the observation that

$$\sigma_{zz}(r, 0^+) = -\sigma_{zz}(r, 0^-) \tag{13.119}$$

The equation of equilibrium gives

$$P + 2\pi \int_0^a r[\sigma_{zz}(r, 0^+) - \sigma_{zz}(r, 0^{-1})] \, dr = 0 \tag{13.120}$$

The integral expression for $\sigma_{zz}(r, 0^+)$ is given by

$$\sigma_{zz}(r, 0^+) = -\frac{8\mu \widetilde{\Delta}_0 (1 - \nu)}{\pi (3 - 4\nu) a} \int_0^\infty \sin(\beta) J_0(\beta \rho) \, d\beta; \quad 0 < r < a \tag{13.121}$$

The evaluation of the integral in Eq. (13.121) yields

$$\sigma_{zz}(r, 0^+) = -\frac{8\mu \widetilde{\Delta}_0 (1 - \nu)}{\pi (3 - 4\nu)[a^2 - r^2]^{1/2}}; \quad 0 < r < a \tag{13.122}$$

The expressions above, Eq. (13.122), and the relationship in Eq. (13.119) can be used in Eq. (13.120) to develop the load-displacement relationship for the embedded rigid disc indentor; the resulting closed form expression is given by (see also Collins (1962), Kanwal and Sharma (1996) and Selvadurai (1976, 1994))

$$P = \frac{32\mu \widetilde{\Delta}_0 a(1 - \nu)}{(3 - 4\nu)} \tag{13.123}$$

13.4.2 Internal indentation of a halfspace

We now consider the axisymmetric problem related to the internal indentation of an elastic halfspace by a rigid circular disk shaped indentor. The disk indentor is embedded in bonded contact with the elastic halfspace region (Figure 13.9). This problem, which was

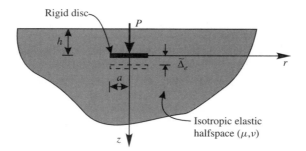

Figure 13.9   Axial internal identation of an isotropic elastic halfspace region

investigated by Selvadurai (1993), can be regarded as a more generalized problem, from which the solutions to problems involving either a deeply embedded rigid disk indentor or a rigid circular indentor which is bonded to the surface of the halfspace can be recovered as special cases. For the formulation of the mixed boundary value problem associated with the internal indentation problem, we utilize the representation of the solution in terms of two harmonic functions $f(r, z)$ and $g(r, z)$, which satisfy

$$\nabla^2 f = 0; \quad \nabla^2 g = 0 \tag{13.124}$$

The displacement and stress components relevant to the formulation of the embedded indentor problem can be expressed in terms of $f(r, z)$ and $g(r, z)$ as follows:

$$u_r = \frac{\partial f}{\partial r} - z \frac{\partial g}{\partial r}$$

$$u_z = (3 - 4v)g + \frac{\partial f}{\partial z} - z \frac{\partial g}{\partial z} \tag{13.125}$$

and

$$\sigma_{zz} = 2G \left[2(1 - v) \frac{\partial g}{\partial z} + \frac{\partial^2 f}{\partial z^2} - z \frac{\partial^2 g}{\partial z^2} \right]$$

$$\sigma_{rz} = 2G \frac{\partial}{\partial r} \left[(1 - 2v)g + \frac{\partial f}{\partial z} - z \frac{\partial g}{\partial z} \right] \tag{13.126}$$

The boundary conditions governing the embedded indentor problem can be specified in relation to the layer region (denoted by superscript[1] and occupying $0 \le r < \infty; -h \le z \le 0$) and the halfspace region (denoted by superscript[2] and occupying $0 \le r < \infty; 0 \le z < \infty$). The boundary and continuity conditions governing the problem are

$$u_z^{(1)} = u_z^{(2)} = \tilde{\Delta}_e; \quad z = 0; \quad r\varepsilon(0, a) \tag{13.127}$$

$$u_r^{(1)} = u_r^{(2)} = 0; \quad z = 0; \quad r\varepsilon(0, a) \tag{13.128}$$

$$u_r^{(1)} = u_r^{(2)}; \quad z = 0; \quad r\varepsilon(a, \infty) \tag{13.129}$$

$$u_z^{(1)} = u_z^{(2)}; \quad z = 0; \quad r\varepsilon(a, \infty) \tag{13.130}$$

$$\sigma_{zz}^{(1)} = \sigma_{zz}^{(2)}; \quad z = 0; \quad r\varepsilon(a, \infty) \tag{13.131}$$

$$\sigma_{rz}^{(1)} = \sigma_{rz}^{(2)}; \quad z = 0; \quad r\varepsilon(a, \infty) \tag{13.132}$$

$$\sigma_{zz}^{(1)} = 0; \quad z = -h; \quad r\varepsilon(a, \infty) \tag{13.133}$$

$$\sigma_{rz}^{(1)} = 0; \quad z = -h; \quad r\varepsilon(a, \infty) \tag{13.134}$$

The solutions of Eq. (13.124) which are applicable in the layer region[(1)] and the halfspace region[(2)], respectively, have the general forms

$$f^{(1)}(r, z) = \int_0^\infty \xi[A(\xi)e^{-\xi z} + B(\xi)e^{\xi z}]J_0(\xi r)\, d\xi$$

$$g^{(1)}(r, z) = \int_0^\infty \xi[C(\xi)e^{-\xi z} + D(\xi)e^{\xi z}]J_0(\xi r)\, d\xi \tag{13.135}$$

and

$$f^{(2)}(r, z) = \int_0^\infty \xi F(\xi)e^{-\xi z}J_0(\xi r)\, d\xi$$

$$g^{(2)}(r, z) = \int_0^\infty \xi G(\xi)e^{-\xi z}J_0(\xi r)\, d\xi$$

where $A(\xi)$, $B(\xi)$, etc. are arbitrary functions. Selvadurai (1993) has shown that by using the representations in Eqs. (13.135) and (13.136), the mixed boundary conditions can be effectively reduced to the solution of a pair of coupled Fredholm integral equations of the form

$$\Omega_1^*(t) + \int_0^a \Omega_1^*(u)K_{11}(u, t)\, du + \int_0^a \Omega_2^*(u)K_{12}(u, t) = 0; \qquad 0 \le t \le a \tag{13.137}$$

$$\Omega_2^*(t) + \int_0^a \Omega_2^*(u)K_{21}(u, t)\, du + \int_0^a \Omega_2^*(u)K_{22}(u, t) = -\frac{4\tilde{\Delta}_e}{\pi}; \quad 0 \le t \le a \tag{13.138}$$

where $\Omega_1^*(t)$ and $\Omega_2^*(t)$ are unknown functions, and $K_{ij}(i, j = 1, 2)$ are kernel functions which take the forms

$$K_{11}(u, t) = -\frac{4t}{(3 - 4v)\pi} \int_0^\infty \frac{\cos(\xi u)\{1 - \cos(\xi t)\}}{\{1 + \coth(\xi h)\}} \left[-\frac{(3 - 4v)}{2}[1 + \coth(\xi h)] \right.$$

$$+ (\xi h + 2v - 1)(\xi h \coth(\xi h) + 1 - 2v)$$

$$\left. + \{2(1 - v) - \xi h\}\{2(1 - v)\coth(\xi h) + \xi h\} \right] \tag{13.139}$$

$$K_{12}(u, t) = \frac{4t}{(3 - 4v)\pi} \int_0^\infty \frac{\cos(\xi u)\{1 - \cos(\xi t)\}}{\xi\{1 + \coth(\xi h)\}} [(1 + \xi h - 2v)\{2(1 - v)\coth(\xi h) + \xi h\}$$

$$- \{\xi h + 2(1 - v)\}\{\xi h \coth(\xi h) + 1 - 2v\}] \tag{13.140}$$

etc.

Considering the equilibrium of the embedded indentor, it can be shown that

$$P = \frac{8\pi\mu(1-\nu)}{(3-4\nu)} \int_0^a \Omega_2(t)\, dt \tag{13.141}$$

The system of coupled integral equations can be solved numerically to obtain the load-displacement relationship for the embedded rigid disk indentor. These numerical results can be calibrated via exact solutions for the indentor load-displacement relationships which are applicable to limiting cases as either $(h/a) \to \infty$ or $(h/a) \to 0$.

In the case when $(h/a) \to \infty$, the problem reduces to that of the rigid disk indentor which is in bonded contact with an isotropic elastic medium of infinite extent. The exact solution for the load-displacement relationship is given by Eq. (13.123). In the specific case when $(h/a) \to 0$, the problem reduces to that of the indentation of the surface of the halfspace by a *bonded* rigid indentor of radius a. This problem was examined by Mossakovskii (1954) and Ufliand (1956) (see also Gladwell, 1980), by solving the mixed boundary value problem

$$u_z(r, 0) = \Delta_0; \quad 0 \le r \le a \tag{13.142}$$

$$u_r(r, 0) = 0; \quad 0 \le r \le a \tag{13.143}$$

$$\sigma_{zz}(r, 0) = 0; \quad a < r < \infty \tag{13.144}$$

$$\sigma_{rz}(r, 0) = 0; \quad a < r < \infty \tag{13.145}$$

Therefore, in contrast to mixed boundary value problem defined by Eqs. (13.1) to (13.3), the last condition (Eq. (13.3)) applicable to σ_{rz} in the range $r\varepsilon(0, \infty)$ is replaced by a mixed set of boundary conditions applicable to u_r and σ_{rz} in the ranges $r\varepsilon(0, a)$ and $r\varepsilon(a, \infty)$, respectively. The solution of the mixed boundary value problem developed by Mossakovskii (1954) and Ufliand (1956) incorporates the oscillatory form of the stress singularity at the boundary of the rigid indentor. The exact closed form solution based on the Hilbert problem is given by

$$\frac{P}{4\mu\Delta_0 a} = \frac{\ln(3-4\nu)}{(1-2\nu)} \tag{13.146}$$

It may be noted that the mixed boundary value problem defined by Eqs. (13.142) to (13.145) can also be solved by using a Hankel transform approach, in which the oscillatory stress singularities are suppressed and the problem is effectively reduced to the solution of a single Fredholm integral equation of the second-kind. Selvadurai (1989) has shown that the incorporation of the oscillatory form of the stress singularity does not significantly influence the axial stiffness of the rigid circular indentor bonded to a halfspace region. Comparison of the results obtained by the two schemes indicate that the maximum difference between the two sets of solutions does not exceed 0.5% when $\nu = 0$ and when $\nu \to 1/2$, the oscillatory form of the stress singularity disappears and Eq. (13.146) reduces to the classical result applicable to the indentation problem defined by Eqs. (13.1) to (13.3). Figure 13.10 illustrates the results for the axial stiffness of the indentor which is located at an arbitrary depth (h/a) beneath the surface of the halfspace region. The influence of the free surface $z = -h$ becomes important as $(h/a) \to 0$. The stiffness of the disk indentor increases as (h/a) increases, and approaches the limit for the rigid disk indentor embedded in an elastic infinite space region when $(h/a) \simeq 7$.

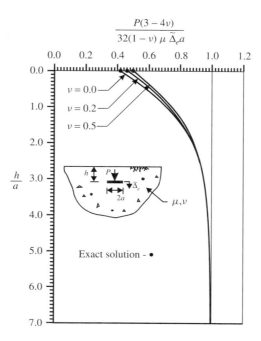

Figure 13.10   Influence of the depth of embedment on the axial stiffness of the embedded rigid circular indentor

13.5   Concluding Remarks

The classical theory of elasticity is a mathematical idealization which has had a powerful influence in the development of the engineering sciences. One of the attractions of the application of classical elasticity to engineering problems is the ability to develop results which have a certain lasting value in the estimation of the response in the working load range. The modeling of the deformable medium as a linearly elastic solid has definite limitations in terms of its applicability to natural geologic materials such as soils and rocks, which invariably have much more complicated constitutive responses which are both nonlinear and time-dependent. Despite these constraints, the linear elastic continuum model continues to serve a useful function in identifying performance of interactive problems in geomechanics. The present contribution focuses on a very limited class of classical problems in geomechanics which deal with indentation of elastic geomaterials with rigid indentors with a circular planform. The regions of the geomaterial are restricted to half-space and infinite space regions which are isotropic. While extensions can be made to reduce these constraints, the primary objective of the study is to present the basic methodology associated in modeling a physical problem in geomechanics by its mathematical equivalent. It is shown that, within the context of a uniquely defined problem, several approaches can be used to solve the mathematical problem. This is illustrated by appeal to the classical problem of the indentation of a halfspace region by a rigid smooth indentor with a flat base.

The presentation also discusses the extension of this classical contact problem to include interaction effects of loads that are located exterior to the rigid indentor. The closed form nature of these results of engineering interest is of particular value in their applicability to geomaterials. The presentation also discusses the application of the elastostatic modeling to the analyses of problems where the embedded nature of the indentor results in a problem where the solution to even the elastostatic problem can be attempted only via numerical techniques. The results of interest to geomechanics can, however, be obtained in convenient forms. The subject matter presented in this chapter has also been extended to cover the class of geomaterials which exhibit anisotropic and inhomogeneous elastic properties that are frequently encountered in the modeling of stratified geomaterials.

Acknowledgements

The work described in this chapter was supported by research grants awarded to the author by the Natural Sciences and Engineering Research Council of Canada over the past two decades. The author is grateful to a number of graduate students and post-doctoral research fellows who have contributed to the geomechanics research program at Carleton University, Ottawa, Canada and, to a limited extent, also at McGill University.

References

Agbezuge, L. K. and Deresiewicz, H. The consolidation settlement of a circular footing. *Isr. J. Tech.*, **13**, (1975) 264–269.

Boussinesq, J. *Applications des potentiels a l'étude de l'equilibre et du mouvement des solides elastique*. Gauthier-Villars, Paris, (1885).

Bycroft, G. N. Forced vibrations of a rigid circular plate on a semi-infinite elastic space and on an elastic stratum. *Phil. Trans. Roy. Soc.*, **A 248**, (1956) 327–368.

Collins, W. D. Some axially symmetric stress distributions in elastic solids containing penny-shaped cracks, I. Cracks in an infinite solid and a thick plate. *Proc. Roy. Soc.*, **A 203**, (1962) 359–386.

Copson, E. T. On the problem of the electrified disc. *Proc. Edin. Math. Soc.*, **8**, (1947) 14–19.

Chiarella, C. and Booker, J. R. The time-settlement behaviour of a rigid die resting on a deep clay layer. *Quart. J. Mech. Appl. Math.*, **28**, (1975) 317–328.

Davis, R. O. and Selvadurai, A. P. S. *Elasticity and Geomechanics*. Cambridge University Press, Cambridge, (1996).

de Pater, A. D. and Kalker, J. J. (eds.). *The Mechanics of Contact Between Deformable Bodies: Proc. IUTAM Symposium*, Delft University Press, Enschede, (1975).

Desai, C. S. and Christian, J. T. (eds.). *Numerical Methods in Geotechnical Engineering*. Wiley, London, (1977).

Florence, A. L. Two contact problems for an elastic layer. *Quart. J. Mech. Appl. Math.*, **14**, (1961) 453–459.

Fung, Y. C. *Foundations of Solid Mechanics*. Prentice-Hall, Englewood Cliffs, NJ, (1964).

Galin, L. A. *Contact Problems in the Theory of Elasticity*. (English Translation edited by I. N. Sneddon). Technical Report G16447, North Carolina State College, Raleigh, NC, (1961).

Gibson, R. E. The analytical method in soil mechanics, Rankine Lecture. *Geotechnique*, **24**, (1974) 115–140.

Gladwell, G. M. L. *Contact Problems in the Classical Theory of Elasticity*. Sijthoff and Noordhoff, The Netherlands, (1980).

Green, A. E. and Zerna, W. *Theoretical Elasticity, 2nd Ed.* Clarenden Press, Oxford, (1968).

Gurtin, M. E. The linear theory of elasticity. In: *Mechanics of Solids II, Encyclopedia of Physics*, Flugge, S. (ed.). Vol. VIa/2, 1-295. Springer-Verlag, Berlin, (1972).

Harding, J. W. and Sneddon, I. N. The elastic stresses produced by the indentation of the plane surface of a semi-infinite solid by a rigid punch. *Proc. Camb. Phil. Soc.*, **41**, (1945) 16–26.

Hetenyi, M. Beams, plates on elastic foundations and related problems. *Appl. Mech. Rev.*, **19**, (1966) 95–102.

Jaeger, J. C. and Cook, N. G. W. *Fundamentals of Rock Mechanics*. Chapman & Hall, London, (1976).

Jeans, J. H. *The Mathematical Theory of Electricity and Magnetism*. Cambridge University Press, London, (1925).

Johnson, K. L. *Contact Mechanics*. Cambridge University Press, London, (1985).

Kanwal, R. P. and Sharma, D. L. Singularity methods for elastostatics. *J. Elasticity*, **6**, (1976) 405–418.

Korenev, B. G. Structures resting on elastic foundations. In: *Structural Mechanics in the USSR*, (Rabinovich, I. M. (ed.)). Pergamon Press, Oxford, (1960) 160–190.

Krynine, D. P. *Soil Mechanics*. McGraw-Hill, New York, (1941).

Little, R. W. *Elasticity*. Prentice-Hall, Englewood Cliffs, NJ, (1973).

Love, A. E. H. *A Treatise on the Mathematical Theory of Elasticity*. Clarendon Press, Oxford, (1927).

Mindlin, R. D. Force at a point in the interior of a semi-infinite solid. *Physics*, **7**, (1936) 195–202.

Mossakovskii, V. I. The fundamental mixed boundary problem of the theory of elasticity for a halfspace with a circular line separating the boundary conditions. *Prikl. Math. Mekh.*, **18**, (1954) 187–196.

Muki, R. Asymmetric problems of the theory of elasticity for a semi-infinite solid and a thick plate. In: *Progress in Solid Mechanics*, Sneddon, I. N. and Hill, R. (eds.). North Holland, Amsterdam, Vol. 1, (1960) 339–349.

Noble, B. Certain dual integral equations. *J. Math. Phys.*, **37**, (1958) 128–136.

Poulos, H. G. and Davis, E. H. *Elastic Solutions for Soil and Rock Mechanics*. Wiley, New York, (1975).

Selvadurai, A. P. S. The load-deflexion characteristics of a deep rigid anchor in an elastic medium. *Géotechnique*, **26**, (1976) 603–612.

Selvadurai, A. P. S. The interaction between a rigid circular punch on an elastic halfspace and a Mindlin force. *Mech. Res. Comm.*, **5**, (1978) 57–64.

Selvadurai, A. P. S. *Elastic Analysis of Soil-Foundation Interaction. Developments in Geotechnical Engineering, Vol. 17*. Elsevier, Amsterdam, (1979a).

Selvadurai, A. P. S. The displacement of a rigid circular foundation anchored to an isotropic elastic halfspace. *Géotechnique*, **29**, (1979b) 195–202.

Selvadurai, A. P. S. Elastic contact between a flexible circular plate and a transversely isotropic elastic halfspace. *Int. J. Solids Structures*, **16**, (1980a) 167–176.

Selvadurai, A. P. S. On the displacement induced in a rigid circular punch on an elastic halfspace due to an external force. *Mech. Res. Comm.*, **7**, (1980b) 351–358.

Selvadurai, A. P. S. Betti's reciprocal relationships for the displacements of an elastic infinite space bounded internally by a rigid inclusion. *J. Struct. Mech.*, **9**, (1981) 199–210.

Selvadurai, A. P. S. The additional settlement of a rigid circular foundation on an isotropic elastic halfspace due to multiple distributed external loads. *Géotechnique*, **32**, (1982) 1–7.

Selvadurai, A.P.S. Fundamental results concerning the settlement of a rigid foundation on an elastic medium due to an adjacent surface load. *Int. J. Num. Analytical Meth. Geomech.*, **7**, (1983) 209–223.

Selvadurai, A. P. S. The influence of boundary fracture on the stiffness of a deeply embedded anchor plate. *Int. J. Num. Analytical Mech. Geomech.*, **13**, (1989) 159–170.

Selvadurai, A. P. S. The axial loading of a rigid circular anchor plate embedded in an elastic halfspace. *Int. J. Num. Analytical Meth. Geomech.*, **17**, (1993) 343–353.

Selvadurai, A. P. S. The settlement of a rigid circular foundation resting on a halfspace exhibiting a near surface elastic non-homogeneity. *Int. J. Num. Analytical Meth. Geomech.*, **20**, (1996) 351–364.

Selvadurai, A. P. S. Analytical methods for embedded flat anchor problems in geomechanics. *Proc. IACMAG '94*, (H. J. Siriwardane and M. M. Zaman, (eds)), Morgantown, West Virginia. Balkema, vol. 1, (1994) 305–321.

Selvadurai, A. P. S. *Partial Differential Equations; Modelling and Applications in Mechanics.* Vol II Springer-Verlag, Berlin, (2000)

Selvadurai, A. P. S. and Yue, Z. Q. On the indentation of a poroelastic layer. *Int. J. Num. Anal. Meth. in. Geomech.*, **18**, (1994) 161–175.

Sneddon, I. N. *Fourier Transforms.* McGraw-Hill, New York, (1951).

Sneddon, I. N. The relationship between load and penetration in the axisymmetric Boussinesq problem for a punch of arbitrary profile. *Int. J. Engrg. Sci.*, **3**, (1965) 47–57.

Sneddon, I. N. and Lowengrub, M. *Crack Problems in the Classical Theory of Elasticity.* Wiley, New York, (1969).

Stagg, K. G. and Zienkiewicz, O. C. (eds.). *Rock Mechanics in Engineering Practice.* Wiley, New York, (1968).

Sternberg, E. On some recent developments in the linear theory of elasticity. In: *Structural Mechanics: Proc. 1st Symposium on Naval Structural Mechanics*, Goodier, J. N. and Hoff, N. J. (eds.). Pergamon Press, Oxford, (1960) 48–72.

Terzaghi, K. *Theoretical Soil Mechanics.* Wiley, New York, (1943).

Timoshenko, S. P. and Goodier, J. N. *Theory of Elasticity.* McGraw-Hill, New York, (1970).

Truesdell, C. Invariant and complete stress functions for general continua. *Arch. Rat. Mech. Anal.*, **4**, (1959) 1–29.

Tschebotarioff, G. P. *Foundations, Retaining and Earth Structures.* McGraw-Hill, New York, (1973).

Ufliand, Ia. S. The contact problem of the theory of elasticity for a die, circular in its plane, in the presence of adhesion. *Prikl. Math. Mekh.*, **20**, (1956) 578–587.

Ufliand, Ia. S. *Survey of Articles on the Applications of Integral Transforms in the Theory of Elasticity* (translation edited by I. N. Sneddon). Technical Report 65-1556, North Carolina State College, Raleigh, NC, (1965).

Westergaard, H. M. *Theory of Elasticity and Plasticity.* Harvard Monographs in Applied Science, No. 3. Wiley, New York, (1952).

Yue, Z. Q. and Selvadurai, A. P. S. Contact problems for saturated poroelastic acid, J. Engineering Mechanics, ASCE, **121**, (1995a) 502–512.

Yue, Z. Q and Selvadurai, A. P. S. On the mechanics of a rigid disc inclusion embedded in a fluid-saturated poroelastic medium, *Int. J. Engineering Science*, **33**, (1995b) 1633–1662.

Zienkiewicz, O. C. and Taylor, R. L. *The Finite Element Method.* McGraw-Hill, New York, (1989).

14

Application of the Theory of Classical Plasticity to the Analysis of the Stress Distribution in Wedges of a Perfectly Frictional Material

John R. Booker and **X. Zheng**
The University of Sydney, Sydney, NSW, Australia

14.1 Introduction

Soil and rock are complex materials with a complex behavior. For this reason, geotechnical engineers have often adopted approximate and empirical methods to try and analyse their behavior. These methods, such as the slip-circle analysis or the sliding wedge analysis, were often very restricted in their range of application, and could only be applied with confidence to a narrow range of situations. In recent years, the theory of plasticity has been used to characterize the behavior of soil and rock. This sound theoretical basis has made it possible to examine many practical situations by formulating them as boundary value problems within a well defined theory, and using modern numerical and analytical methods to analyse these cases.

There are two important methods of analysis used in the theory of plasticity: the most powerful of these is the load path method, which examines the behavior of the material under consideration at every stage of loading, usually by determining the increment of response to each increment of load. This is a very powerful technique, and is often associated with a finite element approximation of the governing equations. It can be used to examine the complete response of the material as the loads are increased, and ultimately to examine the collapse or failure of the material, viz. the point at which the material undergoes very large deformations even when there is no appreciable increase in loads. Often it is only this collapse load which is of interest, as the engineer or designer is primarily concerned with ensuring that there is an adequate safety factor against collapse.

In such cases, it is possible to use the classical theory of plasticity to determine the collapse load without following the details of the complete loading path.

In this chapter, it will be shown that there are a number of important practical problems, for a non-cohesive soil, which may be formulated in terms of the plastic equilibrium of a wedge of such material. This problem will be examined in detail using the classical theory of plasticity, and the method of characteristics and a comprehensive theory will be developed. The theory will then be used to develop solutions to a number of important practical problems, such as determining the bearing capacity of trapezoidal granular embankments.

14.2   Elements of the Theory of Plasticity

14.2.1   Failure

When an element of a perfectly plastic material is loaded, it will initially respond elastically. If the loads are further increased, the element will reach a stage where very large deformations occur even though there is very little change in the applied loads, at this stage the element is said to be in a state of failure, and the element stresses are said to be in a failure state. If the collection of all possible failure states is plotted in stress space they form a failure surface. States of stress within the surface are said to be in the elastic range or safe, while states of stress on the failure surface are said to be failure states. In this chapter, attention will be restricted to a perfectly plastic material which is deforming under conditions of plane strain in the x-y plane. Under such circumstances, the failure condition may be written in the form

$$f(\sigma_{xx}, \sigma_{yy}, \sigma_{xy}) = 0 \qquad (14.1)$$

where $\sigma_{xx}, \sigma_{yy}, \sigma_{xy}$ denote the Cartesian stress components, and where the usual geomechanics convention of regarding compressive stresses as positive is adopted.

For the purposes of this discussion, it will be convenient to introduce an alternative set of variables:

$$X = \tfrac{1}{2}(\sigma_{xx} - \sigma_{yy}) = R\cos(2\Omega)$$

$$Y = \sigma_{xy} \qquad\qquad\quad = R\sin(2\Omega)$$

$$p = \tfrac{1}{2}(\sigma_{xx} + \sigma_{yy}) \qquad\qquad (14.2)$$

where it will be recognized that p is the mean stress, R is the radius of the Mohr circle and Ω is the angle between the major principal stress direction and the x axis.

In terms of these variables, the failure surface may be expressed in the form

$$f(\sigma_{xx}, \sigma_{yy}, \sigma_{xy}) = R - F(p, \Omega) = 0 \qquad (14.3)$$

Failure in the soil can often be represented by the conventional Coulomb failure relationship, so that failure occurs as soon as the shear stress (τ) reaches a critical value, related to the normal stress (σ_n), on any plane:

$$|\tau| = c + \sigma_n \tan(\phi) \qquad (14.4a)$$

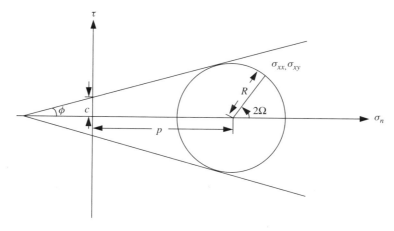

Figure 14.1   Mohr Coulomb failure

where c is the cohesion and ϕ is the angle of internal friction. In future discussions, it will prove useful to introduce the angle μ defined as follows:

$$\mu = \frac{\pi}{4} - \frac{\phi}{2} \tag{14.4b}$$

The Coulomb relation also implies that the Mohr circle of stress must just touch the Coulomb envelope, as shown in Figure 14.1, and that Coulomb failure is achieved along two planes inclined at an angle of $\pm\mu$ to the principal stress direction. It follows from Figure 14.1 that

$$R = \sin(\phi)(p + q) \tag{14.4c}$$

where

$$q = c\cot(\phi) \tag{14.4d}$$

14.2.2   Stress equations

It will be assumed that when an element is in either an elastic state or a plastic state, it must also be in equilibrium, and thus:

$$\frac{\partial \sigma_{xx}}{\partial x} + \frac{\partial \sigma_{xy}}{\partial y} = \gamma_x \tag{14.5a}$$

$$\frac{\partial \sigma_{xy}}{\partial x} + \frac{\partial \sigma_{yy}}{\partial y} = \gamma_y \tag{14.5b}$$

where γ_x and γ_y denote the components of body force in the x and y directions, respectively.

Equations (14.4) and (14.5) constitute a set of hyperbolic equations which can be used to determine the stress state in a plastic region. These equations have been examined in detail by Sokolvski (1960), Hill (1950), Salencon (1977), and Booker and Davis (1977).

The canonical form of these equations is

$$- \sin(2\mu)\frac{\partial p}{\partial s_\alpha} + 2R\frac{\partial \Omega}{\partial s_\alpha} + Q_\alpha = 0$$

$$+ \sin(2\mu)\frac{\partial p}{\partial s_\beta} + 2R\frac{\partial \Omega}{\partial s_\beta} + Q_\beta = 0 \qquad (14.6a)$$

with

$$Q_\alpha = \frac{\partial x}{\partial s_\alpha}[\gamma_x \sin(2\mu) - \gamma_y \cos(2\mu)] + \frac{\partial y}{\partial s_\alpha}[\gamma_x \cos(2\mu) + \gamma_y \sin(2\mu)]$$

$$Q_\beta = \frac{\partial x}{\partial s_\beta}[-\gamma_x \sin(2\mu) - \gamma_y \cos(2\mu)] + \frac{\partial y}{\partial s_\beta}[\gamma_x \cos(2\mu) - \gamma_y \sin(2\mu)]$$

The expressions $\partial/\partial s_\alpha$, $\partial/\partial s_\beta$ indicate derivatives with respect to arc length along the α, β characteristics, respectively. These characteristics are defined by the following relationships:

$$\frac{\partial y}{\partial s_\alpha} = \tan(\Omega - \mu)\frac{\partial x}{\partial s_\alpha}$$

$$\frac{\partial y}{\partial s_\beta} = \tan(\Omega + \mu)\frac{\partial x}{\partial s_\beta} \qquad (14.6b)$$

Equations (14.6b) show that the α, β characteristics are inclined to the principal stress at the angles $-\mu$ and $+\mu$, respectively, as shown in Figure 14.2.

It is interesting that the characteristics thus coincide with the planes on which the Coulomb failure condition is mobilized (Eq. (14.4)). The characteristic lines for bearing capacity failure induced by a smooth rigid strip footing of width B on a deep homogeneous deposit are shown in Figure 14.3 for the particular case in which $\gamma B/c = 8.4$ and $\phi = 30°$.

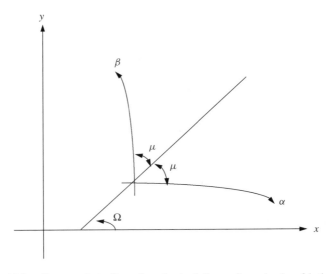

Figure 14.2   Characteristic lines for plastic failure of a cohesive frictional soil

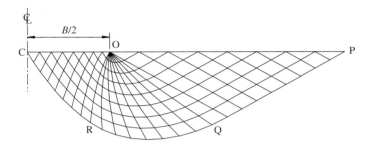

Figure 14.3 Characteristics for the bearing capacity failure of a smooth rigid strip footing on a deep homogeneous soil

The characteristics can be used to determine the stress distribution, as outlined in the appendix. Some important properties of characteristics are also discussed in that section.

14.2.3 Plastic strain rates

In the theory of plasticity, strains are considered as the sum of two distinct components, an elastic component, indicated by the superscript E, and a plastic component, indicated by the superscript P. Because elastic-plastic behavior is best considered as incremental, the relationships are usually expressed in terms of stress rates and strain rates, and thus

$$\dot{\varepsilon} = \dot{\varepsilon}^E + \dot{\varepsilon}^P \tag{14.7a}$$

where the elastic strain component is defined as the strain which would occur if the material responded elastically to the applied stress increases, and thus

$$\dot{\varepsilon}^E = \mathbf{D}^{-1}\dot{\sigma} \tag{14.7b}$$

where $\mathbf{D}$ is the matrix of elastic moduli.

Equation (14.7a) then defines the plastic strain rate. It is found, for a perfectly plastic material, that the actual value of the plastic strain rate cannot be determined from the stress state alone. However, it is found that the ratio of strain rate components is uniquely defined, and depends only upon the stress state. This is often written in the form

$$\dot{\varepsilon}^p_{ij} = \Lambda \frac{\partial g}{\partial \sigma_{ij}} \tag{14.8}$$

where g is a function of stress which is called the plastic potential, and Λ is a one-signed multiplier. It can be seen that the plastic strain rates are normal to the plastic potential when plotted in stress space. If the plastic potential and the failure surface coincide, the material is said to have an associated flow rule. Under such circumstances, it is possible to prove the limit theorems (Drucker Prager and Greenberg, 1952), which show that a statically admissible stress field provides a lower bound to the collapse load, while a kinematically admissible velocity field leads to an upper bound to the collapse load. Some care is necessary in using Eq. (14.8), as it treats the stress components σ_{ij} and σ_{ji} as being independent.

When a body has reached its collapse load it will undergo very large displacement at constant load. This also implies that the stress state will remain constant, and thus that the elastic strain rate is zero. As a consequence, at collapse an elastic-plastic material responds exactly in the same way as a rigid-plastic material, since the actual strain rate is identical to the plastic strain rate. For a Mohr Coulomb material with an associated flow rule, the failure condition can be written as

$$f = 2(p\sin(\phi) + c\cos(\phi) - R) \tag{14.9a}$$

and it is found that the expressions for the strain rates are

$$\frac{\partial v_x}{\partial x} = \Lambda(\cos(2\mu) - \cos(2\Omega)) \tag{14.9b}$$

$$\frac{\partial v_y}{\partial y} = \Lambda(\cos(2\mu) + \cos(2\Omega)) \tag{14.9c}$$

$$\frac{\partial v_x}{\partial y} + \frac{\partial v_y}{\partial x} = -2\Lambda(\sin(2\Omega)) \tag{14.9d}$$

where $\Lambda > 0$ and v_x and v_y are the x and y components of velocity, respectively.

If Λ is eliminated from Eq. (14.8) and the resulting equations are expressed in terms of differentiation along the stress characteristics, it is found that

$$\frac{\partial v_x}{\partial s_\alpha}\cos(\Omega - \mu) + \frac{\partial v_y}{\partial s_\alpha}\sin(\Omega - \mu) = 0 \tag{14.10a}$$

$$\frac{\partial v_x}{\partial s_\beta}\cos(\Omega + \mu) + \frac{\partial v_y}{\partial s_\beta}\sin(\Omega + \mu) = 0 \tag{14.10b}$$

It can be seen that Eqs. (14.10a) and (14.10b) are in canonical characteristic form, and thus the velocity characteristics are identical to the stress characteristics. Once the stress characteristics have been determined the velocity field can be determined in exactly the same manner as the stress field, i.e. as outlined in the appendix. It can be shown using the properties of characteristic equations that a line separating a plastic deforming region from a rigid region must always be a velocity characteristic.

It is sometimes convenient to introduce velocity components (v_α) tangential to the α characteristic and the velocity components (u_α) normal to the α characteristic with similar definitions for v_β, u_β:

$$V_\alpha = V_x\cos(\Omega - \mu) + v_y\sin(\Omega - \mu) \tag{14.11a}$$

$$\mu_\alpha = V_x\sin(\Omega - \mu) + v_y\cos(\Omega - \mu) \tag{14.11b}$$

$$V_\beta = V_x\cos(\Omega + \mu) + v_y\sin(\Omega + \mu) \tag{14.11c}$$

$$V_\beta = -V_x\sin(\Omega + \mu) + v_y\cos(\Omega + \mu) \tag{14.11d}$$

It can be shown by considering a transition zone in which there is a rapid change in velocity that the quantity v_β is continuous across an α characteristic, and similarly, that the quantity v_α is continuous across a β characteristic. In terms of the above variables,

Eqs. (14.10a) and (14.10b) become

$$\frac{\partial v_\alpha}{\partial s_\alpha} = u_\alpha \frac{\partial \Omega}{\partial s_\alpha} \tag{14.12a}$$

$$\frac{\partial v_\beta}{\partial s_\beta} = u_\beta \frac{\partial \Omega}{\partial s_\beta} \tag{14.12b}$$

It is important to check that the quantity Λ is positive; this may be done by evaluating the expression

$$\Lambda = -\frac{\dfrac{\partial v_\alpha}{\partial s_\beta} - u_\alpha \dfrac{\partial \Omega}{\partial s_\beta} + \dfrac{\partial v_\beta}{\partial s_\alpha} - u_\beta \dfrac{\partial \Omega}{\partial s_\alpha}}{2 \sin(2\mu)^2} \tag{14.12c}$$

The quantities $u_\alpha, v_\alpha, u_\beta v_\beta$ are not independent, and thus that the quantities u_α, u_β can be expressed in terms of the quantities v_α, v_β as follows:

$$u_\alpha = \frac{v_\beta - \cos(2\mu)v_\alpha}{+\sin(2\mu)} \tag{14.12d}$$

$$u_\beta = \frac{v_\alpha - \cos(2\mu)v_\beta}{+\sin(2\mu)} \tag{14.12e}$$

so that the entire computation of the velocity field can be expressed in terms of the variables v_α, v_β.

14.3 Analysis of a Granular Wedge

The problem of the distribution of stress in an infinite granular (non-cohesive) wedge is a fundamental problem of the theory of plasticity (see Sokolvsky, 1960).

Such fields occur in many applications; for example, a characteristic field of the type shown in Figure 14.3 can be used to determine the stress distribution necessary to cause plastic collapse beneath a smooth rigid footing. The variation of this distribution of stress (q) is plotted against the semi-width (L) of the footing in a dimensionless form in Figure 14.4 for the particular case of $\phi = 30°$.

It can be seen that as c approaches zero, the stress distribution becomes linear varying from a value of zero at the edge to a maximum value at the center of the footing. It can also be seen from Eq. (A14.2) in the appendix that the stress distribution in the Rankine passive zone outside the footing is radially proportional, in the sense that the stress state at any point on a radial line passing through the edge of the footing is proportional to the distance from the edge of the footing. It may be verified that this is also true throughout the entire plastic region.

For this reason, it is of interest to consider the behavior of a wedge of purely frictional material which is subjected to radially proportional tractions, as depicted in Figure 14.5.

For the situation depicted in Figure 14.5, it is easy to establish that the stress field must be radially proportional, and thus

$$\sigma_{rr} = rS_{rr}(\theta) \tag{14.13a}$$

$$\sigma_{\theta\theta} = rS_{\theta\theta}(\theta) \tag{14.13b}$$

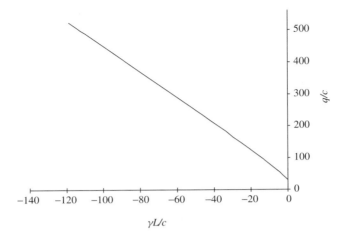

Figure 14.4   Stress distribution beneath a smooth rigid footing

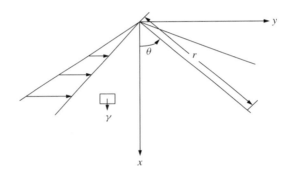

Figure 14.5   Infinite granular wedge

$$\sigma_{r\theta} = rS_{r\theta}(\theta) \tag{14.13c}$$

where σ_{rr}, $\sigma_{\theta\theta}$, $\sigma_{r\theta}$ denote the stress components in polar coordinates.

The stress field must satisfy the equations of equilibrium, which in polar coordinates are

$$\frac{\partial \sigma_{rr}}{\partial r} + \frac{1}{r}\frac{\partial \sigma_{r\theta}}{\partial r} + \frac{\sigma_{rr} - \sigma_{\theta\theta}}{r} = \gamma \cos(\theta) \tag{14.14a}$$

$$\frac{\partial \sigma_{r\theta}}{\partial r} + \frac{1}{r}\frac{\partial \sigma_{\theta\theta}}{\partial r} + \frac{2\sigma_{r\theta}}{r} = -\gamma \sin(\theta) \tag{14.14b}$$

where γ is the unit weight of the material and the direction of gravity is assumed to be in the x direction.

The stress state must also lie on the failure surface, and thus introducing Mohr's representation (Eq. (14.2)) and the failure condition for a purely frictional material, it follows that

$$\sigma_{rr} = rs[1 + \sin(\phi)\cos(2\chi)] \tag{14.15a}$$

$$\sigma_{\theta\theta} = rs[1 - \sin(\phi)\cos(2\chi)] \tag{14.15b}$$

$$\sigma_{r\theta} = r\sin(\phi)\sin(2\chi) \tag{14.15c}$$

where

$$p = rs(\theta) \tag{14.15d}$$

$$\chi = \Omega(\theta) - \theta \tag{14.15e}$$

It follows from Eqs. (14.14) and (14.15) that the partial differential equations of plasticity (Eq. (14.6a)) can be reduced to a pair of ordinary differential equations:

$$\frac{d\Omega}{d\theta} = \frac{\gamma[\cos(\theta) - \cos(2\mu)\cos(2\chi + \theta)] - s\sin^2(2\mu)}{2s\cos(2\mu)[\cos(2\chi) - \cos(2\mu)]} \tag{14.16a}$$

$$\frac{ds}{d\theta} = \frac{s\sin(2\chi) - \gamma\sin(2\chi + \theta)}{[\cos(2\chi) - \cos(2\mu)]} \tag{14.16b}$$

where $\mu = \pi/4 - \phi/2$.

It is interesting to observe that in polar coordinates, the characteristics satisfy the following equations:

$$\frac{d\theta}{dr} = \frac{\tan(\chi - \mu)}{r} \quad \text{on an } \alpha \text{ line} \tag{14.17a}$$

$$\frac{d\theta}{dr} = \frac{\tan(\chi + \mu)}{r} \quad \text{on a } \beta \text{ line} \tag{14.17b}$$

Thus, once the solution of Eq. (14.16) is known, Eqs. (14.17a) and (14.17b) can be integrated to give the α and β lines passing through the point $r = r_0$, $\theta = \theta_0$:

$$r = r_0 \exp\left(\int_{\theta_0}^{\theta} \cot(\chi(\theta') - \mu)\, d\theta'\right) \quad \alpha \text{ line} \tag{14.17c}$$

$$r = r_0 \exp\left(\int_{\theta_0}^{\theta} \cot(\chi(\theta') + \mu)\, d\theta'\right) \quad \beta \text{ line} \tag{14.17d}$$

It can be seen that the α lines of the wedge field are all similar, as are the β lines.

Equation (14.16) have a singularity when $\cos(2\chi) = \cos(2\mu)$. It follows immediately from Eqs. (14.17a) and (14.17b) that the singularities of Eq. (14.16) correspond to straight line characteristics.

It can be shown that, at such a singular point, that if one numerator of the right-hand side of Eq. (14.16) vanishes, then both vanish. There are thus two situations to consider:

(a) Branch points where both the numerators of the right-hand side of Eq. (14.16) vanish.

(b) Wall points where the numerators of the right-hand side of Eq. (14.16) do not vanish, but the denominators do.

The reason for this nomenclature will emerge as the properties of the singular points are developed.

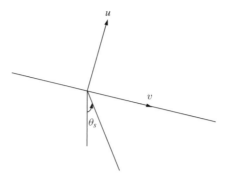

Figure 14.6   Infinite slope of cohesionless soil

In general, it does not appear to be possible to find exact solutions to Eq. (14.16). There is, however, one notable exception. Consider the single stress free slope (180° wedge) shown in Figure 14.6.

In terms of the coordinates u,v shown in Figure 14.6, the equilibrium equations can be written as

$$\frac{\partial \sigma_{uu}}{\partial u} + \frac{\partial \sigma_{uv}}{\partial v} = \gamma \cos(\theta_s) \tag{14.18a}$$

$$\frac{\partial \sigma_{uv}}{\partial u} + \frac{\partial \sigma_{vv}}{\partial v} = -\gamma \sin(\theta_s) \tag{14.18b}$$

Clearly, the solution is independent of the coordinate v, and thus introducing Mohr's representation

$$\sigma_{uv} = p[\cos(2\mu)\sin(2\Omega - 2\theta_s)] = \gamma \cos(\theta_s)v \tag{14.19a}$$

$$\sigma_{vv} = p[1 - \cos(2\mu)\cos(2\Omega - 2\theta_s)] = -\gamma \sin(\theta_s)v \tag{14.19b}$$

It now follows that

$$\cos(2\Omega - \theta_s) = \frac{\cos(\theta_s)}{\sin(\phi)} \tag{14.19c}$$

This equation shows, as is well known, that it is only possible to have a stress free slope for the range $-\phi < \varepsilon < \phi$, and that there are two possible constant values of Ω, one corresponding to passive failure, the other corresponding to active failure (Terzaghi, 1943). This leads to the well-known Rankine solution:

$$\Omega = \Omega_R = \text{constant} \tag{14.20a}$$

$$\chi = \chi_R = \Omega_R - \theta \tag{14.20b}$$

$$p = p_R = \frac{\gamma r[\cos(\theta) - \cos(2\mu)\cos(2\chi_R + \theta)]}{\sin^2(2\mu)} \tag{14.20c}$$

The characteristics for this solution are sets of straight lines. The α characteristics are inclined at an angle $\Omega_R - \mu$ to the x axis, while the β characteristics are inclined at the angle $\Omega_R + \mu$ to the x axis.

It can be verified that this solution, viz.

$$\chi = \chi_R = \Omega_R - \theta \tag{14.21a}$$

$$s = s_R = \frac{\gamma[\cos(\theta) - \cos(2\mu)\cos(2\chi_R + \theta)]}{\sin^2(2\mu)} \tag{14.21b}$$

satisfies Eq. (14.16). The numerator of the second equation of this set vanishes identically for this solution. On either of the straight line characteristics $\theta = \Omega_R - \mu$, $\theta = \Omega_R + \mu$, it is found that $\cos(2\chi) = \cos(2\mu)$ and $s = \gamma \sin(2\chi + \theta)/\sin(2\chi)$, and so these are branch points of the differential Eq. (14.16).

It thus follows that, corresponding to any solution with a branch point, there is an associated Rankine solution which is identical to that solution at the branch point, and represents one extension of that solution beyond the branch point.

14.3.1 Solution in the neighborhood of a branch point

Consider a branch point where

$$\theta = \theta_0 \tag{14.22a}$$

$$\chi = \chi_0 \tag{14.22b}$$

$$\Omega = \Omega_0 = \chi_0 + \theta_0 \tag{14.22c}$$

$$s = s_0 = \frac{\gamma \sin(2\chi_0 + \theta_0)}{\sin(2\chi_0)} \tag{14.22d}$$

with

$$\cos(2\chi_0) = \cos(2\mu) \tag{14.22e}$$

As mentioned previously, at a branch point, one possible continuation is always the associated Rankine solution:

$$\Omega = \Omega_R = \Omega_0 \tag{14.23a}$$

$$\chi = \chi_R = \Omega_0 - \theta \tag{14.23b}$$

$$s = s_R = \frac{\gamma r[\cos(\theta) - \cos(2\chi_0)\cos(2\Omega_0 - \theta)]}{\sin^2(2\chi_0)} \tag{14.23c}$$

Thus, to investigate the solution in the neighborhood of a branch point, it is convenient to introduce

$$\Omega = \Omega_R + \Delta\Omega \tag{14.24a}$$

$$s = s_R + \Delta s \tag{14.24b}$$

$$\theta = \theta_0 + \Delta\theta \tag{14.24c}$$

where Δs, $\Delta\Omega$ both vanish when $\Delta\theta = 0$. In terms of these variables, the differential Eq. (14.16) can be approximated, in the neighborhood of the branch point, by the equations

$$\frac{d(\Delta s)}{d(\Delta\theta)} = +\frac{\Delta s - 2s_0\cot(2\chi_0)\Delta\Omega}{2(\Delta\theta - \Delta\Omega)} \tag{14.24d}$$

$$\frac{d(\Delta\Omega)}{d(\Delta\theta)} = -\frac{\Delta s - 2s_0\cot(2\chi_0)\Delta\Omega}{4s_0\cot(2\chi_0)(\Delta\theta - \Delta\Omega)} \tag{14.24e}$$

It thus follows that

$$\Delta s = -2s_0\cot(2\chi_0)\Delta\Omega \tag{14.25a}$$

which leads to the equation

$$\frac{d(\Delta\Omega)}{d(\Delta\theta)} = +\frac{\Delta\Omega}{(\Delta\theta - \Delta\Omega)} \tag{14.25b}$$

and hence

$$\Delta\theta = -\Delta\Omega(\ln|\Delta\Omega| + X) \tag{14.25c}$$

where X is a constant.

As a consequence of Eq. (14.25c), it can be seen that there is a family of possible solutions, defined by the parameter X, emanating from a branch point.

14.3.2   Solution in the neighborhood of a wall point

Consider a wall point where

$$\theta = \theta_0 \tag{14.26a}$$

$$\chi = \chi_0 \tag{14.26b}$$

$$\Omega = \Omega_0 = \chi_0 + \theta_0 \tag{14.26c}$$

$$s = s_0 \ne s_{R0} \tag{14.26d}$$

where

$$\cos(2\chi_0) = \cos(2\mu) \tag{14.26e}$$

$$s_{R0} = \frac{\gamma\sin(2\chi_0 + \theta_0)}{\sin(2\chi_0)} \tag{14.26f}$$

To investigate the solution in the neighborhood of a branch point, it is convenient to introduce

$$\Omega = \Omega_R + \delta\Omega \tag{14.27a}$$

$$s = s_R + \delta_s \tag{14.27b}$$

$$\theta = \theta_0 + \delta\theta \tag{14.27c}$$

It can then be established that in the neighborhood of the wall point, that to sufficient accuracy;

$$\frac{d(\delta s)}{d(\delta\Omega)} = -2s_0\cot(2\chi_0) \tag{14.28a}$$

$$\frac{d(\delta\theta)}{d(\delta\Omega)} = \eta(\delta\theta - \delta\Omega) \tag{14.28b}$$

where

$$\eta = \frac{4s_0 \cot(2\chi_0)}{s_{R0} - s} \tag{14.28c}$$

It thus follows that

$$\delta s = -2s_0 \cot(2\chi_0)\delta\Omega \tag{14.28d}$$

$$\delta\theta = \delta\Omega + \frac{1}{\eta}(1 - e^{\eta\delta\Omega}) \tag{14.28e}$$

This last equation (Eq. (28e)) can be expanded as a Taylor's series to show that to sufficient accuracy

$$\delta\theta = -\eta\delta\Omega^2 \tag{14.29}$$

Thus, at a particular wall point there exists a family of possible solutions, depending on the parameter s_0. It is interesting to note that if $\cot(2\chi_0)$ is positive, then all solutions must be in the negative θ direction whenever $s_0 > s_{R0}$, and thus the line $\theta = \theta_0$ constitutes a barrier or wall. Likewise, the solutions must be in the positive θ direction whenever $s_0 < s_{R0}$. Corresponding results of course hold when $\cot(2\chi_0)$ is negative. This is a specific case of a much more general theorem due to Bonneau (see Salencon, 1977).

14.4 A Property of Wedge Fields

It follows from the discussion in the previous sections that continuous solutions to Eq. (14.16) can always be assumed to be such that the quantity $\cos(2\chi) - \cos(2\mu)$ does not change sign. If this quantity were to change sign when $\theta = \theta_0$, then that point would either be a branch point, where there are many possible continuations of the solution, and more information would be necessary to proceed, or it would be a wall point and no continuation of the solution would be possible.

This observation leads to the result that for such a field, Ω must be either a strictly increasing or strictly decreasing function of θ, unless it is a Rankine solution, in which case Ω is constant. To establish this, consider the first member of Eq. (14.16) over the interval $\theta_L < \theta < \theta_R$:

$$\frac{d\Omega}{d\theta} = \frac{\gamma[\cos(\theta) - \cos(2\mu)\cos(2\chi + \theta)] - s\sin^2(2\mu)}{2s\cos(2\mu)[\cos(2\chi) - \cos(2\mu)]} \tag{14.30}$$

As was discussed above, it will be assumed that the factor $\cos(2\chi) - \cos(2\mu)$ can be assumed to be of constant sign and non-zero, except possibly at the bounding points $\theta = \theta_L$ and $\theta = \theta_R$. It will also be assumed that s is positive, since negative values of s violate the Mohr Coulomb failure condition for a purely frictional material. It thus follows that the denominator of equation is non-zero, except possibly at the end points of the interval of definition. Now suppose that the solution of Eq. (14.30) is not a Rankine solution, but that numerator of Eq. (14.30) vanishes at the point $\theta = \theta_M$, and that at this point $s = s_M$ and $\Omega = \Omega_M$.

In summary

$$\theta = \theta_M$$

$$\Omega = \Omega_M$$

$$\chi = \chi_M = \Omega_M - \theta_M$$

$$s = s_M = \frac{\gamma[\cos(\theta_M) - \cos(2\mu)\cos(2\chi_M + \theta_M)]}{\sin^2(2\mu)}$$

Now consider the Rankine solution:

$$\Omega = \Omega_M$$

$$s = \frac{\gamma[\cos(\theta) - \cos(2\mu)\cos(2\Omega_M - \theta)]}{\sin^2(2\mu)}$$

It is clear that this solution takes the values $s = s_M$ and $\Omega = \Omega_M$ at the point $\theta = \theta_M$ and satisfies Eq. (14.16) and thus, because this point is an ordinary point of these differential equations, this must be the solution throughout the entire range $\theta_L < \theta < \theta_R$ contrary to the assumption that the solution was not a Rankine solution. It thus follows that the numerator of Eq. (14.30) must not change sign throughout the interval $\theta_L < \theta < \theta_R$. Thus, neither the numerator nor the denominator of the right-hand side of Eq. (14.30) changes sign, and so $d\Omega/d\theta$ is either strictly positive or strictly negative, thus ensuring that Ω is either strictly increasing or strictly decreasing with θ.

14.5 Bearing Capacity of a Footing on a Trapezoidal Embankment

Figure 14.7 shows a symmetric trapezoidal embankment POO'P' which has a crest width of B, and which is made of a purely frictional material. The sides of the embankment are inclined at an angle $\theta = \theta_s$ to the vertical, and since it is assumed that the material is purely frictional, the inclination of the side of the embankment must lie in the range $\pi/2 - \phi < \theta_s < \pi/2 + \phi$. It will be assumed that the top of the embankment is loaded by a rigid footing of width B.

Discussion of the solution for the bearing capacity of a purely frictional trapezoidal embankment will be motivated by an examination of the characteristic field for a smooth

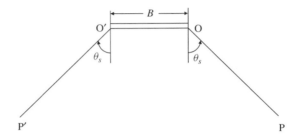

Figure 14.7   Trapezoidal embankment

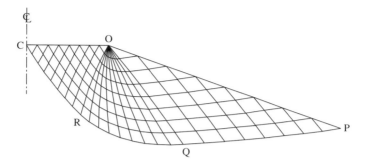

Figure 14.8 Characteristics for bearing capacity failure of a smooth strip footing on a cohesive frictional trapezoidal embankment

footing loading a trapezoidal cohesive frictional embankment, shown in Figure 14.8. These characteristics are for the particular case of $\phi = 30°$, $\theta_s = 70°$ and $\gamma B/c = 7.95$.

Referring to Figure 14.8, it can be seen that, if the embankment is in a state of failure, then on the stress free boundary OP:

$$\Omega = \theta_s \tag{14.31a}$$

$$p = \frac{c \cos(\phi)}{1 - \sin(\phi)} \tag{14.31b}$$

This line is a non-characteristic, and thus as described in the appendix, defines the solution in the region bounded by the α characteristic OQ and the β characteristic PQ.

As indicated in the appendix, there is a point β characteristic located at O and on this characteristic:

$$\Omega = \theta + \mu \tag{14.32a}$$

$$p + c \cot(\phi) = \frac{c \cot(\phi)}{1 - \sin(\phi)} \exp[2 \tan(\phi)(\theta_s - \mu - \theta)] \tag{14.32b}$$

It has been assumed that the rigid footing loading the embankment is smooth, and it thus follows that $\Omega = 0$ on OC. Thus Eqs. (14.32a) and (14.32b) are valid in the range $0 < \Omega < \theta_s$, and thus in the range $-\mu < \theta < \theta_s - \mu$.

It is shown in the appendix that the point β characteristic O and the α characteristic OQ define the solution in the region OQR, where OR is the α characteristic tangential to $\theta = -\mu$ at O and QR is a β characteristic.

Since the footing was assumed to be smooth, it follows that

$$\Omega = 0 \quad \text{on OC} \tag{14.33}$$

This condition, together with the known solution on the α characteristic OR, constitute a mixed problem (appendix) and define the solution in the region OCR where CR is a β characteristic.

Clearly, the solution is symmetric about a vertical line passing through the center of the footing (C), and this defines the solution in the regions O'P'Q', O'Q'R' and O'R'C, completing the solution in the plastic region, where O',P',Q',R' are the images of the points O,P,Q,R reflected in a vertical plane through C.

It can be shown that it is possible to find a velocity field in the plastic region, and also to find a statically admissible stress field in the non-plastic regions, and thus show that the solution is both an upper and a lower bound, and is thus exact.

This solution is valid for all values of cohesion, and so it is possible to consider the case in which $c \to 0$. It is not difficult to show that in the region OPQ the solution becomes

$$\Omega = \Omega_0 \tag{14.34a}$$

$$p = \frac{\gamma r[\cos(\theta) - \sin(\phi)\cos(2\Omega_0 - \theta)]}{\cos^2(\phi)} \tag{14.34b}$$

where Ω_0 can be determined from the relationships

$$\cos(2\Omega_0) = \frac{\cos^2(\theta_s) - \sin(\theta_s)\sqrt{\cos^2(2\mu) - \cos^2(\theta_s)}}{\cos(2\mu)} \tag{14.34c}$$

$$\sin(2\Omega_0) = \frac{\cos(\theta_s)[\sin(\theta_s) + \sqrt{\cos^2(2\mu) - \cos^2(\theta_s)}]}{\cos(2\mu)} \tag{14.34d}$$

It will be recognized that this is a Rankine solution, and that the α characteristic OQ is the straight line $\theta = \Omega_0 - \mu$, and on this line

$$\chi = \mu \qquad \text{on OQ} \tag{14.35a}$$

$$p = \frac{\gamma r \sin(2\mu + \theta)}{\sin(2\mu)} \qquad \text{on OQ} \tag{14.35b}$$

As was observed earlier, on the line OC

$$\Omega = 0 \tag{14.35c}$$

It can be seen that Eqs. (14.35a), (14.35b) and (14.35c) define a wedge field which has a branch point at $\theta = \Omega_0 - \mu$. The solution can thus be found by choosing the member of the family of solutions (see Eq. (14.25c)), emanating from this branch point such that $\Omega = 0$ when $\theta = -\pi/2$. Alternatively, the solution may be found by considering the family of solutions defined by the condition that $s = s_F$ and $\Omega = 0$ when $\theta = -\pi/2$, and choosing the member that has a branch point at $\theta = \Omega_0 - \mu$. The set of characteristics found in this way is shown in Figure 14.9 for the case of an embankment with sides inclined at $20°$ to the horizontal, $\theta_s = 70°$, and an angle of internal friction $\phi = 30°$.

When the footing is rough the solution is derived in a similar fashion. There are two modes of failure: one is for relatively narrow footing (smaller values of $\gamma B/c$); and one is for broader footings (larger values of $\gamma B/c$). A description for the case in which the footing is resting on a homogeneous half space ($\theta_s = 90°$) is given by Davis and Booker (1971) and Bolton and Lau (1993).

The characteristic field for a narrower footing is shown in Figure 14.10 for the particular case of $\phi = 30°$, $\theta_s = 70°$ and $\gamma B/c = 1.86$.

As discussed in the case of the smooth footing, the non-characteristic line OP defines the solution in the region bounded by the α characteristic OQ, and the β characteristic PQ. There is also there is a point β characteristic located at O, and on this characteristic

$$\Omega = \theta + \mu \tag{14.36a}$$

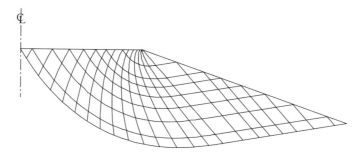

Figure 14.9 Characteristics field for the bearing capacity failure of a smooth strip footing on a purely frictional trapezoidal embankment

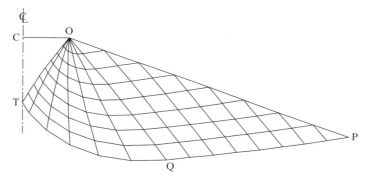

Figure 14.10 Characteristics for the bearing capacity failure of a narrow rough strip footing on a cohesive frictional trapezoidal embankment

$$p + c \cot(\phi) = \frac{c \cot(\phi)}{1 - \sin(\phi)} \exp[2 \tan(\phi)(\theta_s - \mu - \theta)] \qquad (14.36b)$$

This characteristic has an angular span $-\pi/2 < \theta < \theta_s - \mu$, which means that Ω has the range $-(\pi/4 + \phi/2) < \Omega < \theta_s$. The extent of the fan is determined by the condition that the solution is symmetric about the vertical line CT, and thus that $\Omega(T) = 0$ again O′,P′,Q′,R′ denote the images of the points O,P,Q,R reflected in a vertical plane through C. When the velocity field associated with the collapse is determined, it is found that the region OTO′ moves down with the same rigid body movement as the rough footing.

For wider footings the characteristic field is as shown schematically in Figure 14.11. In this case, the non-characteristic line OP defines the solution in the region OPQ; there is a point β characteristic at O which, together with the α characteristic OQ, defines the solution in the fan. The final α characteristic OR of the fan is tangential to the footing at O. Considerations of the velocity solution show that all the α characteristics in the region OSTR are tangential to the footing, and thus $\Omega = -(\pi/4 + \phi/2)$ on OS. This condition, together with the known solution on the α characteristic OR, define a mixed problem as described in the appendix, and thus define the solution in the region OSTR. The point T is chosen so that $\Omega(T) = 0$, and the solution is completed by using the symmetry of the solution about the vertical line CT. Again, O′,P′,Q′,R′ are the images of the points

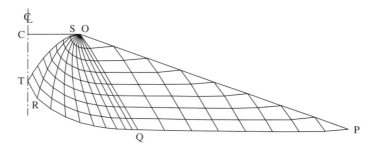

Figure 14.11   Characteristics for the bearing capacity failure of a wide rough strip footing on a cohesive frictional trapezoidal embankment

O,P,Q,R reflected in a vertical plane through C. The characteristic field for the particular case of $\phi = 30°$, $\theta_s = 70°$ and $\gamma B/c = 44.8$ is shown in Figure 14.11.

In the case of a wide footing it is found that only a portion of the soil STS' moves down as a rigid body with the footing, and that there are regions OS, O'S' at the edge of the footing where slip between the footing and the soil takes place. Such regions are velocity discontinuities, and as such must be characteristics, thus confirming the assumption that α characteristics such as OR and ST were tangential to the footing.

The solution is valid for all values of cohesion, and so it is possible to consider the case in which $c \rightarrow 0$. It is not difficult to show that in the region OPQ, the solution is precisely that described by Eqs. (14.34a) and (14.34b). As in the case of a smooth footing, it will be recognized that this is a Rankine solution, and that the α characteristic OQ and the stress state on this line is given by Eqs. (14.35a) and (14.35b). It was shown in previous discussion that on the line OS,

$$\Omega = - \left(\frac{\pi}{4} + \frac{\phi}{2} \right)$$

(14.37a)

and thus that

$$\chi = \mu$$

(14.37b)

It can be seen that Eqs. (14.35a), (14.35b), (14.36a) and (14.36b) define a wedge field which has a branch point at $\theta = \Omega_0 - \mu$. The solution can thus be found by choosing the member of the family of solutions (see Eq. (14.25c)), emanating from this branch point such that $\Omega = -(\pi/2 + \phi/2)$ when $\theta = -\pi/2$. Interestingly Eq. (14.37b) shows that this is a wallpoint. Alternatively, the solution may be found by considering the family of solutions defined by the condition that $s = s_F$ and $\Omega = -(\pi/2 + \phi/2)$ when $\theta = -\pi/2$, that is emanating from a wall point at $\theta = -\pi/2$ (see Eq. (14.29)), and choosing the member that has a branch point at $\theta = \Omega_0 - \mu$. The set of characteristics found in this way is shown in Figure 14.12 for the case of an embankment with sides inclined at 20° to the horizontal, $\theta_s = 70°$, and an angle of internal friction $\phi = 30°$.

It is interesting to observe that in the region at the edge of the footing:

$$\Omega = 0 \qquad \text{for a smooth footing} \qquad (14.38a)$$

$$\Omega = - \left(\frac{\pi}{2} + \frac{\phi}{2} \right) \quad \text{for a rough footing} \qquad (14.38b)$$

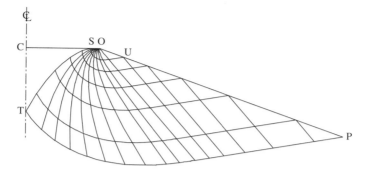

Figure 14.12 Characteristics for the bearing capacity failure of a rough strip footing on a purely frictional trapezoidal embankment

These boundary conditions represent particular cases of a more general boundary condition. In the Mohr–Coulomb failure condition. (Eq. (14.4a)), there are two components of strength; a cohesive part c; and a frictional part $\sigma_n \tan(\phi)$. Now suppose that at the soil footing interface that each component is reduced by a factor m so that

$$|\tau| = c_a + \sigma_n \tan(\delta) \tag{14.38c}$$

where

$$c_a = mc \tag{14.38d}$$

$$\tan(\delta) = m \tan(\phi) \tag{14.38e}$$

The factor m has a value between zero and one, with $m = 0$ corresponding to a smooth footing and $m = 1$ corresponding to a rough footing.

For the case described by Eqs. (14.38a) and (14.38b), it can be shown that on OS:

$$\Omega = -\frac{\alpha + \delta}{2} \tag{14.38f}$$

where

$$\sin(\alpha) = \frac{\sin(\delta)}{\sin(\phi)} \tag{14.38g}$$

Thus, for a smooth footing $\delta = 0$ and $\alpha = 0$, while for a rough footing $\delta = \phi$ and $\alpha = \pi/2$.

This observation may be used to examine footings where the strength of the interface between footing and underlying soil is less than that of the soil itself (see Figure 14.13).

For a wide footing on a trapezoidal embankment, the characteristics are similar to those for a wide rough footing (see Figure 14.10) except that the characteristics ST are no longer tangential to the footing, but inclined at an angle λ given by

$$\lambda = \frac{\pi}{4} + \frac{\phi - \delta - \alpha}{2}$$

so that for perfectly smooth footings $\lambda = \pi/4 + \phi/2$, for perfectly rough footings $\lambda = 0$, and for footings of intermediate roughness λ takes an intermediate value between these two extremes.

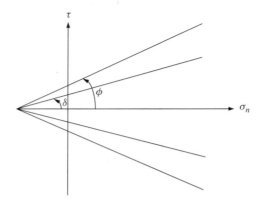

Figure 14.13   Failure conditions at soil footing interface

It can be seen that when $c \to 0$ the solution can be found in a similar fashion to that used to determine the solution for a smooth footing and for a rough footing. Again, it is found that in the region OPQ the solution is precisely that described by Eq. (14.34b), that is a Rankine solution, and that the α characteristic OQ and the stress state on this line is given by Eq. (14.35b). It was shown in the previous discussion that on the line OS

$$\Omega = \lambda - \left(\frac{\pi}{4} + \frac{\phi}{2} \right) \tag{14.39}$$

It can be seen that Eqs. (14.35b) and (14.39) define a wedge field which has a branch point at $\theta = \Omega_0 - \mu$. The solution can thus be found by choosing the member of the family of solutions emanating from this branch point such that $\Omega = -(\pi/4 + \phi/2 - \lambda)$ when $\theta = -\pi/2$. Alternatively, the solution may be found by considering the family of solutions defined by the condition that $s = s_F$ and $\Omega = -(\pi/4 + \phi/2 - \lambda)$ when $\theta = -\pi/2$, and choosing the member that has a branch point at $\theta = \Omega_0 - \mu$. The set of characteristics found in this way for the case of an embankment with sides incline at 20° to the horizontal, $\theta_s = 70°$, and an angle of internal friction $\phi = 30°$ and an angle of interface friction $\delta = 15°$ is shown in Figure 14.14.

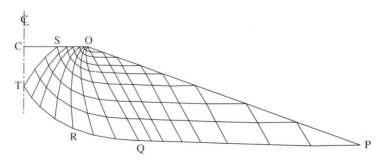

Figure 14.14   Characteristics for a bearing capacity failure of a partially rough footing loading a
purely frictional trapezoidal embankment

Referring to Figure 14.14, it follows from a consideration of equilibrium of the triangle OTO' that the average pressure at failure (q_F) on the footing is given by

$$q_F = \frac{\gamma B}{2} N_\gamma \tag{14.40a}$$

where

$$N_\gamma = \frac{s(T)(1 + \sin(\phi))}{2\gamma \cos(\xi)} - \frac{\tan(\xi)}{2} \tag{14.40b}$$

and $s(T)$ denotes the value of s at T (the point where $\Omega = 0$) and ξ denotes the angle $O'OT$. It can be seen from an examination of Eqs. (14.16a) and (14.16b) that s is proportional to γ, and thus N_γ is dimensionless and depends only upon the strength properties of the embankment, the footing φ and δ and the geometric parameter θ_s.

14.6   Results

The theory developed in the previous section can be used to calculate the bearing capacity of a strip footing on a trapezoidal embankment.

Figure 14.15 shows the variation of bearing capacity for the particular case of an angle of internal friction $\phi = 30°$ for a range of embankment geometries $\pi/2 - \phi < \theta_s < \pi/2 + \phi$, and for a range of angle of friction between the footing and the embankment $0 < \delta < \phi$.

It can be seen that, in general, the greater the inclination of the slope to the vertical, the greater the bearing capacity of a strip footing placed on its crest. Similarly, if the footing is place at the base of a trench with an upward slope on either side $\theta_s > 90°$, the presence of the slopes induces additional stability against bearing capacity failure.

It can also be seen that for all geometries, the bearing capacity of a smooth footing $(\delta = 0)$ is significantly smaller than for the corresponding completely rough footing $(\delta = \phi)$. However, it can also be seen that a small reduction in interface friction, say from $\delta = 20°$ to $\delta = 15°$, leads to a relatively small reduction in bearing capacity.

Figures 14.16 and 14.17 provide similar solutions for the cases in which the angle of internal friction is $\phi = 30°$ and $\phi = 40°$, respectively, and lead to similar observations.

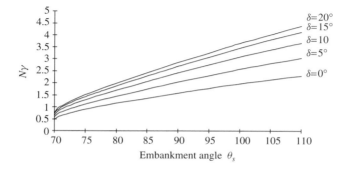

Figure 14.15   Variation of bearing capacity factor N_γ with embankment slope and angle of friction of soil-footing interface for an angle of internal friction $\phi = 20°$

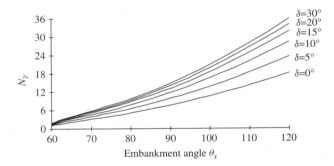

Figure 14.16   Variation of bearing capacity factor N_γ with embankment slope and angle of friction of soil-footing interface for an angle of internal friction $\phi = 30°$

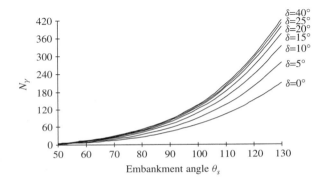

Figure 14.17   Variation of bearing capacity factor N_γ with embankment slope and angle of friction of soil-footing interface for an angle of internal friction $\phi = 40°$

14.7   Velocity Field

To show that the solution is an upper bound, it is necessary to show that it is possible to find a velocity field. This is a straightforward task, and will be illustrated by reference to one of the many cases which were checked.

Figure 14.18 shows the velocity field for a trapezoidal embankment for a cohesionless soil with an angle of internal friction $\phi = 30°$ and with a side slope inclined at $70°$ to the vertical. The footing OO′ is assumed to be rigid, and moves down with a vertical velocity v. At failure the stress state is constant, and thus there is no variation of the strains in the region outside the characteristics TP and TP′ which separate the plastic region from the elastic region. There is a velocity discontinuity at the bounding β characteristic; it was shown earlier that the quantity v_α must be continuous across this line, and thus $v_\alpha = 0$ on TP. Similarly, it can be shown that $v_\beta = 0$ on TP′. The footing OO′ moves down as a rigid body with $v_x = v$ and $v_y = 0$. This ensures that the region STS′ moves down as a rigid body with the same velocity. The quantity v_β must be continuous across the characteristic ST and, similarly, the quantity v_α must be continuous across S′T. It is thus possible to show that:

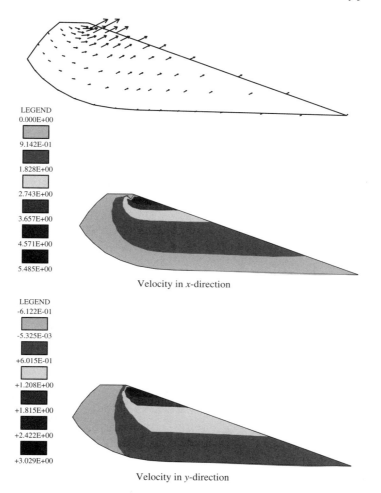

Figure 14.18 Velocity field for bearing capacity failure of a strip footing on a trapezoidal embankment

$$v_x = v \qquad \text{in STS}' \qquad (14.41\text{a})$$

$$v_y = 0 \qquad \text{in STS}' \qquad (14.41\text{b})$$

$$v_\alpha = v\cos(\Omega - \mu) \qquad \text{in STS}' \qquad (14.41\text{c})$$

$$v_\beta = v\cos(\Omega + \mu) \qquad \text{in STS}' \qquad (14.41\text{d})$$

$$v_x = v - \frac{v\sin(\Omega + \mu)}{2\sin(\mu)}\exp(-\cot(2\mu)\Omega) \qquad \text{on ST} \qquad (14.41\text{e})$$

$$v_y = \frac{v\cos(\Omega + \mu)}{2\sin(\mu)}\exp(-\cot(2\mu)\Omega) \qquad \text{on ST} \qquad (14.41\text{f})$$

$$v_\alpha = v(\cos(\Omega - \mu) - \cos(\mu))\exp(-\cot(2\mu)\Omega) \qquad \text{on ST} \qquad (14.41\text{g})$$

$$v_\beta = v\cos(\Omega + \mu) \qquad \text{on ST} \qquad (14.41\text{h})$$

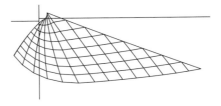

Figure 14.19   A deformed mesh of characteristics for bearing capacity failure of a strip footing on a trapezoidal embankment

$$v_x = -\frac{v\sin(\Omega - \mu)}{2\sin(\mu)}\exp(+\cot(2\mu)\Omega) \quad \text{on TP} \qquad (14.41\text{i})$$

$$v_y = \frac{v\cos(\Omega - \mu)}{2\sin(\mu)}\exp(+\cot(2\mu)\Omega) \quad \text{on TP} \qquad (14.41\text{j})$$

$$v_\alpha = 0 \qquad\qquad\qquad\qquad\qquad \text{on TP} \qquad (14.41\text{k})$$

$$v_\beta = -v\cos(\mu)\exp(+\cot(2\mu)\Omega) \qquad \text{on TP} \qquad (14.41\text{l})$$

The known velocity on the characteristics ST and TP defines the solution in the region STPU, where SU is the β characteristic through S. The known velocity on SU together with the condition that the vertical velocity is v on SO is sufficient to determine the remaining velocity solution.

The velocity field for the particular case of $\phi = 30°$ and $\theta_s = 70°$ is shown in Figure 14.18, both as velocity vectors and as contours of horizontal and vertical velocity.

A deformed mesh of characteristics is shown in Figure 14.19. It was found that the quantity Λ evaluated using Eq. (14.12c) was always positive, and so the velocity solution provided a rigorous upper bound to the collapse load.

14.8   Extension Field

To establish that the solution is also a rigorous lower bound, it is necessary to establish that it is possible to find a statically admissible stress field in the elastic regions. To do this for the region STS', it is merely necessary to assume that it is in a state of plastic failure, and to evaluate the stress distribution within the region from the known values on the characteristics ST and S'T. This is shown in Figure (14.20(a)) for the particular case of a purely frictional material with an angle of internal friction $\phi = 30°$ and for a trapezoidal embankment with a side slope $\theta_s = 70°$.

The extension beyond the characteristics TP and TP' is found in two stages. First, the stress state on the two characteristics TP and TP' are used to generate the plastic field shown in Figure (14.20(b)). Next, the principal stress trajectory passing through P and P' is drawn. It is assumed that one beyond this principal stress trajectory one family of principal stress trajectories is the family of straight lines normal to the original principal stress trajectory PP', and thus that this principal stress trajectory must be convex. The remaining principal stress trajectories are, of course, orthogonal to this family of straight lines, and are shown in Figure (14.20(c)). It is now possible to determine the exact distribution of stress in the extension zone, and to verify that the failure condition is not

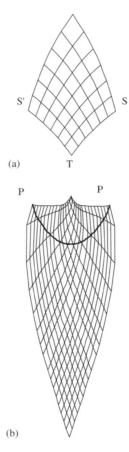

(a)   T

P   P

(b)

Figure 14.20   (a) Stage 1 of the Extension Field; (b) stage 2 of the extension field

violated. This was done for a wide range of geometric and physical parameters, and it was always found that the extension field was statically admissible, and thus also provided a lower bound to the collapse load. The contours of the ratio of available strength to mobilized strength $p\sin(\phi)/R$ are shown for the particular case of a purely frictional embankment with $\phi = 30°$ and $\theta_s = 70°$, are shown in Figure (14.20(c)).

14.9   Conclusions

In this chapter the theory of plasticity has been used to examine the plastic failure of a wedge of perfectly plastic purely frictional material. It was then shown how this solution could be used to develop solutions for the bearing capacity of trapezoidal granular embankments. The bearing capacity of such embankments was examined for a number of embankment geometries, and for a range of physical properties. The solution was determined by first finding the characteristic field in the plastic region. It was then shown that it was possible to determine a velocity field in this region, and hence that solution was

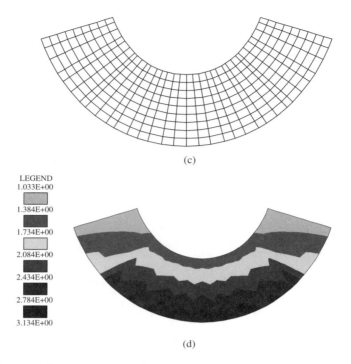

LEGEND
1.033E+00

1.384E+00

1.734E+00

2.084E+00

2.434E+00

2.784E+00

3.134E+00

(d)

Figure 14.20 (*continued*)   (c) contours of available strength/mobilized strength; (d) contour of safety factor in extension field

kinematically admissible and so gave an upper bound to the failure load. Finally, it was demonstrated that the solution could be extended into the elastic regions without violating equilibrium or failure, and was thus statically admissible, and hence gave a lower bound to the collapse load, establishing that the solution was exact.

Appendix: Numerical Calculations Using Characteristics

The α, β characteristics can be used to determine the stress state in a plastic region as follows. Suppose that the stress state is known at two points A,B, as shown in Figure A14.1, and suppose that the α line through A intersects the β line through B at the point P.

Equations (14.6a,b) can be approximated as follows:

$$y_P - y_A = \tan\left(\frac{\Omega_P + \Omega_A}{2} - \mu\right)(x_P - x_A) \tag{A14.1a}$$

$$y_P - y_B = \tan\left(\frac{\Omega_P + \Omega_B}{2} + \mu\right)(x_P - x_B) \tag{A14.1b}$$

$$-\sin(2\mu)(p_P - p_A) + (R_P + R_A)(\Omega_P - \Omega_A) + F_\alpha = 0 \tag{A14.1c}$$

$$+\sin(2\mu)(p_P - p_B) + (R_P + R_B)(\Omega_P - \Omega_B) + F_\beta = 0 \tag{A14.1d}$$

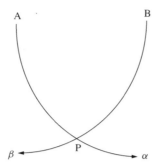

Figure A14.1   Basic calculation using α β characteristics

and

$$F_\alpha = [\gamma_x \sin(2\mu) - \gamma_y \cos(2\mu)](x_P - x_A)$$
$$+ [\gamma_x \cos(2\mu) + \gamma_y \sin(2\mu)](y_P - y_A) \qquad \text{(A14.1e)}$$
$$F_\beta = [-\gamma_x \sin(2\mu) - \gamma_y \cos(2\mu)](x_P - x_B)$$
$$+ [\gamma_x \cos(2\mu) - \gamma_y \sin(2\mu)](y_P - y_B) \qquad \text{(A14.1f)}$$

and where the subscripts P, A, B indicate the value of the particular quantity at that point.

Equations (A14.1a) through (A14.1d) represent a set of nonlinear equations which cannot be solved directly; however, if initial approximations to the values of x_P, y_P, p_P, Ω_P at the point P are known, then Eqs. (A14.1a) and (A14.1b) can be used to determine new approximations to x_P, y_P, and then Eqs. (A14.1c) and (A14.1d) can be used to provide new approximations to p_P, Ω_P. This process can be continued iteratively and usually converges rapidly.

The above basic calculation can be used to solve the fundamental problems. For example, suppose that the stress state is known on the non-characteristic arc AB shown in Figure A14.2. The stress state will be known at the point $P_{11}, P_{22}, \ldots.P_{n,n}$ on the arc, and thus the basic calculation outlined above can be used to calculate the stress state at the points $P_{12}, P_{23}, \ldots.P_{n-1,n}$. Once the stresses at these points are known, they can be used to calculate the stress at points $P_{13}, P_{23}, \ldots, P_{n-2,n}$. The computation can be continued in a similar fashion until the stress field is determined in the region bounded by the original arc $P_{11}, P_{22}, \ldots, P_{nn}$ and the α line $P_{11}, \ldots.P_{1n}$, and the β line $P_{nn}, \ldots.P_{1n}$.

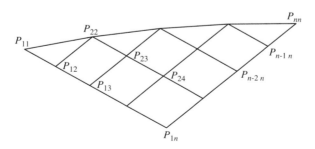

Figure A14.2   Cauchy or initial value problem

The determination of this solution is called the Cauchy or initial value problem. Referring to Figure A14.2 it can be seen that the solution for the region OPQ immediately beneath the stress free surface OP constitutes the solution of a Cauchy problem. It can be shown that in this case, the region OPQ is in a Rankine passive state, and so that

$$p = \frac{\gamma x + c \cos(\phi)}{1 - \sin(\phi)} \qquad (A14.2a)$$

$$\Omega = \pi \qquad (A14.2b)$$

The solution to the Cauchy problem outlined above is, of course, an approximate numerical solution which becomes more accurate as the distance separating the points on the non-characteristic line becomes smaller and smaller. A rigorous justification to the uniqueness of the solution of the Cauchy problem is given by Courant and Hilbert (1965), who also show that if the derivatives of the field quantities are continuous on the non-characteristic line, then they are continuous throughout the domain of definition of the solution. It can be seen by allowing the non-characteristic arc to approach a characteristic, that a solution cannot be extended beyond a characteristic without some additional information. It is also possible to show that a characteristic line is the only line where a discontinuity in the derivative of the field quantities can exist.

Another fundamental problem occurs when the stress field is known on two intersecting characteristics such as shown in Figure A14.3. In this case, the stress field is known at the points $P_{11}, P_{12}, \ldots, P_{1m}$ on the initial α line and at the point P_{21}, of the initial β line the basic calculation described above can then be used to calculate the solution at the points $P_{22}, P_{23}, \ldots P_{2m}$. The solution can now be regarded as being known on the second α line, and a repetition of the calculation described in the previous sentence can be used to calculate the stress field on the third α line, and so on until the solution is known for the region bounded by the first and 'mth' α lines and the 1st and the 'nth' β lines. This solution is known as the solution of the Goursat problem. A rigorous justification to the uniqueness of the solution of the Goursat problem is given by Courant and Hilbert (1965).

An example of the Goursat problem can be seen in Figure A14.3 in the determination of the solution in the 'fan'-like region OQR. It can be shown that there is a point β characteristic at the point O. The stress field at O must satisfy the equation

$$\sin(2\mu)\frac{\partial p}{\partial s_\beta} + 2R\frac{\partial \Omega}{\partial s_\beta} = 0 \qquad (A14.3a)$$

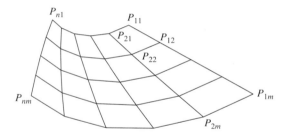

Figure A14.3   Solution of the Goursat problem

where
$$R = p\sin(\phi) + c\cos(\phi) \tag{A14.3b}$$

It then follows that

$$p + c\cot(\phi) = \frac{c\cot(\phi)}{1 - \sin(\phi)}\exp[-2\tan(\phi)(\Omega - \pi/2)] \tag{A14.4a}$$

$$r = 0 \tag{A14.4b}$$

$$\Omega = \mu + \theta \tag{A14.4c}$$

The stress field is thus known on the α characteristic OQ (see Eqs. (A14.2a) and (A14.2b)), and the β characteristic at O, and so the solution can be determined throughout the entire region OQR.

A third fundamental boundary problem is called a mixed problem. In this case, the solution is known on a characteristic line and a relationship between the stress components is known on a specified line. The simplest such problem is one in which Ω is specified on a straight line, as shown in Figure A14.4. It will be assumed for the sake of definiteness that the known characteristic is an α line. To solve the mixed problem, another basic calculation (B) is needed.

Referring to Figure A14.4, and supposing that the stress state is known at the point B (P_{12}) and that Ω is known on the line $P_{11}P_{mm}$, then using the approximations developed in Eq. (A14.1), it is found that

$$y_P - y_B = \tan\left(\frac{\Omega_p + \Omega_B}{2} + \mu\right)(x_P - x_B) \tag{A14.5}$$

Since Ω_P is known, this is the equation of a straight line and, consequently, its point of intersection with the specified straight line OD can be found. Thus, Ω_P, x_P, y_P can all be considered as known. It therefore remains to determine p_P. This can be done by returning to Eq. (A14.1) and observing that

$$\sin(2\mu)(p_P - p_B) + (p_P\sin(\phi) + c\cos(\phi) + R_B)(\Omega_P - \Omega_B) = F_\beta \tag{A14.6}$$

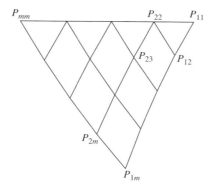

Figure A14.4   Simplified mixed boundary value problem

since all the quantities occurring in F_β are known, it is a simple task to solve Eq. (A14.6) for p_P.

The complete mixed problem can now be solved. Referring to Figure A14.3, calculation of B can be used to calculate the values of stress at point P_{22} from the known value at P_{12}, then calculation A can be used to calculate the values at P_{23} from the known values at P_{22} and P_{13} calculation, as can now be repeated until values at all points on the second α line have been found. The calculation of the points on the third α line commence with calculation B to determine the field quantities at point P_{33} and calculation A to determine all other. The procedure is continued until the solution is known in the region bounded by the first α characteristic $P_{11}....P_{1m}$, the 'mth' β line $P_{1m}....P_{mm}$ and the specified line $P_{11}....P_{mm}$. An example of such a calculation is provided by the determination of the stress field in the region OSR beneath the smooth footing ($\Omega = 0$) shown in Figure 14.3.

References

Bolton, M. D. and Lau, C. K. Vertical bearing capacity factors for circular and strip footings on Mohr–Coulomb soil. *Canad. Geotech. J.*, **30**, (1993) 1024–1033.

Booker, J. R and Davis, E. H. Stability analysis by plasticity theory. In: *Numerical Methods in Geotechnical Engineering*, Desai, C. S. and Christian, J. T. (eds.). McGraw-Hill, New York, (1977).

Courant, R. and Hilbert, D. *Methods of Mathematical Physics*. Interscience, New York, (1965).

Davis, E. H. and Booker, J. R. The bearing capacity of footings from the stand point of plasticity theory. *Proceedings of the First Australian New Zealand Conference in Geomechanics*, Melbourne, Australia, (1971) 276–282.

Drucker, D. C., Prager, W. and Greenberg, H.J. Extended limit design theorems for continuous media. *Quart. J. Mech. Appl. Math.*, **9**, (1952) 381.

Hill, R. *The Mathematical Theory of Plasticity*. Clarendon Press, Oxford, (1950).

Meyerhoff, G. G. The ultimate bearing capacity of foundations. *Geotechnique*, **2**(4), (1951) 301–342.

Salencon, J. *Applications of the Theory of Plasticity in Soil Mechanics*. Wiley, New York, (1977).

Sokolovski, V. V. *Statics of Soil Media*. Butterworths, London, (1960).

Terzaghi, K. *Theoretical Soil Mechanics*. Wiley, New York, (1943).

15

Limit Analysis of Foundation Systems with Offshore Applications

James D. Murff
Exxon Production Research Co., Houston, Texas, USA

15.1   Introduction

Offshore structures are commonly supported on systems of individual foundation elements such as shallow footings or piles. The behavior and particularly the capacity of individual foundation elements has been the focus of a great deal of research, and is clearly a major source of uncertainty. However, it is also of great interest to understand the behavior of the foundation system. As the offshore industry moves toward limit state design and reliability assessments of existing structures become more important, it is logical that system performance takes on added significance.

For the purposes of this discussion we will focus on foundation system capacity, a key element of performance. Figure 15.1 shows a schematic of a typical pile-founded, offshore jacket structure in various stages of installation. This structure is most often used for oil and gas production. It is constructed onshore in a horizontal orientation and is subsequently loaded onto a barge and towed to the site, where it is pushed off the barge or 'launched'. The structure typically comes to equilibrium, floating in a horizontal position. By controlled ballasting operations, tubular members are flooded and the structure is rotated into a vertical position. With the help of a floating crane vessel it is gently set on bottom. Ballasting continues as some of the structure's weight is transferred to spread footings on the base of the structure called mudmats. These footings are used for temporary support pending installation of piles. As such, they must withstand the combined effects of buoyant weight as well as wind, wave, and current which cause horizontal and overturning loads. Mudmat designs must account for contingencies such as sudden weather changes so that the design loading can be significant. In this configuration, the foundation engineer is concerned with the capacity of the system of shallow foundations. The piles are usually installed through the legs (main piles) and through tubular sleeves structurally attached around the platform base (skirt piles). Once the piles are in place, the deck and production

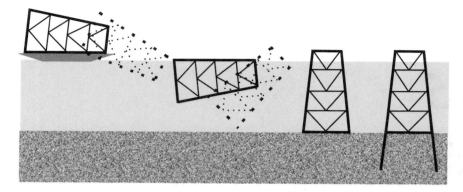

Figure 15.1   Stages of installation of an offshore oil production platform

facilities are installed and platform operations commence. These structures may have a design life of twenty to thirty years or longer and hence the pile foundation system must be designed to withstand very large operating and environmental loads. The above description illustrates a typical case in which a single structure must rely on two different foundation systems during its lifetime.

There are also a number temporary structures supported on systems of individual foundation elements used in offshore exploration. These structures are usually mobile such that they can be floated to a site and the foundation activated to fix the structure to the seabed during drilling operations. Subsequent to drilling the exploration well, the structure is removed. The jackup rig, typically with three independent legs, is a good example of such a structure. As shown schematically in Figure 15.2, the jackup is towed to a site and moored in place while its legs are jacked down to contact the seafloor. Jacking continues until the footings or 'spud cans' penetrate the soil and the bearing capacity is sufficient to support the deck and facilities, which are jacked up out of the wave zone. During this process a preload of ballast water is taken on board storage vessels integrated into the deck. This preload water is discharged after installation, in effect providing a proof test of the foundation against anticipated storm load levels. In strong soils, such as dense sands or heavily overconsolidated clays, the spud cans' penetration may be minimal resulting in a shallow foundation system. In weak soils, such as normally or underconsolidated clays,

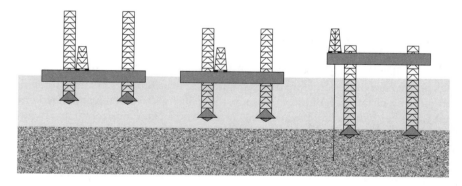

Figure 15.2   Installation of a mobile jackup drilling rig

the spud cans may penetrate thirty meters or more, forming a hybrid system of deep foundations – neither piles nor spread footings.

In the sections immediately following, we consider existing methods of characterizing individual foundation element capacity, including their multi-axial response. Subsequently, Prager's method of generalized plastic limit analysis (Prager, 1959) will be reviewed as it applies to the problem under consideration. This discussion is followed by a general description of how Prager's method is used to develop bounds on typical foundation system capacity. Details of limit analysis implementation for systems of shallow foundations and piles are then developed. Finally, a discussion of the method for practical design and analysis is presented. It is useful to first consider some of the types of foundations that are of particular interest.

15.2   Bearing Capacity of Foundation Elements

15.2.1   General

Offshore foundation elements such as footings or piles are typically subjected to multi-axial loading. The capacity of such a foundation element is, in general, a function of all the loads imposed on it. For the purposes of this discussion, we will focus our attention on planar loading conditions, i.e. resultant lateral loads and moments in the same vertical plane. This condition significantly simplifies the analytical development and is a valid idealization for most loading environments encountered. For other situations, the principles outlined below remain valid but the complexity of formulation and solution may be significantly increased.

15.2.2   Shallow foundations

Figure 15.3 shows a planar shallow foundation subject to multi-axial loading. A general expression for the capacity of this foundation is

$$f(V, H, M/B) = 0 \tag{15.1}$$

where V and H are the resultant vertical and horizontal loads, and M/B is the resultant moment divided by the appropriate base width or other characteristic length to maintain dimensional consistency. Equation (15.1) thus represents a load-interaction or yield surface. In this discussion we will assume ideal, rigid-plastic behavior of the soil such that the yield surface and the failure surface are identical. Equation (15.1) also depends upon other parameters such as the soil characteristics, footing geometry, depth of embedment, and footing inclination with respect to the soil surface.

Equation (15.1) has been extensively studied both experimentally and analytically for both drained and undrained loading conditions. A limited set of analytical solutions has been found for planar loading conditions using the method of characteristics (Prandtl (1921), Prager (1959) and Shield (1954)). These solutions have been generalized using numerical methods by Sokolovskii (1965), Houlsby and Wroth (1982), and many others. Using analytical and/or numerical solutions as starting points, Terzaghi (1943), Meyerhof (1953), Hansen (1970), Vesic (1975) and others have developed explicit versions of Eq. (15.1) by incorporating experimental results and heuristic arguments. The latter are

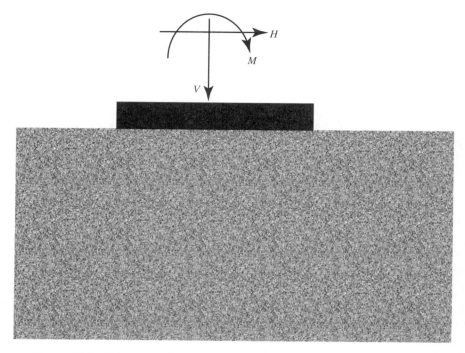

Figure 15.3   Schematic of a shallow foundation subject to multi-axial loading

collectively referred to as 'classical' bearing capacity equations. A typical form of the 'classical' equation is (Terzaghi, 1943)

$$V = \left(cN_cK_c + qN_qK_q + \tfrac{1}{2}\gamma'BN_\gamma K_\gamma\right)A' \tag{15.2}$$

where V = vertical load at failure, c = soil cohesion, N_c, N_q and N_γ are dimensionless functions of soil friction angle, ϕ, K_c, K_q and K_γ are empirical functions of load inclination, footing shape, footing embedment, and footing inclination with respect to the soil surface, γ' is the effective soil unit weight, B is the footing dimension in the plane of the loads, and A' is the footing area adjusted for load eccentricity, i.e. a function of M/B.

As an example, consider a strip footing on the surface of a dry sandy soil with cohesion, $c = 0$, friction angle ϕ, and unit weight, γ'. The footing, with width B, is subjected to a vertical load per unit length V, a horizontal load per unit length H, and an overturning moment per unit length M. According to Vesic (1975), Eq. (15.2) simplifies to the following:

$$V = 0.5\gamma'B^2N_\gamma\left(1 - \frac{H}{V}\right)^3\left(1 - 2\frac{M}{BV}\right) \tag{15.3}$$

Normalizing V, H and M/B by $V_m(= 0.5\gamma'B^2N_\gamma)$ gives

$$V' = \left(1 - \frac{H'}{V'}\right)^3\left(1 - 2\frac{M'}{BV'}\right) \tag{15.4}$$

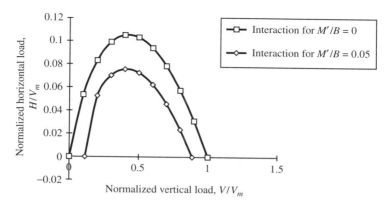

Figure 15.4   Failure-interaction surface for a shallow foundation

where the prime indicates normalized values. Figure 15.4 is a plot of the interaction surface represented by this equation projected on the $H - V$ plane. Similar surfaces can be developed for any specific set of parameters or conditions.

15.2.3   Pile foundations

Methods for developing failure interaction surfaces for pile elements are not as well developed as for shallow foundations. In fact, axial and lateral pile loads are usually considered independently. For axial loads, designs are typically based on a specified factor of safety against axial soil failure. The possible effect of lateral loads on axial capacity is usually ignored. For lateral loading the typical design is based on working stress methods. In this approach, the soil is rather crudely modeled as a continuous 'bed' of nonlinear springs and the pile is modeled as a beam-column. This approach may or may not include the effects of axial load on lateral response. Highly simplified analyses are conducted to insure the pile stresses remain below allowable values. Thus, it is not a standard practice to estimate the pile head lateral load or bending moment capacity of the pile element. However, while such methods are not in common usage for offshore design, the problem of pile capacity under combined loading has been addressed by a number of authors, for example, Broms (1964), Meyerhof (1995) and Murff (1987).

As an example, consider a steel pipe pile with diameter D, wall thickness t, and yield stress σ_y, subjected to combined axial, lateral and moment loading (V, H and M) as shown schematically in Figure 15.5. Consider the case where the pile is sufficiently long that the failure mechanism is the development of a plastic hinge at some depth l, rather than rigid pile rotation with the 'kicking out' of the pile tip. It is assumed that the soil skin friction resistance (axial) in the upper part of the pile (down to the plastic hinge) is negligible such that the axial load at the plastic hinge is equal to the axial load at the pile top V. For this example, the lateral resistance of the soil per unit length of the pile p is assumed to be constant with depth. The moment at the pile hinge is a local maximum, thus the shear at the plastic hinge is zero. Moment and horizontal load equilibrium thus provide the following equations (Murff, 1987):

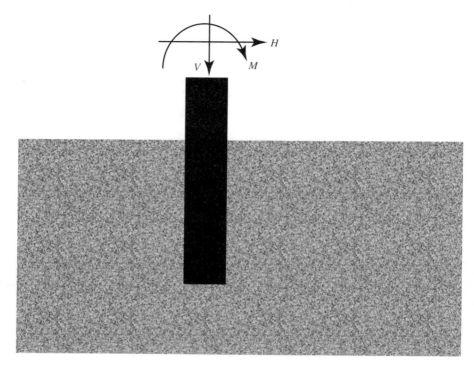

Figure 15.5   Schematic of a pile foundation subject to multi-axial loading

$$\Sigma M = 0 : Hl + M = M_p + \frac{pl^2}{2} \tag{15.5a}$$

$$\Sigma H = 0 : H = pl \tag{15.5b}$$

where

$$M_p = M_y \cos \left(\frac{\pi}{2} \frac{V}{V_y} \right) \tag{15.6a}$$

$$M_y = \frac{\sigma_y}{6} [D^3 - (D - 2t)^3] \tag{15.6b}$$

and

$$V_y = \pi (D - t) t \sigma_y \tag{15.7}$$

Note that M_y is the plastic moment capacity of a tubular member of wall thickness t with no axial load, and V_y is the axial yield capacity of a tubular member with no moment. The cosine term in Eq. (15.6) shows the effect of axial load on moment capacity. Simplifying these equations and solving for the depth to the plastic hinge gives,

$$l = \sqrt{\frac{2(M_p - M)}{p}} \tag{15.8}$$

and substituting in the second of Eq. (15.5) gives

$$H = \sqrt{2(M_p - M)p} \qquad (15.9)$$

Combining Eqs. (15.6), (15.7) and (15.9) provides the equation for the pile head load-interaction surface. Equation (15.9) is written in a normalized form as follows:

$$H' = \sqrt{\cos\left(\frac{\pi}{2}V'\right) - M'} \qquad (15.10)$$

where

$$H' = \frac{H}{\sqrt{2M_y p}} \qquad (15.11a)$$

$$V' = \frac{V}{V_y} \qquad (15.11b)$$

$$M' = \frac{M}{M_y} \qquad (15.11c)$$

which is shown plotted in Figure 15.6. This result can also be obtained by applying the upper bound method.

15.2.4   Other approaches

As mentioned above, there are other approaches for defining interaction surfaces. For example, numerical analyses such as the finite element method (Desai and Able, 1972), using advanced soil constitutive equations (Desai and Siriwardane, 1984), could be used to determine a discrete set of solutions (failure load combinations) for a specific non-dimensional set of parameters. A simpler but more approximate approach is the use of detailed limit analysis methods (Murff and Miller, 1977a,b). A third option is to conduct physical model or field tests to assess discrete failure load combinations as reported, for example, by Georgiadis and Butterfield (1988), Osborne *et al.* (1991), Nova and Montrasio (1991), Murff *et al.* (1992), Meyerhof (1995) and Butterfield and Gotardi (1994). Load combinations determined in this manner will normally have to be fit with an analytical function, splines, or some other piecewise continuous surface.

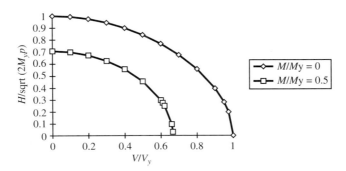

Figure 15.6   Failure-interaction surface for a pile foundation

With this rather simple background, we now have the basic tools to apply the method of generalized plastic limit analysis to a system of foundation elements. The following section describes the theoretical basis for this approach.

15.3   Generalized Theory of Limit Analysis

If a system of forces characterizes the stress state in a continuum, the forces can be considered generalized stresses. The treatment here follows Prager (1959), where the term 'generalized stresses' is used to emphasize the fact that the variables which characterize the stress state do not necessarily have the dimensions of stress. For a given set of generalized stresses $Q_1 \cdots Q_n$, the generalized strain rates $\dot{q}_1 \cdots \dot{q}_n$ are the work rate conjugates of the stresses, i.e.

$$\dot{W} = Q_1\dot{q}_1 + \cdots + Q_n\dot{q}_n \tag{15.12}$$

where $\dot{W}$ is the work rate of the generalized stresses on the strain increments.

As an example, consider the case of a single footing of width B, subjected to vertical load, V, horizontal load H, and moment M, with deformations characterized by vertical displacement v, horizontal displacement h, and rotation m. We can choose the following stress and strain variables,

Generalized stresses: V, H, M/B
Generalized strains: v, h, mB

Although this set is very intuitive, it is not unique. Note that the moment is divided by B and the rotation is multiplied by B to make the force and displacement variables dimensionally consistent while maintaining work conjugacy

$$\dot{W} = V\dot{v} + H\dot{h} + (M/B)(\dot{m}B) \tag{15.13}$$

For a rigid plastic body obeying associated flow, the yield limit can be expressed as (Prager, 1959)

$$f(Q_1 \cdots Q_n) = 0 \tag{15.14}$$

This function is then a convex surface which encloses the coordinate origin in generalized stress space and the generalized strain rates are normal to it, i.e. the surface is a plastic potential. The function is defined such that for stress states within the surface, the function is negative and for those outside, it is positive. As such, it is then possible to apply the upper and lower bound theorems of limit analysis using the generalized definitions of stress and strain and the yield condition in the form of Eq. (15.14). For the purposes here we will primarily employ the upper bound method, that is,

> "If an estimate of the plastic collapse load of a body is made by equating internal rate of dissipation of energy to the rate at which external forces do work in any postulated (kinematically admissible) mechanism of deformation of the body, the estimate will be either high or correct." (Calladine, 1969)

Application of the method requires that a collapse mechanism be postulated and that a method for calculating energy dissipation rate within the deforming material be available. It follows then that the least upper bound solution found is the best solution, i.e. closest to the exact solution.

The associated flow rule provides a means for calculating the generalized plastic strain rates of a deforming plastic body (Calladine, 1969)

$$\dot{q}_i = \lambda \frac{\partial f}{\partial Q_i} \tag{15.15}$$

where λ is a positive scalar. Equations (15.12), (15.14) and (15.15) provide the generalized relationships to calculate energy dissipation rates for a postulated collapse mechanism. It is useful to consider a specific example to demonstrate the upper bound principles.

For the purpose of illustration, consider a strip footing of width B, supported on purely cohesive soil with undrained strength s_u, as shown schematically in Figure 15.7(a). The footing is subjected to vertical load V per unit length, and horizontal load H per unit length. Using the method of characteristics, the kinematic and static solutions can be

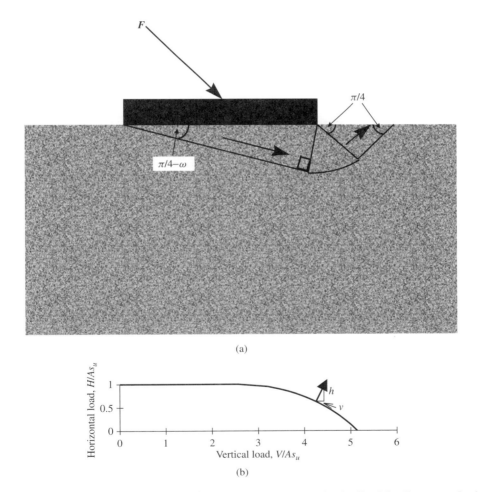

(a)

(b)

Figure 15.7 (a) Failure mechanism for shallow foundation under inclined loading on cohesive soil; (b) failure-interaction surface for shallow foundation under inclined loading on cohesive soil

determined. Also shown in Figure 15.7(a) is the failure mechanism or kinematic solution. The static solution in terms of the vertical load capacity can be determined in terms of the undrained strength and the horizontal load (Chen, 1975). The vertical stress exerted by the footing at failure σ is

$$\sigma = s_u(1 + \pi - 2\omega + \cos 2\omega) \tag{15.16}$$

where $\omega = 1/2 \sin^{-1}(\tau/s_u)$, and $\tau = $ shear stress at failure. Note that $\tau = H/B$ and $\sigma = V/B$. Rewriting Eq. (15.16) in terms of V and H as a yield function gives

$$f = V' - 1 - \pi + \sin^{-1}(H') - \sqrt{1 - H'} = 0 \tag{15.17}$$

where $V' = V/Bs_u$ and $H' = H/Bs_u$. Equation (15.15) gives the following results:

$$\dot{v}' = \lambda \frac{\partial f}{\partial V} = \lambda \frac{1}{Bs_u} \tag{15.18}$$

and

$$\dot{h} = \lambda \frac{\partial f}{\partial H} = \frac{\lambda(1 + H')}{Bs_u \sqrt{1 - H'^2}} \tag{15.19}$$

Equations (15.18) and (15.19) then provide a means for uniquely calculating the ratio of the generalized strain increments (the scalar, λ, cancels). Figure 15.7(b) shows the yield surface in generalized stress space with example strain rate vectors (magnitude is arbitrary). It is also straightforward to demonstrate that strain increment directions are consistent with the kinematic solution shown in Figure 15.7(a).

  The above is simply an illustrative example of the principles of plastic limit analysis for a well known problem framed in terms of generalized stresses and strains. Before proceeding with applications of the theory developed here, it is useful to consider conventional approaches to such a problem.

15.4   Conventional Solution Procedures

There are many possible approaches to assessing the capacity of a foundation system. These vary from a three-dimensional, nonlinear, finite element analysis of the structure and foundation to a highly simplified estimate of foundation loads with subsequent analysis of individual footings. Below, we will briefly discuss these two extremes and allude to some intermediate approaches.

15.4.1   *Three-dimensional finite element analysis*

The most rigorous approach would be to model the structure and the soil continuum using three-dimensional finite element analysis with nonlinear material properties. The design loads would be applied to the structure and the loads would then be scaled up according to some logical rule. For example, the gravity loads might be maintained constant and the environmental loads, such as wind and wave loading, would be increased until convergence of the numerical algorithm is no longer possible. If care is taken, this lack of numerical convergence can provide a good estimate of the system capacity. In practice, this approach is rarely taken. It is expensive, time consuming, and, in light of the uncertainties involved, it is often debatable whether the results are significantly improved over more approximate methods.

15.4.2   *Simplified individual foundation analysis*

In this approach the structure may be treated as a rigid body and the foundation elements as a set of soil springs. The design loads are applied to the structure and the loads on individual foundation elements are estimated based on equilibrium equations. Individual footings are then independently analysed for the relative load combinations acting on that footing. For example, the loads may be scaled up to failure and the footing sized to achieve a specified safety factor for this condition. This approach is quite often taken for simple structure-foundation systems. For example, see Fjeld and Andreassen (1982). A major drawback with this approach is that it does not account for the system nonlinearity. That is, the relative force distribution on a given footing may change dramatically from a nominal elastic response to the ultimate system capacity.

15.4.3   *Intermediate analysis approaches*

For important, complex structures a compromise between the above-mentioned approaches is often made. For example, an offshore structure and its pile foundation may be modeled as an assemblage of nonlinear beam column elements with the axial and lateral soil resistance on the piles modeled as beds of uncoupled, nonlinear springs (Arnold *et al.*, 1977). In as much as the soil spring characterization is reasonable, the nonlinearity of the system is properly accounted for by this approach and a good estimate of system capacity can be made. Of course, the analysis also provides much more information about the system including insights regarding the initiation and development of mechanisms. Analyses of this type are frequently carried out with the specific purpose of determining the system capacity. In fact, the foundation capacity alone is often the target of such investigations.

These analyses are much simpler than the full three-dimensional FEM analyses and have become commonplace in offshore practice. Nonetheless, they are still time consuming and expensive. As will be discussed in the following sections, it is possible to use plastic limit analysis methods to obtain accurate estimates of foundation system capacity with much less effort. This approach is particularly effective when used to complement the more detailed analysis of the structure-foundation system.

15.5   Limit Analysis Solution Procedures

With tools such as those discussed above for individual foundation elements, it is possible to use limit analysis to estimate foundation system capacity. For the purposes here, the upper bound method will be used. We will consider planar failure mechanisms but a simple way for generalizing the mechanism to consider out of plane failures will be described later.

It has been shown previously that individual elements for both shallow foundations and pile foundations can be characterized as yield surfaces in terms of horizontal load, vertical load, and in-plane moment. Consider simple rigid structures, as shown schematically in Figures 15.8. and 15.9, which are supported by individual foundation elements (footings

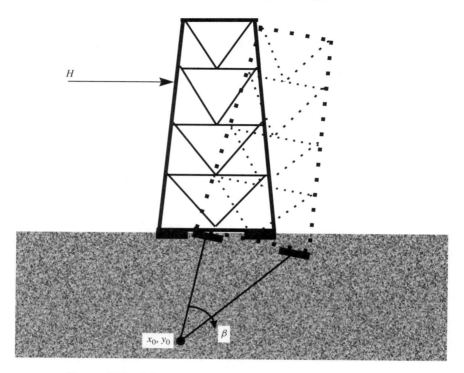

Figure 15.8   Schematic of rigid structure on shallow foundation

or piles). The structure is subjected to a complex set of loads and moments in any directions. Now assume a failure mechanism in which the structure translates and rotates in a particular vertical plane. For an upper bound analysis, only in-plane loads and moments need be considered since these are the only ones that contribute external work. Any such mechanism can be fully described as a rigid body rotation of the structure about an axis perpendicular to the plane of failure (instantaneous center of rotation). For most practical cases the axis of rotation can be assumed to be below the foundation as shown in Figures 15.8 and 15.9 (for battered piles, the center of rotation may be above the foundation as will be discussed in a subsequent section).

Since footings and piles have been characterized as yield surfaces in V, H and M generalized stress space, the analysis method is essentially identical for both systems. For battered piles it is necessary to use definitions of vertical and horizontal loads and displacements that are consistent with the development of the yield surfaces; that is, for the pile element, vertical load is defined as axial load (parallel to the pile axis) and horizontal load is lateral load (perpendicular to the pile axis). Thus, the system load definitions may differ from the element ones. This will be illustrated in the examples following this section.

As shown in Figures 15.8 and 15.9, a general collapse mode of the foundation system involves a rigid body rotation of the structure about a horizontal axis normal to the vertical failure plane. The coordinates of the axis of rotation, x_o and y_o, are optimization parameters in the upper bound analysis. That is, we search for the location of the axis of rotation that will give the minimum upper bound to the collapse load. The known in-plane loads can be resolved into single vertical and horizontal loads and a moment.

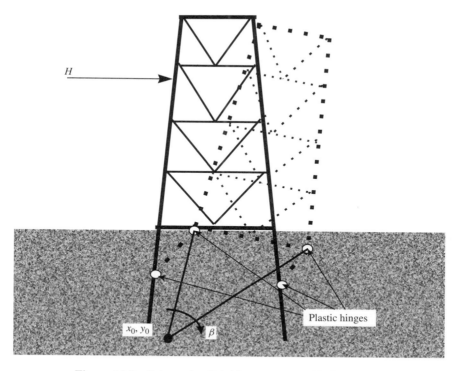

Figure 15.9   Schematic of rigid structure on pile foundation

A single unknown load or scale factor for a set of loads can be determined for any particular mechanism. Typically gravity loads and other well defined loads are considered known and the most uncertain load is considered the unknown. For offshore structures the unknown load is typically taken as the resultant wave, ice or earthquake load. A line of action for the load is estimated based on the geometry (and perhaps the dynamic response) of the structure and the nature of the load mechanism. The foundation capacity is then estimated for a load acting on that line of action using the upper bound method.

Assuming an in-plane failure of the structure and foundation, the upper bound method of plastic limit analysis provides a straightforward means of estimating the foundation capacity. As shown in Figures 15.8 and 15.9, the structure is assumed to rotate rigidly about a horizontal axis perpendicular to the vertical plane of failure. The structure is assumed to undergo a virtual rotation rate $\dot{\beta}$, about the axis of rotation. This, in turn, prescribes the relative velocities and rotation at each foundation element in terms of the location of the axis of rotation and the element coordinates. These relative velocities and the associated flow rule provide two equations in terms of the two unknown forces and the moment for each foundation element. A third equation is provided by the yield surface for each element. In summary, we then have

$$\frac{\dot{v}}{\dot{m}} = \frac{\dfrac{\partial f}{\partial V}}{\dfrac{\partial f}{\partial M}} \tag{15.20a}$$

$$\frac{\dot{h}}{\dot{m}} = \frac{\dfrac{\partial f}{\partial H}}{\dfrac{\partial f}{\partial M}} \qquad\qquad (15.20b)$$

$$f(V, H, M) = 0 \qquad\qquad (15.20c)$$

Note that the ratios of the velocities (left-hand side) are known from the problem geometry and the ratios of the partial derivatives can be determined using the associated flow rule (the scale factor λ cancels in the ratios). The three equations can then be solved (at least in principle) for the unknown forces, V and H and the moment M. The rate of internal energy dissipation $\dot{I}$ is then

$$\dot{I} = V\dot{v} + H\dot{h} + M\dot{m} \qquad\qquad (15.21)$$

From the problem geometry it is also possible to calculate the work rate of the known external forces $\dot{E}$ as

$$\dot{E} = F_{vk}(x_o - x_{vk})\dot{\beta} + F_{hk}(y_o - y_{hk})\dot{\beta} + M_k\dot{\beta} \qquad\qquad (15.22)$$

where the subscripts k and u refer to the known loads and unknown loads, respectively, and v and h refer to vertical and horizontal loads and their coordinates, respectively. The work rate of the unknown external force is

$$\dot{E}_u = F_u \cos(\alpha)(y_o - y_u)\dot{\beta} + F_u \sin(\alpha)(x_o - x_u)\dot{\beta} \qquad\qquad (15.23)$$

where α is the unknown load inclination angle with the horizontal. Equating the external work rates to the internal energy dissipation rates and simplifying gives

$$F_u = \frac{\displaystyle\sum_{i=1}^{N}\{V_i(x_o - x_{fi}) + H_i(y_o - y_{fi}) + M_i\} - F_{vk}(x_o - x_{vk}) - F_{hk}(y_o - y_{hk}) - M_k}{(y_o - y_u)\cos(\alpha) + (x_o - x_u)\sin(\alpha)}$$

$$(15.24)$$

where the subscript fi refers to the ith foundation element. For a given location of the axis of rotation (x_o, y_o), the unknown force can then be determined. All of the above equations are derived in detail in Appendices A–C at the end of this chapter. It follows from the upper bound theorem, that the best solution is the one that minimizes F_u. Furthermore, since the rotation of the structure about some instantaneous center of rotation includes all possible planar motions then the true minimum upper bound should be the exact solution for the planar mechanism assumed. Thus, the system so determined should be in equilibrium. Since we compute the forces and moment on each foundation element, this can be checked.

It is useful to comment here on a common misperception regarding limit analysis methods, in general, and the approach used herein, in particular. The assumption of associated flow is often questioned and consequently this approach may be overlooked by those who favor the so-called limit equilibrium method. The latter is an ad hoc method in which solutions are sought which approximately satisfy the failure condition and equilibrium. In general, upper bound methods are merely simpler ways to find such approximate solutions. As pointed out above, the method used here rigorously satisfies these conditions. Hence, while associated flow arguments can be important in an absolute sense, they are rendered irrelevant in comparisons of the proposed method with limit equilibrium approaches (which in the author's view are typically more difficult to apply).

The above process does not guarantee a globally correct solution since it does not consider out of plane failures. One method for improving this situation is simply to consider many different vertical failure planes. This will enable one to determine which planar failure mechanism is most critical. An example will be discussed in the following section. This approach is restricted to planar mechanisms and does not address situations where torsional failure about the vertical axis is significant. Although torsional modes are usually not major contributors for offshore structures, care is in order. This underscores the value of calibrating the procedure with more rigorous nonlinear structure-foundation analyses.

In summary, then, the following steps are carried out in estimating the foundation failure load:

(1) Idealize the problem as a rigid structure on yielding foundation supports. For a selected vertical failure plane (typically start with a plane containing a principal axis of the structure broadside, for example), select an initial location for the axis of rotation.

(2) For the selected plane determine the resultant known loads and moment in the plane. Determine a point on the line of action of the resultant unknown load.

(3) For the selected axis of rotation calculate the forces and moment on each foundation element using Eqs. (15.20a), (15.20b) and (15.20c).

(4) Evaluate the unknown force F_u as given in Eq. (15.24).

(5) Systematically vary the location of the axis of rotation and repeat Steps (3) and (4) to find the minimum failure load.

(6) Check equilibrium. If the structure is not in equilibrium the minimum is a local value. Search a new region for the axis of rotation.

(7) Finally consider mechanisms in different vertical planes to ensure an out of plane mechanism is not more critical.

Parametric studies considering the effects of known loads (gravity load, for example), pile batter, and foundation geometry, using this approach can be extremely useful in developing insight as to foundation performance. In the following section, specific examples will be discussed to enhance the above description of the methodology.

15.6   Implementation with Examples

15.6.1   *Shallow foundations*

Many mobile offshore structures and a few permanent structures are supported by shallow foundations. The foundation system geometry may range from a single gravity base to a tripod system of jack-up spud cans. Foundations must resist large vertical loads from structure weight, deck equipment, and drilling loads, as well as lateral loads and overturning moments due to environmental effects such as wind and waves. For the purposes here, we will consider a highly simplified foundation system but the principles involved can be readily applied to most shallow foundation system configurations. A discussion of generalization of the model to more complex conditions is included at the end of this section.

Problem description

For the purposes of illustration, we will consider a simple planer structure supported by two shallow foundations. As previously stated, the superstructure is assumed rigid and rigidly attached to the shallow footings. The footings then comply with the structure as it translates and rotates. This mechanism includes all possible planar foundation failure mechanisms. For this example, we will consider a simple case of a planar structure supported by two footings. The yield surface selected to characterize the footing failure mechanism is an empirical function first suggested by R. G. James (Osborne *et al.*, 1991) which includes moment, horizontal load and vertical load as follows:

$$\sqrt{(M/D)^2 + \Lambda_1 H^2} + \Lambda_2 \left(\frac{V^2}{V_c} - V \right) = 0 \tag{15.25}$$

where V_c is the compression capacity of the footing under pure vertical load, D is footing diameter, and Λ_1 and Λ_2 are constants. Equation (15.25) mimics Eq. (15.4) (the general bearing capacity equation), as shown in Figure 15.10, but is somewhat more tractable.

Figure 15.8 shows a schematic of the structure-foundation system and Table 15.1 gives the details of the geometry and component strengths. The vertical load is assumed to be known and the unknown lateral load is strictly horizontal. For the purposes here, we will examine the effects of the vertical load on the structure and the point of application of the resultant unknown horizontal load.

Results

Consider the simple structure as shown in Figure 15.8. For a particular vertical load, we can obtain a set of solutions where the lateral load is applied at different elevations above the mudline. Cases with the load applied near the mudline could represent shallow water structures and those applied at higher points could represent progressively deeper water situations. Solutions for a set of such cases can be represented as interaction diagrams as

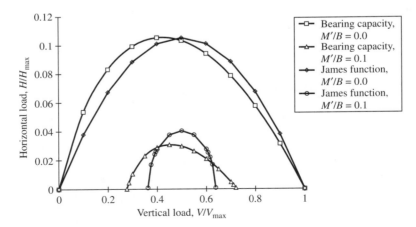

Figure 15.10 Comparison of interaction surfaces for bearing capacity equation and James empirical equation ($\lambda_1 = 1.06$, $\lambda_2 = 0.43$)

Table 15.1　Data for shallow foundation analysis

Property description	Value
Number of footings	2
Footing spacing, center to center	30 m
Footing diameter	6 m
Footing compression capacity	20 MN
Footing tension capacity	0
Capacity parameter, λ_1	1.10
Capacity parameter, λ_2	0.43
Horizontal load height	varies
Vertical load	varies

shown in Figure 15.11. In this figure each curve represents a locus of failure points for a constant vertical load with varying heights of lateral load application which give rise to the failure interaction plot of lateral load vs. overturning moment. Each point represents the failure conditions for particular values of the vertical load, lateral load height, and structure geometry. The unknown lateral load, which would cause failure for those conditions, is found by minimizing the right-hand side of Eq. (15.24) with respect to the coordinates of the assumed axis of rotation. In each plot, for lateral loads applied at the mudline, the overturning moment is zero and the lateral load capacity is maximum. As the lateral load elevation increases, the increasing overturning moment gradually reduces the lateral load capacity until the moment capacity is a maximum and the lateral capacity approaches zero (load at infinite height). As the vertical load increases, the failure interaction plots expand until the vertical load is one-half of the combined vertical capacity of the footings, indicating that, to this point, the lateral capacity is increased by the vertical load. At larger vertical loads, the failure-interaction plots contract until they degenerate into a point where the vertical load equals vertical capacity of the system.

The failure mechanisms for each point provide useful insight as to the foundation system behavior. As an example, Figure 15.12 shows a plot of the coordinates of the axis of rotation of the critical mechanism as the elevation of the lateral load changes. Each curve represents a different vertical load. The lateral load direction is from left to right and the leeward edge of the leeward footing is at the origin. At very low load elevations

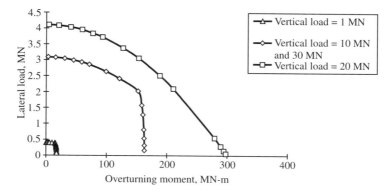

Figure 15.11　Interaction surface for example shallow foundation system

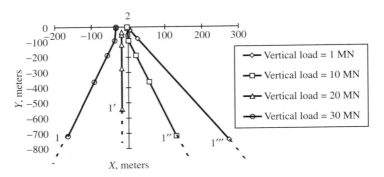

Figure 15.12 Locus of critical centers of rotation with varying load heights for example shallow foundation system

the center of rotation is essentially at an infinite depth below the structure (indicated by dashed lines). This means that the mechanism is pure translation of the structure. As the load elevation increases the center of rotation moves upward toward the structure (past point 1 and ultimately to point 2), and a combined translation-overturning mechanism is apparent. At very high load elevations and vertical loads less than the capacity of one footing, the center of rotation moves to a point near the center of the leeward footing. The mechanism is overturning about that point, with lift-off of the windward footing. At very high load elevations and vertical loads greater than the capacity of one footing, the center of rotation moves to a point near the center of the windward footing and the mechanism is overturning about that point, with plunging of the leeward footing.

Generalization

The examples above are relatively simple and for most practical problems more complex situations must be modeled. This complexity includes the spatial variability of soil resistance and the three-dimensional aspects of a structure and its foundation system.

For spatial variations in soil properties, a convenient closed form solution for the failure surface of the footing (like Eq. (15.25)) may not be available. In these cases it may be possible to use published solutions for special non-homogeneous conditions such as described by Davis and Booker (1973) or to develop a suite of solutions for special cases. It may then be possible to fit an Equation similar to the form of Eq. (15.25) to the numerical results.

It is also possible to incorporate fully three-dimensional mechanisms. For simple overturning, assuming the structure rotates rigidly about a horizontal axis normal to the vertical plane of failure (which is normal to the horizontal axis of rotation), the velocities at each footing are prescribed. In effect, the footings can be projected onto the assumed vertical plane of failure and the system treated as a two-dimensional mechanism. One can then vary the horizontal load orientation to develop a failure interaction surface in terms of orthogonal lateral loads as discussed by Murff (1994). It is also possible to develop a mechanism which includes torsion about the vertical axis of the structure, however, this can become quite complicated and the simplicity of the approach begins to be lost.

15.6.2 Pile foundations

Most permanent offshore structures are founded on steel pipe piles. The foundation system geometry may vary considerably from one of small diameter, vertical piles, distributed densely and uniformly over the foundation area for simple template type structures, usually restricted to shallow water, to one of very large diameter, battered piles distributed sparsely around the periphery of the structure base, more typical of deep water structures. Foundations must resist large vertical loads from structure weight, deck equipment, and drilling loads as well as lateral loads and overturning moments due to environmental effects such as wind and waves. For the purposes of illustration, we will consider a highly simplified foundation system, but the principles involved can be readily applied to most pile system configurations. A discussion of generalization of the model to more complex conditions is included at the end of this section.

Problem description

For the purposes of illustration, we will consider a simple planer structure supported by two piles. As previously stated, the superstructure is assumed rigid, whereas the piles are assumed to develop a two-hinge mechanism, one hinge at the soil surface and one at some depth below as shown in Figure 15.9. This mechanism is quite robust and includes the majority of foundation failure mechanisms. For this example, we will consider a simple case of uniform lateral soil resistance and piles of uniform diameter and wall thickness. It is assumed that the piles are long enough that a plastic hinge forms below the mudline rather than kick-out of the pile toe, and that the depth of the second plastic moment is sufficiently shallow that the axial resistance of the soil between the two plastic moments can be neglected. For most offshore piles, this condition is satisfied to a high degree of approximation. The yield surface which characterizes the pile failure mechanism at the pile top is then given by Eq. (15.9) and implicitly includes the two hinge mechanism. It should be mentioned that Eq. (15.9) can be derived from an upper bound formulation as well as the equilibrium argument previously used (and is therefore exact) and is hence consistent with the approach taken here.

Figure 15.9 shows a schematic of the pile founded structure and Table 15.2 gives the details of the geometry and component strengths. The vertical load is assumed to be

Table 15.2 Data for pile foundation analysis

Property description	Value
Number of piles	2
Pile spacing, center to center	30 m
Pile diameter	1 m
Pile wall thickness	2.5 cm
Pile yield strength	250 Mpa
Lateral soil resistance	0.2 MN/m
Pile batter (Hor.:Vert.)	0 and 1:8
Horizontal load height	varies
Vertical load	varies

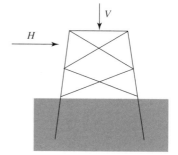

known and the unknown lateral load is strictly horizontal. For the purposes here we will examine the effects of the vertical load on the structure, the point of application of the resultant unknown horizontal load, and the pile batter.

Results

First consider the simple structure with vertical piles. For a particular vertical load we can obtain a set of solutions where the lateral load is applied at different elevations above the mudline. Cases with the load applied near the mudline could represent shallow water structures and those applied at higher points could represent progressively deeper water situations. Solutions for a set of such cases can be represented as interaction diagrams as shown in Figure 15.13. In this figure each curve represents a locus of failure points for a constant vertical load with varying heights of lateral load application which give rise to the failure interaction plot of lateral load vs. overturning moment. Each point represents the failure conditions for particular values of the vertical load, lateral load height, and structure geometry. The unknown lateral load which would cause failure for those conditions is found by minimizing the right-hand side of Eq. (15.24) with respect to the coordinates of the assumed axis of rotation. In each plot, for vertical piles and lateral loads applied at the mudline, the overturning moment is zero and the lateral load capacity is a maximum. As the lateral load elevation increases the increased overturning moment gradually reduces the lateral load capacity until the moment capacity is a maximum and the lateral capacity approaches zero (infinitely high load). As the vertical load increases, the failure interaction plots shrink indicating that the lateral capacity is reduced by the vertical load. Initially this effect is relatively small but as the axial load becomes a significant portion of the axial capacity of the piles it is more pronounced.

The failure mechanisms for each point provide useful insight as to the foundation system behavior. As an example, Figure 15.14(a) shows a plot of the coordinates of the axis of rotation of the critical mechanism as the elevation of the lateral load changes. Each curve represents a different vertical load. The lateral load direction is from left to right and the center of the leeward pile is at the origin. Figure 15.15 shows schematics of possible pile system failure mechanisms. At very low load elevations the center of rotation is essentially at an infinite depth below the structure (indicated by dashed lines). This means that the mechanism is pure translation of the structure due to a shearing

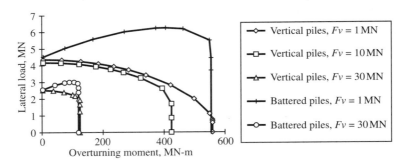

Figure 15.13   Interaction surface for example pile foundation system

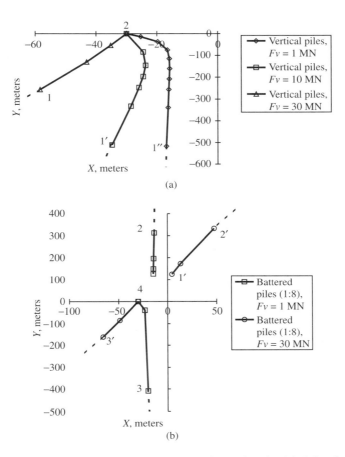

(a)

(b)

Figure 15.14 Locus of critical centers of rotation with varying load heights for example pile foundation system. (a) Vertical piles; (b) battered piles

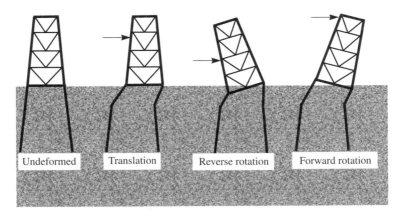

Undeformed   Translation   Reverse rotation   Forward rotation

Figure 15.15 Schematic of pile foundation system failure mechanisms

mechanism in the piles, as shown in Figure 15.15. As the load elevation increases the center of rotation moves upward toward the structure (to point 1 and beyond) and a combined shear-overturning mechanism is apparent. At very high load elevations the center of rotation moves to the top of the windward pile (point 2) and the mechanism is pure overturning about that point. For larger vertical loads, the overturning mechanism is reached at lower lateral load elevations because the system is initially closer to plunging under vertical load.

Next consider the battered pile case. As shown in Figure 15.13, the failure interaction diagrams have a clearly different character. At low lateral load elevations, the lateral capacity is only slightly greater than for the comparable unbattered case. This increase is due to the horizontal component of the axial pile capacity. However, as the load elevation and hence overturning moment increases the lateral capacity actually increases. Eventually the lateral capacity reaches a maximum and the effect of the overturning moment begins to reduce the lateral capacity. The maximum overturning capacity is slightly less than that for the vertical pile case due to the reduction in the vertical component of the axial pile capacity because of pile batter.

Again the failure mechanisms provide insight into the system response. Figure 15.14(b) shows the coordinates of the critical centers of rotation as the lateral load elevation is varied. Figure 15.15 shows the possible failure mechanisms involved. At low lateral load elevations the critical center of rotation is actually above the structure (point 1) indicating that the structure is actually rotating backwards. Of course, it is necessary that the center of rotation is always above the lateral load such that the point of application of the lateral load always has a net translation in the load direction. However, in these cases, the base of the structure is translating faster than the point of load application. Note also that the critical center of rotation, for loads applied at the structure base, is in the neighborhood of the point where projections of the piles upward would intersect. In fact, as the lateral soil resistance goes to zero, the critical center of rotation approaches this point exactly. In this limiting case, the velocity of the pile head would be normal to each pile axis. As the load elevation increases, the velocity of the unknown lateral load is slowed by the reverse rotation, causing the capacity to increase. Eventually, the tendency for reverse rotation is overcome by the increasing overturning moment and the lateral load reaches a maximum with the structure undergoing pure horizontal translation. At this point, the critical center of rotation approaches an infinite elevation (indicated by the dashed lines above point 2 in Figure 15.14(b)), and instantly shifts to a point at infinite depth below the structure (indicated by the dashed lines below point 3 in Figure 15.14(b)). Note that both limiting cases give the same pure translation. As the lateral load elevation continues to increase, the critical center of rotation moves upward toward the structure as it did in the vertical pile case, ultimately approaching the top of the windward pile (point 4).

Generalization

The examples above are relatively simple and for most practical problems more complex situations must be modeled. This complexity includes the variability of lateral soil resistance, pile wall thickness changes and the three-dimensional aspects of a structure and its foundation system.

For most soil and pile variations a convenient closed form solution for the failure surface at the pile head (like Eq. (15.9)) may not be available. One approach that has been successful for these more complicated situations is to include both pile hinges in the global failure mechanism explicitly. Murff and Wesselink (1986) provide a detailed description of this analysis procedure but a brief description here may be helpful. It was found that a simple lateral analysis of each pile alone is sufficiently accurate to locate the depth of the second plastic hinge so that this step can be performed independently before the system is analysed. Assuming that the first plastic hinge occurs at the mudline, the depth of the second plastic hinge can be found by minimizing the lateral load at collapse with respect to that depth. Using numerical integration and relatively crude minimization techniques, this method is capable of dealing with varying soil resistance and pile wall thickness very efficiently (see Murff and Wesselink (1986) for a more detailed discussion). Given the location of the second plastic hinge, the system can be analysed by including this depth in the mechanism. Thus, it is only necessary to minimize the system lateral capacity with respect to the coordinates of the center of rotation of the system mechanism. For a given center of rotation, the velocities at the pile heads are prescribed. Knowing the depth to the second plastic hinge the dissipation rate for each pile mechanism can be determined and hence the system capacity can be calculated using an equation analogous to Eq. (15.24).

It is also possible to incorporate fully three-dimensional mechanisms. For simple overturning, assuming the structure rotates rigidly about a horizontal axis, the velocities at each pile head are prescribed and these simply have to be resolved into axial and lateral components at each pile head. For piles with relatively small batter, the pile profile can be projected on the assumed vertical plane of failure (normal to the horizontal axis of rotation) and the system treated as a two-dimensional mechanism. For small batter angles, this approximation is very accurate and much easier to implement. One can then vary the horizontal load orientation to develop a failure interaction surface in terms of two orthogonal lateral loads as discussed by Murff and Wesselink (1986). It is also possible to develop a mechanism which includes torsion about the vertical axis of the structure however this can become quite complicated and the simplicity of the approach begins to be lost.

Closing comments

As can be seen from the results given above, the characteristics of the foundation systems for shallow foundations and piles are, not surprisingly, very different in their response to such parameters as foundation geometry, vertical load and lateral load location. The methods presented here can efficiently provide very clear insight into these behaviors so that these features can be effectively exploited in the selection and/or design of offshore foundation systems.

15.7   Applications to Practical Design and Analysis

To use the methods described above in practical design and analysis it is important to understand their strengths and weaknesses. It has been the author's experience that their most effective use is to complement more rigorous procedures. For example, a limited number of detailed nonlinear structural analyses (including foundation) can be conducted and the limit analysis methods described here can be used parametrically to assess design or operational sensitivities. This provides a benchmark for the more approximate limit analysis methods.

There are several conditions that the analyst/designer should be particularly aware of:

(1) Limit analysis bounds are not strictly valid where elements of the system exhibit strain or displacement softening.

(2) In this approach we are considering only foundation failures. In many cases the structure may be the weak link. Furthermore, it is possible that the critical collapse mode may be a combined foundation and structure failure. In the latter case, foundation capacity may overestimate system strength.

(3) The assumption of a planar failure mechanism is a kinematic constraint on the system It is possible that an out of plane failure mechanism may be more critical. This is particularly likely where the system is particularly strong (or weak) in one direction. This possibility can be explored to some degree by analysing the system for a range of loading directions as shown schematically in Figure 15.16. If an out of plane load component would cause failure in a different direction before the imposed in-plane failure occurs, the former is more critical and the failure envelope is effectively truncated as shown in Figure 15.16.

(4) In systems where torsion is important, the simple planar mechanism may not be adequate.

(5) The details of the assumed mechanism should be borne in mind. For example, it was assumed that a plastic hinge forms in the piles at the mudline. If a reduction in pile cross section occurs right below the mudline, the plastic moment formation in the lower section may be more critical.

The above cautions simply reiterate the need to understand the analysis and to reconcile unexpected results. If caution is exercised, the method can be an effective complement to the analysis of many offshore structural/foundation systems.

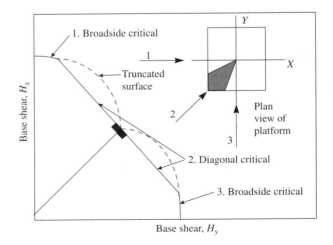

Figure 15.16   Truncation of failure interaction surface for pile foundation system

15.8   Conclusions

Upper bound limit analysis methods for systems of footings can be developed which do not require one to invent the detailed kinematics of foundation failure. Rather, the multi-dimensional interaction diagrams of individual footings can be used as macroscopic yield surfaces following Prager's method of generalized stresses and strains (Prager, 1959). These yield surfaces can then be used in an application of the upper bound method in which the footing displacements play the role of generalized strains which are normal to the yield surface (associated flow). This surprisingly simple analysis method can be used effectively in the evaluation of relatively complex foundation systems, both shallow footings and piles. It can be implemented on a spread sheet with minimal input data required.

Appendix A

The assumed rotation rate $\dot{\beta}$ of a footing or pile head about a specified horizontal axis, as shown in Figures 15.8 and 15.9, prescribes the magnitudes of vertical velocity $\dot{v}$, horizontal velocity $\dot{h}$, and angular velocity $\dot{m}$ as follows (based on geometric constraints):

$$\dot{v} = \dot{\beta}(y_o - y_{fi}) \qquad (A15.1)$$

$$\dot{h} = \dot{\beta}(x_o - x_{fi}) \qquad (A15.2)$$

$$\dot{m} = \dot{\beta} \qquad (A15.3)$$

where x_o, y_o are the coordinates of the axis of rotation, and x_{fi}, y_{fi} are the coordinates of foundation element i.

Each foundation element has a yield function described in load space as

$$f_i(V_i, H_i, M_i) = 0 \qquad (A15.4)$$

where V_i = vertical load, H_i = horizontal load, and M_i = moment.

The loads and plastic displacements associated with yielding of the foundation element are treated as generalized stresses and generalized plastic strains as discussed by Prager (1959). The kinematic variables can then be determined from the associated flow rule as

$$\dot{v} = \lambda \frac{\partial f_i}{\partial V_i} \qquad (A15.5)$$

$$\dot{h}_i = \lambda \frac{\partial f_i}{\partial H_i} \qquad (A15.6)$$

$$\dot{m}_i = \lambda \frac{\partial f_i}{\partial M_i} \qquad (A15.7)$$

where λ is a positive scalar. Note that Eqs. (A15.1)–(A15.7) allow one to calculate the relative velocities, e.g. $\dot{v}_i/\dot{h}_i$ in terms of V_i, H_i and M_i. Eliminating $\dot{v}_i$, $\dot{h}_i$, $\dot{m}_i$, λ and $\dot{\beta}$ among Eqs. (A15.1)–(A15.3) and (A15.5)–(A15.7) gives two independent equations for V_i, H_i, and M_i in terms of geometric parameters. For example, any two of the following equations provide two independent equations:

$$\frac{\dot{v}_i}{\dot{m}_i} = \frac{\partial f_i/\partial V_i}{\partial f_i/\partial M_i} \tag{A15.8}$$

$$\frac{\dot{h}_i}{\dot{m}_i} = \frac{\partial f_i/\partial H_i}{\partial f_i/\partial M_i} \tag{A15.9}$$

$$\frac{\dot{h}_i}{\dot{v}} = \frac{\partial f_i/\partial H_i}{\partial f_i/\partial V_i} \tag{A15.10}$$

A third independent relationship is the yield function itself, Eq. (A15.4). In general, these three equations are nonlinear, but can be solved simultaneously for V_i, H_i and M_i for a prescribed axis of rotation. In other words, the kinematics that result from the assumed collapse mechanism are associated with the direction of a normal to the yield surface and hence are uniquely related to a specific load combination on the yield surface. These loads are only the equilibrium set for the exact solution. The total energy dissipation rate, D_i, can then be determined for all foundation elements as follows:

$$D_t = \sum_{i=1}^{N} V_i v_i + H_i h_i + M_i m_i$$

$$= \sum_{i=1}^{N} V_i (x_o - x_{fi})\dot{\beta} + H_i (y_o - y_{fi})\dot{\beta} + M_i \dot{\beta} \tag{A15.11}$$

The above provides the necessary relationships to calculate the total energy dissipation. Next, it is necessary to determine the work rate of the external loads. It is assumed that known loads F_{vk} and F_{hk} are applied at known locations x_{vk}, y_{vk} and x_{hk}, y_{hk}. The unknown load F_u is applied at some point x_u, y_u at an angle α to the horizontal. The external work rate, E_t, is then

$$E_t = F_{vk}(x_o - x_{vk})\dot{\beta} + F_{hk}(y_o - y_{hk})\dot{\beta}$$

$$+ F_u[(y_o - y_u)\cos\alpha + (x_o - x_u)\sin\alpha]\dot{\beta} \tag{A15.12}$$

Equation external work rate to internal dissipation rate, canceling $\dot{\beta}$s, and solving for the unknown force, F_u gives

$$F_u = \frac{\displaystyle\sum_{i=1}^{N}\{V_i(x_o - x_{fi}) + H_i(y_o - y_{fi}) + M_i\}}{(y_o - y_u)\cos\alpha + (x_o - x_u)\sin\alpha}$$

$$- \frac{+F_{vk}(x_o - x_{vk}) + F_{hk}(y_o - y_{hk})}{(y_o - y_u)\cos\alpha + (x_o - x_u)\sin\alpha} \tag{A15.13}$$

The value of F_u closest to the true solution is found by minimizing F_u with respect to x_o and y_o.

The above equations apply in general to any admissible yield surfaces.

Appendix B

A yield surface of particular interest (Osborne, 1991; Nova and Montrasio, 1991) will be used below as a specific example of the application. A generalized form of this yield surface for a circular footing is

$$f(V, H, M) = \left[\left(\frac{M}{D} \right)^2 + \Lambda_1 H^2 \right]^{\frac{1}{2}} + \Lambda_2 \left[\frac{V^2}{V_c} - V \right] = 0 \qquad (A15.14)$$

where Λ_1 and Λ_2 are constants, V_c is the vertical compression capacity (positive) of the footing and D is the footing diameter.

From Eqs. (A15.5)–(A15.6), we find

$$\dot{v} = \lambda \frac{\partial f}{\partial V} = \lambda \left\{ \Lambda_2 \left[\frac{2V}{V_c} - 1 \right] \right\} \qquad (A15.15)$$

$$\dot{h} = \lambda \frac{\partial f}{\partial H} = \lambda \left\{ \frac{\Lambda_1 H}{[(M/D)^2 + \Lambda_1 H^2]^{1/2}} \right\} \qquad (A15.16)$$

$$\dot{m} = \lambda \frac{\partial f}{\partial M} = \lambda \left\{ \frac{M/D^2}{[(M/D)^2 + \Lambda_1 H^2]^{1/2}} \right\} \qquad (A15.17)$$

From Eqs. (A15.8), (A15.15) and (A15.17), we obtain

$$\Lambda_2 \left\{ \frac{2V}{V_c} - 1 \right\} + \left\{ \frac{M/D^2}{[(M/D)^2 + \Lambda_1 H^2]^{1/2}} \right\} (x_o - x_f) = 0 \qquad (A15.18)$$

and from Eqs. (A15.9), (A15.16) and (A15.17), we obtain

$$\Lambda_1 D^2 H - M(y_o - y_f) = 0 \qquad (A15.19)$$

where the subscript f refers to the footing.

Equations (A15.18) and (A15.19) are combined with the yield function, Eq. (A15.14), to obtain expressions for V, H and M. Solving for H in terms of M from Eq. (A15.19) and substituting into Eq. (A15.18) gives an explicit solution for V, in terms of geometry alone,

$$V^* = \frac{V_c}{2} \left\{ 1 + \frac{(x_o - x_f)/\Lambda_2}{\left[D^2 + \frac{1}{\Lambda_1}(y_o - y_f)^2 \right]^{1/2}} \right\} \qquad (A15.20)$$

This in turn can be used with Eqs. (A15.14) and (A15.19) to solve for M in terms of geometry alone:

$$M^* = \frac{-D^2 \Lambda_2 \left(\frac{V^{*2}}{V_c} - V^* \right)}{\left[D^2 + \frac{1}{\Lambda_1}(y_o - y_f)^2 \right]^{1/2}} \qquad (A15.21)$$

where V^* refers to the expression in Eq. (A15.20). Finally, Eq. (A15.21) is substituted into Eq. (A15.19) to give

$$H^* = \frac{M^*}{\Lambda_1 D^2}(y_o - y_f)$$ (A15.22)

where M^* refers to the expression in Eq. (A15.21). Thus, somewhat surprisingly, the values of V, H and M implied by the footing kinematics can be obtained in closed form. For a general yield surface, numerical methods appropriate for solution of simultaneous, nonlinear equations are required such as Newton–Raphson.

Appendix C

A generalized yield surface for piles under axial and lateral load and moment is given by Eq. (15.9) in the main text. It is assumed here that a plastic hinge forms at the pile top (as well as some depth below), thus substituting $-M_p$ for M in Eq. (15.9) gives the required yield surface:

$$f(V, H, M_p) = H - 2\sqrt{M_p p} = 0$$ (A15.23)

where

$$M_p = M_y \cos\left(\frac{\pi V}{2V_y}\right)$$ (A15.24)

$$M_y = \frac{\sigma_y}{6}[D^3 - (D - 2t)^3]$$ (A15.25)

and

$$V_y = \pi(D - t)t\sigma_y$$ (A15.26)

In these equations the pile is assumed to be a tubular section with a wall thickness t, and a yield strength σ_y.

The associated flow rule provides the following relationships:

$$\dot{h} = \lambda\frac{\partial f}{\partial H} = \lambda$$ (A15.27)

$$\dot{v} = \lambda\frac{\partial f}{\partial V} = \frac{\lambda\pi M_y p \sin\left(\frac{\pi V}{2V_y}\right)}{2V_y\left[pM_y \cos\left(\frac{\pi V}{2V_y}\right)\right]^{\frac{1}{2}}}$$ (A15.28)

Since $M(= -M_p)$ is known at the outset, it is only necessary to find H and V. From Eqs. (A15.1), (A15.2), (A15.10), (A15.27) and (A15.28), we obtain

$$\frac{\dot{h}}{\dot{v}} = \frac{2V_y\left[pM_y \cos\left(\frac{\pi V}{2V_y}\right)\right]^{\frac{1}{2}}}{\pi pM_y \sin\left(\frac{\pi V}{2V_y}\right)} = \frac{y_o - y_f}{x_o - x_f}$$ (A15.29)

Since the right-hand side is known for a specified axis of rotation (x_o, y_o), Eq. (A15.29) is a quadratic equation, and can be solved for V

$$V^* = \frac{2V_y}{\pi} \cos^{-1}\left[\frac{-C + \sqrt{C^2 + 4}}{2}\right] \tag{A15.30}$$

where

$$C = \frac{4V_y^2}{\pi^2 pM_y}\left(\frac{x_o - x_f}{y_o - y_f}\right)^2 \tag{A15.31}$$

It is then straightforward to determine H from Eqs. (A15.1), (A15.21) and (A15.22) as

$$H^* = 2\sqrt{pM_y \cos\left(\frac{\pi V^*}{2V_y}\right)} \tag{A15.32}$$

V^*, H^* and $M^*(= M_p)$ can then be substituted into Eq. (A15.24) to determine the estimate of F_u for the specified axis of rotation.

For battered piles, the velocities determined at the pile head must be resolved in directions along the pile axis (v) and perpendicular to it in the assumed failure plan (h) before applying the above equations. Likewise, the forces calculated $(V$ and $H)$ are forces along and perpendicular to the pile axis, respectively.

References

Arnold, P., Bea, R. G., Idriss, I. M., Reimer, R. B., Beebe, K. E. and Marshall, P. W. A study of soil-pile-structure systems in severe earthquakes. *Proc. Offshore Technology Conference*, Houston, TX, (1977).

Broms, B. B. Lateral resistance of piles in cohesive soil. *J. Soil Mech. Found. Div.*, **90**(2), (1964).

Butterfield, R. and Gotardi, G. A complete three dimensional failure envelope for shallow footings on sand. *Geotechnique*, **44**(1), (1994) 181–184.

Chen, W. F. *Limit Analysis and Soil Plasticity*, Elsevier, Amsterdam, The Netherlands, (1975).

Calladine, C. R. *Engineering Plasticity*. Pergamon Press, Oxford, (1969).

Chen, W. F. and Liu, X. L. *Limit Analysis in Soil Mechanics*. Elsevier, Amsterdam, (1975).

Davis, E. H. and Booker, J. R. The effect of increasing strength with depth on the bearing capacity of clays. *Geotechnique*, **23**(4), (1973) 551–563.

Desai, C. and Able, J. F. *Introduction to the Finite Element Method*. Van Nostrand Reinhold, New York, (1972).

Desai, C. S. and Siriwardane, H. J. *Constitutive Laws for Engineering Materials*. Prentice Hall, Englewood Cliffs, NJ, (1984).

Fjeld, S. and Andreassen, B. Soil-structure interaction of multifooting gravity structures. *Proc. Offshore Technology Conference*, Houston, TX, (1982).

Georgiadis, M. and Butterfield, R. Displacements of footings on sand under eccentric and inclined loads. *Canad. Geotechnical J.*, **25**, (1988) 192–212.

Hansen, J. B. *Bulletin 28: A Revised and Extended Formula for Bearing Capacity*. Copenhagen, Danish Geotechnical Institute, (1970).

Houlsby, G. T. and Wroth, C. P. Direct solutions of plasticity problems in soils by the method of characteristics. *Proc. 4th Int. Conf. on Numerical Methods in Geomechanics*, **3**, (1982) 1059–1072.

Lloyd, J. R. and Clawson, W. C. Reserve and residual strength of pile founded offshore platforms. *Proc. Int. Symp. on the Role of Design, Inspection, and Redundancy in Marine Structural Reliability*. Washington, DC, (1984).

Matlock, H. Correlations for design of laterally loaded piles in soft clay. *Proc. Offshore Technology Conference*, Houston, TX, (1970).

Meyerhof, G. G. The bearing capacity of footings under eccentric and inclined loads, *Proc. Third ICSMFE*, (1953) 440–445.

Meyerhof, G. G. Behavior of pile foundations under special loading conditions. *Canad. Geotechnical J.*, **32**, (1995) 204–222.

Murff, J. D. Plastic collapse of long piles under inclined loading. *Int. J. Num. Anal. Meth. in Geomech.*, **11**, (1987) 185–192.

Murff, J. D. Limit analysis of multi-footing foundation systems. *Proc. Eighth Int. Conf. on Computer Methods and Advances in Geomechanics*, Morgantown, W. VA, (1994) 233–244.

Murff, J. D. and Miller, T. W. Foundation stability on non-homogeneous clays. *J. Geotechnical Engrg. Div.*, **103**(GT10), (1977) 1083–1095.

Murff, J. D. and Miller, T. W. Stability of offshore gravity structure foundations using the upper bound method. *Proc. Offshore Technology Conference*, Houston, TX, (1977).

Murff, J. D., Prins, M. D., Dean, E. R. T., James, R. G. and Schofield, A. N. Jackup rig foundation modeling. *Proc. Offshore Technology Conference*, Houston, TX, (1992).

Murff, J. D. and Wesselink, B. D. Collapse analysis of pile foundations. *Proc. Third Int. Conf. on Numerical Methods in Offshore Piling*, Nantes, France, (1986) 445–459.

Nova, R. and Montrasio, L. Settlements of shallow foundations. *Geotechnique*, **41**, (1991) 243–256.

Osborne, J., Trickey, J, Houlsby, G. T. and James, R. G. Findings from a joint industry study on foundation fixity of jackup units. *Proc. Offshore Technology Conference*, Houston, TX, (1991).

Prager, W. *An Introduction to the Theory of Plasticity*. Addison-Wesley, Reading, MA, (1959).

Prandtl, L. Eindringungsfestigkeit und festigkeit von schneiden. *Zeit. f. Angew. Math. u. Mech.*, **1**, (1921) 15.

Randolph, M. F. and Houlsby, G. T. The limiting pressure on a circular pile loaded laterally in cohesive soil. *Geotechnique*, **34**(4), (1984) 613–623.

Reese, L. C., Cox, W. R. and Koop, F. D. Field testing and analysis of laterally loaded piles in stiff clay. *Proc. Offshore Technology Conference*, Houston, TX, (1970).

Shield, R. T. Plastic potential theory and Prandtl bearing capacity solution. *J. Appl. Mech.* (1954) 193–194.

Skempton, A. W. The bearing capacity of clays. *Proc. Building Res. Congress*, London, (1951) 180–189.

Sokolovskii, V. V., *Statics of Granular Media*, Pergamon Press, New York, (1965).

Terzaghi, K. *Theoretical Soil Mechanics*. Wiley, New York, (1943).

Vesic, A. S. Bearing capacity of shallow foundations. In: *Foundation Engineering Handbook*, Winterkorn, H. F. and Fang, H. Y. (eds.). van Nostrand Reinhold, New York, (1975).

16

Modeling of Interfaces and Joints

Eric Drumm[1], **Kenneth Fishman**[2] and **Musharraf Zaman**[3]

[1]*University of Tennessee, Knoxville, TN, USA*
[2]*McMahon & Mann Consulting Engineer, Buffalo, NY, USA*
[3]*University of Oklahoma, Norman, OK, USA*

16.1   Introduction

Most problems encountered in geomechanics involve systems that are neither homogeneous nor continuous. These systems usually involve two or more different materials, each with different properties. For example, a concrete friction pile may be much stiffer and stronger than the surrounding clay. As the applied load approaches the ultimate capacity, failure occurs as slip at the contact or interface between the pile and soil. Alternatively, two regions of similar materials may be separated by discontinuity (e.g. bedding plane in a rock mass). When the strength of the discontinuity is much lower than that of the surrounding materials, slip or separation along the discontinuity may control the response. In both of these examples, the properties of the contact or interface between the different materials will affect and sometimes govern the response of the system. Figure 16.1 presents several examples in geomechanics in which the behavior of the interface or contact may have a significant effect on the system response.

The behavior of contacts and interfaces is important when dynamic loads are present. However, the static response of many systems can also be highly dependent upon the interface behavior, particularly if there are adjacent materials with a significant difference in stiffness or if the potential for slip or separation exists. There are certain problems in which a large number of discontinuities may exist, and continuum methods may not be appropriate. Examples include a highly jointed rock mass, where the mechanical behavior of the rock mass is largely governed by the discontinuities (Einstein *et al.*, 1996), or where the contact between individual particles or clusters of particles may govern the response (Selvadurai and Boulon, 1995). In problems such as these, it is important that the interface be represented or modeled properly in a numerical analysis. While not within

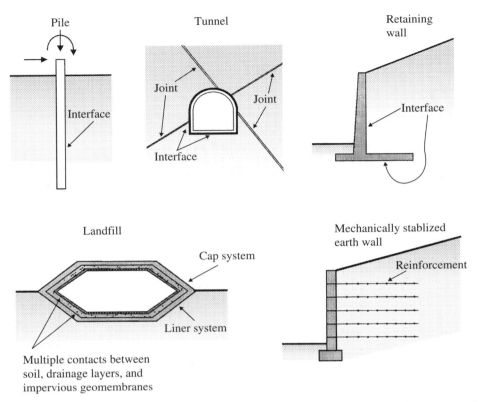

Figure 16.1   Some examples in geomechanics where interface behavior may be important to the overall system response

the scope of this chapter, some of these problems may be best modeled by methods such as the discrete element method (Cundall, 1971; Cundall and Strack, 1979), Discontinuous Deformation Analysis (DDA) (Shi, 1993), block theory (Goodman and Shi, 1985; Mauldon and Goodman, 1990) or micro-mechanics (Chang *et al.*, 1989; Bazant *et al.*, 1990). This chapter is restricted to 'continuum' type approaches to geomechanics problems in which the location and orientation of the interfaces or joints is explicitly known and specified in the model. The words 'interfaces' and 'joints' are used interchangeably here.

16.2   Review of Interface Models

The subjects of contacts, friction, interfaces and joints involve extensive research and publications. Comprehensive reviews are available in the literature (e.g. see Bowden and Tabor, 1950; Rabinowicz, 1965; Desai and Christian, 1977; Desai, 1981; Kikuchi and Oden, 1981). Only a limited review is included here, with emphasis on the development by Desai and co-workers.

Historically, the pursuit for understanding the behavior of interfaces and joints started almost from the beginning of scientific thinking. The most common laws to define interface behavior are the Amonton's (1699) and Coulomb's (1785) law of friction:

$$F = \mu N \tag{16.1}$$

where F = tangential or shear force parallel to the plane of contact, μ = coefficient of friction, and N = force normal to the contact with a nominal area A_0. This law is valid for dry friction between rigid bodies, assumes μ to be constant, and provides a pointwise description leading to definition of gross sliding of one body relative to the other (Desai, 1999). In reality, however, the contact involves mated or non-mated asperities and bodies in contact can be deformable. As a result, the pointwise or local definition is not sufficient, and consideration needs to be given to nonlocal effects due to factors such as nonuniform properties and lack of complete contact. To include elastic component of response, Archard (1958) modified Eq. (16.1) as follows:

$$F = \mu N^m \tag{16.2}$$

where m is the exponent whose value varies between 2/3 and 1; for purely elastic behavior $m = 2/3$, and for ductile contacts $m = 1$. Oden and Pives (1983) discussed the limitations of Coulomb friction law, in particular its local and pointwise character, and presented variational principles by treating the contacting materials as elastic. The friction coefficient μ in Eq. (16.1) was expressed as $\mu = s/p_0$, where s = the average shear strength of the interface and p_0 = an approximately constant local plastic yield pressure. In the general areas of friction and contact mechanisms, models based on strength, limit equilibrium, elastic and classical elastoplastic theories have been presented by many researchers. A review of the subject is given by Desai (1981, 1999) and Kikuchi and Oden (1981).

During the past two decades, many investigators have proposed modified laws for contact problems to account for factors such as asperities, dilation, residual friction angle, high normal stresses, and joint roughness based on empirical/semi-empirical considerations (Zaman, Desai and Selvadurai, 1992). With the advent of computer methods (finite element, boundary element), various interface or joint models have been proposed to characterize 'discrete' interfaces and joints; they include zero and finite thickness elements with linear, nonlinear and plasticity-based models. The formulations here are cast in matrix element equations for incorporation in the computer methods. Models including shear and normal forces as constraint conditions have also been proposed. A review of these and other models is given by Selvadurai and Boulon (1995) and by Zaman (1981).

Among the early models, Ngo and Scordelos (1967) developed a one-dimensional linkage element to study the bond between reinforcement and surrounding concrete in a reinforced concrete beam. Goodman, Taylor and Brekke (1968) extended the idea of linkage element and derived the stiffness matrix for a more general joint element to model the behavior of jointed rocks. Figure 16.2 is a simple one-dimensional line element consisting of four nodes, having two nodes from each material in contact, and zero-thickness. The energy stored in this element in the loading/deformation process is assumed to be due to the relative displacements between top (T) and bottom (B) surfaces. Relative

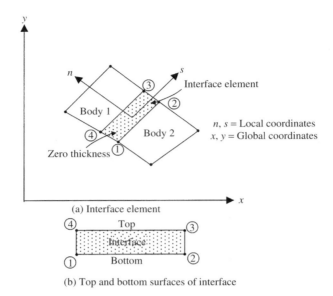

(a) Interface element

(b) Top and bottom surfaces of interface

Figure 16.2   Interface element with zero-thickness

displacements, $u_i{}^r$, is defined as

$$u_i{}^r = u_i{}^T - u_i{}^B \tag{16.3}$$

where $u_i{}^T$ and $u_i{}^B$ are the displacements of the two adjacent points, on the top and the bottom surfaces, respectively. The constitutive relation matrix $[C_j]$ was characterized by the joint stiffness per unit length in the normal (k_{nn}) and tangential (k_{ss}) directions.

$$[C_j] = \begin{bmatrix} k_{nn} & 0 \\ 0 & k_{ss} \end{bmatrix} \tag{16.4}$$

The stiffness matrix was derived by minimizing the potential energy II_p expressed in terms of $[C_j]$ and $\{q_r\}$, where $\{q_r\}$ is the relative displacement vector (obtained from $u_i{}^r$). Applications of this type of zero-thickness joint element have been reported by many researchers including Clough and Duncan (1971), Desai (1977) and Goodman *et al.* (1968, 1977), among others.

An extension of zero-thickness element to three-dimensional analyses was addressed by Mahtab and Goodman (1970). In the context of soil-structure interaction problems, this type of element has been used by Desai and Appel (1976), Phan (1979), and others. Among the main drawbacks of zero-thickness joint elements, adjacent solid elements can penetrate each other and thus violate the kinematic constraints; appropriate computer algorithms can be developed to overcome such problems (Zaman, 1981).

Following a different approach, Zienkiewicz, Best and Dullage (1970) proposed an interface element with non-zero thickness and uniform strain along the thickness directions. The nodal displacements were treated as the primary unknowns. Several problems, including a barrage problem with weak layers of soil and complex loading conditions, were solved using this element. Numerical ill-conditioning was observed in the finite element equations when gap occurs at the interface. Ghaboussi, Wilson and Isenberg (1973) proposed an interface element with relative displacements as independent degrees-of-freedom; however, conceptually it was similar to the joint model proposed by Goodman *et al.* (1968). Desai (1977) extended this concept to develop an axisymmetric interface element for soil-structure interaction problems such as soil-pile interaction. Heuze and Barbour (1982) reported an axisymmetric element for modeling of rock joints and discontinuities. Unlike most previous models, this model was capable of including the effect of dilatancy due to shear. Also, this model has the ability to calculate the increase in joint normal stress due to an increment of shear displacement.

In most of the foregoing interface elements, usually a high value is assigned to normal stiffness (k_{nn}) to prevent interpenetration of adjacent solid elements. When an element fails in shear, with the interface still in compression, the values of shear stiffness (k_{ss}) are reduced to a negligible value keeping the normal stiffness high. Similarly, when tensile stresses develop at the interface, both normal and shear stiffnesses are assigned arbitrarily small values. Also, the cross-diagonal terms of $[C_j]$ are assumed to be zero, so the effect of dilatancy cannot be accounted for in a rational manner. To overcome some of these difficulties, Katona *et al.* (1976) proposed an interface element based on the equilibrium approach. In this model, the nodes of adjacent solid elements forming the interface are connected by an 'imaginary' one-dimensional element (in a two-dimensional space). These imaginary elements play the role of kinematic constraints such as a Lagrangian multiplier constraint. A modified principle of virtual work was utilized in formulating the stiffness matrix. Many other investigators used similar models for solving interface or contact problems in structural and geomechanics (e.g. see Francavilla and Zienkiewicz, 1975; Alvappillai, 1992; Zaman and Alvappillai, 1995; and Selvadurai and Boulon, 1995).

16.2.1   Thin-layer element

Desai and his co-workers (e.g. see Zaman *et al.*, 1984) proposed a simple 'thin-layer element' that allows for various deformation modes such as no-slip, slip, debonding and rebonding for simulating interface behavior between structural and geologic media subjected to static and dynamic loading. The underlying idea of the thin-layer element is based on the assumption that the behavior near the interface (Figure 16.1) involves a finite thin zone (Figure 16.3), rather than a zero thickness zone. The behavior of this thin zone or layer can be significantly different from the behavior of the surrounding structural and geologic materials. This behavior is achieved by adopting appropriate constitutive law for the element.

The constitutive relation matrix $[C_i]$ of the thin-layer element is written in a generic form as

$$[C_i] = \begin{bmatrix} [C_{nn}]_i & [C_{ns}]_i \\ [C_{sn}]_i & [C_{ss}]_i \end{bmatrix} \tag{16.5}$$

where $[C_{nn}]$ = normal component, $[C_{ss}]$ = shear component, and $[C_{ns}]$ and $[C_{sn}]$ represent coupling effect between normal and shear behaviors. In the context of soil-structure interaction problems, since the interface is surrounded by the structural and geological materials, its normal properties during the deformation process should be dependent upon the characteristics of the thin interface zone, as well as the state of stress and properties of the surrounding elements. The shear part $[C_{ss}]$ of the interface constitutive relation was

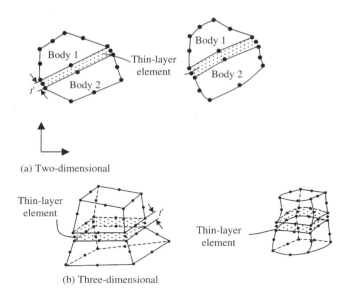

(a) Two-dimensional

(b) Three-dimensional

Figure 16.3   Schematic of thin-layer (interface) element

obtained using the results of direct shear tests (static and cyclic) of interfaces. The shear strain (θ) of the interface was defined approximately as $\theta = u_r/t'$, t' being the thickness of the interface element. The interface (tangent) shear modulus G_i was obtained from $G_i \cong \partial\tau/\partial\theta$ in which $\tau =$ (average) total shear stress at the interface. For two-dimensional (plain strain) problem, $[C_{ss}]_i$ is equal to G_i. For a general soil-structure interaction problem, the value of G_i may depend upon various factors such as amplitude of relative displacement, interface thickness, shear and normal stresses, and roughness. In case of cyclic loading, G_i will also depend upon the number of cycles.

Various modes of deformation (stick, slip, debonding, and rebonding; Figure 16.4) have been accounted for by Desai and co-workers (e.g. see Zaman *et al.*, 1984) in thin-layer element by using an iterative approach based on the redistribution of stresses and kinematic constraints. It is assumed that all interface elements are initially in the stick mode and excitations are applied incrementally with a time increment Δt. For static problems Δt can be viewed as load increment. Based on the configuration of the system at time t, a trial response at $t + \Delta t$ is evaluated by solving the global equations of dynamic equilibrium (or static equilibrium in case of static problems). Based on the updated stesses and displacements (in local coordinate system), the interface conditions are checked to see if any constraints have been violated and if the assumed mode needs to be revised. Appropriate measures (e.g. redistribution of stresses) are taken accordingly. Details of this procedure are given by Zaman *et al.* (1984).

The thin-layer element model was subsequently extended to include various features such as dilatancy, pore pressure, and cyclic effects and was implemented in finite element programs for analysis of complex soil-structure interaction problems (e.g. see Desai and Nagaraj, 1988; Desai and Fishman, 1991; Navayogarajah, Desai and Kiousis, 1990; Sharma and Desai, 1992; Desai and Rigby, 1997). More recently, Desai and his co-workers have reformulated the thin-layer element model within the framework of the 'disturbed state concept', and used it for analysing several real-life interaction problems, including

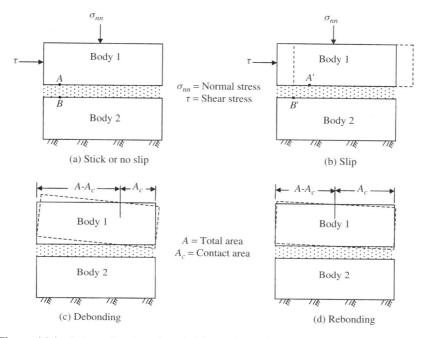

Figure 16.4 Schematic of modes of deformation at interface: two-dimensional case

interface problems associated with electronic packaging and soil-structure interaction (Desai, 1999; Desai *et al.*, 1997, 1998; Desai, Basaran and Zhang, 1997; Desai and Zhang, 1988).

16.3 Constitutive Relationships for Interfaces

Just as there is a wide range of constitutive relationships to describe the stress-strain response of continua, a wide range of models have been used to describe interface response. The primary difference is that most of these models are formulated to describe the shear stress-shear displacement response, since shear strain at the interface is usually the dominant response and is difficult to define. No attempt is made here to include a comprehensive review of the existing constitutive models for interfaces and joints, rather we focus on three specific classes of models that have been used rather widely. These models are: the Ramberg-Osgood nonlinear elastic model, the Mohr–Coulomb elastic-perfectly plastic model and the Hierarchical Single Surface (HiSS) elastoplastic model. Each of these models is briefly described in this section.

16.3.1 Ramberg–Osgood nonlinear elastic model

The non-linear shear stress-deformation response observed at a soil-structure contact may be simulated using a modified Ramberg–Osgood (R-O) nonlinear elastic model (Ramberg and Osgood, 1943; Jennings, 1964). This model has been used to describe the cyclic variation of sand-concrete interfaces (Drumm and Desai, 1984), and follows the approach used by Idriss *et al.* (1978) to represent the cyclic stress-strain behavior of a marine clay. However, rather than expressing the R-O model in terms of shear modulus, the cyclic interface response is expressed in terms of the shear stiffness, K_s,

$$K_s = \frac{\tau}{u_r} \qquad (16.6)$$

where τ = interface shear stress and u_r = the relative displacement at the interface. A number of modifications were required to include the variation of K_s with loading cycles, N.

In the R-O interface model, the stress-displacement or loading curve (Figure 16.5(a)) is assumed to coincide with the backbone curve, which is the path traced by the tips of the hysteresis loops with coordinates u_i, τ_i, (Figure 16.5(b)). The backbone curve is used to characterize the cyclic behavior by employing the Masing criterion. This approach assumes that the unloading-reloading curve (Figure 16.5(b)) is defined by the same function as the backbone curve, except that the stress and deformation axes are expanded by two, and the origin is translated to the point of loading reversal, u_i, τ_i. The initial loading or backbone curve is defined as

$$u_r = u_y \left(\frac{\tau}{K_i u_y} \right) \left(1 + \alpha \left| \frac{\tau}{K_i u_y} \right|^{R-1} \right) \qquad (16.7)$$

and the unloading-reloading portion of the curve is defined as

$$u_r \pm u_i = u_y \left(\frac{\tau - \tau_i}{K_i u_y} \right) \left(1 + \frac{2\alpha}{2^R} \left| \frac{\tau - \tau_i}{K_i u_y} \right|^{R-1} \right) \qquad (16.8)$$

where u_r = the relative displacement of the interface, τ = interface shear stress, and (u_i, τ_i) = the relative displacement and shear stress, respectively, at the point of loading

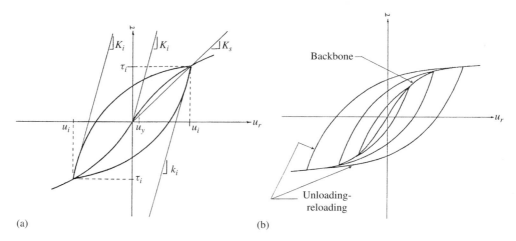

(a)   (b)

Figure 16.5   Ramberg–Osgood relationship for cyclic interface response. (a) Model parameters;
(b) Loading-unloading cycle

reversal, as shown in Figure 16.5(a). The remaining terms in Eqs. (16.7) and (16.8) are: K_i initial shear stiffness, $u_y =$ a reference displacement, and material parameters α and R. The tangent stiffness at the point of load reversal is assumed to be equal to K_i, as indicated in Figure 16.5(a).

To describe either the increase or decrease in stiffness with loading cycles, the concept of a degradation index (Idriss *et al.*, 1978) can be generalized as the cycle index, δ

$$\delta = \frac{(K_S)_N}{(K_S)_1} = N^t \tag{16.9}$$

where $(K_S)_N$ and $(K_S)_1 =$ secant stiffness during cycle N and cycle 1, respectively, and $t =$ the cyclic parameter. For a positive value of t, a stiffening behavior is described, whereas a negative value of t describes degradation behavior. The effect of number of cycles, N, is included in the Ramberg–Osgood formulation through a modification of the initial stiffness, K_i. If it is assumed that the ratio of initial stiffness at cycles N and 1 is equal to the ratio of secant stiffness at cycles N and 1,

$$\frac{(K_i)_N}{(K_i)_1} = \frac{(K_s)_N}{(K_s)_1} = \delta \tag{16.10}$$

then the effect of number of cycles can be included in Eqs. (16.7) and (16.8) through a modification of K_i as

$$(K_i)_N = \delta(K_i)_1$$

Since there are two stress reversals per load cycle, N increases in increments of 0.5 cycle. By utilizing a non-integer value of N, the initial stiffness can be modified at every half cycle. This results in an improved approximation of the open hysteresis loops which may be observed when the stiffness changes significantly with loading cycle. Thus, the effect of loading cycles can be represented by rewriting Eq. (16.7) as

$$u_r = u_y \left(\frac{\tau}{\delta K_i u_y} \right) \left(1 + \alpha \left| \frac{\tau}{\delta K_i u_y} \right|^{R-1} \right) \tag{16.11}$$

for the backbone curve, and rewriting Eq. (16.8) as

$$u_r \pm u_i = u_y \left(\frac{\tau - \tau_i}{\delta K_i u_y} \right) \left(1 + \frac{2\alpha}{2^R} \left| \frac{\tau - \tau_i}{\delta K_i u_y} \right|^{R-1} \right) \tag{16.12}$$

for the unloading-reloading curve. The interface secant stiffness may then be expressed as

$$K_s = \frac{\delta K_i}{1 + \alpha \left| \dfrac{\tau_m}{\delta K_i u_y} \right|^{R-1}} \tag{16.13}$$

In this manner, the cyclic shear stress-displacement response can be described by five parameters: (1) initial stiffness, K_i; (2) parameter R; (3) reference displacement, u_y; (4) parameter α; and (5) cyclic index δ.

16.3.2   Mohr–Coulomb model

Models based on the Mohr–Coulomb and the Drucker–Prager failure criteria have been widely used in the past to represent behavior of engineering materials (e.g. see Desai and Siriwardane, 1984). In elastic-perfectly plastic models a material is usually assumed to behave elastically at all stress levels except at failure. The behavior at the onset of failure is assumed to be governed by one of the assumed failure criteria. A similar approach could be used in modeling interface behavior. Zaman (1981) employed the Mohr–Coulomb criterion to incorporate the effect of dilatancy in the thin-layer element. Xiurun (1981) used a similar approach for predicting dilatancy using a conventional zero-thickness element. The Mohr–Coulomb failure envelope F was defined in the form

$$F = c_a + \sigma_{nn}{}^0 \tan \delta - |\tau^0| \qquad (16.14)$$

where c_a and δ are the apparent cohesive strength and friction angle of the interface, respectively. $\sigma_{nn}{}^0$ and τ^0 are the total normal and shear stresses, respectively. In a conventional Mohr–Coulomb relation, the coefficients c_a and $\tan \delta$ are associated with the ultimate strength of a material. In the context of thin-layer element, however, they are not ultimate strength parameters. As discussed by Zaman *et al.* (1984), for a given $\sigma_{nn}{}^0$, $u_r{}^0$ and N (N being the number of cycles for cyclic problems), a set of c_a and $\tan \delta$ can be determined from appropriate laboratory (or field) tests. This set of c_a and $\tan \delta$ represents the shearing stresses at the interface corresponding to prescribed values of $\sigma_{nn}{}^0$, $u_r{}^0$ and N. In terms of finite element implementation, stresses computed in the global coordinate system at element integration points are transformed into the local coordinate system. The transformed stresses are used to check if the Mohr–Coulomb criterion is satisfied:

$$dF = (\partial F / \partial \sigma_{ij}{}^0)\, d\sigma_{ij} = 0 \qquad (16.15)$$

where $d\sigma_{ij}$ is the incremental stress tensor. Assuming that the incremental strain tensor $d\varepsilon_{ij}$ can be decomposed into elastic and plastic components, it is possible to write

$$d\varepsilon_{ij}{}^e = d\varepsilon_{ij} - d\varepsilon_{ij}{}^p \qquad (16.16)$$

where the superscripts e and p denote elastic and plastic components, respectively. Following the procedure analogous to the theory of plasticity, incremental stress $(d\sigma_{ij})$ and incremental strain $(d\varepsilon_{ij})$ tensors at the interface can be written as

$$d\sigma_{ij} = \left[C_{ijk\ell} - \frac{C_{ijgh}\dfrac{\partial F}{\partial \sigma_{gh}}\dfrac{\partial F}{\partial \sigma_{mn}}C_{mnk\ell}}{\dfrac{\partial F}{\partial \sigma_{pq}}C_{pqrs}\dfrac{\partial F}{\partial \sigma_{rs}}} \right] d\varepsilon_{k\ell} \qquad (16.17)$$

In matrix notation

$$\{d\sigma\} = [C^{ep}]_i\{d\varepsilon\} \qquad (16.18)$$

where $\{d\sigma\}$ and $(d\varepsilon)$ are the incremental stress and strain tensors, respectively, and $[C^{ep}]_i$ is the elastic-plastic constitutive relation matrix for the thin-layer element. Equation (16.15) can be specialized for a two-dimensional case, and F can be assumed as the Mohr–Coulomb failure criterion.

16.3.3   *Hierarchical single surface (HiSS) model*

The HiSS model (discussed in Chapter 10) can be specialized to describe the behavior of rock joints, as described by Desai and Fishman (1991). The yield function in this formulation is written as

$$F = \tau^2 + \alpha\sigma_n{}^n - \gamma\sigma_n{}^2 = 0 \qquad (16.19)$$

where τ and σ_n are the shear and normal stress acting on the joint plane, γ is a strength parameter related to the asymptote to the shear stress-relative tangential displacement curve at very large relative tangential displacement, n is a parameter describing the shape of the yield function, and α is the growth or hardening function. The parameters for the yield function may be found as follows.

According to the HiSS model, at the ultimate state $\alpha = 0$, Eq. (16.19) reduces to

$$\tau_u{}^2 - \gamma\sigma_n{}^2 = 0 \qquad (16.20)$$

where τ_u = asymptotic stress from tests with different σ_n which is often about 10% higher than the peak shear stress. The constant γ is found by using a least square procedure based on the ultimate envelope.

The parameter n is found by considering the stress state during shear where the change in relative displacement normal to the interface is zero (Figure 16.6). The shear stress at this point is denoted as τ_{zs}. If the plastic strain increment is assumed to be normal to the yield surface, then at this point $\partial F/\partial\sigma_n = 0$. Then from Eq. (16.19):

$$\partial F/\partial\sigma_n = n\alpha\sigma_n{}^{n-1} - 2\gamma\sigma_n = 0 \qquad (16.21)$$

and

$$\sigma_n = (2\gamma/n\alpha)^{1/n-2} \qquad (16.22)$$

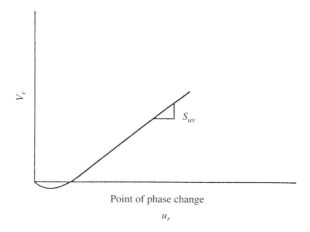

Point of phase change

u_r

Figure 16.6   Typical plot of relative normal, and tangential displacement, U_r, during shear showing point of phase change where change in relative displacement normal to the interface is zero (Desai and Fishman, 1991)

If M_{zs} is defined as the slope of the line representing the locus of points on a plot of τ_{zs} vs. σ_n, then

$$\tau_{zs}^2 / \sigma_n^2 = M_{zs}^2 = -\alpha\sigma^{n-2} + \gamma = \gamma(1 - 2/n) \tag{16.23}$$

The value of n is determined from this expression once γ and M_{zs} are known.

The growth or hardening function can be defined in terms of internal state variables. For rock joints it was found that plotting α vs. the total trajectory of plastic strains, $\xi = \int [(v_r^P)^2 + (u_r^P)^2]^{1/2}$, yielded the most consistent trends. A typical plot of α vs. ξ from testing simulated rock joints is shown in Figure 16.7. The shape of this plot indicates a relationship of the form:

$$\alpha = a\xi^b \tag{16.24}$$

where a and b are material constants. The constant b is negative such that α become smaller as ξ increases. Thus, constants a and b are related to the shear response and can be found by plotting $\ln \alpha$ vs $\ln \xi$ from shear tests under different σ_n.

Nonassociative behavior

The plastic potential function is introduced to account for the nonassociative behavior of joints. It is based on a correction to the yield function and introduced through modification of the growth function α to α_Q as follows:

$$\alpha_Q = \alpha + \kappa(\alpha_1 - \alpha)(1 - r_v) \tag{16.25}$$

such that

$$Q = \tau^2 + \alpha_Q\sigma_n^n - \gamma\sigma_n^2 \tag{16.26}$$

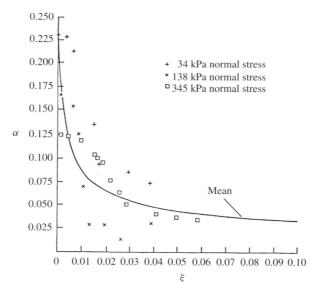

Figure 16.7   Typical plot of growth function, α, vs. total trajectory of plastic strain (Desai and Fishman, 1991)

where κ = material parameter, α_1 = value of α at the initiation of nonassociative response, and r_v = ratio ξ_v/ξ.

The nonassociative parameter κ is determined by considering the relation between the normal and tangential displacements; as a simplification the slope near the ultimate condition can be used. At the ultimate condition ($\alpha = 0$) and Eq. (16.25) reduces to

$$\alpha_Q = \kappa\alpha_1(1 - r_v) \tag{16.27}$$

Using the flow rule (Desai and Siriwardane, 1984), the following relationship is obtained:

$$(dv_r^P/du_r^P)_u = S_{uv} = [(n\alpha_Q\sigma_n^{n-1} - 2\gamma\sigma_n)/2\tau)]_u \tag{16.28}$$

and

$$\kappa\alpha_1(1 - r_v)_u = (2S_{uv}\tau_u + 2\gamma\sigma_n)/n\sigma_n^{n-1}. \tag{16.29}$$

Here the subscript u denotes the ultimate condition and S_{uv} = slope of the v_r vs. u_r curve near the ultimate. Equation 16.29 permits computation of κ.

Desai and Fishman (1991) demonstrate the procedure for determination of parameters and perform back predictions with respect to a series of direct shear tests performed on artificial rock joints with constant asperity angles of $0°, 5°, 7°$ and $9°$. The accuracy of HiSS model was verified by comparing back predictions of test data used to find parameters, and also of test data not used to determine parameters. Quasi-static monotonic loading, unloading and load reversal were treated in the study. The study revealed that the HiSS model is capable of describing rock joint behavior very well, but a nonassociative flow rule is necessary to accurately portray the normal relative displacement during shear.

Correlation of HiSS model parameters to empirical joint model   Barton's model (Barton and Choubey, 1977; Bandis *et al.*, 1981; Barton, 1986; Barton and Bandis, 1987) describing the shear behavior of rock joints is an empirical relationship which involves three parameters determined from simple index tests. The relationship between peak shear stress and normal stress is written as

$$\tau_p = \sigma_n \tan(JRC^* \log_{10}(JCS/\sigma_n) + \phi_b) \tag{16.30}$$

where JRC is the joint roughness coefficient, JCS is joint wall compressive strength and ϕ_b is the basic friction angle of the rock joint.

For practical purposes ϕ_b corresponds to the residual strength of the rock joint during shearing. If the joint is completely unweathered then JCS will be equal to the unconfined compressive strength of unweathered parent material. In general, rock joint walls are weathered to some extent and JCS will be lower than the unconfined compressive strength of the unweathered rock. The joint roughness coefficient is often estimated empirically based on traces of typical joint surfaces rated on a scale from 0 to 20 as proposed by Barton and Choubey (1977).

The mobilized shear stress corresponding to a particular state during deformation can also be expressed as

$$\tau_{mob} = \sigma_n \tan(JRC_{mob}^* \log_{10}(JCS/\sigma_n) + \phi_b) \tag{16.31}$$

where the subscript *mob* means mobilized.

In the HiSS model the parameter γ describes an ultimate condition, extrapolated from the observed peak shear strength. By introducing a proportionality parameter, m, the following relationship between γ and the peak strength parameters of Barton's model is proposed:

$$\sqrt{\gamma} = m \tan(I_p + \phi_r) \qquad (16.32)$$

where $I_p = JRC^* \log_{10}(JCS/\sigma_n)$ and ϕ_r is residual friction angle. The parameter m depends on the level of normal stress. Wun (1992) studied three different sets of rock joint test data and corresponding values of the parameter m. Test data included results from direct shear testing saw cut granite and limestone joints described by Hutson (1987a,b) and simulated rock joints as reported by Fishman (1988). Based on this limited parametric study over a range of normal stress between 0.035 MPa and 6.5 MPa, the parameter m ranged from 0.748 to 1.285, with an average value $m = 1.071$.

The parameter n in the HiSS model is related to the point of phase change between compression and dilation during shear and at this point

$$dv_r/du_r = 0 = i_{mob} \qquad (16.33)$$

where $i_{mob} = JRC_{mob}^* \log_{10}(JCS/\sigma_n)$. Equation (16.33) infers that at the point of phase change JRC_{mob} is zero leading to the following result:

$$\tau_{mob} = \tau_{zs} = \sigma_n \tan(\phi_r) \qquad (16.34)$$

and n is estimated from

$$\tan^2(\phi_r) = \gamma(1 - 2/n) \qquad (16.35)$$

As an aid in estimating the hardening function parameters a and b the shear stress-relative displacement curve is approximated as bilinear. The slope of the first branch of the bilinear relation shown in Figure 16.8 has a slope equal to the initial shear stiffness. Point Y is located at the knee of the stress-relative displacement curve which for rock joints generally occurs at relative tangential displacements between 0.25 mm and 1 mm. Point U represents the peak shear strength of the rock joint and is estimated from Barton's equation. Based on the bilinear shear stress, relative displacement relationship

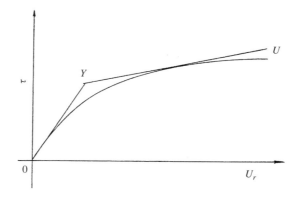

Figure 16.8   Bilinear approximation to the shear stress, α, vs. relative displacement, U_r relationship of rock joints (Wun, 1992)

and Barton's equation for τ_{mob} the term $i_{mob} = JRC_{mob}* \log_{10}(JCS/\sigma_n)$, and hence the relative normal displacement during shear can be obtained numerically at a number of points. With τ, σ_n, u_r and v_r obtained from the synthetic history of stress versus deformation, the $\ln(\alpha)$ vs. $\ln(\zeta)$ relation can be plotted and parameters a and b determined as described before.

16.4   Back Predictions

16.4.1   Back predictions with Ramberg–Osgood model

The Ramberg–Osgood interface model has been used to predict the cyclic response of sand-concrete interfaces, at both low ($D_r = 15\%$, very loose) and high ($D_r = 80\%$, dense) relative density (Drumm and Desai, 1984). The model response was compared with measurements obtained with a large (300 mm by 300 mm) direct shear device. Satisfactory results were obtained up to 100 loading cycles. Details of the parameter determination process are also described by Drumm and Desai (1984).

16.4.2   Back predictions with Mohr–Coulomb model

A thin-layer element with the Mohr–Coulomb model has been validated rigorously by comparing predicted response with laboratory and/or field measurements, and with results obtained from other methods (e.g. see Zaman, 1981; Zaman *et al.*, 1984; Desai, 1987). Only one example is included here.

Figure 16.9 shows the finite element mesh of a buried pipe-soil system that was analysed by Katona (1981). Eight thin-layer elements (Elements 9 through 16) are used to model the interface between the buried pipe (represented by elements 1 through 8), and the surrounding soil medium (represented by elements 17 through 42). The soil medium and pipe material are assumed to be homogeneous and linearly elastic so that the results could be compared with analytical solutions. Three cases of interface friction are considered; $\tan \delta = 0.001$, 2.0 and 0.25 to represent approximately frictionless slip (no bond); stick (bonded or no-slip); and frictional slip conditions, respectively. These values of $\tan \delta$ are used to define the slip function in accordance with the Mohr–Coulomb criterion (Eq. (16.15)), as discussed previously. For frictionless and bonded (stick) cases Burns and Richard (1964) have obtained closed-form elasticity solutions. For frictional case, however, no analytical or closed-form solutions are available. The following material properties are used:

Soil: Young's modulus = 3000 psi (20,670 kPa); Poisson's ratio = 0.333

Pipe: Young's modulus = 2.2×10^6 psi (15.2 Gpa); Poisson's ratio = 0.25

Thin-layer element: Shear modulus = 128.8 t psi;

t' = element thickness = $0.01(\pi R/8)$ (R = average pipe radius in inch)

Interface shear stress, $\tau^0 = 128.8(u_r{}^0) - 250(u_r{}^0)^2$

where $u_r{}^0$ represents the relative displacement at the interface. The soil in the finite element region is assumed to be weightless (because of relatively small height). The overburden pressure is applied incrementally. Typical finite element results for this problem has been

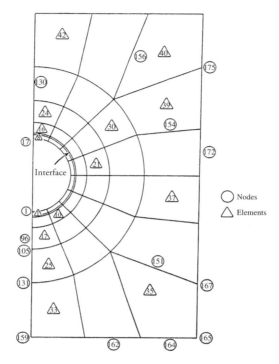

Figure 16.9   Finite element mesh: buried pipe problem

reported by Desai *et al.* (1982), and only selected results are presented in Figures 16.10(a) and (b). The response (radial stress along the spring line and the crown) predicted by the thin-layer element compares favorably with that reported by Burns and Richard (1964) for the extreme interface conditions (bonded and frictionless).

16.4.3   *Back predictions with HiSS model*

Back predictions using parameters determined from test data, and estimated from empirical relations/data, are compared with the observed response. The shear response of rock joints prepared in limestone and tested by Hutson (1987a) are investigated. A computer controlled diamond saw was used to prepare artificial joints in samples of limestone. Rock joint surfaces were carved having a constant asperity angle of 11°. The joint surface had an unconfined compressive strength (*JCS*) of 22.4 Mpa, a basic friction angle $\phi_b = 38.8°$, and a joint roughness coefficient (*JRC*) of eight.

Direct shear tests were performed on the rock joint surfaces under three levels of normal stress, σ_n, including 1, 2.8 and 5.5 Mpa. The normal stiffness, k_n, of the rock joint was determined to be 38.0 Mpa/mm and the shear stiffness, k_s, increased with increasing levels of normal stress such that $k_s = 11.0$, 18.0, and 21.0 Mpa/mm for normal stress levels of 1.0, 2.5 and 5.5 Mpa, respectively. Parameters for the HiSS model were determined from the test data and are presented in Table 16.1, along with those estimated from Barton's empirical model of joint behavior. A value of *m* equal to 1.14 was used in Eq. (16.) to estimate γ.

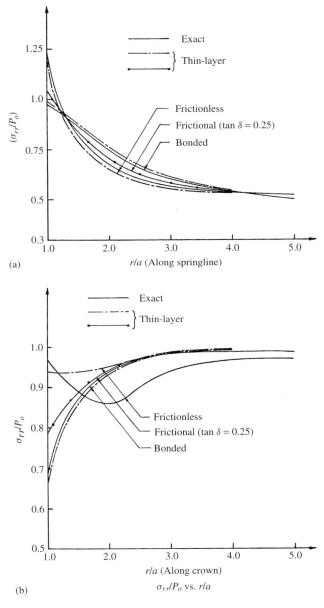

Figure 16.10 Comparison between computed (radial stress, σ_{rr}) results and closed-form solutions for buried pipe problem

Estimated parameters are in reasonable agreement with those determined from actual test data. Figure 16.11 compares back predictions performed with the estimated parameters to the observed responses. Good agreement between predicted and observed shear stress-relative displacement curves is realized for normal stress levels of 1.0 and 2.8 MPa. However, the agreement at $\sigma_n = 5.5$ Mpa is not as good. At this level of normal

Table 16.1 HiSS model parameters for Hutson's (1987a) Limestone Rock Joints

Parameter	From test data	Estimated
γ	1.44	1.46
n	3.03	3.58
a	0.055	0.081
b	−0.498	−0.447
κ	0.517	0.478

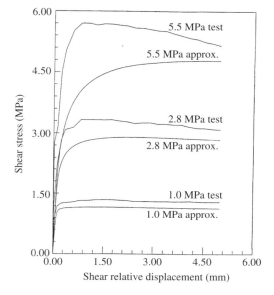

Figure 16.11 Comparison of laboratory test results and back prediction of shear response of rock joints using estimated model parameters (Wun, 1992)

stress strain softening behavior is observed and this was not incorporated into the back predictions. Strain softening behavior can be incorporated within the context of the HiSS model by using the theory of mixtures as described by Frantziskonis and Desai (1987a,b), and the disturbed state concept based on the work of Desai and Ma (1992).

Figure 16.12 compares normal displacement predicted with HiSS model parameters *determined from the test data* to observations. Dilation is overpredicted by the model that employ an associative flow rule, however, an excellent agreement is obtained with the nonassociative flow rule. Figure 16.13 displays normal relative displacement *predicted with estimated parameters*. By comparing results presented in Figures 16.12 and 16.13 it is evident that there is very good agreement between predictions *with determined* and *estimated parameters* when an associative flow rule is employed. For the case of a non-associative flow rule, the response computed *with estimated parameters* is not as accurate as that portrayed *using parameters determined from test data*. However, back predictions of normal response are very sensitive to the parameter κ. The parameter κ can be optimized,

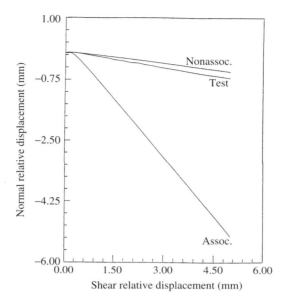

Figure 16.12 Comparison of rock joint normal displacement during shear predicted with HiSS model parameters determined from test data to observations (Wun, 1992)

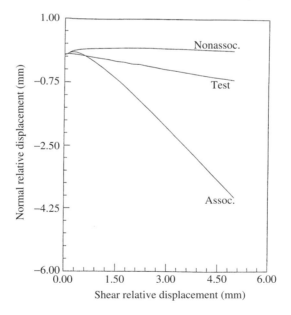

Figure 16.13 Comparison of rock joint normal displacement during shear predicted with estimated HiSS model parameters to observations (Wun, 1992)

improving the accuracy of the plasticity model, using data generated for the joint from Barton's empirical equations.

16.5   Applications

Several applications are described in which interface behavior was included in the analysis of a boundary value problem. These applications include a series of verification problems that are simple, well defined problems used to verify the interface models discussed in the preceding sections, and to compare the results with analytical solutions, laboratory tests, or classical theories. These verification problems are followed by several field problems in which the results are compared with field observations.

16.5.1   Verification problems

Tunnel in a jointed material

The case of a circular opening embedded in a jointed rock mass is investigated, as described by Fishman *et al.* (1991). In this study, results from numerical analysis are compared with the observed response of a small model constructed in the laboratory. The model is made of concrete having dimensions of approximately $0.5 \times 0.5 \times 0.3$ m with a 0.075 m diameter opening in the center. Smooth horizontal joints are included in the model, as depicted in Figure 16.14.

   A surface traction is applied corresponding to a nearly uniform vertical surcharge of 2.1 MPa near the top of the concrete model. Measurements made during loading of the model include an array of strain gages placed along a horizontal line intersecting the spring line, and along a vertical line above the crown of the opening.

   Material properties for the concrete and smooth simulated rock joints are shown in Table 16.2. All material behavior is assumed to be linear elastic. For this study joints are modeled with a high normal stiffness uncoupled from an inherently low shear stiffness.

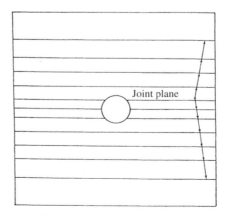

Joint plane

Figure 16.14   Schematic of tunnel model in jointed rock (Fishman *et al.*, 1991)

Table 16.2 Summary of material properties for tunnel model (1 psi = 6.89 kPa)

	Material			
Property	Steel	Rubber	Concrete	Joints
Young's modulus (ksi)	30,000	0.5	3,000	30,000
Poisson's ratio	0.3	0.499	0.3	0.3
Shear modulus (psi)				100

Since smooth joints are considered it is desired to limit the normal deformations in the analysis, and Young's modulus for the joint plane is taken to be 10 times more than that of the surrounding solid rock (concrete).

The finite element mesh used in the numerical solution is shown in Figure 16.15. Finite element analyses were performed using thin-layer interface elements as described by Desai *et al.* (1984), and the zero-thickness interface element described by Goodman *et al.* (1968) to represent the interface (rock joint) behavior. A limited parametric study was performed by Fishman and Ahmad (1997) to demonstrate the effect that varying interface model parameters has on the computed results. Interfaces were modeled having low and very high shear stiffness. Results from finite element analyses are presented in Figures 16.16 and 16.17; low and high G_i or K_s refer to results obtained with the thin-layer and zero-thickness interface elements, respectively. Also shown for reference are results, referred to as *solid model*, considering an isotropic, homogeneous, continuous rockmass, i.e. no joints. One should expect that results obtained using interface elements with high shear stiffness will be close to results obtained with a solid model.

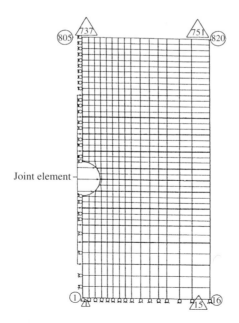

Figure 16.15 Finite element mesh used in analysis of tunnel model in jointed rock (Fishman *et al.*, 1991)

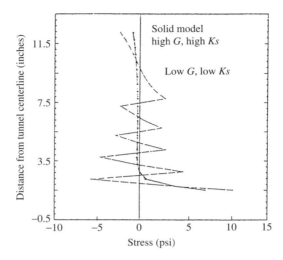

Figure 16.16 Results from finite element analysis of tunnel model in jointed rock showing variation of lateral stress component along the tunnel centerline at the crown of the opening (Fishmen and Ahmad, 1991)

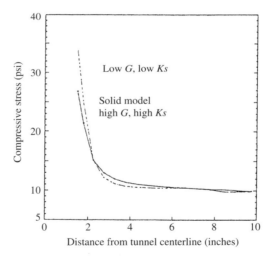

Figure 16.17 Results from finite element analysis of tunnel model in jointed rock showing variation of vertical stress component along the tunnel centerline at the springline of the opening (Fishmen and Ahmad, 1991)

Figures 16.16 and 16.17 indicate that results obtained with either of the interface models cannot be distinguished from one another. As expected, results from analyses incorporating joints with very high shear stiffness are in very close agreement with those from the solid model.

Furthermore, results with interfaces having low shear stiffness show the profound effect the presence of planes with low shear resistance can have on the response of the rockmass surrounding the opening. The jointed rock appears to be acting as a series slabs spanning the opening. The stresses are tensile near the bottom of the slabs and compressive near the top. This is consistent with the behavior of tunnels constructed in laminated rock as discussed by Obert and Duvall (1967). The study demonstrates that the use of thin-layer element in a finite element model provides a simple, yet effective, means for including the effects of rock joints on the response.

Seismic induced sliding of retaining wall

Richards and Elms (1979) applied the concept of the sliding block model described by Newmark (1965) to develop an analytical procedure to estimate seismic-induced permanent deformations of free standing gravity retaining walls. The analysis is pseudo-static and assumes that kinematic interaction between the wall and surrounding soil is negligible. Also, it is assumed that sliding will occur before the wall topples, or failure of the foundation soil occurs.

During seismic loading lateral earth pressure and inertia forces acting on a retaining wall increase in proportion to the level of ground acceleration. Only elastic deformation occurs until the wall yields at a level of acceleration referred to as the threshold acceleration, or the seismic resistance of the retaining wall. At the threshold acceleration, the shear resistance at the base of the wall is consumed by the horizontal component of the thrust exerted on the wall by the backfill, and the inertia force acting on the wall, such that further increments in acceleration cannot be sustained by the wall.

As shown in Figure 16.18 during seismic loading the ground and wall acceleration are assumed to be equal until the threshold acceleration is reached, after which the wall acceleration remains constant until the relative velocity between the ground and the wall is zero, at which point the ground and wall move together again. Relative displacement is determined by integrating the relative acceleration twice with respect to time. Due to the transient nature of accelerations during seismic loading, the threshold acceleration is exceeded a finite number of times during an earthquake, and for a duration depending on

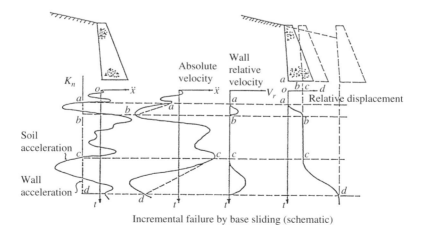

Incremental failure by base sliding (schematic)

Figure 16.18   Schematic of ground and wall motion during an earthquake (Richards and Elms, 1979)

the frequency of the ground motion. Seismically-induced permanent deformation occurs in increments during an earthquake and the total accumulation depends upon the severity, duration, and frequency content of the earthquake.

High levels of acceleration are required to overcome the passive resistance of the back-fill and the analysis only considers permanent displacement to occur when accelerations are directed into the fill and the backfill is in an active condition. Active thrust and the angle of the failure plane for the case of seismic loading may be determined using the Mononabe–Okabe equations (1928). The analysis assumes that the failed region of soil within the backfill behind the retaining wall acts as a rigid body and moves together with the wall during sliding. The sliding surface includes the failure plane within the backfill and the interface between the base of the wall and the foundation soil as depicted in Figure 16.19.

The analytic method proposed by Richards and Elms (1979) has been verified based on results of shaking table tests performed on model retaining walls and from results of finite element analysis (Nadim and Whitman, 1983). In the finite element analysis, the sliding surface is known *a priori,* and represented by interface elements. The interface is modeled as a rigid plastic material (Mohr–Coulomb), and sliding is initiated whenever the shear traction along the interface exceeds the shear resistance at the interface. A dynamic analysis is performed in the time domain and ground motions are input to the finite element model.

An example of the application of the finite element model including interface elements is described with respect to a series of shaking table tests performed on model retaining walls at the State University of New York at Buffalo (Kutschke, 1995). The finite element code ANSYS was used in the study. Figure 16.20 shows the model retaining wall and the discretization used in the analysis (Kutschke, 1995).

The model retaining wall is contained in a test chamber that is 4.5 m long, 0.91 m wide and 1.22 m high. The test chamber is constructed with rigid side walls such that a plane strain condition exists. The test chamber is filled with sand to a depth of 0.91 m and the model retaining wall is 0.45 m high with a base width of 0.15 m. The model retaining wall is a bridge abutment, and includes a bridge deck with a free connection detail at the top of the abutment. The bridge abutment weighs 0.29 kN/m and carries a 0.85 kN/m deck load.

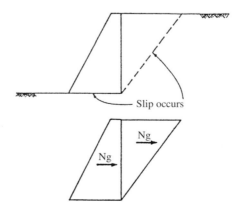

Figure 16.19   Illustration of sliding surfaces for wall and backfill system (Nadim and Whitman, 1984)

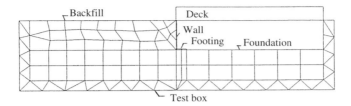

Figure 16.20   Finite element mesh for model retaining wall including sliding surfaces (Kutschke, 1995)

The finite element mesh includes eight-noded quadrilateral elements to model the backfill, foundation and retaining wall model; four-noded beam elements to model the bridge deck, six-node triangular elements to model the test box frame; and contact elements (Bathe, 1982) to model interfaces between the backface of the wall and the backfill, the base of the wall and the foundation soil, and the assumed failure plane within the backfill. Elastic perfectly plastic behavior was assumed for soils and rigid plastic behavior for interfaces. Cohesion was taken as zero and the friction angles and unit weights used to describe the different soils and interfaces in the analyses are shown in Table 16.3.

A single haversine pulse of acceleration was applied to the base of the test box with a frequency of 3 Hertz and maximum amplitude of $0.2g$, where g is the acceleration due to gravity. The finite element results indicate a tendency for the wall to rotate with respect to the top coupled with a sliding mode of failure. Although not reflected in the observations at $0.2g$, at higher accelerations the wall was observed to undergo this mode of deformation. Finite element results indicate that the sliding displacement at the base of the wall due to the input ground motion was approximately 2.54 mm. This is consistent with observations. Sliding predicated with the sliding block method by Richards and Elms (1979) is approximately 1.27 mm. The reason for the discrepancy observed in the result from the sliding block model is due to a phase difference between the ground motion and backfill not considered in the sliding block analysis. The finite element results presented here and by Whitman and Nadim (1984) demonstrate that the simple analytic method proposed by Richards and Elms (1979) yields reasonable results for a sliding mode of failure. The study is also a good example of the use of interface elements in finite element analysis.

Table 16.3   Finite element parameters for retaining wall model

Backfill friction, ϕ_b	30°
Foundation friction, ϕ_f	38°
Backfill wall-soil friction, δ_b	20°
Foundation footing-soil friction, δ_f	20°

16.5.2   Application to problems with field observations

Drilled shaft socketed into rock

Dilled shaft foundations are a commonly used deep foundation method in karst areas. Karst areas are characterized by solution cavities in the limestone bedrock, with a residual soil overburden that often exhibits an increase in water content and decrease in shear strength with depth. An irregular zone of highly weathered or inconsistent rock often exists between the residual soil and the sound or consistent limestone. Because of these very poor subsurface conditions, drilled shafts are usually designed to extend into the consistent rock. Since the skin friction capacity of the residual soil is small, and the thickness and strength of the inconsistent rock are very uncertain, many designs are based on end bearing in the consistent rock, neglecting skin friction in the rock socket.

Field measurements of in-service shafts (Tang *et al.*, 1994) found that less than 20% of the applied load was actually transferred to the tip, with the majority of the load transferred through skin friction to the sides of the socket in the consistent rock. This observation was based on the variation in unit skin friction along the shaft, which was calculated from the vertical stresses measured in the shaft. To investigate the distribution of skin friction in rock socketed shafts, an axisymmetric finite element analysis, Figure 16.21 (Tang, 1995), was performed with the commercial code ABAQUS (HKS, 1995). A typical idealization is shown in Figure 16.22. Initial test computations conducted with elastic-plastic models indicated that under the service loads there was very little yielding in the system. Therefore, linear elastic material models were used for the concrete, soil, inconsistent rock and consistent rock. The concrete/soil and concrete rock interfaces were represented with zero thickness contact elements, and a Mohr–Coulomb criterion was assumed. Details of the material properties used for the analysis are provided by Tang (1995). The analysis was performed both with a smooth rock socket with zero-dilation, and a rough socket with dilation. The results suggested that for the relatively small tip displacements encountered under the service loading, modeling the dilation was not important. However, even for relatively smooth sockets, significant load was transferred into the socket through skin friction.

Figure 16.23(a) compares the unit side friction along the shaft determined from the instrumented shafts, with that obtained from the FE analysis. The results are shown for

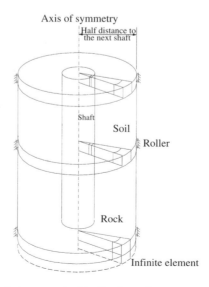

Figure 16.21   Axisymmetric idealization of rock socketed drilled shaft

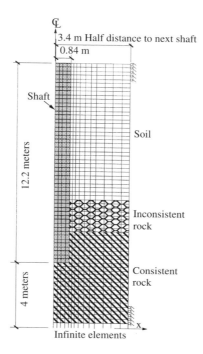

℄

3.4 m Half distance to next shaft

0.84 m

Shaft

12.2 meters

Soil

Inconsistent rock

Consistent rock

4 meters

x

Infinite elements

Figure 16.22   FE idealization for rock socketed drilled shaft

three different dates during construction, corresponding to the completion of the fifth floor, the seventh floor, and six months after completion of the building (eighth floor). The FE results agree favorably with the results from the instrumented shafts. The increase in the unit skin friction within the consistent rock (depth from 9.5 m to 12 m) is evident, as is the negligible amount of friction mobilized in the soil layer (depth from 0 to 6 m). Figure 16.23(b) compares the variation in the FE computed axial stress with that measured in the shaft. The numerical results provide a good depiction of the variation in vertical stress, the rate of which is equal to the unit skin friction. The results indicate that the significant unit skin friction in the competent rock socket causes very low stresses to be transferred to the tip of the shaft, which is consistent with the field observations. Although a substantial portion of the applied load is carried through skin friction (about 80%), only service loadings were applied. At higher loadings approaching shaft capacity, the distribution of skin friction and tip capacity may well change. These results suggest that more economical shafts might be constructed if the rock socket friction were considered in design.

Soil nailed slope

Prior to 1977 when the US Surface Mining Act mandated surface mine reclamation, the extraction of coal from the southern Appalachian mountains of the United States often left unstable or marginally stable waste-rock slopes. These slopes were created when waste-rock was dumped down the hillside below the mine operations. This mine spoil is often

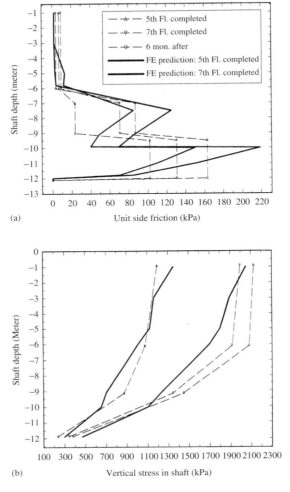

Figure 16.23   Measured and computed response of drilled shaft: (a) unit side friction with depth; (b) vertical stress with depth

comprised of random size material ranging from clay size to boulder size, and generally exists at low density. Many of these slopes are still active, with down slope movement carrying trees and vegetation. This movement may occur very slowly over periods of years, or may occur rapidly corresponding to periods of high rainfall or construction activities. Access to these slopes is often difficult, making conventional repair methods involving movement of large volumes of material demanding and expensive. An investigation (Drumm *et al.*, 1998) was conducted to evaluate the application of established soil nailing techniques to mine waste slopes.

An existing, marginally stable 1H:1V slope about 8 m high was stabilized with nails. The lower 4 m of the slope was steepened to 1H:3.7V (75°), thus creating a 'cut-slope' requiring support (Figure 16.24). The slope was divided into three sections with different nail patterns and lengths as described by Drumm *et al.* (1996), but the section discussed

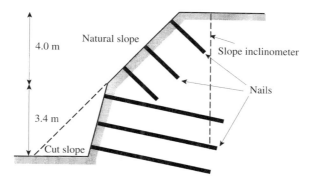

Figure 16.24   Section through soil nailed slope with excavated cut

here utilized 3 m long nails in the upper natural slope and 5.8 m nails in the lower cut slope (Figure 16.24). The design specified 1.5 m horizontal and vertical nail spacing. The 25 mm steel nails were grouted into 200 mm diameter holes augered into the mine spoil. Slope inclinometers were installed to measure the deformations in the nailed slope. Figure 16.24 depicts the approximate location of the inclinometer casing.

A numerical analysis (Yang and Drumm, 2000) was conducted to model the slope deformations and to evaluate the role of bending in the soil nails. A 3-D section of the slope was idealized as shown in Figure 16.25, with a model width equal to the 1.5 m horizontal nail spacing. The steel/grout nails were modeled as a single uniform material with equivalent linear elastic properties. The nails were square in cross-section, with an area equal to that of the grout hole. Zero thickness contact elements surrounded the nails, with the shear strength as determined from nail pull-out tests. The mine spoil was represented by a Drucker–Prager elastic-plastic model, and the analysis was performed with the commercial code ABAQUS (HKS, 1998). The material properties used in the analysis are described by Yang and Drumm (2000). The initial state was first established

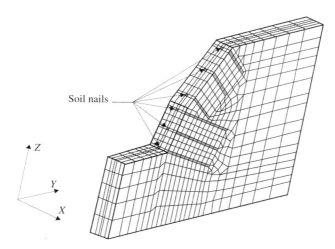

Figure 16.25   Three-dimensional FE idealization of soil nailed slope, with the X dimension equal to the horizontal nail spacing

by fixing all the nodes, and applying the gravitational loading. The nodes were then released, and the slope allowed to deform. The lower slope was then excavated by the incremental removal of elements at the lower portion of the slope (Figure 16.25).

The results of the analysis suggested that the plastic strains were very small throughout the mine waste, and that while the nails developed tension as the lower slope was excavated, only very small bending moments were developed in the nails. The interface elements around the nails ensured that the stresses were nearly zero at the ends of the nails, and maximum near the middle, as would be expected in the actual slope. The observed deformation profiles from the inclinometer measurements are compared in Figure 16.26 with the displacements predicted along a vertical line about 1.5 meters behind the edge of the slope, corresponding to the location of the inclinometer casing (Figure 16.24). The maximum computed displacement at the surface was about 1.2 mm which was comparable with the 0.8 mm measured at the end of construction (about 60 days after the start of construction). However, the numerical results indicate displacements extending to a much greater depth, with a displacement of nearly one mm at the depth corresponding to the bottom of the inclinometer. This would suggest that the bottom of the inclinometer casing may have translated because it was not installed deep enough. If it is assumed that the bottom of the inclinometer casing translated by an amount equal to the computed displacement of one mm, a 'corrected' inclinometer profile is obtained as shown in Figure 16.26. Also shown is the corrected profile corresponding to 800 days after construction. A comparison of the profile at the end of construction with that at 800 days after construction suggests that very little long-term deformation occurred in the lower cut slope with the 5.8 m nails, whereas about 3 mm of additional

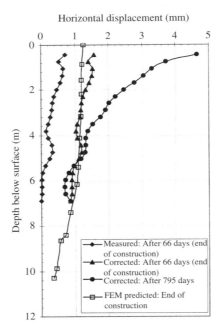

Figure 16.26 Measured and predicated inclinometer displacements along crest of soil nailed slope

local deformation occurred at the top of the upper 4 m of the slope. During the design of the nail system, the shorter 3 m nails installed in the upper 4 m of the 'natural' where not considered to contribute to the overall stability, but were installed to control local deformations. The observed deformations suggest that the these short nails did not control the local deformation, but the 5.8 m nails in the lower slope effectively limited slope deformation.

Additional evidence of the effectiveness of the lower nails was obtained when an adjacent section of the slope, without any nails, was excavated to the same profile as that shown in Figure 16.24. Within about five hours of excavating the lower cut slope, the slope failed. To determine if the FE model of the nailed slope could represent conditions approaching failure, an additional surcharge of 70 kPa surcharge was placed on the surface of the slope. This resulted in significant computed plastic strains (Figure 16.27) along a surface similar to the critical failure surface that would be predicted using limiting equilibrium methods.

This is an example of a problem in geomechanics which could not have been very well approximated without the use of interface elements. Without slip taking place along the nails, very high stresses would be generated at the ends of the nails, and tension produced in the soil. This would have led to an unrealistic stress distribution and a poor representation of the field behavior. In the following example, it will be shown that while interface response may improve the analysis results, for some problems the overall response is not significantly affected unless the stress state is such that significant relative displacement occurs at the contacts.

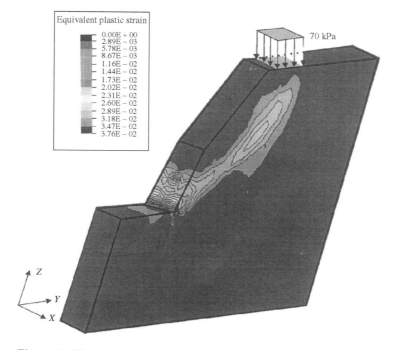

Figure 16.27   Contours of plastic strain in nailed slope with surcharge

Box culvert

A cast-in-place concrete box culvert, overlain with about 12 meters of backfill, was instrumented to measure the vertical and horizontal earth pressures during both the construction and the service loadings. To reduce the earth pressures on the culvert, compaction equipment was prohibited from working within two meters of the culvert, resulting in a zone of low density material around the culvert (Figure 16.28). To evaluate the effects of this low density material on the vertical and lateral earth pressures, a series of finite element analyses was conducted (Yang *et al.*, 1997).

A two-dimensional (plane strain) analysis was conducted, taking advantage of a vertical plane of symmetry through the centerline of the box culvert. The soil was represented by a Drucker–Prager model, the concrete idealized as linear elastic, and the interface was modeled using a zero thickness contact. The analysis was performed using the ABAQUS code (HKS 1998), and details of the material properties used are provided by Yang (2000). Figure 16.29 compares the field measured earth pressures with those obtained from the numerical analysis. Numerical results are shown for three cases: zero interface friction (perfectly smooth), interface friction corresponding to that of the well graded gravel, and no interface behavior (no slip between backfill and concrete). As expected, in the region over the culvert roof where little relative slip expected, the results from all three analyses are nearly identical, and good agreement with the field measurements is obtained. However, because slip would be expected along the upper portion of the culvert wall, the vertical stress distribution over the edge of the roof is somewhat dependent upon the interface friction. Higher vertical earth pressures were measured in this area, and correspond to the increased rigidity of the culvert wall. Without the interface slip, the zone of maximum vertical stress is shifted slightly outside the region immediately above the culvert. The predicted distribution of lateral earth pressures on the culvert wall were somewhat dependent upon the interface friction, with the pressures slightly greater at the bottom when no interface is included.

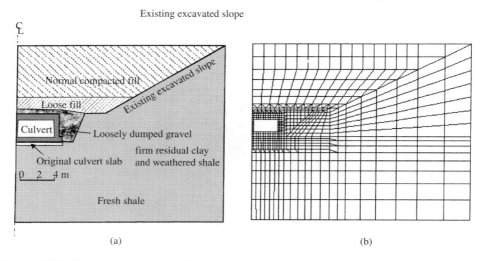

Existing excavated slope

(a)   (b)

Figure 16.28   Instrumented cast-in-place concrete box culvert: (a) subsurface and backfill conditions; (b) 2-D plane strain idealization

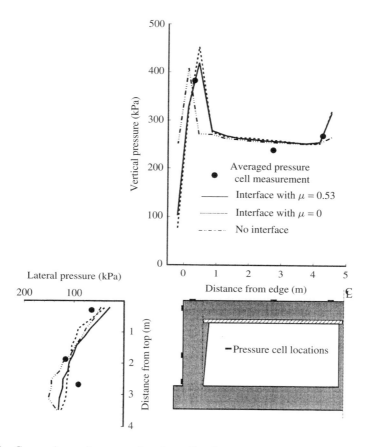

Figure 16.29 Comparison of measured and predicted earth pressures on cast-in-place concrete box culvert

A parametric study of the effect of backfill compaction (Yang *et al.*, 1997), as manifested in backfill modulus, revealed that the vertical pressures were not significantly affected by the degree of compaction. However, the distribution of lateral pressure was significantly affected, with high density backfill resulting in larger pressures at the bottom of the wall. Under the 12 m of backfill, yielding was predicted in only very limited regions around the corner of the culvert. Because the stresses in this area are dependent upon the interface friction, it could be concluded that for greater embankment heights and larger vertical stresses, the interface behavior may be more important than it was under the service loadings. Under the service loading, a reasonable result could probably be obtained without the interface elements, resulting in significantly shorter computational times.

16.6 Summary and Conclusions

Several examples were provided of problems in geomechanics in which the behavior of material contacts/interfaces govern or significantly influence the system response. Due to the important role of these interfaces, it is important that relative slip or separation at interfaces be represented or modeled properly in a numerical analysis. A general review of frequently implemented interface or contact models was given, and a number of newer approaches were mentioned. A discussion of constitutive relationships for interfaces was provided, with details for three different classes of models and some back-predictions of interface behavior. A number of numerical applications are described, with distinction made between verification problems, in which the numerical solution is compared with a well defined analytical solution or classical theory, and field problems, in which the numerical results are compared with field observations. The examples include a tunnel in jointed rock, the seismic sliding of a retaining wall, a rock-socketed drilled shaft foundation, a soil nailed slope, and a cast-in-place concrete culvert. Results from these field examples suggest that interfaces can be very effective when the otherwise continuous system can be divided into several different domains, separated by a relatively small number of discontinuities which dominate the response. However, systems that are highly jointed or include a large number of blocks may be better modeled by other methods. For many practical problems, particularly those well below design capacity or failure, satisfactory results may be obtained with linear elastic material models provided the interfaces are properly represented. It is a good practice to model the system both with and without the interface response to verify that the interface model does not significantly affect the computed quantities in regions of the problem where interface slip or separation is not important. The examples suggest that the incorporation of interface response into the numerical analysis can permit the satisfactory modeling of complex geomechanical systems that could not be adequately modeled otherwise.

References

Alvappillai, A. *Dynamic Analysis of Rigid Pavements Considering Pavement-Subgrade-Aircraft Interaction*. PhD, Dissertation, University of Oklahoma, Norman, (1992).

Amontons, G. De La Resistance Caus'ee les Machines. In: *Memories de l'Academie Royale*, Chez, A. Gerard Kuyper, Amsterdam, (1699) 257–282.

Archard, J. I. Elastic deformation and the laws of friction. *Proc. Roy. Soc. Lond., Series A*, **243**, (1959) 190–205.

Bandis, S., Lumsden, A. C. and Barton, N. R. Experimental studies of scale effects on the shear behavior of rock joints. *Int. J. Rock Mechanics and Mining Sci. and Geomechanics Abstracts*, **18**, (1981) 1–21.

Barton, N. R. Review of a new shear strength criterion for rock joints. *Engrg. Geology*, **7**, (1973) 579–602.

Barton, N. R. Deformation phenomena in jointed rock. *Geotechnique*, **36**, (1986) 147–167.

Barton, N. R. and Bandis, S. Rock joint model for analysis of geologic discontinua. *Constitutive Laws for Engineering Materials; Theory and Applications, Proceedings of the Second International Conference on Constitutive Laws for Engineering Materials*, Tucson, AZ. Elsevier, New York, (1987) 993–1002.

Barton, N. R. and Choubey, V. The shear strength of rock joints in theory and practice. *Rock Mechanics*, **10**, (1977) 1–54.

Bathe, K. J. *Finite Element Procedures in Engineering Analysis*. Prentice Hall, Englewood Cliffs, NJ, (1982).

Bazant, Z. P. and Mazars, J. France-U.S. workshop on strain localization and size effect due to cracking and damage. *J. Engrg. Mech.*, **116**(6), (1990) 1412–1424.

Bowden, F. P. and Tabor, D. *The Friction and Lubrication of Solids, Part I.* Clarendon Press, Oxford, (1950).

Burns, J. Q. and Richard, R. M. Attenuation of stresses for buried cylinders. *Proc. Symp. on Soil-Structure Interaction*, University of Arizona, Tucson, AZ, (1964).

Chang, C. S., Weeraratne, S. P. and Misra, A. Slip mechanism-based constitutive model for granular soils. *J. Engrg. Mech.*, **115**(4), (1989) 790–807.

Clough, R. W. and Duncan, J. M. Finite element analysis of retaining wall behavior. *J. Soil Mech. Fdn. Div.*, **97**(12), (1971) 1657–1673.

Cundall, P. A. A computer model for simulating progressive, large scale movements in blocky rock system. *Proc. Symp. Int. Soc. Rock Mech.*, Nancy, France, **2**(8), (1971).

Cundall, P. A. and Strack, O. D. L. A discrete numerical model for granular assemblies. *Geotechnique*, **29**(1), (1979) 47–65.

Cundall, P. A. and Hart, R. D. Numerical modeling of discontinua. In: *Comprehensive Rock Engineering, Vol 2.* Pergamon, New York, (1993) 231–243.

Desai, C. S. Soil-structure interaction and simulation problems. In: *Finite Elements in Geotechnical Engineering*, Gudehus, (ed.). (1977) 209–250.

Desai, C. S. A Dynamic Multi-Degree-of-Freedom Shear Device. *Report No. 8–36*, Department of Civil Engineering, Virginia Technical College, Blacksburg, VA, (1980).

Desai, C. S. Behavior of interfaces between structural and geologic media. A state-of-the-art paper. *Proc. Int. Conf. on Recent Advances in Geotech. Earthquake Eng. and Soil Dyn.*, St. Louis, MO, (1981).

Desai, C. S. Static and cyclic response of interfaces for analysis and design of soil-structure interaction problems. In: *Geotechnical Modeling and Applications*, Sayed, S. M. (ed.). Gulf Publishing, (1987).

Desai, C. S. DSC for interfaces and joints. In: *Mechanics of Materials and Interfaces*. CRC Press, Boca Raton, FL, (1999).

Desai, C. S. and Appel, G. C. 3-Dimensional analysis of laterally loaded structures. *Proc. Second Int. Conf. Num. Met. in Geomech.*, ASCE, New York, (1976).

Desai, C. S., Basaran, C. and Zhang, W. Numerical algorithms for mesh dependence in the disturbed state concept. *Int. J. Num. Met. in Engrg.*, **40**, (1997) 3059–3083.

Desai, C. S. and Christian, J. T. (eds.). *Numerical Methods in Geotechnical Engineering*. McGraw-Hill, New York, (1977).

Desai, C. S., Drumm, E. C. and Fishman, K. L. Testing and modeling of joints and interfaces under static and cyclic loading. *Proc. 25th Symposium on Rock Mechanics*, Northwestern University, Evanston, IL, (1984) 806–815.

Desai, C. S., Drumm, E. C. and Zaman, M. M. Cyclic testing and modeling of interfaces. *J. Geotech. Engrg.*, **111**, (1985) 793–815.

Desai, C. S. and Fishman, K. L. Constitutive model for rocks and discontinuities. *Rock Mechanics, Proc. 28th U.S. Symposium on Rock Mechanics*, University of Arizona, Tucson, Az. Balkema, Boston, (1987) 609–619.

Desai, C. S. and Fishman, K. L. Plasticity based constitutive model with associated testing for joints. *Int. J. Rock Mechanics and Mining Sci. and Geomechanics Abstracts*, **28**(1), (1991) 15–26.

Desai, C. S. and Hashmi, Q. S. E. Analysis, evaluation and implementation of a nonassociative model for geologic materials. *Int. J. Plasticity*, **5**, (1989) 397–420.

Desai, C. S. and Ma, Y. Modelling of joints and interfaces using disturbed state concept. *Int. J. Num. Anal. Methods in Geomech.*, **16**(9), (1992) 623–663.

Desai, C. S. and Nagaraj, B. K. Modelling for normal and shear behavior at contacts and interfaces. *J. Eng. Mech.*, **114**(7), (1988).

Desai, C. S. and Nagraj, B. K. Modeling of cyclic normal and shear behavior of interfaces. *J. Engrg. Mech. Div.*, **114**, (1988) 1998–1217.

Desai, C. S. and Rigby, D. B. Cyclic interface and joint shear device including pore pressure effects. *J. Geotech. Engrg. Div.*, **123**, (1997) 568–579.

Desai, C. S., Zaman, M. M., Lightner, J. G. and Siriwardane, H. J. Thin-layer element for interfaces and joints. *Int. J. Numer. Methods Geomech.*, **8**, (1984) 19–43.

Desai, C. S. and Zhang, W. Computational aspects of disturbed state constitutive models. *Int. J. Comp. Met. in Appl. Met. and Engrg.*, **151**, (1988) 361–376.

Drumm, E. C. and Desai, C. S. Determination of parameters for a model for the cyclic behaviour of interfaces. *Earthquake Engrg. & Structural Dynamics*, **14**, (1986) 1–18.

Drumm, E. C., Mauldon, M. and Tant, C. R. Stabilization of coal mine waste with soil nails. *Ground Improvement*, Thomas Telford, UK, Vol. 2, (1998) 147–156.

Drumm, E. C., Tant, C. R., Mauldon, M. and Berry, R. M. Instrumentation and monitoring of a nailed mine-waste slope. *Proc. Int. Symposium on Earth Reinforcement*, Fukuoka, Kyushu, Japan, (1996) 741–746.

Einstein, H. H., Halabe, V. B., Dudt, J. P. and Descoeudres, F. Geologic uncertainties in tunneling. *Geotechnical Special Publication No. 58, Uncertainty in the Geologic Environment: From Theory to Practice. Proceedings of Uncertainty '96*, Shackelford, C. D., Nelson, P. P. and Roth, M. J. S. (eds.). Madison Wisconsin, Vol 1, (1996) 239–253.

Fakharian, J. and Evgin, E. An automated apparatus for three-dimensional monotonic and cyclic testing of interfaces. *Geotechn. Testing J.*, **19**(1), (1996) 22–31.

Fishman, K. L. *Constitutive Modeling of Idealized Rock Joints Under Quasi-Static and Cyclic Loading*. PhD, Dissertation, Faculty of the Department of Civil Engineering and Engineering Mechanics, University of Arizona, Tucson, AZ (1988).

Fishman, K. L. and Ahmad, S. Verification for using thin layer finite elements to model rock joints. *Constitutive Laws for Engineering Materials; Recent Advances and Industrial and Infrastructure Applications, Proceedings of the Third International Conference on Constitutive Laws for Engineering Materials*, Tucson, AZ. ASME Press, New York, (1991) 557–560.

Fishman, K. L., Derby, C. W. and Palmer, M. C. Verification study for numerical modeling of jointed rock mass. *Int. J. Num. Anal. Methods in Geomech.*, **15**(1), (1991) 61–70.

Fishman, K. L. and Desai, C. S. A constitutive model for hardening behavior of rock joints. *Constitutive Laws for Engineering Materials; Theory and Applications, Proceedings of the Second International Conference on Constitutive Laws for Engineering Materials*, Tucson, AZ, Elsevier, New York, (1987) 1043–1050.

Fishman, K. L. and Desai, C. S. Measurement of compression/dilation of interfaces/joints during shear. *Geotech. Engrg. Testing J.*, **12**(4), (1989) 297–301.

Fishman, K. L., Drumm, E. C., Ben-Hassine and Ahmad, S. Influence of interface/joint behavior in the response of structural systems. *8th Int. Conf. Association for Computer Methods and Advances in Geomechanics*, University of West Virginia, Morgantown, WV, May 22–28, (1994).

Francavilla, A. and Zienkiewicz, O. C. A note on numerical computation of elastic contact problems. *Int. J. Num. Methods in Engrg.*, **9**, (1975) 913–924.

Frantziskonis, G. and Desai, C. S. Constitutive model with strain softening. *Int. J. Solids and Structures*, **23**(6), (1987) 733–750.

Frantziskonis, G. and Desai, C. S. Elastoplastic model with damage for strain softening geomaterials. *Acta Mechanica*, **68**, (1987) 151–170.

Gässler, G. Full scale test on a nailed wall in consolidated clay. *Proc. Int. Symposium on Earth Reinforcement Practice*, Fukuoka, Japan, (1992) 475–480.

Ghaboussi, J., Wilson, E. L. and Isenberg, J. Finite element for rock joint and interfaces. *J. Soil Mech. and Fdn. Div.*, **99**(10), (1973) 833–848.

Goodman, R. E. *Introduction to Rock Mechanics, 2nd Ed.* Wiley, New York, (1989).

Goodman, R. E. and Christopher, S. J. Finite element analysis for discontinuous rocks. In: *Numerical Methods in Geotechnical Engineering*, Desia, C. S. and Christian, J. T. (eds.). McGraw-Hill, New York, (1977).

Goodman, R. E., Taylor, R. L. and Brekke, T. L. A model for the mechanics of jointed rock. *J. Soil Mech. and Fn. Div.*, **94**(3), (1968) 637–659.

Goodman, R. E. and Shi, G. *Block Theory and its Application to Rock Engineering*. Prentice-Hall, NJ, (1985).

Hibbitt, Karlsson & Sorensen, Inc. *ABAQUS/Standard User's Manual*. Vols. 1, 2. Ver. 5.7, (1998).

Heuze, F. E. and Barbour, T. G. New models for rock joints and interfaces. *J. Geotech. Engrg. Div.*, **108**(5), (1982) 757–776.

Hutson, R. W. Limestone Rock Joint: Direct Shear Data. *Internal Report*, Northwestern University, Evanston, Il., (1987a).

Hutson, R. W. Granite Rock Joint: Direct Shear Data. *Internal Report*, Northwestern University, Evanston, Il., (1987b).

Idriss, I. M., Dobry, R. and Sign, R. D. Nonlinear behavior of soft clays during cyclic loading. *J. Geotech. Engrg. Div.*, **104**, (1978) 1427–1447.

Jennings, P. C. Periodic response of a general yielding structure. *J. Engrg. Mech. Div.*, **90**(2), (1964) 131–166.

Katona, M. G., Smith, J. M., Odello, R. S. and Allgood, J. R. CANDE: A Modern Approach for the Structural Design and Analysis of Buried Culverts. *Report FHWA-RD-77*, Prepared for the Federal Highway Administration, Washington, D.C., (1976).

Kikuchi, N. and Oden, J. T. *Contact Problems in Elasticity*. SIAM, Philadelphia, (1981).

Kutschke, W. G. *Bridge Abutment Rotations and Displacements Due to Seismic Loads*. Master's Thesis, submitted to the Faculty of the Graduate School of the State University of New York at Buffalo, (1995).

Mahtab, M. A. and Goodman, R. E. Three dimensional F.E. analysis of jointed rock slopes. *Proc. 2nd Conf. Int. Soc. For Rock Mech. (ISRM)*, **3**, (1970) 353–360.

Mauldon, M, and Goodman, R. E. Rotational kinematics and equilibrium of blocks in a rock mass. *Int. J. Rock Mechanics Min. Sci. & Geomech. Abstr.*, **27**(4), (1990) 291–301.

Mononobe, N. Earthquake-proof construction of masonry dams. *Proc. World Engrg. Congress*, **9**, (1929) 275.

Nadim, F. and Whitman, R. V. Seismically induced movement of retaining walls. *J. Geotech. Engrg.*, **109**(7), (1983) 915–931.

Navayogarajah, N., Desai, C. S. and Kiousis, P. D. Hierarchical single-surface model for static and cyclic behavior of interfaces. *J. Engrg. Mech.*, **118**(5), (1992) 990–1011.

Newmark, N. M. Effects of earthquakes on dams and embankments. *Geotechnique*, **14**(2), (1965) 139–160.

Ngo, D. and Scordelos, A. C. Finite element analysis of reinforced concrete beams. *ACI J.*, **64**(3), (1967) 152–163.

Obert, L. and Duvall, W. *Rock Mechanics and the Design of Structures in Rock*. Wiley, New York, (1967).

Oden, J. T. and Pives, T. B. Nonlocal and nonlinear friction laws and variational principles for contact problems in elasticity. *J. Appl. Mech.*, **50**, (1983) 67–75.

Okabe, S. General theory of earth pressure. *J. Japanese Soc. Civil Eng.*, **12**(1), (1926).

Paikowsky, S. G., Player, C. M. and Connors, P. J. A dual interface apparatus for testing unrestricted friction of soil along solid surfaces. *Geotech. Testing J.*, **19**(4), (December 1996) 446–451.

Phan, H. V. *Geometric and Material Nonlinear Analysis of Three-Dimensional Soil-Structure Interaction*. PhD Dissertation, VPI&SU, Blacksburg, (1979).

Rabinowicz, E. *Friction and Wear of Materials*. Wiley, New York, (1965).

Ramberg, W. and Osgood, W. R. Description of stress-stain curves by three parameters. *Technical Note 902*, National Advisory Committee for Aeronautics, Washington, DC, (1943).

Richards, R. and Elms, D. Seismic behavior of gravity retaining walls. *J. Geotech. Engrg. Div.*, **105**(4), (1979) 449–464.

Samtani, N. C., Desai, C. S. and Vulliet, L. An interface model to describe viscoplastic behavior. *Int. J. Num. Anal. Methods in Geomech.*, **20**, (1996) 231–252.

Selvadurai, A. P. S. and Boulon, M. J. *Contact Mechanics for Geomaterials*. Studies in Applied Mechanics 42, Elsevier, Amsterdam, (1995).

Sharma, K. G. and Desai, C. S. An analysis and implementation of thin-layer element for interfaces and joints. *J. Engrg. Mech. Div.*, **118**, (1992) 545–569.

Shi, Gen-hua, *Block System Modeling by Discontinuous Deformation Analysis*, Topics in Engineering, Vol. 11, edited by C. A. Brebbin and J. J. Connor, Computational Mechanics Publications, Boston, USA, 209p. (1993).

Stark, T. D., Williamson, T. A. and Eid, H. T. HDPE gemomembrane/geotextile interface shear strength. *J. Geotech. Engrg.*, **122**(3), (1996).

Tang, Q. *Load Transfer Mechanisms of Drilled Shaft Foundations in Karstic Limestone–Behavior under Working Load*. PhD dissertation, The University of Tennessee, Knoxville, TN, (1995).

Tang, Q., Drumm, E. C. and Bennett, R. M. Response of drilled shaft foundations in Karst during construction loading. *Proceedings FHWA International Conference on Design and Construction of Deep Foundations*, Orlando, FL, (1994) 1296–1309.

Whittle, A. J., Hashash, Y. M. A. and Whitman, R. V. Analysis of deep excavation in Boston. *J. Geotechn. Engrg.*, **119**(1), (1993) 69–90.

Wun, R. B. *A Study On Plasticity Based Constitutive Model For Rock Joints*. Master's Thesis, Faculty of the Graduate School, State University of New York at Buffalo, (1992).

Yang, M. Z. *Evaluation of Factors affecting the Earth Pressures on Concrete Box Culverts*. PhD dissertation, The University of Tennessee, Knoxville, TN, (2000).

Yang, M. Z. and Drumm, E. C. "Numerical analysis of the load transfer and deformation in a soil nailed slope." *Proceedings, ASCE Geo-Institute Conference*, Denver, Co., USA (2000).

Yang, M. Z., Drumm, E. C., Bennett, R. M. and Mauldon, M. Influence of compactive effort on earth pressures around a box culvert. *Ninth International Conference of the Association for Computer Methods and Advances in Geomechanics*. The Chinese Academy of Sciences, Wuhan, China, November 2–7, (1997) 2021–2026.

Yang, M. Z., Drumm, E. C., Bennett, R. M. and Mauldon, M. Measurement of Earth pressures on concrete box culverts under highway embankments. *a: Field Instrumentation for Soil and Rock, ASTM STP 1358*, Durham, G. N. and Marr, W. A. (eds.). American Society for Testing and Materials, (1999) 87–100.

Zaman, M. M. *Influence of Interface Behavior in Dynamic Soil-Structure Interaction Problems*. PhD Dissertation, University of Arizona, Tucson, AZ, (1981).

Zaman, M. and Alvappillai, A. Contact-element model for dynamic analysis of jointed concrete pavements. *J. Transp. Engrg. Div.*, **121**(5), (1995) 425–433.

Zaman, M. M., Desai, C. S. and Drumm, E. C. An interface model for dynamic soil-structure interaction. *J. Geotech. Engrg. Div.*, **110**, (1984) 1257–1273.

Zaman, M., Desai, C. S. and Selvadurai, A. P. S *US-Canada Workshop: 'Recent Accomplishments and Future Trends in Geomechanics in the 21st Century*, Norman, OK, (1992).

Zienkiewicz, O. C., Best, B., Dullage, C. and Stagg, K. G. Analysis of nonlinear problems in rock mechanics with particular reference to jointed rock systems. *Proc. Second Congress of the Int. Soc. For Rock Mech. (ISRM)*, Belgrade, Yugoslovia, Vol. 3, (1970) 501–509.

17

Dynamic Soil-Structure Interaction During Seismic Events

Kanthasamy K. Muraleetharan[1], Changfu Wei[1] and **Kyran D. Mish[2]**
[1]*University of Oklahoma, Norman, OK, USA*
[2]*Lawrence Livermore National Laboratory, Livermore, CA, USA*

17.1   Introduction

Dynamic soil-structure interaction during seismic events is a complex problem. Soil, the embedded structural foundation, and the above ground structure undergo a series of complex interactions during earthquakes. The complexity is further increased by the three-phase nature of soils. Soils are nonlinear materials and in general consist of a soil skeleton made of solid grains, a pore fluid (generally water) and a pore gas (generally air). Saturated soils are two-phase media consisting of a soil skeleton and a pore fluid. Other nonlinear behavior such as gapping between foundation elements such as piles and soils and concrete cracking also have been observed during seismic loading of structures. However, the widely used design procedures currently available for seismic design of structures are somewhat over-simplified and do not take into account the complex interactions mentioned above.

The current design methodology involves several disparate, often disjointed or poorly joined, analysis procedures; for example, see Hadjian *et al.* (1992). A typical analysis procedure might involve free field (i.e. assuming no structures are present) dynamic analysis of the foundation soils performed by geotechnical engineers followed by a dynamic analysis of the above ground structure performed by structural engineers. Structural engineers utilize free field surface motion provided by geotechnical engineers to excite the above ground structure with a fixed base assumption. Finally, geotechnical engineers complete the design by performing a static analysis of the foundation using maximum shear forces and moments predicted by structural engineers at the base of the structure. Rudimentary assumptions for foundation soils such as a single phase medium and elastic behavior are not too uncommon in the analyses carried out by geotechnical engineers.

Hadjian *et al.* (1992) conducted a detailed survey of state-of-the-practice among design professionals, and concluded the decoupling of functions between geotechnical and structural engineers described above is an important contributor to the oversimplified analyses carried out in practice.

As the Sylmar earthquake in Southern California inspired a rethinking of structural design methodologies for hospitals and other critical institutional facilities during the decade of the 1970s, more recent seismic disasters, such as the 1989 Loma Prieta earthquake in California and 1995 Hyogoken-Nambu (Kobe) earthquake in Japan, have underscored the importance of preventing large-scale failure of causeways, bridges, and viaducts and focused attention again on the seismic design of structures. Although some of the design codes have begun to recommend alternate, more comprehensive analysis procedures, current practice, in general, still involves poorly coupled analysis procedures. While the current approaches to earthquake design of soil-structure systems have been found to be generally satisfactory, engineers often are unable to predict, with confidence, whether these approaches will result in conservative designs or not, especially for large magnitude earthquakes. Soil related phenomena, such as liquefaction, often observed in large earthquakes increase the lack of confidence in the current methodology. A fundamental, fully coupled analysis procedure that takes into account the complex interactions during seismic loading of soil-structure systems, including the multi-phase nature of the soils, will help to eliminate the uncertainties involved in the seismic design of structures. Although more and more practicing engineers are becoming aware of these fundamental analysis procedures, they are reluctant to use them in practice mainly for two reasons: (i) they do not know how to interpret the quantities calculated by the fully-coupled methods; and (ii) validation of the fully coupled methods are lacking. In addition, the code-based practice can be an impediment to better analysis techniques being used in design. The codes absolve the engineer of responsibility for use of advanced analysis techniques.

To introduce readers to the current practice for seismic design of structures, first a latest seismic design criterion for bridges is reviewed. These design criteria entitled, 'Improved Seismic Design Criteria for California Bridges: Provisional Recommendations' (ATC-32, Applied Technology Council, 1996) were produced following the 1989 Loma Prieta earthquake in a study funded by the California Department of Transportation (Caltrans). Although, these criteria have not been fully implemented into practice, they can be considered the latest thinking in seismic design of bridges. Next, several available analysis procedures with varying degrees of coupling between soil, foundation, and structure are reviewed. An example problem and analyses of this problem using various methods are then presented. The example problem consists of a structural frame supported on piles embedded in a soft clay. The analysis procedures include simplified procedures and a fully coupled procedure. The simplified procedures include free field dynamic analysis, elastic structural analysis assuming a fixed base, and pseudo static pile analysis. The fully coupled analysis is performed using a dynamic finite element computer code, DYSAC2 (Muraleetharan *et al.*, 1988, 1997b). The results are compared and discussed. Finally, conclusions are drawn from the results presented and some recommendations to practicing engineers are provided.

17.2   State of Practice

The current methods for analyzing the interactions among a soil mass, an embedded foundation, and a structure ridge have proven to be useful in a wide variety of cases, but of uncertain accuracy in many important practical situations (such as saturated granular deposits susceptible to liquefaction). The present approach for modeling such soil-structure interactions is shown in Figure 17.1, where a schematic overview of the process of decomposing the analysis into various subtasks is diagrammed. In this representation of one typical aspect of current foundation analysis practice, it can be seen that a complex procedure is utilized in order to replace the actual coupled soil-foundation-structure system with a sequence of individual component analyses. The purpose of this complex decomposition is to avoid performing a full-blown computational analysis of the combined system, because, until recently, such a complete and accurate computer simulation would have required excessive computational resources.

It is essential to note in the discussions below that existing approaches have worked well for a variety of common foundation types, and that their limitations are primarily due to assumptions required to implement these schemes on computer architectures of the 1970s, where memory was scarce, computer performance was slow, and computer effort was appropriately expensive. It is also worth recognizing that these limitations are rapidly

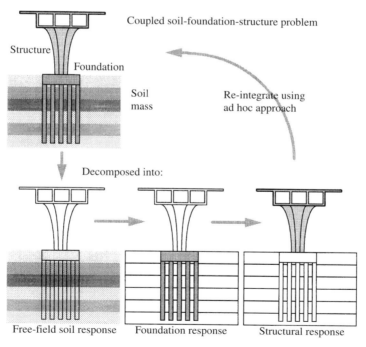

Figure 17.1   Schematic of uncoupled analysis techniques used in current seismic design practice

disappearing today, and hence there is a potential for alternatives more attuned to present and future computer architectures.

Of primary importance in all seismic foundation investigations is the underlying soil mass. In current practice, this soil mass is initially assumed to contain no embedded foundation structural components, and to consist of a series of horizontal soil layers of infinite lateral extent, each of which can be modeled using a linear or near-linear elastic constitutive model that relates the total stress state in the soil to the strain fields induced by the shaking of an earthquake ground motion. These assumptions permit a one-dimensional geometric idealization of the soil mass, as well as a very simple soil constitutive model implementation. The cumulative effect of these simplifications is to permit numerical simulation of the soil dynamic response with little overall computational effort or resources. Unfortunately, the attendant effect of this decoupled scheme is to run the risk of divorcing the simplified model from physical reality, where actual soil masses are three-dimensional in geometry, heterogeneous in distribution, and where soil constitutive response depends upon effective or intergranular stress instead of total stress.

Once the response of the individual soil layers has been approximated, in relatively sophisticated analyses, this idealization of soil behavior is imposed on the embedded soil foundation by assuming that the foundation does not substantially influence the dynamic response of the soil. However, often the foundation-soil interaction is evaluated in a pseudo-static sense only for the purpose of obtaining some representative foundation spring constants for the dynamic analysis of the structure or in the final phase of the design to propagate forces obtained from a fixed base structural analysis back into the soil. The practical result of this simplification is that the potentially complicated coupled response of the soil mass with its embedded foundation is either replaced with a substantially simpler uncoupled problem in which the soil mass imposes its motions on the foundation, or with one where the inertial effects are largely ignored in performing a pseudo-static analysis.

Once the foundation's response history to this imposed motion field or appropriate foundation spring constants has been determined, the earthquake event can be further propagated in a similar fashion to the above ground structural components. At this point, various models for structural response are used to determine member forces in the structure, and depending upon the particular state or local transportation agency's code of practice, as well as the level of importance of the structure, some type of iteration may be performed based upon propagating these structural loads back into the foundation and the soil mass, towards the goal of verifying the overall response of the combined system.

Existing programs such as SHAKE (Schnabel *et al.*, 1972), LPILE (Ensoft, Inc., 1997), SPASM (Matlock *et al.*, 1978), and SEISAB (Imbsen, 1993) are used to perform various components of the soil-pile-structure design task, including (respectively) free-field seismic response, pseudo-static lateral pile analysis, elementary dynamic lateral soil-pile analysis with soil behavior modeled by nonlinear springs and dashpots, and above ground structural analysis. The two computer codes SHAKE and LPILE will be utilized later in the analyses of the example problem and are briefly described below.

17.2.1   Computer code SHAKE

SHAKE (Schnabel *et al.*, 1972) is a simplified one-dimensional total stress code that solves the wave propagation problem for a horizontally layered system in the frequency domain. Constitutive behavior of soil is modeled using an equivalent linear viscoelastic model. This equivalent linear model captures the nonlinear behavior of soils in an approximate sense, by reducing the elastic modulus and increasing the damping as the strain level experienced by the soil increases. Due to strain level dependence of soil properties, SHAKE solves the wave propagation problem using an iterative technique.

17.2.2 Computer code LPILE

LPILE (Ensoft, Inc., 1997) solves the soil-pile interaction problem by treating the pile as a beam on an elastic foundation. Soil reaction to pile deflection is captured through a nonlinear load-deflection or $p - y$ curve. A secant modulus approach is used to relate the pile deflection to soil reaction. However, the pile material is considered to behave elastically. The $p - y$ curves for various soil types have been established mainly through field testing of piles to lateral loading. Shear forces, moments, and axial loads can be considered as pile head loads. Loading of the pile caused by lateral movement of the soil surrounding the pile can be considered in addition to the pile head loads. The program predicts pile deformations and shear force and moment distributions along the pile.

17.2.3 A current seismic code: ATC-32

Following the 1989 Loma Prieta, California, earthquake, which resulted in failure of number of bridges, the Applied Technology Council (ATC) proposed a project to Caltrans to review and revise existing seismic design of bridges. This project was funded by Caltrans and resulted in the report ATC-32 entitled 'Improved Seismic Design Criteria for California Bridges: Provisional Recommendations' (Applied Technology Council, 1996). Although Caltrans has not yet made a decision to implement the recommendations in ATC-32, this report does reflect the current state of practice in seismic design. Analyses recommended by ATC-32 for important bridges are briefly summarized here. ATC-32 recommends for:

(1) An important bridge with simple structural configuration (Type I), for both functional and safety evaluations, equivalent static analysis (A) or elastic dynamic analysis (B).

(2) An important bridge with complex structural configuration (Type II) for functional evaluation elastic dynamic analysis (B) and for safety evaluation elastic dynamic analysis (B) or inelastic static analysis (C), substitution of inelastic dynamic analysis for inelastic static analysis is considered acceptable.

Damage to important bridges during functional evaluation design earthquake should be minimal and the bridges should be usable immediately. Damage to important bridges during safety evaluation design earthquake should be reparable; however, the bridge should be usable immediately. Caltrans currently defines the safety evaluation earthquake as the Maximum Credible Earthquake (MCE) at a site. This earthquake only has a small probability of occurring during a useful life of a bridge. Alternatively, the safety evaluation earthquake may be probabilistically defined as one with a 1000–2000 year return period. The functional evaluation earthquake is defined as an event which has an approximate probability of 40% of occurring during the life time of the bridge. This probability translates into an earthquake with a return period of approximately 300 years (Zechlin and Chai, 1998).

Equivalent static analysis allows an equivalent static force to be applied to the structure. The magnitude of this force is determined from the value of the design spectra at the structure's fundamental period of vibration. Elastic dynamic analysis can be performed using a multi-modal response spectrum analysis using a lumped-mass 'stick' model or using several base motion histories. The structural members are considered elastic and some consideration for soil response, if possible, through the use of elastic springs is recommended. Foundation spring constants are obtained by pseudo static analyses of piles. Inelastic static analysis or 'push-over' analysis is carried out by applying loads incrementally until the structure has reached the ultimate displacements. Inelasticity of structural member, such as plastic hinging, is considered in this analysis. Inelastic dynamic analysis may be substituted for inelastic static analysis. In inelastic dynamic analysis, the structure is subjected to a number of base motion histories and the response is studied. Again, plastic hinging and soil spring constants are considered in the analysis.

17.3   State-of-the-Art

A number of researchers in the last decade or so have performed more advanced soil-pile-structure interaction analyses than currently practiced or recommended in the codes. Cai *et al.* (1995) presented a three-dimensional subsystem approach for soil-pile-structure interaction problems. The soil-pile subsystem was analysed using an elastoplastic constitutive model for soils and structural elements for piles. This subsystem was analysed by solving the equations of motions. However, pore water pressure generation and dissipation in the soil was not considered. The pile head motions obtained through the analysis of the soil-pile subsystem were fed into a space frame structural subsystem and the structural analysis was carried out.

Abghari and Chai (1995) analyzed soil-pile-structure interaction by considering a single pile in the foundation system and approximating the above ground structure as a lumped mass on top of a column. Free-field displacement-time histories were given to the pile through dashpots and springs and the whole system was analyzed together.

Anandarajah *et al.* (1995) performed a truly fully coupled analysis of a soil-pile-structure interaction system by solving the coupled dynamic governing equations of the whole system using a computer code HOPDYNE (Anandarajah, 1990). Structure and piles were modeled as beam elements and an elastoplastic constitutive model was used to represent the clay soil. Soil response was still modeled under undrained conditions and therefore pore pressure dissipation and redistribution were not considered in the analysis.

DYNAFLOW (Prevost, 1987) and DYSAC2 (Muraleetharan *et al.*, 1988, 1997b) are two other computer codes that are capable of performing fully coupled analysis of soil-pile-structure systems. The computer DYSAC2 is used in the analyses presented in this chapter, and a brief description is provided below.

17.3.1   A fully coupled analysis: Computer code DYSAC2

The fully coupled analysis procedure used in this chapter is based on the finite element solution of the dynamic governing equations for a saturated porous media (soil skeleton and pore fluid). The details of this formulation and numerical implementation are given by Muraleetharan *et al.* (1994b). The two-dimensional numerical implementation of the formulation resulted in the computer code DYSAC2 (Muraleetharan *et al.*, 1988, 1997b). Four-noded isoparametric elements with reduced integration for the fluid bulk modulus terms are used in DYSAC2. Nodal variables per node are two soil skeleton and two fluid displacements. A three-parameter time integration scheme called the Hilber–Huges–Taylor α-method (Hilber *et al.*, 1977) is used, together with a predictor/multi-corrector algorithm, to integrate the spatially discrete finite element equations. This time integration scheme provides quadratic accuracy and desirable numerical damping characteristic to damp the high frequency spurious modes.

In DYSAC2, stress-strain behavior of the soil skeleton can be described by an isotropic linear elastic model and various bounding surface elastoplastic models. Two constitutive models based on the bounding surface plasticity theory are used in DYSAC2: one for cohesive soils (Dafalias and Herrmann, 1986) and another for non-cohesive soils (Yogachandran, 1991). Only the constitutive model for cohesive soil is used in this chapter and briefly described below. Predictions made by DYSAC2 have been verified using centrifuge model test results (Muraleetharan and Arulanandan, 1991; Arulanandan and Manzari, 1992; Muraleetharan, 1993; Muraleetharan *et al.*, 1994a) and DYSAC2 has been used in the design of waterfront structures subjected to seismic loading (Wittkop, 1993; Muraleetharan *et al.*, 1997a).

Constitutive model

The constitutive model for cohesive soils used in DYSAC2 is based on bounding surface plasticity theory applied to isotropic cohesive soils within the framework of critical state soil mechanics. The major drawback of classical plasticity theory based models for soils is that it allows no plastic strains to occur inside the yield surface. This feature poses a difficulty in describing the behavior of overconsolidated soils and the undrained cyclic loading of soils within the yield surface. This problem is alleviated in the bounding surface plasticity models by allowing plastic strains to occur for stress points within the bounding surface. These plastic strains are proportional to the proximity of the actual stress point to an image stress point on the surface defined according to a certain mapping rule.

The bounding surface and radial mapping rule used in the current model are illustrated in Figure 17.2. The quantities I and J shown in Figure 17.2 are the standard stress invariants, and α is the Lode angle. The bounding surface consists of three surfaces with continuous tangents at their connecting points:

(1)   An ellipse 1 defined by parameter R.
(2)   A hyperbola defined by the parameter A.
(3)   An ellipse 2 (tension region) defined by the parameter T.

In addition, an elastic nucleus inside the bounding surface, described by the parameter S, is also considered. Within the elastic nucleus the behavior is assumed to be purely elastic. While the plastic strains within the bounding surface (outside the elastic nucleus) during loading are taken into account, unloading behavior is still treated as elastic. An associated flow rule is used to describe the direction of plastic strains.

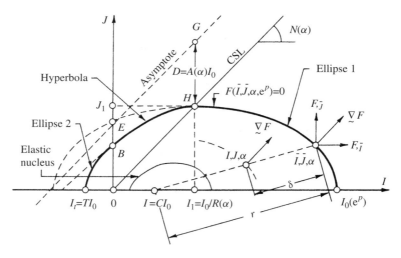

Figure 17.2 Schematic illustration of the bounding surface used for cohesive soils in DYSAC2 (from Dafalias and Herrmann, 1986)

Table 17.1 Properties of Speswhite Kaolin

Property		Value
Liquid limit		69
Plasticity index		31
Specific gravity		2.62
Traditional model parameters		
Slope of isotropic consolidation line on $e - \ln p'$ plot	λ	0.25
Slope of elastic rebound line on $e - \ln p'$ plot	κ	0.05
Slope of critical state line in $q - p'$ space (compression)	M_c	0.88
Shear modulus	G	34 768 kPa
Bounding surface configuration parameters		
Ratio of extension to compression value of M	M_e/M_c	1.0
Value of parameter defining the ellipse 1 in compression	R_c	2.4
Value of parameter defining the hyperbola in compression	A_c	0.01
Parameter defining the ellipse 2 (tension zone)	T	0.01
Projection center parameter	C	0.0
Elastic nucleus parameter	S	1.0
Ratio of triaxial extension to compression value of R	R_e/R_c	0.92
Ratio of triaxial extension to compression value of A	A_e/A_c	1.20
Hardening parameters		
Shape hardening parameter in triaxial compression	h_c	3.0
Ratio of triaxial extension to compression value of h	h_e/h_c	1.0

Notes: e = void ratio; p' and q are stress variables in triaxial space, $p' = (\sigma_1' + 2\sigma_3')/3$, $q = \sigma_1' - \sigma_3'$, σ_1' and σ_3' are the principal effective stresses.

All the significant bounding surface parameters are listed in Table 17.1 (values given are for speswhite kaolin, the soil later used in the example problem). Model parameter values are obtained directly from the laboratory test results and by calibrating the model against the laboratory results. To obtain the values of the parameters without any assumptions seven laboratory tests are required, namely,

(1) A single isotropic or anisotropic (K_o) consolidation or drained compression test with loading and unloading phases.

(2) Consolidated-undrained (preferable) or drained triaxial compression and extension tests (with pore pressure measurements) on specimens in the normally, lightly, and heavily overconsolidated regions (a total of six tests).

If the differences between the values of certain model parameters in triaxial compression and extension are neglected or assumed, only triaxial compression tests are necessary for the model calibration.

With a single set of parameter values, the model simulates the behavior of soils at all degrees of overconsolidation, subjected to either monotonic or cyclic compression and/or extension loading, under either drained or undrained conditions.

17.4   Example Problem and Analyses

The example site and the structure chosen to carry out various analyses are shown in Figure 17.3. The subsurface consists of a 50 m thick speswhite kaolin clay layer underlain by a competent bedrock. The structure is a single storey concrete frame consisting of two 0.6 m diameter, 10 m tall columns and a 2 × 2 m and 15 m long beam. Such a frame is a 2-D representation of a typical highway bridge bent. The two columns are supported on 0.6 m diameter, 35 m long concrete piles. Various properties of the structure and soil are summarized in Table 17.2. Structure, soil, and foundation conditions described in Figure 17.3 and Table 17.2 are not uncommon in the design of bridges (for example, see Abghari and Chai, 1995).

Three types of analyses were carried out. First free field analyses were carried out using SHAKE and DYSAC2. Secondly, the frame was analysed assuming a fixed base and using surface motions obtained from the free field analyses (Elastic Dynamic Analysis, ATC-32). The piles are then analysed using moments and shear forces obtained from the fixed based frame analyses. Finally, a fully coupled analysis of the entire soil-pile-structure system is performed using the computer code DYSAC2.

17.4.1   Free field analyses

Bedrock motions

The input base acceleration-time histories used in the analyses are shown in Figure 17.4. The input motions are East-West and Up-Down components of the 1995 Hyogoken-Nambu earthquake recorded at the Kobe Port Island at a depth of 83 m below ground surface (Iwasaki and Tai, 1996). The response spectra of the base motions are shown in Figure 17.5, from which it can be seen that the predominant periods are 0.36 s and 0.07 s for the horizontal and vertical base motions, respectively.

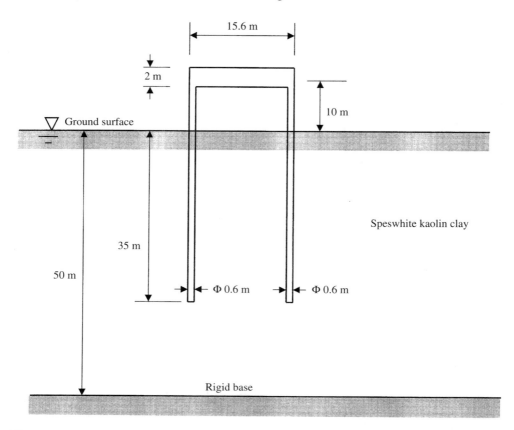

Figure 17.3   Geometry of the site and the structure used in the example soil-pile-structure interaction analyses

Table 17.2   Properties of the structure and the soil

Structure	Soil
Modulus of elasticity = 3.172×10^7 kPa	Void ratio = 1.478
Mass = 2.404 Mg/m^3	Dry unit weight = 10.37 kN/m^3
	Submerged unit weight = 6.41 kN/m^3
	Shear wave velocity = 145 m/s
Beam	Hydraulic conductivity = 1.7×10^{-9} m/s
Cross sectional area, $A = 4$ m^2	Bulk modulus of soil grains and
Second moment of area, $I = 1.333$ m^4	pore fluid = 3.69×10^6 kPa
Columns and Piles	Overconsolidation ratio = 6.0
Cross sectional area, $A = 0.2827$ m^2	Coefficient of earth pressure at rest,
	$K_o = 0.45$
Second moment of area, $I = 6.362 \times 10^{-3}$ m^4	

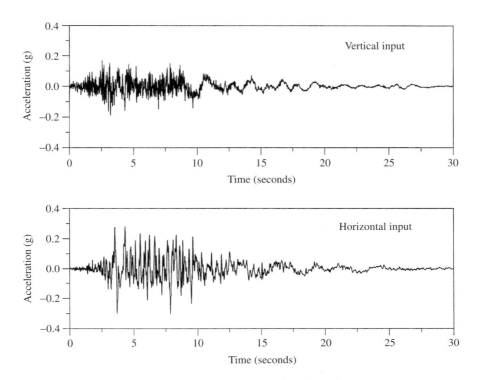

Figure 17.4   Input base acceleration time histories

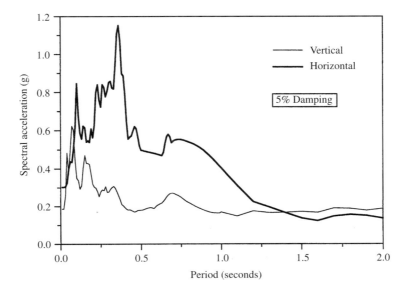

Figure 17.5   Response spectra of base motions

SHAKE analysis

The modulus and damping curves used in the equivalent linear SHAKE analyses are shown in Figure 17.6. These are typical curves for clays suggested by Seed and Idriss (1970). The shear wave velocity of the soil deposit is 145 m/s (Table 17.2). Shear wave velocity of 145 m/s corresponds to a small strain shear modulus (G_{max}) of 34 768 kPa. The shear modulus value of 34 768 kPa was also used in the DYSAC2 analysis to represent the elastic response in the elastoplastic model, making the SHAKE and DYSAC2 analyses consistent. The bedrock was assumed to have a shear wave velocity of 2438 m/s (8000 ft/s), and a unit weight of 23.6 kN/m³. Horizontal base motion was specified at the bed rock level and the surface motion was obtained. Horizontal surface acceleration-time history predicted by SHAKE and the corresponding response spectrum are shown in Figures 17.7 and 17.8, respectively. As can be seen from Figures 17.7 and 17.8, bedrock motion is substantially attenuated by SHAKE. A peak in the surface response spectrum predicted by SHAKE at 0.86 s suggests that, according to the equivalent linear analysis, the clay layer has a fundamental period of 0.86 s.

DYSAC2 analysis

The two-dimensional finite element mesh used in the DYSAC2 free field analysis is shown in Figure 17.9. Some of the key elements and nodes are also indicated in Figure 17.9. A free field analysis with DYSAC2 can also be performed using a column of elements. However, a true 2-D analysis was performed for the dual purposes of obtaining the free field response, as well as to study the influence of the piles and the structure on soil motions. Exactly the same soil mesh was used in the soil-pile-structure interaction analysis. In DYSAC2 analyses, the semi-infinite nature of the soil deposit is simulated by slaving the displacements of the nodes at the left and right boundaries to corresponding

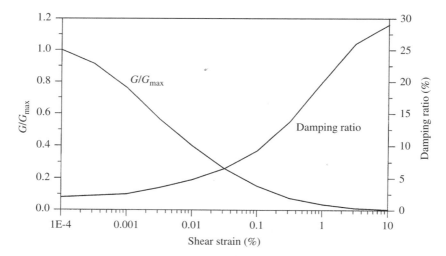

Figure 17.6   Modulus and damping curves used in the SHAKE analysis (after Seed and Idriss, 1970)

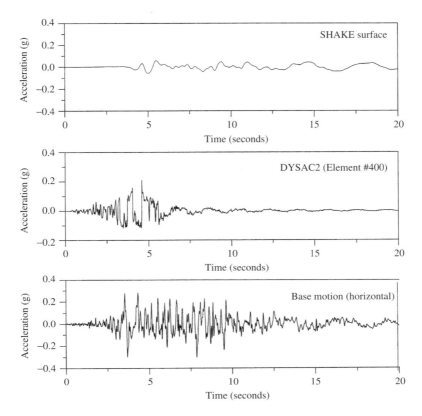

Figure 17.7   Horizontal surface acceleration time histories from SHAKE and DYSAC2

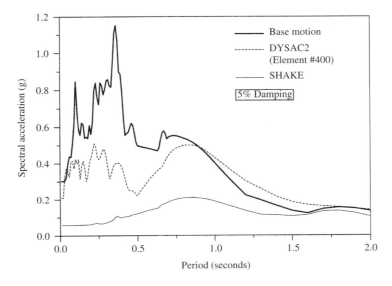

Figure 17.8   Comparison of SHAKE and DYSAC2 horizontal surface acceleration response spectra

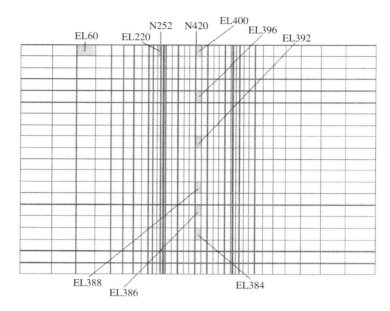

Figure 17.9 Finite element mesh used in the DYSAC2 free field analysis

adjacent nodes. In addition, the horizontal fluid displacements of these columns of nodes are slaved to horizontal solid displacement at the same nodes to restrict the fluid flow to the vertical direction only. No further boundary conditions were specified at the left and right boundaries. In other words, these boundaries were allowed to displace vertically and horizontally. Some results are shown in this section to indicate the effectiveness of such an approach in simulating a horizontal soil deposit.

A DYSAC2 analysis was performed using both the horizontal and vertical input base motions. The base of the DYSAC2 finite element mesh was assumed to be rigid. Surface horizontal acceleration-time history predicted by DYSAC2 at the center of the mesh (Element #400) is shown in Figure 17.7. The response spectrum corresponding to this motion is shown in Figure 17.8. DYSAC2 predicted considerably higher peak surface acceleration (0.21 g) than SHAKE (0.06 g). Furthermore, in addition to producing a peak in the response spectrum around 0.86 s, similar to SHAKE, DYSAC2 also predicted several high frequency peaks with periods less than 0.5 s. One could argue that another modulus and damping curve in SHAKE would have produced a maximum surface acceleration closer to that of DYSAC2. However, it is impossible for the equivalent linear SHAKE to match the frequency response of a fully coupled, elastoplastic, DYSAC2 analysis.

Excess pore water pressure time histories predicted by DYSAC2 along a vertical column of elements are shown in Figure 17.10. Pore water pressure responses along other columns are similar to the one shown in Figure 17.10. Maximum excess pore water pressures of 20–25% of the initial effective vertical stresses were produced throughout the clay layer during shaking. Pore water pressures produced during shaking can be seen to redistribute slowly following the earthquake. Again it is possible to model the change in soil behavior resulting from this 20–25% pore pressure increase using some sort of modulus reduction scheme in an equivalent linear model, however, it is nearly impossible to replicate dilation and contraction behavior predicted by the elastoplastic constitutive

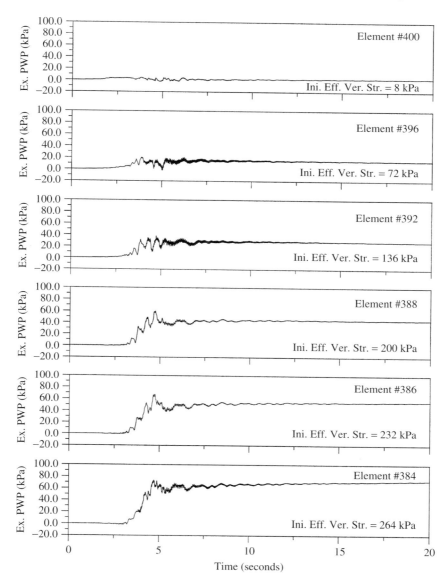

Figure 17.10 Excess pore water pressure (Ex. PWP)-time histories predicted by DYSAC2 without the structure (Ini. Eff. Ver. Str. = Initial effective vertical stress)

model used in DYSAC2 by utilizing an equivalent linear model. Dilation and contraction behavior, as seen in the cyclic pore pressure time histories, is the most likely reason for the higher frequencies seen in the surface response spectra predicted by DYSAC2 (Figure 17.8).

Horizontal and vertical motions along the surface of the soil layer predicted by DYSAC2 are compared in Figures 17.11 and 17.12, respectively, and the corresponding response

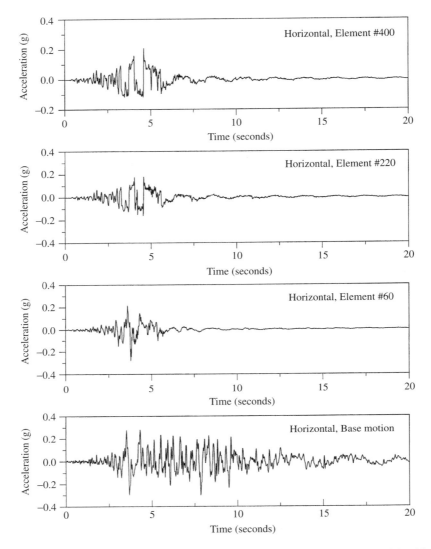

Figure 17.11 Horizontal surface acceleration-time histories predicted by DYSAC2 without the structure

spectra are shown in Figures 17.13 and 17.14, respectively. Although not identical, horizontal and vertical motions at Element Nos. 220 and 400 (in the middle of the finite element mesh) are substantially similar. Horizontal motion at Element #60 deviates from motions at Element Nos. 220 and 400 in the high frequency range (periods less than 0.2 s) and the vertical motion deviates in the low frequency range (periods greater than 0.3 s). These comparisons indicate that the above referenced technique to model a horizontal deposit of soil by slaving nodal degrees of freedom at the left and right boundaries is sufficiently efficient for regions some distance away from the boundaries.

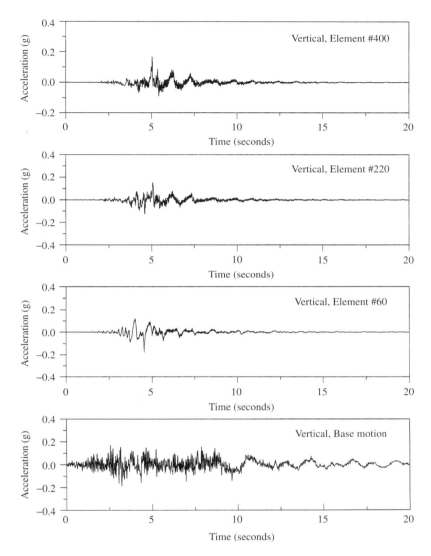

Figure 17.12   Vertical surface acceleration-time histories predicted by DYSAC2 without the structure

17.4.2   Uncoupled analyses of the structure and the foundation

First, an equivalent static analysis of the structure was performed using the surface response spectra obtained through a SHAKE analysis. Secondly, an elastic dynamic analyses of the structure assuming a fixed base was performed using both the SHAKE and the DYSAC2 predicted surface horizontal motions. Maximum shear and moments at the base of the structure obtained in all these analyses were used to study the behavior of the piles using the computer code LPILE.

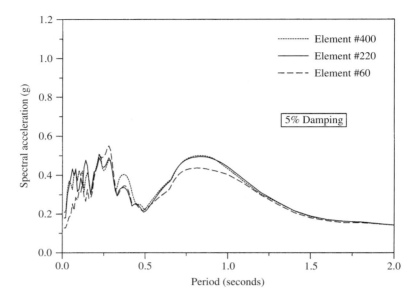

Figure 17.13   Horizontal surface acceleration response spectra from the DYSAC2 analysis without the structure

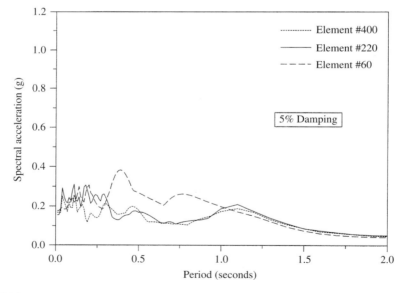

Figure 17.14   Vertical surface acceleration response spectra from the DYSAC2 analysis without the structure

Equivalent static analysis

The first natural period of the structure can be calculated as 1.11 s using lumped mass techniques and assuming the second moment of area (I) of the beam is much larger than that of the columns, the base of the columns are fixed, and neglecting the axial strains. Spectral acceleration for this natural period from the SHAKE surface response spectrum (Figure 17.8) is 0.162 g. A spectral acceleration of 0.162 g and a mass of 151 Mg (mass of the beam plus half the mass of each columns) will result in a base shear of 240 kN or 120 kN per column, and a moment of 600 kN.m at the base of each column. Although ATC 32 specifies that the base shear should be at least 0.4W (593 kN), where W is the weight of the structure, subsequent pile analysis is carried out using pile head shear and moment of 120 kN and 600 kN.m, respectively, for comparison with other analyses.

Elastic dynamic analyses

Elastic dynamic analyses of the structure were carried out using a frame analysis program based on beam elements and consistent mass techniques; however, axial strains in the beams and the columns were still neglected. Bases of the columns were also assumed fixed. Therefore, the three degrees of freedom considered by the frame analysis program were the translation of the beam and two rotations at the beam-column connections. The frame analysis program provided the natural periods of the structure as 1.11 s, 0.06 s, and 0.01 s. Note that substantial energy can be seen in the DYSAC2 predicted surface response spectrum (Figure 17.8) near two of these natural periods (1.11 s and 0.06 s), whereas SHAKE predicts a peak near only the 1.11 s natural period.

Both the SHAKE and DYSAC2 surface horizontal motions shown in Figure 17.7 were used as input base motions in the frame analysis program and shear and moments at the base of the columns were calculated utilizing a 5% modal damping (same in all three modes). All the results including those from the equivalent static analysis are summarized in Table 17.3. As it can be seen from Table 17.3, for the simple structure considered, equivalent static analysis predicts column loads with sufficient accuracy in comparison to an elastic dynamic analysis. It can also be noticed that when DYSAC2 surface motion is used in the elastic dynamic analysis, column loads are twice those corresponding to the SHAKE motion.

Table 17.3   Shear forces and bending moments at the base of a column through fixed base structural analyses

Type of analysis	Column base shear (kN)	Column base moment (kN.m)
Equivalent static w/ SHAKE response spectrum	120	600
Elastic dynamic-SHAKE motion (5% damping)	107	602
Elastic dynamic-DYSAC2 motion (5% damping)	239	1341

Table 17.4   Soil layering and properties used in the LPILE analyses

Soil layer No.	Depth range (m)	Undrained shear strength (kPa)	Soil modulus parameter, k (kPa/m)	Values of strain at 50% of the maximum stress (ε_{50})
1	0 to 1	4.2	8140	0.02
2	1 to 35	4.2 to 128.5	136 000	0.007
3	35 to 50	128.5 to 175.3	271 000	0.005

Table 17.5   Pile head displacements and maximum pile shear forces using uncoupled LPILE analyses

Type of structural analysis used for pile head forces	Pile head displacement (m)	Maximum shear force in the pile (kN)
Equivalent static w/SHAKE response spectra	0.176	201
Elastic dynamic-SHAKE motion (5% damping)	0.161	191
Elastic dynamic-DYSAC2 motion (5% damping)	0.550	400

Pseudo-static pile analyses

The column base loads shown in Table 17.3 were used as pile head loads and pseudo-static analyses of piles were performed using the computer code LPILE. Soil layering and properties used in LPILE analyses are summarized in Table 17.4. Shear strengths of the soil given in Table 17.4 are consistent with the strengths predicted by the bounding surface elastoplastic model used in DYSAC2. Maximum pile head displacements and shear forces predicted by LPILE are listed in Table 17.5. Deformed shapes of the pile are shown in Figure 17.15.

17.4.3   Fully coupled analysis using DYSAC2

A two-dimensional fully coupled finite element analysis of the soil-pile-structure system was performed using DYSAC2. The only type of element available in the current version of DYSAC2 is a four-noded isoparametric element with two solid and two fluid degrees of freedom per node. This type of element was used to model both the soil and the structure, however, as discussed in the next section, the stiffness values of the structural elements were adjusted to represent beam elements. Furthermore, a very low fluid compressibility was used to avoid pore pressure generation within structural elements. Structural elements used for the piles had the same hydraulic conductivity as the clay, and therefore did not impede the pore pressure redistribution around the piles. No interface elements were used in between the structure and the soil and therefore no gapping was considered in the analysis presented here.

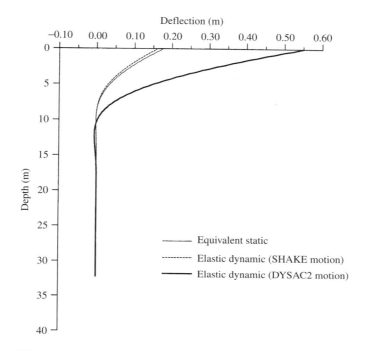

Figure 17.15   Deformed shapes of the pile predicted by LPILE

Calibrating the stiffness values of structural elements

Results from the LPILE analysis and the simple bending theory were used to calibrate the stiffness values of the 2-D isoparametric elements used in DYSAC2 for the piles and the columns, respectively. Displacements obtained from these two analyses were provided to 2-D finite element meshes of piles and columns and the modulus of elasticity was reduced until the shear stresses along the pile or the columns were matched. The static counterpart to DYSAC2, the computer code SAC2 (Herrman and Mish, 1983) was used in these analyses. The meshes used in SAC2 analyses are exactly the same ones later used in the soil-pile-structure interaction analysis utilizing DYSAC2.

Results of these two analyses are presented in Figures 17.16 and 17.17. Note that the LPILE and the simple bending results shown in Figures 17.16 and 17.17, respectively, correspond to the equivalent static analysis described in a previous section. The actual modulus of elasticity of the concrete (3.172×10^7 kPa) was reduced by a factor of 23.4 in SAC2 to obtain the pile response shown in Figure 17.16 and by a factor of 6.42 to obtain the column response shown in Figure 17.17. Several other pertinent properties of the 2-D structural elements used in these SAC2 analyses and subsequent DYSAC2 analyses are summarized in Table 17.6. It can be seen from Figures 17.16 and 17.17 that SAC2 analyses match shear forces obtained from the beam analyses except near the top. This level of matching was considered sufficient to represent the structural behavior.

As a final check of validity of using 2-D continuum element to model beam elements, a dynamic analyses of the structure was performed using DYSAC2 assuming a fixed base

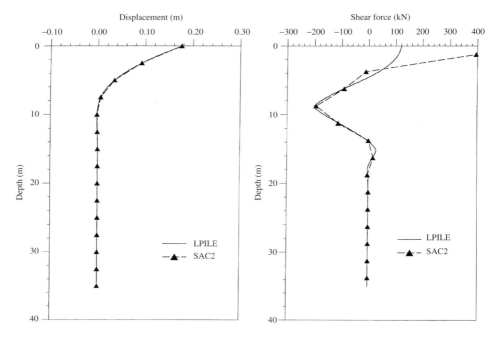

Figure 17.16 Comparison of displacements and shear forces in the pile using LPILE and 2-D continuum finite elements (SAC2)

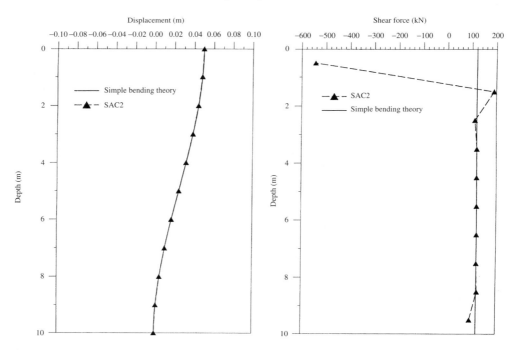

Figure 17.17 Comparison of displacements and shear forces in the columns using the simple bending theory and 2-D continuum finite elements (SAC2)

Table 17.6   Properties of the structural elements used in SAC2 and DYSAC2 analyses

Property	Column	Pile
Young's modulus (kPa)	4.941×10^6	1.358×10^6
Poisson's ratio	0.2	0.2
Dry unit weight (kN/m³)	22.69	22.69
Void ratio	0.1	0.1
Hydraulic conductivity (m/s)	9.8×10^{-12}	1.7×10^{-9}
Bulk modulus of pore fluid (kPa)	2.2	2.2

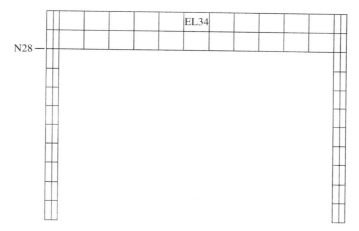

Figure 17.18   Finite element mesh used in the DYSAC2 fixed base analysis

and the results were compared to those from a frame analysis program. No additional structural damping was introduced in this DYSAC2 analysis or the soil-pile-structure analysis described next. The SHAKE surface motion shown in Figure 17.7 is used as the input base motion for all the analyses. The frame analysis program was run with two damping values of 5% and 1%. The finite element mesh used in the DYSAC2 analysis is shown in Figure 17.18. Horizontal displacement and acceleration-time histories of the beam are compared in Figures 17.19 and 17.20, respectively. Response spectra of acceleration-time histories are compared in Figure 17.21. As expected, linear elastic DYSAC2 analysis behaves in an undamped mode. In terms of the maximum quantities, it can be considered that the damping present in 2-D continuum elements is between 1% and 5%. DYSAC2 predicted a softer response by vibrating at slightly larger first natural period of 1.3 s (Figure 17.21) compared to 1.11 s predicted by the frame analysis program. Maximum shear forces in the columns were only 36 kN as compared to 107 kN obtained through the elastic dynamic analysis (Table 17.3). It turned out that the main cause of this softer response was the accidental use of a reduced modulus of elasticity value (same as that of the columns, Table 17.6) for the beam. When the DYSAC2 analysis was repeated using the actual modulus of elasticity of concrete (3.172×10^7 kPa) for the beam, the natural period became 1.2 s and the maximum column shear force became 73 kN. Still somewhat softer response of the structure could be due to the differences

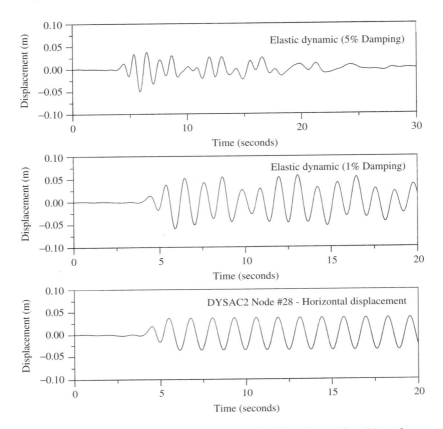

Figure 17.19 Comparison of horizontal displacement-time histories predicted by a frame analysis
program and DYSAC2

in static and dynamic stiffness values of the 2-D continuum elements. Note that the
stiffness values were calibrated under static conditions. The soil-pile-structure analysis
reported below was not repeated with the actual modulus of elasticity for the beam and
therefore the structure used in the analysis is softer than the actual structure. However, the
conclusions made regarding the influence of the structure on the overall behavior should
still be valid.

Soil-pile-structure analysis

The finite element mesh used in the fully coupled soil-pile-structure interaction analysis is
shown in Figure 17.22. Certain key elements and nodes are also indicated in Figure 17.22.
Horizontal acceleration-time histories predicted by DYSAC2 along the surface of the
soil during the soil-pile-structure interaction analysis are compared in Figure 17.23. The
response spectra of these motions are shown together with response spectra of horizontal motions from exactly the same locations but without the structure (Figure 17.11)
in Figures 17.24 through 17.26. Vertical acceleration-time histories along the surface of
the soil are shown in Figure 17.27 and the response spectra are compared in Figures 17.28

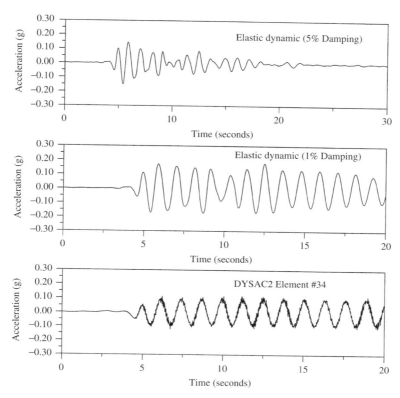

Figure 17.20   Comparison of horizontal acceleration-time histories predicted by a frame analysis program and DYSAC2

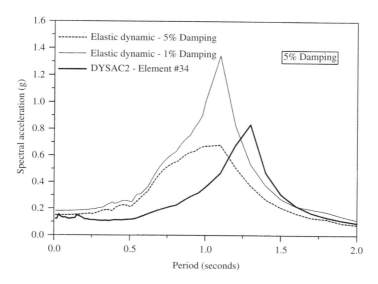

Figure 17.21   Comparison of horizontal acceleration response spectra predicted by a frame analysis program and DYSAC2

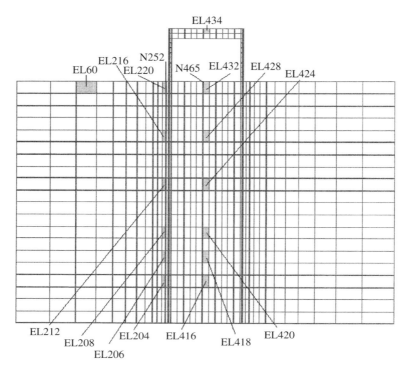

Figure 17.22   Finite element mesh used in the DYSAC2 soil-pile-structure interaction analysis

through 17.30. At some distance from the structure (Element #60) the main effect of the structure is to reduce the vertical acceleration felt by the soil. Since this element is close to the boundary, it is hard to say whether this is a true effect of the structure or an artifact of the boundary. Near the columns (Element #220) the main effect of the structure is to add significant high frequency components (periods less than 0.1 s) to both the horizontal and vertical motions. Stiffening of the soil caused by the presence of the structure is the most likely reason for these high frequency components. In between the columns (Element #400 or #432) the main effect of the structure is to reduce the horizontal motions of the soil and to increase the vertical motions. Again the stiffening effect of soil caused by the piles will force the soil in between the structure to undergo more rocking motions in unison with the structure. These comparisons of acceleration-time histories show that the presence of even a simple structure, such as that considered here, can substantially change the free field soil surface motion casting doubt on the practice of using free field motions to analyze a structure in an uncoupled manner.

Pore pressure response of soils predicted during the soil-pile-structure analysis along two columns of elements, one near the piles and another in between the piles, are shown in Figures 17.31 and 17.32, respectively. Comparing these figures with Figure 17.10, it can be seen that substantial dilation is predicted within the depth of embedment of piles (0–35 m depth) during shaking. This dilation is more prominent near the piles than in between the piles. It can also been seen that following shaking pore pressures are

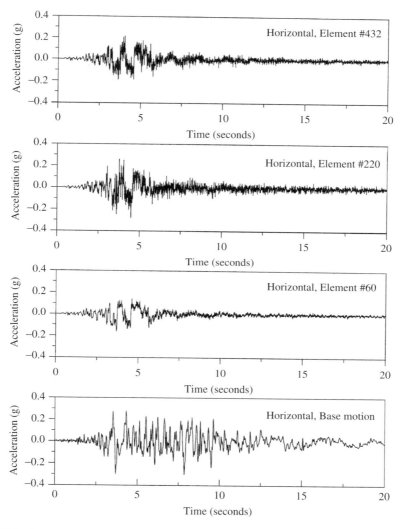

Figure 17.23 Horizontal soil surface acceleration-time histories predicted by DYSAC2 with the structure

redistributing towards equilibrium values similar to those seen without the structure. The most likely reason for the dilation is the pull of piles on soil elements. If gapping between soil and the piles is allowed, this dilation might be less, however, it would still be present in clayey soils such as the one considered here. The dilation of soil will also strengthen the soil, and would have partly contributed to the high frequency response of soil near the piles.

Horizontal and vertical displacement-time histories predicted by DYSAC2 near the structure are compared in Figures 17.33 and 17.34, respectively. It can be seen from these figures that, while the horizontal displacements are unchanged by the presence of the structure, vertical movement of soil near the structure is substantially reduced.

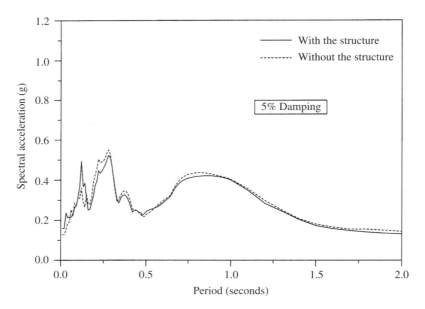

Figure 17.24   Comparison of horizontal soil motions at Element #60 with and without the structure

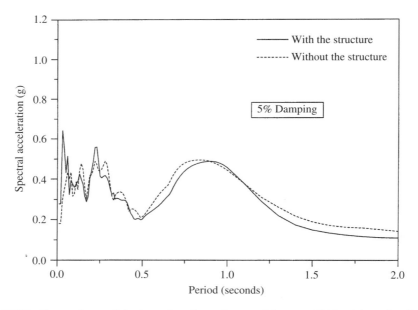

Figure 17.25   Comparison of horizontal soil motions at Element #220 with and without the structure

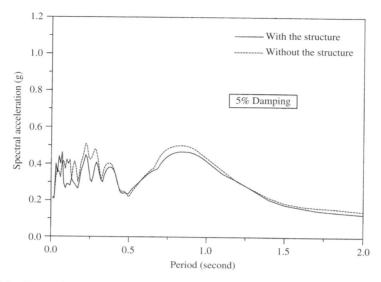

Figure 17.26   Comparison of horizontal soil motions at Element #400 (without the structure) and Element #432 (with the structure)

Deformed shape of the piles and columns are shown in Figure 17.35, at $t = 5.30$ s, when the left pile head displacement reaches its maximum value of 0.113 m. At this instance the maximum shear force in the left pile is 24 kN. These values are substantially lower than those from the elastic dynamic analyses shown in Figure 17.15 and Table 17.5 (DYSAC2 motion), again likely due to a softer structure considered in the analysis. To further investigate the causes, an additional LPILE analysis was performed. A pile head shear force of 36 kN (maximum column shear from the DYSAC2 fixed base structure analysis) and maximum horizontal soil displacements near the piles obtained from the DYSAC2 analysis without the piles were input into LPILE and the behavior of the pile was analyzed. The deformed shape of the pile predicted by LPILE is also shown in Figure 17.35. LPILE predicted a maximum pile shear of 56 kN. The LPILE deformation shape shown in Figure 17.35 is more similar to the DYSAC2 predicted deformation shape than what is shown in Figure 17.15. Therefore it appears that the soil movement might be the main mechanism of pile loading when the shear transmitted through pile head is small.

Acceleration-time histories of the beam predicted by DYSAC2 during the soil-pile-structure interaction analysis are shown in Figure 17.36 and the response spectra are compared in Figures 17.37 and 17.38. It can be seen from Figures 17.37 and 17.38 that substantial amplification of vertical base motion and amplification of horizontal motions in the high frequency range (periods less than 0.09 s) are predicted by DYSAC2 during the soil-pile-structure interaction analysis. For comparison, also shown in Figure 17.36 is the horizontal acceleration-time history of the beam obtained by a frame analysis program (elastic dynamic analysis) with 1% damping and using horizontal surface motion predicted by DYSAC2 (Fig. 17.11, Element 400). In Figure 17.39, horizontal acceleration response spectra of elastic dynamic analysis and soil-pile-structure DYSAC2 analysis are compared. Although the frame analysis program predicted the maximum beam acceleration of about 0.4 g, the frequency response is completely different from the DYSAC2

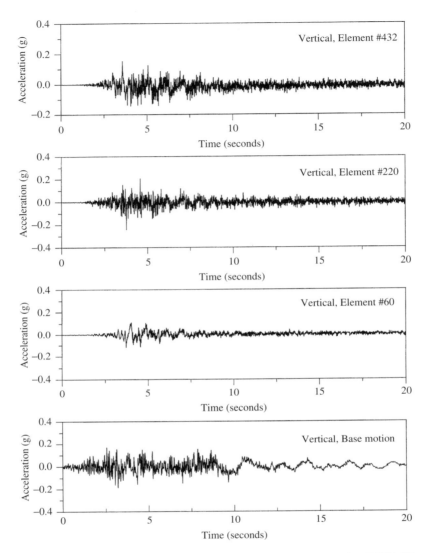

Figure 17.27   Vertical surface soil acceleration-time histories predicted by DYSAC2 with the structure

analysis. Only a fully coupled analysis can produce the complex frequency response resulting from various interactions taking place during a seismic excitation of a soil-pile-structure system.

17.5   Conclusions and Recommendations

Based on the analyses performed for the example problem, following general conclusion can be made:

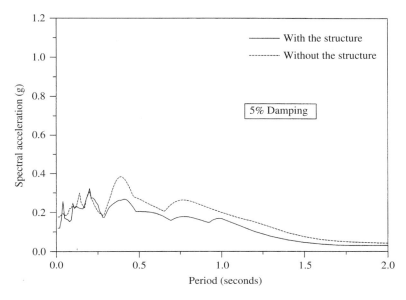

Figure 17.28   Comparison of vertical soil motions at Element #60 with and without the structure

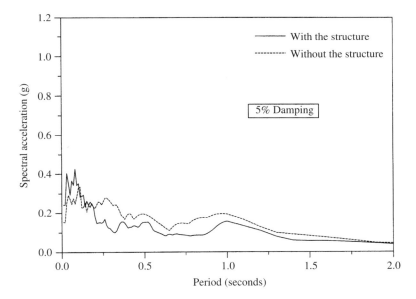

Figure 17.29   Comparison of vertical soil motions at Element #220 with and without the structure

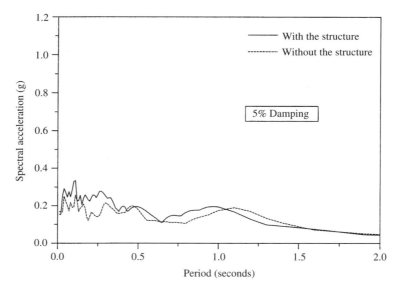

Figure 17.30 Comparison of vertical soil motions at Element #400 (without the structure) and Element #432 (with the structure)

- In a free field analysis, although equivalent linear analysis such as SHAKE might be capable of producing maximum surface accelerations correctly, it is unlikely to predict the frequency response of the surface motion. Only a detailed description of the constitutive behavior of soil, such as through an elastoplastic model, is likely to capture the frequency response correctly.

- Presence of a structure can substantially alter the soil motions. In general, presence of a structure will add high frequency components to soil motions near the columns/piles and reduce horizontal motions and increase vertical motions in between the columns. These effects are mainly related to stiffening of the soil caused by the structure.

- Since free field motions are likely to be altered by the structure, these motions should be used with extreme caution in structural analyses.

- Only a fully coupled analysis is capable of capturing the complex interactions occurring in a soil-pile-structure system during seismic events and the resulting complex frequency response and vertical vibrations. It is also important that the fully coupled analysis consider pore water pressure generation and dissipation in soils during and after shaking.

Therefore, it is recommended that, whenever possible, practicing engineers should use a fully coupled analysis for the seismic design of soil-pile-structure systems. If it is not feasible to perform detailed laboratory testing to calibrate an elastoplastic constitutive

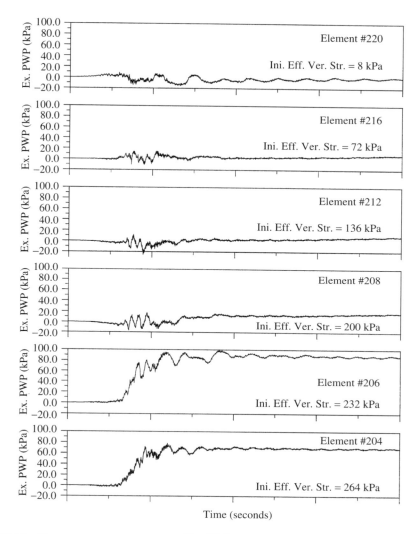

Figure 17.31 Excess pore water pressure (Ex. PWP)-time histories predicted by DYSAC2 near the piles (Ini. Eff. Ver. Str. = Initial effective vertical stress)

model such as that presented in this chapter, it is still better to perform a fully coupled analysis using a simple elastoplastic constitutive model, such as a Mohr–Coulomb type model, rather than performing an uncoupled analysis as in the current practice. Engineers are more likely to have better insight into the behavior of a structure through a fully coupled analysis that includes pore pressure generation dissipation than any other analysis.

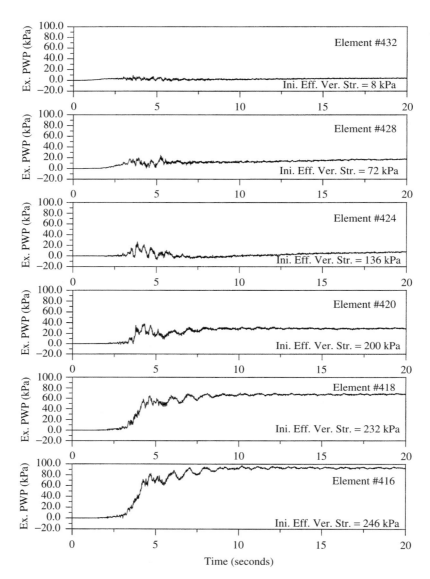

Figure 17.32 Excess pore water pressure (Ex. PWP)-time histories predicted by DYSAC2 in between the piles (Ini. Eff. Ver. Str. = Initial effective vertical stress)

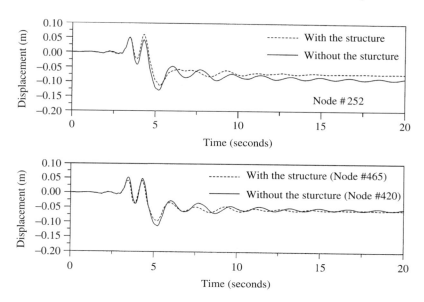

Figure 17.33 Comparison of horizontal displacement-time histories predicted by DYSAC2 with and without the structure

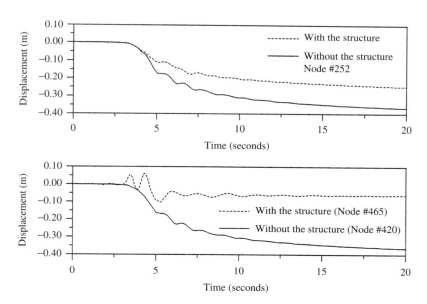

Figure 17.34 Comparison of vertical displacement-time histories predicted by DYSAC2 with and without the structure

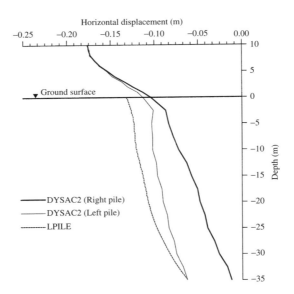

Figure 17.35   Deformed shape of the piles and the columns at $t = 5.30$ s predicted by DYSAC2 in comparison with LPILE predictions when the soil movement was considered

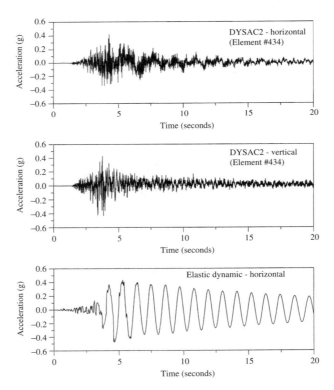

Figure 17.36   Acceleration-time histories of the beam predicted by DYSAC2 and a frame analysis program

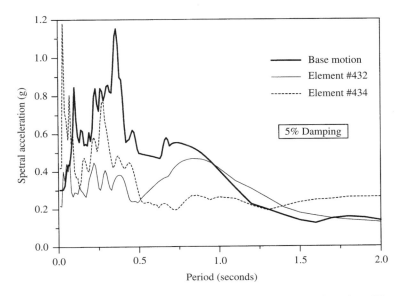

Figure 17.37 Horizontal acceleration response spectra of the base, the soil surface (Element #432), and the beam (Element #434) predicted by DYSAC2

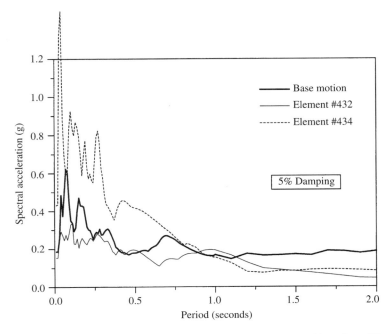

Figure 17.38 Vertical acceleration response spectra of the base, the soil surface (Element #432), and the beam (Element #434) predicted by DYSAC2

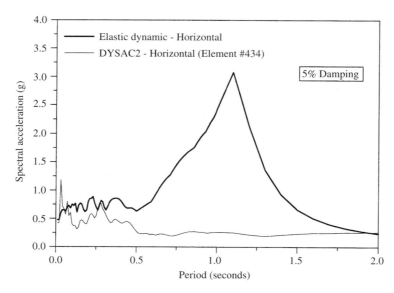

Figure 17.39 Comparison of horizontal acceleration response spectra of the beam predicted by a frame analysis program and DYSAC2

References

Abghari, A. and Chai, J. Modeling of soil-pile-superstructure interaction for bridge foundations. In: *Performance of Deep Foundations*, Turner, J. P. (ed.), *Geotechnical Special Publication No. 51*, ASCE, New York, (1995) 45–59.

Anandarajah, A. HOPDYNE–A finite element computer program for the analysis of static, dynamic, and earthquake soil and soil-structure systems. *Civil Engineering Report*, The John Hopkins University, Baltimore, MD, (1990).

Anandarajah, A., Rashidi, H. and Arulanandan, K. Elasto-plastic finite element analyses of soil-structure system under earthquake excitations. *Comput. and Geotechnics*, **17**, (1995) 301–235.

Applied Technology Council. *Improved seismic design criteria for California bridges: Provisional recommendations*. ATC-32, Redwood City, CA, (1996).

Arulanandan, K. and Manzari, M. T. Liquefaction analysis of stratified soil layers: A verification of coupled stress-flow approach. *Proceedings 10th World Conference on Earthquake Engineering*, Madrid, Spain, (1992) 1349–1353.

Cai, Y. X., Gould, P. L. and Desai, C. S. Investigation of 3-D nonlinear seismic performance of pile-supported structures. In: *Performance of Deep Foundations*, Turner, J. P. (ed.), *Geotechnical Special Publication No. 51*, ASCE, New York, (1995) 17–31.

Dafalias, Y. F. and Herrmann, L. R. Bounding surface plasticity II: Application to isotropic cohesive soils. *J. Engrg. Mech. Div.*, ASCE, **112**(12), (1986) 1263–1291.

Ensoft, Inc. *LPILE Plus version 3.0m: A Program for the analysis of piles and drilled shafts under lateral load*, User's Manual, Austin, TX, (1997).

Hadjian, A. H., Fallgren, R. B. and Tufenkjian, M. R. Dynamic soil-pile-structure interaction: State-of-practice. In: *Piles Under Dynamic Loads*, Prakash, S. (ed.), *Geotechnical Special Publication No. 34*, ASCE, New York, (1992) 1–26.

Herrmann, L. R. and Mish, K. D. User's manual for SAC2: A two-dimensional, nonlinear, time dependent, soil analysis code using the bounding surface plasticity model. *Technical Report*, Department of Civil Engineering, University of California, Davis, CA, (1983).

Hilber, H. M., Hughes, T. J. R. and Taylor, R. L. Improved numerical dissipation for time integration algorithms in structural dynamics. *Earthquake Engrg. and Structural Dynamics*, **5**, (1977) 283–292.

Imbsen, R. A. SEISAB. *Computer Code*, Imbsen and Associates, Inc., Sacramento, CA, (1993).

Iwasaki, Y. and Tai, M. Strong motion records at Kobe Port Island. *Soils and Foundations*, 'Special Issue on Geotechnical Aspects of the January 17 1995 Hyogoken-Nambu Earthquake', Japanese Geotechnical Society, Tokyo, Japan, (1996) 29–40.

Matlock, H., Foo, S. H. C., Tsai, C. F. and Lam, I. SPASM 8: A dynamic beam-column program for seismic pile analysis with support motions. *Report to Chevron Oil Field Research Co.*, Fugro Inc., Long Beach, CA, (1978).

Muraleetharan, K. K. Numerical prediction for model no. 1. *Proceedings, International Conference on the Verification of Numerical Procedures for the Analysis of Soil Liquefaction Problems*, A. A. Balkema, Netherlands, **1**, (1993) 187–196.

Muraleetharan, K. K. and Arulanandan, K. Dynamic behavior of earth dams containing stratified soils. *Proceedings International Conference, Centrifuge 1991*, A. A. Balkema, Netherlands, (1991) 401–408.

Muraleetharan, K. K., Arulmoli, K., Jagannath, S. V., Wittkop, R. C. and Foxworthy, J. E. Validation of a computer code for the analysis of dike retaining structures. *Proceedings International Conference Centrifuge 1994*, A. A. Balkema, Netherlands, (1994a) 203–208.

Muraleetharan, K. K., Mish, K. D. and Arulanandan, K. A fully-coupled non-linear dynamic analysis procedure and its verification using centrifuge test results. *Int. J. Num. Anal. Methods in Geomechanics*, **18**, (1994b) 305–325.

Muraleetharan, K. K., Mish, K. D., Yogachandran, C. and Arulanandan, K. DYSAC2 (Version 1.0): Dynamic soil analysis code for 2-dimensional problems. *Computer Code*, Department of Civil Engineering, University of California, Davis, CA, (1988).

Muraleetharan, K. K., Arulmoli, K., Wittkop, R. C. and Foxworthy, J. E. Use of centrifuge and numerical modeling in the design of Port of Los Angeles' Pier 400. *Transportation Research Record No. 1582*, Transportation Research Board, National Research Council, (1997a) 13–21.

Muraleetharan, K. K., Mish, K. D., Yogachandran, C. and Arulanandan, K. User's manual for DYSAC2 (Version 7.0): Dynamic soil analysis code for 2-dimensional problems. *Technical Report*, School of Civil Engineering and Environmental Science, University of Oklahoma, Norman, OK, (1997b).

Prevost, J. H. DYNAFLOW: A nonlinear transient finite element analysis program. *Report No. 81-SM-1*, Department of Civil Engineering, Princeton University, Princeton, NJ, (1987).

Schnabel, P. B., Lysmer, J. and Seed, H. B. SHAKE: A computer program for earthquake response analysis of horizontally layered sites. *Report No. UCB/EERC-72/12*, Earthquake Engineering Research Center, University of California, Berkeley, CA, (1972).

Seed, H. B. and Idriss, I. M. Soil moduli and damping factors for dynamic response analysis. *Report No. EERC 70-10*, Earthquake Engineering Research Center, University of California, Berkeley, CA, (1970).

Wittkop, R. C. Applications of VELACS philosophy to Port of Los Angeles Pier 400 project. *Proceedings International Conference on the Verification of Numerical Procedures for the Analysis of Soil Liquefaction Problems*, A. A. Balkema, Netherlands, **2**, (1993) 1647–1655.

Yogachandran, C. Numerical and centrifuge modeling of seismically induced flow failures. *PhD Dissertation*, University of California, Davis, CA, (1991).

Zechlin, E. T. and Chai, J. Nonlinear dynamic analysis of large diameter pile foundations for the bay bridge. *Proceedings (Geotechnical Special Publication No. 75), Specialty Conference, Geotechnical Earthquake Engineering and Soil Dynamics III*, Geo-Institute, ASCE, (1998) 1223–1234.

18

Dynamics of Rigid Pavements Including Vehicle-Pavement Interaction Effects

Musharraf Zaman
University of Oklahoma, Norman, OK, USA

18.1   Introduction

Study of the dynamic response of pavements due to moving loads such as vehicles and aircraft has received significant attention in recent years because of its relevance to the reliable design of pavements and airport runways. Although the importance of more accurate dynamic analyses of pavements has been realized, analytical solutions for such problems are available only for some simple cases partly due to the mathematical complexities involved. Most of the available analytical solutions represent the pavement by an infinitely long beam or plate, and neglect the dynamic interaction between the moving loads and the pavement (e.g. see Achenbach and Sun, 1965; Kenney, 1954; Thompson, 1963). Some of these limitations of the analytical solutions have been largely overcome by the development of high-speed computers and efficient numerical techniques such as the Finite Element Method (FEM). By using the Finite Element (FE) technique, concrete pavements can be efficiently and accurately modeled including the dynamic interaction between moving loads and the pavement and a realistic representation of the suspension system.

The moving load problems were first encountered in the design of railroad bridges. By assuming that the bridge can be modeled by simply supported beams, several researchers (e.g. see Ayre *et al.*, 1951) solved this problem analytically. Yoshida and Weaver (1971) presented a numerical solution technique based on the FEM to analyse dynamic response of bridges to moving loads. They used both beam and plate elements separately to model the bridge. Subsequently, Taheri and Ting (1990) presented solution algorithms for bridges based on the structural impedance and finite element methods. In their study, a simplified spring-dashpot suspension system was used to represent the moving vehicles.

Analysis of continuously supported beams and plates subjected to moving loads has been addressed by many researchers during the past few decades. In most cases, solutions were obtained by neglecting interaction effects due to moving loads. For example,

Kenney (1954) presented an analytical steady state solution for an infinite beam resting on a Winkler foundation due to a moving force. The classical (thin) beam theory was used in this analysis. Crandall (1957) analysed the same beam problem utilizing the Timoshenko beam theory and moving force approach. The foundation damping was not included. Achenbach and Sun (1965) extended this analysis by including the foundation damping. Thompson (1963) carried out a similar investigation assuming the pavement as an infinitely long thin plate.

The analytical solutions discussed above cannot be effectively applied to pavement design, since they are obtained for an infinitely long beam or plate and neglect the dynamic vehicle-pavement interaction. Ledesma (1988) and Taheri (1986) reported the results of a more comprehensive study involving the dynamic analysis of concrete pavements subjected to moving vehicular and aircraft loads, including vehicle-pavement interaction. These studies were based on the finite element method, and used thin plate models to idealize the pavement. The foundation was modeled by discrete viscoelastic springs. The thin plate model does not consider the effect of transverse shear deformations that may be important for pavements, particularly for airport pavements. Further, the discrete viscoelastic spring model is not very realistic because the actual soil medium is continuous.

This chapter presents an improved solution algorithm based on the Finite Element Method (FEM) to analyse rigid pavements under moving vehicular or aircraft loads. In the FEM algorithm, the concrete pavement is discretized by thick plate elements that account for the transverse shear deformation and bending. The underlying soil medium is modeled by continuous linear elastic spring and dashpot systems. The dynamic interaction between the moving load and the pavement is considered by modeling the vehicle or the aircraft suspension system by a simplified spring-dashpot unit (Figure 18.1). Graphical results are presented to demonstrate the significance of dynamic interaction between the pavement and the vehicle/aircraft on pavement response.

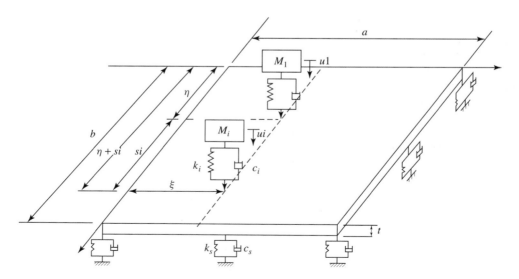

Figure 18.1   Vehicle/aircraft-pavement model

18.2 Finite Element Model

The finite element model adopted for pavement is based on the classical theory of thick plates resting on a continuous Winkler foundation that accounts for the transverse shear deformation of the plate. The formulation is based on the assumptions that deflections are small compared with the thickness of the plate, and that a normal to the middle surface of the undeformed plate remains straight, but not necessarily normal to the middle surface of the deformed plate. Also, the stresses normal to the middle surface are considered negligible. The words 'vehicle' and 'aircraft' are used interchangeably, because the same methodology can be used to model the vehicle-pavement and aircraft-pavement interaction effects.

Four noded, isoparametric rectangular elements with three degrees of freedom per node (namely, the transverse displacement w, rotation about x-axis θ_x, and rotation about y-axis θ_y) are considered in the development of the finite element formulation. Independent bilinear shape functions are assumed for displacement and rotational degrees of freedom.

Following Bathe (1982), the strain energy of an isotropic, linear elastic plate including transverse shear deformation effects can be expressed in the form

$$U(w, \theta_x, \theta_y) = \frac{1}{2} \int_A [H^T][C_b][H] \, dA + \frac{1}{2} \int_A [\gamma]^T [C_s][\gamma] \, dA - \int_A wq(x, y) \, dA \quad (18.1)$$

$$[H] = \begin{bmatrix} \dfrac{\partial \theta_x}{\partial x} \\[2mm] \dfrac{-\partial \theta_y}{\partial y} \\[2mm] \dfrac{\partial \theta_x}{\partial y} - \dfrac{\partial \theta_y}{\partial x} \end{bmatrix} \quad (18.2a)$$

$$[\gamma] = \begin{bmatrix} \dfrac{\partial w}{\partial y} - \theta_y \\[2mm] \dfrac{\partial w}{\partial x} + \theta_x \end{bmatrix} \quad (18.2b)$$

$$[C_b] = \frac{Et^3}{12(1 - v^2)} \begin{bmatrix} 1 & v & 0 \\ v & 1 & 0 \\ 0 & 0 & \dfrac{1-v}{2} \end{bmatrix} \quad (18.3a)$$

$$[c_S] = \frac{Etk}{2(1 + v)} \begin{bmatrix} 1 & 0 \\ 0 & 1 \end{bmatrix} \quad (18.3b)$$

In the above equations, q is the intensity of the load per unit area, E is the Young's modulus, v is the Poisson's ratio, k is the shear correction factor, t is the plate thickness and A is the element area. The first integral in Eq. (18.1) represents the bending energy, and the second represents the shear energy. Using independent shape functions N_i, the nodal variables w, θ_x and θ_y can be written in matrix form as

$$w(x, y) = [N]_w[d] \quad (18.4a)$$

$$\theta_x(x, y) = [N]_{\theta_x}[d] \quad (18.4b)$$

$$\theta_y(x, y) = [N]_{\theta_y}[d] \quad (18.4c)$$

where

$$[N]_w = [N_1\,0\,0\,N_2\,0\,0 \ldots N_n\,0\,0] \tag{18.5a}$$

$$[N]_{\theta_x} = [0\,N_1\,0\,0\,N_2 \ldots 0\,N_n\,0] \tag{18.5b}$$

$$[N]_{\theta_y} = [0\,0\,N_1\,0\,0\,N_2 \ldots 0\,0\,N_n] \tag{18.5c}$$

$$[d] = [w_1\,\theta_{x_1}\,\theta_{y_1} \ldots w_n\,\theta_{x_n}\,\theta_{y_n}] \tag{18.5d}$$

and n represents the number of nodes of an element. Substituting Eqs. (18.4a) to (18.4c) into Eqs. (18.2a) and (18.2b) yields

$$[H] = [B]_b[d] \tag{18.6a}$$

$$[\gamma] = [B]_s[d] \tag{18.6b}$$

where

$$[B]_b = \begin{bmatrix} [N]_{\theta_x}^x \\ -[N]_{\theta_y}^y \\ [N]_{\theta_x}^y - [N]_{\theta_y}^x \end{bmatrix} \tag{18.7a}$$

$$[B]_s = \begin{bmatrix} [N]_w^y - [N]_{\theta_y} \\ [N]_w^x + [N]_{\theta_x} \end{bmatrix} \tag{18.7b}$$

and the superscripts x and y in Eqs. (18.7a) and (18.7b) represent the derivatives with respect to x and y, respectively.

In view of Eqs. (18.6) and (18.7), Eq. (18.1) can be expressed in a simplified form as follows:

$$\begin{aligned}
U(d) &= \frac{1}{2} \int_{-1}^{+1} \int_{-1}^{+1} [d]^T [B]_b^T [C_b][B]_b[d]\,|J|\,dr\,ds \\
&+ \frac{1}{2} \int_{-1}^{+1} \int_{-1}^{+1} [d]^T [B]_s^T [C_s][B]_s[d]\,|J|\,dr\,ds \\
&- \int_{-1}^{+1} \int_{-1}^{+1} [d]^T [N]_w^T q\,|J|\,dr\,ds
\end{aligned} \tag{18.8}$$

where $|J|$ denotes the determinant of the so-called Jacobian matrix and r, s represent the local coordinates. The term $U(d)$ in Eq. (18.8) contains the strain energy due to bending and transverse shear deformation of the plate and the potential energy due to the applied external loads. The total strain energy of the pavement-foundation system can be obtained by adding the strain energy U_f of the foundation which supports the pavement. By treating the underlying support system as a Winkler medium (distributed springs), and by assuming full contact between pavement and subgrade, the strain energy due to foundation (U_f) can be written as

$$U_f(w) = \frac{1}{2} \int_A w^T k_s w\,dA \tag{18.9}$$

where A is the surface area of the plate element and k_s is the modulus of subgrade reaction. By substituting Eq. (18.4a) into Eq. (18.9), and performing the desired integrations in the local coordinates, we get

$$U_f(d) = \frac{1}{2} \int_{-1}^{+1} [d]^T [N]_w^T k_s [N]_w [d] \, |J| \, dr \, ds \tag{18.10}$$

The total strain energy U_t of the plate-foundation system is thus given by

$$U_t = U + U_f \tag{18.11}$$

By equating the first variation of Eq. (18.11) to zero, the force-deflection equation for the plate-foundation element can be expressed as

$$[[k]_b + [k]_s + [k]_f][d] = [Q] \tag{18.12}$$

where

$$[k]_b = \text{bending stiffness} = \int_{-1}^{+1} \int_{-1}^{+1} [B]_b^T [C_b][B]_b |J| \, dr \, ds \tag{18.13a}$$

$$[k]_s = \text{shear stiffness} = \int_{-1}^{+1} \int_{-1}^{+1} [B]_s^T [C_s][B]_s |J| \, dr \, ds \tag{18.13b}$$

$$[k]_f = \text{foundation stiffness} = \int_{-1}^{+1} \int_{-1}^{+1} [N]_w^T k_s [N]_w |J| \, dr \, ds \tag{18.13c}$$

$$[Q] = \text{force vector} = \int_{-1}^{+1} \int_{-1}^{+1} [N]_w^T q |J| \, dr \, ds \tag{18.13d}$$

18.2.1 Dynamic interaction between pavement-moving vehicle/aircraft

The dynamic interaction between pavement and a moving vehicle/aircraft is taken into account by treating the vehicle/aircraft and the pavement as an integrated system, as shown in Figure 18.1. In this model, a simplified idealization involving a concentrated load supported by a linear spring-dashpot system is assumed to represent the vehicle or the aircraft. For the idealized system in Figure 18.2, the load $q(x, y)$ acting on the element due to dynamic vehicle-pavement interaction can be expressed as

$$q(x, y) = F_i \delta(x - \xi, y - \eta_i) + mg - m\ddot{w} - c_s \dot{w} \tag{18.14}$$

where F_i is the pavement-vehicle/aircraft interaction force, mg is the plate weight, $m\ddot{w}$ is the plate inertia force, and $c_s \dot{w}$ is the force due to foundation damping. In the above equation, $\delta(x - \xi, y - \eta_i)$ denotes a Dirac-delta function which equals unity at $x = \xi$ and $y = \eta_i$, and zero elsewhere, and the over-dot denotes derivative with respect to time. Applying the d'Alembert principle for the vehicle/aircraft suspension system shown in Figure 18.2 gives

$$M_i \ddot{u}_i(t) + c_i [\dot{u}_i(t) - \dot{w}(\xi, \eta_i)] + k_i [u_i(t) - w(\xi, \eta_i)] = M_i g \tag{18.15}$$

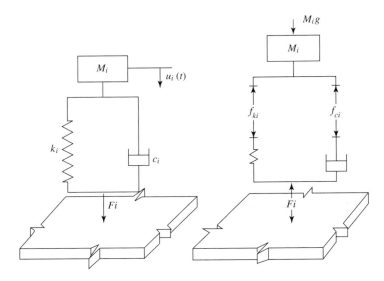

Figure 18.2 Free body diagram of the suspension

where M_i is the vehicle/aircraft mass, u_i is the vertical displacement of the suspended mass, c_i is the damping coefficient of the suspension system, k_i is the spring constant of the suspension system, and (ξ, η_i) denotes the location of the vehicle/aircraft.

In view of Figure 18.2 and Eqs. (18.14) and (18.15), the dynamic interaction force, F_i, can be expressed as

$$F_i = M_i g - M_i \ddot{u}_i \tag{18.16}$$

Substituting Eqs. (18.13d), (18.15) and (18.16) into Eq. (18.12) leads to

$$[[K]_b + [k]_s + [k]_f][d] = \int_{-1}^{+1} \int_{-1}^{+1} [N]^T \{[mg - m\ddot{w} - c_s \dot{w}]$$

$$+ [M_i g - M_i \ddot{u}_i] \delta(x - \xi, y - \eta_i)\} |J| \, dr \, ds \tag{18.17}$$

By expressing the derivatives of plate deflection w in the right-hand side of Eq. (18.17) in terms of nodal variables $[d]$, and by assembling the individual element matrices, the dynamic equilibrium equation for the pavement-subgrade system can be expressed in the form

$$[k][e] = [w] - [M]\frac{d^2}{dt^2}[e] - [c]\frac{d}{dt}[e] + [\tilde{N}(\xi, \eta_i)]^T M_i g - [\tilde{N}(\xi, \eta_i)]^T M_i \ddot{u}_i \tag{18.18}$$

where

$$[k] = \sum \{[k]_b + [k]_s + [k]_f\} \tag{18.19a}$$

$$[W] = \sum \int_{-1}^{+1} \int_{-1}^{+1} [N]^T mg \, |J| \, dr \, ds \tag{18.19b}$$

$$[M] = \sum \int_{-1}^{+1} \int_{-1}^{+1} [N]^T m[N] |J| \, dr \, ds \tag{18.19c}$$

$$[c] = \sum \int_{-1}^{+1} \int_{-1}^{+1} [N]^T c_s[N] |J| \, dr \, ds \tag{18.19d}$$

In Eq. (18.18), $[e]$ is the global displacement vector, and the tilde above $[N]$ denotes that the shape functions are evaluated for a specific element where the vehicle or aircraft masses are located. Also, the sign Σ in the above equations represents the assembly of the individual element matrices. Rearranging Eq. (18.18) yields

$$[M]\frac{d^2[e]}{dt^2} + [c]\frac{d[e]}{dt} + [k][e] = [w] + [\tilde{N}(\xi, \eta_i)]^T M_i g - [\tilde{N}(\xi, \eta_i)]^T M_i \ddot{u}_i \tag{18.20}$$

Similarly, the dynamic equation of equilibrium for the aircraft suspension system given by Eq. (18.15) can be transformed into the following form:

$$M_i \ddot{u}_i + c_i \dot{u}_i + k_i u_i = M_i g + c_i [\tilde{N}(\xi, \eta_i)]\frac{d}{dt}[e] + c_i [\tilde{N}(\xi, \eta_i)]_t[e] + k_i [\tilde{N}(\xi, \eta_i)][e] \tag{18.21}$$

where t in $[\tilde{N}(\xi, \eta_i)]_t$ denotes the derivative with respect to time. Equations (18.20) and (18.21) are the two sets of required coupled equations from which the nodal displacement vector $[e]$ and the vehicle or aircraft displacements u_i can be obtained.

18.2.2 Solution scheme

Before solving the equations of dynamic equilibrium given by Eqs. (18.20) and (18.21), it is convenient to change the time variable t in $[e(t)]$ and $u_i(t)$ by using the position ξ of the vehicle or aircraft mass as a pseudo-time variable as proposed by Taheri *et al.* (1990). The displacements and its derivatives can be expressed in terms of ξ as follows:

$$[e(t)] = [d(\xi)] \tag{18.22a}$$

$$\frac{d}{dt}[e(t)] = \dot{\xi}\frac{d[e(\xi)]}{d\xi} \tag{18.22b}$$

$$\frac{d^2}{dt^2}[e(t)] = (\dot{\xi})^2\frac{d^2}{d\xi^2}[e(\xi)] + \ddot{\xi}\frac{d}{d\xi}e(\xi) \tag{18.22c}$$

Similar expressions can be written for $u_i(t)$, $\dot{u}_i(t)$ and $\ddot{u}_i(t)$ also. The displacements at each time step can be evaluated by applying a direct time integration procedure. The Newmark–Beta integration scheme is used here (Newmark, 1959).

By using the relationships in Eqs. (18.22) and applying the Newmark–Beta method for time integration, Eq. (18.20) is transformed into the following:

$$\left\{ [k] + \left[\frac{1}{h^2\beta}v_m^2 + \frac{\alpha}{h\beta}a_m \right][M] + \frac{\alpha}{h\beta}v_m[c] \right\} e_j + \left(\frac{1}{h^2\beta}v_m^2 + \frac{\alpha}{h\beta}a_m \right)[\tilde{N}(\xi, \eta_i)]^T M_i\{u_i\}_j$$

$$= [\tilde{N}(\xi, \eta_i)]^T M_i \left[v_m^2 \left(\frac{1}{h^2\beta}u_i + \frac{1}{h\beta}\dot{u}_i + \left(\frac{1}{2\beta} - 1 \right)\ddot{u}_i \right) \right.$$

$$+a_m \left(\frac{\alpha}{h\beta} u_i + \left(\frac{\alpha}{\beta} - 1 \right) \dot{u}_i + \frac{h}{2} \left(\frac{\alpha}{\beta} - 2 \right) \ddot{u}_i \right) \Bigg]_{j-1}$$

$$+ v_m^2 [M] \left(\frac{1}{h^2\beta} \underline{e} + \frac{1}{h\beta} \underline{\dot{e}} + \left(\frac{1}{2\beta} - 1 \right) \underline{\ddot{e}} \right)_{j-1}$$

$$+ (a_m[M] + v_m[c]) \left(\frac{\alpha}{h\beta} \underline{e} + \left(\frac{\alpha}{\beta} - 1 \right) \underline{\dot{e}} + \frac{h}{2} \left(\frac{\alpha}{\beta} - 2 \right) \underline{\ddot{e}} \right)_{j-1}$$

$$+ [w] + [\tilde{N}(\xi, \eta_i)]^T M_i g \tag{18.23}$$

Similarly, Eq. (18.21) yields

$$\left(-v_m c_i [\tilde{N}(\xi, \eta_i)]_x - \frac{\alpha}{h\beta} v_m c_i [\tilde{N}(\xi, \eta_i)] - k_i [\tilde{N}(\xi, \eta_i)] \right) \underline{e}_j$$

$$+ \left[k_i \pm \left(\frac{1}{h^2\beta} v_m^2 + \frac{\alpha}{h\beta} a_m \right) M_i + \frac{\alpha}{h\beta} v_m c_i \right] \{u_i\}_j$$

$$= M_i g + v_m^2 M_i \left[\frac{1}{h^2\beta} u_i + \frac{1}{h\beta} \dot{u}_i + \left(\frac{1}{2\beta} - 1 \right) \ddot{u}_i \right]_{j-1}$$

$$+ (a_m M_i + v_m c_i) \left[\frac{\alpha}{h\beta} u_i + \left(\frac{\alpha}{\beta} - 1 \right) \dot{u}_i + \frac{h}{2} \left(\frac{\alpha}{\beta} - 2 \right) \ddot{u}_i \right]_{j-1}$$

$$- c_i v_m [\tilde{N}(\xi, \eta_i)] \left[\frac{\alpha}{h\beta} \underline{e} + \left(\frac{\alpha}{\beta} - 1 \right) \underline{\dot{e}} + \frac{h}{2} \left(\frac{\alpha}{\beta} - 2 \right) \underline{\ddot{e}} \right]_{j-1} \tag{18.24}$$

In Eqs. (18.23) and (18.24), v_m and a_m denote velocity and acceleration of the moving vehicle or aircraft, respectively. α and β are the parameters which determine the accuracy and stability of the time-integration scheme employed. The values of 0.5 and 0.25 are used here for α and β, respectively. The unconditional stability is generally achieved for $\beta \geq 0.25$. For convenience, Eqs. (18.23) and (18.24) can be represented in matrix form as:

$$\begin{bmatrix} [S_{11}][S_{12}] \\ [S_{21}][S_{22}] \end{bmatrix}_j \begin{bmatrix} \underline{e} \\ \underline{u}_i \end{bmatrix}_j = \begin{bmatrix} [R_1] \\ [R_2] \end{bmatrix}_j \tag{18.25}$$

where

$$[S_{11}]_j = [k] + \left(\frac{1}{h^2\beta} v_m^2 + \frac{\alpha}{h\beta} a_m \right) [M] + \frac{\alpha}{h\beta} v_m [c] \tag{18.26a}$$

$$[S_{12}]_j = \left(\frac{1}{h^2\beta} v_m^2 + \frac{\alpha}{h\beta} a_m \right) [\tilde{N}(\xi, \eta_i)]^T M_i \tag{18.26b}$$

$$[S_{21}]_j = -v_m c_i [\tilde{N}(\xi, \eta_i)]_x - \frac{\alpha}{h\beta} v_m c_i [\tilde{N}(x, \eta_i)] - k_i [\tilde{N}(\xi, \eta_i)] \tag{18.26c}$$

$$[S_{22}]_j = \left[k_i + \left(\frac{1}{h^2\beta} v_m^2 + \frac{\alpha}{h\beta} a_m \right) M_i + \frac{\alpha}{h\beta} v_m c_i \right] \tag{18.26d}$$

$$[R_1]_j = [w] + [\tilde{N}(\xi, \eta_i)]^T M_i g$$

$$+ [\tilde{N}(\xi, \eta_i)]^T M_i \left[v_m^2 \left(\frac{1}{h^2\beta} u_i + \frac{1}{h\beta} \dot{u}_i + \left(\frac{1}{2\beta} - 1 \right) \ddot{u}_i \right) \right.$$

$$\left. + a_m \left(\frac{\alpha}{h\beta} u_i + \left(\frac{\alpha}{\beta} - 1 \right) \dot{u}_i + \frac{h}{2} \left(\frac{\alpha}{\beta} - 2 \right) \ddot{u}_i \right) \right]_{j-1}$$

$$+ v_m^2 [M] \left(\frac{1}{h^2\beta} \underline{e} + \frac{1}{h\beta} \underline{\dot{e}} + \left(\frac{1}{2\beta} - 1 \right) \underline{\ddot{e}} \right)_{j-1}$$

$$+ (a_m[M] + v_m[c]) \left(\frac{\alpha}{h\beta} \underline{e} + \left(\frac{\alpha}{\beta} - 1 \right) \underline{\dot{e}} + \frac{h}{2} \left(\frac{\alpha}{\beta} - 2 \right) \underline{\ddot{e}} \right)_{j-1} \quad (18.26e)$$

$$[R_2] = M_i g + v_m^2 M_i \left[\frac{1}{h^2\beta} u_i + \frac{1}{h\beta} \dot{u}_i + \left(\frac{1}{2\beta} - 1 \right) \ddot{u}_i \right]_{j-1}$$

$$+ (a_m M_i + v_m c_i) \left[\frac{\alpha}{h\beta} u_i + \left(\frac{\alpha}{\beta} - 1 \right) \dot{u}_i + \frac{h}{2} \left(\frac{\alpha}{\beta} - 2 \right) \ddot{u}_i \right]_{j-1}$$

$$- c_i v_m [\tilde{N}(\xi, \eta_i)] \left[\frac{\alpha}{h\beta} \underline{e} + \left(\frac{\alpha}{\beta} - 1 \right) \underline{\dot{e}} + \frac{h}{2} \left(\frac{\alpha}{\beta} - 2 \right) \underline{\ddot{e}} \right]_{j-1} \quad (18.26f)$$

It may be noted that the system of equations given by Eq. (18.25) cannot be solved easily because the stiffness matrix $[S]_j$ is unsymmetric, since $[S_{21}]$ is not equal to $[S_{12}]^T$. However, the submatrix $[S_{11}]$ is a banded symmetric matrix which contains most of the elements of $[S]_j$. By taking advantage of the symmetry and bandedness of $[S_{11}]$, the following mixed iterative and direct elimination scheme is used to solve for the unknown displacements $[e]_j$ and $[u_i]_j$.

The matrix Eq. (18.25) can be separated into two matrix equations as follows:

$$[S_{11}]_j \underline{e}_j + [S_{12}]_j [u_i]_j = [R_1]_j \quad (18.27a)$$

$$[S_{21}]_j \underline{e}_j + [S_{22}]_j [u_i]_j = [R_2]_j \quad (18.27b)$$

From the above two equations, the matrix $[u_i]_j$ is eliminated by static condensation, and the resulting single matrix equation is rearranged in the form

$$[S_{11}]_j \underline{e}_j = [R_1]_j - [S_{12}]_j [S_{22}]_j^{-1} [R_2]_j + [S_{12}]_j [S_{22}]_j^{-1} [S_{21}]_j \underline{e}_j \quad (18.28)$$

Equation (18.28) is solved by using an iterative procedure in which the unknown nodal displacement vector $\underline{e}_j$ is first approximated by $\underline{e}_{j-1}$, and then the resulting system of equation is solved by the direct elimination scheme. After computing $\underline{e}_j^1$ from first iteration, it is substituted into the right-hand side of the equation, and a new system of nodal displacement vector $\underline{e}_j^2$ corresponding to the second iteration is computed. This process is repeated until the displacements converge to a specified tolerance. It is observed that, in most of the cases analysed, the displacements converge within five iterations. Once the nodal displacement vector $\underline{e}_j$ is obtained, the vehicle or aircraft displacements u_i can be evaluated from Eq. (18.27b). It should also be noted that the corresponding solution for moving force problems can be obtained by dropping the terms corresponding to the inertia of the moving vehicle or aircraft (Alvappillai *et al.*, 1993).

18.3   Modeling of Jointed Pavements

Since the pavement joints are an integral part of the pavement system, they should be accurately represented in the analysis to obtain a realistic response. Huang and Wang (1973) developed a solution scheme based on the FEM to analyse a two-slab system connected with dowel bars at the joint. The effect of dowel joints was taken into account by specifying the joint efficiency across a joint. In the subsequent development, the imaginary shear transfer (vertical) springs were used along the joint between two adjacent slabs to indirectly account for the load transfer devices. This type of model has been widely used for aggregate interlock and keyed joints (Tabatabaie and Barenberg, 1978; Larralde, 1984) and for dowel joints (Chou, 1981), because of the model's simplicity and more realistic approach than the model specifying the joint load transfer efficiency. In this model, the moment transfer efficiency is assumed to be zero. In an attempt to include the moment transfer ability of pavement joints, Tia *et al.* (1987) developed a model that represents the joint by a series of shear and torsional springs. Looseness of the dowel bars is accounted for by specifying a slip distance such that the shear and moment stiffnesses become fully effective only when the slip distance is overcome. An important disadvantage of this method is that it is difficult to obtain a proper spring constant without adequate field-test data.

For dowel joints, the stiffness of the model spring can be approximately determined based on the analytical study that assumes the dowel bar encased in concrete as a beam resting on a Winkler-type elastic foundation (Friberg, 1938). The modulus of the Winkler foundation is quantified by a parameter known as the modulus of dowel support (K), which characterizes the dowel-concrete interaction effects. The accurate prediction of K may require the experimental investigation of doweled joints. Ioannides and Korovesis (1990, 1992) presented a methodology to estimate K from the load-deflection data.

In other related studies (Tabatabaie and Barenberg, 1978; Tayabji and Colley, 1983; Larralde, 1984), the dowel joints were modeled by deep beam elements in which the effects of shear deformation of the dowel bar were included. The length of the beam element is taken as the width of the joint opening. The relative deformation between the dowel bar and the concrete is represented by a vertical spring that extends between the surrounding concrete and the dowel bar. If the dowel-concrete interaction is neglected, the dowel bar is assumed to be fully embedded into the pavement. In such cases, a large value for the spring stiffness is assumed to ensure no relative deformation between the dowel and the pavement.

Although the above-mentioned studies presented the analysis of discontinuous pavements, they are limited to static cases only. The dynamic interaction between moving vehicles/aircraft and the pavement is neglected for convenience. The majority of the literature available on the dynamic analysis of pavements treats the pavement system as an infinitely long beam/plate (e.g. see Kerr, 1981). Thus, the results cannot be effectively utilized for discontinuous pavements with load transfer devices. In a number of recent studies (Kukreti *et al.*, 1992; Zaman *et al.*, 1991; Taheri *et al.*, 1990) involving the dynamic analysis of rigid pavements to moving vehicle or aircraft loads, some aspects of vehicle/aircraft-pavement-subgrade interaction have been taken into account. For example, Kukreti, Taheri and Ledesma (1992) used a non-conforming thin plate element to model the pavement, and the discontinuities or pavement joints were modeled by vertical spring elements connecting the two nodes forming a joint.

In this section, a procedure based on the FEM is presented to analyse the dynamic response of jointed concrete pavements to moving vehicular or aircraft loads. A special joint element is employed to accurately model the dowel pavement joints, based on the contact theory. This joint element provides an improved numerical model for the pavement dowel joint and considers the effects of dowel looseness and dowel embedment.

In the finite-element idealization (Figure 18.3), the concrete pavement is modeled by four-noded rectangular thin-plate elements, instead of thick plate idealization used in the previous section, and the underlying soil medium is represented by continuous Winkler springs and dashpots. Doweled joints are considered for pavement transverse joints. The dowel bars at the transverse joints are represented by massless plane frame elements. The dowel-pavement interaction effects are represented by employing contact elements between the dowel bar and the pavement. The aggregate interlock or keyways are assumed in the longitudinal joints and are represented by the vertical spring elements. A spring-dashpot system is used to represent the landing gear of the aircraft. It is assumed that the vehicle or aircraft travels along a straight line with a specified initial horizontal velocity and acceleration.

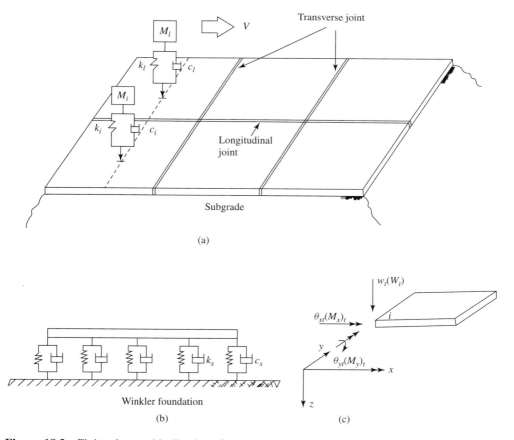

Figure 18.3 Finite-element idealization of pavement components. (a) Aircraft-pavement subgrade system; (b) subgrade idealization; (c) plate element; (d) doweled joint; (e) aggregate interlock and key joint

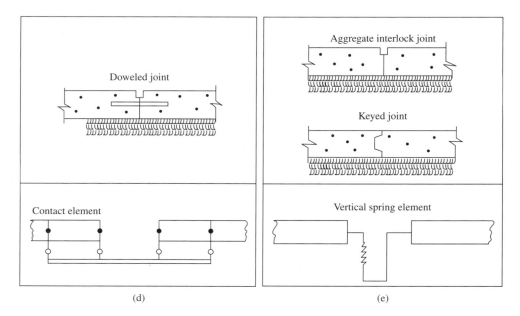

(d)   (e)

Figure 18.3   *(continued)*

18.3.1   *Contact element model for dowel joints*

When a load is applied near the dowel joint of a slab, the loaded slab must first deflect an amount equal to the dowel looseness before it transfers any load to the adjacent slab. In other words, the load transfer occurs only after the slab comes in contact with the dowel. This clearly explains that the pavement-dowel behavior is essentially a contact problem.

A considerable amount of work has been done in the past to develop various analysis techniques to solve contact problems based on the finite-element approach, and no effort is made to review them here. Interested readers can refer to, for example, Francavilla and Zienkiewicz (1975), Sachdeva and Ramakrishnan (1981), Bathe and Chaudhary (1985), Nour-Omid and Wriggers (1986), Chen and Tsai (1986), Kikuchi and Song (1981), Campos *et al.* (1982), Zaman *et al.* (1984), and Desai and Nagaraj (1988).

The contact element employed here is based on the Lagrange-multiplier method. This method is chosen because of its relative simplicity and appropriateness. The constraint conditions for the contact problem are satisfied by introducing Lagrange parameters. Both the nodal variables and the Lagrange parameters are treated as independent unknown variables. The introduction of Lagrange parameters to enforce the constraint condition increases the number of unknowns. Further, the resulting stiffness matrix is indefinite and contains zero diagonal terms, which require special treatment during the solution procedure, such as row interchange during the Gaussian elimination.

Formulation of contact problem

The contact between the pavement and the dowel bar is induced by the moving vehicular or aircraft loads. The algorithm described here considers the friction between the dowel bar and the pavement. Other features of the model are given by Alvappillai (1992) and Zaman and Alvappillai (1995).

In the finite-element discretization, the dowel bar is idealized by the two-noded mass-less plane frame elements that include the effects of shear deformation. The plane frame element has three DOFs per node, namely: vertical deflection (w), rotation about *x-axis* (θ_x), and axial deformation (u). Contact elements are employed at all possible contact points between the pavement and the dowel bar. An additional imaginary node (j) is introduced between the corresponding nodes on the pavement (i) and the dowel (k) (see Figure 18.4). Nodes i and k have DOFs of ($w_i, \theta_{x_i}, \theta_{y_i}$) and ($w_k, \theta_{x_k}, u_k$), respectively. The DOFs corresponding to node j are the pavement-dowel interaction forces. The contact element employed at the free end of the dowel bar includes the following contact forces: the normal (λ_n) and the tangential (λ_h) forces. The node i on the pavement does not have the degree of freedom corresponding to the horizontal displacement u. This is because the pavement is assumed to have a very large axial stiffness and, therefore, a negligible axial deformation. At the fixed end of the bar, an additional force (moment) (λ_{M_x}) is considered, to enforce perfect contact in terms of rotation about the x-axis (Figure 18.4).

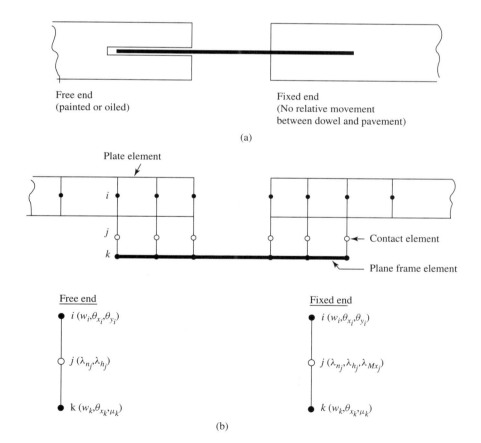

(a)

(b)

Figure 18.4 Doweled pavement joints. (a) Conventional representation; (b) finite-element representation

In the finite-element formulation, the equilibrium state of the pavement system is obtained by minimizing the total potential energy (e.g. see Eqs. (18.11) and (18.12)). Appropriate constrain conditions are added to the potential energy to account for the contact state. The equilibrium for the total system that satisfies the contact conditions can be represented in the following form:

$$
\begin{bmatrix}
[K]_{pav} & [B]_{ij} & 0 \\
[B]_{ji} & [B]_{jj} & [B]_{jk} \\
0 & [B]_{kj} & [K]_{dow}
\end{bmatrix}
\left\{
\begin{array}{c}
d_i^{pav} \\
\lambda_j \\
d_k^{dow}
\end{array}
\right\}
=
\left\{
\begin{array}{c}
\{R\}_i \\
\{\gamma\}_j \\
\{R\}_k
\end{array}
\right\}
\tag{18.29}
$$

where $[K]_{pav}$ = stiffness matrix of the pavement subgrade system, and $[K]_{dow}$ = stiffness matrix of the dowel bar. The matrix $[B]$ in Eq. (18.29) represents the contact kinematics, and its specific terms can be obtained by considering it for different contact conditions (see Alvappillai, 1992).

18.3.2 Modeling of longitudinal joints

The aggregate interlock of keyed joint (Figure 18.3) is modeled by a series of vertical springs joining two adjacent slabs at the joints. The stiffness matrix of a spring element that has two degrees of freedom (vertical deflections w_i and w_j) can be written as follows:

$$
\begin{bmatrix}
k_s & -k_s \\
-k_s & k_s
\end{bmatrix}
\tag{18.30}
$$

where k_s = spring stiffness that may be specified as an input. The moment transfer efficiency of the aggregate interlock or keyed joint is neglected here.

Dynamic vehicle/aircraft-pavement interaction

The dynamic interaction force $q(x, y)$ in this case can be expressed in the form (Zaman et al., 1991)

$$
q(x, y) = (M_i g - M_i \ddot{u}^a)\delta^a(x - \xi, y - \eta) + mg - m\ddot{w} - c_s \dot{w}
\tag{18.31}
$$

where M_i = vehicle or aircraft mass, u = vertical displacement of the suspended mass, mg = plate or slab weight, $m\ddot{w}$ = inertia force due to plate or slab mass, $c_s \dot{w}$ = force due to foundation damping, δ^a = Dirac delta function, and (ξ, η_i) = position of the jth vehicle or aircraft wheel load. The overdots denote derivative with respect to time. Following a procedure similar to that described in section 18.2, the dynamic force-deflection equation for the pavement and the suspended vehicle or aircraft mass can be obtained for this case (see Zaman and Alvappillai (1995) for details). The final equation will be essentially of the same form as Eq. (18.25).

18.3.3 Solution scheme for contact algorithm

The contact algorithm discussed above to represent stick, slip, rebounding conditions between the dowel and the pavement uses an iterative procedure. At first, the stiffness matrix and the force vector [[S] and {R}, as in Eq. (18.25)], are formed by assuming that there is no contact between the dowel bar and the pavement along the dowel joints. Because of the introduction of constraint conditions, {e} in Eq. (18.25) now includes the unknown contact forces ({λ}) and [S₁₁] and {R} are modified accordingly. The sub-matrices $[S_{12}]$, $[S_{21}]$ and $[S_{22}]$ in Eq. (18.25) remain unchanged during contact iterations. The resulting equilibrium equations are solved to obtain the unknown displacement field. If there is no interpenetration of nodes, the computations will continue for the next time step. If there is penetration, however, the appropriate constraint conditions are imposed at the corresponding nodal pairs, and the stiffness matrix and force vector are modified before solving for the next contact iterative step. These iterations are continued until the displacements and the contact forces at all the nodes satisfy the constraint conditions.

After the displacement field and contact forces have converged for a particular time step, the computations are continued for the next time step. The introduction of constraint conditions preserves the bandedness of the stiffness matrix but not the symmetry, all the time. In addition, the stiffness matrix becomes indefinite and has zero diagonal terms. The Gaussian elimination method with partial pivoting, the rows of the coefficient matrix are interchanged whenever a small number in the diagonal terms are identified. This method is capable of solving the system of equations that result from the imposition of contact conditions (Alvappillai, 1992).

18.4   Verification

To verify the accuracy of the finite algorithm presented in the preceding sections, numerical results for a number of problems were compared with the existing analytical and experimental results. Satisfactory correlations were obtained for all cases. Details of verification are discussed by Alvappillai (1992), Alvappillai *et al.* (1993) and Zaman, Taheri and Alvappillai (1991).

18.5   Application to Pavement Without Joints

The finite element model discussed in section 18.2 was used to analyse the dynamic response of pavements including the interaction between pavement and moving vehicle/aircraft. A parametric analysis was conducted to evaluate the effects of selected parameters on the dynamic response of concrete pavements. The results presented in this section are limited to a single pavement slab traversed by a moving suspension unit representing an aircraft wheel load.

The following properties for slab and suspension system are assumed in the analysis:

Slab dimensions: 762×381 cm,

Thickness $(h) = 15.2$ cm,

Young's modulus $(E) = 2.48 \times 10^7$ N/m²,

Poisson's ratio $(v) = 0.15$,

Mass density $= 2325$ kg/m³,

Modulus of subgrade reaction $(k_s) = 81.4\,\text{MN/m}^3$,

Foundation damping ratio $= 5\%$.

Weight $(P) = 343\,\text{kN}$,

Spring stiffness $= 175\,\text{kN/cm}$,

Damping ratio $= 0.5\%$,

Velocity of the moving mass $(v) = 161\,\text{km/hr}$.

The results of this analysis are presented in Figures 18.5–18.10 in a nondimensional form. The effect of slab thickness on the dynamic response of the pavement due to a moving wheel load is presented in Figures 18.5–18.8. The results are obtained for different slab thicknesses ranging from 15–50 cm, while the other parameters are kept constant. Figure 18.5 shows the maximum nondimensional deflection $(\overline{w} = wD/Pl^2)$ at the center of the pavement for cases in which the soil medium is idealized by continuous and discrete springs (note that the discrete spring idealization is a special case of the continuous spring idealization). It can be seen that the continuous soil springs give larger deflections than discrete springs for small slab thickness. The difference in results for the continuous and discrete soil spring decreases with the increase in slab thickness. Figures 18.6–18.8 compare the thick plate, thick plate with reduced integration, and thin plate solutions for static, moving force and moving mass problems. These results are obtained with continuous soil spring model.

Figure 18.9 presents the variation of maximum central deflection for varying subgrade modulus. The solutions presented are obtained by using the thick plate elements with continuous soil springs. As expected, the dynamic response decreases with increasing

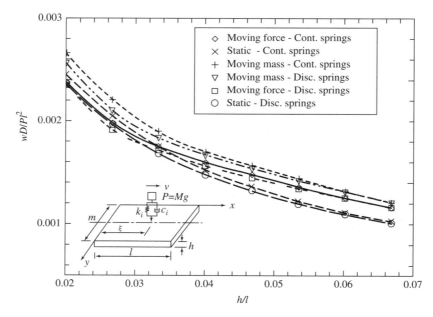

Figure 18.5 Effect of slab thickness on the maximum center deflection – comparison of continuous and discrete foundation models

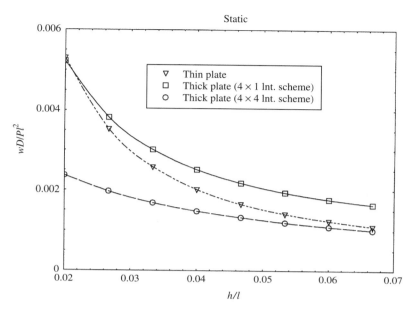

Figure 18.6 Effect of slab thickness on the maximum center deflection – comparison of static thick and thin plate solutions

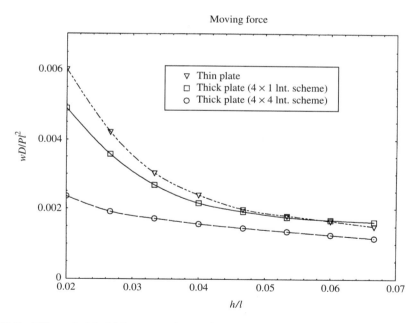

Figure 18.7 Effect of slab thickness on the maximum center deflection – comparison of moving force solutions by thick and thin plate elements

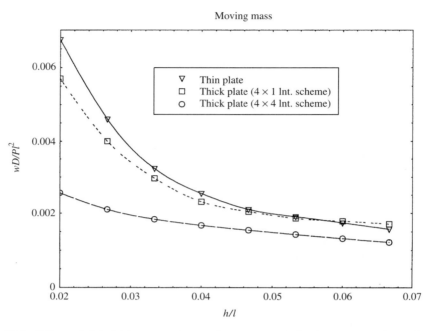

Figure 18.8   Effect of slab thickness on the maximum center deflection – comparison of moving mass solution by thick and thin plate elements

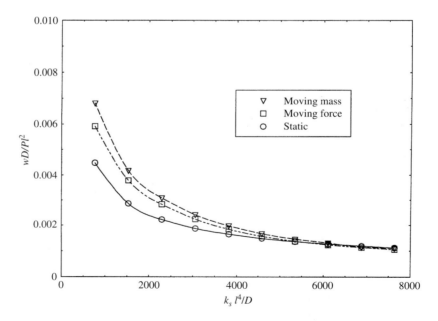

Figure 18.9   Effect of subgrade modulus on the maximum center deflection

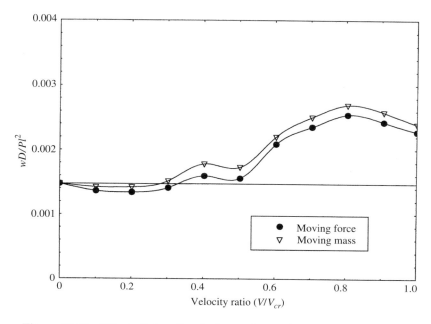

Figure 18.10   Effect of aircraft velocity on the maximum center deflection

subgrade modulus. It is also noted that the static, moving force and moving mass solutions produce the same maximum response for large values of subgrade modulus.

The effect of vehicle velocity ratio (velocity/critical velocity or v/v_{cr}) on the maximum central deflection of the slab is presented in Figure 18.10. It can be seen that the moving force and moving mass solutions yield smaller deflections than the corresponding static solutions for smaller velocity ratios of the moving load. For increasing load velocity ratios, there exist two critical velocity ratios (0.3 and 0.8 in this case) at which the slab deflections attain peak values. However, the first critical response is less important compared with that of the second.

18.6   Application to Jointed Pavements

Some representative results of finite element analyses of a jointed pavement system due to a moving aircraft are presented in this section. The pavement system used for the analysis consists of three discrete slabs in the longitudinal direction and two slabs in the transverse direction, as shown in Figure 18.11. Keyed joint and doweled joints are assumed for the longitudinal and transverse joints, respectively.

The aircraft considered in the analysis is the B-727 model, which has twin landing gear with a maximum gross weight of 751.7 kN (169 kips). It is assumed that the main landing gear carries 90% of the total aircraft weight, and that the remaining 10% is carried by the nose gear. Therefore, each gear in the main twin assembly carries approximately 342.5 kN (77 kips). For the sake of simplicity, only the main landing gear represented by two suspended moving masses is considered in the analysis. Figure 18.11 shows the aircraft-pavement model used in the FE analysis.

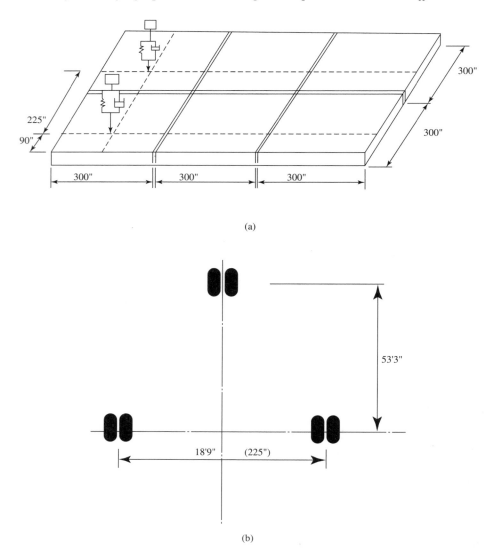

(a)

(b)

Figure 18.11 Aircraft-pavement model assumed for parametric study. (a) Model; (b) aircraft
(B727) wheel arrangement

The following geometric and material properties for the pavement-subgrade system are
assumed, based on the typical values recommended by the Federal Aviation Administration
(1978).

● *Pavement-subgrade properties*
  Six 762 cm (300 in.) by 762 cm (300 in.) slabs with pavement thickness = 30.48 cm
  (12 in.), Young's modulus (E) = 24.9 × 10^6 kPa (3.6 × 10^6 psi), Poisson's ratio (ν) =
  0.15, mass density (ρ) = 237 kg· (s^2/m^4 (0.0002174 lb· (s^2/in.^{4}), modulus of subgrade
  reaction (k) = 815 kPa/cm (300 pci), and subgrade damping = 5%.

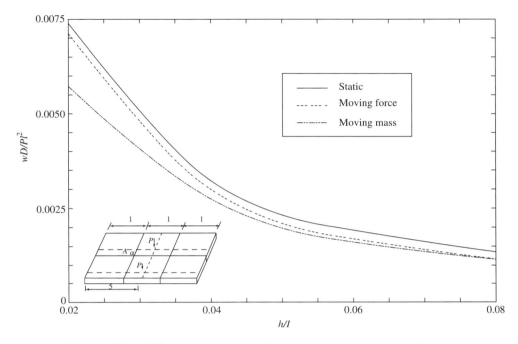

Figure 18.12 Effect of pavement thickness on maximum corner deflection

- *Joint properties*
  Transverse doweled joint: dowel bar diameter $(d) = 2.54$ cm $(1.0$ in.) with spacing $(s) = 38.1$ cm $(15$ in.), dowel looseness $(\gamma) = 0.0127$ mm $(0.0005$ in.), dowel length $= 91.4$ cm $(36$ in.), joint width $= 0.64$ cm $(0.25$ in.), Young's modulus of dowel bar $= 2 \times 10^8$ kPa $(29 \times 10^6$ psi), and shear modulus of dowel bar $= 7.6 \times 10^7$ kPa $(11 \times 10^6$ psi). It is assumed that one end of the dowel bar is fully embedded in the pavement, while the other end is free to move vertically and horizontally.
  Longitudinal joint: the longitudinal keyed joint is represented by a very stiff vertical spring element.

- *Aircraft properties*
  The spring stiffness of the aircraft suspension $= 1.75 \times 10^7$ N/m $(1 \times 10^5$ lb/in.) with damping of aircraft suspension $= 0.5\%$, and aircraft velocity $= 161$ k/h $(100$ mph).

The results of the FE analysis are presented for static, moving force, and moving mass loading conditions in a dimensionless form. The effect of pavement thickness (h/l) on the maximum pavement deflection (wD/Pl^2) is shown in Figure 18.12. As expected, the pavement deflection decreases with an increasing pavement thickness. The moving force and moving mass solutions give less deflection for this particular case in which the aircraft is assumed to travel at a constant speed of 161 k/h $(100$ mph).

Figure 18.13 shows the effect of dowel looseness or initial gap (γ) on the dowel joint efficiency (Alvappillai, 1992). The joint efficiency is observed to decrease (approximately) linearly with an increasing dowel looseness. About 99% of this joint efficiency is achieved when the dowel has no looseness. The joint efficiency decreases to 80%, 72%, and 70% in the cases of static, moving force, and moving mass loading conditions, respectively, when the dowel looseness increases from 0 to 0.001 in. ($\gamma/w = 0.04$, where w is the joint width). This suggests that the dowel-pavement interaction effects due to dowel looseness are more pronounced in cases of dynamic loadings, than in static cases.

The variation of the maximum pavement deflections with dowel looseness is presented in Figure 18.14. As observed, the deflection increases with an increasing dowel looseness in a (approximately) linear manner. When γ/w increases from 0 to 0.04 (dowel looseness from 0 to 0.0254 mm (0.001 in.)), the maximum deflections increase by about 6%.

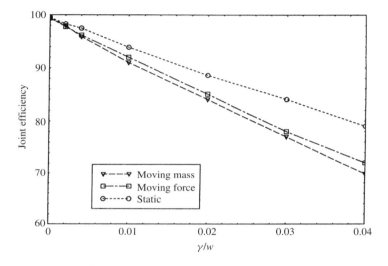

Figure 18.13   Effect of dowel looseness on joint efficiency

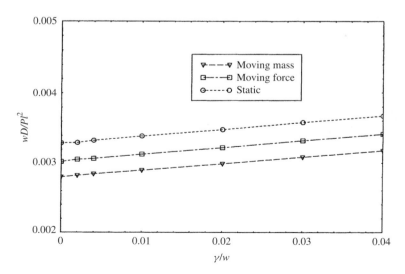

Figure 18.14   Variation of maximum pavement corner deflection with dowel looseness

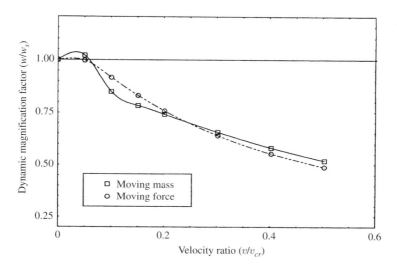

Figure 18.15   Effect of aircraft velocity on maximum corner deflection

The effect of aircraft velocity (v/v_{cr}) on the maximum pavement deflection, represented by the dynamic magnification factor (w/w_s), is depicted in Figure 18.15. The critical aircraft velocity v_{cr}, can be given by the following expression:

$$v_{cr} = \left[\frac{4kD}{(\rho h)^2} \right]^{1/4}$$

(18.32)

where D = plate rigidity = $Eh^3/12(1 - v^2)$, and ρ = plate mass density. The moving mass solution is observed to give a larger deflection than the static solution for small velocity ratios $(v/v_{cr} < 0.05)$. A maximum dynamic magnification factor of about 1.05 is observed at a velocity ratio of 0.05. The moving force and moving mass solutions give smaller deflections than the static solutions for large velocity ratios.

18.7   Conclusions

The dynamic response of concrete pavements subjected to moving loads is discussed with emphasis on the finite element modeling. Modeling of two specific cases is considered: (a) pavements without joints; and (b) jointed pavements. Emphasis is given to modeling the dynamic interaction between the pavement and the moving vehicle or aircraft. The first finite element model uses thick plate elements including the effects of transverse shear deformation to idealize the concrete pavements. The underlying soil medium supporting the pavement is idealized by continuous elastic spring-dashpot system. The dynamic vehicle-pavement interaction effects are included by using an idealized model of the suspension unit. It can be seen from the numerical results that the continuous soil spring model yields larger deflections than that of the discrete soil spring model, particularly for softer subgrades.

The second finite element model can accurately analyse the dynamic response of multiple, jointed concrete pavements to moving vehicular or aircraft loads. This model considers the dowel-pavement interaction effects and includes the effects of dowel loose-ness and dowel embedment on the pavement response.

Numerical results indicate the importance of dowel-pavement interaction effects on the pavement response. Increased pavement deflection is obtained when the dowel-pavement interaction effects are considered. Dowel looseness caused by repeated loads and other factors decreases the joint efficiency of the pavement joints.

It can be seen that the static loading condition is generally more critical in determining the pavement thickness. However, the dynamic analysis may be important in the design of pavement joints.

Acknowledgements

The materials included in this chapter have been contributed by a number of former students and colleagues of the author. The author is particularly indebted to Dr A. Alvappillai, Dr A. R. Kukreti, and Dr M. R. Taheri for their technical contribution. Thanks are also extended to Mr R. H. Ledesma.

References

Achenbach, J. D. and Sun, C. T. Moving load on a flexibly supported Timoshenko beam. *Int. J. Solids Structures*, **1**, (1965) 353–370.

Alvappillai, A. Dynamic analysis of rigid pavements considering pavement-subgrade-aircraft inter-action. *PhD Dissertation*, School of Civil Engineering and Environmental Science, University of Oklahoma, Norman, OK, (1992).

Alvappillai, A., Zaman, M. M. and Taheri, M. R. Dynamic response of concrete pavements resulting in viscoelastic foundation to moving loads. *Euro. J. Mechanics, A: Solids*, **12**, (1993) 73–93.

Ayre, R. S., Jacobson, L. S. and Hsu, C. S. Transverse vibration of one- and two-span beam under the action of a moving mass load. *Proc. 1st US National Congress of Applied Mechanics* (1951) 81–90.

Bathe, K. J. *Finite Element Procedures in Engineering Analysis*. Prentice-Hall, Englewood Cliffs, NJ, (1982).

Bathe, K. J., and Chaudhary, A. A solution method for planar and axisymmetric contact problems. *Int. J. Num. Methods in Engrg.*, **21**(1), (1985) 65–88.

Campos, L. T., Oden, J. T. and Kikuchi, N. A numerical analysis of a class of contact problems with friction in elastostatics. *Comp. Methods Appl. Mech. Engrg.*, **34**, (1982) 821–845.

Chen, W. H. and Tsai, P. Finite element analysis of elastodynamic sliding contact problems with friction. *Computers and Struct.*, **22**(6), (1986) 925–938.

Chou, Y. T. Structural analysis computer programs for rigid multicomponent pavement structures with discontinuities. WESLI-QID and WESLAYER. *Technical Report 1, 2, and 3*, US Army Engrg. Wtrwy. Experiment Station, Vicksburg, MI, (1981).

Crandall, S. H. The Timoshenko beam on an elastic foundation. *Proc. Third Midwestern Conference on Solid Mechanics*, Ann Arbor, MI, (1957) 146–159.

Desai, C. S. and Nagaraj, B. K. Modeling for cyclic normal and shear behavior of interfaces. *J. Engrg. Mech.*, **114**(7), (1988) 1198–1217.

Federal Aviation Administration. Airport pavement design and evaluation. *FAA Advisory Circular*, AC 150/5320-6c, Washington, DC, (1978).

Francavilla, A. and Zienkiewicz, O. C. A note on numerical computation of elastic contact prob-lems. *Int. J. Num. Methods in Engrg.*, **9**(4), (1975) 913–924.

Friberg, B. F. Load and deflection characteristics of dowels in transverse joints in concrete pavements. *Proc., HRB, Vol. 18, Part 1*, Hwy. Res. Board, Washington, DC, (1938) 140–154.

Huang, W. H. and Wang, S. T. Finite-element analysis of concrete slabs and its implications for rigid pavement design. *HRB, HRR 466*, Hwy. Res. Board, Washington, DC, (1973) 55–69.

Ioannides, A. M. and Korovesis, G. T. Backcalculation of joint related parameters in concrete pavements. *Proc. 3rd Int. Conf. on Bearing Capacity of Roads and Airfields*, Vol. 1, (1990) 549–558.

Ioannides, A. M. and Korovesis, G. T. Analysis and design of doweled slab-on-grade pavement systems. *J. Transp. Engrg.*, **118**(6), (1992) 745–768.

Kenney, J. T. Steady-state vibration of beam on elastic foundation for moving load. *J. Appl. Mech.*, **21**, (1954) 359–364.

Kerr, A. D. Continuously supported beams and plates subjected to moving load – a survey. *SM Archives*, No. 6 (1981) 401–449.

Kikuchi, N. and Song, Y. J. Penalty/finite element approximations of a class of unilateral problems in linear elasticity. *Quart. Appl. Math.*, **39**(1), (1981) 1–22.

Kukreti, A. R., Taheri, M. R. and Ledesma, R. H. Dynamic analysis of rigid airport pavements with discontinuities. *J. Transp. Engrg.*, **118**(3), (1992) 341–359.

Larralde, J. Structural analysis of rigid pavement with pumping. *PhD Dissertation*, Purdue University, West Lafayette, IN, (1984).

Ledesma, R. H. Vehicle-guideway interaction in rigid airport pavements with discontinuities. *Master Thesis*, University of Oklahoma, OK, (1988).

Newmark, N. M. A method of computation for structural dynamics. *J. Eng. Mech. Div.* (1959) 67–94.

Nour-Omid, B. and Wriggers, P. A two-level iteration method for solution of contact problems. *Comp. Methods in Appl. Mech. Engrg.*, **54**(1), (1986) 131–144.

Sachdeva, T. D. and Ramakrishnan, C. V. A finite element solution for the two-dimensional elastic contact problems with friction. *Int. J. Num. Methods in Engrg.*, **17**, (1981) 1257–1271.

Tabatabaie, A. M. and Barenberg, E. J. Finite element analysis of jointed or cracked concrete pavements. *Transportation Res. Rec. 671*, Transportation Research Board, Washington, DC, (1978) 11–19.

Taheri, M. R. Dynamic response of plates to moving loads: Structural impedance and finite element methods. *PhD Dissertation*, Purdue University, LA, (1986).

Taheri, M. R. and Ting, T. C. Dynamic response of plates to moving loads: Finite element method. *Comput. Struct.*, **34**(3), (1990) 509–521.

Taheri, M. R., Zaman, M. and Alvappillai, A. Dynamic response of concrete pavements to moving aircraft. *Mathematical Modelling*, **14**(11), (1990) 562–575.

Tayabji, S. D. and Colley, B. E. Improved rigid pavement joints. *Transportation Research Record 930* (1983) 69–93.

Thompson, W. E. Analysis of dynamic behavior of roads subject to longitudinally moving loads. *HRB*, **39**, (1963) 1–24.

Tia, M., Armaghani, J. M., Wu, C. L., Lei, S. and Toye, K. L. FEACONS III computer program for analysis of jointed concrete pavements. *Transp. Res. Rec. 1136*, Transport Research Board, Washington, DC, (1987) 12–22.

Yoshida, D. M. and Weaver, W. Finite element analysis of beams and plates with moving loads. *Publ. Int. Assoc. Bridge Struc. Engr.*, **31**, (1971) 179–195.

Zaman, M. M., Desai, C. S. and Drumm, E. C. Interface model for dynamic soil-structure interactions. *J. Geotech. Engrg.*, **110**(9), (1984) 1257–1273.

Zaman, M., Taheri, M. R. and Alvappillai, A. Dynamic analysis of a thick plate on viscoelastic foundation to moving loads. *Int. J. Num. Anal. Methods in Geomech.*, **15**, (1991) 627–647.

Zaman, M. M. and Alvappillai, A. Contact-element algorithm for dynamic analysis of concrete pavements. *ASCE J. Transp. Div.*, **121**(5), (1995) 425–433.

19

A Practical Modeling Technique for Assessing Potential Contaminant Impacts Due to Landfills

R. Kerry Rowe[1] and **John R. Booker**[2]

[1]*The University of Western Ontario, London, Ontario, Canada*
[2]*The University of Sydney, NSW, Australia*

19.1 Introduction

Despite the widespread development of standard barrier system design for landfills, there is a growing recognition that minimum standard design may not necessarily provide the required level of groundwater protection (e.g. see Rowe *et al.*, 1996; Rowe, 1997). As a consequence, there is also a growing recognition of the need to perform modeling on a case-by-case basis to assess whether a given design will meet a performance standard (e.g. as per USEPA, 1994; MoE, 1998). This approach can minimize the potential for predictable unacceptable impacts, and hence minimize the liability of both designers and regulators (Estrin and Rowe, 1997).

Modeling of landfill barrier systems provides a considerable challenge to conventional numerical technique (e.g. finite element; finite difference) due to the large variability in parameters both spatially and temporally. For example, modeling impact may require consideration of plastic (geomembrane) liners as thin as 1–2 mm, geosynthetic clay liners 6–11 mm thick, compacted clay liners 0.6–1.5 m thick, natural deposits 1–10 m + thick and landfill dimensions that may be of the order of hundreds to thousands of metres. At the same time, the hydraulic conductivity and diffusion coefficients of different components of the barrier systems may vary over more than five orders of magnitude. Also, the dominant transport mechanism may change from diffusion (i.e. the movement of contaminants due to a concentration gradient; see Rowe *et al.* (1995) for details regarding diffusion as a transport mechanism) and advection (i.e. the movement of contaminants with flowing water or leachate). However, these conditions can be readily dealt with by adopting semi-analytic (finite layer) techniques, as initially described by Rowe and Booker (1984), and more recently extended by Rowe *et al.* (1995) and Rowe and Booker (1995, 1998).

Since its introduction in 1984, the finite layer technique has been widely adopted for modeling contaminant impact due to landfills with engineered barrier systems (e.g. Franz and Rowe, 1993; Panigrahi *et al.*, 1993; Manaserio *et al.*, 1997). This chapter draws together and summarizes the latest advances in the development of the finite layer technique. The application of the technique is then illustrated with respect to a hypothetical landfill.

19.2   Finite Layer Theory

The basic finite layer theory for 1-D and 2-D conditions (Rowe and Booker, 1984) has been extended to consider modeling fractured systems (Rowe and Booker, 1989, 1990), and more recently to model complex time histories and variations in parameters with time due to factors such as clogging of a leachate collection system or failure of a geomembrane liner once its service life is reached (Rowe and Booker, 1995, 1997, 1998). Under these circumstances, there is a change in boundary conditions or layer properties (e.g. diffusion coefficient of a contaminant through an HDPE geomembrane as the geomembrane ages, change in advective velocity as a leachate collection system clogs and a leachate mound develops) involves modeling one set of conditions for a period of time t^*, storing the concentration history at time t^*, and then using this as initial conditions for modeling the next period of time, $t' = t - t^*$, with the changed conditions.

We seek a solution to the advection-dispersion equation

$$nD_{xx}\frac{\partial^2 c}{\partial x^2} + nD_{zz}\frac{\partial^2 c}{\partial z^2} - nv_x\frac{\partial c}{\partial x} - nv_z\frac{\partial c}{\partial z}$$

$$= (n + \rho K_d)\frac{\partial c}{\partial t} + n\lambda c \tag{19.1}$$

where D_{xx}, D_{zz} are the coefficient of hydrodynamic dispersion in the x and z direction, v_x, v_z are the groundwater velocities in the x and z directions, n is the porosity, ρ is the dry density, K_d is the partitioning/distribution coefficient, λ is the first order decay coefficient, and c is the concentration at position (x, z) at time t, such that we can change boundary conditions and layer properties at discrete time t^*.

19.2.1   1-D conditions

Considering 1-D conditions first, for a given set of boundary conditions and layer properties, the modeling of contaminant migration through a layer is pursued generally as described by Rowe and Booker (1985). However, if at some time t^* these conditions change, then the concentrations in the layer at time t^* just prior to the change can be approximated by the relationship

$$c^* = Xe^{\Omega z} \tag{19.2a}$$

and represented in terms of the concentrations at the top (c_p^*) and bottom (c_q^*) of the layer, such that

$$\Omega = \frac{\ln(c_q^*) - \ln(c_p^*)}{z_q - z_p} \tag{19.2b}$$

and

$$\ln X = \frac{z_q \ln(c_p^*) - z_p \ln(c_q^*)}{z_q - z_p} \tag{19.2c}$$

where z_p, z_q are the co-ordinates at the top and bottom of the layer, respectively ($z_p \le z \le z_q$).

We now introduce the time t', which is the time that has elapsed after a change in condition (at time t^*) has occurred, such that

$$t' = t - t^* \tag{19.3a}$$

and then introducing the notation

$$c'(t') = c(t) = c(t' + t^*) \tag{19.3b}$$

$$f'(t') = f(t) = f(t' + t^*) \tag{19.3c}$$

where $c(t), f(t)$ are the concentration and mass flux at some depth z at time t.

Taking a Laplace transform of the 1-D version of Eq. (19.1), it is then possible to establish a relationship between the transformed nodal fluxes $\overline{f'}_p, \overline{f'}_q$ and the transformed nodal concentration $\overline{c'}_p, \overline{c'}_q$ for any layer p:

$$\begin{bmatrix} \overline{f'}_p \\ -\overline{f'}_q \end{bmatrix} = -\begin{bmatrix} A_p \\ B_p \end{bmatrix} + \begin{bmatrix} Q_p & R_p \\ S_p & T_p \end{bmatrix} \begin{bmatrix} \overline{c'}_p \\ \overline{c'}_q \end{bmatrix} \tag{19.4}$$

where the superior bar $(\overline{c}, \overline{f})$ denotes that this is a transformed function; Q_p, R_p, S_p and T_p are functions of the current layer properties (see Rowe and Booker, 1985; Rowe *et al.*, 1995) for $t \ge t^*$; and A_p and B_p are functions of both the current layer properties and the hereditary data (c_p^*, c_q^*), and can be derived for both unfractured and fractured layers, as described by Rowe and Booker (1995).

One of the most useful boundary conditions involves considering the finite mass of contaminant, such that the decrease in source concentration with time, as mass is removed from the landfill either by leachate collection or contaminant migration. However, the conditions relating to this boundary condition can also change at some time t^* (e.g. due to an increase in mass due to a landfill expansion, or a change in the operation of the leachate collection). Thus,

$$c_1'(t') = c_1(t^*) - \frac{1}{H_r'} \int_0^{t'} f_1'(t')dt' - \frac{q_c'}{H_r'} \int_0^{t'} c_1'(t')dt' \tag{19.5a}$$

where H_r' is the current 'reference height of leachate', and is a measure of the leachable mass of contaminant per unit area of landfill (see Rowe *et al.* (1995) for details on how to calculate H_r for a given landfill) and q_c' is the volume of leachate collected per unit area ($t^* < t$) and $c_1(t^*)$ is the concentration of contaminant in the landfill at time t^*.

Taking the Laplace transform of Eq. (19.5a), it can be rearranged into the form

$$-\overline{f'}_1 = T_0 \overline{c'}_1 + B_0 \tag{19.5b}$$

where

$$T_0 = sH'_r + q'_0 \tag{19.5c}$$

$$B_0 = -H'_r c(t^*) \tag{19.5d}$$

and s is the Laplace transform parameter.

The base boundary conditions for a landfill of length L parallel to the direction of groundwater flow and an aquifer of thickness h_b having a porosity n_b and horizontal Darcy velocity v'_b at $t > t^*$ can be written in the form

$$\overline{f'}_{m+1} = Q_{m+1}\overline{c'}_{m+1} - A_{m+1} \tag{19.6a}$$

where

$$Q_{m+1} = n_b h_b \left[s + \frac{v'_b}{L} \right] \tag{19.6b}$$

$$A_{m+1} = n_b h_b c_{m+1}(t^*) \tag{19.6c}$$

For a multilayered deposit, consideration of continuity of flux and concentration at the layer boundaries and incorporating the boundary conditions given by Eqs. (19.5) and (19.6) gives

$$
\begin{bmatrix}
T_0 + Q_1 & R_1 & 0 \\
S_1 & T_1 + Q_1 & R_1 \\
 & & \cdot \\
 & & \cdot \\
 & & \cdot \\
 & & T_m + Q_m & R_m \\
 & & S_m & T_m + Q_{m+1}
\end{bmatrix}
\begin{bmatrix}
\overline{c'}_1 \\
\overline{c'}_2 \\
\cdot \\
\cdot \\
\cdot \\
\overline{c'}_{m+1}
\end{bmatrix}
=
\begin{bmatrix}
B_0 + A_1 \\
B_1 + A_2 \\
\\
B_m + A_{m+1}
\end{bmatrix}
\tag{19.7}
$$

where the coefficients $T_k, Q_k, S_k, R_k, A_k, B_k (k = 1, ..., m)$ can be determined as described by Rowe and Booker (1995) for either fractured or unfractured layers, T_o, B_o are obtained from Eq. (19.5), and A_{m+1}, Q_{m+1} are obtained from Eq. (19.6).

19.2.2   2-D conditions

The approach described above can be generalized to two-dimensional conditions (Rowe and Booker, 1998). In this case, the concentration at time t^* in a layer can be described by a generalized form of Eq. (19.2) as:

$$\Phi = E e^{\epsilon z} \tag{19.8a}$$

where

$$\Phi = R \int_{-\infty}^{\infty} c^*(x, z)^{i\xi x} dx \tag{19.8b}$$

$$R = 1 + \rho K_d/n \tag{19.8c}$$

$$\epsilon = \frac{\ln(\Phi_q) - \ln(\Phi_p)}{z_q - z_p} \tag{19.8d}$$

$$\ln(E) = \frac{z_q \ln(\Phi_p) - z_p \ln(\Phi_q)}{z_q - z_p} \tag{19.8e}$$

and

$$\Phi_p = \Phi(z_p) \tag{19.8f}$$

$$\Phi_q = \Phi(z_q). \tag{19.8g}$$

By taking Laplace and Fourier transforms of Eq. (19.1), the relationship between concentration and vertical flux for any layer can be written in a form similar to Eq. (19.3), replacing $\overline{f'}_p$, $\overline{f'}_q$, $\overline{c'}_p$ and $\overline{c'}_q$ by terms that reflect that a Fourier transform has also been used $(\overline{F'}_p, \overline{F'}_q, \overline{C'}_p, \overline{C'}_q)$, and where A_k, B_k, Q_k, R_k, S_k and T_k are a function of the layer properties (D_{zz}, D_{xx}, etc.), and A_k, B_k are also functions of the hereditary information at time $t^*(\Phi_{k-1}, \Phi_k, \in)$ (see Rowe and Booker (1998) for details).

For the two-dimensional case, an aquifer can be modeled in a manner similar to that for 1-D (Eq. (19.6a)) viz.

$$\overline{F'}_{m+1} = Q_{m+1}\overline{C'}_{m+1} - A_{m+1} \tag{19.9a}$$

where

$$Q_{m+1} = hn_b\left(s + D_H\xi^2 + \frac{iv_b\xi}{n_b} + \lambda_b\right) \tag{19.9b}$$

and

$$A_{m+1} = hn_b \int_{-\infty}^{\infty} c_{b_o}(x)e^{-i\xi x}dx \tag{19.9c}$$

and ξ is the Fourier transform parameter. By invoking continuity of flux and concentration and applying Eq. (19.9), one can establish a system of equations:

$$\begin{bmatrix} Q_1 & R_1 & 0 & . & . & . \\ S_T & & & & & \\ 0 & K & & & & \\ . & & & & & \\ . & & & & & \\ . & & & & & \end{bmatrix} \begin{bmatrix} \overline{C'}_0 \\ \overline{C'}_1 \\ . \\ . \\ . \\ \overline{C'}_n \end{bmatrix} = \begin{bmatrix} A_1 + \overline{F}_T \\ W \end{bmatrix} \tag{19.10a}$$

where

$$K = \begin{bmatrix} Q_2 + T_1 & R_2 & 0 & \ldots & 0 \\ S_2 & Q_3 + T_2 & R_3 & \ldots & 0 \\ 0 & S_3 & Q_4 + T_3 & \ldots & 0 \\ . & & & & \\ . & & & & \\ 0 & & Q_n + T_{n-1} & & R_n \\ 0 & & S_n & & T_n + Q_{m+1} \end{bmatrix} \tag{19.10b}$$

$$W^T = [A_2 + B_1, \ldots, A_{n+1} + B_n] \tag{19.10c}$$

The surface flux $\overline{F}_T$ can be written in terms of the surface concentration (Rowe and Booker, 1998) as

$$\overline{F}_T = \psi\overline{C}'_0 - \Theta \tag{19.11a}$$

where

$$\psi = P_1 - Q_1 R_1 [e_1^T K^{-1} e_1] \tag{19.11b}$$

$$\Theta = A_1 - Q_1 [e_1^T K^{-1} w] \tag{19.11c}$$

and

$$e_1^T = [1 \quad 0 \quad \ldots \quad 0 \quad 1] \tag{19.11d}$$

Eq. (19.11) and the Fourier inversion theorem can then be used to calculate the flux distribution at the base of the landfill, viz.

$$\overline{f}_T = \frac{1}{2\pi} \int_{-\infty}^{\infty} [\psi\overline{C}'_0 - \Theta] e^{ix\xi} d\xi \tag{19.12}$$

and hence the average concentration in the landfill can be deduced from the mass balance equation

$$\left(s + \lambda + \frac{q'_c}{H'_r}\right) \overline{c}_0^*(S) = c_0^* - \frac{1}{LH} \int_{r-L/2}^{L/2} \overline{f}_T(x, s) dx \tag{19.13}$$

19.3 Application of 1-D Finite Layer Theory to Examine the Impact Due to a Hypothetical Landfill

To illustrate the theory given in the previous section, consideration will be given to the potential impact that could arise due to a moderately sized landfill (15 m average thickness of waste) that has a barrier system (as shown in Figure 19.1) consisting of a primary leachate collection system, a 1.2 m thick compacted clay liner, a 0.3 m thick granular 'hydraulic control layer', and 5.7 m thick natural aquitard overlying a 0.6 m thick aquifer. The base contours have been selected so that after consideration of any lowering of water levels due to the hydraulic shadow of the landfill (see Rowe *et al.*, 1995), the potentiometric surface is above the design leachate level in the landfill, and hence there would be a small inward flow of water from the aquifer into the land-fill. Thus, the landfill is designed as a 'hydraulic trap' (see Rowe *et al.*, 1995), so that the inward flow of groundwater provides some hydraulic containment of contam-inants.

Under normal operating conditions, the Hydraulic Control Layer (HCL) will be saturated, and will adopt a water level that is in equilibrium with the water level in the aquifer and lechate. Under these conditions, the system is totally passive (no water is either added or removed by man after initial saturation). However, as discussed in the following subsections, this unit can be operated to control contaminant impact in the event of some premature or unexpected change in either leachate levels in the landfill (e.g. due to clogging of the leachate collection system), or a change in groundwater levels in the underlying aquifer. To be definitive, it is assumed that the landfill is approximately

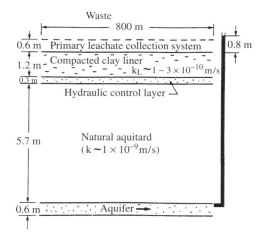

Figure 19.1 Schematic showing landfill designed with hydraulic containment

Table 19.1 Assumed leachate characteristics for example case

parameter	Initial source concentration c_o	Mass ratio p	K_{oc} (mL/g)	Half-life $t_{1/2}$ (years)
Chloride	1600–2500 (mg/L)	1.8 g/kg	–	–
Benzene	20 (µg/L)	0.014 mg/kg	862	25
Dichloro-methane	3300 (µg/L)	2.3 mg/kg	345	10
Vinyl Chloride	55 (µg/L)	0.039 mg/kg	375	25

800 m long in the direction parallel to groundwater flow. Analyses were performed for chloride, benzene, dichloromethane (DCM) and vinyl chloride. The assumed contaminant parameters are given in Table 19.1. The organic carbon content was assumed to be $f_{oc} = 0.36\%$, and hence the value of the partitioning coefficient K_d could be deduced from the K_{oc} values given in Table 19.1 such that

$$K_d = K_{oc} \cdot f_{oc} \tag{19.14}$$

The natural aquitard was assumed to have a hydraulic conductivity of 1×10^{-9} m/s. The compacted clay liner was assumed to have a specified hydraulic conductivity of 3×10^{-10} m/s (or less), but that the actual, as constructed, hydraulic conductivity was 1×10^{-10} m/s.

The contaminant transport modeling reported in the following sections was conducted using the finite layer theory outlined in Section 19.2 for 1-D conditions as implemented in Program POLLUTE v6 (Rowe and Booker, 1996).

19.3.1   Base case

The base case used the flow modeling along the section shown in Figure 19.1 and assumed that the compacted clay liner had a maximum specified design hydraulic conductivity of 3×10^{-10} m/s. This gave rise to an average (upward) vertical Darcy velocity from the aquifer to the landfill of -0.00048 m/a, and an average horizontal flow in the aquifer of $0.32 \, m^3/a/m$. These flows were used to model the potential impact of the landfill on the concentrations of the contaminants in the aquifer and hydraulic control layer.

19.3.2   Better than specified liner

As noted above, the base case assumes that the liner has a hydraulic conductivity of 3×10^{-10} m/s. This represents the specified design value for the liner and the use of this value is 'conservative' (i.e. gives a worst case) for predicting the potential drawdown due to the 'shadow effect', and the consequent potential effect of changes in groundwater flow conditions on groundwater chemistry. However, a liner is likely to have a lower hydraulic conductivity than specified and a value of 1×10^{-10} m/s is assumed here. Making this assumption, the calculated Darcy velocity is reduced to -0.0004 m/a (upward). The difference between this and the value $(-0.00048 \, m/a)$ obtained for the specified liner is not very significant. The quite small effect can be attributed to the fact that the low hydraulic conductivities of the aquitard provide the primary hydraulic control on vertical flows from the aquifer.

19.3.3   Predicted impacts: Chloride

Analyses performed for the migration of chloride from the landfill considering both the 'base case' and 'better than specified case', source concentrations of 1600 and 2500 mg/L and infiltration of 0.1 and 0.2 m/a through the cover (the latter covering the expected infiltration range). For this range of conditions, the calculated increase in chloride concentrations (above background) in the aquifer under operating conditions was between 15–35 mg/L after periods of the order of 530–650 years.

The chloride concentration in the hydraulic control layer was assumed to be initially the same as the average value for the aquifer. The increase in concentration above this initial value was then calculated and the range of values is bounded by the two curves given in Figure 19.2. These curves correspond to a range in likely increase of chloride in the aquifer of between 15 and 35 mg/L. Thus, using a figure such as Figure 19.2, one could, by monitoring the hydraulic control layer, assess whether the landfill is performing as expected.

Organic compounds

Analyses performed for dichloromethane, vinyl chloride and benzene all indicated that there would be negligible impact on the aquifer, and that any increase in concentration should be below current detection limits. In the hydraulic control layer, the peak increase in concentration is likely to occur at between 60 and 120 years, but is expected to be less than 5 µg/L, and is probably undetectable.

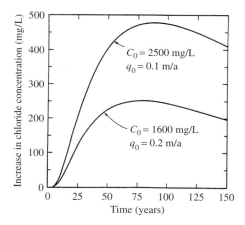

Figure 19.2   Calculated increase in chloride concentration (above background) with time in the Hydraulic Control Layer for limiting cases considered

Monitoring and control implications

The Hydraulic Control Layer (HCL) may be used as a monitoring point, and the subsequent potential impact on groundwater quality in the aquifer may be related to the observed concentration in the HCL. Any consistent increase in concentration in the HCL greater than that given in Figure 19.2 for chloride, or greater than 5 μg/L for the organic contaminants, would warrant some investigation to identify the reason for the unexpected behavior.

Chloride is likely to be the most troublesome parameter. Figure 19.3 shows a relationship between the increase in concentration in the HCL (above the background value) to the subsequent increase in the aquifer. The curve represents the upper bound, and the

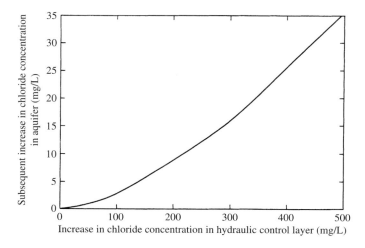

Figure 19.3   Trigger plot obtained by modeling and relating the measured increase in chloride concentration in the Hydraulic Control Layer to the subsequent expected increase in chloride concentration in the receptor aquifer

impact in the aquifer is likely to be less than that predicted by this curve. The peak impact on the HCL is expected to occur between 40 and 90 years after completion of the landfill cell. It would take about another 400 years for the corresponding peak impact to be observed in the aquifer, assuming that the hydraulic trap is maintained over this period. The implication of prior termination of pumping of leachate, and the consequent loss of the hydraulic trap will be examined in the next subsection.

In the event that concentrations measured in the HCL signal a potential unacceptable future impact on the aquifer, the HCL could be designed so that it can be used in an active mode. For example, by introducing 'clean' water with negligible chloride into the manholes at one end of the HCL and then removing water from the other end of the HCL, it would be possible to control the concentration in the HCL. Once the chloride concentration in the HCL increases significantly (e.g. by 100 mg/L) above the nominal background value, calculations indicate that an exchange of water of about 10 l/min (5200 m^3/a) would be likely to control impact in the aquifer to effectively zero.

19.3.4   *Contaminant transport modeling: failure of primary leachate collection system (PLCS) (Case 2)*

A series of analyses was performed to examine the potential impact on the aquifer in the event of a failure of the primary leachate collection system at times ranging from 20 to 150 years after landfilling. The leachate mound was calculated based on mounding between perimeter drains and conditions of continuity of flow for infiltration through the cover of both 0.1 and 0.2 m/a. It was assumed that the HCL was pumped such that the head in the HCL was maintained at a specified value (0.25 m above the top of the compacted clay liner) once a failure in the PLCS has been detected (e.g. by the observation of leachate mound more than 1 m above the nominal liner). For the range of cases considered, the increase in chloride concentration (above background levels) in the aquifer would be less than 40 mg/L. Under the same conditions, the increase in the concentration of organic compounds (dichloromethane, vinyl chloride, benzene) in the aquifer would be negligible (below current detection limits).

The foregoing calculations indicate that the HCL has the potential to control impact on the aquifer even if there is a failure of the PLCS after 20 years operation. The operation modeled simply involved pumping to control the head in the HCL such that there was still upward flow from the aquifer to the HCL. This resulted in maintaining a hydraulic trap with respect to the aquifer, and hydraulically decoupling the effects of leachate mounding in the landfill from the flow in the natural aquitard and aquifer. The impact could be further reduced by the addition of clean water as previously discussed.

19.3.5   *Contaminant transport modeling: termination of operation of PLCS and HCL (Case 3)*

In this case, it was assumed that at some time pumping from both the PLCS and, if relevant, the HCL will be terminated. As discussed in the previous section, the impacts of a failure of the PLCS can be mitigated by pumping from the HCL and ensuring a hydraulic trap between the aquifer and the HCL. However, if pumping from both the PLCS and HCL is terminated, then the leachate mound will cause an increased flow to the aquifer, which will accelerate and accentuate the impact on the aquifer. For the assumed conditions ($c_o = 1600\,\text{mg/L}$, $q_o = 0.2\,\text{m/a}$, $p = 1.8\,\text{g/kg}$), the concentration in the HCL increases to a peak value of about $250\,\text{mg/L}$ at 80 years (see Figure 19.2), and then begins to decrease as chloride migrates down through the underlying aquitard towards the aquifer. When the operation of the PLCS/HCL ceases, the increased leachate mound causes downward flow and hence removes contaminants from the HCL, and causes it to migrate towards the aquifer, giving a peak impact of $80\,\text{mg/L}$ at about 320 years, assuming termination of operation of the engineered system at 200 years.

Whether this increase is acceptable or not would depend upon the background concentration in the aquifer. By varying the time at which the operation of the engineered collection system is terminated one can assess the contaminating lifespan (i.e. the period of time when active maintenance and pumping is required) by comparing the predicted impacts with allowable values. For the conditions examined here, it may be concluded that with respect to impact on the aquifer, the likely contaminating lifespan is of the order of several decades for organic contaminants. For chloride, the contaminating lifespan of the landfill could range between 100 and 600 years depending on the allowable increase in concentration in the aquifer.

19.4 Conclusions

Considerable advances have been made in the development of theoretical techniques that can be readily applied to modeling the potential impact due to landfills. This chapter has summarized finite layer theory that can be used to model landfills involving layers with very different properties as well as a change in the properties with time due to factors such as (a) changes in the operation of a leachate collection system, (b) failure of a geomembrane liner, (c) expansion of a landfill, etc..

The application of the theory has been illustrated with respect to a hypothetical landfill that is operated to provide hydraulic containment. The effect of assumed conditions and failure of the primary leachate collection system were examined, and it is shown that a hydraulic control layer (below the primary clay liner) can effectively control impact on the underlying aquifer.

Acknowledgement

Funding of the program of research into the clogging of leachate collection systems came from Collaborative Grant CPG0163097 provided by the Natural Sciences and Engineering Research Council of Canada.

References

Estrin, D. and Rowe, R. K. Legal liabilities of landfill design engineers and regulators. *Proceedings Sardinia '97: Sixth International Landfill Symposium*, Vol. 5, S. Margherita di Pula, Cagliari, Italy, (October 1997) 65–76.

Franz, T. J. and Rowe, R. K. Simulation of contaminant transport at a landfill site using models. *Int. J. Num. Anal. Methods in Geomech.*, **17**(7), (1993) 435–455.

Manassero, M., Di Molfetta, A., Pacitti, M. and Veggi, S. The case history of a landfill for hazardous industrial wastes. *Proceedings of GeoEnvironment '97*, A. A. Balkema, Rotterdam, (1997) 3–29.

MoEE. *Ontario Ministry of Environment Guideline B-7*. Incorporation of the reasonable use concept into MoEE groundwater management activities, (1994).

MoE. Landfill standards: A guideline on the regulatory and approval requirements for new or expanding landfill sites. *Ontario Ministry of Environment* May 1998, Queens Printer for Ontario, PIBS 3651E.

Panigrahi, B. K., Pandit, A., Hebson, C. S. and Rowe, R. K. Input parameter requirements for groundwater models. *Joint CSCE-ASCE National Conference on Environmental Engineering*, Montreal, July 2 (1993) 1265–1272.

Rowe, R. K. The design of landfill barrier systems: should there be a choice? *Ground Eng.*, (August 1997) 36–39.

Rowe, R. K. and Booker, J. R. A novel technique for the analysis of 1-D pollutant migration. *Proceedings of the International Conference on Numerical Methods for Transient and Coupled Problems*, Venice, Italy, (1984) 699–709.

Rowe, R. K. and Booker, J. R. 1-D pollutant migration in soils of finite depth. *J. Geotech. Eng.*, **111**(GT4), (1985) 479–499.

Rowe, R. K. and Booker, J. R. A semi-analytic model for contaminant migration in a regular two or three dimensional fracture network: Conservative contaminants. *Int. J. Num. Anal. Methods in Geomech.*, **13**, (1989) 531–550.

Rowe, R. K. and Booker, J. R. Contaminant migration in a regular two or three dimensional fracture network: Reactive contaminants. *Int. J. Num. Anal. Methods in Geomech.*, **14**, (1990) 401–425.

Rowe, R. K. and Booker, J. R. A finite layer technique for modelling complex landfill history. *Canad. Geotech. J.*, **32**(4), (1995) 660–676.

Rowe, R. K. and Booker, J. R. *POLLUTE v. 6–1-D Pollutant Migration Through a Non-Homogeneous Soil.* ©1983, 1990, 1994, 1996. Geotechnical Research Centre, University of Western Ontario, London N6A 5B9, Canada, (1996).

Rowe, R. K. and Booker, J. R. Recent advances in modelling contaminant impact due to clogging. Keynote paper, *9th International Conference of the Association for Computer Methods and Advances in Geomechanics*, Vol. 1, Wuhan, China, November (1997) 43–56.

Rowe, R. K. and Booker, J. R. Modelling impacts due to multiple landfill cells and clogging of leachate collection systems. *Canad. Geotech. J.* **35**(1), (1998) 1–14.

Rowe, R. K., Hrapovic, L. and Armstrong, M. D. Diffusion of organic pollutants through HDPE geomembrane and composite liners and its influence on groundwater quality. *Geosynthetics: Applications, Design and Construction: Proceedings 1st European Geosynthetics Conference*, Maastricht, Netharlands (October 1996) 737–742.

Rowe, R. K., Quigley, R. M. and Booker, J. R. *Clayey Barrier Systems for Waste Disposal Facilities*. E & FN Spon (Chapman & Hall), London, UK, (1995).

USEPA. Office of the Federal Register, National Archives and Records Administration, Code of Federal Regulations, Vol. 40, *Protection of the Environment*, Parts 190–259, Revised as of July 1 (1994).

20

Finite Element Analysis of Free Surface Seepage Flows

Annamaria Cividini and **Giancarlo Gioda**
Politecnico di Milano, Milan, Italy

20.1 Introduction

The analysis of unconfined seepage flows is often required in geotechnical engineering, for instance when dealing with the design of earth dams, with the stability analysis of natural slopes, etc. They belong to a broad class of free boundary value problems (Cryer, 1977) involving differential equations on a domain part of whose boundaries (the so-called 'free surface', or 'phreatic line' in the present context) is unknown, and has to be determined, together with the distribution of hydraulic head, as part of the solution.

Depending upon the problem at hand, steady or transient analyses could be required. The former is characterized by a free surface the shape of which is constant with time, while in the latter, the free surface moves with time, eventually reaching a steady state configuration.

Owing to their intrinsic geometrical nonlinearity, these analyses are frequently carried out using numerical, computer-oriented techniques. Among them, the finite element method (e.g. see Zienkiewicz *et al.*, 1966; Desai, 1972; France, 1974; Desai, 1977) and, more recently, the boundary integral equation method (e.g. see Liggett, 1977; Banerjee *et al.*, 1981; Liggett and Liu, 1983) are most widely used.

Confining our attention to the finite element approaches, they can be grouped into two main categories, which are referred to in the following as 'variable' and 'fixed' mesh procedures.

The techniques of the first group (e.g. see Taylor and Brown, 1967; Desai, 1972; Chen and Li, 1973) consist of iterative processes that modify the geometry of a finite element mesh and, in particular, of its elements adjacent to the free surface, until a sufficient approximation of the correct shape of the flow domain is reached. One of the free surface boundary conditions is directly imposed on it at the beginning of each iteration, while the second condition is used to define a more accurate position of the free surface for the subsequent iteration.

The procedures of the second group do not change the geometry of the finite element mesh during the iterative solution process (e.g. see Desai, 1976; Bathe and Khoshgoftaar,

1979; Desai and Li, 1983; Lacy and Prevost, 1987), and determine the shape of the phreatic line in a way similar to that used in nonlinear stress analyses for defining the boundary between 'elastic' and 'plastic' regions.

Both constant and variable mesh procedures present advantages and shortcomings. In particular, the variable mesh methods are usually more accurate than those operating with a constant geometry. On the other hand, they might present stability problems during the iterative solution process, which in some cases lead to apparent non-unique solutions. This effect is observed when the intersection between the free surface and a pervious boundary exposed to the atmosphere (or seepage face) has to be determined.

In the following, the equations governing the seepage flows and the relevant boundary conditions are first recalled both for the continuum problem and for its finite element discretization. Then, some variable and fixed mesh procedures for steady state and transient analyses are outlined. Some comments are also presented on solution approaches based on the so-called 'conformal mapping' technique.

Finally, the sources of instability which might affect the variable mesh approaches are discussed, and a procedure is discussed aimed at avoiding these negative effects.

A matrix notation is used in this chapter. Capital and lower case boldface letters denote matrices and column vectors, respectively. A superscript T means transpose and a superimposed dot means time derivative.

20.2   Governing Equations and Boundary Conditions

In a Cartesian reference system x_1, x_2, x_3, the isothermal seepage flow of a compressible liquid through a deformable porous medium is governed by the following equation of continuity:

$$\sum_{i}^{3} \frac{\partial}{\partial x_i}(\rho v_i) - \rho \bar{q} = \frac{\partial}{\partial t}(\rho S_r n) \tag{20.1}$$

where v_i are the components of the fluid discharge velocity, ρ is the density of the liquid, $\bar{q}$ is the assigned flux per unit volume (positive in the case of a source), S_r is the degree of saturation, n is the porosity, and t is the time. The right-hand side term in Eq. (20.1) represents the rate of accumulation of the liquid mass per unit total volume.

Assuming that the soil skeleton is undeformable and perfectly saturated, and that both soil grains and pore liquid are incompressible, the rate of accumulation of the liquid vanishes and Eq. (20.1) reduces to the following simplified form:

$$\sum_{i}^{3} \frac{\partial v_i}{\partial x_i} - \bar{q} = 0 \tag{20.2}$$

Since the majority of seepage phenomena in soil mechanics involve laminar flow conditions, Darcy's law can be adopted as a linear 'constitutive' relationship between the components v_i of the discharge velocity vector $\mathbf{v}$ and those of the gradient $\mathbf{i}$ of the hydraulic head h,

$$\mathbf{v} = \begin{Bmatrix} v_1 \\ v_2 \\ v_3 \end{Bmatrix} = -\mathbf{Ki} \tag{20.3a}$$

where

$$\mathbf{i} = \left\{ \begin{array}{c} i_1 \\ i_2 \\ i_3 \end{array} \right\} = \left\{ \begin{array}{c} \partial h/\partial x_1 \\ \partial h/\partial x_2 \\ \partial h/\partial x_3 \end{array} \right\} \tag{20.3b}$$

In Eqs. (20.3a) and (20.3b), $\mathbf{K}$ is the matrix of the permeability coefficients, and the following well known definition holds for h, where the kinetic term depending on the square of the discharge velocity is neglected,

$$h = \frac{p}{\gamma} + x_2 \tag{20.4}$$

Here x_2 is the upward vertical coordinate, p is the pressure of the pore liquid and γ its unit weight.

If the principal directions of permeability coincide with the Cartesian directions, $\mathbf{K}$ becomes a diagonal matrix as follows:

$$\mathbf{K} = \begin{bmatrix} k_1 & 0 & 0 \\ 0 & k_2 & 0 \\ 0 & 0 & k_3 \end{bmatrix} \tag{20.5}$$

Under this last hypothesis, by substituting Eqs. (20.3a) and (20.3b) into Eq. (20.2), and taking into account Eq. (20.5), the following final relationship governing the problem at hand is arrived at:

$$\sum_{i=1}^{3} \frac{\partial}{\partial x_i} \left(k_i \frac{\partial h}{\partial x_i} \right) - \bar{q} = 0 \tag{20.6}$$

It can be seen that Eq. (20.6) holds for both steady state and transient seepage problems, in the latter case neglecting the inertial effects.

As to the boundary conditions to be associated with Eq. (20.6), let us consider the two-dimensional transient problem shown in Figure 20.1, in which the water levels in the upper and lower reservoirs are known functions of time t. In this case, the following conditions hold where a dash characterizes the known quantities:

Pervious (constrained) boundaries:

$$h = \bar{h}_1(t) \quad \text{on } 1 - 2 \tag{20.7a}$$

and

$$h = \bar{h}_2(t) \quad \text{on } 4 - 5 \tag{20.7b}$$

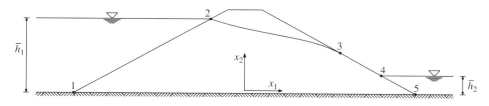

Figure 20.1   Unconfined flow through an homogeneous earth dam

Impervious boundary normal to the x_2 axis:

$$\frac{\partial h}{\partial x_2} = v_2 = 0 \quad \text{on } 5 - 1 \tag{20.8}$$

The general expression of Eq. (20.8) in the three-dimensional case is given by

$$\sum_{i}^{3} \alpha_i \frac{\partial h}{\partial x_i} = 0 \tag{20.9}$$

where α_i are the direction cosines of the vector normal to the impervious boundary. The above boundary conditions are those commonly met in confined problems. Additional conditions are present in the unconfined case:

'Wet' boundary, or seepage face:

$$h = x_2 = \bar{g}(x_1) \quad \text{on } 3 - 4 \tag{20.10}$$

Free surface (transient problems):

$$h = x_2 = F(x_1, t) \quad \text{and} \quad v_F = v_n/m \quad \text{on } 2 - 3 \tag{20.11a, b}$$

where

$$v_n = v_1\beta_1 + v_2\beta_2 \tag{20.11c}$$

Equation (20.10) requires that the hydraulic head on the seepage face coincides with the elevation expressed by function $\bar{g}$, which is a known function describing the shape of the downstream slope of the dam.

The two conditions on the free surface require that the hydraulic head be equal to its unknown elevation expressed by function F (Eq. (20.11a)) and that the velocity v_F of the free surface coincides with the fluid velocity normal to it, v_n, divided by the effective porosity m (Eq. (20.11b)). Equation (20.11c) expresses the flow velocity normal to the free surface in terms of the velocity components in the x_1 and x_2 directions, and of the direction cosines β_1 and β_2 of the vector normal to the free surface.

Note that the effective porosity m in Eq. (20.11b) is, in general, different from the 'standard' porosity n. In fact, while n is defined as the ratio between the volume of voids and the total volume of a soil element, m is the ratio between the area of void and the total area of a portion of the moving free surface. Hence, even though m is often assumed to coincide with n, its value depends upon the direction cosines β, while n is a scalar quantity independent on the direction.

In steady state conditions, the functions of time in Eqs. (20.7a) and (20.7b) become constants and the boundary conditions in Eqs. (20.11a), (20.11b) and (20.11c) reduce to the form below:

Free surface (steady state problems):

$$h = x_2 = F(x_1) \quad \text{and} \quad v_F = v_n = 0 \quad \text{on } 2 - 3 \tag{20.12a, b}$$

It should be observed that the portions of the boundary having a known geometry are characterized by only one condition, expressed either in terms of the free variable h (as in the case of a pervious boundary exposed to the atmosphere (Eqs. (20.7, 20.10)), or of its space derivatives (as in the case of an impervious boundary (Eq. (20.9)). On the contrary, both conditions are simultaneously present on the free surface (e.g. see Eqs. (20.12a) and (20.12b)).

This apparent redundancy of the number of boundary conditions is due to the fact that the free surface has an unknown geometry. As a consequence, an additional condition is necessary with respect to the remaining portions of the contour in order to evaluate its unknown shape during solution.

20.3 Finite Element Formulation of the Confined Problem

When the finite element method is adopted in a confined seepage analysis, the flow domain, which has a known shape, is subdivided into elements within which the distribution of hydraulic head $h(x_1, x_2, x_3)$ depends upon the corresponding nodal values **h** through a vector **b** of interpolation functions

$$h(x_1, x_2, x_3) = \mathbf{b}(x_1, x_2, x_3)^\mathrm{T} \mathbf{h} \tag{20.13}$$

In addition to the vector of nodal hydraulic heads **h**, each element is characterized by the vector of nodal fluxes **q**. In terms of standard stress analysis, these vectors correspond to the nodal displacement and to the nodal force vectors, respectively.

The relationship governing the flow problem for each finite element can be derived by writing an equation equivalent, in hydraulics terms, to the principle of virtual works based on Green's theorem used in stress analysis.

In particular, let us consider a finite element characterized by a given vector of nodal fluxes **q** and to the corresponding internal distribution of flow velocity **v**. Let us apply to this element a 'virtual' variation of nodal hydraulic heads, collected in vector $\delta\mathbf{h}$ which induces a virtual variation of the hydraulic gradient $\delta\mathbf{i}$ within the element, according to Eq. (20.3b).

On the basis of the above-mentioned principle, the following relationship can be established:

$$(\delta\mathbf{h})^\mathrm{T} \mathbf{q} = \int_V (\delta\mathbf{i})^\mathrm{T} \mathbf{v} \, dV \tag{20.14}$$

where V represents the element volume.

Taking into account Eqs. (20.3) and (20.13), vectors **v** and $\delta\mathbf{i}$ can be expressed in terms of the nodal hydraulic head vector **h** and of its virtual variation $\delta\mathbf{h}$:

$$\mathbf{v} = -\mathbf{K} \begin{bmatrix} (\partial\mathbf{b}/\partial x_1)^\mathrm{T} \\ (\partial\mathbf{b}/\partial x_2)^\mathrm{T} \\ (\partial\mathbf{b}/\partial x_3)^\mathrm{T} \end{bmatrix} \mathbf{h} \tag{20.15a}$$

$$(\delta\mathbf{i})^\mathrm{T} = (\delta\mathbf{h})^\mathrm{T} \begin{bmatrix} \dfrac{\partial\mathbf{b}}{\partial x_1} & \dfrac{\partial\mathbf{b}}{\partial x_2} & \dfrac{\partial\mathbf{b}}{\partial x_3} \end{bmatrix} \tag{20.15b}$$

By substituting Eqs. (20.15a) and (20.15b) into Eq. (20.14), and by eliminating vector $\delta\mathbf{h}$ from the right and left side terms, a linear relationship is obtained between nodal hydraulic heads **h** and assigned nodal fluxes **q**:

$$\mathbf{Mh} = \mathbf{q} \tag{20.16}$$

The so-called 'flow' matrix **M** in Eq. (20.16) is expressed by the following integral over the volume V of the element:

$$\mathbf{M} = \int_V \left[\frac{\partial \mathbf{b}}{\partial x_1} \quad \frac{\partial \mathbf{b}}{\partial x_2} \quad \frac{\partial \mathbf{b}}{\partial x_3} \right] \mathbf{K} \begin{bmatrix} (\partial \mathbf{b}/\partial x_1)^\mathrm{T} \\ (\partial \mathbf{b}/\partial x_2)^\mathrm{T} \\ (\partial \mathbf{b}/\partial x_3)^\mathrm{T} \end{bmatrix} dV \qquad (20.17)$$

where **K** is the permeability matrix (Eq. (20.5)).

The system of linear equations governing the solution of the discrete problem is arrived at by assembling Eq. (20.17) written for all the elements of the mesh. The intrinsic singularity of the assembled flow matrix is eliminated by 'constraining' the values of the hydraulic head of the nodes belonging to pervious boundaries (Eqs. (20.7a) and (20.7b)).

In general, internal sources (equivalent in stress analysis terms to body forces) are absent in the majority of seepage problems of interest in geomechanics. However, if an internal source $\bar{q}$ is present within a finite element of volume V, the equivalent (in the finite element sense) nodal fluxes **q** can be evaluated through the following integral and added to the data vector in Eq. (20.16):

$$\mathbf{q} = \int_V \mathbf{b}^\mathrm{T} \bar{q} \, dV \qquad (20.18)$$

The impervious boundaries through which no water flow takes place correspond, in terms of stress analysis, to portions of the contour where no external tractions are applied. Hence, no external 'loads' or constraints have to be applied to these traction free boundaries.

The finite element analysis of confined seepage problem (i.e. problems in which all boundaries of the flow domain have known geometry) does not present particular diffi-culties. In fact, it consists of the linear analysis of the discretized system with only one free variable in each node.

On the contrary, unconfined analyses are, in general, more complicated since, in addi-tion to the distribution of hydraulic head, the location of the free surface (and, hence, the shape of the domain within which the seepage flow takes place) is unknown. In the following sections, some variable and fixed mesh techniques will be summarized.

20.4   Some Variable Mesh Techniques for Unconfined Analysis

20.4.1   Steady state case

The variable mesh techniques modify the geometry of the finite element grid during the solution process until the shape of the free surface is described with a sufficient accuracy.

A relatively simple iterative procedure of this kind was proposed by Taylor and Brown (1967) for steady state problems. At the beginning of each iteration, a confined analysis is performed, on the basis of the current geometry of the mesh, in which the free surface is treated as an impervious boundary (Eq. (20.12b)). If the shape of the phreatic line is inaccurate, the calculations lead to hydraulic heads at the free surface nodes different from their elevation. To fulfil the second free surface condition (Eq. (20.12a)), the nodes on the free surface are moved so that their elevation becomes equal to the corresponding hydraulic head previously computed. Then, the modified mesh is used in the subsequent iteration.

This procedure presents an ambiguity in defining the movement of the node at the intersection between the free surface and the seepage face or 'seepage point' (i.e. point 3 in Figure 20.1). In fact, even though this node belongs to the phreatic line, it has in general a non-vanishing flux related to the flow of water through the portion of the seepage face adjacent to it. Consequently, for this node the pervious boundary condition (Eq. (20.12b)) has to be adopted during the confined analysis and, hence, the procedure for displacing the other free surface nodes cannot be applied to it.

This difficulty can be minimized by reducing the mesh size adjacent to the seepage point, or by relating the position of this node to that of the node next to it on the free surface through some interpolation function.

A slightly more complex solution method, requiring two confined analyses for each iteration, was proposed for steady problems by Neuman and Witherspoon (1970), who subsequently extended it to the transient case (Neuman and Witherspoon, 1971).

In the first analysis, the free surface is considered as a pervious boundary, and the hydraulic head of its nodes is set equal to their elevation (Eq. (20.12a)). From this calculation, the fluxes at the nodes located on the seepage face (line 3-4 in Figure 20.1) are evaluated. Then a second confined analysis is performed in which the free surface is treated as an impervious contour and the nodal fluxes previously computed are imposed on the seepage face. Finally, the mesh is modified by making the elevation of the free surface nodes equal to the corresponding hydraulic heads obtained from the second analysis.

Also, this approach presents a difficulty in treating the seepage point. In fact, the flux evaluated at this node by the first confined analysis is due to the contributions of both the free surface (which is a pervious boundary in this calculation) and the seepage face. In the second analysis, however, only the part of flux through the seepage face should be imposed on the seepage point. To overcome this problem, it was suggested to impose to the seepage point a flux equal to one half of the flux obtained at the adjacent node on the wet boundary.

20.4.2 Transient conditions

Among various approaches for transient analyses, that proposed by Taylor *et al.* (1973) is considered here. The technique that in its original version constrained the free surface nodes to move along vertical lines during the transient process, is extended to the case of nodal movement along non-vertical directions.

Consider the side S of a three-dimensional isoparametric finite element which coincides with a portion of the phreatic surface (cf. Figure 20.2). It can be easily shown that the nodal fluxes equivalent to a distributed fluid velocity v_n normal to side S are given by the following integral:

$$\mathbf{q} = \int_S \mathbf{f}(\zeta, \eta) v_n(\zeta, \eta) \, dS \tag{20.19}$$

where $\mathbf{q}$ is the vector collecting the fluxes at the $\mathbf{N}$ nodes of the side, $\mathbf{f}$ is the vector of the interpolation functions of the same nodes:

$$\mathbf{f} = \{f_1(\zeta, \eta) \quad f_2(\zeta, \eta) \cdots f_N(\zeta, \eta)\}^{\mathrm{T}} \tag{20.20}$$

and ζ, η are the curvilinear coordinates on the surface S.

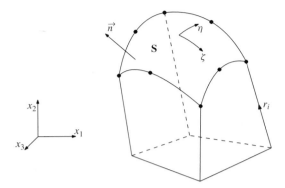

Figure 20.2 Finite element side discretizing a portion of the free surface in a three dimensional problem

The fluid velocity v_n is related to the velocity of the free surface v_F by the boundary condition (Eq. (20.11b)), which is rewritten here for convenience:

$$v_n(\zeta, \eta) = m v_F(\zeta, \eta) \tag{20.21}$$

In turn, v_F can be expressed in terms of its Cartesian components $(\dot{x}_1, \dot{x}_2, \dot{x}_3)$ and of the direction cosines of the normal to S grouped in vector $\boldsymbol{\alpha}_n$:

$$v_F(\zeta, \eta) = \boldsymbol{\alpha}_n^T(\zeta, \eta) \left\{ \begin{array}{c} \dot{x}_1(\zeta, \eta) \\ \dot{x}_2(\zeta, \eta) \\ \dot{x}_3(\zeta, \eta) \end{array} \right\} \tag{20.22}$$

In the isoparametric formulation of the finite element method, the coordinates, and hence their time derivatives, are interpolated in the same manner as other variables. Consequently, a relation can be established between the components of the free surface velocity and the time derivative of the nodal coordinates $\dot{\mathbf{x}}$, through the matrix $\mathbf{F}$ of 'shape' functions of the N nodes on side S:

$$\left\{ \begin{array}{c} \dot{x}_1(\zeta, \eta) \\ \dot{x}_2(\zeta, \eta) \\ \dot{x}_3(\zeta, \eta) \end{array} \right\} = \mathbf{F}(\zeta, \eta)\dot{\mathbf{x}} \tag{20.23a}$$

where

$$\mathbf{F}(\zeta, \eta) = \begin{bmatrix} F_1(\zeta, \eta) & 0 & 0 & & F_N(\zeta, \eta) & 0 & 0 \\ 0 & F_1(\zeta, \eta) & 0 & \cdots & 0 & F_N(\zeta, \eta) & 0 \\ 0 & 0 & F_1(\zeta, \eta) & & 0 & 0 & F_N(\zeta, \eta) \end{bmatrix} \tag{20.23b}$$

Assume that, during the transient process, the nodes of side S move along given lines, and denote with $\mathbf{r}$ the vector of the linear abscissae along these lines (cf. Figure 20.2). Then, it is straightforward to relate the nodal velocity components $\dot{\mathbf{x}}$ to $\dot{\mathbf{r}}$ through a matrix $\mathbf{A}$ containing the direction cosines $\boldsymbol{\alpha}_i$ of the N lines of movement,

$$\dot{\mathbf{x}} = \mathbf{A}\dot{\mathbf{r}} \tag{20.24a}$$

where

$$\dot{\mathbf{r}} = \{\dot{r}_1 \dot{r}_2 \dots \dot{r}_N\}^{\mathrm{T}} \tag{20.24b}$$

and

$$\mathbf{A} = \begin{bmatrix} \alpha_1 & & & \\ & \alpha_2 & & \mathbf{0} \\ & & \ddots & \\ & \mathbf{0} & & \alpha_N \end{bmatrix} \tag{20.24c}$$

Finally, substitution of Eqs. (20.21–20.24) into Eq. (20.19) leads to the following relationship between the nodal fluxes $\mathbf{q}$ and the nodal velocities in the directions of movement $\dot{\mathbf{r}}$:

$$\mathbf{q} = \mathbf{R}\dot{\mathbf{r}} \tag{20.25a}$$

where

$$\mathbf{R} = m \int_S [\mathbf{f}(\zeta, \eta)\boldsymbol{\alpha}_n^T(\zeta, \eta)\mathbf{F}(\zeta, \eta)\mathbf{A}]\,dS \tag{20.25b}$$

By suitably assembling Eq. (20.25a) written for all the element sides belonging to the phreatic surface, a global relationship is arrived at, which is used in the time marching procedure for transient analysis.

At the beginning t of the generic time increment Δt, a confined analysis is first carried out based on the current geometry of the finite element grid. This calculation, in which the free surface is treated as a pervious boundary, yields the flux vector $\mathbf{q}$ at the free surface nodes. Then, the velocities $\dot{\mathbf{r}}(t)$ of these nodes along their directions of movement are determined by solving the system of linear equations (20.25a). If forward Euler (explicit) time integration scheme is adopted, the coordinates of the free surface nodes at the end of the time increment Δt can be directly evaluated:

$$\mathbf{x}(t + \Delta t) = \mathbf{x}(t) + \Delta t \mathbf{A}\dot{\mathbf{r}}(t) \tag{20.26}$$

Alternatively, if an implicit time integration scheme is chosen based, for instance, on the following relationship:

$$\mathbf{x}(t + \Delta t) = \mathbf{x}(t) + \Delta t \mathbf{A}\{\beta\dot{\mathbf{r}}(t) + (1 - \beta)\dot{\mathbf{r}}(t + \Delta t)\} \quad (0 \leq \beta \leq 1) \tag{20.27}$$

then an iterative process has to be carried out for each time increment.

This approach for transient analyses presents an ambiguity in determining the movement of the node corresponding to the seepage point. In fact, if the entire flux computed by the confined analysis at this node is introduced in Eq. (20.25a), the velocity of the seepage point is likely to be incorrectly estimated because no account is taken of the part of flux, leaving the mesh through the seepage face. To avoid this problem, only a certain percentage of the flux at the seepage point should be considered in evaluating its velocity.

In spite of this provision, spurious oscillations of the phreatic surface may show up during the time integration, which depends upon the mentioned ambiguity in the movement of the seepage point. To constrain them, the position of the seepage point may be related to those of the adjacent node on the free surface through a suitable interpolation function.

20.5   'Fixed' Mesh Approaches

The techniques of this group operate on finite element meshes, the geometry of which remains constant during the iterative solution process. Consequently, the mesh should contain a part of the porous medium larger than that in which the seepage flow takes place.

These procedures are, in general, less accurate than those based on a variable mesh in defining the correct shape of the free surface. In fact, their results are markedly influenced by the size of the finite elements. On the other hand, they offer some appreciable advantages. In particular, they can be adopted in 'coupled' seepage and stress analyses (Li and Desai, 1983), which would not be a straightforward extension for the variable mesh approaches.

Among these techniques, two will be recalled here which involve solution procedures similar to those customarily adopted for nonlinear stress analyses.

20.5.1   *Variable permeability approach*

When applied in steady state conditions, this procedure replaces the intrinsic geometrical nonlinearity of free surface problems with a nonlinear relationship between pore pressure and the coefficient of permeability (Bathe and Khoshgoftaar, 1979; Desai and Li, 1983).

The real value of the permeability is adopted in the calculations if the pore pressure is positive, while a small (close to zero) value is assumed when the pore pressure becomes equal to, or decreases below, the atmospheric pressure (conditions that take place on the free surface and in the elements above it). Figure 20.3 shows a possible idealization of the relationship between pore pressure and coefficient of permeability that can be adopted in the calculations. Note that, for negative pore pressure, the permeability is slightly greater than zero, to avoid possible numerical instabilities during solution.

The solution of the nonlinear material problem can be based on the well known modified Newton scheme. This leads to an iterative procedure in which the assembled flow matrix has to be evaluated and triangularized only once, with a consequent reduction of the computational cost.

A confined analysis is first performed that yields the initial trial distribution of nodal hydraulic heads. In this initial analysis, in which the real values of the permeability coefficients is adopted for all elements, the contour of the mesh which is likely to be above the free surface is considered as an impervious boundary.

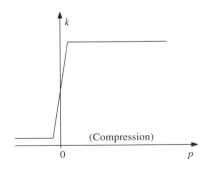

Figure 20.3   Schematic relationship between pore pressure p and coefficient of permeability k

On the basis of the initial nodal heads and of the element 'shape functions', the hydraulic head, and hence the pore pressure, is evaluated through Eq. (20.13) at the integration points within each element used in the numerical integration of the flow matrix (cf.Eq. (20.17)). The permeability at each integration point is determined according to the material law schematically shown in Figure 20.3, and the corresponding flow matrix $\mathbf{M}'$ of the assembled mesh is worked out. Then, the variation of the nodal hydraulic heads $\Delta\mathbf{h}_i$ for the next (i-th) iteration is determined:

$$\mathbf{M}\Delta\mathbf{h}_i = \mathbf{q} - \mathbf{q}_i \tag{20.28}$$

which depends, through the nodal fluxes $\mathbf{q}_i$, upon the flow matrix and on the nodal heads computed at the end of the previous iteration:

$$\mathbf{q}_i = \mathbf{M}'_{i-1}\mathbf{h}_{i-1} \tag{20.29}$$

The hydraulic head vector is finally updated and another iteration is performed:

$$\mathbf{h}_i = \mathbf{h}_{i-1} + \Delta\mathbf{h}_i \tag{20.30}$$

The above solution technique was extended to the analysis of transient free surface problems by Desai and Li (1983).

The described constant mesh procedure presents an ambiguity, similar to that affecting the variable mesh methods, in determining the position of the seepage point. In fact, since on the surface of seepage the hydraulic head is equal to the elevation, the previously described scheme for determining the integration points above the free surface cannot be adopted. To reduce this drawback, a flow-deflecting zone can be introduced at the seepage face (Bromhead, 1977), consisting of a thin layer of elements having a permeability markedly higher than that of the porous material.

20.5.2   Residual flow procedure

This solution procedure allows the free surface to pass through the elements by applying to the relevant nodes of the mesh 'residual' fluxes determined in such a way that the two boundary conditions on the free surface are simultaneously satisfied, at least in an average sense.

Consider first the steady state case (Desai, 1976). As in the previous constant mesh method, the iterative solution process is initiated by performing a confined analysis, in which the mesh contour above the extimated free surface is treated as an impervious boundary.

The nodal hydraulic heads obtained by this calculation are used to find the elements crossed by the line characterized by a vanishing pore pressure (denoted by S in Figure 20.4) that represents a first approximation of the free surface. The position of this line (on which only Eq. (20.12a) is fulfilled) within the elements is determined by means of Eqs. (20.13) and (20.4).

To enforce also the second free surface condition (Eq. (20.12b)), it is necessary to evaluate the flow velocity v_n normal to the free surface segments. Taking into account Eqs. (20.3) and (20.13), the following relationship is obtained, where ϑ represents the

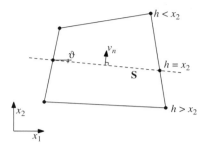

Figure 20.4   Approximation of the free surface S crossing a finite element

angle between the free surface segment and the x_1 direction (cf. Figure 20.4):

$$v_n = - \begin{bmatrix} \sin \vartheta \\ \cos \vartheta \end{bmatrix}^{\mathrm{T}} \mathbf{K} \begin{bmatrix} (\partial \mathbf{b}/\partial x_1)^{\mathrm{T}} \\ (\partial \mathbf{b}/\partial x_2)^{\mathrm{T}} \\ (\partial \mathbf{b}/\partial x_3)^{\mathrm{T}} \end{bmatrix}_S \mathbf{h} \qquad (20.31)$$

The flux associated with v_n crossing the free surface can be seen, using a terminology suited for stress analysis problems, as a 'distributed load' on segment S. This load has to be reduced to zero to fulfil the boundary condition (Eq. (20.12b)). This can be done by applying to the nodes of the corresponding element a set of 'residual' nodal fluxes $\mathbf{q}$ that produce across segment S a flux equal to the one associated to v_n, but with opposite sign:

$$\mathbf{q} = - \int_S \mathbf{b} v_n \, dS \qquad (20.32)$$

By applying the residual fluxes $\mathbf{q}$, evaluated for all the elements crossed by the free surface, to the relevant nodes and by solving again Eq. (20.16), a new vector of nodal hydraulic heads is determined. This leads to more refined approximation of the free surface geometry, and to a new residual flow vector. The iterative process is continued until the changes in geometry of the free surface becomes negligible.

When dealing with transient seepage problems (Bathe *et al.*, 1982), the flow velocity normal to the free surface is related to the velocity of the free surface itself by Eq. (20.11b) that, taking into account Eq. (20.11a), can be rewritten in the following form:

$$v_n - m\dot{h} \cos \vartheta = 0 \qquad (20.33)$$

where $\dot{h}$ is the derivative with respect to time of the hydraulic head.

To eliminate the flux through line S in excess to that associated with Eq. (20.33), the following residual flux vector q can be applied to the nodes of the elements crossed by the free surface:

$$\mathbf{q} = - \int_S \mathbf{b}(v_n - m\dot{h} \cos \vartheta) \, dS \qquad (20.34)$$

Taking into account Eqs. (20.31) and (20.13), the above equation can be re-written in the following equivalent form:

$$\mathbf{q} = \left\{ \int_S \mathbf{b} \begin{bmatrix} \sin \vartheta \\ \cos \vartheta \end{bmatrix}^{\mathrm{T}} \mathbf{K} \begin{bmatrix} (\partial \mathbf{b}/\partial x_1)^{\mathrm{T}} \\ (\partial \mathbf{b}/\partial x_2)^{\mathrm{T}} \\ (\partial \mathbf{b}/\partial x_3)^{\mathrm{T}} \end{bmatrix} \, dS \right\} \mathbf{h} + \left\{ \int_S m\mathbf{b}\mathbf{b}^{\mathrm{T}} \cos \vartheta \, dS \right\} \dot{\mathbf{h}} \qquad (20.35)$$

Adopting Euler's backward scheme for time integration, the solution for each time step can be obtained through an iterative procedure analogous to that adopted for steady state analyses.

Various applications of the residual flow procedure, like those involving three-dimensional conditions and the simultaneous flow of different fluids, were discussed by Desai *et al.* (1983), Sugio and Desai (1987) and Desai and Baseghi (1988).

Due to its rather general characteristics, this method can be also applied to other free surface problems. For instance, its extension to the determination of the frozen-unfrozen boundary during the artificial freezing of cohesionless soils is discussed by Gioda *et al.* (1994).

20.6   Some Alternative Solution Approaches

Various numerical techniques for free surface analyses have been presented in the literature which cannot be placed in the two previously defined categories. Only some of them are mentioned in the following, without discussing the details of their solution procedures.

One of the early numerical approaches for unconfined analyses (Desai and Sherman, 1971), particularly oriented towards problems involving the flow through river banks, was based on a finite difference scheme for solving the diffusion problem and on the approximate 'method of fragments' for determining the variation of the free surface during time. The combined use of the finite element method and of the method of fragments was subsequently suggested (Desai, 1973) to reduce the computational effort required for the solution of the nonlinear problem.

Another approximated solution method for transient problems was presented by Cividini and Gioda (1984). This procedure, based on the finite element method, operates on meshes of constant geometry, and models the free surface by means of a series of segments which coincide with sides of the elements. Therefore, only a discrete number of different positions can be reached by the free surface during the transient process. Figure 20.5 shows the numerical solution of the transient flow through a zoned dam induced by a 'rapid' filling of the upper reservoir.

Transient unconfined problems were also studied by Burghignoli and Desideri (1985), who proposed a method based on a mapping technique and on a finite difference integration scheme. This procedure requires a bijective mapping to be introduced between the physical domain, the shape of which is *a priori* unknown and varies with time for transient problems, and a fixed integration domain, having a conveniently simple (square) shape. In other words, a relationship has to be established such that every point of the fixed domain corresponds to a unique point of the physical domain, and *vice versa*. Due to its simple shape, a finite difference integration scheme can be adopted on the fixed domain. The solution is mapped on the physical domain, obtaining both the hydraulic head distribution and the shape of the flow region. Some comments on this method can be also found in Gioda and Desideri (1988).

A different mapping technique (Morland and Gioda, 1990), limited to the analysis of steady state seepage through homogeneous trapezoidal dams, is based on a general approach for diffusion problems with a moving boundary (Morland, 1982). In this case, the bijective mapping is established between the physical domain and a fixed domain having

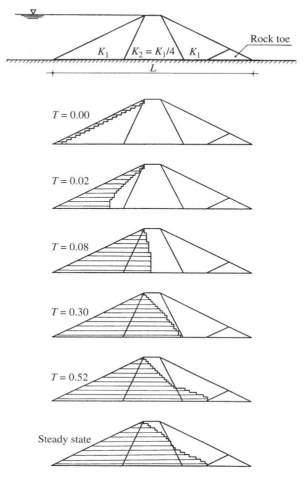

Figure 20.5 Approximated analysis of the transient flow caused in a zoned dam by the filling of the upper reservoir. The non-dimensional time T is defined as $T = tK/mL$ (where t = time, K = coefficient of permeability, m = porosity)

a triangular shape. The unconfined analysis in the physical domain is made equivalent to the problem of determining the unknown distribution of permeability in the fixed domain, under the transformed boundary conditions. The problem in the fixed domain is solved through a finite element analysis equivalent to that adopted for a confined problem. A simple 'backward' transformation recovers the distribution of hydraulic head in the physical domain and the shape of the free surface.

This mapping technique showed problems in determining the correct position of the seepage point similar to those affecting the variable mesh techniques. To remove them, it was imposed in the discretized problem the condition that the free surface is tangent to the downstream side of the dam. Note that, in standard finite element analyses, it would not, in general, be possible to impose this condition unless higher order isoparametric elements are used.

20.7   Position of the Seepage Point in Variable Mesh Analyses

It was previously mentioned that variable mesh techniques present some problems in determining the position of the seepage point, i.e. of the intersection between the surface of seepage and the wet boundary. To obtain an insight into this aspect of the numerical solution let us consider the steady state seepage flow through the homogeneous and isotropic rectangular block dam shown in Figure 20.6.

The numerical analyses are based on Taylor and Brown's (1967) technique. Since the solution procedure does not provide a criterion for determining the elevation of the seepage point *P*, a series of analyses was carried out by choosing *a priori* the position of this point, and then applying the previously outlined method of solution. The results of these analyses are shown by dashed lines in Figure 20.6, while a solid line represents the 'optimal' shape of the free surface evaluated according to a modified procedure that will be subsequently outlined.

Two aspects of these solutions should be pointed out. First, no stability problems were observed during the analyses, regardless of the difference between the initial trial shape of the free surface and the final results. Secondly, all solutions are correct 'in the model'; in fact they exactly fulfil the boundary condition defined by Eq. (20.12a) at the first and last nodes of the free surface (that also belong to pervious boundaries) and Eqs. (20.12a) and (20.12b) at the remaining free surface nodes. However, the only correct solution is the one in which the 'correct' position of the seepage point is assumed.

In other words, the finite element variable mesh technique provides an infinite number of stable solutions that fulfil the governing equation within the flow domain and all relevant boundary conditions. This indicates that the numerical analysis is affected by a non-uniqueness of solution, which is not present in the continuum problem, and that is apparently introduced by the discretization process.

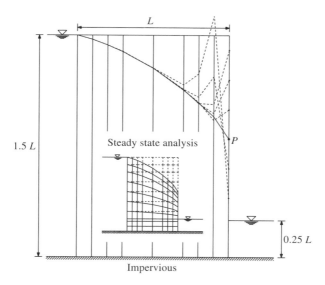

Figure 20.6   Influence of the chosen position of the seepage point *P* on the numerically evaluated free surface from the variable mesh analysis of the steady state flow through a rectangular block dam

Similar effects can also be observed when solving transient free surface problems (Cividini and Gioda, 1989). In this case, the numerical results indicate that the solution process is stable during time. In fact, no appreciable changes in the numerical results were observed by varying the value of the time increment and the integration technique from explicit to implicit. However, the iterative process may converge to inaccurate solutions for the intermediate time steps, and it may also lead to an inaccurate final (steady state) geometry of the phreatic line.

To eliminate the observed 'oscillations' of the free surface close to the seepage point, various provisions have been suggested in the literature. In some cases, it is attempted to constrain them by describing the shape of the phreatic line by means of *a priori* chosen functions or splines. In other instances, the use of very fine meshes close to the seepage face is suggested, so that the oscillations are confined in a narrow zone having a limited influence on the overall results.

Conditions imposed directly on the node corresponding to the seepage point or in its vicinity are also used. For instance, by assuming that the flux at the seepage node is equal to half the flux evaluated at the neighboring node on the wet boundary, or by imposing a given slope of the phreatic line at the intersection with the seepage face.

Even though these provisions, if properly used, may lead to satisfactory results, they do not appear to have a general validity.

A technique initially developed for avoiding these drawbacks in steady state analyses (Gioda and Gentile, 1987), and subsequently extended to transient conditions (Cividini and Gioda, 1989), will be briefly recalled here.

The most straightforward way to eliminate the observed non-uniqueness of numerical results is to establish a criterion for choosing among them the one closest to the correct physical solution. Since this effect is not present in the continuum problem, it seems preferable to work out a criterion involving quantities which are meaningful only for the discrete model, i.e. the nodal variables in the finite element context. In addition, in the spirit of the finite element method the criterion should not be of a 'local' nature (i.e. involving nodal variables only at one specific location), but it should be expressed by integrals or summations over a significant portion of the finite element mesh.

Finally, from the numerical viewpoint, the influence of this criterion should become weaker and weaker with increasing refinement of the mesh, and it should virtually disappear when tending to the continuum problem.

It can be observed that this criterion cannot involve the nodal variables of the assembled finite element mesh. In fact, they should obey only the governing equation and the relevant boundary conditions. Consequently, additional relationships imposed among them would lead to inaccurate solutions 'in the model'.

Hence, the remaining alternative is to base the criterion on the nodal variables at the element level, in particular, on the nodal fluxes leaving and entering the element facing the free surface at the free surface nodes. A pictorial representation of these quantities is shown in Figure 20.7. Note that the nodal fluxes are scalar variables, hence they are represented in Figure 20.7 by means of arrows only for graphical reasons. These fluxes are evaluated by multiplying the flow matrix (cf. Eq. (20.17)) of each element adjacent to the phreatic line by the corresponding nodal hydraulic heads.

The impervious boundary condition imposed on the free surface implies the continuity of the inter-element nodal fluxes at the free surface nodes, e.g. $q_{2A} = q_{1B}$ (because of this condition, they are referred to in the following as fluxes 'tangent' to the free surface), but

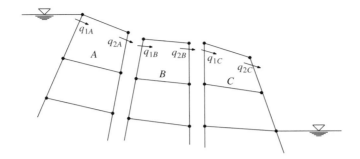

Figure 20.7   Pictorial representation of the inter-element fluxes at the free surface nodes

no constraint is imposed between the fluxes at the nodes of the same element, e.g. q_{1B} and q_{2B}.

Now, considering that in the correct solution of the unconfined steady state problem the streamlines in the vicinity of the phreatic surface are almost parallel to it, it seems reasonable to choose among the infinite possible numerical solutions the one corresponding to the most regular variation of the nodal fluxes entering and leaving the element facing the free surface.

This criterion can be translated into numerical terms by choosing the solution that minimizes the following function Q:

$$\min_{x_{2P}}\{Q^2 = \sum_i [q_{1i} + q_{2i} + \cdots + q_{ni}]^2\} \tag{20.36}$$

Here, x_{2P} is the elevations of the seepage points that represents the free variables of the minimization problem. The index i includes the elements facing the free surface and n represents, for each of these elements, the number of nodes belonging to the phreatic surface. Note that, since inward and outward fluxes have opposite signs, the function Q tends to become zero if a constant flux takes place through the free surface nodes. Equation (20.36) can be also adopted in the presence of several seepage points.

Many standard algorithms are applicable to the minimization of the above function. Among them, the so-called direct search techniques seem preferable, since they reach the minimum through a sequence of evaluations of the function without requiring the determination of its gradient (e.g. see Himmelblau, 1972). This characteristic leads to a simple iterative solution procedure.

At each iteration the minimization program modifies the free variable vector (i.e. the elevations of the seepage points), and an unconfined analysis is carried out, obtaining (in addition to the shape of the phreatic line corresponding to the current position of the seepage nodes) the nodal fluxes and the value of function Q. On the basis of this value, the minimization program defines new elevations of the seepage points and the subsequent iteration is carried out. The process continues until a convergence criterion is fulfilled.

The heavy solid line in Figure 20.6 represents the shape of the steady state free surface obtained by the modified solution procedure for the rectangular block dam problem.

The extension of this procedure to transient problems, in two- and three-dimensional conditions, can be found in Cividini and Gioda (1989).

20.8   Concluding Remarks

Some finite element approaches for the solution of free surface seepage problems have been discussed, subdividing them into variable mesh and fixed mesh procedures.

The techniques of the first group modify the geometry of a finite element mesh during an iterative process until the correct shape of the free surface is modeled with a sufficient approximation. Those in the second group operate on meshes the shape of which does not change, and present some similarities with the methods used for nonlinear elastic or elasto-plastic stress analyses.

Also, some alternative solution strategies have been reviewed. Some of these strategies introduce particular simplifying assumptions to reduce the computational effort. Other approaches introduce a mapping 'rule' between the physical domain and a fixed domain having a known, simple geometry. The problem is integrated over the fixed domain, and the shape of the free surface is subsequently determined through the mapping.

The results of the numerical studies based on the described techniques suggest comments on the numerical analysis of unconfined seepage flows.

Some approximate procedures actually reduce the computational time with respect to the non-approximate ones. However, the high speed of modern computers perhaps reduce this advantage from a practical viewpoint.

The mapping techniques, being limited to simple geometries (e.g. trapezoidal dams) or to the case of homogeneous media, have an application potential smaller than that of other numerical procedures. On the other hand, they are characterized by an accuracy higher than that of other fixed mesh methods.

From the programming point of view, the fixed mesh methods present some advantages with respect to the variable mesh procedures. In fact, they are similar to the methods commonly used for solving elasto-plastic stress analyses and, consequently, their implementation can be based on already available finite element codes for nonlinear stress analysis. This leads to a non-negligible reduction of the programming effort with respect to the variable mesh procedures that usually require the preparation of *ad hoc* codes. In addition, some programming difficulties may be encountered when applying the variable mesh methods to layered or non-homogeneous media.

As to the accuracy of results, it appears that variable mesh methods define the shape of the free surface with a precision higher than that of the fixed mesh methods, when both are applied to the same finite element grid. On the other hand, it should be recognized that variable mesh approaches are, in general, affected by stability and convergence problems that are less pronounced in the fixed mesh analyses.

For steady state analyses, these stability problems can be particularly evident if the initial trial shape of the free surface is markedly different from the final solution. A similar effect might show up also in transient analyses, if the adopted value of the time increments is too large, i.e. when a marked difference exists between the shapes of the free surface at two subsequent solution steps.

Another case in which variable mesh methods could present convergence problems, that in some instances lead to an apparent non-unique solution, is when the point of intersection between the free surface and a pervious boundary has to be determined. A possible way to overcome this difficulty consists of introducing a criterion for choosing among the possible solutions the most accurate one. This is apparently more effective than to enforce local conditions directly on the seepage point or on the elements in its vicinity.

It should be noted, however, that this provision requires the use of a mathematical programming algorithm for function minimization. Consequently, its introduction leads to an increase in programming effort with respect to standard methods.

From the computational point of view, the variable mesh approaches are, in general, heavier than fixed mesh methods. This is due to the fact that variable mesh methods modify the finite element grid during the solution process and, consequently, reassemble at least a part of the flow matrix and triangularize it at each iteration. This is not required by fixed mesh methods using the modified Newton scheme.

In choosing between the analysis techniques, one should also consider that the variable mesh methods are restricted only to the solution of seepage flow problems. On the contrary, fixed mesh methods can be also used in coupled analyses, i.e. in the consolidation problems when the influence of the seepage flow on the deformation of the soil skeleton cannot be neglected.

From the above considerations its seems possible to conclude that fixed mesh methods are in general more 'robust', and perhaps more 'flexible', than variable techniques. They are particularly suitable for unconfined seepage analyses when a limited approximation of the free surface is required. If more refined results are needed, then a variable mesh analysis could be necessary. Some of the mentioned stability problems can be avoided in this case if the input data of the variable mesh analysis are defined on the basis of the geometry of the flow region obtained from a preliminary fixed mesh calculation.

Acknowledgements

Some of the examples presented in this chapter are part of studies supported by the Ministry of University and Research of the Italian Government.

References

Banerjee, P. K., Butterfield, R. and Tomlin, G. R. Boundary element methods for two-dimensional problems of transient ground water flow. *Int. J. Numer. Anal. Methods in Geomech.*, **5**, (1981) 15–31.

Bathe, K. J. and Khoshgoftaar, M. R. Finite element free surface seepage analysis without mesh iteration. *Int. J. Numer. Anal. Methods in Geomech.*, **3**(13), (1979) 13–22.

Bathe, K. J., Sonnad, V. and Domigan, P. Some experiences using finite element methods for fluid flow problems. *Proc. 4th Int. Conf. on Finite Element Methods in Water Resources*, Hannover, (1982).

Bromhead, E. N. Discussion of *Finite Element Scheme for Unconfined Flow.*, by C. S. Desai. *Int. J. Numer. Methods in a Engrg.*, **11**(5), (1977) 908–910.

Burghignoli, A. and Desideri, A. Free surface flow in soils. *Proc. 5th Int. Conf. on Numerical Methods in Geomechanics*, Nagoya, (1985).

Chen, R. T. S. and Li, C.Y. On the solution of transient free-surface flow problems in porous media by the finite element method. *J. Hydrology*, **20**, (1973) 49–63.

Cividini, A. and Gioda, G. An approximate F. E. analysis of seepage with a free surface. *Int. J. Numer. Anal. Methods in Geomech.*, **8**, (1984) 549–566.

Cividini, A. and Gioda, G. On the variable mesh finite element analysis of unconfined seepage problems. *Geotechnique*, **39**, (1989) 251–267.

Cryer, C. W. A bibliography of free boundary problems. *Technical Report No. 1793*, University of Wisconsin, Mathematics Research Center, (1977).

Desai, C. S. and Sherman, W. C. Unconfined transient seepage in sloping banks. *J. Soil Mech. Found. Div.*, **97**(SM2), (1971) 357–373.

Desai, C. S. Seepage analysis of earth banks under drawdown. *J. Soil Mech. Found. Div.*, **98**(SM11), (1972) 1143–1162.

Desai, C. S. Approximate solution for unconfined seepage. *J. Irrigation and Drainage Div.*, **99**(IR1), (1973) 71–87.

Desai, C. S. Finite element residual schemes for unconfined flow. *Int. J. Numer. Methods in Engrg.*, **10**, (1976) 1415–1418.

Desai, C. S., Flow through porous media. In *Numerical Methods in Geotechnical Engineering*, Desai, C. S. and Christian J. T. (eds.). McGraw-Hill, New York, (1977).

Desai, C. S. and Li, C. G. A residual flow procedure and application for free surface flow in porous media. *Int. J. Adv. Water Resources*, **6**, (1983) 27–35.

Desai, C. S., Lightner, J. G. and Somasundaram, S. A numerical procedure for three-dimensional transient free surface seepage. *Adv. Water Resources*, **6**, (1983) 175–181.

Desai, C. S. and Baseghi B. Theory and verification of residual flow procedure for 3-D free surface seepage. *Adv. Water Resources*, **11**, (1988) 192–203.

France, P. W. Finite element analysis of three-dimensional groundwater flow problems. *J. Hydrology*, **21**, (1974) 381–398, .

Gioda, G. and Gentile, C. A nonlinear programming analysis of unconfined steady state seepage. *Int. J. Numer. Anal. Methods in Geomech.*, **11**, (1987) 283–305.

Gioda, G. and Desideri, A. Some numerical techniques for free surface seepage analysis. *Proc. 6th Int. Conf. on Numerical Methods in Geomechanics*, Innsbruck, (1988).

Gioda, G., Locatelli, L. and Gallavresi, F. A numerical and experimental study of the artificial freezing of sand. *Canad. Geotech. J.*, **31**(1), (1994).

Himmelblau, D. M. *Applied Non-linear Programming*. Mc-Graw-Hill, New York, (1972).

Lacy, S. J. and Prevost, J. H. Flow through porous media: a procedure for locating the free surface. *Int. J. Numer. Anal. Methods in Geomech.*, **11**(6), (1987) 585–601.

Li, G. C. and Desai, C. S., Stress and seepage analysis of earth dams. *J. Geotech. Engrg.*, **109**(7), (1983) 946–960.

Liggett, J. A. Location of free surface in porous media. *J. Hydraulic Div.*, **103**(HY4), (1977) 353–365.

Liggett, J. A. and Liu, P. L. F. *The Boundary Integral Equation Method for Porous Media Flow*. George Allen and Unwin, London, (1983).

Morland, L. W. A fixed domain method for diffusion with a moving boundary. *J. Engrg. Math.*, **16**, (1982) 259–269.

Morland, L. W. and Gioda G. A mapping technique for steady state unconfined seepage analysis. *Int. J. Numer. Anal. Methods in Geomech.*, **14**, (1990) 303–323.

Neuman, S. P. and Witherspoon, P. A. Finite element method of analyzing steady seepage with a free surface. *Water Resources Res.*, **6**(3), (1970) 889–897.

Neuman, S. P. and Witherspoon, P. A. Analysis of nonsteady flow with a free surface using the finite element method. *Water Resources Res.*, **7**(3), (1971) 611–623.

Taylor, R. L. and Brown, C. B. Darcy flow solution with a free surface. *J. Hydraulic Div.*, **93**(HY2), (1967) 25–33.

Taylor, C. J., France, P. W. and Zienkiewicz O. C. Some free surface transient flow problems of seepage and irrotational flow. In *The Mathematics of Finite Elements and Applications*, Whiteman, J. R. (ed.). Academic Press, London, (1973).

Sugio, S. and Desai, C. S. Residual flow procedure for sea water intrusion in unconfined aquifers. *Int. J. Numer. Methods in Engrg.*, **24**, (1987) 1439–1450.

Zienkiewicz, O. C., Mayer, P. and Cheung, Y. K. Solution of anisotropic seepage by finite elements. *J. Engrg. Mech. Div.*, **92**(EM1), (1966) 111–120.

21

Simulating Fully 3-D Hydraulic Fracturing

B. J. Carter[1], J. Desroches[2], A. R. Ingraffea[1] and P. A. Wawrzynek[1]
[1]*Cornell University, Ithaca, NY, USA*
[2]*Schlumberger Well Services, Houston, TX, USA*

21.1   Introduction

Hydraulic fracturing, the process of initiation and propagation of a crack by pumping fluid at relatively high flow rates and pressures, is one of several techniques for creating cracks in rock. Fractures in the earth's crust are desired for a variety of reasons, including enhanced oil and gas recovery, re-injection of drilling or other environmentally sensitive wastes, measurement of *in situ* stresses, geothermal energy recovery, and enhanced well water production. These fractures can range in size from a few meters to hundreds of meters, and their cost is often a significant portion of the total development cost. In locations where the *in situ* stress field, including the directions, is known and the wellbore is aligned with one of the far-field principal stresses, the hydraulic fracture geometry can be predicted and controlled with reasonable accuracy. For those wellbores that are not aligned with such a direction (deviated wells), the hydraulic fracture geometry is usually more complex and more difficult to model, especially close to the wellbore where the local stress field is significantly different from the far-field stresses. Field data from hydraulic fracturing operations exist primarily in the form of pressure response curves. It is difficult to define the actual hydraulic fracture geometry from this data alone, however. Therefore, numerical simulations are used to evaluate and predict the location, direction and extent of these hydraulic fractures.

Simulations range from two- to fully three-dimensional depending on the degree of complexity of the wellbore and fracture geometries, the capability of the available simulator, and the required accuracy of the predictions. Numerous 2-D, pseudo-3-D, and planar 3-D hydraulic fracturing simulators exist, and these simulators work very well in many cases where the geometry of the fracture is easily defined and constrained to a single plane. However, there are instances where a fully 3-D simulator is necessary for more accurate modeling. For example, fractures from deviated wellbores are generally non-planar with arbitrary crack front shapes. Most hydrofracturing simulators simply ignore the near-wellbore effects of deviated wells and assume a planar starting crack that has extended beyond this region. The problem with this approach is that most of the difficulties and failures in hydraulic fracturing of deviated wellbores can be attributed to restricted flow in the near-wellbore region. Restricted flow usually is due to fracture reorientation and interaction with other fractures. Hydraulic fracturing of deviated wells, where the fractures reorient as they propagate, clearly requires fully 3-D simulation capabilities and accurate modeling of the near-wellbore geometry.

Efficient numerical simulation of fully 3-D hydraulic fracturing requires at least two key components. The first is a capability for representing and visualizing the complex wellbore and fracture geometries. The second is a method for solving the highly nonlinear coupling between the equations for the fluid flow in the fracture and the deformation and propagation of the fracture. The first component can be partitioned into several sub-components, including geometrical and topological solid modeling tools, routines to model geometry and topology changes for fracture propagation, automated meshing and remeshing capabilities, visualization of response information, and analysis control information (i.e. input for a stress analysis program). The second component consists of a stress analysis procedure, fluid flow simulation capabilities, and a method for coupling the structural response with the fluid flow, including rules for determining hydraulic fracture propagation direction and extent. The authors have developed a simulator that includes all of these components for modeling multiple, non-planar, fully 3-D, hydraulic fracture propagation. This simulator treats hydraulic fracturing as a quasi-static process, and the solution consists of a series of 'snapshots in time' of the fracture geometry, fluid pressures, and crack opening displacements.

A brief history of hydraulic fracturing and the development of hydrofracturing simulators are discussed in the next section. Then the key components of the new simulator are described in some detail, including the software framework for model representation, the coupled elasticity and lubrication theory, the finite element implementation of this theory, and the iterative solution procedure. The simulator has been verified for a radial (penny-shape) and a slot-like (part-through) fracture by comparing results to those from a robust and accurate 2-D simulator, Loramec (Desroches and Thiercelin, 1993). Further simulations that compare well with experimental results show the ability of the system to model more realistic hydraulic fracture geometries. This chapter concentrates on the development of the simulator rather than actual simulations, however. Detailed simulations of hydraulic fracturing from cased, perforated, deviated wellbores is possible now, but is computationally intensive and is left as the subject of future publications.

21.2   Background

The process of hydraulic fracturing is not new. Nature has produced many such fractures in the earth's crust (see, for example, Bahat, 1991). The first recorded application of hydraulic fracturing for enhancing oil recovery that the authors are aware of was in 1947 in Kansas (see Howard and Fast, 1970). The 'Hydrafrac' concept was formalized first by Clark (1949), although others had recognized previously that 'pressure parting' could occur during acid treatment and water injection, a phenomenon considered to be closely related to the 'Hydrafrac' concept (Howard and Fast, 1970). The Hugotan field in western Kansas was the site of the first hydraulic fracturing operation, and by the mid-1960s, hydraulic fracturing had become the dominant method of stimulation in this and many other fields, including geomechanics (Howard and Fast, 1970).

During this early period of hydraulic fracturing, two simple models were proposed to try to predict the shape and size of a hydraulic fracture based on the rock and fluid properties, the pumping parameters, and the *in situ* stresses (Khristianovic and Zheltov, 1955; Geertsma and de Klerk, 1969; Perkins and Kern, 1961; Nordgren, 1972). The models are known as the KGD and PKN models, and their description can be found in the above references and in many other summaries or texts (e.g. Geertsma, 1989; Mendelsohn, 1984a,b). Perkins and Kern (1961) and Geertsma and de Klerk (1969) also derived a model for radial hydraulic fracturing. The radial, KGD, and PKN models are essentially two-dimensional plane strain formulations with fluid flow only along the length (or radius) of the fracture. The fracture width and shape are related to the fluid pressure distribution in the fracture; the KGD model has a constant height and constant width through the height, while the PKN model has a constant height and an elliptical vertical cross-section.

The 2-D models are not able to simulate both vertical and lateral propagation. Therefore, pseudo-3-D models were formulated by removing the assumption of constant and uniform height (Settari and Cleary, 1986; Morales, 1989). The height in the pseudo-3-D models is a function of position along the fracture as well as time. The major assumption is that the fracture length is much greater than the height, and an important difference between the pseudo-3-D and the 2-D models is the addition of a vertical fluid flow component. The pseudo-3-D models have been used to model fractures through multiple rock layers with differing stresses and properties. These models are simple, fast, and relatively effective. Warpinski *et al.* (1994) recently provided brief descriptions and a comparison of predictions for a number of simulators, including 2-D and pseudo-3-D models.

Pseudo-3-D models cannot handle fractures of arbitrary shape and orientation, however; fully 3-D models are required for this purpose. The literature contains numerous references to fully 3-D simulators; the majority of these are limited to planar fracture surfaces, however. These are called planar 3-D simulators in this chapter to differentiate them from true fully 3-D simulators which can model out-of-plane fracture growth. Planar 3-D simulators have been developed by Clifton and Abou-Sayed (1979), Barree (1983), Touboul *et al.* (1986), Morita *et al.* (1988), Advani *et al.* (1990), and Gu and Leung (1993). Out of plane 3-D hydraulic fracture growth has been modeled by Lam *et al.* (1986), Vandamme and Jeffrey (1986), and Sousa *et al.* (1993). To the authors' knowledge, only Carter *et al.* (1994), using the predecessor of the simulator described herein, have modeled 3-D fracture in the near-wellbore region of a cased, perforated, and deviated wellbore.

The increasing use of deviated wellbores implies that fully 3-D hydraulic fracture simulators are vital to the petroleum industry. Hydrofracturing is often less effective for deviated wellbores as compared to traditional vertical wells. Some of the problems have been attributed to a poor understanding of the mechanics of fracture initiation and propagation from a deviated wellbore. The complex state of stress which is generated around an inclined wellbore (Yew and Li, 1988; Ong and Roegiers, 1995) means that the fracture propagates with a complex geometry (Behrmann and Elbel, 1990; Hallam and Last, 1991; Weijers and de Pater, 1992; Abass *et al.*, 1996). The complex stress state and fracture geometry can limit the fracture width at the wellbore and hinder the injection of proppant into the fracture leading to premature screenout (Hallam and Last, 1991; Soliman *et al.*, 1996). Nevertheless, the advantages of drilling inclined wellbores are significant. For example, the ability to drill several wells from a single location minimizes production infrastructure and impact on the environment. Therefore, the ability to model hydraulic fracturing from deviated wells is of ever increasing importance.

In addition to inadequate modeling of the fracture geometry, many of the current hydraulic fracturing simulators do not predict the correct wellbore fluid pressure or fracture geometry even for planar fractures. The proposed reasons for this are numerous (Medlin and Fitch, 1983; Warpinski, 1985; Shlyapobersky *et al.*, 1988; Jeffrey, 1989; Palmer and Veatch, 1990; Johnson and Cleary, 1991; Gardner, 1992; Papanastasiou and Thiercelin, 1993; de Pater *et al.*, 1993 and van den Hoek *et al.*, 1993). The simulator developed here addresses this problem by properly modeling the near crack tip behavior (SCR, 1993). Furthermore, the simulator is ideal for modeling non-planar hydraulic fracturing, and is able to model multiple branching, intersecting, and merging fractures as will be shown in the following section.

21.3 Model Representation

The first key component of an efficient, fully 3-D, hydraulic fracture simulator is the geometric representation of the model. Representation implies computer storage and visualization of the model topology and geometry. This portion of the hydraulic fracture simulator is actually a general purpose, fully 3-D, fracture analysis code, called FRANC3D, under development at Cornell University since 1987 (Martha, 1989; Wawrzynek, 1991; Potyondy, 1993). FRANC3D is capable of modeling multiple, arbitrary, non-planar, 3-D cracks in complex structures, and has pre-and post-processing capabilities for both finite and boundary elements. It relies on a boundary surface representation of the model and a radial edge data structure for storing and accessing topological and geometrical information. It has the ability to do fully automatic or fully user-controlled crack growth simulations, including post-processing of the response information, modifying the geometry, remeshing, and updating the boundary conditions for each stage of crack growth. The complex geometry associated with perforated, cased, and deviated wellbores with multiple non-planar evolving 3-D cracks requires a sophisticated, but easy to use simulation capability. FRANC3D has these capabilities, and some of its individual components are described briefly in the following sections.

21.3.1 *Representational model of fracture propagation*

Crack growth simulation in FRANC3D is an incremental process, where a sequence of operations is repeated for a progression of models (Figure 21.1). Each step in the

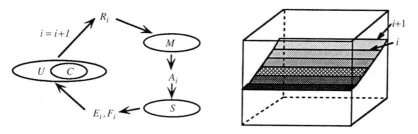

Figure 21.1 Incremental crack growth simulations; *i* denotes the increment of crack growth

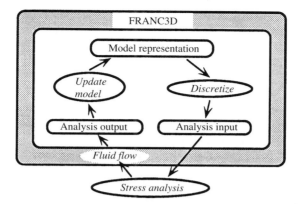

Figure 21.2 FRANC3D encompasses most of the conceptual crack growth simulation model

process relies on previously computed results and represents one crack configuration. There are four primary collections of data, or databases, required for each step. The first is the representational database, denoted R_i (where the subscript identifies the step). The representational database contains a description of the solid model geometry, including the cracks, the boundary conditions, and the material properties. The representational database is transformed by a discretization (or meshing, M) process to a stress analysis database A_i. The analysis database contains a complete, but approximate description of the body, suitable for input to a solution procedure (S), usually a finite or boundary element stress analysis program.

The solution procedure is used to transform the analysis database to an equilibrium database E_i which consists of field variables, such as displacements and stresses, that define the equilibrium solution for the analysis model A_i. The equilibrium model should contain field variables and material state information for all locations in the body, and in the context of a crack growth simulation, should also contain values for stress-intensity factors, or other fracture parameters F_i for all crack fronts. The equilibrium database is used in conjunction with the current representational database to update ($U(C)$) the representational model R_{i+1} including the increment of crack growth as governed by the fracture parameters F_i and the crack growth function C. This process is performed repeatedly (Figure 21.1) until a suitable termination condition is reached.

FRANC3D encompasses all components of this conceptual model except for the stress analysis procedure (Figure 21.2). The individual components consist of unique databases and functions that operate on the databases, some of which are described in more detail in the following sections.

21.3.2 Solid modeling for crack growth simulations

Simulation of crack growth is more complicated than many other applications of computational mechanics because the geometry and topology of the structure evolve during the simulation. For this reason, a geometric description of the body that is independent of any mesh needs to be maintained and updated as part of the simulation process. The geometry database should contain an explicit description of the solid model including the crack.

The three most widely used solid modeling techniques, boundary representation (B-rep), Constructive Solid Geometry (CSG), and Parametric Analytical Patches (PAP) (Hoffmann, 1989; Mäntylä, 1988; Mortenson; 1985), are capable of representing uncracked geometries. A B-rep modeler stores surfaces and surface geometries explicitly. If explicit topological adjacency information (as defined in the next section) is available as well, two topologically distinct surfaces can share a common geometric description. Cracks, for instance, consist of two surfaces that have the same geometric description; for this reason, among others, a boundary representation was found to be the most suitable of the three modeling techniques for modeling cracks.

21.3.3   Computational topology as a framework for crack growth simulation

Explicit topological information is an essential feature of the representational database for crack growth simulations. The topology of an object is the information about relationships, proximity, and order among features of the geometry – incomplete geometric information. These are the properties of the actual geometry that are invariant with respect to geometric transformations; the geometry can change, but the topology remains the same (Figure 21.3). A topology framework serves as an organizational tool for the data that represents an object and the algorithms that operate on the data.

There are several reasons for using a topological representation for crack growth simulation: (1) topological information, unlike geometrical information, can be stored exactly with no approximations or ambiguity; (2) there are formal and rigorous procedures for storing and manipulating topology data (Mäntylä, 1988; Hoffmann, 1989; Weiler, 1986); (3) any topological configuration can represent an infinite number of geometrical configurations; and (4) topology generally changes much less frequently than the geometry during crack propagation. Investigations into the use of data structures for storing information needed for crack propagation simulations (Wawrzynek and Ingraffea, 1987a,b) showed that topological databases were a convenient and powerful organizing agent, and efficient topological adjacency queries make this data structure ideally suited for interactive modeling.

Explicit topological information is used as a framework for the representational database R_i and aids in implementation of the meshing function M and the updating function

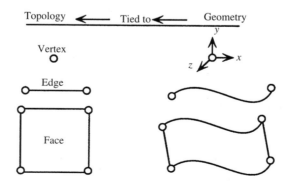

Figure 21.3   Relationship between topology and geometry

$U(C)$. In particular, by using a topological database in conjunction with a B-rep modeler, topological entities can serve as the principal elements of the database with geometrical descriptions and all other attributes (such as boundary conditions and material properties) accessed through the topological entities.

Several topological data structures have been proposed for manifold objects: the winged-edge (Baumgart, 1975), the modified winged-edge, the face-edge, the vertex-edge (Weiler, 1985), and the half-edge (Mäntylä, 1988) data structures. However, these data structures cannot be used for modeling fully 3-D hydraulic fracturing because features like bi-material interfaces create non-manifold topologies. Weiler (1986) presented another edge-based data structure for storing non-manifold objects, called the radial-edge, and outlined the corresponding generalized non-manifold Euler operators. The basic topological entities used for modeling are vertices, edges, faces, and regions. An internal crack, for example, consists of vertices, edges, and faces with a null volume region between the crack surfaces. The edge entity is the object through which topological relationships are maintained and queried (Figure 21.4). As the name implies, the edge uses are ordered radially about the edge; each face has two face uses and each face use has a corresponding edge use on the given edge. The radial ordering allows for efficient storage, querying and manipulation of the model topology. As shown in Figure 21.5, this data structure, in addition to bi-material interfaces, is clearly able to represent model topologies consisting of branching or intersecting cracks; both are important features when modeling hydraulic fracturing in a layered rock mass from a cased and deviated wellbore.

A crack is defined within this representational database by both geometry and topology. It consists of multiple surfaces in order to represent the evolving geometry as well as the possibility of intersecting, branching, and merging cracks. Crack surfaces are arranged in pairs (main and mate surface; see Figure 21.5), and each surface is composed of faces, edges and vertices. The edges and vertices are further classified based on their location on the crack surface. For instance, crack front edges represent the leading edge of the crack within the solid. Note that crack growth involves modifying the model topology and geometry to represent newly created fracture surfaces.

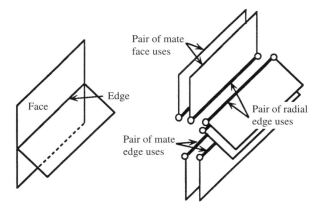

Figure 21.4 The radial-edge database relies on the radial ordering of edge uses about each edge. A face has two face uses and the edge has a use with respect to each face use (after Weiler, 1986)

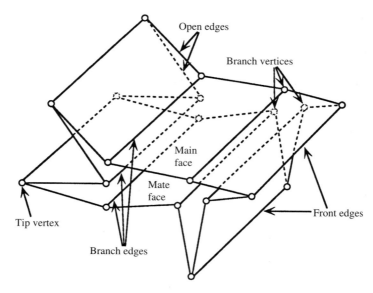

Figure 21.5   The topology of a typical branching crack

21.3.4   *Meshing, crack growth and model update functions*

The combination of the boundary representation solid model and the radial edge topological database comprises the representational model R_i. Complex 3-D models of deviated, perforated, and cased wellbores including multiple non-planar fractures can be built fairly quickly and easily using this representational model. The other components of the abstract model, such as the discretization (meshing), crack growth and model updating, post-processing, and visualization also take advantage of the topological database. The meshing capabilities (Potyondy *et al.*, 1995) and the crack growth and model update functions (Martha *et al.*, 1993; Carter *et al.*, 1997) in FRANC3D have been described elsewhere. For completeness however, a brief description of both functions is needed here as they relate to modeling hydraulic fracturing.

FRANC3D maintains a consistent geometric representation of the model at each step of propagation. During fracture propagation, the previous crack surface geometry remains the same; new fracture surface is simply added to the model to represent the crack growth (Figure 21.6). There are some exceptions when it is necessary to rebuild the entire fracture geometry, but this is beyond the scope of the present discussion. Therefore, the mesh that is attached to the existing geometric crack surfaces is unaffected by fracture growth because the existing geometry does not change. In truth, the mesh is removed from the geometric crack surface during propagation, but an identical mesh can be regenerated on that surface. A new mesh is attached to the new crack surface. Thus, the process of modeling crack propagation involves neither a 'fixed' nor a 'moving' mesh as described in other hydraulic fracturing literature. The mesh on the existing crack surface can remain fixed or it can be modified, but a new mesh must be added to the new crack surface. Mapping of information from the previous step of propagation to the current step is discussed later.

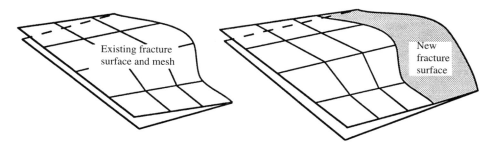

Figure 21.6   Fracture propagation is modeled by adding new crack surface geometry. The mesh which is attached to the geometry can remain the same for the existing surface

21.4   The Physics and Mechanics of Hydraulic Fracturing

The second important component of a robust and accurate, fully 3-D, hydraulic fracture simulator is the ability to properly model the fluid flow coupled with the fracture deformation and propagation. The following discussion is restricted to the framework of linear elasticity and lubrication theory.

21.4.1   Elastic stress analysis

BES is a linear elastic, 3-D, boundary element program (Lutz, 1991). It is based on a direct formulation and uses special hypersingular integration techniques and non-conforming elements on and around the crack surfaces. It is capable of handling multiple loading cases, specifically generating basic solutions (displacements and tractions) for unit tractions at points on the crack surface. Unit tractions are applied to each node on the discretized crack surface. The traction is distributed according to the shape functions of the incident elements, starting at unity at the given node and vanishing to zero at all adjacent nodes. The displacements at all nodes in the structure are evaluated for each unit traction loading case, providing a matrix of solutions whose generic element K_{ij} is the displacement at node i due to a unit traction at node j.

The set of basic solutions is combined to build a single influence matrix which then is used along with the equilibrium fluid pressures to determine the overall structural response due to both the far field boundary conditions and the fluid pressure in the crack. The displacements in the structure can be computed by multiplying each of the basic solutions, obtained for a unit traction at a node, by the fluid pressure at that node. This means that the stress analysis, which is the most time consuming process of the entire simulation, can be performed once for a specific model or crack geometry. Various fluid properties and flow parameters then can be used in the hydraulic fracture simulator based on this single stress analysis.

21.4.2   Fluid flow

A peculiarity of hydraulic fracturing consists of the strong nonlinear coupling between fluid flow and solid deformation, particularly in the vicinity of the fracture front. Proper coupling, as derived by the Geomechanics Group at Schlumberger Cambridge Research (SCR, 1993, 1994), yields an analytical model for pressure and width near the crack front which corresponds to a stress singularity that is different from the usual Linear Elastic Fracture Mechanics (LEFM) solution. A new term has been coined, Linear Elastic Hydraulic Fracturing (LEHF), to reflect the difference in the solutions. The use of the LEHF analytical model provides an elegant solution to the numerical problem otherwise associated with modeling the front of a hydraulic fracture. (Note that a 3-D crack front becomes a crack tip in 2-D.)

Review of 2-D LEHF solution

Following an approach similar to Spence and Sharp (1985), the Geomechanics Group at SCR (SCR, 1993) has shown that a fluid lag often develops near the crack tip, and this fluid lag negates the influence of the rock fracture toughness. However, by assuming that the fluid reaches the crack tip, a particular singularity develops in both the fluid pressure and the stress field ahead of the crack tip which is unique to hydraulic fracturing (SCR, 1994). This yields an intermediate asymptotic solution for the width and pressure in the crack which is independent of fracture toughness, provided that more energy is dissipated in the fluid than in creating new fracture surface. It is intermediate in the sense that there exists a small region at the very tip of the fracture where a fluid lag develops and LEFM holds, but has little effect on the rest of the solution and is not taken into account.

To obtain the general solution for fluid flow in the vicinity of the tip of a hydraulic fracture propagating in an impermeable solid, the following assumptions are made: crack propagation is self-similar and steady-state, the rock mass is a linear elastic solid in plane strain, lubrication theory is valid, and the fluid is incompressible with a power-law shear-thinning consistency. The boundary conditions include the far-field minimum principal stress, the fracture width at the crack tip, which must be zero, and the fluid velocity at the crack tip where the tip is a moving boundary. Details of how the solution is obtained can be found in SCR (1994). The final expressions for the pressure p and crack opening w in the vicinity of the crack tip, are as follows:

$$p - \sigma_3 = p_h \left[\left(\frac{2\sqrt{2}(2+n')}{\pi(2-n')} \right) \left(\frac{L_h}{L} \right)^{n'/(2+n')} - \left(\frac{L_h}{\xi} \right)^{n'/(2+n')} \right] \qquad (21.1)$$

$$w = \xi^{2/(2+n')} L_h^{n'/(2+n')} \left[c_1(n') - c_2(n') \left(\frac{\xi}{L} \right)^{\frac{2+3n'}{4+2n'}} \right] \qquad (21.2)$$

where

$$L_h = V \left(\frac{K'}{E'} \right)^{1/n'} \qquad (21.3)$$

is a characteristic length particular to hydraulic fracturing, and

$$p_h = E' \left[\left(\frac{\cos((1-\alpha)\pi)}{\sin(\alpha\pi)} \right)^{1+n'} \left(\frac{2n'+1}{n'^2(2+n')} \right)^{n'} \right]^{1/(2+n')} \qquad (21.4)$$

is a characteristic pressure with $\alpha(n') = 2/(2 + n')$. V is the crack tip speed which is equal to the fluid velocity at the tip. L is the fracture half length and ξ is the position from the crack tip. σ_3 is the far-field minimum stress. $E' = E/(1 - v^2)$ is the effective Young's modulus where E is the Young's modulus and v is Poisson's ratio. K' is the consistency index and n' is the power-law exponent. c_1 and c_2 are constants (SCR, 1994) that are evaluated best numerically; for a Newtonian fluid, they are $c_1 \cong 7.21$ and $c_2 \cong 3.17$. Note that the flow rate at the tip q_b is a function of the speed and the crack opening, $q_b = V \cdot w = f(V)$.

From these equations, the fluid pressure at the crack tip is found to be singular; for a Newtonian fluid, $n' = 1$ and the order of the singularity is 1/3. The stress at the crack tip has the same order of singularity. The order of the singularity depends on the fluid properties only and, within the assumptions made here, is always weaker than the 1/2 obtained from linear elastic fracture mechanics.

A similar solution was developed for the permeable case and can be found in Lenoach (1995). It involves a supplementary length scale because of the leak-off process and yields a solution which exhibits yet another singularity, also weaker than that of linear elastic fracture mechanics.

Extension of the LEHF solution to 3-D

The behavior of a hydraulic fracture in the vicinity of its propagating front is easily described by the LEHF solution, provided two additional assumptions are made: the crack front is considered to be locally under plane strain conditions, and the local fluid flow parallel to the crack front is negligible. We shall restrict ourselves to the case of a Newtonian fluid, but the process can easily be extended to power-law fluids. The width w in the vicinity of the fracture front is then described by the LEHF solution along any normal to the crack front:

$$w = (2)(3^{7/6}) \left(\frac{\mu V}{E'} \right)^{1/3} \rho^{2/3} = \beta V^{1/3} \tag{21.5}$$

where $\beta = (2)(3^{7/6}) \left(\frac{\mu}{E'} \right)^{1/3} \rho^{2/3}$, and ρ is the curvilinear distance, measured on the fracture surface, between any point and the fracture front. μ is the fluid viscosity.

The equation describing the mass conservation within the bulk of the fracture can then be transformed in order to accommodate the introduction of the LEHF solution. The fracture is considered as a surface in 3-D space, F. Consider a subdomain Ω of the fracture, characterized by a boundary Γ, a pressure field p, an aperture field w and fluid flux field q. t is the time. If one considers any closed domain A included in Ω (Figure 21.7), fluid enters and leaves A through its boundary ∂A. Fluid is stocked by an increase of the fracture width w with time t.

Considering also a source term at a point O included in A (injection or sink of value $Q(t)$), one then can write for an incompressible fluid:

$$\int_{\partial A} q \cdot n \, d\partial A + \int_A \frac{\partial w}{\partial t} dA = Q(t)_{at \, O} \tag{21.6}$$

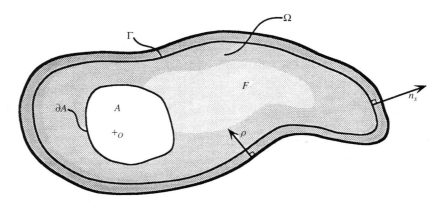

Figure 21.7   Fracture geometry for fluid flow analysis (showing F, Γ, Ω, A and ∂A)

After transformation of the contour integral into a surface integral, one gets

$$\int_A \left(\frac{\partial w}{\partial t} + \mathrm{div}\, q - Q(t)\delta(O) \right) dA = 0 \qquad (21.7)$$

where δ is the Dirac operator. This equation must hold whatever the chosen domain A, which yields the mass conservation equation for any point inside the considered domain Ω:

$$\frac{\partial w}{\partial t} + \mathrm{div}\, q = Q(t)\delta(O) \qquad (21.8)$$

One can write a variational form for Equation (21.8) by considering an auxiliary pressure field δp:

$$\int_\Omega \delta p \frac{\partial w}{\partial t} d\Omega + \int_\Omega \delta p \,\mathrm{div}\, q \, d\Omega = Q(t)\delta P(O) \qquad (21.9)$$

Using s as the curvilinear abscissa along the boundary Γ and n_s as the outward normal to Γ at point s (Figure 21.7), one can apply the divergence theorem to the second integral of Eq. (21.9), to get

$$\int_\Omega \delta p \,\mathrm{div}\, q \, d\Omega = -\int_\Omega q \cdot \mathrm{grad}(\delta p)\, d\Omega + \int_\Gamma \delta p (q_\Gamma \cdot n_s)\, ds \qquad (21.10)$$

The lubrication approximation for a Newtonian fluid of viscosity μ gives

$$q = -\frac{w^3}{12\mu}\, \mathrm{grad}\, p \qquad (21.11)$$

Setting $q_\Gamma \cdot n_s = q_b$ (Figure 21.7) and plugging the lubrication equation into Eq. (21.9), it becomes

$$\int_\Omega \delta p \frac{\partial w}{\partial t} d\Omega + \int_\Omega \left(\frac{w^3}{12\mu}\,\mathrm{grad}\, p \right) \cdot (\mathrm{grad}\, \delta p)\, d\Omega + \int_\Gamma \delta p\, q_b\, d\Gamma = Q(t)\delta p(O) \qquad (21.12)$$

As illustrated in Figure 21.7, it is convenient to choose Ω so as to describe the bulk of the fracture, whereas $F - \Omega$ corresponds to the domain of the fracture in the vicinity of the crack front that is described by the LEHF solution. As Γ now is the boundary between Ω and the front region of the fracture where the asymptotic solution holds, then

$$q_b = Vw \tag{21.13}$$

where V is the velocity of the crack and w is now a known function of V and the distance ρ from the crack front. Therefore, $q_b = \beta V^{4/3}$, which gives the final fluid flow equation to be solved outside the vicinity of the crack front:

$$\int_\Omega \delta p \frac{\partial w}{\partial t} d\Omega + \int_\Omega \frac{w^3}{12\mu} (\text{grad} \, p \cdot \text{grad} \delta p) \, d\Omega + \int_\Gamma \delta p \beta V^{4/3} d\Gamma = Q(t)\delta p(O) \tag{21.14}$$

The first two integrals correspond to the usual hydraulic fracture solution. The third integral is a contour integral that incorporates the asymptotic solution for the fluid flow. The right-hand side accommodates the boundary condition of a source term at any point and can be extended for the case of a line source.

21.5 Finite Element Implementation of Fluid Flow Coupled with the Structural Response

The finite element method is used to solve the fluid flow equation coupled with the structural response. The special form of the fluid flow equation developed in Section 21.4 allows one to separate the fracture surface in two regions: the bulk of the fracture Ω where width and pressure are to be determined and regular shell finite elements are used, and the near vicinity of the crack front where special tip elements adjacent to the crack front are used. Thanks to the form of Eq. (21.14), the special tip elements only represent the LEHF solution, and nodal width and pressure are never computed, except at the boundary Γ between the regular elements and the tip elements.

21.5.1 Finite element formulation of fluid flow equation

Considering a set of n shape functions N over the domain Ω, the pressure can be approximated in the finite element sense by $p = \sum_{i=1}^{i=n} N_i \hat{p}_i$, where $\hat{p}_i$ is the pressure at node i (or in matrix form $p = N^T \hat{p}$). The small increment in pressure δp is approximated in a similar way $\delta p = \sum_{i=1}^{i=n} N_i \delta \hat{p}_i$. These two expressions, when introduced into (Eq. 21.14), lead to

$$\int_\Omega \left(\sum_{i=1}^{i=n} N_i \delta \hat{p}_i \right) \frac{\partial w}{\partial t} d\Omega + \int_\Omega \frac{w^3}{12\mu} \text{grad} \left(\sum_{j=1}^{j=n} N_j \hat{p}_j \right) \cdot \text{grad} \left(\sum_{i=1}^{i=n} N_i \delta \hat{p}_i \right) d\Omega$$

$$+ \int_\Gamma \left(\sum_{i=1}^{i=n} N_i \delta \hat{p}_i \right) \beta V^{4/3} d\Gamma = Q(t) \left(\sum_{i=1}^{i=n} N_i(O)Q(t) \right)$$

$$\tag{21.15}$$

Requiring stationarity of Eq. (21.15) leads to a set of n differential equations. Introducing the finite element approximation for the width w together with the time discretization, the *i*th equation can be written as:

$$\int_{\Omega} N_i \left(\sum_{j=1}^{j=n} N_j \frac{\hat{w}_j(t_{n+1}) - \hat{w}_j(t_n)}{t_{n+1} - t_n} \right) d\Omega$$

$$+ \frac{1}{12\mu} \int_{\Omega} \left(\sum_{k=1}^{k=n} N_k \hat{w}_k \right)^3 \mathrm{grad} N_i \cdot \mathrm{grad} \left(\sum_{j=1}^{j=n} N_j \hat{p}_j \right) d\Omega \qquad (21.16)$$

$$+ \int_{\Gamma} N_i \beta V^{4/3} d\Gamma = Q(t) N_i(O)$$

where $\hat{w}_j$ is the value of the width w at node j.

For simplicity, the *i*th shape function is associated with the node at which it takes a value of 1. It is worth noting that a shape function N_i is equal to zero when the point considered for the integration is not part of an element containing the associated *i*th node. Therefore, the integrals in Eq. (21.16) can be reduced to integrals over a single element Ω_i:

$$\int_{\Omega_i} N_i \left(\sum_{j=1}^{j=n} N_j \frac{\hat{w}_j(t_{n+1}) - \hat{w}_j(t_n)}{t_{n+1} - t_n} \right) d\Omega$$

$$+ \frac{1}{12\mu} \int_{\Omega_i} \left(\sum_{k=1}^{k=n} N_k \hat{w}_k \right)^3 \mathrm{grad} N_i \cdot \mathrm{grad} \left(\sum_{j=1}^{j=n} N_j \hat{p}_j \right) d\Omega \qquad (21.17)$$

$$+ \int_{\Gamma_i} N_i \beta V^{4/3} d\Gamma = Q(t) N_i(O)$$

where Γ_i is the portion of the boundary Γ which intersects the element Ω_i. Two cases can be considered, depending upon the location of the element Ω_i with respect to the boundary Γ.

21.5.2 Element not adjacent to Γ

If Γ_i is void (the element is not adjacent to the boundary), the computation of the various integrals over Ω_i is straightforward and the integral over Γ_i is zero. The *i*th equation can be reordered in the following way:

$$Q(t) N_i(O) = \int_{\Omega_i} N_i \left(\sum_{j=1}^{j=n} N_j \frac{\hat{w}_j(t_{n+1}) - \hat{w}_j(t_n)}{t_{n+1} - t_n} \right) d\Omega$$

$$+ \sum_{j=1}^{j=n} \hat{p}_j \left(\int_{\Omega_i} \frac{1}{12\mu} \left(\sum_{k=1}^{k=n} N_k \hat{w}_k \right)^3 \mathrm{grad} N_i \cdot \mathrm{grad} N_j \right) d\Omega \qquad (21.18)$$

21.5.3   *Element adjacent to* Γ

If Γ_i is not void (the element is adjacent to the boundary), the second and third term of Eq. (21.17) are to be modified. The pressure $\hat{p}_i$ at each node belonging to the boundary Γ_i satisfies

$$\hat{p}_i = \hat{P}_i - E'^{2/3} \left(\frac{\mu}{3\hat{\rho}_i} \right)^{1/3} \hat{V}_i^{1/3} \tag{21.19}$$

Note that this equation simply means that the derivative of the pressure normal to the boundary is known. Assuming that there are n nodes in the considered element, m of which are not on the boundary, the second term I_i of Eq. (21.17) can be transformed into:

$$I_i = \int_{\Omega_i} \frac{1}{12\mu} \left(\sum_k N_k \hat{w}_k \right)^3 \mathrm{grad} N_i \cdot \mathrm{grad} \left(\sum_{j=1}^n N_j \hat{p}_j \right) d\Omega$$

$$= \int_{\Omega_i} \frac{1}{12\mu} \left(\sum_k N_k \hat{w}_k \right)^3 \mathrm{grad} N_i \cdot \mathrm{grad} \left(\sum_{j=1}^m N_j \hat{p}_j + \sum_{j=m+1}^n N_j (\hat{P}_j - \hat{B}_j \hat{V}_j^{1/3}) \right) d\Omega$$

$$\tag{21.20}$$

where $\hat{B}_j = E'^{2/3} \left(\dfrac{\mu}{3\hat{\rho}_j} \right)^{1/3}$.

Finally,

$$I_i = \int_{\Omega_i} \frac{1}{12\mu} \left(\sum_k N_k \hat{w}_k \right)^3 \mathrm{grad} N_i$$

$$\cdot \mathrm{grad} \left(\sum_{j=1}^m N_j \hat{p}_j + \sum_{j=m+1}^n N_j \hat{P}_j - \sum_{j=m+1}^n \hat{B}_j \hat{V}_j^{1/3} \right) d\Omega \tag{21.21}$$

For convenience, $\hat{P}_j$ can be lumped with $\hat{p}_j$ (note that $\hat{p}_j$ is not the true pressure at the nodes sitting on Γ_i), leading to a more compact notation

$$I_i = \int_{\Omega_i} \frac{1}{12\mu} \left(\sum_k N_k \hat{w}_k \right)^3 \mathrm{grad} N_i \cdot \mathrm{grad} \left(\sum_{j=1}^n N_j \hat{p}_j \right) d\Omega$$

$$- \int_{\Omega_i} \frac{1}{12\mu} \left(\sum_k N_k \hat{w}_k \right)^3 \mathrm{grad} N_i \cdot \mathrm{grad} \left(\sum_{j=m+1}^n N_j \hat{B}_j \hat{V}_j^{1/3} \right) d\Omega$$

$$\tag{21.22}$$

Assuming l nodes along Γ_i, one can also define a finite element approximation for the flow q_b of the fluid normal to Γ_i. For the case of a Newtonian fluid:

$$q_b = \sum_{k=1}^{k=l} M_k \hat{B}_k \hat{V}_k^{4/3} \tag{21.23}$$

where $\hat{V}_k$ is the speed at node k, $\hat{\beta}_k$ is the evaluation of β at node k, and M is a set of shape functions. Then, the third term of Eq. (21.17) becomes:

$$\int_{\Gamma_i} N_i q_b V^{4/3} d\Gamma = \int_{\Gamma_i} N_i \left(\sum_{k=1}^{k=l} M_k \hat{\beta}_k \hat{V}_k^{4/3} \right) d\Gamma \qquad (21.24)$$

where $\hat{\rho}_k$ is the curvilinear distance from the fracture tip to node k.

If the various nodal speeds are known, the ith equation can now be reordered as follows:

$$\int_{\Omega_i} N_i \left(\sum_{j=1}^{j=n} N_j \frac{\hat{w}_j(t_{n+1}) - \hat{w}_j(t_n)}{t_{n+1} - t_n} \right) d\Omega$$

$$+ \sum_{j=1}^{j=n} \hat{p}_j \int_{\Omega_i} \frac{1}{12\mu} \left(\sum_{k=1}^{k=n} N_k w_k \right)^3 \mathrm{grad} N_i \cdot \mathrm{grad} N_j \, d\Omega$$

$$- \int_{\Omega_i} \frac{1}{12\mu} \left(\sum_k N_k \hat{w}_k \right)^3 \mathrm{grad} N_i \cdot \mathrm{grad} \left(\sum_{j=m+1}^{n} N_j \hat{B}_j \hat{V}_j^{1/3} \right) d\Omega$$

$$+ \int_{\Gamma_i} N_i \left(\sum_{k=1}^{l} M_k \hat{\beta}_k \hat{V}_k^{4/3} \right) d\Gamma \qquad (21.25)$$

$$= Q(t) N_i(O)$$

Equations (21.18) and (21.25) were implemented in FRANC3D in order to model fluid flow in the fractures while accounting for the near crack tip behavior.

21.5.4   *Coupling with the structural response of the rock*

To adequately consider the coupling between elasticity and fluid flow, both the structural and the fluid flow equations are solved at the same time for both the nodal widths and fluid pressures.

Linear elasticity provides the following linear relation between the fracture apertures w and the fluid pressure p

$$w = w_p + w_o$$
$$w = G \, p + w_o \qquad (21.26)$$

where w_o is the contribution from the external stresses and G is the influence function relating the fracture opening due to the internal fluid pressure (w_p) to the fluid pressure in the fracture. This relation can be expressed as a set of equations using the flexibility matrix K computed by BES (see previous section). For the nodes which are located neither on the crack front nor on Γ, the ith equation of this set can be written as

$$\sum_j K_{i,j} \hat{p}_j - \hat{w}_i + w_{oi} = 0 \qquad (21.27)$$

where w_{oi} is the nodal width induced by the external stresses. For nodes located on the crack front, the width is set to zero. The fluid pressure is set to zero also in order to conform with the assumed existence of a fluid lag; therefore this relation disappears at the crack front. For the pressure at the nodes located on Γ, there is an extra term to be taken into account and the equation becomes:

$$\sum_j K_{i,j} \hat{p}_j - \hat{w}_i = \sum_k^{\Gamma \ nodes} K_{i,k} \hat{B}_k \hat{V}_k^{1/3} - \hat{w}_{oi} \tag{21.28}$$

When solving for both nodal width and pressure, a strong non-linearity arises in the set of fluid flow equations due to the cubic power of the crack opening displacement w^3. This is handled by considering that, for every term in which w^3 arises, the value of w will be fixed and equal to either the value of the width at the previous iteration or, for the first iteration of any time stage, the value of the width at the previous time stage.

One can then reorder the ith equation of the set of fluid flow equations as

$$\sum_{j=1}^{j=n} \frac{\hat{w}_j}{\Delta t} \int_{\Omega_i} N_i N_j d\Omega + \sum_{j=1}^{j=n} \hat{p}_j \int_{\Omega_i} \frac{1}{12\mu} \left(\sum_{k=1}^{k=n} N_k w_k \right)^3 \mathrm{grad} N_i \cdot \mathrm{grad} N_j \, d\Omega$$

$$= \sum_{j=1}^{j=n} \frac{\hat{w}_j(t_n)}{\Delta t} \int_{\Omega_i} N_i N_j \, d\Omega - \int_{\Gamma_i} N_i \left(\sum_{k=1}^{1} M_k \hat{B}_k \hat{V}_k^{4/3} \right) d\Gamma + Q(t) N_i(O) \tag{21.29}$$

$$+ \int_{\Omega_i} \frac{1}{12\mu} \left(\sum_k N_k \hat{w}_k \right)^3 \mathrm{grad} N_i \cdot \mathrm{grad} \left(\sum_{j=m+1}^{n} N_j \hat{B}_j \hat{V}_j^{1/3} \right) d\Omega$$

This results in a set of n fluid flow equations and n structural equations with n unknown nodal widths and n unknown nodal fluid pressures. The solution process is described in more detail in Section 21.7.

21.6   Extensions to the Model and Formulation

Current trends in research and in field applications are pushing development of the simulator towards being able to model indirect vertical fracturing (with fracture containment, pinching, closure and fluid recession as corollaries), sand control (with screen-out as a corollary) and steam injection. It could also be used as a tool to study the need for special completions (with the ability to handle multiple fractures from cased and inclined wellbores as a corollary). Some of the features that are needed to model these problems include: (1) partial propagation of a fracture front across material interfaces; (2) simultaneous solution of crack surface contact tractions and fluid flow pressures during fracturing of layered formations with differing stresses and stiffnesses; (3) branching, intersecting, and merging cracks (for example, consider T-shaped fractures that can form at material interfaces); (4) proppant transport; and (5) bi-material interface slip and pressurization. As a step towards those developments, BES has been modified recently to run on parallel computers such as the IBM SP2 and the SG Power Challenge (Shah *et al.*,

1997), allowing larger more complex problems to be modeled with increased accuracy and efficiency.

Unlike the subjects mentioned above, each a research topic in itself, a fairly simple extension to the simulator can be made in order to model hydraulic fracture propagation in the presence of a porous and permeable rock mass where fluid is able to leak out of the fracture into the surrounding rock.

21.6.1 Introduction of leak-off

Carter's 1-D formulation (Carter, 1957), which uses a single lumped leak-off coefficient C_l, is the standard model in oil field applications for taking into account fluid leaking away from the fracture into the formation. It is a simple model of fluid diffusion through the fracture walls. The underlying assumptions are that the outward flow is normal to the direction of fracture propagation and that variations in the difference between the fluid pressure in the fracture and the pore pressure remain small.

The Carter leak-off model takes the form:

$$u_l = \frac{2C_l}{\sqrt{\tau(x)}} \tag{21.30}$$

where u_l is the fluid velocity of the leaking fluid, and τ is the exposure time at point x. Note that the factor 2 in front of C_l corresponds to fluid leaking through the two faces of the fracture, where C_l is an empirical constant related to the porosity and permeability of the rock.

This introduces an extra term in the mass balance equation, which now becomes

$$\frac{\partial w}{\partial t} + \operatorname{div} q + \frac{2C_l}{\sqrt{\tau(x)}} = Q(t)\delta(O) \tag{21.31}$$

This impacts the finite element formulation of the fluid flow equation described in Section 21.5. An extra term has to be added on the right-hand side of Eq. (21.29). For the ith equation, this supplementary term is:

$$- \int_{\Omega_i} N_i \left(\frac{2\sum_j N_j C_{lj}}{\sqrt{t_n + \Delta t - \sum_j N_j \theta_j}} \right) d\Omega \tag{21.32}$$

where C_{lj} is the value of the leak-off coefficient at node j and θ_j is the time at which the fluid reached node j.

The presence of leak-off also modifies the behavior in the vicinity of the tip. The pressure and width tip fields are now governed by a new series of equations (see Lenoach, 1995):

$$w = 6.6687 \left(\frac{\mu C_l}{E'} \right)^{1/4} V^{1/8} \rho^{5/8} + 2.2545 \left(\frac{\mu}{C_l E'} \right)^{1/2} V^{3/4} \rho^{3/4} \tag{21.33}$$

$$p(\rho) = p_o - 0.43 E'^{3/4} (\mu C_l)^{1/4} V^{1/8} \rho^{-3/8} - 0.42 E'^{1/2} \left(\frac{\mu}{C_l} \right)^{1/2} V^{3/4} \rho^{-1/4} \tag{21.34}$$

with similar notations as in Section 21.4. p_o is a constant dependent upon fracture geometry.

This changes the expression for the flow $q_b = V \cdot w = f(V)$ across the boundary Γ between regular elements and tip elements. Introducing the variational formulation for the fluid flow equation, the right-hand side term corresponding to the flow across the boundary still is

$$\int_\Gamma \delta p(q_\Gamma \cdot n_s)ds = \int_\Gamma \delta p\, q_b\, ds \tag{21.35}$$

Introducing the new expression for the width as given by Eq. (21.33) and the finite element discretization, this becomes for the ith equation:

$$\int_{\Gamma_i} N_i(\alpha V^{9/8} + \gamma V^{7/4})d\Gamma = \int_{\Gamma_i} N_i \left[\left(\sum_j N_j\alpha_j \right) \left(\sum_k N_k\hat{V}_k \right)^{9/8} \right.$$

$$\left. + \left(\sum_j N_j\gamma_j \right) \left(\sum_k N_k\hat{V}_k \right)^{7/4} \right] d\Gamma \tag{21.36}$$

$$\alpha_i = 6.6687 \left(\frac{\mu C_l}{E'} \right)^{1/4} \rho_i^{5/8}$$

$$\gamma_i = 2.255 \left(\frac{\mu}{C_l E'} \right)^{1/2} \rho_i^{3/4} \tag{21.37}$$

where E' and C_l are also evaluated at point i.

It is clear that α tends to zero, whereas γ tends to infinity when the leak-off coefficient tends to zero. In short, the permeable solution does not tend in a clear manner towards the impermeable solution when the leak-off coefficient tends to zero. It is, however, clear physically and verified numerically that the width field corresponding to a very small amount of leak-off is extremely similar to that described by the impermeable solution (Eq. (21.5)).

A criterion, therefore, is needed to switch from the permeable to the impermeable solution when the amount of leak-off is small. The permeable solution actually relies on a parameter η to be small (Lenoach, 1995), where η is given by

$$\eta = \left(\frac{\mu}{E'} \right)^{1/3} \left(\frac{V}{4C_l} \right)^{2/3} \tag{21.38}$$

As the exponent in ρ is different for the two solutions (impermeable and permeable), no absolute criterion can easily be established. It was determined empirically that as soon as η was greater than $1/20$, one could safely switch from the permeable solution to the impermeable solution. In order to be able to mix the two solutions, two nodal switch functions $G(x_i)$ and $H(x_i)$ were introduced: so that $G = 0$ and $H = 1$ if $\eta(x_i) < 1/20$ (permeable case), and $G = 1$ and $H = 0$ otherwise (impermeable case).

A mixed sink term is then

$$\int_{\Gamma_i} N_i \left(\left(\sum_j H_j N_j\alpha_j \right) \left(\sum_k H_k N_k\hat{V}_k \right)^{9/8} + \left(\sum_j H_j N_j\gamma_j \right) \left(\sum_k H_k N_k\hat{V}_k \right)^{7/4} \right.$$

$$\left. + \left(\sum_j G_j N_j\beta_j \right) \left(\sum_k G_k N_k\hat{V}_k \right)^{4/3} \right) d\Gamma \tag{21.39}$$

$$\alpha_i = 6.6687 \left(\frac{\mu C_l}{E'} \right)^{1/4} \rho_i^{5/8}$$

$$\beta_i = 7.206 \left(\frac{\mu}{E'} \right)^{1/3} \rho_i^{2/3}$$

$$\gamma_i = 2.255 \left(\frac{\mu}{C_l E'} \right)^{1/2} \rho_i^{3/4} \tag{21.40}$$

This approach was implemented and was found to function well. The only drawback it presents is the necessity to gradually increase the leak-off coefficient from zero to its desired value during the solution process for the first time stage in order to preserve reasonably quick convergence of the algorithm.

21.7 Solution Procedure

Hydraulic fracture propagation is a nonlinear, time-dependent, moving boundary problem that involves simultaneous satisfaction of solid deformation, fluid flow and fracture mechanics. Following the discretization in both time and space, the solution consists of a series of 'snapshots' that correspond to unique instances in time and crack shape. Two approaches can be followed to obtain the next term in such a series: one can either fix the time step and look for the corresponding geometry, or fix the geometry and look for the corresponding time. Although the first approach is more intuitive, the latter scheme was chosen as it minimizes the amount of computation. Note that the crack initiation process is not modeled; rather a model with an initial starting crack is the beginning point for the simulation. The mechanics of fracture initiation and breakdown pressure are beyond the scope of this chapter.

21.7.1 Computing the solution for a particular model

The special case of the initial solution for the starting crack will be detailed in section 21.7.2. For subsequent time stages (actually fracture geometry stages), it is assumed that the solution at the previous time stage t_n is known. It consists of crack geometry $\Omega^*(t_n)$, crack volume $Vol(t_n)$, a set of nodal widths $\hat{w}_j(t_n)$, a set of nodal pressures $\hat{p}_j(t_n)$, and a set of nodal crack front speeds $\hat{V}_j(t_n)$. Assuming the geometry of the current model was computed from the previous solution, the process which leads to the solution of the current model is complicated and has been separated into three subsections for clarity.

Meshing and updating information

As described in section 21.3, crack propagation involves a change in the model geometry. The mesh that is attached to the crack surface is merely a mathematical representation of the true geometry. For a given stage of fracture propagation, the previous, $\Omega^*(t_n)$, and current, $\Omega^*(t_{n+1})$, fracture geometries are known. The current fracture surface contains the previous surface plus some new surface representing the propagation. A mesh is attached to the geometry surface; in general, the mesh on the portion of the current fracture surface that was the previous fracture surface can remain fixed or can be modified (to reduce the number of elements for example). The set of nodal widths and pressures from the previous solution must be mapped onto the mesh of the current fracture. This is a trivial procedure if the mesh has remained fixed. If the mesh has been modified, the nodal widths and pressures for the new mesh must be interpolated from the previous mesh. Efficient algorithms to find the location of the node on the previous mesh and to interpolate the response have been written for this purpose (see Potyondy, 1993), but will not be discussed here.

Computation of the flexibility matrix

From the fracture geometry $\Omega^*(t_{n+1})$, the elastic properties of the model and the corresponding boundary conditions, BES is used to compute the flexibility matrix as described in section 21.4.2. Using the flexibility matrix, one can convert the nodal pressures inside the fracture into nodal fracture widths. This reflects both the influence of the geometry of the fracture, the geometry of the model, the elastic properties of the model, and the boundary conditions on the model. The computation of the elastic flexibility matrix is by far the most time consuming part of the analysis, and guided the earlier choice of fixing the geometry instead of the time.

Iterative solution of the fluid flow

The results from the previous time stage, t_n, are used as starting values for the solution process. Given an initial set of values for crack opening displacement, fluid pressure and crack front speed, the iterative scheme proceeds at two levels. First, the set of nonlinear equations describing the fluid flow inside the fracture coupled to the structural response of the model is solved in an iterative manner. Once convergence is achieved based on the crack opening displacements, the global mass balance and the crack front speeds are considered in the following way to determine the absolute time corresponding to the current geometry. The total volume of fluid injected should be equal to the volume of the fracture minus the volume of fluid that has leaked away into the formation. Also, the fluid speed at each point of the crack front should equal the crack propagation speed, i.e. the crack advance divided by the time step. Note that, within the framework used here, these two relationships both express satisfaction of global mass balance. If these two relationships are not satisfied, the time step is adjusted and the fluid flow equation is solved iteratively again. This process continues until the solution has converged on both the nodal values and the time, thus satisfying both elasticity and fluid flow.

Fracture propagation

Before a crack can be propagated, the total model displacements must be determined. This is done by combining the displacements due to the far-field applied loading with the displacements produced by the fluid pressurization in the fracture.

The process of propagating a crack in FRANC3D has been described elsewhere (Martha *et al.*, 1993; Carter *et al.*, 1997). The only difference in this case is that the crack is

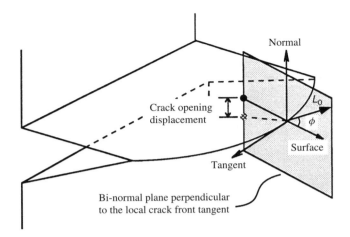

Figure 21.8   Propagation of a hydraulic fracture

propagated using an *ad hoc* propagation criterion for hydraulic fracturing (Figure 21.8). It is worth describing this criterion in more detail as the authors believe that more research is needed in this area.

The local extension of the fracture propagation along the crack front is proportional to the speed of the fluid at that point. It is scaled by a maximum propagation length L_0 (input by the user) so that the point where the maximum speed is attained is propagated by L_0. Indeed, a reasonable value for L_0 will depend upon the model and crack geometry as well as the desired accuracy but, because of the linear interpolation in time, should not be greater than 10% of the fracture penetration into the formation.

The direction of fracture propagation at each point is determined according to the maximum circumferential stress criterion. The criterion is evaluated at discrete points along the front in a plane normal to the crack front tangent (Figure 21.8). The angle ϕ is computed using the mode I and II stress intensity factors K_1 and K_2. Quasi-static propagation of the fracture requires that the mode I stress intensity factor be equal to the fracture toughness of the rock K_{1c} at each point. The assumption underlying the special solution used here at the fracture tip is that, although not explicitly taken into account, a fluid lag would exist and automatically adjust its length to satisfy this condition (SCR, 1993). Therefore, $\phi(x)$ is taken to satisfy

$$K_{1c} \sin(\phi(x)) + K_2(x)(3 \cos(\phi(x)) - 1) = 0 \tag{21.41}$$

Once the crack has been propagated, the model can be remeshed for the next stage of analysis.

21.7.2   Special case for the first time stage

As described in the previous section, the solution process requires the results of a previous step. This is achieved for the first model through approximate analytical solutions. These solutions are used to produce the complete description of a 'dummy' crack, smaller than the initial crack, which is then considered as the previous model by the iterative solver.

This provides the initial values to be used for the real fracture geometry, and the solution scheme then proceeds as described in Section 21.7.1.

Two solutions have been developed covering most initial crack geometries in hydraulic fracturing, one for a slot crack and one for a penny-shaped crack (or radial crack). Following the approach of the SCR Geomechanics Group (SCR, 1994), the main assumption behind these solutions is that the pressure profile near the tip is governed by a singular field, and that the resulting stress intensity factor is zero. This allows one to obtain a pressure profile from which all other quantities are derived. These solutions, therefore, produce time, crack opening displacements, crack front speeds and fluid pressure inside the crack as a function of fracture extension, elastic parameters, and pumping parameters.

21.8   Verification and Validation

One of the most important aspects in the development of any numerical simulator is the verification of the approach as well as the computer software. To verify this 3-D simulator, one must analyse problems for which the solution is known. In this case, the fully 3-D simulator is used to model simple 2-D fractures: a KGD and a radial hydraulic fracture. The results from the 3-D simulator are compared with those of a previously verified 2-D simulator. Furthermore, the 3-D simulator is used to model some experiments that were performed at Delft University which provides additional validation and shows the capability of the simulator to model fully 3-D hydraulic fractures in the near-wellbore region.

21.8.1   Modeling hydraulic fractures of simple geometries

FRANC3D has been compared with a hydraulic fracturing model (Desroches and Thiercelin, 1993) which can only model simple geometries such as KGD and radial, but considers full coupling of fluid flow, elastic deformation and fracture propagation. It includes a complete description of the fracture tip, in particular the effect of fracture toughness and the possibility for the fracturing fluid not to reach the tip of the fracture ('fluid lag'). This model, referred to as Loramec, has been validated using laboratory experiments (de Pater *et al.*, 1996). Although the formulation of these two models is radically different, the underlying physical assumptions are similar so they should yield similar results for a similar fracture geometry.

To simulate a plane strain geometry with FRANC3D, a small slot crack was introduced in a cube (Figure 21.9). The fracture was propagated with an evenly distributed line source. Data for comparison with Loramec were taken in the middle of the crack. The relevant parameters were: Young's modulus of 24.1 GPa, Poisson's ratio of 0.02, fracture toughness of 0.25 MPa$\sqrt{\text{m}}$, fluid viscosity of 130 000 cp, flow rate/height of 5.92×10^{-9} m^3/s and closure stress of 9.7 MPa.

Figure 21.10(a) shows the fracture extension versus time for Loramec and FRANC3D together with a profile of pressure (Figure 21.10(b)) and width (Figure 21.10(c)) from the fracture mouth to the fracture tip. Excellent agreement between FRANC3D and Loramec is observed. Note that the FRANC3D model used only six linear elements along the fracture extension in addition to the last tip element. This demonstrates how well the tip element captures the essence of the solution, allowing one to use a coarse mesh away from the tip.

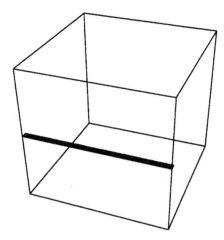

Figure 21.9   Modeling a plane strain fracture with FRANC3D

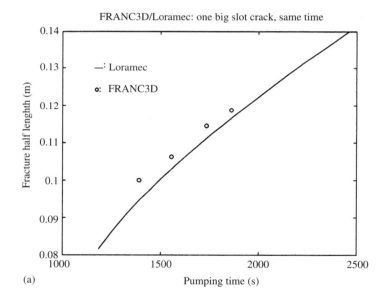

(a)

Figure 21.10   Comparison between FRANC3D and Loramec: plane strain case

To simulate an axisymmetric (radial) geometry, a penny-shaped crack was introduced in the center of a cube (Figure 21.11). The size of the crack was two orders of magnitude smaller than that of the cube to simulate an infinite medium. The fracture was propagated using a point source at the center of the crack. Data for comparison with Loramec were taken along a crack radius. The relevant parameters were similar to those for the KGD geometry except for the injection rate of $2.96 \times 10^{-9} \mathrm{m}^3/\mathrm{s}$. These parameters correspond to those of one of the laboratory experiments that was used to validate the Loramec model. Figure 21.12(a) shows the fracture radius versus time for Loramec and FRANC3D,

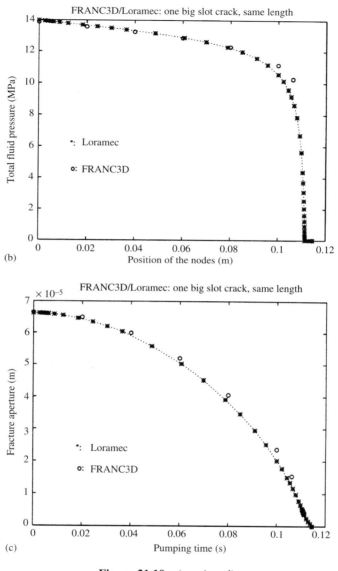

Figure 21.10 (*continued*)

together with a profile of pressure (Figure 21.12(b)) and width (Figure 21.12(c)) from the fracture center to the fracture tip. There is excellent agreement between the two solutions for this model as well.

21.8.2   *Modeling experimental data*

To study the 3-D geometry of hydraulic fractures, a matrix of hydraulic fracturing experiments using a true triaxial loading frame was carried out on physical models at

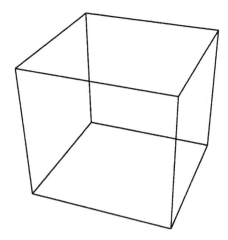

Figure 21.11   Modeling an axisymmetric fracture with FRANC3D

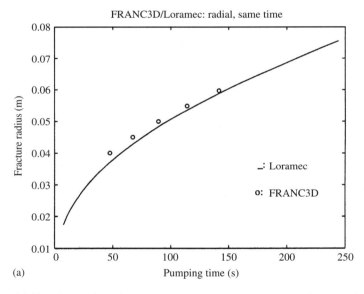

Figure 21.12   Comparison between FRANC3D and Loramec: axisymmetric case

the University of Delft (Weijers, 1995). The simplest configuration, an open-hole model was chosen for comparison with FRANC3D here so that the extra complexity due to the perforations and casing could be avoided. Simulations of the other test configurations are on-going (Shah *et al.*, 1997), but cannot be described here due to time and space limitations.

The geometry of the chosen configuration, COH10, is presented in Figure 21.13. A cylindrical wellbore of radius 1 cm was drilled in a 30 cm edge cube. The wellbore was drilled parallel to the direction of the minimum stress, perpendicular to the preferred fracturing plane. The experimental parameters were: Young's modulus E = 35 GPa, Poisson's

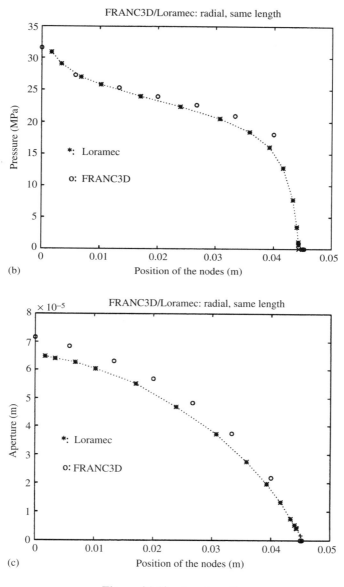

Figure 21.12 (*continued*)

ratio $\nu = 0.15$, fracture toughness $K_{1c} = 0.6\,\mathrm{MPa}\sqrt{\mathrm{m}}$, fluid viscosity $\mu = 100\,000\,\mathrm{cp}$, injection rate $Q = 0.088\,\mathrm{cc/min}$, $\sigma_{1,2,3} = 23$, 12.1 and 9.7 MPa, respectively. Splitting of the blocks after the experiment showed that a single fracture initiated parallel to the wellbore and slowly turned to propagate perpendicularly to the wellbore. The pressure in the wellbore was recorded during the experiment. Two displacement transducers *DIS2* and *DIS3* were set $90°$ apart in the middle of the wellbore to measure the deformation of the wellbore. *DIS2* measures the deformation in the direction of the maximum principal

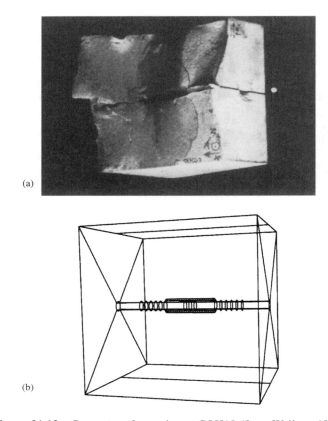

(a)

(b)

Figure 21.13   Geometry of experiment COH10 (from Weijers, 1995)

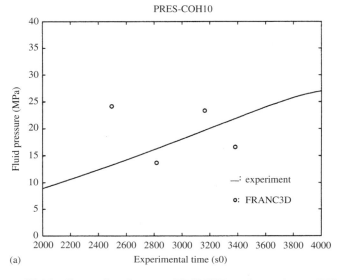

(a)

Figure 21.14   Comparison between FRANC3D and experiment COH10

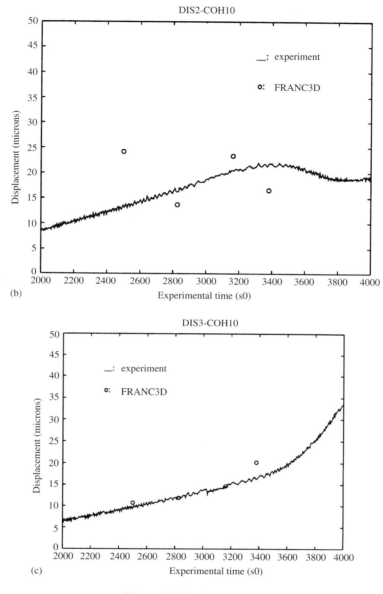

Figure 21.14 (*continued*)

stress. The output of FRANC3D was compared against both this data and the radius of reorientation of the fractures, which was measured after dismantling the blocks.

Although categorized as 'open-hole', there was still a borehole assembly in this experiment, which consisted of two 10 cm long packers glued at the two extremities of the wellbore, leaving an open-hole section of 10 cm at the center of the wellbore. To simulate the wellbore assembly that was used during the experiment, the cylinder representing the wellbore was split into three regions; the middle section corresponds to the open-hole section whereas the other two correspond to the packer arrangement. As the packer assembly was glued to the wellbore, it was considered as a rigid inclusion and was modeled by a zero normal displacement boundary condition on the packer zone. In the open-hole section, two initial slot fractures were added, 180° apart. Their location was taken as indicated by the experiments; the fracture initiation process was not modeled.

Figure 21.14 shows the comparison between the data obtained from the experiment and the output of FRANC3D. The comparison is made in terms of wellbore pressure (Figure 21.14(a)) and displacements of the wellbore wall as measured by the two LVDT sensors *DIS2* (Figure 21.14(b)) and *DIS3* (Figure 21.14(c)). Apart from numerical noise, the pressure predicted by FRANC3D is comparable to that measured in the laboratory and the two horizontal displacements are well predicted. As *DIS3* is an indirect measurement of the fracture aperture, this indicates that the fracture aperture at the wellbore is well predicted by the model. The only discrepancy between the experiment and the output of the model is the radius of reorientation of the fracture; the experiment showed that the fracture was slow to turn (about 10 cm for the radius of reorientation (Weijers, 1995)), whereas the simulated fracture turned much more quickly towards the preferred fracturing plane (a radius of reorientation of about 2 cm was computed). It is proposed that this is due to some influence of the packer assembly that is not taken into account by our modeling (de Pater, personal communication, 1996).

21.9   Conclusions

It is necessary to have a fully 3-D simulator to correctly model some of the problems regularly encountered in the field during hydraulic fracturing, such as the near wellbore fractures of a deviated well. If a robust and accurate simulator is available, improvements in oil/gas recovery are possible as well as savings in time and money during completion and re-injection.

A simulator that is capable of modeling multiple, fully 3-D, non-planar hydraulic fractures has been built. It maintains a consistent geometric representation of the fracture geometry throughout the analysis and properly couples the fluid flow in the fracture with the fracture deformation. It is verified using previous solutions for simple 2-D crack geometries as well as 3-D experimental results. Enhancements and additions are in progress to allow more complex problems to be modeled and to increase the speed, efficiency, and accuracy of the simulator.

An additional advantage of the software is that it is easily adapted to other problems of fracture in geomechanics, such as pressure grouting of cracks in dams (Ingraffea *et al.*, 1995), cohesive crack propagation (Carter *et al.*, 1993), and compression induced fracture from inclined flaws in brittle materials (Germanovich *et al.*, 1996).

Acknowledgements

The authors wish to acknowledge Schlumberger for permission to publish this work as well as for their intellectual and financial support during the development of the simulator. The additional support of the Cornell Fracture Group and the Cornell National Supercomputing Facility is also hereby acknowledged.

The additional support of the Cornell Fracture Group, the Cornell Theory Center, and the National Science Foundation (CMS grant 9625406) is also hereby acknowledged.

References

Abass, H. H., Hedayati, S. and Meadows, D. L. *Nonplanar fracture propagation from a horizontal wellbore: experimental study.* SPEPE, (August 1996) 133–137.

Advani, S. H., Lee, T. S. and Lee, J. K. Three dimensional modeling of hydraulic fractures in layered media: Finite element formulations. *J. Energy Res. Tech.*, **112**, (1990) 1–18.

Bahat, D. *Tecton-fractography.* Springer-Verlag, (1991).

Barree, R. D. A practical numerical simulator for three-dimensional hydraulic fracture propagation in heterogeneous media. *SPE Paper 12273, SPE Symp. Reservoir Simulation*, San Francisco, (November 1983) 15–18.

Baumgart, B. G. A polyhedron representation for computer vision. *AFIPS Proc.*, **44**, (1975) 589–596.

Behrmann, L. A. and Elbel, J. L. Effect of perforations on fracture initiation. *JPT* (May 1991) 608–615.

Carter, B. J., Ingraffea, A. R. and Bittencourt, T. N. Topology-Controlled Modeling of Linear and Non-Linear 3D-Crack Propagation in Geomaterials. In: *Fracture of Brittle Disordered Materials: Concrete, Rock and Ceramics: Proc. IUTAM Conf.*, Brisbane, Australia, E&FN Spon Publishers, (1993) 301–318.

Carter, B. J., Wawrzynek, P. A., Ingraffea, A. R. and Morales, H. Effect of casing on hydraulic fracture from horizontal wellbores. *1st North American Rock Mechanics Symposium*, Austin, TX, (June 1994) 185–192.

Carter, B. J., Chen, C-S., Ingraffea, A. R. and Wawrzynek, P. A. Recent advances in 3D computational fracture mechanics. *Ninth International Conference on Fracture*, Sydney, (April 1997).

Carter, R. D. (1957) Derivation of the general equation for estimating the extent of the fractured area. Appendix to: *Optimum Fluid Characteristics for Fracture Extension*, Howard, G. C. and Fast, C. R. (eds.). Drilling and Production Practices, API, (1957) 261–270.

Clark, J. B. A hydraulic process for increasing the productivity of wells. *Trans. AIME*, **186**, (1949) 1.

Clifton, R. J. and Abou-Sayed, A. S. On the computation of the three-dimensional geometry of hydraulic fractures. *SPE Paper 7943*, (1979).

Desroches, J. and Thiercelin, M. Modelling the propagation and closure of micro-hydraulic fractures. *Int. J. of Rock Mech. & Geomech. Abstr.*, **30**, (1993) 1231–1234.

Emermann, S. H., Turcotte, D. L. and Spence, D. A. Transport of magma and hydrothermal solutions by laminar and turbulent fluid fracture. *Physics of the Earth and Planetary Interiors*, **41**, (1986) 249–259.

Gardner, D. C. High fracturing pressures for shales and which tip effects may be responsible. *Proc. 67th SPE Annual Tech. Conf. and Exhib.*, Washington, DC, (1992) 879–893.

Geertsma, J. Two-dimensional fracture propagation models, in "Recent Advances in Hydraulic Fracturing". *Monograph Series*, **12**, SPE, Richardson, TX, (1989) 81–94.

Geertsma, J. and de Klerk, F. A rapid method of predicting width and extent of hydraulically induced fractures. *JPT* (1969) 1571–1581.

Germanovich, L. N., Carter, B. J., Dyskin, A. V., Ingraffea, A. R. and Lee, K. K. Mechanics of 3D crack growth in compression. In: *Tools and Techniques in Rock Mechanics: 2nd North American Rock Mechanics Symposium, NARMS'96*, Montreal, Canada, (June 1996) 1151–1160.

Gu, H. and Leung, K. H. 3D numerical simulation of hydraulic fracture closure with application to minifracture analysis. *JPT* (March 1993) 206–211.

Hallam, S. D. and Last, N. C. Geometry of hydraulic fractures from modestly deviated wellbores. *JPT* (June 1991) 742–748.

Hoffmann, C. M. *Geometric and Solid Modeling: An Introduction.* Morgan Kaufmann, San Mateo, CA, (1989).

Howard, G. C. and Fast, C. R. Hydraulic Fracturing. *Monograph Series,* **2**, SPE, Richardson, TX, (1970).

Ingraffea, A. R., Carter, B. J. and Wawrzynek, P. A. Application of computational fracture mechanics to repair of large concrete structures. In: *Fracture Mechanics of Concrete Structures: Proc. 2nd Int. Conf. Fracture Mech. of Concrete Structures (FRAMCOS2)*, Zurich, Switzerland, (1995).

Jeffrey, R. G. The combined effects of fluid lag and fracture toughness on hydraulic fracture propagation. *Proc. Joint Rocky Mountains Regional Meeting and Low Permeability Reservoir Symposium*, Denver, CO, (1989) 269–276.

Johnson, E. and Cleary, M. P. Implications of recent laboratory experimental results for hydraulic fractures. *Proc. Rocky Mountain Regional Meeting and Low-Permeability Reservoirs Symposium*, Denver, CO, (1991) 413–428.

Khristianovic, S. A. and Zheltov, Y. P. Formation of vertical fractures by means of highly viscous fluid. *Proc. 4th World Petroleum Congress*, Rome, Paper 3 (1955) 579–586.

Lam, K. Y., Cleary, M. P. and Barr, D. T. A complete three dimensional simulator for analysis and design of hydraulic fracturing. *SPE Paper 15266*, (1986).

Lenoach, B. Hydraulic fracture modelling based on analytical near-tip solutions. In: *Computer Methods and Advances in Geomechanics*, Siriwardane and Zaman, (eds.). Balkema, Rotterdam, (1995) 1597–1602.

Lutz, E. E. Numerical methods for hypersingular and near-singular boundary integrals in fracture mechanics. *PhD Thesis,* Cornell University, Stanford, CA, (1991).

Mäntylä, M. *An Introduction to Solid Modeling.* Computer Science Press, Rockville, MD, (1988).

Martha, L. F. A topological and geometrical modeling approach to numerical discretization and arbitrary fracture simulation in three-dimensions. *PhD Thesis*, Cornell University, Stanford, CA, (1989).

Martha, L. F., Wawrzynek, P. A. and Ingraffea, A. R. Arbitrary crack representation using solid modeling. *Engrg. with Comput.*, **9**, (1993) 63–82.

Medlin, W. L. and Fitch, J. L. Abnormal treating pressures in MHF treatments. *Proc. SPE Annual Technical Conference and Exhibition,* San Fransisco, CA, (1983).

Mendelsohn, D. A. A review of hydraulic fracture modeling – I: General concepts, 2D models, motivation for 3D modeling. *J. Energy Res. Tech.*, **106**, (1984a) 369–376.

Mendelsohn, D. A. A review of hydraulic fracture modeling – II: 3D modeling and vertical growth in layered rock. *J. Energy Res. Tech.*, **106**, (1984b) 543–553.

Morales, R. H. Microcomputer analysis of hydraulic fracture behavior with a pseudo-three dimensional simulator. *SPEPE,* (February 1989) 69–74.

Morita, N., Whitfill, D. L. and Wahl, H. A. Stress-intensity factor and fracture cross-sectional shape predictions from a three-dimensional model for hydraulically induced fractures. *JPT*, (October 1988) 1329–1342.

Mortenson, M. E. *Geometric Modeling.* Wiley, New York, (1985).

Nordgren, R. P. Propagation of a vertical hydraulic fracture. *SPEJ* (1972) 306–314.

Ong, S. H. and Roegiers, J.-C. Fracture initiation from inclined wellbores in anisotropic formations. *JPT* (July 1996) 612–619.

Palmer, I. D. and Veatch R. W. Abnormally high fracturing pressures in step rate tests. *SPE Production Engrg.* (1990) 315–323.

Papanastasiou, P. and Thiercelin, M. Influence of inelastic rock behaviour in hydraulic fracturing. *Int. J. Rock Mech. Min. Sci. & Geomech. Abstr.*, **30**, (1993) 1241–1247.

de Pater, C. J., Weijers, L., Savic, M., Wolf, K. A. A., van den Hoek, P. J. and Barr, D. T. Experimental study of non-linear effects in hydraulic fracture propagation. *SPE Production and Facilities*, **9**, (1994) 239–246.

de Pater, C. J., Desroches, J., Groenenboom, J. and Weijers, L. Physical and numerical modeling of hydraulic fracture closure. *SPE Production and Facilities*, (May 1996) 122–127.

de Pater, C. J. (1996) Personal Communication.

Perkins, T. K. and Kern, L. R. Widths of hydraulic fractures. *J. Petrol. Technol.* (1961) 937–949.

Potyondy, D. O. A software framework for simulating curvilinear crack growth in pressurized thin shells. *PhD Thesis,* Cornell University, Ithaca, NY, (1993).

Potyondy, D. O., Wawrzynek, P. A. and Ingraffea, A. R. An algorithm to generate quadrilateral or triangular element surface meshes in arbitrary domains with applications to crack propagation. *Int. J. Numer. Meth. in Engrg.*, **38**, (1995) 2677–2701.

SCR Geomechanics Group. On the modelling of near tip processes in hydraulic fractures. *Int. J. Rock Mech. Min. Sci. & Geomech. Abstr.*, **30**, (1993) 1127–1134.

SCR Geomechanics Group. The crack tip region in hydraulic fracturing. *Proc. Royal Society, Mathematical and Physical Sciences, Series A*, **447**, (1994) 39–48.

Settari, A. and Cleary, M. P. Development and testing of a pseudo-three-dimensional model of hydraulic fracture geometry. *SPE, Trans. AIME*, **281**, (1986) 449–466.

Shah, K., Carter, B. J. and Ingraffea, A. R. Simulation of hydrofracturing in a parallel computing environment. *36th U.S. Rock Mech. Symposium*, Columbia University, NY, (1997).

Shlyapobersky, J., Wong, G. K. and Walhaug, W. W. Overpressure calibrated design of hydraulic fracture stimulations. *Proc. SPE Annual Tech. Conf. and Exhib.*, Houston, TX, (1988) 133–148.

Shlyapobersky, J., Walhaug, W. W., Sheffield, R. E. and Huckabee, P. T. Field determination of fracturing parameters for overpressure calibrated design of hydraulic fracturing. *Proc. 63rd Annual Tech. Conf. and Exhib.*, Houston, TX, SPE Paper 18195, (1988).

Spence, D. A. and Sharp, P. Self-similar solutions for elastodynaic cavity flow. *Proc Roy. Soc. Lond. A*, **400**, (1985) 289–313.

Soliman, M. Y., Hunt, J. L. and Azari, M. Fracturing horizontal wells in gas reservoirs. SPE Paper 35260, Mid-Continent Gas Symposium, Amarillo, TX, (1996).

Sousa, J. L. S., Carter, B. J. and Ingraffea, A. R. Numerical simulation of 3D hydraulic fracture using Newtonian and power-law fluids. *Int. J. Rock Mech. Min. Sci. & Geomech. Abstr.*, **30**, (1993) 1265–1271.

Touboul, E., Ben-Naceur, K. and Thiercelin, M. Variational methods in the simulation of three-dimensional fracture propagation. *27th US Rock Mech. Symp.*, (1986) 659–668.

Vandamme, L., Jeffrey, R. G. and Curran, J. H. Pressure distribution in three-dimensional hydraulic fractures, *SPE Production Engrg.*, **3**, (1988) 181–186.

van den Hoek, P. J., van den Berg, J. T. M. and Shlyapobersky, J. Theoretical and experimental investigation of rock dilatancy near the tip of a propagating hydraulic fracture. *Int. J. Rock Mech. Min. Sci. & Geomech. Abstr.*, **30**, (1993) 1261–1264.

Warpinski, N. R. Measurement of width and pressure in a propagating hydraulic fracture. SPEJ (February 1985) 46–54.

Warpinski, N. R., Moschovidis, Z. A., Parker, C. D. and Abou-Sayed, I. S. Comparison study of hydraulic fracturing models-Test case: GRI staged filed experiment No. 3. *SPE Production & Facilities.* (February 1994) 7–16.

Wawrzynek, P. A. Discrete modeling of crack propagation: theoretical aspects and implementation issues in two and three dimensions. *PhD. Thesis*, Cornell University, (1991).

Wawrzynek, P. A. and Ingraffea, A. R. Interactive finite element analysis of fracture processes: an integrated approach. *Theor. & Appl. Fract. Mech.*, **8**, (1987a) 137–150.

Wawrzynek, P. A. and Ingraffea, A. R. An edge-based data structure for two-dimensional finite element analysis. *Engrg. with Comput.*, **3**, (1987b) 13–20.

Weijers, L. and de Pater, C. J. Fracture reorientation in model tests. *SPE Paper 23790, Formation Damage Control Conference*, Lafayette, LA, (1992).

Weijers, L. The near-wellbore geometry of hydraulic fractures initiated from horizontal and deviated wells. *PhD Dissertation*, Delft University of Technology, (1995).

Weiler, K. Edge-based data structures for solid modeling in curved-surface environments. *IEEE Comp. Graph & Appl.*, **5**, (1985) 21–40.

Weiler, K. Topological structures for geometric modeling. *PhD Dissertation*, Rensselaer Polytechnic Institute, Troy, NY, (1986).

Yew, C. H. and Li, Y. Fracturing of a deviated well. *SPEPE*, (November 1988) 429–437.

22

The Theory of Anisotropic Poroelasticity with Applications

Younane Abousleiman[1] and Lizheng Cui[2]

[1]*Lebanese American University, Byblos, Lebanon*
[2]*The University of Oklahoma, Norman, OK, USA*

22.1 Introduction

Geomaterials such as rocks and soils, in general, exhibit anisotropy either due to layered or microstructural characteristics, meaning that their properties vary with direction. Many authors have devoted considerable efforts to the problems of material anisotropy (Lekhnitskii, 1963; Berry and Fairhurst, 1966; Barla and Wane, 1970; Amadei, 1983). However, most of the previous studies treated the materials as linear elastic. In recent years, more attention is being directed towards the role of pore fluid in saturated geomaterials and the theory of linear poroelasticity of Biot (1941), in explaining the crucial role of the fluid pore pressure.

The constitutive relations for anisotropic poroelasticity were first presented by Biot (1955), and later extended by Biot and Willis (1957), to describe the various mechanical properties of an anisotropic fluid saturated porous medium. In their work, the experimental procedures outlined for identification of the directional mechanical properties may not be easy to realize. Lately, Thompson and Willis (1991), following the interpretation of the poroelastic material constants of Rice and Cleary (1976), reformatted the anisotropic poroelastic constitutive relations. The outcome of their reformation is a set of constitutive relations in which the parameters that appear are given explicitly in terms of drained and undrained compliances, and the components of a generalized Skempton's pore pressure coefficient tensor.

Sedimentary rocks typically have laminated structure; as such, their directional mechanical properties exhibit the same poroelastic properties in all horizontal directions, but different properties are exhibited vertically. This form of transverse isotropy will be elaborated on in this chapter. In the following, the continuum mechanics approach is first utilized to rewrite the poroelastic constitutive equations and discuss the relevant coefficients. Secondly, the micromechanics approach is reformulated and the various micromechanical parameters are identified. Thirdly, the solutions for the governing equations of transversely isotropic media with various boundary and loading conditions are presented.

22.2 Theory of Anisotropic Poroelasticity

The theory of anisotropic poroelasticity was presented by Biot (1955) using the continuum mechanics approach. To gain more physical insight of the poroelastic constituents, the micromechanical approach was also adopted to study the constitutive relations (Carroll, 1979, 1980; Carroll and Katsube, 1983; Katsube, 1985; Thompson and Willis, 1991). Both approaches are addressed in the following. The porous medium is assumed linear elastic in this chapter. However, the derivations can be easily extended for nonlinear poroelastic media at an incremental level (Nur and Byerlee, 1971; Cui *et al.*, 1996).

22.2.1   Constitutive law

Define the basic dynamic and kinematic variables as σ_{ij} = total stress tensor, p = pore pressure, e_{ij} = overall strain tensor, and ζ = variation of fluid content. The constitutive relations for the general anisotropic poroelasticity can be expressed as (Biot, 1955; Thompson and Willis, 1991)

$$\sigma_{ij} = M_{ijkl} e_{kl} - \alpha_{ij} p \tag{22.1}$$

$$p = M(\zeta - \alpha_{ij} e_{ij}) \tag{22.2}$$

where M_{ijkl}, α_{ij}, and M are material constants. Due to the symmetry of the stress and strain tensor, based on the existence of the functional of an elastic potential (Biot, 1955), it can be shown that

$$M_{ijkl} = M_{jikl} = M_{ijlk} = M_{klij} \tag{22.3}$$

$$\alpha_{ij} = \alpha_{ji} \tag{22.4}$$

In view of the above symmetry, only 21 M_{ij}, 6 α_{ij}, and 1 M are independent. The total number of independent constitutive coefficients for the general anisotropic poroelasticity is therefore 28, as compared with the 21 coefficients for anisotropic elasticity.

22.2.2   Continuum mechanics coefficients

The physical significance of the continuum mechanics coefficients are discussed as follows. First, noting from Eq. (22.1) that under drained condition $p = 0$, the coefficient M_{ijkl} is identified as the modulus tensor for the drained elastic material. Secondly, Eq. (22.1) can be rearranged as

$$\sigma_{ij} + \alpha_{ij} p = M_{ijkl} e_{kl} \tag{22.5}$$

Therefore, it can be observed that there exists an 'effective stress tensor'

$$\sigma'_{ij} = \sigma_{ij} + \alpha_{ij} p \tag{22.6}$$

whose constitutive relation is governed by the elastic law. The coefficients α_{ij} are then referred to as the 'generalized Biot's effective stress coefficients'.

It is also sometimes convenient to write Eq. (22.5) as

$$e_{ij} = C_{ijkl}(\sigma_{kl} + \alpha_{kl} p) \tag{22.7}$$

where C_{ijkl} is the compliance tensor of the drained material, and defined as the inverse of the modulus tensor such that

$$M_{ijkl}C_{klmn} = \tfrac{1}{2}(\delta_{im}\delta_{jn} + \delta_{in}\delta_{jm}) \tag{22.8}$$

To gain further insight into the above system, a few more constitutive coefficients are defined. The relation defined by Eq. (22.1) can be rewritten as

$$\sigma_{ij} = (M_{ijkl} + M\alpha_{ij}\alpha_{kl})e_{kl} - M\alpha_{ij}\zeta \tag{22.9}$$

Under the undrained condition, $\zeta = 0$, Eq. (22.9) can be expressed by

$$\sigma_{ij} = M^u_{ijkl}e_{kl} \tag{22.10}$$

where

$$M^u_{ijkl} = M_{ijkl} + M\alpha_{ij}\alpha_{kl} \tag{22.11}$$

is the undrained modulus tensor which characterizes the apparent elastic modulus of the material under the undrained condition.

Another quantity of interest is the generalized Skempton's pore pressure coefficient tensor, B_{ij}, that relates the pressure rise under the undrained condition due to the stress loading. Rewrite Eq. (22.10) as

$$e_{ij} = C^u_{ijkl}\sigma_{kl} \tag{22.12}$$

where C^u_{ijkl} is the undrained compliance tensor, and

$$C^u_{ijkl} = C_{ijkl} - \frac{M}{1 + M\alpha_{mn}C_{mnop}\alpha_{op}}C_{ijst}\alpha_{st}\alpha_{uv}C_{uvkl} \tag{22.13}$$

Substituting Eq. (22.12) into Eq. (22.2) with $\zeta = 0$, we obtain

$$p = -M\alpha_{ij}C^u_{ijkl}\sigma_{kl} \tag{22.14}$$

It is of interest to note that, unlike the isotropic case, the pore pressure can be generated by pure shear stress under material anisotropy. The generalized Skempton's pore pressure coefficients are defined under the undrained condition, so that

$$p = -\frac{B_{kl}}{3}\sigma_{kl} \tag{22.15}$$

Hence,

$$B_{kl} = 3M\alpha_{ij}C^u_{ijkl} = \frac{3M\alpha_{ij}C_{ijkl}}{1 + M\alpha_{mn}\alpha_{op}C_{mnop}} \tag{22.16}$$

22.2.3 Micromechanical coefficients

The coefficients defined in the preceding sections are 'bulk' parameters. It is often of interest to know their relation with the micromechanical parameters that are properties of the solid grain, porosity, fluid, etc. First, the effective stress coefficient α_{ij} is examined. Defining the Terzaghi's effective stress as

$$\overline{\sigma}_{ij} = \sigma_{ij} + \delta_{ij}p \tag{22.17}$$

the constitutive relation can be expressed in terms of $\overline{\sigma}_{ij}$ and p:

$$e_{ij} = C_{ijkl}\overline{\sigma}_{kl} - c_{ij}p \tag{22.18}$$

where c_{ij} is a compliance tensor whose significance is yet to be identified. Equation (22.18) can be re-organized into:

$$e_{ij} = C_{ijkl}\sigma_{kl} + (C_{ijkk} - c_{ij})p \tag{22.19}$$

From Eqs. (22.19), (22.7), and (22.8), it is found that

$$\alpha_{ij} = \delta_{ij} - M_{ijkl}c_{kl} \tag{22.20}$$

Next, a constitutive relation for the solid grain deformation is addressed. The total strain tensor e_{ij} can be decomposed into a solid part (e^s_{ij}) and a pore part (e^p_{ij}):

$$e_{ij} = (1 - \phi)e^s_{ij} + \phi e^p_{ij} \tag{22.21}$$

where ϕ is the porosity. The total stress tensor can be similarly decomposed as

$$\sigma_{ij} = (1 - \phi)\sigma^s_{ij} - \phi\delta_{ij}p \tag{22.22}$$

A linear constitutive law for the solid grain deformation can be presented as:

$$e^s_{ij} = \frac{1}{1 - \phi}C^s_{ijkl}\sigma_{kl} + \frac{\phi}{1 - \phi}c^s_{ij}p \tag{22.23}$$

or

$$e^s_{ij} = C^s_{ijkl}\sigma^s_{kl} + \frac{\phi}{1 - \phi}(c^s_{ij} - C^s_{ijkk})p \tag{22.24}$$

Two compliance tensors C^s_{ijkl} and c^s_{ij} introduced above relate to the properties of the solid grains.

The fluid deformation can now be examined. The pore strain can be decomposed into two parts:

$$e^p = e^f = e^f_{(1)} + e^f_{(2)} \tag{22.25}$$

where $e_{(1)}^f$ is associated with the dilation of the interstitial fluid, and $e_{(2)}^f$ is due to the fluid squeezed out of the frame. Apparently, $e_{(2)}^f$ is tied to the variation of fluid content ζ in the following fashion:

$$\zeta = \phi e_{(2)}^f \tag{22.26}$$

On the other hand, a constitutive law can be built for $e_{(1)}^f$ as:

$$e_{(1)}^f = -C_f p \tag{22.27}$$

where C_f is the fluid compressibility. Utilizing the definitions in Eqs. (22.21) and (22.25), we can write Eq. (22.26) as the following:

$$\zeta = e - (1 - \phi)e^s - \phi e_{(1)}^f \tag{22.28}$$

Substitution of Eqs. (22.19), (22.22), (22.23), and (22.27) in Eq. (22.28) yields:

$$\zeta = (C_{iikl} - C_{iikl}^s)\sigma_{kl} + [(C_{iikk} - c_{ii}) + \phi(C_f - c_{ii}^s)]p \tag{22.29}$$

Comparison of the first part on the right-hand side of Eq. (22.29) with Eqs. (22.2) and (22.20) reveals that

$$(C_{iikl} - C_{iikl}^s) = \alpha_{ij}C_{ijkl} = (C_{iikl} - c_{kl}) \tag{22.30}$$

It is apparent that

$$c_{ij} = C_{ijkk}^s \tag{22.31}$$

Hence, the compliance coefficients defined so far has been reduced to four independent systems: C_{ijkl}, C_{ijkl}^s, c_{ij}^s and C_f.

With the identity in Eq. (22.31), Eq. (22.20) can be rewritten as:

$$\alpha_{ij} = \delta_{ij} - M_{ijkl}C_{klmm}^s \tag{22.32}$$

In particular, Eq. (22.32) shows that Biot's effective stress coefficient tensor is the property of the skeleton and the porous solid only. It is realized that out of the 28 independent bulk material coefficients for general anisotropy, only one coefficient, namely M, is tied to the fluid property.

Setting $\zeta = 0$ in Eq. (22.29) and comparing with Eq. (22.15), it can be found that

$$B_{ij} = \frac{3(C_{ijkk} - C_{ijkk}^s)}{(C_{llmm} - C_{llmm}^s) + \phi(C_f - c_{ll}^s)} \tag{22.33}$$

Another useful relation is given by

$$M = \frac{1}{(\delta_{ij} - \alpha_{ij})(C_{ijkk} - C_{ijkk}^s) + \phi(C_f - c_{ii}^s)} \tag{22.34}$$

The micromechanical relations presented above seem to be quite complicated. It should be noted that in presenting C_{ijkl}, C_{ijkl}^s, etc., the usual symmetry of a general anisotropic material has been assumed. It may be argued that if the solid grain is microscopically isotropic, and that the global anisotropy is a consequence of the geometric arrangement of the pores, not the microscopic material, the above relations may be further simplified. For example, with the assumption of microisotropy for the solid grain, it can be shown that:

$$\alpha_{ij} = \delta_{ij} - \frac{M_{ijkk}}{3K_s} \tag{22.35}$$

where K_s is the bulk modulus of the solid grain. Applying similar simplification to Eqs. (22.33) and (22.34) yields

$$B_{ij} = \frac{3C_{ijkk} - \delta_{ij}/K_s}{(C_{iijj} - 1/K_s) + \phi(C_f - 1/K_s)} \tag{22.36}$$

$$M = \frac{K_s}{(1 - M_{iijj}/(9K_s)) - \phi(1 - K_sC_f)} \tag{22.37}$$

22.2.4   Unjacketed hydrostatic test

The above constitutive laws were derived without the assumption of microscopic homogeneity. Consider a porous media which is homogeneous at the grain scale. When such a material is subjected to an unjacketed hydrostatic loading, namely that $\sigma_{ij} = -\delta_{ij}P$ on the boundary surface of the element, and $p = P$ everywhere in the pores, the resulting stress corresponds to a uniform confining pressure, P, everywhere in the solid constituent and a zero Terzaghi's effective stress $\bar{\sigma}_{ij}$. The solid constituent, pore and the skeleton will then experience a uniform strain field without shape change, namely,

$$e_{ij} = e_{ij}^s = e_{ij}^p \tag{22.38}$$

Under these conditions, Eqs. (22.19) and (22.24) become

$$e_{ij} = -C_{ijkl}^s P \tag{22.39}$$

$$e_{ij}^s = -C_{ijkl}^s P + \frac{\phi}{1-\phi}(c_{ij}^s - C_{ijkk}^s)P \tag{22.40}$$

Hence, it is concluded that

$$c_{ij}^s = C_{ijkk}^s \tag{22.41}$$

Under the hypothesis of microhomogeneity, the compliance tensor c_{ij}^s is identified by Eq. (22.41).

22.2.5   Transverse isotropy

Most geomaterials exhibit some form of anisotropy. In particular, if the material has isotropic properties in the bedding plane but is anisotropic in the direction perpendicular to it, the transverse anisotropic model is called for. In this case, there are 5 independent M_{ijkl}, 2 α_{ij} and one M, which bring the total number of independent coefficients for poroelasticity to 8. To provide these parameters in their measurable form, a simplification of the presentation for Hooke's law (Boresi and Chong, 1987) is written. The stress and strain parameters are redefined as

$$\sigma_1 = \sigma_{11};\ \sigma_2 = \sigma_{22};\ \sigma_3 = \sigma_{33};\ \sigma_4 = \sigma_{12};\ \sigma_5 = \sigma_{13};\ \sigma_6 = \sigma_{23}$$

$$e_1 = e_{11};\ e_2 = e_{22};\ e_3 = e_{33};\ e_4 = 2e_{12};\ e_5 = 2e_{13};\ e_6 = 2e_{23} \qquad (22.42)$$

The stress-strain relation defined by Eqs. (1) and (2) can be rewritten as:

$$\sigma_i = M_{ij}e_j - \alpha_i p \qquad (22.43)$$

$$p = M(\zeta - \alpha_i e_i) \qquad (22.44)$$

where $i, j = 1, \ldots, 6$ and $M_{ij} = M_{ji}$. For general anisotropy, 21 independent M_{ij}, 6 α_i and 1 M exist. For transverse isotropy, only five of the M_{ij} coefficients are independent. If we select the z-axis to coincide with the axis of rotational symmetry, the five independent coefficients can be expressed in a matrix form as

$$M_{ij} = \begin{bmatrix} M_{11} & M_{12} & M_{13} & 0 & 0 & 0 \\ M_{12} & M_{11} & M_{13} & 0 & 0 & 0 \\ M_{13} & M_{13} & M_{33} & 0 & 0 & 0 \\ 0 & 0 & 0 & (M_{11} - M_{12})/2 & 0 & 0 \\ 0 & 0 & 0 & 0 & M_{55} & 0 \\ 0 & 0 & 0 & 0 & 0 & M_{55} \end{bmatrix} \qquad (22.45)$$

In terms of an engineering representation (Boresi and Chong, 1987), these coefficients can be expressed as

$$M_{11} = \frac{E(E' - Ev'^2)}{(1 + v)(E' - E'v - 2Ev'^2)} \qquad (22.46a)$$

$$M_{12} = \frac{E(E'v + Ev'^2)}{(1 + v)(E' - E'v - 2Ev'^2)} \qquad (22.46b)$$

$$M_{13} = \frac{EE'v'}{(E' - E'v - 2Ev'^2)} \qquad (22.46c)$$

$$M_{33} = \frac{E'^2(1 - v)}{(E' - E'v - 2Ev'^2)} \qquad (22.46d)$$

$$M_{55} = G' \qquad (22.46e)$$

The above relations contain five engineering constants: E and E', the drained Young's moduli in the isotropic plane and in the z-direction, respectively; v and v', the drained Poisson's ratios in the isotropic plane and associated with the z-direction, respectively; and G', the shear modulus associated with the z-direction. It is also noted that Eqs. (22.45) and (22.46) are also valid in the cylindrical coordinate system $r\theta z$, if the $r\theta$ plane coincides with the isotropic plane of the material.

With the assumption of micro-homogeneous and micro-isotropic solid skeleton, α and α', which are, respectively, the Biot effective stress coefficients in the isotropic plane and in the z-direction, can be expressed in terms of M_{ij} and the bulk modulus of the solid grain:

$$\alpha = 1 - \frac{M_{11} + M_{12} + M_{13}}{3K_s} \tag{22.47a}$$

$$\alpha' = 1 - \frac{2M_{13} + M_{33}}{3K_s} \tag{22.47b}$$

Therefore, those drained elastic engineering parameters (which lead to all necessary M_{ij}), M, and K_s are enough to define the transversely isotropic poroelastic constitutive relations (Cheng, 1997). Under this circumstance, similar to the corresponding elastic case, the degree of the material anisotropy can be defined by two anisotropic ratios: $n_E = E/E'$ and $n_v = v/v'$.

22.2.6 Governing equations

In this section, other basic equations in anisotropic poroelasticity are presented in the Cartesian coordinate system for transversely isotropic materials. Again, the z-axis is assumed to coincide with the rotational elastic symmetric axis.

Equilibrium equation

The total stress components must satisfy the equilibrium equation which is the same as its counterpart in continuum mechanics:

$$\sigma_{ij,j} = 0 \tag{22.48}$$

in which the body force term is ignored.

Strain-displacement relation

With the assumption of small strain, strain and displacement components satisfy the kinematic relation:

$$e_{ij} = \tfrac{1}{2}(u_{i,j} + u_{j,i}) \tag{22.49}$$

where u_i is the component of the overall displacement vector of the porous body in the ith-direction.

Navier equation

For many cases, the governing equations expressed in terms of the displacement components may be more convenient. From Eqs. (22.48), (22.43), (22.45), and (22.49), the Navier equation for transversely isotropic materials is obtained:

$$(M_{ii} + M_{ij} + M_{ik})\nabla^2 u_i + M_{ij}u_{j,ij} + M_{ik}u_{k,ik} - \alpha_i p_{,i} = 0 \tag{22.50}$$

In the above, no summation over repeated indices is taken; i, j, and k form a circulant order, i.e. (1, 2, 3), (2, 3, 1), or (3, 1, 2); ∇^2 is the Laplacian operator; $M_{ij} = M_{ji}$, $M_{11} = M_{22}$, and $M_{13} = M_{23}$; and $\alpha_1 = \alpha_2 = \alpha$ and $\alpha_3 = \alpha'$.

Darcy's law

It is further assumed that the orientation of the anisotropic permeabilities is coincident with the elastic anisotropy. Then, the pore fluid flow follows the anisotropic Darcy's law:

$$q_i = -\kappa_i p_{,i} \tag{22.51}$$

where no summation is taken over i, q_i is the component of the specific discharge vector in the ith-direction, $\kappa_i = k_i/\mu$, in which $k_1 = k_2 = k$ is the intrinsic permeability in the isotropic plane, $k_3 = k'$ is the same quantity in the z-direction, and μ is the viscosity of the pore fluid.

Fluid mass balance equation

Ignoring the source and sink, the fluid mass balance requires

$$\frac{\partial \zeta}{\partial t} - \kappa(p_{,11} + p_{,22}) - \kappa' p_{,33} = 0 \tag{22.52}$$

22.3 The Mandel Problem

Mandel, in 1953, presented one of the first solution for the three-dimensional consolidation theory of Biot (1941), demonstrating the non-monotonic pore-water pressure response. Later, Cryer (1963) observed the similar phenomenon of pore pressure development at the center of a sphere consolidating under hydrostatic pressure. This type of non-monotonic pressure response has been referred to as the Mandel–Cryer's effect in the literature (Schiffman *et al.*, 1969; Gibson *et al.*, 1990), and is a distinctive feature of the coupled consolidation theory, versus the traditional uncoupled Terzaghi's theory (Terzaghi, 1943). This physical phenomenon has been confirmed in the laboratory (Gibson *et al.*, 1963; Verruijt, 1965), and in the field (the Noordbergum effect) (Verruijt, 1965; Rodrigues, 1983).

Mandel's solution has been widely used as a benchmark problem for testing the validity of numerical codes of poroelasticity (Christian and Boehmer, 1970; Cheng and Detournay, 1988; Cui *et al.*, 1996). Its geometry also matches with one of the testing configuration of rock or stiff clay samples (Dickey *et al.*, 1968). The original Mandel's solution was limited to the incompressible isotropic materials; and only the pore pressure solution was provided. Cheng and Detournay (1988) extended Mandel's solution to the full poroelastic case, and listed the displacement, stress and pressure expressions. Recently, Abousleiman *et al.* (1996a) and Cui (1995) extended Mandel's solution to account for the anisotropy of the porous media. The analytical solution for Mandel's problem in transversely isotropic materials is presented in this chapter.

22.3.1 Problem definition

As illustrated in Figure 22.1, a transversely isotropic rock specimen of rectangular cross-section is sandwiched between two rigid, frictionless plates. The axis of material rotational

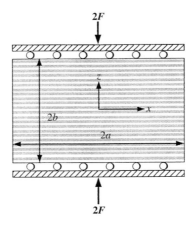

Figure 22.1   Mandel's problem

symmetry is chosen to coincide with the z-axis. It is assumed that the specimen is long in the y-direction (perpendicular to the paper) such that the plane strain (and flow) condition prevails, namely

$$u_x = u_x(x, z, t); \quad u_y = 0; \quad u_z = u_z(x, z, t) \tag{22.53a}$$

$$q_x = q_x(x, z, t); \quad q_y = 0; \quad q_z = q_z(x, z, t) \tag{22.53b}$$

The boundary conditions are the following:

$$\sigma_{xx} = \sigma_{xz} = p = 0, \quad \text{at } x = \pm a; \tag{22.54}$$

$$\sigma_{zx} = q_z = 0, u_z = \text{constant}, \quad \text{and} \quad \int_{-a}^{a} \sigma_{zz} \, dx = -2F, \quad \text{at } z = \pm b \tag{22.55}$$

where F is the force per unit length applied to the rigid plates.

Restricted by the sample geometry and boundary conditions, it is apparent that the displacement component in the z-direction, u_z, is independent of x and y. Therefore,

$$e_{zz} = e_{zz}(t) \tag{22.56}$$

$$q_z = 0 \tag{22.57}$$

The condition defined by Eq. (22.56) implies a planar deformation such that an originally flat, horizontal plane remains un-distorted during consolidation.

Utilizing Eqs. (22.52) and (22.51), it is easily deduced that

$$p = p(x, t) \tag{22.58}$$

$$q_x = q_x(x, t) \tag{22.59}$$

The continuity Eq. (22.52) with the initial condition of an undrained state, $\zeta = 0$ at $t = 0$, shows that

$$\zeta = \zeta(x, t) \tag{22.60}$$

From the above conditions and the constitutive relation defined by Eq. (22.44), it is clear that

$$e_{xx} = e_{xx}(x, t) \tag{22.61}$$

Judging from Eqs. (22.56), (22.58), (22.61), and (22.43), it is concluded that

$$\sigma_{xx} = \sigma_{xx}(x, t) \tag{22.62}$$

$$\sigma_{zz} = \sigma_{zz}(x, t) \tag{22.63}$$

Substituting Eq. (22.63) in the equilibrium Eq. (22.48) yields:

$$\nabla^2 \sigma_{xz} = 0 \tag{22.64}$$

Therefore, the boundary condition of no shear stress (see Eq. (22.55)) leads to

$$\sigma_{xz} = 0 \tag{22.65}$$

Furthermore, utilizing the above conditions, Eq. (22.48) and the boundary condition of no normal stress at the side boundaries require the last simplifying condition:

$$\sigma_{xx} = 0 \tag{22.66}$$

Note that these conditions presented above were also provided by Mandel (1953) for the isotropic case based on physical intuition.

22.3.2 Field equations

Using the above conditions for the Mandel's problem, for the current transversely isotropic case, from Eqs. (22.52), (22.43), and (22.44), it can be derived that

$$\left(\frac{\partial}{\partial t} - c_T \frac{\partial^2}{\partial x^2} \right) (\sigma_{zz} + A_1 p) = 0 \tag{22.67}$$

where c_T is the diffusivity coefficient in the isotropic plane:

$$c_T = \frac{\kappa M M_{11}}{M_{11} + \alpha^2 M} \tag{22.68}$$

and

$$A_1 = \frac{\alpha_1{}^2 M_{33} - 2\alpha_1\alpha_3 M_{13} + \alpha_3{}^2 M_{11}}{\alpha_3 M_{11} - \alpha_1 M_{13}} + \frac{M_{11} M_{33} - M_{13}{}^2}{M(\alpha_3 M_{11} - \alpha_1 M_{13})} \tag{22.69}$$

From Eqs. (22.49), (22.48), (22.43), and (22.44), the Beltrami–Michell compatibility relation in terms of stresses is obtained as follows:

$$\frac{\partial^2}{\partial x^2} (\sigma_{zz} + A_2 p) = 0 \tag{22.70}$$

with

$$A_2 = \frac{\alpha_3 M_{11} - \alpha_1 M_{13}}{M_{11}} \tag{22.71}$$

22.3.3   Solution

The system composed of Eqs. (22.67) and (22.70) is sufficient for the solution of the two variables σ_{zz} and p. First, solving Eq. (22.70) and considering symmetry, $\sigma_{zz}(-x, t) = \sigma_{zz}(x, t)$, and the boundary condition for pressure in Eq. (22.54), it can be shown that

$$\sigma_{zz} + A_2 p = C_1(t) \tag{22.72}$$

Using separation of variable, the solution of Eq. (22.67) can be expressed by

$$\sigma_{zz} + A_1 p = -\frac{F}{a} - \sum_{i=1}^{\infty} D_i \cos \frac{\beta_i x}{a} \exp\left(-\frac{\beta_i^2 c_T t}{a^2}\right) \tag{22.73}$$

The constant term in front of the series is determined as follows: as $t \to \infty$, $p \to 0$, the material behaves elastically and the stress distribution becomes $\sigma_{zz} = -F/a$.

The pore pressure can be obtained from Eqs. (22.72) and (22.73) in the form:

$$p = \frac{1}{A_2 - A_1} \sum_{i=1}^{\infty} D_i \left(\cos \frac{\beta_i x}{a} - \cos \beta_i\right) \exp\left(-\frac{\beta_i^2 c_x t}{a^2}\right) \tag{22.74}$$

where the boundary condition $p = 0$ at $x = \pm a$ has been incorporated. The stress can thus be solved from Eq. (22.73) as the following:

$$\sigma_{zz} = -\frac{F}{a} - \frac{A_2}{A_2 - A_1} \sum_{i=1}^{\infty} D_i \cos \frac{\beta_i x}{a} \exp\left(-\frac{\beta_i^2 c_x t}{a^2}\right)$$

$$+ \frac{A_1}{A_2 - A_1} \sum_{i=1}^{\infty} D_i \cos \beta_i \exp\left(-\frac{\beta_i^2 c_x t}{a^2}\right) \tag{22.75}$$

The stress condition at $z = \pm b$ as in Eq. (22.55) requires:

$$\frac{A_2}{A_2 - A_1} \sum_{i=1}^{\infty} \frac{2D_i a}{\beta_i} \sin \beta_i \exp\left(-\frac{\beta_i^2 c_T t}{a^2}\right) = \frac{A_1}{A_2 - A_1} \sum_{i=1}^{\infty} 2D_i a \cos \beta_i \exp\left(-\frac{\beta_i^2 c_x t}{a^2}\right) \tag{22.76}$$

It is clear that β_i must satisfy the characteristic equation:

$$\frac{\tan \beta_i}{\beta_i} = \frac{A_1}{A_2} \tag{22.77}$$

Constants D_i are determined from the initial condition. At the instant of loading, the system is undrained, i.e. $\zeta = 0$. The stress distribution is again elastic; hence, $\sigma_{zz}(x, 0) = -F/a$. From the constitutive relations in Eqs. (22.43) and (22.44), the pore pressure generated from Skempton's effect is found to be:

$$p(x, 0) = -\frac{\sigma_{zz}(x, 0)}{A_1} = \frac{F}{aA_1} = \frac{1}{A_2 - A_1} \sum_{i=1}^{\infty} D_i \left(\cos \frac{\beta_i x}{a} - \cos \beta_i\right) \tag{22.78}$$

Utilizing the orthogonal properties of the cosine function, Eq. (22.78) can be multiplied by $\cos(\beta_j x/a)$ and integrate from $-a$ to $+a$ to find the coefficients,

$$D_j = \frac{2F(A_2 - A_1)}{aA_1} \frac{\sin \beta_j}{\beta_j - \sin \beta_j \cos \beta_j} \tag{22.79}$$

Finally, the complete solution is presented as follows:

$$
u_x = \left[\frac{F}{a} \frac{M_{13}}{M_{11}M_{33} - M_{13}^2} \right.
$$

$$
- \frac{2F}{a} \frac{\alpha_1\alpha_3 M + M_{13}}{A_1 M(\alpha_3 M_{11} - \alpha_1 M_{13})} \sum_{i=1}^{\infty} \frac{\sin \beta_i \cos \beta_i}{\beta_i - \sin \beta_i \cos \beta_i} \exp\left(-\frac{\beta_i^2 c_T t}{a^2} \right) \left. \right] x
$$

$$
+ 2F \frac{\alpha_1}{A_2 M_{11}} \sum_{i=1}^{\infty} \frac{\cos \beta_i}{\beta_i - \sin \beta_i \cos \beta_i} \sin \frac{\beta_i x}{a} \exp\left(-\frac{\beta_i^2 c_T t}{a^2} \right) \tag{22.80}
$$

$$
u_z = -\frac{F}{a} \frac{M_{11}}{M_{11}M_{33} - M_{13}^2}
$$

$$
\times \left[1 + 2 \left(\frac{A_2}{A_1} - 1 \right) \sum_{i=1}^{\infty} \frac{\sin \beta_i \cos \beta_i}{\beta_i - \sin \beta_i \cos \beta_i} \exp\left(-\frac{\beta_i^2 c_T t}{a^2} \right) \right] z \tag{22.81}
$$

$$
\sigma_{xx} = 0 \tag{22.82}
$$

$$
\sigma_{zz} = -\frac{F}{a} - \frac{2FA_2}{aA_1} \sum_{i=1}^{\infty} \frac{\sin \beta_i}{\beta_i - \sin \beta_i \cos \beta_i} \cos \frac{\beta_i x}{a} \exp\left(-\frac{\beta_i^2 c_T t}{a^2} \right)
$$

$$
+ \frac{2F}{a} \sum_{i=1}^{\infty} \frac{\sin \beta_i \cos \beta_i}{\beta_i - \sin \beta_i \cos \beta_i} \exp\left(-\frac{\beta_i^2 c_T t}{a^2} \right) \tag{22.83}
$$

$$
\sigma_{xy} = 0 \tag{22.84}
$$

$$
p = \frac{2F}{aA_1} \sum_{i=1}^{\infty} \frac{\sin \beta_i}{\beta_i - \sin \beta_i \cos \beta_i} \left(\cos \frac{\beta_i x}{a} - \cos \beta_i \right) \exp\left(-\frac{\beta_i^2 c_T t}{a^2} \right) \tag{22.85}
$$

$$
q_x = \frac{2Fk_x}{a^2 \mu A_1} \sum_{i=1}^{\infty} \frac{\beta_i \sin \beta_i}{\beta_i - \sin \beta_i \cos \beta_i} \sin \frac{\beta_i x}{a} \exp\left(-\frac{\beta_i^2 c_T t}{a^2} \right) \tag{22.86}
$$

$$
q_z = 0 \tag{22.87}
$$

22.3.4 Examples

A set of data reported by Aoki *et al.* (1993) for Trafalgar shale is adopted here:

$$E_u = 22.0 \text{ GPa}; \quad E_u' = 18.8 \text{ GPa}; \quad G' = 7.23 \text{ GPa};$$

$$\nu_u = 0.27; \quad \nu_u' = 0.34; \quad B = 0.51; \quad B' = 0.63$$

where the subscript u indicates the Young's moduli and Poisson's ratios for the undrained materials, B and B' are, respectively, the Skempton's pore pressure coefficients in the isotropic plane and the z-direction. A total of eight coefficients are generally required to define the constitutive relations for transverse isotropy. With the assumption of micro-isotropy and microhomogeneity of the material, the seven coefficients reported are suffi-cient to fully define the bulk continuum behavior. Solving the micromechanical relations

(Cheng, 1997), we obtain the following parameters, which are all defined in a consistent manner:

$$E = 20.6\,\text{GPa};\ E' = 17.3\,\text{GPa};\ \nu = 0.189;\ \nu' = 0.246;$$

$$M_{11} = 24.1\,\text{GPa};\ M_{12} = 6.80\,\text{GPa};\ M_{13} = 7.62\,\text{GPa};\ M_{33} = 21.0\,\text{GPa};$$

$$M_{55} = 7.23\,\text{GPa};\ M_{11}^u = 32.6\,\text{GPa};\ M_{12}^u = 15.3\,\text{GPa};\ M_{13}^u = 16.3\,\text{GPa};$$

$$M_{33}^u = 29.9\,\text{GPa};\ M = 15.8\,\text{GPa};\ \alpha = 0.733;\ \alpha' = 0.749;\ K_s = 48.2\,\text{GPa}$$

The following values for the hydraulic properties are estimated: $k_x = 1 \times 10^{-7}$ darcy, μ (for water) $= 0.001$ Pa·s. Using the above data set, a specimen with the vertical section (Figure 22.1) with dimensions 10 cm $\times$ 10 cm is examined. The force on the specimen is assumed to be $F = 1 \times 10^6$ N/m.

Figure 22.2 illustrates the evolution of pore pressure along the x-axis (at any constant elevation z). At $t = 0^+$, a uniform pressure of the magnitude

$$p(x, 0) = \frac{F}{3a}\left[B_z + B_x \frac{M_{13}^u(M_{11}^u - M_{12}^u)}{M_{11}^u M_{33}^u - M_{13}^{u\,2}}\right] \tag{22.88}$$

arise due to the Skempton's effect. As time progresses we observe that the pore pressure in the center region rise above the initial value, which is shaded in light tone for easy visualization. This is known as the Mandel–Cryer's effect.

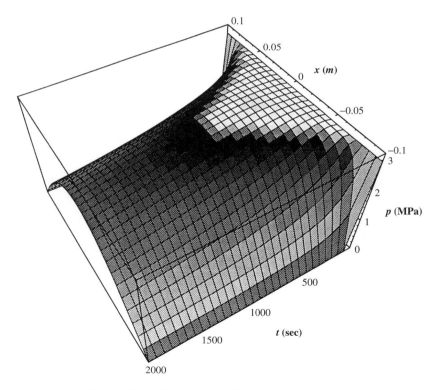

Figure 22.2 Pore pressure history along the x-axis

Next the normal stress $-\sigma_{zz}$ is inspected along the same horizontal section. Figure 22.3 shows that at $t = 0^+$, the uniform stress

$$\sigma_{zz}(x, 0) = -\frac{F}{a} \tag{22.89}$$

exist. Due to the load transferring effect, the stress near the center rises above the initial value, which is again marked in light shade, while that on the edges dips below. As $t \to \infty$, the same uniform stress condition, namely $\sigma_{zz}(x, \infty) = -F/a$, must return.

Figure 22.4 gives the vertical displacement at the top, $-u_z(b, t)$, and the horizontal displacement at the right edge, $u_x(a, t)$. Note that, in the vertical direction, after the initial settlement, the specimen continues to consolidate. In the horizontal direction, the sample instantaneously expands due the Poisson's effect. It then contracts as the effective Poisson's ratio changes from the undrained value to the drained one.

It is of interest to examine the consequence of ignoring anisotropy. For that purpose, the following set of isotropic data for vertical properties is utilized: case 1: $E = 17.3$ GPa, $\nu = 0.189$ $\nu_u = 0.27$, and $B = 0.63$, while, for horizontal properties the following values are used: case 2: $E = 20.6$ GPa, $\nu = 0.246$ $\nu_u = 0.34$, and $B = 0.51$. In Figure 22.5 the pore pressure at the center of the specimen ($x = 0$) is illustrated for the transverse isotropic (solid line), and the two isotropic cases (dashed lines).

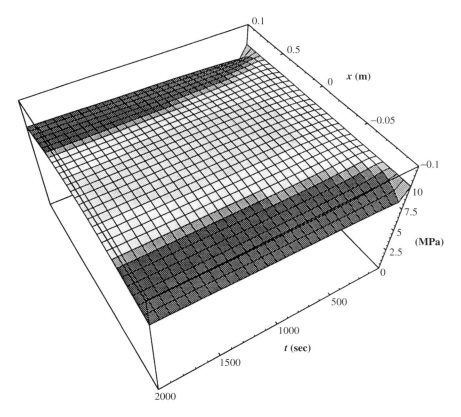

Figure 22.3   Vertical compressive stress ($-\sigma_{zz}$) along the x-axis

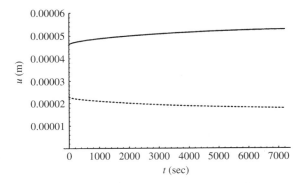

Figure 22.4   Vertical displacement at the top (solid line) and horizontal displacement at the right edge (dashed line)

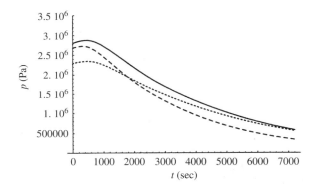

Figure 22.5   Pore pressure history at the center of specimen for case (a) (dashed line) and (b) (solid line)

22.4   Borehole Problem

The stress concentration near an excavated or pressurized borehole is of critical interest to the petroleum industry, since it is a major factor controlling the borehole stability which is of great importance for the borehole drilling. It is common that a borehole is drilled through a fluid saturated formation; therefore, a poroelastic stress analysis of the borehole is necessary. The analytical solutions for stresses and pore pressure around the borehole in isotropic poroelastic media were obtained by Detournay and Cheng (1988) for a plane strain geometry, and Cui *et al.* (1997) for the most general geometry. For transversely isotropic rocks, the analytical poroelastic solution for boreholes was also derived recently by Abousleiman *et al.* (1995), and it is presented below.

22.4.1   Problem description

It is assumed that an infinitely long borehole is drilled perpendicular to the isotropic plane of a transversely isotropic poroelastic formation. The borehole axis is designated as the

z-axis in Figure 22.6(a). The principal axes of the far-field stresses are shown as x'-y'-z' with the three stress components denoted as $S_{x'}$, $S_{y'}$ and $S_{z'}$. The inclination of the borehole is measured by the two angles φ_x and φ_z, as depicted in Figure 22.6(a). Figure 22.6(b) shows the far-field stress components from the vantage-point of the borehole coordinates x-y-z. It is noted that there now exist normal as well as shear stress components which are denoted by S_i and S_{ij}, respectively.

The boundary conditions of the problem can be imposed as: at the far-field, i.e. $r \to \infty$,

$$\sigma_{xx} = -S_x; \quad \sigma_{yy} = -S_y; \quad \sigma_{zz} = -S_z;$$

$$\tau_{xy} = -S_{xy}; \quad \tau_{yz} = -S_{yz}; \quad \tau_{xz} = -S_{xz}; \quad p = p_o \tag{22.90}$$

where p_o is the formation virgin pore pressure; and at the borehole wall, i.e. $r = R$,

$$\sigma_{rr} = -p_w H(t); \quad \tau_{r\theta} = \tau_{rz} = 0; \quad p = p_i H(t) \tag{22.91}$$

in which p_w is the wellbore pressure; p_i is the pore pressure at the borehole wall, and $H(t)$ is the Heaviside step function.

In view of the linearity, the problem can be decomposed into three sub-problems:

Problem I

At the far-field ($r \to \infty$):

$$\sigma_{zz} = -\nu'(S_x + S_y) - (\alpha' - 2\nu'\alpha)p_o; \quad \sigma_{xx} = -S_x; \quad \sigma_{yy} = -S_y;$$

$$\tau_{xy} = -S_{xy}; \quad \tau_{yz} = \tau_{xz} = 0; \quad p = p_o \tag{22.92}$$

At the borehole wall ($r = R$):

$$\sigma_{rr} = -p_w H(t); \quad \tau_{r\theta} = \tau_{rz} = 0; \quad p = p_i H(t) \tag{22.93}$$

Problem II

At the far-field ($r \to \infty$):

$$\sigma_{zz} = -S_z + [\nu'(S_x + S_y) + (\alpha' - 2\nu'\alpha)p_o];$$

$$\sigma_{xx} = \sigma_{yy} = \tau_{xy} = \tau_{yz} = \tau_{xz} = p = 0 \tag{22.94}$$

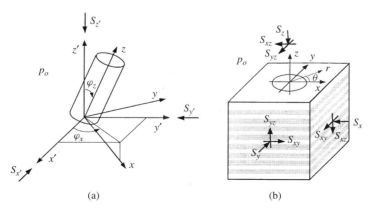

(a)   (b)

Figure 22.6   Inclined borehole in a transversely isotropic material

At the borehole wall $(r = R)$:

$$\sigma_{rr} = \tau_{r\theta} = \tau_{rz} = p = 0 \tag{22.95}$$

Problem III
At the far-field $(r \to \infty)$:

$$\sigma_{xx} = \sigma_{yy} = \sigma_{zz} = \tau_{xy} = p = 0;$$
$$\tau_{yz} = -S_{yz}; \quad \tau_{xz} = -S_{xz} \tag{22.96}$$

At the borehole wall $(r = R)$:

$$\sigma_{rr} = \tau_{r\theta} = \tau_{rz} = p = 0 \tag{22.97}$$

Next, solution for each sub-problem is sought individually.

22.4.2 Solution for Problem I

Problem I is a plane strain problem. Following the same procedure proposed for the isotropic case (Detournay and Cheng, 1988), problem I is decomposed as: initial stress and pore pressure fields; and three different modes which correspond, respectively, three different loading conditions at the borehole wall. Then solution is solved for each mode individually. The solution for Problem I is finally obtained by the superposition.

For all three modes, at infinity $(r \to \infty)$,

$$\sigma_{rr}^{(i)} = \tau_{r\theta}^{(i)} = p^{(i)} = 0 \tag{22.98}$$

However, each mode has different boundary conditions at the borehole wall. More specifically:

Mode 1
At $r = R$,

$$\sigma_{rr}^{(1)} = P_o H(t); \quad \tau_{r\theta}^{(1)} = p^{(1)} = 0 \tag{22.99}$$

Mode 2
At $r = R$,

$$\sigma_{rr}^{(2)} = \tau_{r\theta}^{(2)} = 0; \quad p^{(2)} = -p_o H(t) \tag{22.100}$$

Mode 3
At $r = R$,

$$\sigma_{rr}^{(3)} = -S_o \cos 2\theta H(t); \quad \tau_{r\theta}^{(3)} = S_o \sin 2\theta H(t); \quad p^{(3)} = 0 \tag{22.101}$$

The first two modes are axisymmetric problems. Similar to the isotropic case (Rice and Cleary, 1976; Detournay and Cheng, 1993), from Eqs. (22.52) and (22.44), the irrotational condition (attributed to axisymmetry), the diffusion of pore pressure can be uncoupled from the solid deformation (Cui, 1995):

$$\frac{\partial p}{\partial t} - c_T \left(\frac{\partial^2 p}{\partial r^2} + \frac{1}{r} \frac{\partial p}{\partial r} \right) = 0 \tag{22.102}$$

The axisymmetry also leads to (Cui, 1995):

$$u_r = \frac{A}{r} + \frac{\alpha}{M_{11}} \frac{1}{r} \int_a^r rp\,dr \tag{22.103}$$

where A is an integration constant. In the above, that all variables are finite and $e = 0$ at infinity is implied. Therefore, the stresses can be expressed by:

$$\sigma_r = -2G\frac{A}{r^2} - \alpha\left(1 - \frac{M_{12}}{M_{11}}\right)\frac{1}{r^2}\int_a^r rp\,dr \tag{22.104a}$$

$$\sigma_\theta = 2G\frac{A}{r^2} + \alpha\left(1 - \frac{M_{12}}{M_{11}}\right)\frac{1}{r^2}\int_a^r rp\,dr - \alpha\left(1 - \frac{M_{12}}{M_{11}}\right)p \tag{22.104b}$$

Mode 1 can be proven to be purely elastic since the variation of stresses around the borehole does not disturb the pore pressure field. Hence, the solution is obtained as:

$$\sigma_{rr}^{(1)} = H(t)P_o R^2/r^2 \tag{22.105a}$$

$$\sigma_{\theta\theta}^{(1)} = -H(t)P_o R^2/r^2 \tag{22.105b}$$

$$p^{(1)} = 0 \tag{22.105c}$$

For mode 2, $p^{(2)}$ can be obtained from the diffusion Eq. (22.102) with the condition defined by Eq. (22.100); however, it is solved analytically in the Laplace transform domain (Rice and Cleary, 1976; Detournay and Cheng, 1988). $p^{(2)}$, $\sigma_{rr}^{(2)}$, and $\sigma_{\theta\theta}^{(2)}$ are given below as their Laplace transforms:

$$\tilde{p}^{(2)} = -\frac{P_o}{s}\frac{K_0(\xi)}{K_0(\beta)} \tag{22.106a}$$

$$\tilde{\sigma}_{rr}^{(2)} = -\frac{P_o}{s}\frac{2G\alpha}{M_{11}}\left[\frac{R}{r}\frac{K_1(\xi)}{\beta K_0(\beta)} - \frac{R^2}{r^2}\frac{K_1(\beta)}{\beta K_0(\beta)}\right] \tag{22.106b}$$

$$\tilde{\sigma}_{\theta\theta}^{(2)} = \frac{P_o}{s}\frac{2G\alpha}{M_{11}}\left[\frac{R}{r}\frac{K_1(\xi)}{\beta K_0(\beta)} - \frac{R^2}{r^2}\frac{K_1(\beta)}{\beta K_0(\beta)} + \frac{K_0(\xi)}{K_0(\beta)}\right] \tag{22.106c}$$

where $\tilde{\ }$ denotes the Laplace transform; s, the Laplace transform variable; $\xi = r\sqrt{s/c_T}$, $\beta = R\sqrt{s/c_T}$; and K_n, the modified Bessel function of the second kind of order n.

Again, mode 3 is solved analytically in the Laplace transform domain. From the corresponding Eqs. (22.48), (22.49), (22.43), (22.44), and (22.52) in cylindrical coordinate system, the Navier-type equation and the diffusion equation can be derived (see Cui, 1995):

$$M_{11}\frac{\partial e}{\partial r} - 2G\frac{1}{r}\frac{\partial \omega}{\partial \theta} - \alpha\frac{\partial p}{\partial r} = 0 \tag{22.107}$$

$$M_{11}\frac{1}{r}\frac{\partial e}{\partial \theta} + 2G\frac{\partial \omega}{\partial r} - \alpha\frac{1}{r}\frac{\partial p}{\partial \theta} = 0 \tag{22.108}$$

$$\frac{\partial \zeta}{\partial t} - c_T\left(\frac{\partial^2 \zeta}{\partial r^2} + \frac{1}{r}\frac{\partial \zeta}{\partial r} + \frac{1}{r^2}\frac{\partial^2 \zeta}{\partial \theta^2}\right) \tag{22.109}$$

where

$$\omega = \frac{1}{2} \left(\frac{1}{r} \frac{\partial (r u_\theta)}{\partial r} - \frac{1}{r} \frac{\partial u_r}{\partial \theta} \right) \tag{22.110}$$

Applying the Laplace transform to Eqs. (22.107), (22.108), and (22.109) yields

$$\frac{M_{11} + \alpha^2 M}{2G} \frac{\partial \tilde{e}}{\partial r} - \frac{1}{r} \frac{\partial \tilde{\omega}}{\partial \theta} - \frac{\alpha M}{2G} \frac{\partial \tilde{\zeta}}{\partial r} = 0 \tag{22.111a}$$

$$\frac{M_{11} + \alpha^2 M}{2G} \frac{1}{r} \frac{\partial \tilde{e}}{\partial \theta} + \frac{\partial \tilde{\omega}}{\partial r} - \frac{\alpha M}{2G} \frac{1}{r} \frac{\partial \tilde{\zeta}}{\partial \theta} = 0 \tag{22.111b}$$

$$\frac{\partial^2 \tilde{\zeta}}{\partial r^2} + \frac{1}{r} \frac{\partial \tilde{\zeta}}{\partial r} + \frac{1}{r^2} \frac{\partial^2 \tilde{\zeta}}{\partial \theta^2} - \frac{s}{c_T} \tilde{\zeta} = 0 \tag{22.111c}$$

In view of the asymmetry of mode 3, it can be assumed that

$$(\tilde{\zeta}^{(3)}, \tilde{e}^{(3)}, \tilde{u}_r^{(3)}, \tilde{\sigma}_r^{(3)}, \tilde{\sigma}_\theta^{(3)}, \tilde{p}^{(3)}) = (\tilde{Z}, \tilde{E}, \tilde{U}_r, \tilde{S}_r, \tilde{S}_\theta, \tilde{P}) \cos 2\theta \tag{22.112a}$$

$$(\tilde{\omega}^{(3)}, \tilde{u}_\theta^{(3)}, \tilde{\tau}_{r\theta}^{(3)}) = (\tilde{W}, \tilde{U}_\theta, \tilde{S}_{r\theta}) \sin 2\theta \tag{22.112b}$$

Following Detournay and Cheng (1988), the following expressions of stresses and pore pressure in the Laplace transform domain are then obtained:

$$\tilde{p}^{(3)} = \cos 2(\theta - \theta_r) \frac{S_o}{s} \left[\frac{c_T}{2G\kappa} C_1 K_2(\xi) + A_1 C_2 \frac{R^2}{r^2} \right] \tag{22.113a}$$

$$\tilde{\sigma}_{rr}^{(3)} = \cos 2(\theta - \theta_r) \frac{S_o}{s} \left[A_1 C_1 \left(\frac{1}{\xi} K_1(\xi) + \frac{6}{\xi^2} K_2(\xi) \right) - A_2 C_2 \frac{R^2}{r^2} - 3C_3 \frac{R^4}{r^4} \right] \tag{22.113b}$$

$$\tilde{\sigma}_{\theta\theta}^{(3)} = \cos 2(\theta - \theta_r) \frac{S_o}{s} \left\{ -A_1 C_1 \left[\frac{1}{\xi} K_1(\xi) + \left(1 + \frac{6}{\xi^2} \right) K_2(\xi) \right] + 3C_3 \frac{R^4}{r^4} \right\} \tag{22.113c}$$

$$\tilde{\tau}_{r\theta}^{(3)} = \sin 2(\theta - \theta_r) \frac{S_o}{s} \left[2A_1 C_1 \left(\frac{1}{\xi} K_1(\xi) \right. \right. \tag{22.113d}$$

$$\left. \left. + \frac{3}{\xi^2} K_2(\xi) \right) - \frac{A_2}{2} C_2 \frac{R^2}{r^2} - 3C_3 \frac{R^4}{r^4} \right] \tag{22.113e}$$

where

$$C_1 = \frac{4}{2A_1(D_3 - D_2) - A_2 D_1} \tag{22.114a}$$

$$C_2 = -\frac{4D_1}{2A_1(D_3 - D_2) - A_2 D_1} \tag{22.114b}$$

$$C_3 = \frac{2A_1(D_2 + D_3) + 3A_2 D_1}{3[2A_1(D_3 - D_2) - A_2 D_1]} \tag{22.114c}$$

In the above,

$$A_1 = \frac{\alpha M}{M_{11} + \alpha^2 M} \tag{22.115a}$$

$$A_2 = \frac{M_{11} + M_{12} + 2\alpha^2 M}{M_{11} + \alpha^2 M} \tag{22.115b}$$

$$D_1 = \frac{M_{11}}{2G\alpha} K_2(\beta) \tag{22.115c}$$

$$D_2 = \frac{1}{\beta} K_1(\beta) + \frac{6}{\beta^2} K_2(\beta) \tag{22.115d}$$

$$D_3 = 2\left(\frac{1}{\beta} K_1(\beta) + \frac{3}{\beta^2} K_2(\beta)\right) \tag{22.115e}$$

The final solution for Problem I is thus expressed by:

$$\sigma_{rr}^{(I)} = -P_o + S_o \cos 2(\theta - \theta_r) + \sigma_{rr}^{(1)} + \sigma_{rr}^{(2)} + \sigma_{rr}^{(3)} \tag{22.116a}$$

$$\sigma_{\theta\theta}^{(I)} = -P_o - S_o \cos 2(\theta - \theta_r) + \sigma_{\theta\theta}^{(1)} + \sigma_{\theta\theta}^{(2)} + \sigma_{\theta\theta}^{(3)} \tag{22.116b}$$

$$\sigma_{zz}^{(I)} = v'[\sigma_{rr}^{(I)} + \sigma_{\theta\theta}^{(I)}] - (\alpha' - 2v'\alpha)p^{(I)} \tag{22.116c}$$

$$\tau_{r\theta}^{(I)} = -S_o \sin 2(\theta - \theta_r) + \tau_{r\theta}^{(3)} \tag{22.116d}$$

$$\tau_{rz}^{(I)} = \tau_{\theta z}^{(I)} = 0 \tag{22.116e}$$

$$p^{(I)} = p_o + p^{(2)} + p^{(3)} \tag{22.116f}$$

where

$$P_o = \frac{S_x + S_y}{2} \tag{22.117a}$$

$$S_o = \sqrt{\left(\frac{S_x - S_y}{2}\right)^2 + S_{xy}^2} \tag{22.117b}$$

$$\theta_r = \frac{1}{2} \tan^{-1} \frac{2S_{xy}}{S_x - S_y} \tag{22.117c}$$

Generally, a numerical technique for the inverse of the Laplace transform (e.g. Stehfest method (Stehfest, 1970)) is required to evaluate $\sigma_{rr}^{(2)}$, $\sigma_{rr}^{(3)}$, $\sigma_{\theta\theta}^{(2)}$, $\sigma_{\theta\theta}^{(3)}$, $\tau_{r\theta}^{(3)}$, $p^{(2)}$, and $p^{(3)}$.

22.4.3   Solution for Problem II

For the material anisotropy concerned here, Problem II is part of the initial stress state without the disturbance of the pore pressure field; hence, is purely elastic. Its solution is the same as the one for the corresponding isotropic problem which was studied by Cui et al. (1997). Hence,

$$\sigma_{zz}^{(II)} = -S_z + \left[v'(S_x + S_y) + (\alpha' - 2v'\alpha)p_o\right] \tag{22.118a}$$

$$\sigma_{rr}^{(II)} = \sigma_{\theta\theta}^{(II)} = \tau_{r\theta}^{(II)} = \tau_{rz}^{(II)} = \tau_{\theta z}^{(II)} = p^{(II)} = 0 \tag{22.118b}$$

In fact, these sets of stresses and pore pressure satisfy all the governing equations and boundary conditions; hence, they form an admissible poroelastic solution.

Generally speaking, taking off the core material which originally occupied the borehole geometry disturbs initial stress and pore pressure fields around the borehole and results in the redistribution of stresses and pore pressures. Problem II presents the parts of stress and pore pressure fields which are not disturbed through this process. Therefore, its solution must be uniform and the same as its initial state.

22.4.4   Solution for Problem III

This problem is an anti-shear problem. The anisotropic elasticity (Amadei, 1983) shows that the excavation of the borehole only disturbs the anti-shear stresses. For the transverse isotropy investigated in this chapter, shear deformation is uncoupled with the pore fluid flow. Therefore, Problem III is also purely elastic. The solution is the same as the elastic one. Under this circumstance, the anisotropy does not affect anti-shear stress field (however, the displacements are affected.) Hence, the solution can be obtained as:

$$\tau_{rz}^{(III)} = -(S_{xz}\cos\theta + S_{yz}\sin\theta)\left[1 - H(t)\frac{R^2}{r^2}\right] \tag{22.119a}$$

$$\tau_{\theta z}^{(III)} = (S_{xz}\sin\theta - S_{yz}\cos\theta)\left[1 + H(t)\frac{R^2}{r^2}\right] \tag{22.119b}$$

$$\sigma_{rr}^{(III)} = \sigma_{\theta\theta}^{(III)} = \sigma_{zz}^{(III)} = \tau_{r\theta}^{(III)} = p^{(III)} = 0 \tag{22.119c}$$

22.4.5   Final solution

Finally, the solution for an inclined borehole in a transversely isotropic poroelastic medium is obtained by superposition defined in terms of the following equations:

$$\sigma_{rr} = \sigma_{rr}^{(I)} \tag{22.120a}$$

$$\sigma_{\theta\theta} = \sigma_{\theta\theta}^{(I)} \tag{22.120b}$$

$$\sigma_{zz} = \sigma_{zz}^{(I)} + \sigma_{zz}^{(II)} \tag{22.120c}$$

$$\tau_{r\theta} = \tau_{r\theta}^{(I)} \tag{22.120d}$$

$$\tau_{rz} = \tau_{rz}^{(III)} \tag{22.120e}$$

$$\tau_{\theta z} = \tau_{\theta z}^{(III)} \tag{22.120f}$$

$$p = p^{(I)} \tag{22.120g}$$

In the above equations, components that are identically zero are omitted.

22.4.6   Examples

The far field *in situ* stresses and formation pore pressure gradients adopted for the analysis are the following (Woodland, 1990): $S_{x'} = 29$ MPa/km, $S_{y'} = 20$ MPa/km, $S_{z'} = 25$ MPa/km, and $p_o = 9.8$ MPa/km. The radius of the borehole was assumed to be

$R = 0.1$ m. The deviation of the borehole was defined by $\varphi_x = 30°$ and $\varphi_z = 60°$; and the depth of the formation was assumed to be 1000 m. For convenience, only the case of an excavation was analyzed, i.e. $p_w = p_i = 0$.

It is assumed that $E = 20.6$ GPa, $v = 0.189$, $M = 15.8$ GPa, $K_s = 48.2$ GPa, and $\kappa = \kappa' = 8.64 \times 10^{-6}$ m²/MPa· day. The above data set were used previously for the analyses of the anisotropic Mandel's problem. For given anisotropic ratios of $n_E = E/E'$ and $n_v = v/v'$, E' and v' can be determined; hence, a transversely isotropic material is defined.

In all the results presented below, the effective stress indicates the Terzaghi's effective stress ($\sigma_{ij} + p\delta_{ij}$); and the negative value of the stress, which implies the rock mechanics sign convention for stresses (i.e. positive for compression), is used. All calculated pore pressures and stresses are presented as a function of r/R along $\theta = 90°$.

Figure 22.7 presents the pore pressure profiles with $n_v = 0.5, 1, 2$ for fixed $n_E = 1$ at different times. It is observed that the material anisotropy influences the pore pressure only at small time ($t = 0.001$ day). For a long time duration ($t = 1$ day), all pore pressure distributions for different n_v are identical; indicating that the anisotropy due to Poisson's ratios does not affect pore pressure at large time.

In Figure 22.8 the effective radial stress distribution shows similar results of the anisotropy effect as for the pore pressure. The stress for $n_v = 0.5$ is much more influenced by the anisotropy than for $n_v = 2$. However, unlike the case for the pore pressure, the anisotropy effect on the stress still exists for large time. It seems that the anisotropy effect on the effective radial stress becomes more significant at places far away from the borehole wall for large times.

Figures 22.9 and 22.10 describe, respectively, the effective tangential and axial stresses around the borehole. It can be seen from these figures that at small time ($t = 0.001$ day) for different ratios of n_v, magnitudes of stresses around borehole are quite different, but they are close at the places away from the borehole. When time duration becomes larger ($t = 1$ day), the effect of anisotropy on stresses is transmitted to the places away from

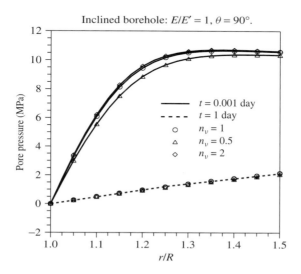

Figure 22.7 Pore pressure near the borehole varying with r/R for $E/E' = 1$ and along the direction of $\theta = 90°$

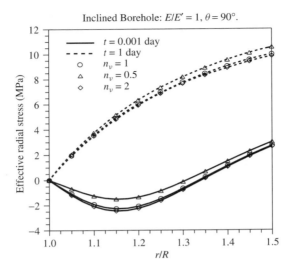

Figure 22.8   Effective radial stress $(-\sigma'_{rr})$ near the borehole varying with r/R for $E/E' = 1$ and along the direction of $\theta = 90°$

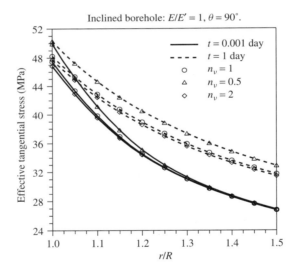

Figure 22.9   Effective tangential stress $(-\sigma'_{\theta\theta})$ near the borehole varying with r/R for $E/E' = 1$ and along the direction of $\theta = 90°$

the borehole. It should also be noted that the material anisotropy effect is significant on effective tangential and axial stresses compared to the pore pressure and the radial stress, thus affecting wellbore stability analyses.

For the cases of varied n_E with fixed $n_v = 1$, analyses showed that material anisotropy effects on the pore pressure and all effective stresses are qualitatively similar to the previous cases of varied n_v, although they are not presented here. Notice that material

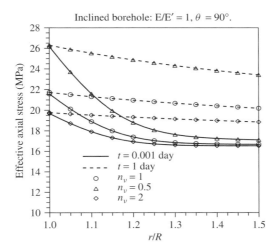

Figure 22.10   Effective axial stress $(-\sigma'_{zz})$ near the borehole varying with r/R for $E/E' = 1$ and along the direction of $\theta = 90°$

anisotropy would have a combination of these two ratios, and the corresponding analysis should be conducted.

22.5   Cylinder Problem

The cylinder is a most widely used sample geometry in laboratory testing in soil/rock mechanics, especially in the petroleum industry. Therefore, the solution for a saturated porous cylinder is very useful for determinations of poromechanical properties, and studies of failure criteria for porous media. In addition, it can also be applied to the strain relaxation method for *in situ* stress measurements (Abousleiman and Cheng, 1996). The poroelastic solutions for isotropic cylinders were obtained by Detournay and Cheng (1993), and Abousleiman *et al.* (1996b) for a plane strain and a generalized plane strain geometry, respectively; the solutions for transversely isotropic cylinders were also derived by Cui (1995), and Abousleiman and Cui (1998) under plane strain and generalized plane strain conditions, respectively. The poroelastic solution for cylinder in transversely isotropic media under both plane strain and generalized plane strain conditions is presented next.

22.5.1   *Problem description*

The length of the cylinder is assumed to be much greater than its radius; the isotropic plane of the material symmetry is also normal to the axis of the cylinder (the z-axis in Figure 22.11); and, the flow in the z-direction is not allowed. The boundary conditions commonly encountered in the laboratory can be summarized at the cylinder's outer boundary, $r = R$ as

$$p = p_o(t); \ \text{or}, \ q_r = Q(t)/2\pi R;$$

$$\sigma_{rr} = -P_o(t) - S_o \cos 2\theta H(t); \ \tau_{r\theta} = S_o \sin 2\theta H(t); \ \tau_{rz} = 0 \qquad (22.121)$$

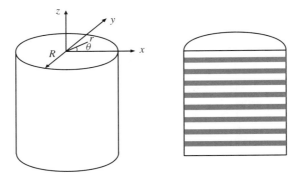

Figure 22.11 Schematic of the cylinder

In addition, a uniform axial strain is imposed which is similar in laboratory tests called a 'stroke control test', i.e.

$$e_{zz} = -e_o(t) \tag{22.122}$$

In the above, p_o is a time-dependent pore pressure applied on the cylindrical surface; $Q(t)$ is the total flow discharge across the cylindrical surface per unit depth; P_o is the normal pressure applied on the surface and it may be a time-dependent loading; S_o is a constant which represents the amplitude of sinusoidal distributions of normal pressure and shear stress applied on the surface; and e_o is the magnitude of the compressive axial strain and may also be time-dependent.

In the light of the linearity, the condition in Eq. (22.121) can be decomposed into the following modes for which combinations (modes superposition) can simulate many laboratory tests configurations:

Mode 1
  At $r = R$,

$$p = p_o(t); \quad \sigma_{rr} = \tau_{r\theta} = 0 \tag{22.123}$$

  with the plane strain condition $e_{zz} = 0$.

Mode 2
  At $r = R$,

$$q_r = Q(t)/2\pi R; \quad \sigma_{rr} = \tau_{r\theta} = 0 \tag{22.124}$$

  with the plane strain condition $e_{zz} = 0$.

Mode 3
  At $r = R$,

$$p = 0; \quad \sigma_{rr} = -P_o(t); \quad \tau_{r\theta} = 0 \tag{22.125}$$

  with the plane strain condition $e_{zz} = 0$.

Mode 4
  At $r = R$,

$$q_r = 0; \quad \sigma_{rr} = -P_o(t); \quad \tau_{r\theta} = 0 \tag{22.126}$$

  with the plane strain condition $e_{zz} = 0$.

Mode 5

At $r = R$,

$$p = 0; \quad \sigma_{rr} = -S_o \cos 2\theta H(t); \quad \tau_{r\theta} = S_o \sin 2\theta H(t) \tag{22.127}$$

with the plane strain condition $e_{zz} = 0$.

Mode 6

At $r = R$,

$$q_r = 0; \quad \sigma_{rr} = -S_o \cos 2\theta H(t); \quad \tau_{r\theta} = S_o \sin 2\theta H(t) \tag{22.128}$$

with the plane strain condition $e_{zz} = 0$.

Mode 7

At $r = R$,

$$p = 0; \quad \sigma_{rr} = \tau_{r\theta} = 0\tau_{rz} = 0 \tag{22.129}$$

with a generalized plane strain condition $e_{zz} = -e_o$.

Mode 8

At $r = R$,

$$q_r = 0; \quad \sigma_{rr} = \tau_{r\theta} = 0\tau_{rz} = 0 \tag{22.130}$$

with a generalized plane strain condition $e_{zz} = -e_o$.

Except for modes 5 and 6, which are asymmetric, all other modes are axisymmetric. Modes 5 and 6 are solved below first. Then, the general solution for the axisymmetric cylinder problem is derived. According to the corresponding boundary conditions, all axisymmetric modes are then solved individually.

22.5.2 Solution for asymmetric problem

Modes 5 and 6 are asymmetric problems. Their corresponding elastic solution can be obtained as:

$$\sigma_{rr} = -S_o \cos 2\theta H(t) \tag{22.131a}$$

$$\sigma_{\theta\theta} = S_o \cos 2\theta H(t) \tag{22.131b}$$

$$\tau_{r\theta} = S_o \sin 2\theta H(t) \tag{22.131c}$$

$$\sigma_{zz} = 0 \tag{22.131d}$$

It is noted that the stress field expressed by Eq. (22.131) does not result in any disturbance of initial pore pressure field (Skempton's effect) and the volumetric strain, and the flow boundary conditions for modes 5 and 6 are trivial. Hence, the pore pressure distribution must be trivial. In other words, modes 5 and 6 are purely elastic. Therefore, the solution for modes 5 and 6 are exactly the same as their elastic counterparts and expressed by

$$\sigma_{rr} = -S_o \cos 2\theta H(t) \tag{22.132a}$$

$$\sigma_{\theta\theta} = S_o \cos 2\theta H(t) \tag{22.132b}$$

$$\tau_{r\theta} = S_o \sin 2\theta H(t) \tag{22.132c}$$

$$\sigma_{zz} = 0 \tag{22.132d}$$

$$p = 0 \tag{22.132e}$$

22.5.3   Solution for axisymmetric problem

For an axisymmetric cylinder problem, such as all other modes except for modes 5 and 6, the axisymmetry leads to

$$u_\theta = 0; \quad q_\theta = 0; \quad \tau_{r\theta} = \tau_{rz} = \tau_{\theta z} = 0 \tag{22.133}$$

Except that u_z is a function of r, z and t, all other variables, such as σ_{rr}, $\sigma_{\theta\theta}$, σ_{zz}, p, q_r and u_r, are functions of the radial distance r and time t only. The general solution for an axisymmetric cylinder problem can be derived analytically in the Laplace transform domain. The solution in the time domain is obtained via the inverse of the Laplace transform through a numerical technique.

Because of axisymmetry, the diffusion Eq. (22.109), which is also valid under the generalized plain strain condition (Cui, 1995), can be written as

$$\frac{\partial \zeta}{\partial t} - c_T \left(\frac{\partial^2 \zeta}{\partial r^2} + \frac{1}{r} \frac{\partial \zeta}{\partial r} \right) = 0 \tag{22.134}$$

Applying the Laplace transform to Eq. (22.134) yields

$$\frac{d^2 \tilde{\zeta}}{dr^2} + \frac{1}{r} \frac{d\tilde{\zeta}}{dr} - \frac{s}{c_T} \tilde{\zeta} = 0 \tag{22.135}$$

Setting $\xi = r\sqrt{s/c_T}$, the following equation can be obtained:

$$\frac{\alpha M}{M_{11} + \alpha^2 M} \tilde{\zeta} = A_1 I_0(\xi) \tag{22.136}$$

Applying the solution for the fluid content variation $\tilde{\zeta}$ and using Eqs. (22.43), (22.44), (22.49), and (22.51), the general solution for the axisymmetric cylinder problem (associated with modes 1, 2, 3, 4, 7, or 8) is derived as:

$$\tilde{u}_r = \frac{A_1}{\sqrt{s/c_T}} I_1(\xi) + A_2 r \tag{22.137a}$$

$$\tilde{p} = \frac{M_{11}}{\alpha} A_1 I_0(\xi) - 2\alpha M A_2 - \alpha' M \tilde{e}_{zz} \tag{22.137b}$$

$$\tilde{q}_r = -\kappa \sqrt{\frac{s}{c_T}} \frac{M_{11}}{\alpha} A_1 I_1(\xi) \tag{22.137c}$$

$$\frac{\tilde{\sigma}_{rr}}{2G} = -A_1 \frac{I_1(\xi)}{\xi} + \frac{M_{11} + M_{12} + 2\alpha^2 M}{2G} A_2 + \frac{M_{13} + \alpha\alpha' M}{2G} \tilde{e}_{zz} \tag{22.137d}$$

$$\frac{\tilde{\sigma}_{\theta\theta}}{2G} = -A_1 \left(I_0(\xi) - \frac{I_1(\xi)}{\xi} \right) + \frac{M_{11} + M_{12} + 2\alpha^2 M}{2G} A_2 + \frac{M_{13} + \alpha\alpha' M}{2G} \tilde{e}_{zz} \tag{22.137e}$$

$$\tilde{\sigma}_{zz} = -\left(\frac{\alpha'}{\alpha} M_{11} - M_{13} \right) A_1 I_0(\xi) + 2(M_{13} + \alpha\alpha' M)A_2 + (M_{33} + \alpha'^2 M)\tilde{e}_{zz} \tag{22.137f}$$

In the above, A_1 and A_2 are arbitrary functions of the Laplace transform variable s; and I_n is the modified Bessel function of the first kind of order n. For individual problems,

A_1 and A_2 can be determined from corresponding loading and boundary conditions. For modes 1, 2, 3, 4, 7, and 8, A_1 and A_2 are obtained below:

Mode 1

$$\tilde{e}_{zz} = 0 \tag{22.138a}$$

$$A_1 = \tilde{p}_o \frac{M_{11} + M_{12} + 2\alpha^2 M}{2G} \frac{\beta}{C_1 I_1(\beta)} \tag{22.138b}$$

$$A_2 = \frac{\tilde{p}_o}{C_1} \tag{22.138c}$$

where $\beta = R\sqrt{s/c_T}$, and C_1 is expressed by:

$$C_1 = \frac{M_{11} + M_{12} + 2\alpha^2 M}{2G} \frac{M_{11}}{\alpha} \frac{\beta I_0(\beta)}{I_1(\beta)} - 2\alpha M \tag{22.139}$$

Mode 2

$$\tilde{e}_{zz} = 0 \tag{22.140a}$$

$$A_1 = -\frac{\tilde{Q}}{2\pi\beta} \frac{\alpha}{\kappa M_{11} I_1(\beta)} \tag{22.140b}$$

$$A_2 = -\frac{\tilde{Q}}{2\pi\beta^2} \frac{2\alpha G}{\kappa M_{11}(M_{11} + M_{12} + 2\alpha^2 M)} \tag{22.140c}$$

Mode 3

$$\tilde{e}_{zz} = 0 \tag{22.141a}$$

$$A_1 = -\frac{\tilde{P}_o}{GC_3} \frac{\alpha^2 M}{M_{11} I_0(\beta)} \tag{22.141b}$$

$$A_2 = -\frac{\tilde{P}_o}{2GC_3} \tag{22.141c}$$

where C_3 is expressed by:

$$C_3 = \frac{M_{11} + M_{12} + 2\alpha^2 M}{2G} - \frac{2\alpha^2 M}{M_{11}} \frac{I_1(\beta)}{\beta I_0(\beta)} \tag{22.142}$$

Mode 4

$$\tilde{e}_{zz} = 0 \tag{22.143a}$$

$$A_1 = 0 \tag{22.143b}$$

$$A_2 = -\frac{\tilde{P}_o}{M_{11} + M_{12} + 2\alpha^2 M} \tag{22.143c}$$

Mode 7

$$\tilde{e}_{zz} = -\tilde{e}_o \tag{22.144a}$$

$$A_1 = -\tilde{e}_o \frac{1}{M_{11}I_0(\beta)} \left(\alpha\alpha'M + 2\alpha^2 M \frac{C_7}{C_3} \right) \tag{22.144b}$$

$$A_2 = -\tilde{e}_o \frac{C_7}{C_3} \tag{22.144c}$$

where C_3 is given by Eq. (22.142) and C_7 is expressed by:

$$C_7 = \frac{\alpha\alpha'M}{M_{11}} \frac{I_1(\beta)}{\beta I_0(\beta)} - \frac{M_{13} + \alpha\alpha'M}{2G} \tag{22.145}$$

Mode 8

$$\tilde{e}_{zz} = -\tilde{e}_o \tag{22.146a}$$

$$A_1 = 0 \tag{22.146b}$$

$$A_2 = \tilde{e}_o \frac{M_{13} + \alpha\alpha'M}{M_{11} + M_{12} + 2\alpha^2 M} \tag{22.146c}$$

It should be noted that A_1 in modes 4 and 8 is zero; hence, no flow takes place in the cylinder. In fact, for modes 4 and 8 the pore pressure is identical everywhere, and the boundary surface is impermeable (jacketed test). Therefore, the pore fluid is not allowed to flow in the cylinder; and the modes represent the undrained state, i.e. solutions for modes 4 and 8 are what we call the 'undrained solutions'.

22.5.4 *Examples*

The radius of the cylinder is set to 0.1 m. It is assumed that a linearly time-dependent compressive axial strain rate ('stroke rate') of 10^{-8} 1/sec is applied, i.e. $e_{zz} = -10^{-8}t$ with a free stress at the cylindrical boundary, $\sigma_{rr} = \tau_{r\theta} = \tau_{rz} = p = 0$. These conditions result in a uniaxial problem for an elastic medium; however, a poroelastic cylinder behaves in a pseudo-three-dimensional geometry. The material properties used for borehole analyses are adopted here to study the effect of material anisotropy on the anisotropic cylinder problems.

Figure 22.12 presents the pore pressure at the center of the cylinder as a function of time for the cases of $n_v = 0.5, 1, 2$ with $n_E = 1$. It is observed that the pore pressure increases from the initial (trivial) level once the axial strain is applied. After a certain time duration, it reaches a constant level as time increases, since the pore pressure field in the cylinder becomes steady. The magnitude of the pore pressure changes significantly with varied ratio of n_v. The greater the n_v, the higher the pore pressure. Similar effects on the tangential stress are also observed in Figure 22.13, which shows the tangential stresses at the boundary surface $(r = R)$ for the cases of a fixed n_E. It is known that for a non-porous elastic cylinder under the same boundary conditions, the problem is uniaxial and the tangential stress is trivial. However, for the poroelastic problem a tensile tangential stress is developed as time progresses. This suggests that the saturated cylinder may be fracturing due to a compressive axial strain.

The total axial stress at the center varying with time for the cases of $n_E = 0.5, 1, 2$ with $n_v = 1$ is displayed in Figure 22.14. In this figure, the negative values of the stress (i.e. positive indicates a compressive stress) varying with time are plotted. The axial stress

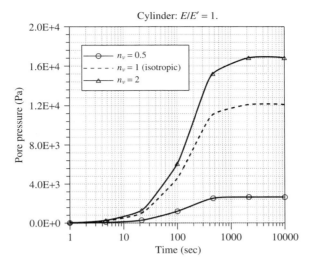

Figure 22.12   Pore pressure at $r = 0$ in the cylinder varying with time for $E/E' = 1$

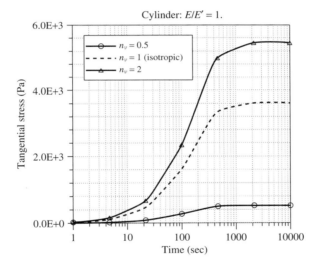

Figure 22.13   Tangential stress at $r = R$ in the cylinder varying with time for $E/E' = 1$

is almost linearly dependent of time. The effect of material anisotropy on the stress is found to be considerable; the magnitude of the stress increases as the ratio n_E decreases. It is noted that the magnitude of the axial stress is much greater than the ones of all other dynamic variables. Considering that the slow axial strain rate and the free stress boundary conditions were applied, these results are expected. Although the results are not presented herein, it is of interest to point out that the effect of variation of n_ν on the axial stress is negligible for $n_E = 1$. This indicates that the difference between two Poisson's ratios may not be a major contributing factor in material anisotropy in terms of the influence on the axial stress.

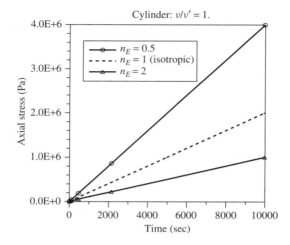

Figure 22.14   Axial stress $(-\sigma_{zz})$ at $r = 0$ in the cylinder varying with time for $v/v' = 1$

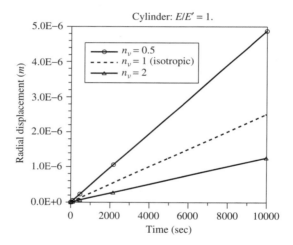

Figure 22.15   Radial displacement at $r = R$ in the cylinder varying with time for $E/E' = 1$

In contrast with the axial stress, the radial displacement at the boundary surface is sensitive to the variation of n_v, but insignificantly influenced by the change of n_E. The displacement varying with time for cases of $n_v = 0.5, 1, 2$ with $n_E = 1$ is presented in Figure 22.15. It is observed that the displacement is also almost linearly dependent of time and a greater ratio n_v leads to a larger displacement.

Finally, unlike the elastic problem, with coupled pore fluid effect the cylinder problem under mode 7 is not uniform any more. A non-trivial pore pressure field results from the compaction of the cylinder. This field is not uniform because of the zero pore pressure enforced at the boundary. Hence, the deformation along the axial direction is resisted by the pore fluid and the resistance is also spatial-dependent. This is basically why the poroelastic problem is pseudo-three-dimensional. The radial and tangential stresses are thus generated in the poroelastic problem. Figure 22.16 presents the variations of the

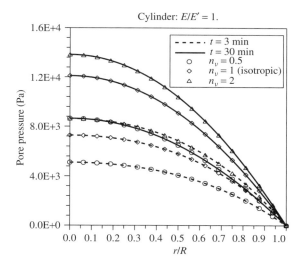

Figure 22.16   Pore pressure in the cylinder varying with r/R at different times for $E/E' = 1$

pore pressure along the radial direction at $t = 3$ min and $t = 30$ min. Non-uniform pore pressure field is observed in the figures. Inhomogeneity of stress and pore pressure fields is affected by the axial strain rate and the anisotropy of Poisson's ratios.

22.6   Conclusion

The theory of anisotropic poroelasticity, especially, the constitutive relations of the anisotropic porous media, was presented in detail. Both continuum mechanics and micromechanics approaches were employed to study the physical significance of poromechanical parameters. The relations between those poromechanical properties required in constitutive equations and some measurable engineering parameters were also derived. In particular, governing equations for transversely isotropic materials were presented for the application of anisotropic poroelasticity to engineering problems. Poroelastic solutions in transversely isotropic media for Mandel's problem, wellbore, and cylinder were then obtained; and material anisotropy effects on these problems were analysed. The results also showed the qualitative difference between anisotropic poroelasticity and elasticity.

Acknowledgements

The work was supported by a National Science Foundation grant to the Rock Mechanics Research Center, the Oklahoma Center for the Advancement of Science and Technology, and the Rock Mechanics Consortium at The University of Oklahoma.

References

Abousleiman, Y., Cui, L., Cheng, A. H-D., Roegiers, J.-C. and Leshchinsky, D. Poroelastic solution of an inclined borehole in an transversely isotropic medium. *Proc. 35th US Symposium on Rock Mechanics*, J. J. K. Daemen and R. Shultz (eds.). (1995) 313–318.

Abousleiman, Y. and Cheng, A. H.-D. Anelastic strain recovery of deep cores with presence of pore pressure. *Engineering Mechanics: Vol. 2, Proc. of the 11th Engineering Mechanics Conference*, Y. K. Lin and T. C. Su (eds.). (1996) 935–938.

Abousleiman, Y., Cheng, A. H.-D., Cui, L., Detournay, E. and Roegiers, J.-C. Mandel's problem revisited. *Géotechnique*, **46**(2), (1996a) 187–195.

Abousleiman, Y., Cheng, A. H.-D., Jiang, C. and Roegiers, J.-C. Poroviscoelastic analysis of borehole and cylinder problems. *Acta Mechanica*, **119**, (1996b) 119–219.

Abousleiman, Y. and Cui, L. Poroelastic solutions in transversely isotropic media for wellbore and cylinder. *Int. J. Solid Structures*, **35**(34–35), (1998) 4905–4929.

Amadei, B. *Rock Anisotropy and the Theory of Stress Measurements* In: *Lecture Notes in Engineering*, C. A. Brebbia and S. A. Orszag (eds.). Springer-Verlag, (1983).

Aoki, T., Tan, C. P. and Bamford, W. E. Effects of deformation and strength anisotropy on borehole failures in saturated rocks. *Int. J. Rock Mech. Min. Sci. & Geomech. Abstr.*, **30**, (1993) 1031–1034.

Barla, G. and Wane, M. T. Stress-relief method in anisotropic rocks by means of gauges applied to the end of a borehole. *Int. J. Rock Mech. Min. Sci. & Abstr.*, **7**, (1970) 171–182.

Berry, D. S. and Fairhurst, C. Influence of rock anisotropy and time dependent deformation on the stress-relief and high-modulus inclusion techniques of in situ stress determination. *Testing Technique for Rock Mechanics*, ASTM STP 402, (1966) 190.

Biot, M. A. General theory of three-dimensional consolidation. *J. Appl. Phys.*, **12**, (1941) 155–164.

Biot, M. A. Theory of elasticity and consolidation for a porous anisotropic solid. *J. Appl. Phys.*, **26**, (1955) 182–185.

Biot, M. A. and Willis, D. G. The elastic coefficients of the theory of consolidation. *J. Appl. Mech.*, **24**, (1957) 594–601.

Boresi, A. P. and Chong, K. P. *Elasticity in Engineering Mechanics*. Elsevier, Amsterdam, (1987).

Carroll, M. M. An effective stress law for anisotropic elastic deformation. *J. Geophys. Res.*, **84**, (1979) 7510–7512.

Carroll, M. M. Mechanical response of fluid-saturated porous materials. In: *Theoretical and Applied Mechanics, 15th Int. Cong. Theoretical & Appl. Mech., Toronto*, F. P. J. Rimrott and B. Tabarrok (eds.), (1980) 251–262.

Carroll, M. M. and Katsube, N. The role of Terzaghi effective stress in linearly elastic deformation. *J. Energy Res. Tech.*, **105**, (1983) 509–511.

Cheng, A. H-D. and Detournay, E. A direct boundary element method for plane strain poroelasticity. *Int. J. Numer. Anal. Methods in Geomech.*, **12**, (1988) 551–572.

Cheng, A. H.-D. Material coefficients of anisotropic poroelasticity. *Int J. Rock Mech. Min. Sci. & Geomech. Abstr.*, **34**(2), (1997) 199–205.

Christian, J. T. and Boehmer, J. W. Plane strain consolidation by finite elements. *J. Soil Mech. Found. Div.*, **96**, (1970) 1435–1457.

Cryer, C. W. A comparison of the three-dimensional consolidation theories of Biot and Terzaghi. *Quart. J. Mech. Appl. Math.*, **16**, (1963) 401–412.

Cui., L. Poroelasticity with application to rock mechanics. *PhD Dissertation*, University of Delaware, (1995).

Cui, L., Abousleiman, Y., Cheng, A. H-D., Kaliakin, V. N. and Roegiers, J.-C. Finite element analysis of anisotropic poroelasticity: a generalized Mandel's problem and an inclined borehole problem. *Int. J. Numer. Anal. Methods in Geomech.*, **20**, (1996) 381–401.

Cui, L., Cheng, A. H-D. and Abousleiman, Y. Poroelastic solution for an inclined borehole. *J. Appl. Mech.*, **64**, (1997) 32–38.

Detournay, E. and Cheng, A. H-D. Poroelastic response of a borehole in a non-hydrostatic stress field. *Int. J. Rock Mech. Min. Sci. & Geomech. Abstr.*, **25**, (1988) 171–182.

Detournay, E. and Cheng, A. H-D. Fundamental of poroelasticity. In: *Comprehensive Rock Engineering: Principles, Practice & Projects, Vol. II, Analysis and Design Method*, C. Fairhurst (ed.). Pergamon Press, (1993) 113–171.

Dickey, J. W., Ladd, C. C. and Rixner, J. J. A plane strain device for testing clays. *Research in earth physics phase report No. 10*, MIT Civil Engineering Research Report R68-3, (1968).

Gibson, R. E., Knight, K. and Taylor, P. W. A critical experiment to examine theories of three-dimensional consolidation. *Proc. European Conf. Soil Mech. Found. Eng.*, **1**, Wiesbaden, Germany, (1963) 69–76.

Gibson, R. E., Gobert, A. and Schiffman, R. L. On Cryer's problem with large displacements and variable permeability. *Géotechnique*, **40**, (1990) 627–631.

Katsube, N. The constitutive theory for fluid-filled porous materials. *J. Appl. Mech.*, **52**, (1985) 185–189.

Lekhnitskii, S. G. *Theory of Elasticity of an Anisotropic Elastic Body*. Holden-Day, San Francisco, CA, (1963).

Mandel, J. Consolidation des sols (étude mathématique). *Géotechnique*, **3**, (1953) 287–299.

Nur, A. and Byerlee, J. D. An exact effective stress law for elastic deformation of rock with fluids. *J. Geophys. Res.*, **76**(26), (1971) 6414–6419.

Rice, J. R. and Cleary, M. P. Some basic stress-diffusion solutions for fluid saturated elastic porous media with compressible constituents. *Rev. Geophys. Space Phys.*, **14**, (1976) 227–241.

Rodrigues, J. D. The Noordbergum effect and characterization of aquitards at the Rio Maior Mining Project. *Ground Water*, **21**, (1983) 200–207.

Schiffman, R. L., Chen, A. T-F. and Jordan, J. C. An analysis of consolidation theories. *J. Soil Mech. & Fndn. Div.*, **95**, (1969) 295–312.

Stehfest, H. Numerical inversion of Laplace transforms. *Comm. ACM.*, **13**, (1970) 47–49, 624.

Terzaghi, K. *Theoretical Soil Mechanics*. Wiley, New York, (1943).

Thompson, M. and Willis, J. R. A reformulation of the equations of anisotropic poroelasticity. *J. Appl. Mech.*, **58**, (1991) 612–616.

Verruijt, A. 'Discussion' *Proc. 6th Int. Conf. on Soil Mechanics and Foundation Engineering, Montreal*, Canada, **3**, (1965) 401–402.

Verruijt, A. Elastic storage of aquifers. In: *Flow Through Porous Media*, R. J. M. DeWiest (ed.). Academic Press, New York, (1969).

Woodland, D. C. Borehole instability in the western Canadian overthrust belt. *SPE Drill. Eng.*, **5**, (1990) 27–33.

23

Modeling Fluid Flow and Rock Deformation in Fractured Porous Media

Mao Bai

The University of Oklahoma, Norman, OK, USA

23.1 Introduction

Fluid flow in fractured porous media has been a subject of intensive study for almost four decades. The first paper on this subject should be attributed to Barenblatt *et al.* (1960), in which a phenomenological dual-porosity model was proposed. Barenblatt *et al.*'s proposal was initially ignored, until Warren and Root (1963) suggested a reservoir simulation model based on a simplification of the Barenblatt *et al.*'s original model. Warren and Root's simplification stimulated substantial interest in reservoir engineering to simulate the behavior of naturally fractured reservoirs. Among numerous publications in the related field, some of the influential contributions include: (a) Kazemi (1969) and deSwaan (1976), where transient interporosity flow between fractures and matrix blocks were accommodated; (b) Crawfort *et al.* (1976), where the temporal pressure slope changes were identified as the dual-porosity behavior in actual well test results (Figure 23.1); and (c) Bourdet *et al.* (1984), where the dual-porosity response in the well tests was further probed using the pressure derivative method. Certain noticeable review papers were written by Kamal (1983), Gringarten (1984) and Chen (1989). Several books are also available to readers having desire to study the subject systematically. Among them, the following have received a great deal of attention in the literature (Auguilera, 1980; Reiss, 1980; Streltsova, 1988; Prat, 1990; Chilingarian *et al.*, 1992).

In comparison, fluid flow in fractured poroelastic media has received less attention due to its 'limited' scope in terms of applications. Up to the present, there is no book contributing to the subject. The only relevant book chapter is by Elsworth (1993), who provided some mathematical and numerical formulations to summarize the dual-porosity poroelastic conceptualization initially proposed by Aifantis (1977, 1980). Aifantis' proposal may be viewed as a natural extension of Biot's single-porosity poroelasticity (Biot, 1941), with the combination of the dual-porosity fluid flow model by Barenblatt *et al.* (1960). The mathematical basis

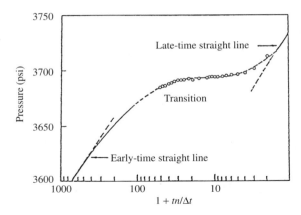

Figure 23.1   Temporal pressure profile of dual-porosity behavior

for this extension can be traced to the mixture theory (Crochet and Naghdi, 1966; Atkin and Craine, 1976), in which any material in a composite medium shows its physical, thermal, hydraulic, and mechanical characteristics distinctly different from these of other intervening materials, and therefore deserves separate description. Practically, the necessity of this extension may be attributed to the inclusion of significant mechanical impacts on solid deformation in fractured or heterogeneous media (Wilson and Aifantis, 1982) on groundwater flow in fractured aquifers (Huyakorn *et al.*, 1983), and in particular, on petroleum production from naturally fractured reservoirs (Bai *et al.*, 1993). Although the analytical solutions are occasionally seen for the applications of the dual-porosity poroelastic theory under the simplified boundary and initial conditions (Bai *et al.*, 1995), numerical methods, preferably the finite element methods, appear to be the dominant tool adapted for any sensible utilizations (Khaled *et al.*, 1984; Elsworth and Bai, 1992; Bai and Elsworth, 1994; Bai *et al.*, 1995). Compared with the single-porosity poroelasticity (Biot, 1941) and the dual-porosity flow (Barenbaltt *et al.*, 1960), the development of the dual-porosity poroelastic models has been restricted primarily due to the complications in the parametric determination, since more parameters are required in the simulation than those in the two former approaches. Physical conceptualization and laboratory determination of these parameters may be referred to Wilson and Aifantis (1982), Bai *et al.* (1993) and Berryman and Wang (1995).

'Coupled processes' are defined as events occurring in a retroactive and simultaneous fashion. There are variety of coupled processes in subsurface. One example related to the present subject is the fluid flow through deformable porous media, where the fluid potential may impose additional force to affect the equilibrium of an elastic system, with a reciprocal consequence that the expansion or contraction of the solid body may exert internal influences over the variations of the fluid flux. This is a typical poroelastic phenomenon. If the porous media are also naturally fractured, fluid flow is not only affected by the rock deformation within matrix blocks and fractures, but also by the interporosity mass exchange between the matrix pores (primary porosity) and the fractures (secondary porosity). This multiple coupled processes are referred as the dual-porosity poroelastic behavior of the fractured porous media, a prime subject of the present study.

The coupled processes may be more general than the flow-deformation system frequently cited. A fully conserved system may maintain the balanced momentum, mass and energy throughout its temporal and spatial evolution. In this aspect, comprehensive coupling is achievable for a poroelastic system if the influence of thermal transport can be considered. Theoretically, other processes may be coupled with this thermoporoelastic response, such as (a) chemical reactions, (b) viscous and plastic behavior of rock constituents, (c) heterogeneities such as fractures, (d) nonlinearities, and (e) multi-phase fluids. In reality, however, meaningful and practical analyses are confined to the processes associated with maximum three parallel processes.

In general, conservation laws hold for momentum, mass and energy. For an isothermal case, stress equilibrium must be maintained for the load-deformation behavior of the reservoir, while fluid mass is concurrently conserved. Reservoir deformation as a result of production is coupled with induced pore pressure change. This effect is most pronounced in areas of large change in total stress, such as in the vicinity of producing wells, with the magnitude of this impact diminishing with distance. In terms of coupling fluid flow with solid deformation, Biot (1941) presented a comprehensive poroelastic formulation in three-dimensional space, with displacement and pressure as primary unknowns. Rice and Cleary (1976) introduced a series of alternative solutions based on the stress function method for the poroelasticity. Using Melan's (1940) solution for a central point dilatation in a semi-elastic space, Segall (1985) solved the coupled flow-deformation problem for a reservoir subjected to fluid extraction. Recent advances extend the traditional dual-porosity approach to encompass the coupled processes which include partial or comprehensive coupling of fluid flow, solid deformation, heat transfer and solute transport (Bai *et al.*, 1993; Bai and Roegiers, 1995). Research efforts have also been focused on identifying local influences such as convective flow (Bai and Roegiers, 1994) or nonlinear flow near a well (Bai *et al.*, 1994) in a dual-porosity medium. Accompanying numerical advances include the development of a three-dimensional finite element model capable of evaluating coupled flow-deformation in poroelastic dual-porosity media (Bai and Meng, 1994). With regard to the numerical treatment of fluid flow in the presence of a free surface frequently encountered in the geotechnical engineering, the residual flow procedure introduced by Desai and his colleagues can be applied (Desai and Sherman, 1971; Desai, 1976; Desai and Li, 1983; Baseghi and Desai, 1990). This residual flow procedure was recently used to examine the coupled process involving chemical transport and stress state of soil related to the slope stability (Ahn *et al.*, 1996).

In view of the modeling of fluid flow and rock deformation in fractured porous media, this chapter is devoted to summarizing the current conceptualizations and mathematical formulations, numerical methodologies, and certain engineering applications of the dual-porosity poroelasticity theory.

23.2   Assumptions

Since multiple complexities are involved in the study of dual-porosity poroelastic response of fractured porous media, it is impossible to accommodate the description of all antic-ipated physical phenomena in the model formulation. As a result, imposing various assumptions may lead to simplified models amendable for analytical and numerical simula-tions. The assumptions consist of embedded physical, mathematical and numerical aspects, as discussed in the following.

- Representative Elementary Volume (REV; Bear, 1972) concept is applied. In terms of scales, REV is substantially smaller than the studied domain, but significantly larger than the microscopic pore scale. Constitutive relations are established at the microscopic point.

- Fluid is Newtonian. Flow is laminar and linear, where Darcy's flow velocity is valid.

- Both rock deformation and fluid compressibility are assumed to be sufficiently small to maintain linear constitutive relationships (e.g. Hooke's law and Darcy's law), linear momentum and mass conservation (e.g. neglecting higher order nonlinear terms), and validity of superposition.

- Fractures and porous matrix are treated as two distinct media, which are perceptive only from microscopic scale. Any activities such as interactive flow and deformation between these two media are viewed as internal.

- Fluid flow in fractures and in porous matrix are considered as separate events, linked only by the leakage terms characterizing the interporosity flow as a result of the pressure difference between the two media. This mathematically 'separate' but physically 'overlapping' system represents the unique dual-porosity conceptualization proposed first by Barenblatt *et al.* (1960).

- As a result of this 'separate' rule, flow parameters are assessed on an individual basis. For example, fracture porosity and matrix porosity are defined as the void fracture volume and void pore volume versus the total volume, respectively. In each flow equation, the porosity represents only a fraction of the respective void volume, instead of the total volume. With respect to corresponding void volumes, matrix porosity should be greater or much greater than fracture porosity.

- Porous matrix is, in general, isotropic. However, anisotropic flow properties are permitted (e.g. allowing permeability tensor). Because flow in fractures is more significant than interstitial flow in matrix pores, fracture permeability should be larger or much larger than matrix permeability.

- There are, in general, two types of the interporosity flow between the fractures and the porous matrix, i.e. transient flow and quasi-steady flow. The latter approach is adopted in the present study.

- The impact of fluid pressures on the solid equilibrium is incorporated in a lumped fashion, envisioning as separate seepage forces acting on the solid grains. While the stresses are retained indiscriminative between the fractures and the porous matrix, the strains are superposed from these two individual media.

- Volumetric strain rate changes affect the flux variations for both fractures and porous matrix.

- In finite element method, one element may compose of numerous fractures and matrix blocks, which has predetermined fracture and matrix characteristics that may differ from those of other elements. In addition to other degrees-of-freedom such as displacements, the fracture and matrix pressures, averaged within the element for fractures and for porous matrix, are both designated at each nodal point. This 'two-pressure at one-point' scenario seems unphysical, but is a typical result of the 'separate' and 'overlapping' system.

- Even though two pressures are resulted from the dual-porosity analysis, usually only fracture pressure is used in the application since the flow is assumed to conduct mainly in the fractures. For this reason, decoupling of the fracture pressure from the matrix pressure using the Laplace transform frequently becomes practical, as proposed by Warren and Root (1963).

23.3 Mathematical Formulation

This section is aimed at providing the mathematical framework for the poroelastic fractured systems, starting from the traditional single-porosity model, and progressively upgrading to the multi-porosity model.

23.3.1 Single-porosity poroelasticity

The study of fluid flow in deformable, saturated porous media as a coupled deformation-dependent flow system was initiated with the work of Terzaghi (1923) in his one-dimensional consolidation model. The Terzaghi theory was extended to a more general three-dimensional form by Biot (1941). In all the development that follows, the porous medium is assumed to possess a continuous distribution of a single type of void space satisfying a single permeability (as shown in Figure 23.2), which may be termed as a single-porosity/single-permeability medium.

The general stress-strain relationship incorporating effective stress effects through pore pressures may be written as

$$\varepsilon_{ij} = \frac{1+\nu}{E}\sigma_{ij} - \frac{\nu}{E}\sigma_{kk}\delta_{ij} - \frac{\alpha}{3H}p\delta_{ij} \tag{23.1}$$

where E is the elastic modulus, ν is the Poisson ratio, α is the pressure ratio factor (or Biot coefficient), H is Biot's constant, σ_{kk} is the total stress (or sum of component stresses), p is fluid pressure, δ_{ij} is the Kronecker delta, and ε_{ij} and σ_{ij} are the strain and stress tensors, respectively. In this analysis, the Greek and Latin subscripts have the values of 1, 2 and 1, 2, 3, respectively; a comma stands for differentiation, and summation is implied over the repeated Latin subscripts.

The equilibrium equation in the absence of body force and inertial effects may be given as

$$\sigma_{ij,j} = 0 \tag{23.2}$$

and the strain-displacement relation is defined as

$$\varepsilon_{ij} = \tfrac{1}{2}(u_{i,j} + u_{j,i}) \tag{23.3}$$

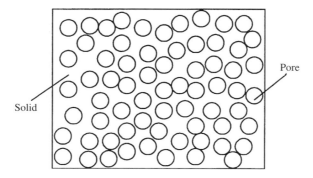

Figure 23.2 Single-porosity medium

where u_i are displacements. The equation governing a solid body deformation is obtained through substitution of Eq. (23.1) and the strain-displacement relation of Eq. (23.3) into the equilibrium condition defined by Eq. (23.2), to yield

$$Gu_{i,jj} + (\lambda + G)u_{k,ki} + \alpha p_{,i} = 0 \tag{23.4}$$

where G is the shear modulus and λ is a Lamé constant.

In the fluid phase, Darcy flow velocity, v_i, can be expressed as

$$v_i = -\frac{k}{\mu} p_{,i} \tag{23.5}$$

where k is permeability and μ is fluid dynamic viscosity. The basic statement of flow continuity requires that the divergence of the flow velocity be equal to the rate of fluid accumulation per unit volume of space, therefore

$$v_{i,i} = \alpha \dot{\sigma}_{kk} - c^* \dot{p} \tag{23.6}$$

where c^* is termed the lumped compressibility. Substituting Eq. (23.5) into Eq. (23.6) gives the governing flow equation as

$$-\frac{k}{\mu} p_{,kk} = \alpha \dot{\varepsilon}_{kk} - c^* \dot{p} \tag{23.7}$$

Equations (23.4) and (23.7) constitute the governing equations for the deformation-dependent behavior of a single-porosity/single-permeability medium. In general, this model is applied to a nonfractured formation with uniform porosity and permeability.

23.3.2 *Dual-porosity poroelasticity*

For a fractured medium, it is generally recognized that the fractures add secondary porosity to the original porosity by breaking the porous medium into blocks. The dual-porosity conceptualization of a fractured medium considers the fluid in fractures and the fluid in matrix blocks as separate and overlapping continua. For the commonly accepted fractured porous media, the fracture and matrix phases are distinctly different in both porosity and permeability. In formal language, these media possess two degrees of porosity and permeability. The model may be best represented by Figure 23.3 where high porosity/low permeability matrix and low porosity/high permeability fractures are typical characteristics of the medium. The governing equation of solid deformation may be expressed as follows:

$$Gu_{i,jj} + (\lambda + G)u_{k,ki} + \sum_{m=1}^{2} \alpha_m p_{m,i} = 0 \tag{23.8}$$

The governing equation for the fluid phase is

$$-\frac{k_m}{\mu} p_{m,kk} = \alpha_m \dot{\varepsilon}_{kk} - c_m^* \dot{p}_m \pm \xi(\Delta p), \tag{23.9}$$

where $m = 1$ and 2 represent matrix blocks and fractures, respectively. k_m is the permeability of phase m, and ξ corresponds to a fluid transfer rate representing the intensity of

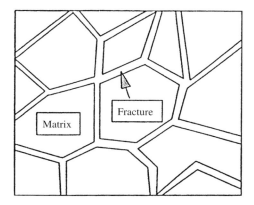

Figure 23.3 Dual-porosity medium

flow between the fractures and matrix driven by the pressure gradient, Δp. A positive sign indicates outflow from the matrix, and a negative sign indicates inflow into the matrix.

Under the assumption of low matrix permeability, a fracture flow mechanism may be incorporated in the formulation. The dual-porosity/dual-permeability model is suitable for the simulation of the fractured formation with low-permeability matrix blocks.

23.3.3 Triple-porosity poroelasticity

For severely fractured media, however, a dual-porosity model may not be appropriate even in the local geometry. An immediate extension of the dual porosity conceptualization is to triple porosity. An example of a triple-porosity model is where a dominant fracture system intercepts a less pervasive and nested fracture system, which in turn is set within a porous matrix.

The term 'triple porosity' is not new in the literature. Abdassah and Ershaghi (1986) considered a reservoir where fractures have homogeneous properties throughout and interact with two groups of separate matrix blocks that have different permeabilities and porosities. It should be pointed out, however, that the Abdassah and Ershaghi approach could still be considered as a dual-porosity approach, particularly in view of the numerical technique where local parametric adjustment may be easily achieved.

For a triple-porosity system, matrix pores are interwoven with percolating fissures, and they interact with open cracks through fluid exchange among different phases (Figure 23.4). The governing equation for the solid phase is given by Bai *et al.* (1993):

$$Gu_{i,jj} + (\lambda + G)u_{k,ki} + \sum_{m=1}^{3} \alpha_m p_{m,i} = 0 \tag{23.10}$$

where $m = 1, 2$, and 3 are the subscripts for matrix, fissures, and cracks, respectively.

For the fluid phase, the field equations can be described by a short form as

$$-\frac{1}{\mu}k_m p_{m,kk} = \alpha_m \dot{\varepsilon}_{kk} - c_m^* \dot{p}_m \pm \xi_{mm_1}(p_{m_1} - p_m) \pm \xi_{mm_2}(p_{m_2} - p_m) \tag{23.11}$$

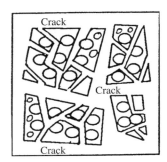

Figure 23.4 Triple-porosity medium

where subscripts m, m_1 and m_2 follow rotational order, i.e. $m = 1, 2, 3$; $m_1 = 2, 3, 1$; and $m_2 = 3, 1, 2$; k_i are the matrix, fissure, and crack permeabilities, respectively. ξ_{ij} are the fluid transfer rates between phase i and phase j. The interporosity flow as a result of pressure differentiation is assumed between all three phases. The triple-porosity model is applicable to the severely fractured reservoir with high permeability.

23.3.4 *Multi-porosity poroelasticity*

By the definition of Aifantis' multi-porosity theory (Aifantis, 1977, 1980), any media that exhibit finite discontinuities in the porosity field are considered to possess a multi-porosity property. However, no unified formulation based on the multi-porosity theory has ever been written. Following the previous derivation, the formulation for a multi-porosity/multi-permeability system is readily available.

In the solid phase, the effect of fluid pressure on the solid deformation within each individual component may be superposed to form (Bai *et al.*, 1993)

$$Gu_{i,jj} + (\lambda + G)u_{k,ki} + \sum_{m=1}^{I} \alpha_m p_{m,i} = 0 \qquad (23.12)$$

where I is the number of distinct porosities of the porous medium.

In the fluid phase, a separate equation must be written for each component of distinct porosity or permeability. For component j,

$$-\frac{1}{\mu}k_j p_{j,kk} = \alpha_j \dot{\varepsilon}_{kk} - c_j^* \dot{p}_j \pm \sum_{i=1(j\neq i)}^{I} \xi_{j(i)}(p_i - p_j) \qquad (23.13)$$

It may be readily demonstrated that the aforementioned single-porosity, dual-porosity, and triple-porosity models are special cases of the unified multi-porosity/multi-permeability formulation exposed by Eqs. (23.12) and (23.13).

23.4 Parametric Analysis

For practical purpose, the dual-porosity model may be sufficient to characterize the behavior of fluid flow and solid deformation in commonly encountered fractured porous media. In this section, the parametric relationship in the single-porosity poroelasticity is presented first as a basis and a prelude for the subsequent discussion on the dual-porosity poroelastic parameters.

23.4.1 Single-porosity poroelasticity

In general, Terzaghi's effective stress law is defined as

$$\sigma^e_{ij} = \sigma_{ij} - p\delta_{ij} \tag{23.14}$$

where σ^e_{ij} is the effective stress tensor and σ_{ij} is the total stress tensor. However, the validity of Terzaghi's effective stress remains questionable, particularly for rock engineering because primary assumptions are of incompressible solid grains. Where grain compressibility is finite, a stress ratio term must be added.

In Biot's theory, the volumetric strain can be written as

$$\theta = \frac{1}{K}\left[\frac{1}{3}(\text{trace }(\sigma_{ij})) - \frac{K}{H}p\right] \tag{23.15}$$

where K is the effective modulus of the skeleton, H is a Biot constant (Biot, 1941), defined as $H = 1/K - 1/K_s$, and K_s is the bulk modulus of solid grains. For isotropic stress conditions, $\theta = 1/K\,\sigma^e$, where effective stress σ^e may be redefined as $\sigma^e = \frac{1}{3}\sigma_{kk} - \alpha p$, where α is termed Biot coefficient and σ_{kk} is the total stress. This equation reduces to Terzaghi's theory only when α is unity. Although α may approach unity for compressible rocks (Kranz *et al.*, 1979; Walsh, 1981), the magnitude may be determined from both the degree of rock fracturing and solid bulk modulus. Geertsma (1957) and Skempton (1960) proposed, on experimental grounds, that

$$\alpha = 1 - \frac{K}{K_s} \tag{23.16}$$

where K_s is the bulk modulus of grains. Nur and Byerlee (1971) verified this relationship theoretically. It is only when the effective compressibility of the dry aggregate is much greater than the intrinsic compressibility of the solid grains ($K \ll K_s$) that the Terzaghi relationship of Eq. (23.14) is valid.

In Biot's approach (1941),

$$c^* = \frac{1}{R} - \frac{\alpha}{H} \tag{23.17}$$

where R is another Biot constant. If Terzaghi's effective stress law is followed, c^* is most widely suggested to be (Verruijt, 1969; Bear, 1972; Huyakorn and Pinder, 1983)

$$c^* = \frac{n}{K_f} \tag{23.18}$$

where n is the porosity and K_f is the fluid bulk modulus.

If the compressibility of the solid grains is not negligible, however, then (Liggett and Liu, 1983)

$$c^* = \frac{n}{K_f} + \frac{1-n}{K_s} \tag{23.19}$$

Geertsma (1957) suggested adding the influence of the compressibility of solid skeleton, then Eq. (23.19) can be modified as

$$c^* = \frac{n}{K_f} + \frac{\alpha - n}{K_s} \tag{23.20}$$

c^* was suggested by Rice and Cleary (1976) as

$$c^* = \frac{n}{K_f} - \frac{1+n}{K_s} + \frac{1}{K} \tag{23.21}$$

If we equate Eqs. (23.17) and (23.19), then R may be derived for Biot's theory as follows:

$$\frac{1}{R} = \frac{n}{K_f} + \frac{1-n}{K_s} + \frac{1}{K}\left(1 - \frac{K}{K_s}\right)^2 \tag{23.22}$$

Biot's field equations in the stress-based method can be expressed as (Cleary, 1977):

$$Gu_{i,kk} + \frac{G}{1 - 2v}u_{k,ki} + \frac{3(v_u - v)}{B(1 + v_u)(1 - 2v)}p_{,i} = 0 \tag{23.23}$$

where v_u and v are the undrained and drained Poisson ratios, respectively. B is Skempton's constant (Skempton, 1954) which is defined as the change in pore pressure per unit change in confining pressure under undrained conditions.

The governing equation in the fluid phase may be written as

$$-\frac{k}{\mu}p_{,kk} = \frac{3(v_u - v)}{B(1 + v_u)(1 - 2v)}\dot{\varepsilon}_{kk} - \frac{9(v_u - v)}{B^2E(1 + v_u)}\dot{p} \tag{23.24}$$

Due to identical unknowns in terms of the fluid phase between Eq. (23.7) for the displacement-based method, and Eq. (23.24) for the stress-based method, the following relationships are readily obtained:

$$\alpha = \frac{3(v_u - v)}{B(1 + v_u)(1 - 2v)} \tag{23.25}$$

$$c^* = \frac{9(v_u - v)}{B^2E(1 + v_u)} \tag{23.26}$$

Biot coefficient α may be modified as:

$$\alpha = \frac{E}{3(1 - 2v)}\left(\frac{1}{K} - \frac{1}{K_s}\right) \tag{23.27}$$

Comparing Eq. (23.25) with Eq. (23.27), the Skempton constant may be expressed as

$$B = \frac{9(v_u - v)}{E(1 + v_u)}\left(\frac{1}{K} - \frac{1}{K_s}\right)^{-2} \tag{23.28}$$

It is noted that, for the case $K \ll K_s$, Eq. (23.28) collapses to Eq. (23.25) with $\alpha = 1$.

The coefficient B can be derived from Eq. (23.26) as:

$$B = \sqrt{\frac{9(v_u - v)}{c^* E(1 + v_u)}} \tag{23.29}$$

Equating Eq. (23.28) to Eq. (23.29), undrained and drained Poisson ratios v_u, v can be described by

$$v_u = \frac{v + \psi}{1 - \psi} \tag{23.30}$$

$$v = v_u - \psi(1 + v_u) \tag{23.31}$$

where

$$\psi = \frac{E}{9c^*} \left(\frac{1}{K} - \frac{1}{K_s} \right)^2 \tag{23.32}$$

It is understood from Eq. (23.32) that $\psi \geq 0$; therefore, $v_u \geq v$. If $\psi = 0$ and $v_u = v$, then $\alpha = 0$ and the fluid flow is fully decoupled from the solid deformation. Since $\frac{1}{2} \geq v_u \geq v$ (Rice and Cleary, 1976), the following relation exists $0 \leq \psi \leq (1 - 2v)/3$. By substituting Eq. (23.30) into Eq. (23.29), Skempton's constant B may reduce to

$$B = \frac{1}{c^*} \left(\frac{1}{K} - \frac{1}{K_s} \right) \tag{23.33}$$

If c^*, proposed by Rice and Cleary (1976), in Eq. (23.21), is used in Eq. (23.33), then

$$B = \left[\frac{1}{K} - \frac{1}{K_s} \right] \left[\frac{n}{K_f} - \frac{1 + n}{K_s} + \frac{1}{K} \right]^{-1} \tag{23.34}$$

When $K \ll K_s$, which is often the case for soil,

$$B = \left[1 + \frac{nK}{K_f} \right]^{-1} \tag{23.35}$$

This is the identical form of B as proposed by Skempton (1954).

Experimental determination of the parameters for the single-porosity model is by no means trivial. It needs careful preparation, setup, and performance to control and measure the interactive behavior of pore pressure, external load, and induced stresses, as well as displacements. In view of the theoretical basis for the parametric relationships, readers can refer to the papers by Biot and Willis (1957) and Geertsma (1957) for these details.

23.4.2 Dual-porosity poroelasticity

The main characteristic of the dual-porosity model is to distinguish between the fracture and intergranular flows. In the general formulation of the dual-porosity model given previously, the possibility exists for flow through both the blocks and the fractures with a transfer function describing the fluid exchange between the two continua.

It is of interest to review the historical development of the effective stress law in dual-porosity models. Wilson and Aifantis (1982) were among the first to provide the effective stress law for the dual-porosity media through extending the analytical evidence by Nur and Byerlee (1971). Their formulation may be expressed as

$$\sigma_{ij}^e = \sigma_{ij} - \sum_{i=1}^{2} \alpha_i p_i \delta_{ij}$$

(23.36)

where

$$\alpha_1 = \frac{K^*}{K}\left(1 - \frac{K}{K_s}\right), \quad \alpha_2 = 1 - \frac{K^*}{K}$$

(23.37)

and K^*, K, and K_s are the bulk moduli of fractured rock, rock without fractures (i.e. porous matrix), and solid grains, respectively. It should be noted that the nomenclature slightly different from the paper by Wilson and Aifantis (1982) is used for the purpose of maintaining consistency.

Tuncay and Corapcioglu (1995), on the other hand, offered a quite similar formulation using the concept of volume average. The only difference in their formulation is the definition for α_i:

$$\alpha_1 = \frac{K^*}{K}\left(1 - \beta_f - \frac{K}{K_s}\right), \quad \alpha_2 = 1 - (1 - \beta_f)\frac{K^*}{K}$$

(23.38)

where β_f is the volume fraction of fractures. If β_f vanishes, Tuncay and Corapcioglu's formulation is identical to Wilson and Aifantis' formulation.

Even though the above formulation shows advantage that the defined parameters can be experimentally determined, there are several drawbacks inherited in the model. First, the relationship between the fractured media and porous matrix is interwoven. As a result, the respective parameters are less independent. The concept becomes intrinsically complex. Secondly, when the pore pressures are identical between the fractures and matrix, the above formulation collapses to the single-porosity one as proposed by Biot (1941). The argument is that, even though the pressures are identical in the fractures and the matrix, the medium itself should still reflect the two-phase characteristics. In other words, the dual-porosity media merge to the single-porosity media only when the fracture space vanishes. The following conceptualization, also extended from Nur and Byrlee's model (1971), may overcome the weakness in the above-mentioned models.

In Nur and Byrlee's model, two-stage loading is employed. The first stage relates the pure pressure load to the compressibility of solid grains, while the second stage imposes the pure total stress to the compressibility of porous medium skeleton. The effective volume variation can be expressed as

$$\frac{\Delta V}{V} = \frac{p}{K_s} + \frac{\sigma - p}{K}$$

(23.39)

where V is the total volume, and K is the bulk modulus of porous media, σ is the total stress. The effective stress σ^e can be described by

$$\sigma^e = K\frac{\Delta V}{V} = \sigma - \alpha p$$

(23.40)

where

$$\alpha = 1 - \frac{K}{K_s} \tag{23.41}$$

Assuming the principle of superposition in which the pore pressure in the first stage of loading is the sum of the pressures of the matrix and fractures, while the two pressures are subtracted in the second stage of loading from the total stress, the effective volume change for a dual-porosity medium can be given as follows:

$$\frac{\Delta V}{V} = \frac{p_1}{K_s} + \frac{p_2}{K_{fr}} + \frac{\sigma - p_1 - p_2}{K^{\#}} \tag{23.42}$$

where p_1 and p_2 are the pore pressures in matrix and fractures, $K^{\#}$, K_s, and K_{fr} are the bulk moduli of fractured porous media, porous matrix media, and fractured media, respectively. As a result, the effective stress for the dual-porosity media σ^e can be deduced as

$$\sigma^e = K^{\#} \frac{\Delta V}{V} = \sigma - \alpha_1 p_1 - \alpha_2 p_2 \tag{23.43}$$

where

$$\alpha_1 = 1 - \frac{K^{\#}}{K_s}, \quad \alpha_2 = 1 - \frac{K^{\#}}{K_{fr}} \tag{23.44}$$

Since each medium possesses its own characteristic coefficient α_i, this formulation can be easily extended to the multi-porosity cases. One remaining question is the laboratory determination of the fracture bulk modulus K_{fr}. In fact, test of such a medium may be represented by the test of a typical fractured rock such as a fractured granite where the porous matrix contributes little to the total fluid flow due to the exceptionally small matrix porosity and permeability. If such a test can be properly worked out, the advantage of using the previous conceptualization (e.g. Wilson and Aifantis (1982) model where the test of the fractured rock has been avoided) is largely overshadowed. If the pore pressures in matrix and fractures are identical (i.e. $p_1 = p_2 = p$), the effective stress in Eq. (23.43) reduces to

$$\sigma^e = \sigma - \left[2 - K^{\#} \left(\frac{1}{K_s} + \frac{1}{K_{fr}} \right) \right] p \tag{23.45}$$

It may be noted that Eq. (23.45) is not converted to the Biot's single-porosity formulation, in contrast to other formulations (e.g. Eq. (23.36) when $p_1 = p_2 = p$). This is physically correct because the dual-porosity media still exist although the pressures are equalized between the two media. However, if the fractures vanish, which is equivalent to letting $K_{fr} = K^{\#} = K$, or letting $\alpha_2 = 0$ in Eq. (23.44), then Eq. (23.45) reduces to

$$\sigma^e = \sigma - \left(1 - \frac{K}{K_s} \right) p \tag{23.46}$$

which is identical to Biot's single-porosity model (Biot, 1941).

Instead of naming it Biot coefficient, α for a dual-porosity medium is labelled as the pressure ratio factor. The matrix pressure ratio factor α_a can be evaluated by Eq. (23.16) (Nur and Byerlee, 1971). Determination of the fracture pressure ratio factor α_2 appears to be more difficult, because of its nonlinear dependence on the stress history. Robin (1973) suggested that

$$\alpha_2 = 1 - v_p \beta_s \frac{\partial v_p}{\partial p} \tag{23.47}$$

where v_p is the pore volume, β_s is the rock compressibility in fractured media and $(\partial v_p)/(\partial p)$ is the rate of change in pore volume with applied hydrostatic pressure for a joint with no pore fluid.

Based on an experimental study, Walsh (1981) suggested that α_2 varies between 0.5 and 1. In this work, it was determined that $\alpha_2 = 0.9$ for joints with polished surfaces, and $\alpha_2 = 0.56$ for a joint made from a tension fracture.

Kranz *et al.* (1979) expressed their results in a similar manner, proposing that α_2 should be less than 1 for jointed rock, and α_2 approximates 1 for whole rock. They pointed out that the stress dependence of α_2 is a function of both surface roughness and ambient pressure.

It may be worth mentioning that the effective stress law, as discussed so far, does not consider the volume proportionality between fluids and solids. Suklje (1969) may be the first to propose an effective stress law with the inclusion of the porosity factor, as described below:

$$\sigma^e = \sigma - \left[1 - (1 - n)\frac{K}{K_s} \right] p \tag{23.48}$$

where n is the porosity. To consider the influence of fluid compressibility, the effective stress law may be further modified as

$$\sigma^e = \sigma - \left\{ 1 - K \left[\frac{1-n}{K_s} + \frac{n}{K_f} \right] \right\} p \tag{23.49}$$

The validity of Eq. (23.49) needs to be further examined through laboratory experiments.

To determine α_1 and α_2 in the present chapter, Suklje's proposal (Suklje, 1969) is extended for dual-porosity media. As a first approximation, the difference between the bulk modulus of the fractured rock and that of the rock without fractures is neglected. The volumetric strain can be expressed as

$$\theta = \frac{\sigma_{kk}}{3K} - \sum_{m=1}^{2} \frac{p_m}{K} \left[1 - (1 - n_m)\frac{K}{K_s} \right] \tag{23.50}$$

where n_m is the porosity of phase m. Note that the volumetric strain, $\theta = \varepsilon_{kk}$, yields

$$\alpha_i = 1 - (1 - n_i)\frac{K}{K_s} \quad (i = 1, 2) \tag{23.51}$$

In general, $n_1 > n_2$; therefore, $\alpha_1 > \alpha_2$. It is readily confirmed (Kranz *et al.*, 1979; Walsh, 1981) that α increases with a decrease in fracturing.

Relative compressibility, c^*, may be expressed as

$$c_m^* = \frac{\alpha_m - n_m}{\Lambda_m} + \frac{n_m}{K_f} \tag{23.52}$$

or

$$c_m^* = \frac{\alpha_m - n_m}{\Lambda_m} + \frac{n_m}{K_f} \qquad (23.53)$$

where $\Lambda_1 = K_s$ for $m = 1$, and $\Lambda_2 = K_n s$ for $m = 2$ in which K_n is the fracture normal stiffness and s is the fracture spacing.

Matrix permeability, k_1, may be estimated if the hydraulic radius, d, of the capillaries or pores is known or can be estimated. A suitable capillary equation can be used to calculate the permeability (Bear, 1972) as

$$k_1 = \frac{n_1^3}{(1 - n_1)^2} \frac{d^2}{180} \qquad (23.54)$$

where n_1 is the matrix porosity, and d is the mean grain size or hydraulic radius. As an alternative, pumping tests may be conducted in the porous medium alone to yield values of the matrix permeability, k_1.

To evaluate fracture permeability k_2, we assume an idealized, regularly spaced, fracture system. If the fracture opening is estimated either from direct measurements or through pressure/flow relationships, then the fracture permeability in the direction parallel to each fracture set may be calculated directly from parallel plate analogy (Snow, 1968; Louis, 1969) as follows:

$$k_2 = \frac{b^3}{12s} \qquad (23.55)$$

where b is the fracture opening and s is the fracture spacing. Fracture permeability may also be obtained from *in situ* measurements such as pumping tests, provided that the flux contribution from the matrix may be verified as minimal.

It should be noted that the fracture and porous medium permeabilities described are initial values. Since fracture apertures and capillary diameters are influenced by stress changes, the initial magnitudes of permeability will be modified as the stress field varies. The stress-dependent matrix permeability can be derived via the elastic contact theory, shown as follows (Bai and Elsworth, 1994):

$$k_1 = k_0 \left\{ 1 \mp \frac{1}{2} \left[\frac{9(1 - v^2)^2}{2} \left(\frac{\pi \Delta \sigma}{E} \right)^2 \right]^{1/3} \right\}^2 \qquad (23.56)$$

where k_0 is the absolute permeability of the porous medium, $\Delta \sigma$ is the stress change, E is the Young's modulus, and v is the Poisson ratio. A negative sign in the equation refers to compressional loading while a positive sign corresponds to dilational loading.

In contrast, the stress-dependent fracture permeability can be written as (Bai and Elsworth, 1994)

$$k_2 = \frac{1}{12s} \left[b + \frac{(s + b)\Delta \sigma}{s K_n + E} \right]^3 \qquad (23.57)$$

where b is the fracture aperture.

Equations (23.56) and (23.57) are to be implemented in the steady state numerical models.

On the basis of dimensional analysis and the assumption of quasi-steady flow conditions, Barenblatt *et al.* (1960) defined that the rate of fluid mass transfer ξ from the porous matrix blocks to the fractures is given as

$$\xi = \xi_0(k_1 S_s^2) \tag{23.58}$$

where ξ_0 is a leakage constant, k_1 is the matrix permeability and S_s is the specific surface of the fractures, i.e. the surface area of fractures per unit volume of the porous medium.

In general, ξ can be expressed as

$$\xi = \frac{k_1}{\mu}\alpha^* \tag{23.59}$$

where α^* is the shape factor defined by Warren and Root (1963) as

$$\alpha^* = \frac{4j(j+2)}{\varsigma^2} \tag{23.60}$$

where $j = 1, 2$ or 3 in terms of the number of orthogonal sets of fractures, ς is a characteristic dimension of the matrix blocks, and can be written as

$$\varsigma = \frac{3a^*b^*c^*}{a^*b^* + b^*c^* + c^*a^*} \tag{23.61}$$

For three mutually orthogonal fracture sets, $j = 3$, $a^* = b^* = c^* = s$ in which s is the fracture spacing. As a result,

$$\alpha^* = \frac{60}{s^2} \tag{23.62}$$

For a dual-porosity medium, parametric determination through experimental testing becomes substantially more difficult than that for a single-porosity medium, as a result of increased number of parameters and difficulty involved in the separation of influences between the fracture and the matrix. For the conceptual design of this type of experiment, readers may wish to refer to the paper by Berryman and Wang (1995).

23.5   Numerical Schemes

For modeling coupled phenomena with complicated domain geometries or boundary and initial conditions, numerical schemes may be the only tools that can cope with the complexities. In this chapter, the finite element method is chosen for the numerical analysis. The finite element modeling schemes can be designed according to the accuracy requirements for the coupling between rock deformation and fluid flow in the governing poroelastic formulation.

23.5.1   Steady state analysis

It appears to be a common perception that the poroelastic effects are manifested in the porous media only during the transient period of fluid flow. This is true for the fully coupled poroelastic phenomena where the compressibilities of porous media and interstitial fluid are controlled by the effective volumetric strain merely during the transient mass storage changes, and this behavior retroactively affects the change of solid deformation such as grain expansion and contraction via a fluid seepage force in a time-dependent fashion. However, the poroelastic behavior may remain in effect during the steady state flow, but in a less direct manner. Because the flow rate is maintained at the level irrelevant to the time changes, the pore pressure variation no longer affects the spatial changes of porous spaces and solid skeleton. As a result, the fluid flow in the saturated rock may be influenced by the geometric variations of void spaces and/or grain sizes, characterized through the changes in original permeability, as a result of application of external load. In particular, this type of coupling occurs in a sequential, instead of interactive fashion. It should be emphasized that applying external load in a steady state manner which allows the rearrangement of porous media structure may result in the nonequilibrium of fluid mass equivalent to that in the transient flow.

Uncoupled modeling

In this case, the rock deformation and fluid pressure do not interact with each other. The coupling is recovered only through strain-permeability relationship in the section discussing the dual-porosity parameters. The matrix form of the finite element method can be written as

$$\begin{bmatrix} \mathbf{A} & 0 \\ 0 & \mathbf{G} \end{bmatrix} \begin{bmatrix} \mathbf{u} \\ \mathbf{p} \end{bmatrix} = \begin{bmatrix} \mathbf{F} \\ \mathbf{Q} \end{bmatrix} \tag{23.63}$$

where

$$\mathbf{A} = \int_V \mathbf{B}^T \mathbf{DB} \, dV \tag{23.64}$$

$$\mathbf{G} = \frac{1}{\mu} \int_V \nabla \mathbf{N}^T \mathbf{k} \nabla \mathbf{N} \, dV \tag{23.65}$$

$$\mathbf{F} = \int_S \mathbf{N} \mathbf{f} \, dS \tag{23.66}$$

$$\mathbf{Q} = \int_S \mathbf{M} \mathbf{q} \, dS \tag{23.67}$$

where $\mathbf{N}$ and $\mathbf{M}$ are the shape functions for the displacement $\mathbf{u}$ and pressure $\mathbf{p}$, $\mathbf{B}$ is the strain-displacement matrix, $\mathbf{A}$ and $\mathbf{G}$ are stiffness and conductance matrices, $\mathbf{F}$ and $\mathbf{Q}$ are the vectors of applied boundary tractions $\mathbf{f}$ and prescribed nodal fluxes $\mathbf{q}$, V and S are the volume and the surface of the calculated domain, $\mathbf{D}$ and $\mathbf{k}$ are the elasticity and the permeability matrices which can be defined as the constants E and k for a one-dimensional case, respectively.

The numerical procedure can be itemized as:

- Derive Δu and Δp from Eq. (23.63).
- Calculate strain $\Delta \varepsilon$ from the derived displacement Δu.
- Substitute the strain into the permeability-strain relationships given in the previous section to assess the permeability k.

- Determine the change in flow rate using Darcy's law and the derived Δk and Δp, e.g.

$$\Delta Q = -A^* \frac{\Delta k}{\mu} \frac{\Delta p}{L} \qquad (23.68)$$

where μ is the fluid dynamic viscosity, L is the flow length, and A^* is the effective flow cross-section area.

Coupled modeling

To consider the principle of effective stress and interaction of rock stress and pore pressure, the explicit coupled numerical analysis can be formulated. The matrix form of the finite element formulation can be written as

$$\begin{bmatrix} \mathbf{A} & \mathbf{R} \\ \mathbf{0} & \mathbf{G} \end{bmatrix} \begin{bmatrix} \mathbf{u} \\ \mathbf{p} \end{bmatrix} = \begin{bmatrix} \mathbf{F} \\ \mathbf{Q} \end{bmatrix} \qquad (23.69)$$

where the coupling matrix $\mathbf{R}$ is

$$\mathbf{R} = \alpha \int_V \mathbf{B}^T \mathbf{m} \mathbf{N} \, dV \qquad (23.70)$$

and α is Biot's coefficient (Biot, 1941), $\mathbf{m}$ is a one-dimensional vector and can be expressed as (a) unity, (b) $(1\,1\,0)^T$, and $(1\,1\,1\,0\,0\,0)^T$ for 1-D, 2-D and 3-D formulations, respectively.

The numerical procedure is similar to that in the uncoupled approach. In general, the coupled procedure provides a more accurate description of the interactive response between fluid flow and rock deformation. Strictly speaking, the solution for both coupled and uncoupled approaches is nonlinear because the permeability contains the strain component which is a primary unknown. As a result, the iterative solution procedure is required and the solution convergence must be tested for each step of calculation. As an approximation, the solution can be considered as linear if the permeability change is assumed to be ultimate and the strain variation is envisioned at a terminal stage. In other words, the permeability change occurs only after the permanent strain has been reached. This approximation seems justifiable because of the application of steady state loads.

23.5.2 Transient analysis

A general framework for an alternative dual-porosity poroelastic formulation was provided by Elsworth and Bai (1992). By envisioning two separate flows within the two individual but overlapping media, the process of laminar flow can be characterized by Darcy's law as

$$\mathbf{q}_i = \frac{-k_i}{\mu} \nabla (p_i + \gamma z) \qquad (23.71)$$

where subscripts $i = 1$ and 2 represent the matrix and fractures, respectively; k is the permeability, p is the fluid pressure, μ is the fluid dynamic viscosity, γ is the unit weight of the fluid, and z is the elevation of the control volume.

The effective stress law for a dual-porosity medium may be expressed as

$$\partial \sigma_i = \partial \sigma_i^e + \mathbf{m}\alpha_i \partial p_i \tag{23.72}$$

where $\partial \sigma$ are the total stresses, $\partial \sigma^e$ are the effective stresses, α_i is the pressure ratio factor for the individual phase (α_1 is Biot's coefficient), $\mathbf{m}$ is a one-dimensional vector and can be expressed as $\mathbf{m}^T = (1\,1\,1\,0\,0\,0)$ for a three-dimensional formulation.

Further developments on the effective stress law are based on the consideration that the solid deformations of the fractures and the matrix are lumped, such that

$$\partial \varepsilon = \sum_{i=1}^{2} \partial \varepsilon_i \tag{23.73}$$

where $\partial \varepsilon$ are the total strains. The stress compatibility condition is maintained, i.e.

$$\partial \sigma_1 = \partial \sigma_2 = \partial \sigma \tag{23.74}$$

This treatment is different from the case where fluid can migrate between two individual media, spatially and temporally. Because the variation of strain in each individual phase is only due to the change in effective stress

$$\partial \varepsilon_i = \mathbf{C}_i \partial \sigma_i^e \tag{23.75}$$

where $\mathbf{C}_i$ is the compliance matrix. This relationship can be reversed as

$$\partial \sigma_i^e = \mathbf{D}_i \partial \varepsilon_i \tag{23.76}$$

where $\mathbf{D}_i$ is the elasticity matrix.

Combining Eqs. (23.72) through (23.76) results in a modified effective stress law in a dual-porosity system:

$$\partial \sigma = \mathbf{D}_{12} \left[\partial \varepsilon + \sum_{i=1}^{2} \alpha_i \mathbf{C}_i \mathbf{m} \partial p_i \right] \tag{23.77}$$

where the combined elasticity matrix $\mathbf{D}_{12}$ can be defined in a three-dimensional geometry explicitly for an isotropic medium as (Bai and Meng, 1994):

$$\mathbf{D}_{12} = \left(\sum_{i=1}^{2} \mathbf{C}_i \right)^{-1} = |\mathbf{D}_{12}|^{-1} \begin{bmatrix} d_{11} & d_{12} & d_{13} & 0 & 0 & 0 \\ d_{21} & d_{22} & d_{23} & 0 & 0 & 0 \\ d_{31} & d_{32} & d_{33} & 0 & 0 & 0 \\ 0 & 0 & 0 & d_{44} & 0 & 0 \\ 0 & 0 & 0 & 0 & d_{55} & 0 \\ 0 & 0 & 0 & 0 & 0 & d_{66} \end{bmatrix} \tag{23.78}$$

where

$$|\mathbf{D}_{12}| = \left[\frac{1}{K_n s} + \frac{2(1+v)}{E} \right]^3 \left[\left(\frac{1}{K_n s} + \frac{1}{E} \right)^3 - \frac{3v^2}{E^2} \left(\frac{1}{K_n s} + \frac{1}{E} \right) - \frac{2v^3}{E^3} \right] \tag{23.79}$$

$$d_{11} = d_{22} = d_{33} = \left[\frac{1}{K_n s} + \frac{2(1+v)}{E} \right]^3 \left[\left(\frac{1}{K_n s} + \frac{1}{E} \right)^2 - \frac{v^2}{E^2} \right] \tag{23.80}$$

$$d_{12} = d_{21} = d_{13} = d_{31} = d_{23} = d_{32} = \left[\frac{1}{K_n s} \right.$$

$$\left. + \frac{2(1+v)}{E} \right]^3 \left[\frac{v}{E} \left(\frac{1}{K_n s} + \frac{1}{E} \right) + \frac{v^2}{E^2} \right] \tag{23.81}$$

$$d_{44} = d_{55} = d_{66} = |\mathbf{D}_{12}| \left[\frac{1}{K_n s} + \frac{2(1+v)}{E} \right]^{-1} \tag{23.82}$$

and E is the elastic modulus, v is the Poisson ratio, s is the fracture spacing, and K_n is the fracture normal stiffness. For simplicity, it is assumed that orthogonal fractures are aligned parallel to the global axes and of uniform spacing and stiffness. These assumptions considerably simplify the form of the constitutive relation, D_{12}, but are not a requirement of the analysis, as more complex relations may be readily incorporated.

Mapping functions

In the finite element method, interpolation (shape) functions are used to map the element displacements and fluid pressures at the nodal points. For the fluid pressure approximation in phase i, one has

$$\mathbf{p}_i^* = \mathbf{Mp} \tag{23.83}$$

or at the nodal level

$$p_i^* = \sum_{j=1}^{8} M_j p_{ij}$$

where $\mathbf{p}$ is a vector of nodal pressures for the eight-node three-dimensional element. $\mathbf{M}$ is a vector of shape functions for pressure, which can be given in the following long form for three-dimensional geometries:

$$M_j = \tfrac{1}{8}(1 - \varphi\varphi_j)(1 - \eta\eta_j)(1 - \zeta\zeta_j) \tag{23.84}$$

where φ, η, and ζ represent the local coordinates of the bi-unit cube with the origin at the centroid, and are confined by the magnitudes of -1 and 1.

Although frequently used in practice, it is incorrect to assume that the interpolation functions for solid displacements $\mathbf{N}$ and for fluid pressures $\mathbf{M}$ are identical. It is understood that the component of the partial stress tensor is continuously differentiable to the first order. As a result, the polynomial interpolation functions for the pore pressure distribution must be one order lower than that chosen for the displacement field. In this work, a quadratic displacement field and a linear pressure field are chosen. For the choice of a 3-D element, a higher order representation can be accommodated by adding a central node to the element. This internal nodal variable is designed only to achieve a higher order interpolation for the displacement than for the pressure. For a nine-node element (eight corner nodes and one central node), the expressions for the approximation in mapping nodal displacements may be described as

$$\mathbf{u}^* = \mathbf{Nu} \tag{23.85}$$

or at the nodal level

$$u_x^* = \sum_{j=1}^{9} N_j u_{xj}, \quad u_y^* = \sum_{j=1}^{9} N_j u_{yj}, \quad u_z^* = \sum_{j=1}^{9} N_j u_{zj}$$

where $\mathbf{u}$ is a vector of nodal displacements, $\mathbf{u} = u_x$, u_y, and u_z, and $\mathbf{N}$ is a vector of shape functions for displacements. For the first eight nodes, $\mathbf{N}$ are chosen to be identical to $\mathbf{M}$ in Eq. (23.84). For the central node, however, $\mathbf{N}$ is given as follows:

$$N_9 = (1 - \varphi^2)(1 - \eta^2)(1 - \zeta^2) \tag{23.86}$$

This higher order shape function improves the elemental behavior in the displacement and subsequent stress modes. For simplicity, the superscript * for variables in Eqs. (23.83) and (23.85), indicating the finite element approximation, are omitted in the following description.

Strains within a single element may be related to nodal displacements through the derivatives of the shape functions $\mathbf{B}$ as

$$\varepsilon = \mathbf{Bu} \tag{23.87}$$

where the strain-displacement matrix $\mathbf{B}$ may be given explicitly as

$$\mathbf{B} = \begin{bmatrix} N_{i,x} & 0 & 0 \\ 0 & N_{i,y} & 0 \\ 0 & 0 & N_{i,z} \\ N_{i,y} & N_{i,x} & 0 \\ 0 & N_{i,z} & N_{i,y} \\ N_{i,z} & 0 & N_{i,x} \end{bmatrix} \tag{23.88}$$

where a comma represents differentiation in the standard manner with respect to the global x, y, and z coordinates.

Conservation relationships

Substituting the modified effective stress law in Eq. (23.77) into the force equilibrium equation $\sigma_{ij,j} = 0$ where inertial effects are neglected, applying then the variational principle, and dividing by Δt, the momentum balance in finite element form can be expressed as

$$\int_V \mathbf{B}^T \mathbf{D}_{12} \mathbf{B} \, dV \dot{\mathbf{u}} + \sum_{i=1}^{2} \alpha_i \int_V \mathbf{B}^T \mathbf{D}_{12} C_i \mathbf{m} \mathbf{M} \, dV \dot{\mathbf{p}}_i = \int_S \mathbf{N} \, dS \dot{\mathbf{f}} \tag{23.89}$$

where V and S are the volume and surface of the integral domain, respectively, and $\mathbf{f}$ is a vector of applied boundary tractions.

Equating the divergence of Darcy's velocity in Eq. (23.71) to the rate of fluid accumulation due to all sources such as the effect of temporal variation of volumetric strain, applying Galerkin's principle and pressure mapping functions, and neglecting the impact of fluid body force, the mass balance in finite element form may be given for each phase i as

$$\frac{1}{\mu} \int_V \nabla \mathbf{M}^T \mathbf{k}_i \nabla \mathbf{M} \, dV \mathbf{p}_i = \alpha_i \int_V \mathbf{N}^T \mathbf{m} \mathbf{D}_{12} \mathbf{C}_i \mathbf{B} \, dV \dot{\mathbf{u}}$$

$$- c_i^* \int_V \mathbf{M}^T \mathbf{M} \, dV \dot{\mathbf{p}}_i \pm \xi \int_V \mathbf{M}^T \mathbf{M} \, dV \Delta \mathbf{p} + \int_V \mathbf{M}^T \mathbf{M} \, dV \mathbf{q}_i \qquad (23.90)$$

where c^* is the lumped compressibility considering the compression of fluid, solid grains and fractures; $\mathbf{q}$ is a vector of applied boundary fluid sources, ξ is the leakage factor indexing the rate of interporosity flow due to pressure difference between fractures and matrix blocks (Warren and Root, 1963), the *del* operator is defined in the usual manner as $\nabla^T = \{\partial/\partial x, \partial/\partial y, \partial/\partial z\}$ and operates on the matrix of shape functions interpolating fluid pressures M, and the sign $\pm$ is determined from the relative directions of flow between matrix and fractures.

Matrix form finite element solutions

The dual-porosity poroelastic finite element formulation given in Eqs. (23.89) and (23.90) are coupled. The solutions are, therefore, required to be obtained simultaneously from the system of equations. The governing equations for a representative time level of $t + \theta \Delta t$ $(0 \leq \theta \leq 1)$ may be established. When adopting a fully implicit finite difference scheme in the time discretization domain, the final matrix form of the equations may be written as

$$\frac{1}{\Delta t^\phi} \begin{bmatrix} \mathbf{A} & \mathbf{R}_1 & \mathbf{R}_2 \\ \mathbf{R}_1^T & \mathbf{G}_{11} & \mathbf{G}_{12} \\ \mathbf{R}_2^T & \mathbf{G}_{21} & \mathbf{G}_{22} \end{bmatrix} \begin{bmatrix} \mathbf{u} \\ \mathbf{p}_1 \\ \mathbf{p}_2 \end{bmatrix}^\phi = \frac{1}{\Delta t^\phi} \begin{bmatrix} \mathbf{A} & \mathbf{R}_1 & \mathbf{R}_2 \\ \mathbf{R}_1^T & \mathbf{F}_1 & 0 \\ \mathbf{R}_2^T & 0 & \mathbf{F}_2 \end{bmatrix} \begin{bmatrix} \mathbf{u} \\ \mathbf{p}_1 \\ \mathbf{p}_2 \end{bmatrix}^t + \begin{bmatrix} \mathbf{f} \\ \mathbf{q}_1 \\ \mathbf{q}_2 \end{bmatrix}^\phi \qquad (23.91)$$

where superscript T represents the matrix transposition, $\phi = t + \Delta t$, and

$$\mathbf{A} = \int_V \mathbf{B}^T \mathbf{D}_{12} \mathbf{B} \, dV \qquad (23.92)$$

$$\mathbf{R}_1 = \alpha_1 \int_V \mathbf{B}^T \mathbf{D}_{12} \mathbf{C}_1 \mathbf{m} \mathbf{M} \, dV \qquad (23.93)$$

$$\mathbf{R}_2 = \alpha_2 \int_V \mathbf{B}^T \mathbf{D}_{12} \mathbf{C}_2 \mathbf{m} \mathbf{M} \, dV \qquad (23.94)$$

$$\mathbf{G}_{11} = -\frac{\Delta t^\phi}{\mu} \int_V \nabla \mathbf{M}^T \mathbf{k}_i \nabla \mathbf{M} \, dV - (c_1^* + \Delta t \xi) \int_V \mathbf{M}^T \mathbf{M} \, dV \qquad (23.95)$$

$$\mathbf{G}_{12} = \Delta t \xi \int_V \mathbf{M}^T \mathbf{M} \, dV, \quad \mathbf{G}_{21} = \mathbf{G}_{12}^T \qquad (23.96)$$

$$\mathbf{G}_{22} = -\frac{\Delta t^\phi}{\mu} \int_V \nabla \mathbf{M}^T \mathbf{k}_2 \nabla \mathbf{M} \, dV - (c_2^* + \Delta t \xi) \int_V \mathbf{M}^T \mathbf{M} \, dV \qquad (23.97)$$

$$\mathbf{F}_1 = -c_1^* \int_V \mathbf{M}^T \mathbf{M} \, dV \qquad (23.98)$$

$$\mathbf{F}_2 = -c_2^* \int_V \mathbf{M}^T \mathbf{M} \, dV \tag{23.99}$$

$$\mathbf{f} = \int_S \mathbf{N} \mathbf{f}^* \, dS \tag{23.100}$$

$$\mathbf{q}_1 = \int_S \mathbf{M} \mathbf{q}_1^* \, dS \tag{23.101}$$

$$\mathbf{q}_2 = \int_S \mathbf{M} \mathbf{q}_2^* \, dS \tag{23.102}$$

Initial condition and ramp loading

It is generally assumed that the rock displacements $\mathbf{u}$ and pore pressures $\mathbf{p}$ are continuous functions in time. However, this assumption cannot be made when the load is instantaneously applied. As a result, a special treatment of the initial condition is required. Normally, the initial displacement field and fluid pressure distributions can be evaluated through solving the following static undrained governing equation:

$$\begin{bmatrix} \mathbf{A} & \mathbf{R}_1 & \mathbf{R}_2 \\ \mathbf{R}_1^T & \mathbf{G}_{11} & \mathbf{G}_{12} \\ \mathbf{R}_2^T & \mathbf{G}_{21} & \mathbf{G}_{22} \end{bmatrix} \begin{bmatrix} \mathbf{u} \\ \mathbf{p}_1 \\ \mathbf{p}_2 \end{bmatrix} = \begin{bmatrix} \mathbf{f} \\ \mathbf{0} \\ \mathbf{0} \end{bmatrix}$$

However, for slightly compressible fluids, the static governing equation may be ill-conditioned and an alternative method is to use a ramp loading to approximate the initial conditions. In this study, a linear function in the first time step is used to avoid violent oscillation that may result from the step loading.

23.6 Applications

Since its proposal over half century ago, poroelastic theory has been applied to numerous fields including petroleum, hydrogeology, civil and environmental engineering, water resources, geophysics, chemical and mechanical engineering, engineering mechanics, aerospace engineering, and mining. It can be stated that the applications are suitable for analysing any phenomenon involving fluid-solid interactions. However, due to its complexity, the dual-porosity poroelastic theory has incurred only limited applications, mainly in the areas of the subsurface research. For this reason, plus the page restraint, only a couple of applications related to subsurface mining and tunneling are briefly described in the following.

23.6.1 Underground mining

The mine was located in West Virginia. The main aquifer was located approximately 34–40 m below the surface. Decreasing permeabilities recorded with depth suggest the dominance of secondary porosity (fractures) in determining conductivity magnitudes. The rock properties were determined primarily through laboratory technique, and scaled to the in-situ conditions. The coal seam was approximately 216 m deep at the mine. The mine was situated within the Pittsburgh seam with an average extraction thickness of 1.75 m. The mined panel was 183 m wide by 2195 m long. The developing longwall panel was affected by two adjacent panels extracted earlier.

The cross-sectional overburden lithology and geometric description are depicted in Figure 23.5. The corresponding modeling parameters are listed in Table 23.1. The finite element mesh is illustrated in Figure 23.6, where the refined mesh is designed for the mining area. As a first approximation, the steady state dual porosity behavior of poroelastic media is evaluated, where the coupling seepage force and the flow in matrix blocks are neglected (uncoupled numerical scheme as described early).

The subsidence measured 17 months after mining on surveying line is selected for comparison. The substantial subsidence over the seam may be attributed to the impact from adjacent mining. It is noted that the irregular shape of this subsidence curve may be the result of a smaller development pillar on the tailgate side. To match the observed curve by Finite Element Method (FEM), the strata directly overlying the gob are assumed to consist of softer materials than elsewhere due to the impact of caving. The subsidence curve predicted by one of the influence function methods (Liu *et al.*, 1984) is also plotted in Figure 23.7 for comparison. The figure illustrates reasonable agreement between the subsidence predictions from different methods, in particular, over the gob area.

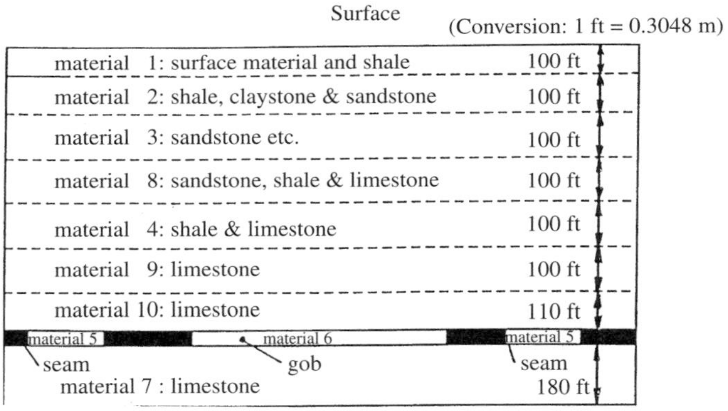

Figure 23.5   Overburden lithology

Table 23.1   Modeling parameters for the mining case

Material	Conductivity (m/s $\times 10^{-7}$)	Modulus (MPa)	Poisson's ratio
1	32.19	7.52	0.28
2	8.02	16.52	0.25
3	2.41	41.56	0.18
4	1.27	22.02	0.22
5	2.36	9.05	0.37
6	1.42	$\ll 1$	0.40
7	1.13	94.99	0.10
8	0.76	41.56	0.18
9	2.55	25.71	0.20
10	5.29	25.71	0.20

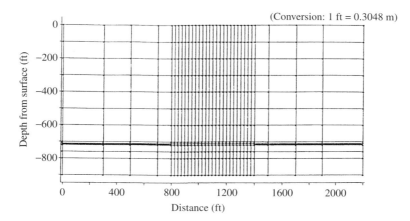

Figure 23.6   Finite element mesh layout (mining)

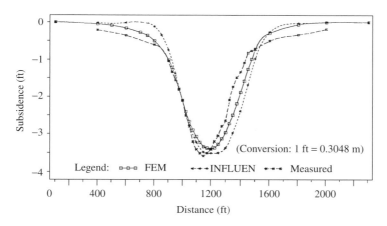

Figure 23.7   Comparison of surface subsidence

Calculated normalized post-mining hydraulic conductivity distribution in the vertical direction is illustrated in Figure 23.8 for an assumed fracture spacing, s, of 0.1 m. As anticipated, the greatest change in hydraulic conductivity occurs in the area immediately over the gob where large deformations result. The variation of hydraulic conductivity is strongly dependent upon the assumed fracture spacing and ascribed elastic parameters. It is apparent from Figure 23.8 that a dramatic increase of vertical conductivity occurs in the surface tension zone in contrast to a decrease in the ground compression zone, reflecting the influence of tensile and compressive strains in the surface region.

23.6.2   *Tunnel stability*

Construction of an underground tunnel in combination with surface pipelines has been considered as a potential means to provide a major potable water supply for the city. The proposed site may be located in fractured and karsified limestones and dolomites

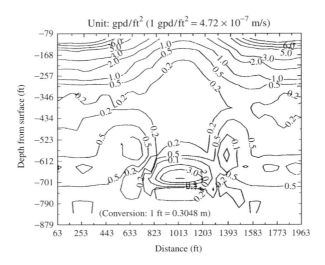

Figure 23.8   Post-mining vertical hydraulic conductivity

overburdened by alluvial deposits. Although the groundwater table is supposed to be well below the tunnel level, difficulties in view of maintaining tunnel stability may be encountered during the service period of the tunnel as a result of the impact of consolidation due to (a) seasonal precipitation; (b) karsified cavity disruption; (c) solid subsidence; (d) storage and conductance interaction between fractures and matrix blocks; and, (d) tectonic stress activation. Common concerns should also consider such aspects as quality and permeability of tunnel lining, ground and water pressures, grouting configuration, and mechanical properties of the surrounding rock mass. In addition, the stress states such as gravitational and horizontal stresses as well as shear stresses due to faulting in the fractured rock mass may also greatly affect the tunnel stability.

The primary objectives of the numerical modeling are to identify some key factors of influence associated with the stability of a pair of waterway tunnels, separated 30 m laterally, and situated in fractured carbonate rock masses at 60 m below surface. The tunnels, each about 3 m in diameter, are subject to both gravitational load and pore pressure due to combined seepage-running water forces as a result of inflow and outflow from the tunnels, respectively. Some field and modeling parameters are described in Table 23.2. In the analysis, displacement and time have been normalized against maximum values for generalization purposes.

For computational efficiency, the analysis was completed using the quartered symmetry finite element mesh shown in Figure 23.9. The interactive effects between the development of rock deformation and dissipation of fluid pressure can be manifested within fractured rock masses due to the large fracture compliance. At 10 m above the tunnel, Figure 23.10 indicates that an increase of rock deformation is accompanied by a decrease of pore pressure in the fractures during the examined period (dimensionless time is the ratio between the examined time and the time beginning the steady state flow). Assuming a fixed proportionality between displacement, strain and stress, this consolidation process will affect the tunnel stability due to the temporal variation of the stress status in the surrounding rock mass.

Table 23.2   Modeling parameters

Parameter	Symbol	Fluid	Fracture	Matrix	Unit
Gravitational load	F_g			1.5083	MN/m^2
Pore pressure load	F_p	0.6033			MN/m^2
Elastic modulus	E			200	GN/m^2
Poisson ratio	ν			0.3	
Porosity	n_1, n_2		0.1	0.4	
Permeability	k_1, k_2		10^{-3}	10^{-5}	m^4/(MN s)
Fracture stiffness	K_n		2.0		MN/m^2/m
Fluid bulk modulus	K_f	1			MN/m^2
Fracture spacing	s^*		0.5		m
Elapsed time	t	10			min

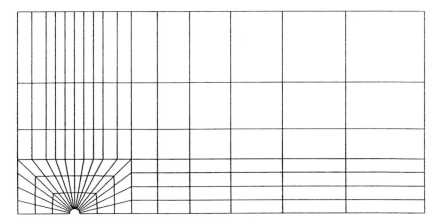

Figure 23.9   Finite element mesh layout

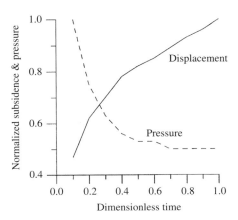

Figure 23.10   Temporal displacement and pressure

At only 3.5 m above the edge of the tunnel, Figure 23.11 reflects the comparison of various subsidences under five different scenarios: (a) conventional elastic media; (b) fractured porous media with outflow from the tunnel when the tunnel fills with water; (c) fractured media with reduced outflow; (d) poroelastic media without flow occurrence at the tunnel; and (e) fractured media with natural inflow. It is of interest to note that the subsidences immediate above the tunnel roof vary substantially. As discussed previously, the dramatic fluid flow counteracts the development of rock deformations, represented by the most conservative subsidence magnitude from the media with outflow when the tunnel is filled with water. An apparent increase in maximum subsidence can be observed for the media with the reduced outflow. The magnitude of subsidence continue to increase for the poroelastic media without any fluid flow from the tunnel as a representation of undrained condition, where the resistance from the pore pressure change against rock deformation is virtually removed. The maximum subsidence can be expected for the media with natural inflow at the tunnel edge, where the consolidation process is most pronounced. The subsidence value for the conventional elastic media appears to accidentally fall between the media with outflow and with the reduced outflow. A similar observation can be made for the calculated surface subsidences from the four different situations, as illustrated in Figure 23.12, except that the minimum subsidence appears to be associated with the elastic media, and all subsidence curves are smooth because the evaluated points are relatively remote from the excavated tunnel.

23.7   Conclusions

The theory of the dual-porosity poroelasticity introduced in this chapter appears to be an effective tool for modeling fluid flow and rock deformation in fractured porous media. However, it is also subject to such an intensive debate on its potential for future development, and the practical applications that what have been described so far may imply the significance but may also appear inconclusive. In comparison with the classical single-porosity poroelasticity, at best, this dual-porosity poroelastic theory may provide us with a more sensible physical intuition and a more accurate mathematical description for the behavior of a coupled subsurface phenomenon with complexity; at worst, it may offer a more flexible tool in matching the experimental results due to the added parameters.

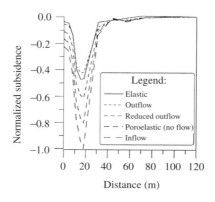

Figure 23.11   Subsidence over the tunnel

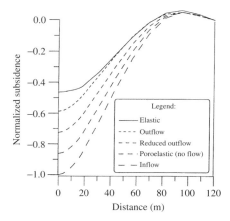

Figure 23.12 Surface subsidence

Although this theory can be modified with the increased complexities in the constitutive relations and in the multiple coupling with other parallel processes, the true virtue of this method perhaps rests on its simplicity of using a linear system to approximate potentially a nonlinear system dominant in reality. Its ability to simulate the unusual changes of a primary unknown in relation to time and space demonstrates this virtue, which cannot be replicated using any simpler equivalent porous medium models. As a result, the dual-porosity poroelastic model may be a natural choice whenever a neutral method between a simple and a complicated systems is sought. Up to the present, the development of this method with regard to the understanding of the related physical process has been severely hindered by the difficulties in the parametric determination using experimental techniques. However, this hurdle should be eventually removed in the foreseeable future because of the improvements in testing technology and experimental layouts currently taking place and expected in the future.

Acknowledgements

The author is most grateful to his former advisor, Dr Derek Elsworth, for his consistent guidance and stimulating inspiration on the subject described in this chapter. Help from the former PhD candidate and colleague, Dr Fanhong Meng is deeply appreciated. Particular gratitude is due to the useful comments and constructive suggestions from Dr Younane Abousleiman. Thanks also go to Dr Musharraf Zaman for his helpful comments in the final version of this chapter. Financial support of the National Science Foundation, Oklahoma State and Industrial Consortium under contract EEC-9209619 and under the S/IUCRC program, along with the administrative support from Dr Jean-Claude Roegiers, are also greatly acknowledged.

References

Abdassah, D. and Ershaghi, I. Triple-porosity system for representing naturally fractured reservoirs. *Soc. Pet. Eng. J.* (1986) 113–127.

Ahn, T. B., Desai, C. S., Armaleh, S. H. and Contractor, D. N. Effect of chemicals on stress and stability analysis. *Proc. 3rd Int. Symp. on Environmental Geotechnology*, San Diego, CA, (1996).

Aifantis, E. C. Introducing a multi-porous medium. *Developments in Mechanical*, **37**, (1977) 265–296.

Aifantis, E. C. On the problem of diffusion in solids. *Acta Mechanical*, **37**, (1980) 265–296.

Atkin, R. J. and Craine, R. E. Continuum theories of mixtures: basic theory and historical development. *Quart. J. Mech. Appl. Math.*, **29**, (1976) 209–244.

Auguilera, R. *Naturally Fractured Reservoirs*. Petroleum, Tulsa, (1980).

Bai, M. and Elsworth, D. Modeling of subsidence and stress-dependent hydraulic conductivity of intact and fractured porous media. *Rock Mech. and Rock Engrg.*, **27**, (1994) 209–234.

Bai, M., Elsworth, D. and Roegiers, J.-C. Multi-porosity/multi-permeability approach to the simulation of naturally fractured reservoirs. *Water Resour. Res.*, **29**, (1993) 1621–1633.

Bai, M., Ma, Q. and Roegiers, J.-C. A nonlinear dual-porosity model. *J. Appl. Math. Modelling*, **18**, (1994) 602–610.

Bai, M. and Meng, F. Study of naturally fractured reservoirs using three-dimensional finite elements. *Report RMRC-94–19*, (1994).

Bai, M. and Roegiers, J.-C. Fluid flow and heat flow in deformable fractured porous media. *Int. J. Engrg. Sci.*, **32**, (1994) 1615–1633.

Bai, M., Roegiers, J.-C. and Elsworth, D. Poromechanical response of fractured-porous rock masses. *J. Pet. Sci. Eng.*, **13**, (1995) 155–168.

Barenblatt, G. I., Zheltov, I. P. and Kochina, N. Basic concepts in the theory of seepage of homogeneous liquids in fissured rocks. *Prikl. Mat. Mekh.*, **24**, (1960) 852–864.

Baseghi, B. and Desai, C. S. Laboratory verification of the residual flow procedure for three-dimensional free surface flow. *Water Resour. Res.*, **26**, (1990) 259–272.

Bear, J. *Dynamics of Fluids in Porous Media*. Elsevier, NY, (1972).

Berryman, J. G. and Wang, H. F. The elastic coefficients of double-porosity models for fluid transport in jointed rock. *J. Geophy. Res.*, **100**, (1995) 24611–24627.

Biot, M. A. General theory of three-dimensional consolidation. *J. Appl. Phys.*, **12**, (1941) 155–164.

Biot, M. A. and Willis, D. G. The elastic coefficients of the theory of consolidation. *J. Appl. Mech.*, **24**, (1957) 594–601.

Bourdet, D., Ayoub, J. A. and Pirard, Y. M. Use of pressure derivative in well test interpretation. *SPE 12777, Cal. Regional Meeting*, (1984) 1–17.

Chen, Z.-X. Transient flow of slightly compressible fluid through double-porosity, double-permeability systems – a state-of-the-art review. *Transport in Porous Media*, **4**, (1989) 147–184.

Chilingarian, G. V., Mazzullo, S. J. and Rieke, H. H. *Carbonate Reservoir Characterization: A Geologic-Engineering Analysis, Part I* Elsevier, Amsterdam, (1992).

Cleary, M. P. Fundamental solutions for a fluid-saturated porous solid. *Int. J. Solids Structures*, **13**, (1977) 785–806.

Crawford, G. E., Hagedorn, A. R. and Pierce, A. E. Analysis of pressure buildup tests in a naturally fractured reservoir. *J. Pet. Tech.* (1976) 1295–1300.

Crochet, M. J. and Naghdi, P. M. On constitutive equations for flow of fluid through an elastic solid. *Int. J. Engrg. Sci.*, **4**, (1966) 383–401.

Desai, C. S. Finite element residual schemes for unconfined flow. *Int. J. Numer. Methods Engrg.*, **10**, (1976) 1415–1418.

Desai, C. S. and Li, G. C. A residual flow procedure and application for free surface flow in porous media. *Adv. Water Resour.*, **6**, (1983) 73–98.

Desai, C. S. and Sherman, W. C. Unconfined transient seepage in sloping banks. *J. Soil Mech. Found. Div. Am. Soc. Civ. Eng.*, **97**, (1971) 357–373.

deSwaan-O. A. Analytical solutions for determining naturally fractured reservoir properties by well testing. *Soc. Pet. Eng. J.* (1976) 117–22.

Elsworth, D. *Computational Methods in Fluid Flow: Comprehensive Rock Engineering:2*. Pergamon Press, (1993) 173–189.

Elsworth, D. and Bai, M. Coupled flow-deformation response of dual porosity media. *J. Geotech. Engrg.*, **118**, (1992) 107–124.

Geertsma, J. The effect of fluid pressure decline on volumetric changes of porous rocks. *Trans. AIME* (1957) 331–340.

Gringarten, A. C. Interpretation of tests in fissured and multilayered reservoirs with double-porosity behavior: theory and practice. *J. Pet. Tech.* (1984) 549–564.

Huyakorn, P. S., Lester, B. H. and Faust, C. R. Finite element techniques for modeling groundwater flow in fractured aquifers. *Water Resour. Res.*, **19**, (1983) 1019–1035.

Huyakorn, P. S. and Pinder, G. *Computational Methods in Subsurface Flow*. Academic Press, NY, (1983).

Javandel, I. and Witherspoon, P. A. Application of the finite element method to transient flow in porous media. *Soc. Pet. Eng. J.*, **8**, (1968) 241–252.

Kamal, M. M. Interference and pulse testing – a review. *Soc. Pet. Eng. J.* (1983) 2257–2270.

Kazemi, H. Pressure transient analysis of naturally fractured reservoirs with uniform fracture distribution. *Soc. Pet. Eng. J.* (1969) 451–61.

Khaled, M. Y., Beskos, D. E. and Aifantis, E. C. On the theory of consolidation with double porosity–3. a finite element formulation. *Int. J. Num. Anal. Meth. in Geomech.*, **8**, (1984) 101–123.

Kranz, R. L., Frankel, A. D. Engelder, T. and Scholz, C. H. The permeability of whole and jointed Barre granite. *Int. J. Rock Mech. Min. Sci. and Geomech. Abstr.*, **16**(3), (1979) 225–234.

Liggett, J. A. and Liu, P. L. F. *The Boundary Integral Equation Method for Porous Media Flow*. George Allen & Unwin, London, (1983).

Liu, T. *et al. Coal Mine Ground Movement and Strata Failure*. China Coal Ministry Publishing Company, (1984).

Louis, C. Groundwater flow in rock masses and its influence on stability. *Rock Mech. Research Report, no.10*, Imperial College, UK, (1969).

Melan, E. Der Spannungszustand der durch eine Einzelkraft im innern beanspruchten Halbscheibe. *Angew. Math. Mech.*, **12**, (1940) 343–346.

Nur, A. and Byerlee, J. D. An exact effective stress law for elastic deformation of rock with fluids. *J. Geophys. Res.*, **76**(26), (1971) 6414–6419.

Pinder, G. F. and Frind, E. O. Application of Galerkin's procedure to aquifer analysis. *Water Resour. Res.*, **8**, (1972) 108–120.

Prat, Da G. *Well Test Analysis for Fractured Reservoir Evaluation*. Elsevier, Amsterdam, (1990).

Reiss, L. H. *The Reservoir Engineering Aspects of Fractured Formations*. Institut Français du Petrole, (1980).

Rice, J. R. and Cleary, M. P. Some basic stress-diffusion solutions for fluid saturated elastic porous media with compressible constituents. *Rev. Geophys. Space Phys.*, **14**, (1976) 227–241.

Robin, P. Y. F. Note on effective pressure. *J. Geophys. Res.*, **78**(14), (1973) 2434–2437.

Segall, P. Stress and subsidence resulting from subsurface fluid withdrawal in the epicentral region of the 1983 Coalinga earthquake. *J. Geophys. Res.*, **B8**, (1985) 6801–6816.

Skempton, A. W. The pore pressure coefficients A and B. *Geotechnique*, **4**, (1954) 143–147.

Skempton, A. W. Effective stress in solid, concrete and rock. *Pore Pressure and Suction in Soils*, Butterworths, London, (1960).

Snow, D. T. Rock fracture spacings, openings, and porosities. *J. Soil Mech. Found. Div., Proc. ASCE*, **94**, (1968) 73–91.

Streltsova, T. D. *Well Testing in Heterogeneous Formations*. Wiley, New York, (1988).

Suklje, L. *Rheological Aspects of Soil Mechanics*. J Wiley, New York, (1969).

Terzaghi, K. V. Die Berechung der Durchassigkeitsiffer des Tones aus dem Verlauf der hydrodynamischen Spannungserscheinungen, Sitzungsber. Akad. Wiss. *Wien Math Naturwiss., Kl. Abt. 2A*, **132**, (1923) 105.

Tuncay, K. and Corapcioglu, M. Y. Effective stress principle for saturated fractured porous media. *Water Resour. Res.*, **31**(12), (1995) 3103–3106.

Verruijt, A. An elastic storage of aquifers. *Flow Through Porous Media*, Academic Press, NY, (1969) 331–376.

Walsh, J. B. Effect of pore pressure and confining pressure on fracture permeability. *Int. J. Rock Mech. Min. Sci. and Geomech. Abstr.*, **18**(3), (1981) 429–435.

Warren, J. E. and Root, P. J. The behavior of naturally fractured reservoirs. *J. Soc. Pet. Eng.*, **3**, (1963) 245–255.

Wilson, R. K. and Aifantis, E. C. On the theory of consolidation with double porosity. *Int. J. Engrg. Sci.*, **20**, (1982) 1009–1035.

24

Models for Uncertainty and its Propagation with Applications to Geomechanics

Roger Ghanem
The Johns Hopkins University, Baltimore, MD, USA

24.1   Introduction

As geomechanics comes to increasingly rely on complex mathematical models for the structure and behavior of geomaterials, along with efficient computational procedures for their resolution, the need becomes pressing to assess the significance of uncertainty on the predictions from such models. Indeed, the perceived increase in accuracy associated with added sophistication and numerical resolution must be consistent with the level of accuracy present in the data used to select and calibrate the appropriate mathematical models. Significant recent developments in computational resources and data acquisition tools make it imperative to develop rational procedures for integrating models of data with models of the physics in order to manage a consistent level of crudeness in the various components of the predictive process. This will hopefully lead to a meaningful coordinated refinement at both the level of data representation, as well as the level of physical and numerical modeling. Choosing a probabilistic framework for data representation provides a rich mathematical context into which to pose the problem.

Mathematical models of physical phenomena usually attempt to address a very well defined aspect in isolation of all others. This is necessitated by the complexity of nature and the restriction of our methods of inquiry to certain forms of logical statements. These isolated aspects of nature usually represent the behavior of a given system in a particular and isolated mode of operation. Since it is generally not possible to isolate and examine a physical system in this fashion, it should be expected that predictions from mathematical models would generally disagree with the observed behavior of the systems they purport to represent. By generalization and analogy, however, models permit us, starting from an observed fact, to follow a certain chain of logic to deduce and predict a number of additional facts. It should be remembered, though, that only the observed fact is certain, and all the others are merely probable. As noted by Poincaré (1905), however, it is far better to predict with uncertainty than to never have predicted at all.

In addition to uncertainties regarding the applicability of a certain model to a particular problem, which is typically associated with the model having captured the physical mechanisms known to be at play, uncertainties are also present in identifying the parameters to be used in that model. These parameters are usually obtained through some fitting procedure. Were the model perfect, some set of parameters would produce a perfect fit between the predictions of the model and the observed data. Given the approximate nature of the model, however, fitting errors will necessarily be incurred in the estimated values of these parameters. An important question can then be stated as one of deducing the confidence to be attached to the predictions of the model given the observed scatter, or uncertainty, in its parameters. Implicit here is the assumption that these parameters are the parameters that produce a best fit for the predictions of the model against experimental data. A closely related and equally important question is that of specifying the maximum level of uncertainty to be tolerated in the data if a decision about accepting or rejecting a particular model is desired with a specified confidence.

In trying to model physical phenomena, certain intrinsic principles have to be satisfied by any mathematical model attempting to rationalize observed behavior. Such principles may consist of some equilibrium or conservation laws whose applicability is not questioned within the confines of the problem at hand. Consider the set of models, denoted by S, consisting of all models that satisfy such consistency requirements. Further, consider a model $\mathcal{M}(\theta_1, \ldots, \theta_p) \in S$ parameterized by the set $(\theta_1, \ldots, \theta_p)$, and denote by $\mathcal{M}_s$ the model whose parameter set is equal to a specific set θ_s indexed by s. Moreover, let $q_s \equiv q(\mathcal{M}_s)$ symbolize the functional dependence, of some predicate of the model, on the parameter set θ. The quantity q_s could, for instance, represent the stress or the acceleration predicted by the model $\mathcal{M}_s$ at some point in the domain of interest. Finally, the symbol $\hat{q}$ will be used to denote the observed value of the physical quantity q_s attempts to predict. It should be noted that to different models $\mathcal{M}_i$ there may correspond different sets of predicates. Quantities that are defined with respect to one model and not with respect to another one could either be viewed as non-measurable with respect to this latter model, or to be constant in its context. Although the parameter set for a given model is uncertain, the relationship between this set and the predicates, q_s, is deterministic, and is completely specified through the functional $\mathcal{M}$. Thus, once the confidence in θ_s has been quantified, the confidence in q_s is uniquely determined. A probabilistic framework will be used to represent the uncertainty in the problem. The practical implication of this framework is that a probabilistic measure is associated with each of the model selection and parameter selection steps, inducing a well defined and computable probabilistic measure on the predicate q. Probabilistic inquiries of the form, $P(Q \in \mathcal{Q})$, will be posed, where a capital letter will be used to denote a random variable and a calligraphic letter will be used to denote a set to which the random variable may belong. We would thus be interested in probabilities of certain events occurring relating to a physical quantity Q. Given the above definitions, the following equalities hold:

$$P_Q(\mathcal{Q}) \equiv P(Q \in \mathcal{Q}) = P(Q(\mathcal{M}(\theta)) \in \mathcal{Q}) = P(\Theta \in \mathcal{T}|\mathcal{M}) \qquad (24.1)$$

where $\mathcal{T}$ denotes a subset of the set of possible parameters, and the last probability is conditioned on having adopted a particular model to represent the phenomenon being predicted. Equation (24.1) relates the probability measure of the predicted values to that of the model parameters, once a model has been selected. This transformation between the two probability measures is uniquely defined by the model $\mathcal{M}$ and can therefore be perceived as a characteristic of the model, in the same way that a frequency transfer function is a characteristic of a dynamical system. Once the expression on the left-hand side of Eq. (24.1) has been evaluated for all relevant sets $\mathcal{Q}$, hypothesis testing for the model given the observations can be readily performed. Thus, the set $\mathcal{Q}_\alpha$ defined by

$$P(Q \in \mathcal{Q}_\alpha) = 1 - \alpha \tag{24.2}$$

refers intuitively to a set into which model prediction would fall with probability $1 - \alpha$. An experimental observation $\hat{q}$ made within this set would corroborate the mathematical model with a confidence of at least $1 - \alpha$. Clearly, the largest such α for a given observation, $\hat{q}$, is of great value in the present context, as it can be used to define the confidence in the model afforded by the data.

Although conceptually straightforward, evaluating the probabilistic characterization of Q requires a significant computational effort, and is typically performed using a Monte Carlo simulation procedure. A framework is presented in this chapter that while simplifying the evaluation of such a probabilistic characterization provides a format that is conducive to posing more complex queries relating to the worth of information and optimal sampling.

The next section presents a brief review of uncertainty in geomechanics and its relevance in a number of application areas. This is followed by a brief review of probabilistic concepts leading to the Hilbert space description of random variables and processes. This will permit the application of deterministic approximation theory to the random case. Following that, the spectral stochastic finite element method is presented along with a number of examples demonstrating its applicability in areas of relevance to geomechanics.

24.2   Uncertainty in Geomechanics

Geomechanics provides the basic engineering and science foundation for a very diverse spectrum of applications. These range from geotechnical engineering to soil mechanics and multiphase flows and transport in porous media including environmental remediation and reservoir simulation. Interest is therefore great in developing methodologies for quantifying the propagation of uncertainty in this branch of science. In addition to quantifying the justifiable confidence in the predictive capability of mechanistic models, in many of the above applications, uncertainty quantification will enable the rationalization of data management and acquisition strategies that accommodate multidisciplinary constraints.

24.2.1   Characterizing uncertainty in geomechanics

In geomechanics, uncertainty can be derived from a number of sources chief among which is the variability in the underlying geological structure. Clearly, this variability is manifested differently at different scales of analysis. Thus, depending on the problem being addressed, models with varying scale resolution will be more or less appropriate.

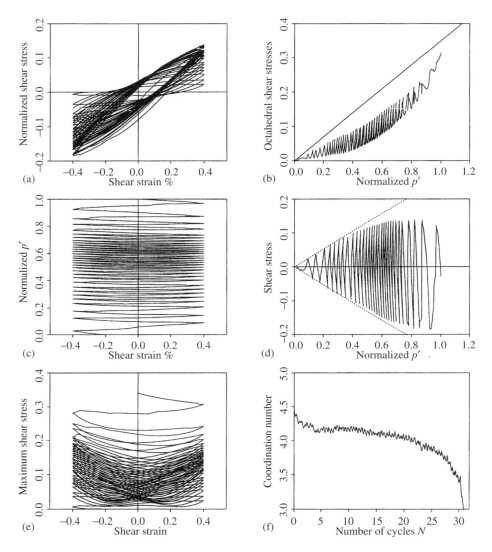

Figure 24.1 Stress-strain curves from three-dimensional discrete element simulations of sand subjected to cyclic load. Void ratio = 0.4, Shear strain amplitude = 0.4%, Simple shear test; Simulation 1

Motivated by increasing computational resources, efforts at bridging these length scales have mushroomed recently. In particular, the Discrete Element Method (DEM) has been used to quantify macro-scale parameters based on the micro-scale description of the soil fabric. While most efforts in this direction have aimed at evaluating deterministic parameters to be used as nominal parameters in macro-scale constitutive models, some research has aimed at providing a probabilistic description of the macro-scale (El-Mestkawy, 1998; Ghanem and El-Mestkawy, 1997, 1998). Figures 24.1 and 24.2 show stress-strain curves associated with two DEM simulations featuring 2000 spheres. These

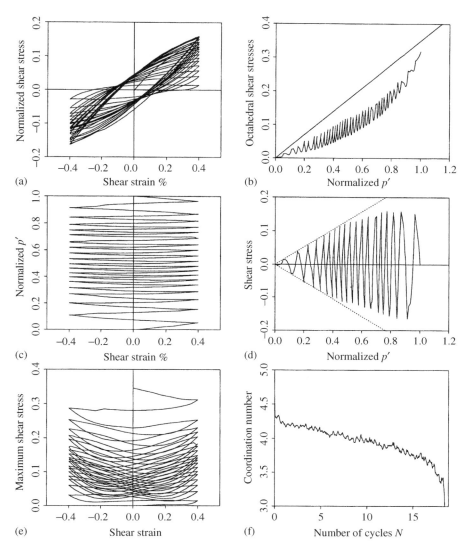

Figure 24.2 Stress-strain curves from three-dimensional discrete element simulations of sand subjected to cyclic load; Void ratio = 0.4, Shear strain amplitude = 0.4%, Simple shear test; Simulation 2

two simulations are associated with similar packings of spheres with the only difference corresponding to their original arrangement. In each case, the diameters of the spheres were drawn from a statistical population representative of Ottawa sand. Figure 24.3 shows the probability density function (pdf) for the normal forces between the spheres taken over the whole simulation domain, and corresponding to one DEM simulation. The different plots in the figure show the evolution of the pdf with the number of cycles. It is noted that while the general shape of the pdf remains the same, the number of contacts is further

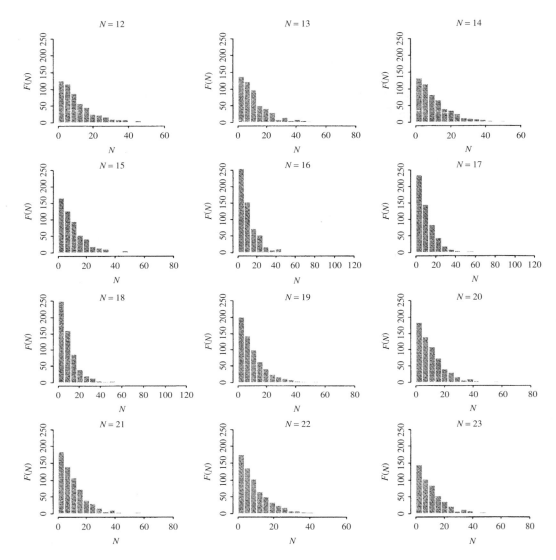

Figure 24.3 Probability density functions for the strength of the normal contacts throughout the spatial extent of one DEM simulation at successive cycles

reduced with each cycle. It is clear that by propagating first physical principles, namely elastic or inelastic interactions between spheres in contact, variability at the macro-scale can be captured. Additional research efforts are needed to quantify this variability, taking into account the corresponding spatial fluctuations.

In addition to computational models such as DEM, other analytical techniques can be used to quantify uncertainty in geomechanics. Those typically consist of postulating some model for the behavior of the geoenvironment, and then computing the statistics of the parameters of that model through data fitting. These models typically refer to energy dissipation and stiffness degradation mechanisms as functions of strain levels or frequency.

Relevant models could also refer to relative permeability as a function of saturation, as well as the various diffusion, adsorption and retardation factors, to name only a few of the parameters used to characterize flows in a porous medium.

The approach presented in this chapter can be used to characterize the variability in each of the above parameters as stochastic processes with random fluctuations in their dependence on the underlying variables. Figure 24.4, for example, refers to a spatially fluctuating interface between two two-dimensional layers, while Figure 24.5

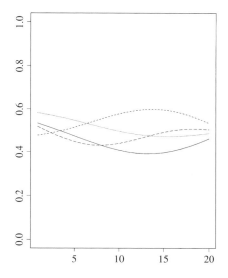

Figure 24.4   Random interface between two layers: four realizations

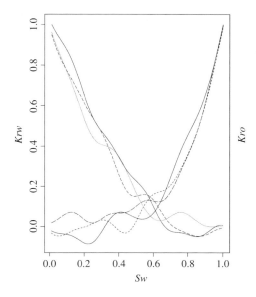

Figure 24.5   Relative permeability curves as stochastic processes: Three realizations

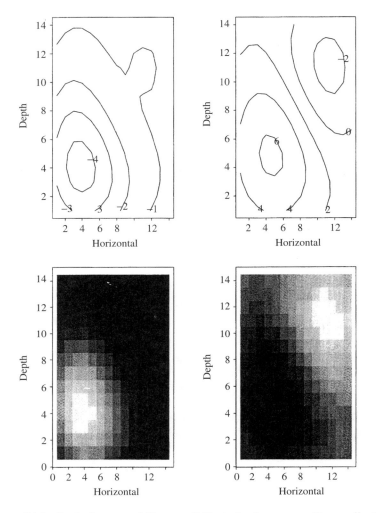

Figure 24.6 Intrinsic permeability as a 2-D stochastic process: Two realization

refers to random relative permeability curves for water and oil flow in a porous medium, and Figure 24.6 shows a random intrinsic permeability varying randomly over a two-dimensional domain.

24.3   Review of Probabilistic Concepts

The development presented in this chapter hinges on the definition of random variables as *measurable functions* from the space of elementary events to the real line. As functions, approximation theory, as developed for deterministic functions, can be applied to random variables. The main question to be addressed, already raised in the previous section, is the characterization of the solution to a physical problem where some parameters of the model have been modeled as stochastic processes. The answer to this question lies in the realization that in the deterministic finite element method, as well as most other numerical analysis techniques, a solution to a deterministic problem is known once its projection on a basis in an appropriate function space has been evaluated. It often happens, in deterministic analysis, that the coefficients in such a representation have an immediate physical meaning, which distracts from the mathematical significance of the solution. Carrying this argument over to the case involving stochastic processes, the solution to the problem will be identified with its projection on a set of appropriately chosen basis functions. A random variable will thus be viewed as a function of a single variable, θ, that refers to the space of elementary events. A stochastic process or field, α, is then a function of $n + 1$ variables where n is the physical dimension of the space over which each realization of the process is defined. Monte Carlo simulation can be viewed as a collocation along this θ dimension. Other approximations along this dimension are possible, and can be associated with different choices of basis functions in the corresponding space of random variables. This theoretical development is consistent with the identification of the space of second order random variables as a Hilbert space with the inner product on it defined as the operation of statistical correlation (Loeve, 1977). Second order random variables are those random variables with finite variance, they are mathematically similar to deterministic functions with finite energy.

24.3.1 *Mathematical characterization of random variables*

As indicated in the introduction to this chapter, a mapping, $\mathcal{M}$, can be conceptually defined between events corresponding to random parameters and events corresponding to measurable outputs. Referring to Figure 24.7, this mapping takes subsets $\mathcal{T}$ into subsets $\mathcal{Q} = \mathcal{M}(\mathcal{T})$. A measure can be associated to each subset $\mathcal{T}$ reflecting a subjective assessment of the likelihood of that event occurring. The corresponding measure on $\mathcal{Q}$ is uniquely computable by the mapping $\mathcal{M}(\cdot)$. To facilitate the analysis, and given that most of measurable events of interest refer to numerical measurements, it is usually convenient to work on measures defined on the real line. Random variables provide a mechanism for effecting that, as they are defined as mappings from the sets of basic events onto the real line. In the figure, each of the events $\mathcal{T}$ and $\mathcal{Q}$ are mapped, through appropriate random variables, into subsets of the real line, the measure of which are identified with the measure of the corresponding event. The mapping $\mathcal{M}(\cdot)$ is now replaced by a new mapping $\mathcal{N}(\cdot)$ between the random variables. It is usually this mapping that is dealt with in the context of mechanistic modeling. As already indicated above, it can be shown that, as mappings go, second order random variables have a number of very interesting properties, so much so that their collection forms a Hilbert Space when endowed with the inner product defined through the operation of statistical correlation. This Hilbert space structure is very convenient as these spaces form much of the foundation of deterministic numerical analysis: projections on subspaces as well as convergent approximations can now be defined. It is clear, from looking at Figure 24.7, that the space of random variables η contains that of the variables γ as a proper subset.

The picture is therefore as follows: The data, modeled as random variables or processes, live in the Hilbert space $\mathcal{H}_\mathcal{G}$. Assuming the data to be well defined in a probabilistic sense provides a full characterization of this space, in which a set of basis functions, ξ, will have to be identified. This will be accomplished in the section below on Karhunen–Loeve expansions. The state of the system, again modeled as a random variable or process,

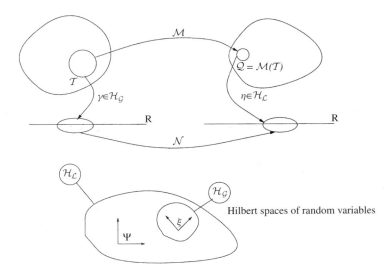

Figure 24.7   Functional dependencies in a probabilistic characterization of the problem

resides in the Hilbert space $\mathcal{H}_\mathcal{L}$. A set of basis functions, ψ, will also be identified in this space, which in general will be different from the basis ξ since this latter one spans only a subset of the space of second order random variables, namely those that characterize the data. Identifying a basis for the space $\mathcal{H}_\mathcal{L}$ will be accomplished in the section below dealing with the Polynomial Chaos expansion.

Karhunen–Loeve expansion

The Karhunen–Loeve expansion (Loeve, 1977) of a stochastic process $\alpha(\mathbf{x}, \theta)$, is based on the spectral expansion of its covariance function $R_{\alpha\alpha}(\mathbf{x}, \mathbf{y})$. Here, $\mathbf{x}$ and $\mathbf{y}$ are used to denote spatial coordinates, while the argument θ indicates the random nature of the corresponding quantity. The covariance function being symmetrical and positive definite, by definition, has all its eigenfunctions mutually orthogonal, and they form a complete set spanning the function space to which $\alpha(\mathbf{x}, \theta)$ belongs. It can be shown that if this deterministic set is used to represent the process $\alpha(\mathbf{x}, \theta)$, then the random coefficients used in the expansion are also orthogonal. The expansion then takes the following form,

$$\alpha(\mathbf{x}, \theta) = \bar{\alpha}(\mathbf{x}) + \sum_{i=1}^{\infty} \sqrt{\lambda_i} \xi_i(\theta) \phi_i(\mathbf{x}) \tag{24.3}$$

where $\bar{\alpha}(\mathbf{x})$ denotes the mean of the stochastic process, and $\{\xi_i(\theta)\}$ forms a set of orthogonal random variables. Furthermore, $\{\phi_i(\mathbf{x})\}$ are the eigenfunctions and $\{\lambda_i\}$ are the eigenvalues of the covariance kernel, and can be evaluated as the solution to the following integral equation:

$$\int_D R_{\alpha\alpha}(\mathbf{x}, \mathbf{y}) \phi_i(\mathbf{y}) d\mathbf{y} = \lambda_i \phi_i(\mathbf{x}) \tag{24.4}$$

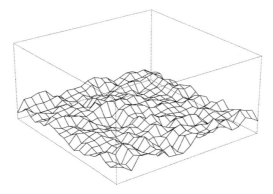

Figure 24.8   Two-dimensional random field generated with the Karhunen–Loeve expansion

where $\mathcal{D}$ denotes the spatial domain over which the process $\alpha(\mathbf{x}, \theta)$ is defined. The most important aspect of this spectral representation is that the spatial random fluctuations have been decomposed into a set of deterministic functions in the spatial variables multiplying random coefficients that are independent of these variables. If the random process being expanded, $\alpha(\mathbf{x}, \theta)$, is Gaussian, then the random variables $\{\xi_i\}$ form an orthonormal Gaussian vector. The Karhunen–Loeve expansion is mean-square convergent irrespective of the probabilistic structure of the process being expanded, provided it has a finite variance (Loeve, 1977). Figure 24.8 shows a typical stochastic process with covariance function given by

$$R_{\alpha\alpha}(\mathbf{x}, \mathbf{y}) = \sigma_\alpha e^{-\dfrac{|x_1 - y_1|}{b_1} - \dfrac{|x_2 - y_2|}{b_2}} \tag{24.5}$$

Figure 24.9 shows the first four eigenmodes, $\phi_1(\mathbf{x})$, $\phi_2(\mathbf{x})$, $\phi_3(\mathbf{x})$, $\phi_4(\mathbf{x})$, obtained through the numerical solution of Eq. (24.4). The abscissa in these figures refer to the node numbers, and the ordinate refers to the magnitude of the normalized eigenmode. Note that for this particular form of the covariance function, as well as for a number of other useful forms, defined over regular geometric domains, an analytical solution of the integral eigenvalue problem has been obtained (Ghanem and Spanos, 1991). Figure 24.10 shows the monotonic decay of the eigenvalues λ_i. The monotony of this decay is guaranteed by the symmetry of the covariance function, and the rate of the decay is related to the correlation length of the process being expanded. Thus, the closer a process is to white noise, the more terms are required in its expansion, while at the other limit, a random variable can be represented by a single term. In physical systems, it can be expected that material properties vary smoothly at the scales of interest in most applications, and therefore only few terms in the Karhunen–Loeve expansion can capture most of the uncertainty in the process.

Polynomial chaos expansion

The covariance function of the solution process is not known a priori, and hence the Karhunen–Loeve expansion cannot be used to represent it. Since the solution process is a function of the material properties, nodal concentrations, $S(\theta)$, can be formally expressed

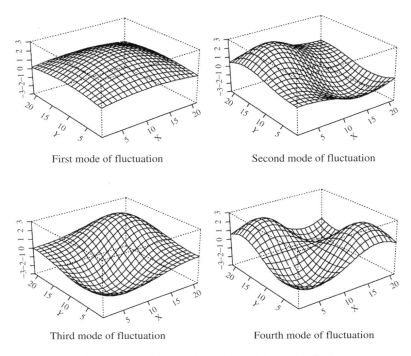

First mode of fluctuation   Second mode of fluctuation

Third mode of fluctuation   Fourth mode of fluctuation

Figure 24.9   The first four scales of fluctuation evaluated from the Karhunen–Loeve expansion

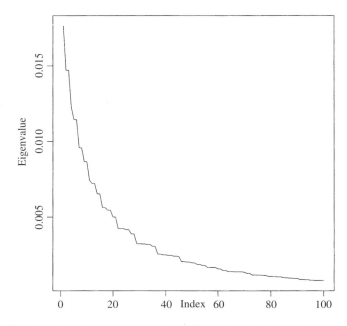

Figure 24.10   Contribution of successive scales of fluctuation in the Karhunen–Loeve expansion

as some nonlinear functional of the set $\{\xi_i(\theta)\}$ used to represent the material stochasticity. It has been shown (Cameson and Mastin, 1947) that this functional dependence can be expanded in terms of polynomials in Gaussian random variables, referred to as Polynomial Chaos. Namely,

$$S(\theta) = a_0 \Gamma_0 + \sum_{i_1=1}^{\infty} a_{i_1} \Gamma_1(\xi_{i_1}(\theta)) + \sum_{i_1=1}^{\infty} \sum_{i_2=1}^{i_1} a_{i_1 i_2} \Gamma_2(\xi_{i_1}(\theta), \xi_{i_2}(\theta)) + \cdots \quad (24.6)$$

In this equation, the symbol $\Gamma_n(\xi_{i_1}, \ldots, \xi_{i_n})$ denotes the Polynomial Chaos (Wiener, 1938; Kallianpur, 1980) of order n in the variables $(\xi_{i_1}, \ldots, \xi_{i_n})$. These are generalizations of the multidimensional Hermite polynomials to the case where the independent variables are functions measurable with respect to the Wiener measure. Introducing a one-to-one mapping to a set with ordered indices denoted by $\{\Psi_i(\theta)\}$ and truncating the Polynomial Chaos expansion after the Pth term, Eq. (24.6) can be rewritten as

$$S(\theta) = \sum_{j=0}^{P} S_j \Psi_j(\theta) \quad (24.7)$$

Table 24.1 shows the explicit expressions for the first few of these polynomials. These polynomials are orthogonal in the sense that their inner product $\le \Psi_j \Psi_k \ge$, which is defined as the statistical average of their product, is equal to zero for $j \neq k$. Moreover, they can be shown to form a complete basis in the space of second order random variables. A complete probabilistic characterization of the process $S(\theta)$ is obtained once the deterministic coefficients S_j have been calculated. A given truncated series can be refined along the random dimension either by adding more random variables to the set $\{\xi_i\}$ or by increasing the maximum order of polynomials included in the Polynomial Chaos expansion. The first refinement takes into account higher frequency random fluctuations of the underlying stochastic process, while the second refinement captures strong nonlinear dependence of the solution process on this underlying process.

Table 24.1 Polynomial Chaoses and their variances. Two terms in the Karhunen–Loeve expansion

ith Polynomial Chaos	Order of the Homogeneous Chaos	Ψ_i	$\langle \Psi_i^2 \rangle$
0	0	1	1
1	1	ξ_1	1
2		ξ_2	1
3	2	$\xi_1^2 - 1$	2
4		$\xi_1 \xi_2$	1
5		$\xi_2^2 - 1$	2
6	3	$\xi_1^3 - 3\xi_1$	6
7		$\xi_1^2 \xi_2 - \xi_2$	2
8		$\xi_1 \xi_2^2 - \xi_2$	2
9		$\xi_2^3 - 3\xi_2$	6
10	4	$\xi_1^4 - 6\xi_1^2 + 3$	24
11		$\xi_1^3 \xi_2 - 3\xi_1 \xi_2$	6
12		$\xi_1^2 \xi_2^2 + \xi_1^2 - \xi_2^2 + 1$	4
13		$\xi_1 \xi_2^3 - 3\xi_1 \xi_2$	6
14		$\xi_2^4 - 6\xi_2^2 + 3$	24

It should be noted at this point that the Polynomial Chaos expansion can be used to represent, in addition to the solution process, stochastic processes that model non-Gaussian material properties. The processes representing the material properties are thus expressed as the output of a nonlinear system to a Gaussian input.

In the next section, the Karhunen–Loeve and the Polynomial Chaos expansions are integrated into the equations governing the behavior of typical geomechanics problems, and procedures are developed for evaluating an expansion of the solution process with respect to the Polynomial Chaos basis.

24.4   The Stochastic Finite Element Method

24.4.1   Overview

In this section, the formulation is presented in terms of generic time-dependent equations that are representative of many problems in geomechanics. The linear problem is treated first to keep the notation concise, with the nonlinear problem treated in a later section. Consider the equation

$$\frac{\partial}{\partial t} S(\mathbf{x}, t, \theta) = \mathcal{Q}_a[S(\mathbf{x}, t, \theta)] + f(\mathbf{x}, t, \theta), \quad x \in \mathcal{B}, t \in [0, T], \quad \theta \in \Omega \tag{24.8}$$

where $\mathcal{Q}_a[\cdot]$ is a linear differential operator whose coefficients $a(\mathbf{x}, t, \theta)$ can be modeled as random fields exhibiting random fluctuations both in space and time, and Ω is the set of elementary events. By virtue of the randomness of the differential operator, the solution to the differential equation will itself be a random process. Separating the random component of $\mathcal{Q}_a[\cdot]$ from its deterministic component results in the following equation:

$$\frac{\partial}{\partial t} S(\mathbf{x}, t, \theta) = (\mathcal{L} + \mathcal{R}_\alpha)[S(\mathbf{x}, t, \theta)] + f(\mathbf{x}, t, \theta) \tag{24.9}$$

where now $\mathcal{L}[\cdot]$ indicates a deterministic linear operator whose coefficients represent the average values of the random coefficients, and $\mathcal{R}_\alpha[\cdot]$ denotes a random linear operator whose coefficients $\alpha(\mathbf{x}, t, \theta)$ are zero mean random fields. In the above, it is assumed that the random coefficients $a(\mathbf{x}, t, \theta)$ appear as multiplicative constants in the explicit expression of the operator $\mathcal{Q}_a[\cdot]$. Extensions to more complicated situations have already been carried out (Ghanem and Dham, 1998; Ghanem, 1998; Ghanem and Red-Horse, 1999). For clarity of presentation, however, the simpler, linear case is used in this section.

Conceptually, given the joint probabilistic information about $\mathcal{R}_\alpha[\cdot]$ and $f(\mathbf{x}, t, \theta)$, the solution is completely determined when the joint probabilistic information about $\mathcal{R}_\alpha[\cdot]$, $f(\mathbf{x}, t, \theta)$ and $S(\mathbf{x}, t, \theta)$ is obtained. Traditionally, a number of simplifying assumptions are introduced and a partial solution of the problem, restricted to the second order statistics of the solution, is sought. The process $\alpha(\mathbf{x}, t, \theta)$ can be thought of as representing the stochastic fluctuations about its mean of some property of the system described by the operator $\mathcal{R}_\alpha$. In general, the process $\alpha(\mathbf{x}, t, \theta)$ is continuous with respect to $\mathbf{x}$ although the form of the corresponding functional dependence is not explicitly known. This fact presents a major difficulty in relation to the numerical treatment of the problem, specially when these processes are integrated with respect to $\mathbf{x}$, as is customary in finite element techniques. The difficulty may be overcome by using the Karhunen–Loeve expansion

(Loeve, 1977). Accordingly, $\alpha(\mathbf{x}, t, \theta)$ may be represented as

$$\alpha(\mathbf{x}, t, \theta) = \sum_{i=1}^{\infty} \sqrt{\lambda_i} \xi_i(\theta) \psi_i(\mathbf{x}, t) \tag{24.10}$$

where $\xi_i(\theta)$ is a set of zero-mean, orthogonal random variables, λ_i and $\psi_i(\mathbf{x}, t)$ are deterministic quantities representing, respectively, the eigenvalues and eigenfunctions of the covariance kernel of $\alpha(\mathbf{x}, t, \theta)$. Obviously, the joint distribution of $\{\xi_i(\theta)\}$ depends upon that of $\alpha(\mathbf{x}, t, \theta)$. In the case of a Gaussian process, the set $\xi \equiv \{\xi_i(\theta)\}$ will form a Gaussian vector. Equation (24.10) can be viewed as expressing the random fluctuations $\alpha(\mathbf{x}, t, \theta)$ in terms of deterministic scales $\phi_i(\mathbf{x}, t)$ multiplied by random amplitudes $\sqrt{\lambda_i} \xi_i(\theta)$. In view of Eq. (24.10), Eq. (24.9) becomes

$$\frac{\partial}{\partial t} S(\mathbf{x}, t, \theta) = (\mathcal{L} + \mathcal{R}_\xi)[S(\mathbf{x}, t, \theta)] + f(\mathbf{x}, t, \theta) \tag{24.11}$$

where the subscript ξ on $\mathcal{R}$ indicates that the problem has now been recast in a space spanned by a denumerable set of random variables. Using the deterministic finite element method (Akin, 1982) this partial differential equation is approximated by an ordinary differential equation in R^N of the form

$$\frac{d}{dt} \mathbf{S}(t, \theta) = [\mathbf{L} + \mathbf{R}_\alpha] \mathbf{S}(t, \theta) + \mathbf{f}(\theta) \tag{24.12}$$

where the argument θ for the $N \times N$ matrix $\mathbf{R}$ and the N-dimensional vectors $\mathbf{S}(t, \theta)$ and $\mathbf{f}$ shows their random dependence. Symbolically, the random solution vector may be expressed as

$$\mathbf{S}(t, \theta) = \mathbf{s}_t[\{\xi(\theta)\}] \tag{24.13}$$

where $\mathbf{s}_t[\cdot]$ is a nonlinear vector functional of its argument. Expressed in this form, the solution can be viewed as the output of a nonlinear filter to white noise. This nonlinear filter is physically represented by the transformation, through the porous medium, of the scales of heterogeneity of the hydraulic properties into the scales of heterogeneity of the solution fields. The problem then can be restated as one of finding a convergent expansion, preferably optimal, of the functional $\mathbf{s}_t[\cdot]$.

Next, the solution is expressed through its expansion with respect to the Polynomial Chaos system in the form

$$S(\theta) = \sum_{i=0}^{\infty} S_i \Psi_i(\theta) \tag{24.14}$$

where S_i is some constant independent of θ, and $\{\Psi_i(\theta)\}_{i=1}^{\infty}$ is a basis in Θ. Using the following notation:

$$\mathbf{R}_0 = \mathbf{L} \quad \text{and} \quad \xi_0(\theta) = 1 \tag{24.15}$$

Eq. (24.11) can be rewritten as

$$\frac{d}{dt} \mathbf{S}(t, \theta) + \left[\sum_{m=0}^{M} \xi_m(\theta) \mathbf{R}_m \right] \mathbf{S}(t, \theta) = \mathbf{f}. \tag{24.16}$$

In this equation, $\mathbf{R}_m$ can be thought of as the stiffness or conductance matrix associated with material property given by the mth scale of fluctuation.

Truncating Eq. (24.14) at the Pth polynomial, and substituting for $\mathbf{S}(t, \theta)$, Eq. (24.16) yields an expression for the resulting error

$$\varepsilon = \mathbf{f} - \sum_{i=0}^{P} \Psi_i(\theta) \frac{d}{dt} \mathbf{S}_i(t) + \left[\sum_{m=0}^{M} \sum_{i=0}^{P} \xi_m(\theta) \Psi_i(\theta) \mathbf{R}_m \, \mathbf{S}_i(t) \right] \tag{24.17}$$

This error results from truncating the series in Eq. (24.14) after a finite number of terms, as well as from using a finite number of elements from the infinite set $\{\xi(\theta)\}_{i=1}^{\infty}$. The error, as expressed by Eq. (24.17), is minimized by requiring it to be orthogonal to the solution space, with respect to the inner product defined by the statistical averaging operator. A procedure similar to the one used in the previous section, for the deterministic finite element method, can be used here. Mathematically, this is equivalent to

$$(\varepsilon, \Psi_s(\theta)) = 0 \quad s = 0, 1, \dots \tag{24.18}$$

The orthogonality constraint results in a set of algebraic equations that can be solved for the vector coefficients $\mathbf{S}_i(t)$. From these coefficient, the spectral expansion of the solution vector is determined with respect to the Polynomial Chaos basis. From Eq. (24.14), it can be seen that all the probabilistic information concerning the random vector $\mathbf{S}(t, \theta)$ is contained in the expansion coefficients. Namely, the mean of the solution is equal to $\mathbf{S}_0$, and the covariance of the solution is given by $\mathbf{S}_i \mathbf{S}_i^T \sum_i \langle \Psi_i^2 \rangle$. So once these coefficients have been computed, the probability distribution of the solution vector $\mathbf{S}(t, \theta)$ can be determined.

Table 24.1 shows explicit expressions for the first ten two-dimensional Polynomial Chaoses and their variances. Repetitive application of the orthogonality requirement given by Eq. (24.18) for successive Polynomial Chaoses results in the matrix equation

$$\frac{d\mathbf{S}}{dt} + \mathbf{A}\mathbf{S} = \mathbf{F} \tag{24.19}$$

where $\mathbf{S}$ is the $(P + 1)N$-dimensional vector of coefficients, $\mathbf{A}$ is a $(P + 1)N \times (P + 1)N$ matrix constituted of block submatrices, where the (i, j) block is an $N \times N$ matrix given by

$$\mathbf{A}_{ij} = \sum_{m=0}^{M} \mathbf{R}_m \langle \Psi_i(\theta) \Psi_j(\theta) \xi_m(\theta) \rangle, \quad i, j = 0, \dots, b - 1 \tag{24.20}$$

and $\mathbf{F}$ is a $(P + 1)N$ dimensional vector given by

$$\mathbf{F}_i = \langle \mathbf{f}\Psi_i(\theta) \rangle, \quad i = 0, \dots, b - 1 \tag{24.21}$$

Table 24.2 shows the value of the coefficients $\langle \Psi_i(\theta)\Psi_j(\theta)\xi_m(\theta) \rangle$ for $m = 2$. In this table, the order M of the Karhunen–Loeve expansion and the order P of the polynomial chaos are equal to 2 and 3 respectively. The symbolic manipulation program Mathematica was used to develop similar tables for arbitrary values of M and P.

24.4.2 Connection to Monte Carlo simulation and hybrid implementations

The formalism developed in the previous section can be seen to encompass the Monte Carlo simulation technique as a special case. In particular, if the functions $\Psi_i(\theta)$,

Table 24.2   Values of the coefficient in Eq. (24.20), $\langle \Psi_i(\theta)\Psi_j(\theta)\xi_m(\theta)\rangle$. Two terms in the K–L expansion

$m = 2$ i	0	1	2	3	4	5	6	7	8	9
j										
0	0	0	0	1	0	0	0	0	0	0
1	0	0	0	0	2	0	0	0	0	0
2	0	0	0	0	0	0	2	0	0	0
3	1	0	0	0	0	0	0	2	0	0
4	0	2	0	0	0	0	0	0	6	0
5	0	0	0	0	0	0	0	0	0	0
6	0	0	2	0	0	0	0	0	0	0
7	0	0	0	2	0	0	0	0	0	0
8	0	0	0	0	6	0	0	0	0	0
9	0	0	0	0	0	0	0	0	0	0

instead of referencing the Polynomial Chaos basis, referred to delta functions $\delta(\theta - \theta_i)$, a collocation-like procedure is obtained along the random dimension θ. The coefficients of the delta functions in this expansion correspond to individual realizations of the process being represented; these realizations being sampled at the coordinates $\{\theta_i\}$. Implementing this expansion into the formulation presented above results in a set of algebraic equations to be solved, at every iteration of every time step, of the following general form:

$$
\begin{bmatrix}
\mathbf{K}_1 & 0 & 0 & 0 & 0 & 0 \\
0 & \mathbf{K}_2 & 0 & 0 & 0 & 0 \\
0 & 0 & \cdot & 0 & 0 & 0 \\
0 & 0 & 0 & \cdot & 0 & 0 \\
0 & 0 & 0 & 0 & \cdot & 0 \\
0 & 0 & 0 & 0 & 0 & \mathbf{K}_{nsimul}
\end{bmatrix}
\left\{
\begin{matrix}
\mathbf{S}_1 \\ \cdot \\ \cdot \\ \cdot \\ \cdot \\ \mathbf{S}_{nsimul}
\end{matrix}
\right\}
=
\left\{
\begin{matrix}
\mathbf{f}_1 \\ \cdot \\ \cdot \\ \cdot \\ \cdot \\ \mathbf{f}_{nsimul}
\end{matrix}
\right\}
\tag{24.22}
$$

The off-diagonal submatrices, in these equations, are all zero, due to the fact that the coefficients d_{ijk}, in this case, are all zero unless $i = j = k$. It is clear, then, that in this case, the number of unknown subvectors, *nsimul*, is equal to the number of independent simulations to be performed in the context of a Monte Carlo simulation procedure. The corresponding equation associated with the Polynomial Chaos basis has the general form

$$
\begin{bmatrix}
\mathbf{K}_{00} & \mathbf{K}_{01} & \cdots & \mathbf{K}_{0P} \\
\mathbf{K}_{10} & \mathbf{K}_{11} & \cdots & \mathbf{K}_{1P} \\
\cdot & \cdot & \cdots & \cdot \\
\cdot & \cdot & \cdots & \cdot \\
\mathbf{K}_{P0} & \mathbf{K}_{P1} & \cdots & \mathbf{K}_{PP}
\end{bmatrix}
\left\{
\begin{matrix}
\mathbf{S}_0 \\ \cdot \\ \cdot \\ \cdot \\ \mathbf{S}_P
\end{matrix}
\right\}
=
\left\{
\begin{matrix}
\mathbf{f}_0 \\ \cdot \\ \cdot \\ \cdot \\ \mathbf{f}_P
\end{matrix}
\right\}
\tag{24.23}
$$

where coupling occurs between the various submatrices, and the index P refers to the number of terms retained in the Polynomial Chaos expansion. The above discussion can be used as a vehicle to assess the relative efficiency of the procedure described above in comparison to a traditional Monte Carlo simulation. Noting that, in general, *nsimul* is much larger that P suggests that, depending on the solution procedure used (matrix storage algorithm, iterative solution procedure, etc.), there may be a threshold level for which one procedure may be more efficient than the other. Given that the number of iterations in an iterative solution scheme depends upon the level of the uncertainty (as measured by the coefficient of variation of material properties), and on the level of nonlinearity in the system, it is not clear at this stage that a simple identification of this threshold level is possible. An important outcome of the above discussion, however, is the new perspective on Monte Carlo simulation procedures, and their embedding as a special case of the procedure presented in this chapter. Indeed, this new perspective paves the way for the development of the hybrid stochastic finite element method, detailed in the next section.

Using a P term Polynomial Chaos expansion for the solution process, and expanding the residual vector in terms of a basis set $\Phi(\theta)$, results in

$$\mathbf{S} = \sum_{i=0}^{P} \mathbf{S}_i \Psi_i + \sum_{i=0}^{M} \mathbf{S}_i^* \Phi_i \tag{24.24}$$

In the following, the expansion for the residual will be associated with a Monte Carlo simulation procedure. This implies that the set Φ_i is identified with the sequence of delta function, $\delta(\theta - \theta_i)$. Thus, the inner product of any random variable with a function Φ_i is given by

$$\langle f(\theta)\Phi_i(\theta)\rangle = f(\theta_i) \tag{24.25}$$

and results in the realization $f(\theta_i)$ of f. Substituting Eq. (24.24) into Eq. (24.36), and taking the inner product with Ψ_k and Φ_k, respectively, results in the following systems of equations:

$$\sum_{j=0}^{P}\sum_{i=0}^{L}\langle\Psi_i\Psi_j\Psi_k\rangle\mathbf{K}_i\mathbf{S}_j = \langle\Psi_k\mathbf{f}\rangle - \sum_{j=0}^{M}\sum_{i=0}^{L}\mathbf{K}_i\langle\Psi_i\Phi_j\Psi_k\rangle\mathbf{S}_j^* \quad k = 0, \ldots \tag{24.26}$$

and

$$\langle\mathbf{KS}^*\Phi_k\rangle = \langle\mathbf{f}\Phi_k\rangle - \sum_{j=0}^{P}\sum_{i=0}^{L}\langle\Psi_i\Psi_j\Phi_k\rangle\mathbf{K}_i\mathbf{S}_j \quad k = 0, \ldots \tag{24.27}$$

Equation (24.26) is associated with Eq. (24.23), while Eq. (24.27) can be associated with Eq. (24.22). A new equation, similar to Eq. (24.22) and (24.23), can be written down that symbolically captures the hybrid stochastic finite element procedure presented herein. Specifically, Eq. (24.26) and (24.27) can be grouped into the following equation:

$$
\begin{bmatrix}
\mathbf{K}_1 & \cdots & 0 & & & & \\
0 & \cdots & 0 & & [\mathbf{K}_{\text{hybrid}}] & & \\
0 & \cdots & \mathbf{K}_{\text{nsimul}} & & & & \\
& & & \mathbf{K}_{00} & \cdot & \cdot & \cdot & \mathbf{K}_{0P} \\
& & & \mathbf{K}_{10} & \cdot & \cdot & \cdot & \mathbf{K}_{1P} \\
& & & \cdot & & & \cdot \\
& [\mathbf{K}_{\text{hybrid}}] & & \cdot & \cdot & \cdot & \cdot & \cdot \\
& & & \cdot & & & \cdot \\
& & & \mathbf{K}_{P0} & \cdot & \cdot & \cdot & \mathbf{K}_{PP}
\end{bmatrix}
\begin{Bmatrix}
\mathbf{S}_1^* \\ \cdot \\ \cdot \\ \mathbf{S}_{\text{nsimul}}^* \\ \mathbf{S}_0 \\ \cdot \\ \cdot \\ \cdot \\ \mathbf{S}_P
\end{Bmatrix}
=
\begin{Bmatrix}
\mathbf{f}_1^* \\ \cdot \\ \cdot \\ \mathbf{f}_{\text{nsimul}}^* \\ \mathbf{f}_0 \\ \cdot \\ \cdot \\ \cdot \\ \mathbf{f}_P
\end{Bmatrix}
\tag{24.28}
$$

where $\mathbf{K}_{\text{hybrid}}$ represents a coupling term between the two components of the solution. Equations (24.26) and (24.27) are indeed coupled through the terms, $\langle\Psi_i\Psi_j\Phi_k\rangle\mathbf{K}_i\mathbf{S}_j$ and $\langle\Psi_i\Psi_k\rangle\mathbf{S}_j^*$. This last average is evaluated as the arithmetic means of the solution of

Eq. (24.27) for different values of k. The double summation on the right-hand side of Eq. (24.26) can be rewritten as

$$\sum_{j=0}^{M}\sum_{i=0}^{L}\mathbf{K}_i\langle\Psi_i\Phi_j\Psi_k\rangle\mathbf{S}_j^* = \Psi_k\sum_{i=0}^{L}\mathbf{K}_i\Psi_i\sum_{j=0}^{M}\mathbf{S}_j^* \tag{24.29}$$

where all the terms on the right-hand side of Eq. (24.23) refer to a particular realization of the corresponding random quantities (associated with the θ_j). Assuming that the mean value of the solution process has been well represented by the Polynomial Chaos approximation, the quantity $\sum\mathbf{S}_j^*$ is approximately zero. The interaction between Eq. (24.26) and (24.27) can thus be neglected and Eq. (24.26) can be rewritten as

$$\sum_{j=0}^{P}\sum_{i=0}^{L}\langle\Psi_i\Psi_j\Psi_k\rangle\mathbf{K}_i\mathbf{S}_j = \langle\Psi_k\mathbf{f}\rangle \quad k = 0, \ldots \tag{24.30}$$

This decoupling eliminates the need for an iterative solution process whereby the values of the Polynomial Chaos coefficients, $\mathbf{S}_j$, are updated following a certain number of simulations, $\mathbf{S}_j^*$. From the orthogonality property of the Polynomial Chaos,

$$\mathbf{R}_{uu} = \sum_{i=1}^{P}\mathbf{S}_i^T\mathbf{S}_i\langle\Psi_i^2\rangle + \mathbf{R}_{u^*u^*} \tag{24.31}$$

It is therefore clear that the process $\mathbf{S}^*$ has a smaller variance than the original solution process $\mathbf{S}$. This clearly indicates that it requires less realizations to achieve a certain accuracy with specified confidence level. Specifically, for a Gaussian random variable, the number of samples required to achieve accuracy greater than δ in the estimating the mean value, given a confidence interval of width Z_α, is given by

$$n = \frac{Z_\alpha^2\sigma^2}{\delta^2} \tag{24.32}$$

where Z_α denotes the width of the confidence interval, and δ denotes a specified tolerance. Clearly, although the solution to a stochastic partial differential equation is not a Gaussian process, the foregoing analysis still provides useful guidelines regarding the required number of realizations. Thus if enough terms are taken in the Polynomial Chaos expansion to reduce the variance of the solution process by a half, it can be expected that four times less samples are required in the simulation of the deviation of the Polynomial Chaos expansion from the exact process. A compromise can thus be achieved between the number of simulations, and the number of terms used in the Polynomial Chaos expansion of the solution.

The foregoing analysis can also be cast in the framework of variance reduction techniques. Denoting the probability of failure of the system whose response is sought by P_f results in

$$P_f = \int_{\mathbf{x}\in\mathcal{D}_f} f_\mathbf{x}(\mathbf{x})d\mathbf{x} \tag{24.33}$$

where $\mathcal{D}_f$ denotes the failure domain of the system, and $f_\mathbf{x}(\mathbf{x})$ denotes the joint density function of the random variables $\mathbf{x}$. Typical variance reduction techniques are based on rewriting the probability of failure in the form

$$P_f = \int_{\mathbf{x}\in\mathcal{D}_f} \frac{f_\mathbf{x}(\mathbf{x})}{f_S(\mathbf{x})} f_S(\mathbf{x})d\mathbf{x} \tag{24.34}$$

where $f_S(\mathbf{x})$ is some new sampling density function, thus effectively weighting the original density function in such a way as to favor the occurrence of rare events. The procedure presented in this chapter, can be viewed, instead, as rewriting the probability of failure in the following form:

$$P_f = \int_{\mathbf{x}+\mathbf{x}^* \in \mathcal{D}_f} f_{\mathbf{x}^*} d\mathbf{x}^* \tag{24.35}$$

In the above, the variables $\mathbf{x}$ represent the part of the solution corresponding to the spectral expansion. This technique can therefore be viewed as enhancing the Monte Carlo simulation procedure by modifying the domain of simulation instead of weighing the current domain as is typical with variance reduction techniques. Of course, more advanced procedures for enhancing Monte Carlo simulation have recently been developed but a review of their performance is outside the scope of this work.

24.4.3   *Material nonlinearities*

Generalizing Eq. (24.8) to nonlinear operators and following through with space and time discretization, leads to equations having the following generic form:

$$\mathbf{K}(\mathbf{S}, \alpha(\theta))\mathbf{S}(\theta) = \mathbf{f}(\theta) \tag{24.36}$$

where the nonlinear dependence of matrix $\mathbf{K}$ on the solution vector $\mathbf{S}$, and the random nature of the various quantities involved are explicitly shown. Moreover, the coefficient matrix $\mathbf{K}$ is random through its dependence on a vector stochastic process, α. Representing this process in terms of its Karhunen–Loeve expansion, and using the Polynomial Chaos expansion (Ghanem and Spanos, 1991) for the solution process, results in the following equation:

$$\mathbf{S} = \sum_{j=0}^{P} \mathbf{S}_j \Psi_j \tag{24.37}$$

$$\sum_{j=0}^{P} \Psi_j \mathbf{K}(\mathbf{S}, \xi(\theta))\mathbf{S}_j = \mathbf{f}(\theta) \tag{24.38}$$

where the vector of uncorrelated random variables, ξ, now replaces the stochastic process α. The error in the above equation is constrained to be orthogonal to the approximating space. This requirement results in the following set of equations:

$$\sum_{j=0}^{P} \langle \Psi_k \Psi_i \mathbf{K}(\mathbf{S}, \xi(\theta)) \rangle \mathbf{S}_j = \langle \Psi_k \mathbf{f}(\theta) \rangle \quad k = 0, 1, \ldots. \tag{24.39}$$

Representing $\mathbf{K}$ in its Polynomial Chaos expansion, Eq. (24.38) becomes

$$\mathbf{K} = \sum_{i=0}^{L} \Psi_i \mathbf{K}_i, \quad \mathbf{K}_i = \frac{\langle \Psi_i \mathbf{K} \rangle}{\langle \Psi_i^2 \rangle} \tag{24.40}$$

results in the following:

$$\sum_{j=0}^{P} \sum_{i=0}^{L} \langle \Psi_k \Psi_i \Psi_m \rangle \mathbf{K}_i \mathbf{S}_j = \langle \Psi_k \mathbf{f}(\theta) \rangle \quad k = 0, 1, \ldots. \tag{24.41}$$

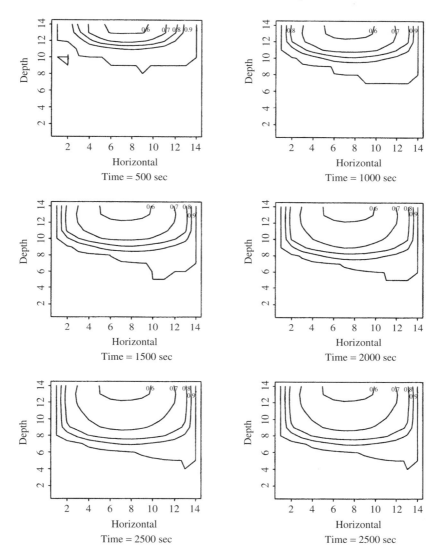

Figure 24.11 Realization of the water saturation levels at successive time steps: Simulation 1

It should be emphasized here that representing **K** in terms of its Polynomial Chaos expansion is neither necessary nor necessarily most efficient. In particular, using Eq. (24.39) in conjunction with such techniques as stratified sampling, or Latin hypercube sampling, may lead to more efficient algorithms. These sampling techniques can be viewed as an efficient way to evaluate the integral implied by the averaging process (which is nothing more than an inner product in the space of random variables).

Figures 24.11 and 24.12 show two realizations of the evolution of water saturation levels corresponding to the flow of oil into an initially water-filled reservoir. The intrinsic permeability of the reservoir was assumed to be a lognormal stochastic process with a specified correlation function. These realizations were obtained by synthesizing the

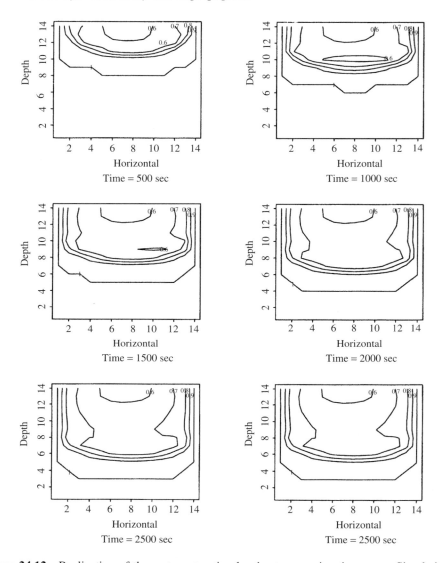

Figure 24.12   Realization of the water saturation levels at successive time steps: Simulation 2

polynomial chaos expansion after its coefficients had been evaluated. Two different seed numbers were used in the synthesis. Once the polynomial chaos coefficients have been computed, simulation of additional realizations is a very efficient process.

24.5   Decision Support

An important feature of the development presented herein is the integration of decision support into the numerical simulation. This can be easily accomplished in the formulation, since a series representation of the solution as a function of all the random quantities in the problem would have been evaluated. In this section, two procedures for utilizing that expansion in a decision-making context are presented.

24.5.1 *Probabilistic characterization of the solution*

The representation of the solution as the Polynomial Chaos expansion, contains a complete probabilistic characterization. Indeed, while the mean value of the solution field is given by the zeroth order term in the expansion, the covariance matrix of the solution is readily obtained as

$$\mathbf{R}_{SS} = \sum_{i=1}^{P} \langle \Psi_i^2 \rangle \mathbf{S}_i \mathbf{S}_i^T \qquad (24.42)$$

Higher order statistics can be obtained in a similarly simple format. More important than these statistical moments, though, is the possibility to efficiently generate a large statistical population of the solution process. For every simulation of the stochastic field representing the porous medium, a corresponding simulation can be obtained for the random variables Ψ_i appearing in the polynomial chaos expansion. This permits the synthesis of the corresponding simulation of the solution process by simply evaluating the expansion. Once a statistically significant population has been generated, probabilities of rare events can be accurately evaluated through standard statistical techniques.

24.5.2 *Optimal sampling locations*

According to the foregoing, the vector solution field $\mathbf{S}(\theta)$ is obtained in the form

$$\mathbf{S}(\theta) = \sum_{i=0}^{P} \mathbf{S}_i \Psi_i(\theta) \qquad (24.43)$$

where Ψ_i is one of the polynomials featured in Table 24.1 or a similar table for the appropriate value of M. Recall that M refers to the number of terms used to represent the spatial randomness of the domain. Taking the derivative of the above equation with respect to $\boldsymbol{\alpha}(t, \theta)$ results in

$$\frac{\partial \mathbf{S}}{\partial \boldsymbol{\alpha}(t, \theta)} = \sum_{i=0}^{M} \frac{\partial \mathbf{S}}{\partial \xi_i} \frac{\partial \xi_i}{\partial \boldsymbol{\alpha}(t, \theta)} \qquad (24.44)$$

It should be noted here that the proper interpretation should be given to the derivatives appearing in Eq. (24.44). Indeed, these are derivatives with respect to random variables, and the usual mean square calculus does not apply to this case. Recalling that random variables are treated in the present context as elements of a Hilbert space, these derivatives should be understood as directional derivatives (Ghanem, 1999). Moreover, it can be shown that based on Eq. (24.3), the following relation holds between ξ_i, $\alpha(\mathbf{x}, t, \theta)$ and $\psi(\mathbf{x}, t, \theta)$:

$$\xi_i(\theta) = \frac{1}{\sqrt{\lambda_i}} \int_{\beta} \alpha(\mathbf{x}, t, \theta) \psi_i(\mathbf{x}, t) d\mathbf{x} \qquad (24.45)$$

Upon discretization, this last equation becomes

$$\xi_i(\theta) = \frac{1}{\sqrt{\lambda_i}} \boldsymbol{\alpha}^T(t, \theta) \psi_i(t, \theta) \qquad (24.46)$$

so that

$$\frac{\partial \xi_i(\theta)}{\partial \boldsymbol{\alpha}(t, \theta)} = \frac{1}{\sqrt{\lambda_i}} \psi_i^T(t, \theta) \qquad (24.47)$$

Moreover, using Eq. (24.43) results in,

$$\frac{\partial \mathbf{S}}{\partial \xi_i} = \sum_{j=0}^{P} \frac{\partial \Psi_j}{\partial \xi_i} . \mathbf{S}_j \qquad (24.48)$$

where the terms $(\partial \Psi_j)/(\partial \xi_i)$ can be readily evaluated based on tables similar to Table 24.1, and $\mathbf{S}_j$ has been evaluated using the SFEM approach described in Section 24.4.1. Finally, Eq. (24.44) is written as

$$\frac{\partial \mathbf{S}(t, \theta)}{\partial \boldsymbol{\alpha}(t, \theta)} = \sum_{j=0}^{P} \sum_{i=0}^{M} \frac{\partial \Psi_j}{\partial \xi_i} \mathbf{S}_j(t) \psi_i^T(t, \theta) \qquad (24.49)$$

In this equation, the quantity $\mathbf{S}_j(t)\psi_i^T(t, \theta)$ is deterministic while the quantity $(\partial \Psi_j)/(\partial \xi_i)$ is a random variable. An element $(\partial \mathbf{S}_i(t, \theta))/(\partial \boldsymbol{\alpha}_j(t, \theta))$ in this Jacobian matrix represents the sensitivity of the solution $\mathbf{S}_i(t, \theta)$ at node i with respect to the hydraulic property $\boldsymbol{\alpha}_j(t, \theta)$ at node j. Given the spatial distribution of the uncertainty in the material property, $\boldsymbol{\alpha}(t, \theta)$, Eq. (24.49) can be used to determine the location of the solution $\mathbf{S}_i(t, \theta)$ that is least sensitive to this uncertainty. Also, spatial locations for sampling the material property $\boldsymbol{\alpha}(t, \theta)$ can be determined in such a manner that they contain the most information regarding the solution process at a given point. These correspond to nodal values associated with a large mean value, and a small variance, of the Jacobian in Eq. (24.49). From Eq. (24.49), statistics and confidence intervals are developed for the optimal sampling locations. This is accomplished by simulating the right-hand side of the equation, a very easy task now that the deterministic coefficients have been calculated. From these simulations, large sample statistics of the sensitivity coefficients can be calculated. Moreover, Eq. (24.49) can be used in an adaptive sampling scheme. Specifically, as predictions of $\mathbf{S}(t)$ are made in time, the location of the optimal samples is likely to change. This change will always be predicted by Eq. (24.49).

The optimal sampling locations, both of the material properties as well as of the solution process, are useful for identifying a reduced model of the overall system.

24.6 Conclusions

As science gets more sophisticated in its probing of nature, either through more efficient computing or more powerful sensing, two key issues come to the fore. The first of these addresses model validation and certification: given the present state of certainty in the data, how much confidence is justified in the prediction of the model? The second question addresses the worth of information: given a certain level of resources, how should they be allocated to maximize their effect on the confidence in the predictability of the model? Having models the confidence in which predictions is quantifiable and controllable is a significant shift in paradigm from deterministic, or even probabilistic, models that could only carry through the forward process of acting on the present state of knowledge and information. This is akin, in some ways, to adaptive mesh refinement in deterministic FEM, whereby the level of approximation is adaptively refined or coarsened as suggested by the analysis. In the present case, mesh refinement would be replaced by data refinement. The format of the solution provided by the Polynomial Chaos expansion is such that analytical manipulations leading, for instance, to optimal sampling strategies, is possible.

Although the methodology presented in this chapter is very computer intensive, it is expected that with the rapid development of computational resources, it will soon become routine to carry through mechanistic-based analyses all the way to the decision-making level.

Acknowledgements

The financial support of the National Science Foundation under Grant No. CMS-9870005 and of the Sandia National Laboratories, NM, under Contract No. BC-4596 is gratefully acknowledged.

References

Akin, J. E. *Application and Implementation of Finite Element Methods*. Academic Press, New York, (1982).

Cameron, R. H. and Martin, W. T. The orthogonal development of nonlinear functionals in series of Fourier–Hermite functionals. *Ann. Math*, **48**, (1947) 385–392.

El-Mestkawy, M. *Discrete Element Simulation of Soil Liquefaction under Cyclic Loading. PhD Dissertation*, Department of Civil Engineering, The State University of New York, Buffalo, NY, (1998).

Ghanem, R. and Spanos, P. *Stochastic Finite Elements: A Spectral Approach*. Springer-Verlag, Berlin, (1991).

Ghanem, R. and El-Mestkawy, M. Discrete element analysis of random packing of randomly shaped granules. *ICOS-SAR'97, Kyoto, Japan, November 24–28 (1997)*.

Ghanem, R. and Dham, S. Stochastic finite element analysis for multiphase flow in heterogeneous porous media. *Transport in Porous Media*, **32**, (1998) 239–262.

Ghanem, R. and El-Mestkawy, M. Efficient statistically equivalent representations of random packings with random shapes. *ASCE 12th Engineering Mechanics Conference*, La Jolla, CA, May 17–20 (1998).

Ghanem, R. Scales of fluctuation and the propagation of uncertainty in random porous media. *Water Resources Res.*, **34**(9), (September 1998) 2123–2136.

Ghanem, R. and Red-Horse, J. Propagation of uncertainty in complex physical systems using a stochastic finite element approach. *Physica D* (1999, to appear).

Ghanem, R. Ingredients for a general purpose stochastic finite elements formulation. *Comput. Methods in Appl. Mechanics and Engrg.*, **168**(1–4), (1999) 19–34.

Kallianpur, G. *Stochastic Filtering Theory*. Springer-Verlag, Berlin, (1980).

Loeve, M. *Probability Theory, 4th ed.* Springer-Verlag, New York, (1977).

Poincaré, H. *Science et Hypothèse* (1905).

Wiener, N. The homogeneous chaos. *Am. J. Math*, **60**, (1938) 897–936.

25

Natural Slopes in Slow Movement

Laurent Vulliet
Swiss Federal Institute of Technology (EPFL), Lausanne, Switzerland

25.1 Introduction

Natural hazards and natural disaster prevention and mitigation are becoming major issues in a world facing a marked change in attitude towards nature. In particular, the degree of acceptance of natural hazards is continuously decreasing in our modern society, and to permit preventive measures to be taken, scientists are asked to produce models which cannot only explain the physical nature of these events, but also predict their occurrence. This is particularly the case in the area of landslides.

In geotechnical engineering research, the main efforts in the domain of natural slopes have been concentrated on *stability analysis*. First with simple kinematic models, then with more sophisticated geometry, but always with a rigid-perfectly plastic body assumption, the question was: What is the global safety factor of that particular slope? The exercise turned into a simple classification problem with two categories: stable or unstable slopes.

With this rather simple approach, it is, however, possible to make some predictions and testing, for example, the influence of boundary conditions and material parameters on global stability.

A notable improvement was offered by *probabilistic methods* that allow consideration of the natural variability of input parameters and the probability of failure. It extends the classes of the above-mentioned classification, and permits the safety level of a slope to be quantified.

Numerical techniques (mostly *finite elements*) and advanced material modeling (*elasto-plasticity*) later came to better model the behavior of soil/rock masses. In the case of landslide, *ad hoc* techniques were to be implemented to assess the global stability of a slope.

In case of *slowly-moving landslides*, however, the type of problem changes completely. Here the notion of stability loses its sense, and the question is not one of evaluating a stability factor, but of modeling a velocity field. Slowly-moving natural slopes show characteristic displacement velocity history, with periods of calm (a few millimeters or centimeters per year) followed by periods of acceleration (up to several meters per day) with, in general, strong but sometimes indirect correlation with rainfall. In some cases, movements degenerate into fast-moving (catastrophic) landslides.

The models pertaining to the prediction of such slow-moving landslides are still rare, and not much effort has been devoted in the geomechanics and geotechnical community to improve the state of the art. These models can be classified into two main categories:

- *mechanical models* using the principles of continuum mechanics to propose well-posed boundary value problems,

- *statistical models* using existing recorded data to propose extrapolation (prediction) algorithms.

In this chapter, three families of model which fit into both of the above-mentioned categories and were developed by the author and co-workers will be reviewed: two nonlinear viscous three-dimensional (3-D) models, and an elasto-viscoplastic two-dimensional (2-D) model concerning the so-called 'mechanical models', and a neural network model concerning the second class of 'statistical models'.

The treatment of the subject covers some of the mathematical analysis, the numerical techniques, the parameter determination and real case studies.

25.2   Landslide as a 3-D Viscous Flow

25.2.1   *Incompressible one-phase body*

Main idea

There is a strong analogy between slow-moving slopes and glaciers: similar shape, similar time history (including 'surges'), same accumulation in the upper part due to the collapse of surrounding materials, same erosion in the lower part due to river erosion, and similar velocity profiles with depth.

As a consequence, the 3-D viscous model presented below was developed by the author, in collaboration with Professor K. Hutter, an extension of existing models in the field of glaciology (Hutter, 1983). The general formulation of the models for natural slopes in slow movement is given in Hutter and Vulliet (1985), Vulliet (1986a), and Vulliet and Hutter (1988a); constitutive modeling is treated more specifically in Vulliet and Hutter (1988b,c,e); the special case of a multilayered sliding mass is presented in Vulliet and Hutter (1988d); some considerations on the numerical models are given in Vulliet (1986b, 1995).

General formulation

In this first model, a nonlinear viscous gravity-driven *incompressible one-phase body* is considered. It corresponds to a 'total stress analysis' in soil mechanics. The associated geometry is shown in Figure 25.1.

Let ρ and v_i be the soil density and the soil velocity, σ_{ij} the total stress tensor, and g_i the gravity vector; then the conservation of mass and momentum equations can be

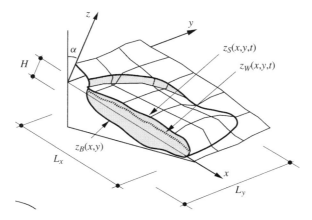

Figure 25.1   Three-dimensional view and section of a landslide, showing the inclined-coordinate system and some dimensional parameters

expressed as

$$v_{i,j} = 0 \qquad (25.1a)$$

$$\sigma_{ij,j} + \rho g_i = 0 \qquad (25.1b)$$

where the incompressibility assumption implies $\rho =$ const. Kinematic equations for the stress-free surface of the terrain are presented in detail in Vulliet and Hutter (1988a). We will only focus here on the constitutive laws.

Postulating a nonlinear viscous behavior for a Reiner–Rivlin isotropic fluid, the stress-strain rate relation can be written in a generic form as

$$D_{ij} = f(J_{2D})t_{ij} \qquad (25.2)$$

where D_{ij} is the stretching tensor, t_{ij} is the deviatoric part of the stress tensor,

$$D_{ij} = \tfrac{1}{2}(v_{i,j} + v_{j,i}), \quad t_{ij} = \sigma_{ij} - \tfrac{1}{3}\sigma_{kk}\delta_{ij} \qquad (25.3)$$

and $f(\cdot)$ is a nonlinear function of J_{2D} the second deviatoric invariant* of t_{ij}:

$$J_{2D} = \tfrac{1}{2}t_{ij}t_{ij} \qquad (25.4)$$

Assuming a constant value of $f(\cdot)$ would yield a Newtonian fluid that does not correspond to the real behavior of soil in general. Expressions of $f(\cdot)$ are given in Table 25.1, based on extensive laboratory tests and *in situ* measurements.

The soil mass can also slide along the base (or 'sliding surface'). The dynamic condition at the base can be expressed by the following constitutive relationship at the base (interface behavior):

$$v_{B_i} = F_B(\tau_B)\tau_{B_i} \qquad (25.5)$$

---

* The invariant notation from Desai and co-workers (1980, 1984, 1986, 1987, 1991, 1995) is used here.

Table 25.1   Creep response functions

Creep law	$f(J_{2D})$, Eq. (25.2)	Note	Value[1]
Newton	$1/\mu$		$\mu = 5.3 \cdot 10^9 \, kNm^{-2} \cdot s$ (1)
			$\mu = 8.06 \cdot 10^9 \, kNm^{-2} \cdot s$ (2)
Bingham	$[1 - \sqrt{J_{2D0}/J_{2D}}]/\mu$	when $J_{2D} > J_{2D0}$	
Norton (power law)	$A(J_{2D})^{(m-1)/2}$		$m = 5.3, A = 6.7 \cdot 10^{-17}$ $(kNm^{-2})^{-5.3}s^{-1}$ (1)
Prandtl–Eyring	$[A\sinh(B\sqrt{J_{2D}})]/\sqrt{J_{2D}}$	$\to AB, if \sqrt{J_{2D}} \to 0$	
Prandtl	$[A\exp(B\sqrt{J_{2D}})]/\sqrt{J_{2D}}$	$\to B/\sqrt{J_{2D}}$, if $\sqrt{J_{2D}} \to 0$	
Newton-Norton	$1/\mu + A(J_{2D})^{(m-1)/2}$	$\to 1/\mu, if J_{2D} \to 0$	$\mu = 1.4 \cdot 10^{10} \, kNm^{-2}s$, $m = 9.6$, $A = 1.2 \cdot 10^{-23}$ $(kNm^{-2})^{-9.6}s^{-1}$ (1)

[1] based on field measurements (landslide names indicated: 1 = Villarbeney, 2 = Chloewena). Values not always related to best fit.

where v_{B_i} is the sliding velocity at the base, and τ_{B_i} is the stress vector tangential to the base and τ_B its norm. Note that, in principle, the sliding velocity in this incompressible case cannot be a function of the normal stress at the base, σ_B. However, to express the effect of consolidation (i.e. overburden pressure) on the sliding properties, σ_B dependence can be implemented. This procedure would be similar to the 'consolidated undrained' approach with undrained strength parameter.

Several explicit expressions for the sliding law $F(\cdot)$ are presented in Table 25.2.

Asymptotic development and numerical model

The 'shallow mass' approximation similar to that used in glaciology or river hydraulics can be used with great advantage here. It has been shown (see Figure 25.2) that, in general, the thickness H of most slowly-moving landslides is much smaller than their length L_x

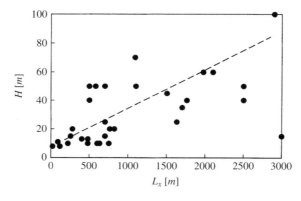

Figure 25.2   Typical dimensions of natural slopes in slow movement (depth H versus length L_x); data from Swiss landslides

Table 25.2 Sliding laws

Sliding law	$F_B(\tau_B)$, Eq. (25.5)	Note	Value[2]
Weertman	$B\tau_B^{n-1}$		$B = 3.24\ 10^{-9}$ ms^{-1}kPa^{-2}, $n = 2$ (1)
Power law with threshold	$B\dfrac{(\tau_B - \tau_{Bo})^n}{\tau_B}$	$\begin{cases} if\ \tau_B > \tau_{Bo} \\ = 0\ \text{otherwise} \end{cases}$	$B = 1.74\ 10^{-6}$ m^{-1}kPa^{-1}, $n = 1$, $\tau_0 = 80$ kPa (1)
Threshold function[1] of σ_B'	$B\dfrac{\Gamma^n}{\tau_B}$	$\begin{cases} if\ \Gamma > 0 \\ = 0\ \text{otherwise} \end{cases}$	
Stress ratio dependance	$B\dfrac{\tau_B^{n-1}}{(c' + \sigma_B' \tan\phi')^n}$		$B = 2.16\ 10^{-8}$ m/s, $n = 15$, $c' = 0$, $\phi' = 9.5°$ (2)
Stress ratio and cohesionless interface	$B\dfrac{\tau_B^{n-1}}{(\sigma_B')^n}$		$B = 1.27\ 10^{-9}$ m/s, $n = 1$ (3) $B = 2.6\ 10^{-6}$ m/s, $n = 2$ (4)
Hyperbolic	$\dfrac{C}{c' + \sigma_B' \tan\phi' - \tau_B} - \dfrac{CD}{\tau_B}$		$C = 1.22\ 10^{-9}$m/s, $D = 5$, $c' = 0$, $\phi' = 9.5°$ (2)

[1] $\Gamma \equiv \tau_B - c' - \sigma_B' \tan\phi'$ in the Mohr–Coulomb case.
[2] Landslide names are: (1) Chloewena, (2) Sallédes, (3) Le Day, (4) La Frasse.

(downslide) or width L_y (perpendicular to the slope). A rigorous treatment of the set of general equations introduces dimensionless quantities that can be evaluated. If lower order terms (smaller than unity) are neglected, the set of equations is drastically reduced and permits a rather simple numerical solution of the 3-D case (Hutter and Vulliet, 1985; Vulliet and Hutter, 1988a).

One main outcome of the asymptotic development is to uncouple the determination of the state of stress and velocity to the free surface evolution. Once stresses and velocities are determined everywhere in the moving soil mass, the free surface evolution equation can be used to determine the evolution of the position of the free surface, $z_S(x, y, z, t)$; this evolution equation takes the form of a kinematic wave equation.

The set of simplified equations is solved using finite differences (VITAL1 programme: Vulliet, 1986a,b). Different time integration schemes (explicit or implicit) are possible with different degrees of numerical stability (see Vulliet, 1995).

Determination of material parameters

Material parameters entering in the explicit forms of the creep law $f(\cdot)$, Eq. (25.2), and the sliding law $F_B(\cdot)$, Eq. 25.5 can be determined from laboratory creep experiments or *in situ* measurement. In the laboratory, only short-term (a few weeks, rarely a few months) creep tests are possible (Vulliet and Hutter, 1988b). It is questionable as to how representative these tests are of the real behavior of natural slopes already in movement for decades or centuries.

In situ measurements of velocity profiles (from inclinometer readings) also permit the determination of material parameters. However, in this case, a strong assumption is needed for the determination of the stress state, as this state can hardly be measured. It is thought, however, that a sound physically-based *estimation* of these quantities does not affect the final result to any higher degree than other simplifications and assumptions (e.g. the geometry of the sliding surface).

As a first approximation, a natural slope at the location of the inclinometer (see the typical geometry in Figure 25.1) can be seen as experiencing simple shear. In this case, Eq. (25.2) reduces to

$$\tfrac{1}{2}\dot{\gamma}(z) = f(\tau)\tau(z) \tag{25.6}$$

where $\dot{\gamma}$ is the shear strain rate obtained from two consecutive readings, and $\tau(z)$ is the shear stress that can be expressed as

$$\tau(z) = \rho g(z_s - z)\sin\alpha_{xy} \tag{25.7}$$

In Eq. (25.7), $(z_s - z)$ represents the depth measured along the normal to the free surface, and α_{xy} is the *local* slope inclination at the location (x, y) of the inclinometer (averaged over a length equal to the landslide thickness).

As a consequence, the functional $f(\cdot)$ can be evaluated for a given slope. Figure 25.3 shows such relations for the landslides of Villarbeney and Chloewena in Switzerland. Some numerical values of the material parameters entering in the creep law are given in Table 25.1.

Along the base, the procedure is similar, and τ_B in Eq. (25.2) is only evaluated along the steepest descent from Eq. (25.7) at the base, i.e. $z = z_B$. Figure 25.4 shows plots of stress level as a function of sliding velocity for the landslides of Sallédes in France and Le Day in Switzerland; it permits the determination of $F_B(\cdot)$. Some numerical values of the material parameters entering in the sliding law for this case are given in Table 25.2.

Case studies

Several real landslides have been studied by this model (see Vulliet 1986; Vulliet and Hutter, 1988a,b,c; Vulliet, 1995; Vulliet and Bonnard, 1996).

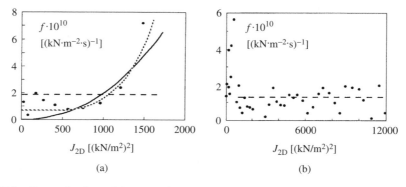

Figure 25.3 Determination of the creep law based on inclinometric measurement from (a) Villarbeney, and (b) Chloewena landslides

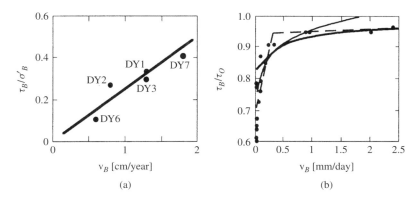

Figure 25.4   Determination of the sliding law based on measurement from (a) Le Day, and (b) Sallédes landslides (the reference stress τ_o is given by a Mohr–Coulomb criterion)

Figure 25.5 shows the result for the case of La Frasse. This landslide lies in the Swiss alps, 60 km east of Lausanne. It covers an area 2000 m long by 400 m wide; it is about 60 m thick and has a mean slope angle of approximately 13°. The geology consists mainly of clayey flisch and marl from the Eocene period. La Frasse landslide has been intensively studied and instrumented so that the geometry of the free surface is known, and the location of the phreatic and basal surfaces has been determined for some points. The complete three-dimensional shape of the base has been constructed by successive interpolation between known points. Here, the long-term (120 years) surface velocity field was used, which was obtained by comparing old and new maps (Engel, 1986). For computation, the domain of the landslide was divided into a rectangular grid with mesh size of $\Delta x = 100$ m and $\Delta y = 50$ m; the total number of nodes amounted to 259. A nonlinear sliding law was employed with phenomenological constants, as listed in Table 25.2, determined by back-calculation on a 2-D profile along the steepest descent. As can be seen in Figure 25.5, the agreement between calculated and measured surface velocities is satisfactory. The direction of the velocity vectors are well reproduced, as are their magnitudes in most cases.

Figures 25.6 and 25.7 present results obtained for the Chlöwena landslide. A reactivation phase occurred in the spring and summer 1994 and dammed the Höllbach River at its toe at the beginning of August 1994. Located near the town of Freiburg, in the Swiss Prealps, this large slide, involving some 40 mio m³ of superficial deposits (mainly from Flysch), formed a small temporary lake which could have increased in volume according to the final size of the natural dam, thus causing a major threat to the industrial town of Marly downstream. The active landslide mass rests in its upper part on Gurnigel Flysch composed of alternate sequences of sandstones and shales and in its lower part on an 18% sloping compacted moraine composed of silt, gravel and Flysch debris. The depth of the slide (max 40 m) has been assessed on the basis of three boreholes equipped with inclinometers. The main goal of the analysis was to predict the progression of the landslide mass with time and give possible dam height values. This is a typical 'class A' prediction. Undrained condition was assumed to better represent the paroxysmic phase of summer 1994. Direct shear tests and *in situ* vane shear tests indicated an undrained cohesion varying between 20 and 80 kPa. Using information given by the inclinometer situated in

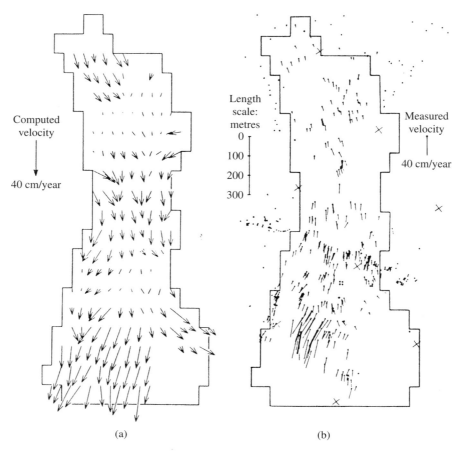

Figure 25.5 Simulation of the La Frasse landslide (a) and comparison with (b) velocity measurements

the middle part of the landslide and a threshold stress $\tau_o = 80\,\text{kPa}$, the parameter B of the sliding law (a power law with threshold, see Table 25.2) was determined for an exponent $n = 1$. The total length of the mesh is 2300 m, with a mesh size of $\Delta x = 100\,\text{m}$ and $\Delta y = 50\,\text{m}$. It can be seen in Figure 25.6 that the velocities at time $t = 0$ are larger in a region situated 500 m above the village of Falli-Hölli; this is what was observed during the initial period of movements. After 40–60 days, the sliding material is following the Höllbach River Bed. The calculated period was 200 days, and although the mass was still moving slightly, no major change in the dam geometry could be seen. Figure 25.7 shows the predicted dam geometry, with a total height of about 40 m, as communicated to the crisis staff headquarters by August 15 1994. At the end of August 1994, the dam started to form, and coarse measurement of its geometry was done by a forest engineer and reported regularly (see Figure 25.7). Considering the crude assumptions used and the fact that the movements continued slightly in 1995, the prediction was acceptable.

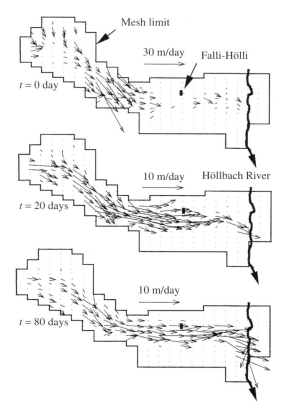

Figure 25.6   Simulation of the Chloewena landslide showing the evolution of the velocity field with time; this computation served for prediction of the geometry of the dam being formed at the bottom of the valley due to the slope movement (total displacement more than 200 m, velocity up to 6 m/day)

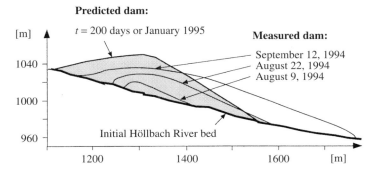

Figure 25.7   Prediction of the geometry of the dam forming at the bottom of the valley of the Chloewena landslide (*a priori* prediction compared with later measurements)

25.2.2 Compressible two-phase body

Generalities

It is well established that landslides are strongly influenced by groundwater seepage conditions. A model that would allow for such influence must be written either in effective stress and one-phase flow, if uncoupling water flow/solid flow is reasonable, or as a mixture of fully-coupled solid and fluid mass. The second approach has been developed as an extension of the previous 3-D one-phase flow model.

General assumptions

In this second model, a nonlinear viscous gravity-driven *compressible-two phase body* is considered. Equations of conservation of mass and momentum are written for both phases, and coupling is produced by Darcy's law and compressibility of the soil skeleton. The true densities of water and mineral grains are assumed to be constant (i.e. water and solid grains are incompressible). However, the mixture is compressible, since porosity changes are possible. The mass balance equations for both constituents given by

$$\frac{\partial n}{\partial t} + (n v_i^w)_{,i} = 0 \tag{25.8a}$$

$$-\frac{\partial n}{\partial t} + [(1-n)v_i^w]_{,i} = 0 \tag{25.8b}$$

while the momentum balance equation can be written as (neglecting inertia terms):

$$-(1-n)p_{w,i} + \sigma'_{ij,j} + \rho_s g_i + \frac{n\rho_w g}{k}(v_i^w - v_i^s) = 0 \tag{25.8c}$$

$$-n p_{w,i} + \rho_w g_i - \frac{n\rho_w g}{k}(v_i^w - v_i^s) = 0 \tag{25.8d}$$

where p_w is the pore water pressure, k the relative (Darcy) coefficient of permeability, ρ the total bulk mass density, and ρ_s and ρ_w are the partial densities of soil grains and water, respectively.

To account for the fact that the sliding mass may deform within its body, several approaches are possible. A reasonable approach is to exclude viscous isotropic deformations, and to relate porosity changes to a non-linear elastic response:

$$D_{ii} = -0.435 C_c (1-n) \frac{\dot{p}'}{p'} \tag{25.9a}$$

$$D_{ij} - \frac{1}{3} D_{kk} \delta_{ij} = b t_{ij} \tag{25.9b}$$

$$b = \tilde{b}(J_{2D}, n) \tag{25.9c}$$

For the sliding at the base, a proposed modified version of Eq. (25.2) is expressed as:

$$v_{B_i} = F'_B(\sigma'_B, \tau_B)\tau_{B_i} \tag{25.10}$$

Some explicit expressions of F' are given in Table 25.2.

Note that the complete formulation here involves two free surfaces: the terrain surface, and the groundwater (see the cited references). Asymptotic development and numerical modeling will not be treated here.

Evaluation of material parameters

The same procedure as described in Section 25.2.1 can be applied here. The main difference is that pore water pressure p_w has to be taken into account to evaluate the local state of stress. The *in situ* measurement of p_w can easily be made by piezometers and pore pressure transducers. Considering again the sliding at the inclinometer position as a simple shear, explicit forms of material functions and values of material parameter can be found based on Eq. (25.9).

Application

Study of landslides in the field, as well as on academic exercise, shows the strong interdependence between groundwater seepage and sliding velocity.

25.2.3 Comments on the viscous flow models

The proposed viscous models permit a three-dimensional(3-D) analysis of time-dependent behavior of natural slopes in slow movement, in the case where the thickness is small in comparison with other dimensions. Multi-sliding surfaces can be treated and groundwater flow is taken into account with a full hydromechanical coupling. The scaling and asymptotic development appears as a very useful tool to reduce the complexity of the problem; with this approach, it is possible to develop lower order models and simple numerical algorithms based on finite differences.

The major limitation of these models is connected with the constitutive law. Considering soil as a nonlinear viscous body, even compressible and nonlinear elastic volumetric behavior, excludes plastic effects. Thus, plastic stress redistribution is not reproduced. Further, the shallowness assumption implies that the calculated velocity field is not adequate close to the lateral limits of the mass; local effects like retaining walls, anchors, etc., cannot be modeled.

25.3 Landslide as an Elasto-Viscoplastic 2-D Body

25.3.1 Overview

To overcome some of the limitations of the viscous flow models, an elasto-viscoplastic model has been developed. Most of the work was accomplished during the author's stay at the University of Arizona, as part of a joint project with Professor C. S. Desai and co-workers.

A version of the Hierarchical Single Surface (HiSS) family of constitutive models is formulated here for characterizing the elastoviscoplastic behavior of soils and interfaces. Various versions, allowing for associative, nonassociative, anisotropic hardening, disturbance or damage and fluid flow pressure have been presented elsewhere (Desai *et al.*, 1986; Desai, 1990).

The present elasto-viscoplastic version, referred to as δ_{vp}, allows for elastic, plastic and viscous creep deformations, continuous yielding or hardening, volume changes, and stress-path effects. The viscous part of the model is based on the theory of elasto-viscoplasticity by Perzyna (1966). First developments of the δ_{vp} are reported by Desai and Zhang (1987), then by Vulliet and Desai (1989), Samtani and Desai (1991), Desai, Samtani and Vulliet (1995), Samtani (1990), Vulliet, Samtani and Desai (1991), and Samtani, Desai and Vulliet (1991, 1994, 1996).

25.3.2   General assumptions and constitutive laws

The usual decomposition of a strain tensor into elastic and viscoplastic components, assuming that strains are small, can be expressed as

$$d\varepsilon_{ij} = d\varepsilon_{ij}^e + d\varepsilon_{ij}^{vp} \tag{25.13}$$

with

$$d\varepsilon_{ij}^e = C_{ijkl}^{e-1} d\sigma_{kl}' \tag{25.14}$$

and

$$d\varepsilon_{ij}^{vp} = \langle\Phi\rangle \frac{\partial F}{\partial \sigma_{ij}'} \tag{25.15}$$

where

$$
\begin{cases}
\Phi = 0 & \text{when } F \leq 0 \tag{25.16a}\\[2mm]
\Phi = \Gamma \left(\dfrac{F}{F_o}\right)^N & \text{when } F > 0 \tag{25.16b}
\end{cases}
$$

where σ_{ij}' is the effective stress tensor, C_{ijkl}^e the elastic constitutive tensor, and Φ is a scalar flow function of its argument F, the yield function. F_o is a normalized constant having the same unit as F, and Γ and N are a material (fluidity) parameter and a material constant, respectively. The angle bracket $\langle\rangle$ has the meaning of a switch-on-switch-off operator.

In Eq. (25.16), the viscoplastic flow rule is expressed as a power law, in a similar way as the Norton law presented in the viscous model. Other laws are possible, for example, an exponential law (Samtani, 1990). The yield surface F is expressed through the HISS model as (Desai *et al.*, 1986):

$$F = J_{2D} - (-\alpha J_1^n + \gamma J_1^2)(1 - \beta S_r)^{-0.5} = 0 \tag{25.17}$$

where $S_r = (\sqrt{27}/2)J_{3D}J_{2D}^{-1.5}$ is a stress ratio, β, γ, n are material parameters, J_1 is the first invariant of the effective stress tensor σ_{ij}', and J_{2D} and J_{3D} the second and third invariant of the deviatoric stress tensor, respectively. Note that the invariants are normalized with respect to the atmospheric pressure, p_a. The hardening function α is expressed as

$$\alpha = \frac{a_1}{(\xi_{vp})^{\eta_1}} \tag{25.18}$$

where $\xi_{vp} = \int (d\varepsilon_{ii}^{vp} d\varepsilon_{kk}^{vp})^{1/2}$ is the length of viscoplastic volumetric strain trajectory, and a_1 and η_1 are the hardening parameters.

A similar model has been developed for interfaces (Samtani, 1990; Samtani, Desai and Vulliet, 1996), where the yield function specializes to (Desai and Fishman, 1991)

$$F_i = \tau^2 + \alpha_i(\sigma'_n)^{n_i} - \gamma_i(\sigma'_n)^2 = 0 \tag{25.19}$$

where τ and σ'_n are the shear and normal effective stress at the interface, respectively, normalized with respect to the atmospheric pressure, p_a. The hardening function is given by

$$\alpha_i = \frac{\gamma_i \exp(-a\xi^v_{vp})}{(1 + \xi^d_{vp})^b} \tag{25.20}$$

where a and b are material parameters for the interface. ξ^v_{vp} and ξ^d_{vp} are the lengths of viscoplastic volumetric and deviatoric strain trajectories, respectively.

Based on laboratory tests, it was found appropriate to characterize the behavior of interfaces using nonassociative plasticity. Hence, the plastic potential function Q_i was defined as

$$Q_i = \tau^2 + \alpha_Q(\sigma'_n)^{n_i} - \gamma_i(\sigma'_n)^2 = 0 \tag{25.21}$$

with α_Q expressed as a function of α for associative plasticity as

$$\alpha_Q = \alpha + \alpha_p\left(1 - \frac{\alpha}{\alpha_i}\right)^\kappa \tag{25.22}$$

where α_i is the value of α at the beginning of shear loading, α_p is the value of α_Q at the peak shear stress, and κ is the nonassociative parameter.

Finally, the viscoplastic strain rate increment for the interface is given by

$$\left\{\begin{matrix} d\varepsilon^n_{vp} \\ d\varepsilon^s_{vp} \end{matrix}\right\} = \langle\Phi_i\rangle \left\{\begin{matrix} \dfrac{\partial Q_i}{\partial \sigma'_n} \\ \dfrac{\partial Q_i}{\partial \tau} \end{matrix}\right\} \tag{25.23}$$

where $\varepsilon^n_{vp} = v_r/d$ and $\varepsilon^s_{vp} = u_r/d$ are the normal and shear viscoplastic strains, respectively, v_r and u_r are, respectively, the normal and tangential relative displacement jump at the interface, and d is the interface thickness. A possible explicit form of the flow law Φ_i is proposed as

$$\begin{cases} \Phi_i = 0 & \text{when } F_i \leq 0 \tag{25.24a} \\[2mm] \Phi_i = \Gamma_i\left(\dfrac{F_i}{F_{i0}}\right)^{N_i} & \text{when } F_i > 0 \tag{25.24b} \end{cases}$$

where Γ_i and N_i are interface viscous parameters.

25.3.3   Evaluation of material parameters

Methods for finding the elastic and plastic parameters of the constitutive model presented above from laboratory experiments are described in detail by Desai *et al.* (1986), Desai (1990), and Samtani (1990). A brief description of the method used for finding the viscous parameters Γ, F_0 and N is given below (Samtani, 1990; Desai, Samtani and Vulliet, 1995).

Squaring the expression for the viscoplastic strain rate increments (Eq. 25.15), the following relation is obtained:

$$\frac{1}{2} d\varepsilon_{ij}^{vp} d\varepsilon_{ij}^{vp} \equiv dI_2^{vp} = (\langle\phi\rangle)^2 \frac{1}{2} \frac{\partial F}{\partial \sigma_{ij}'} \frac{\partial F}{\partial \sigma_{ij}'} \tag{25.25}$$

where dI_2^{vp} is the second invariant of $d\varepsilon_{ij}^{vp}$. Then

$$\langle\Phi\rangle \equiv \chi = \sqrt{\frac{dI_2^{vp}}{\frac{1}{2} \frac{\partial F}{\partial \sigma_{ij}'} \frac{\partial F}{\partial \sigma_{ij}'}}} \tag{25.26}$$

Hence, postulating the power law given Eqs. (25.16a) and (25.16b) yields

$$\Gamma \left(\frac{F}{F_o}\right)^N \equiv \chi \tag{25.27}$$

which, on taking the natural logarithm on both sides, gives

$$\ln \Gamma + N \ln \left(\frac{F}{F_o}\right) = \ln \chi \tag{25.28}$$

The value of F/F_0 and χ are computed from laboratory tests at different points, a plot of $\ln(F/F_0)$ versus $\ln(\chi)$ is obtained, and the values of Γ and N are found from the intercept and the slope of the regression line (see Figure. 25.8).

Clay from the Villarbeney landslide in Switzerland was tested at the EPFL and at the University of Arizona. The material parameters obtained from these tests are given in Table 25.3. Figure 25.9 shows a comparison between the experimental and the model prediction (Desai, Samtani and Vulliet, 1995). Overall, an excellent correlation is observed.

25.3.4   *Numerical model*

The model was implemented in an existing finite element code used by the group of Professor Desai at the University of Arizona (SSTIN programme for soil-structure interaction). Eight-noded isoparametric elements were used to discretize the solid body, while six-noded isoparametric elements were used to represent interfaces. The algorithm is presented in detail in Owen and Hinton (1980), and Desai and Zhang (1987), and Desai, Samtani and Vulliet (1995).

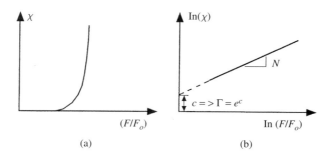

Figure 25.8   Viscous parameters

Table 25.3   Parameters for the elasto-viscoplastic model determined for the Villarbeney clay

Soil			Interface		
Parameter	Symbol	Value	Parameter	Symbol	Value
Modulus	E	10,400 kPa	Normal stiffn.	k_n	$8 \cdot 10^8$ kPa/m
Poisson's ratio	ν	0.35	Shear stiffness	k_s	$2.8 \cdot 10^5$ kPa/m
Ultimate	γ	1.93	Ultimate	γ_i	0.24
Ultimate	β	0.64	–	–	–
Transition	n	2.04	Transition	n_i	2.04
Hardening	a_1	1.47	Hardening	a	143
Hardening	η_1	0.06	Hardening	b	10
Nonassociative	–	–	Nonassociative	κ	0.57
Fluidity	Γ	$1.5 \cdot 10^{-4}$ min^{-1}	Fluidity	Γ_i	$5.7 \cdot 10^{-2}$ min^{-1}
Exponent	N	2.58	Exponent	N_i	3.15

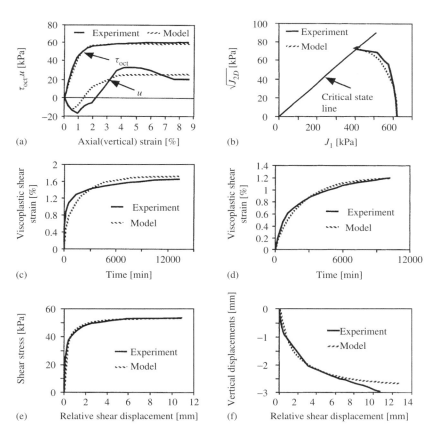

Figure 25.9   Material parameter determination for the elasto-viscoplastic model (Villarbeney clay): RTE undrained test with $\sigma_3 = 200$ kPa (a) and (b), simple shear creep tests with a shear-to-normal-stress ratio of 0.6 and preconsolidation stress of 200 kPa (c), and 400 kPa (d), interface test with a constant normal stress of 100 kPa (e) and (f)

25.3.5 Case studies

Two landslides are studied with the visco-elastoplastic model. Figure 25.10 is an illustration in the case of La Frasse (Vulliet and Desai, 1989), a landslide already presented above in the context of the viscous models. Here a 2-D analysis is performed using 132 isoparametric eight-node elements and a total of 491 nodes. As in Vulliet and Hutter (1988a), the groundwater level is assumed to coincide with the free ground surface. Since the displacement of the soil mass is characterized by strong shear deformations in the vicinity of the sliding surface (base) of the landslide, two materials are considered, a weaker one used for the 5 m thick solid elements used to represent the sliding surface zone. It can be seen in Figure 25.10 that the results are consistent with measured data and previous viscous analysis. In the lower part of the landslide ($x > 1500$ m) the calculated velocities are underestimating the observed behavior, mainly because inhomogeneity and artesian water pressures in the lower part were neglected in the computation.

Figure 25.11 shows the results of the visco-elastoplastic computations in the case of the Villarbeney landslide (Desai, Samtani and Vulliet, 1995; Samtani, Desai and Vulliet, 1996). This landslide is located south of Freiburg in Switzerland, and is mainly composed of clayey soil. Piezometric measurements indicated a groundwater level situated at about 2 m below the ground surface, relatively stable during the time period of about one year considered here. Hence, the influence of steady state seepage forces for depths below 2 m was considered in the finite element analysis as imposed (and uncoupled) body forces. In addition, surface traction on the toe was imposed, assuming the traction to be 75% of the limiting Coulomb passive earth pressure. The profile corresponds to the location of borehole E1 situated in the middle of the landslide, with an average base slope angle of 14° and a depth of movements limited to 17 m. Since no major lateral spreading of the slide can be observed at this location, the assumption of plane strain is acceptable along this section. The total discretized length is 150 m. The mesh consists of 60 eight-noded elements with 202 nodes, including 12 interface elements. Velocity profile obtained after one year of measurements compares relatively well with computational results corresponding to steady viscoplastic strain rates.

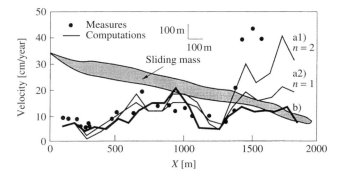

Figure 25.10   Velocity distribution along the axis of the La Frasse landslide: measurements (dots), viscous flow model with two values of the exponent n in the power law (a), elasto-viscoplastic model (b)

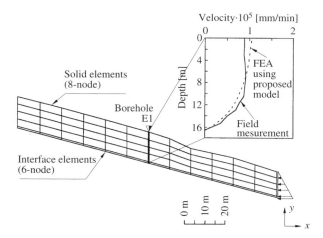

Figure 25.11   Villarbeney landslide: finite element mesh and result of an elasto-viscoplastic computation of velocity profile at borehole E1

25.3.6   *Comments on the elasto-viscoplastic model*

The elasto-viscoplastic model constitutes a useful simulation tool for creeping natural slopes, including sliding at the base (interface). It integrates appropriate constitutive modeling of geomaterials and interfaces, gravity, steady-state seepage, end tractions. It is validated by laboratory and field tests.

Limitations are: (1) the constitutive model can allow for primary and secondary creep, but does not allow for any tertiary creep; (2) the effect of pore fluid pressure is superimposed on the mechanical response, and the full coupling between the fluid and solid phases is not included; (3) it does not allow for a 'catastrophic' landslide with large acceleration and dominant inertia contribution to the momentum; (4) the two-dimensional idealization is valid essentially in the central region of the landslide. (A 3-D extension of the model would, however, be possible); (5) further field measurements are needed to better validate the model.

Compared to the viscous flow model discussed in Section 25.2, the elasto-viscoplastic model offers the advantage of including local effects on the velocity field (anchored wall, existing structure, soil heterogeneity, etc.).

25.4   **Prediction with Neural Networks**

25.4.1   *Main idea*

The above-described models belong to the category of 'hard computing'. They imply a careful deterministic description of the geometry, a choice of constitutive approximations and material parameters, and a good knowledge of existing boundary conditions.

Another approach is presented below that better takes into account the uncertainty of the problem, and treats it as a statistical response of a natural slope to complex solicitation and environment. The following model makes use of neural networks; as such it belongs to the category of 'soft computing'. It is believed that a coupling of both hard and soft computational models can provide a very effective tool for the prediction of slope movements.

Let us consider the velocity evolution of slow-moving landslides as a time series controlled by several parameters: rainfall, pore-water pressures, past displacements and velocities, etc. It is no surprise that statistical models have been developed first using regression analysis, time series or hydrology-derived reservoir models (see, for example, Pouget and Livet, 1994). The performance of such models is good, but the main problem with polynomial functions is to find the mathematical form best suited to a particular site.

Another approach would be to use a *neural network*. If the configuration of the network (i.e. the number of neurons and hidden layers) is be found to be site-independent at least to some extent, what only remain, site-specific is the learning process (determining weights). This makes the model more versatile in terms of predictive capability.

25.4.2 Configuration of the neural network

Detailed explanation of neural networks are given by Haykin (1994). Specific application of the concept by the author and his coworkers is given by Mayoraz *et al.* (1996, 1997).

A multilayer perceptron (MLP) is used here. The structure of the MLP is based on the basic neuron shown in Figure 25.12. A basic neuron with N inputs contains N adjustable weights $w_1, w_2, \ldots, w_N$, and performs a weighted sum of its input, thresholded by a nonlinear function. In MLP algorithms, the threshold function is usually a sigmoid function.

The basic neurons are connected together to form the network of interest, which can have several layers. In our case, only one hidden layer and an output layer of only one neuron are used (see Figure 25.13). The optimal number of hidden neurons is determined empirically, and the minimum number of neurons in the hidden layer was found to be eight.

In this approach, it is assumed that the complete landslide can be seen as a single mass sliding with average velocity v. The input parameters are based on a daily measurement of rainfall, porewater pressure, and velocity. Pre-processing is necessary to obtain the appropriate vector of input parameters, for example as follows:

- sliding average velocity over five days, v_5, at time $t - 1 : v_5(t - 1)$,
- sliding average velocity over five days, v_5, at time $t - 2$ (the day before yesterday):

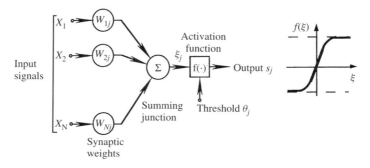

Figure 25.12 Basic neuron i and its threshold sigmoid function f

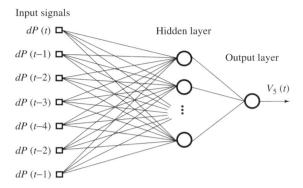

Figure 25.13   Network geometry and input

$v_5(t-2)$,

- net rainfall (after subtracting evapo-transpiration) average over three days at time $t-1$: RA$(t-1)$,
- net average rainfall over three days at time $t-2$: RA$(t-2)$,
- sum of net rainfall over the last week, RS$_7$,
- sum of net rainfall over the last two weeks, RS$_{14}$,
- average pore pressure at time $t-1$: P$(t-1)$,
- average pore pressure at time $t-2$: P$(t-2)$,
- etc.

Different vectors of input parameters have been used in various case studies, showing that sound engineering judgment (i.e. a comprehension of the mechanics of the particular site) may reduce the size of the vector.

25.4.3   *Training and prediction*

To make any prediction, the network must first be trained with an existing data set. This supposes, of course, that such a data set exists, and implies a first limitation of such a neural network to cases where former monitoring is available. This training phase consists of adjusting the weights by minimizing the difference between computed value and desired value of the output. Different algorithms are available for this task, usually based on gradient descent or conjugate gradient methods, known as 'back-propagation algorithms'. Among them the powerful Marquard–Levenberg algorithm (Hagan and Menhaj, 1994), often used in nonlinear least squares problems, is used in this study.

It has been shown that without any other control, the network may overfit the training set. It will reproduce the training set extremely well, but will give a relatively poor prediction for other sets. To avoid such a problem, the existing data set is divided into two sub-sets: the training set, from which the weights are found; and the validation set.

Periodically, the network applies the weights resulting from the training to the validation set; as soon as the global error on this set reaches a minimum, training is stopped. Once the weights are determined, the network can be used for class A prediction.

Different types of predictions have been made, namely:

- prediction of the velocity based on complete input signals;
- prediction of the porewater pressure based on rainfalls;
- prediction of the velocity based on porewater pressure.

A three-day prediction still offered reasonable results. After three days, the quality of the prediction degenerates in most cases (although a five-day prediction of pore pressure was possible). It has been shown that a three-day prediction of the velocity is best obtained with a recurrent network (see Figure. 25.14).

Results

The site of Sallèdes (France) serves as an illustration to the proposed neural network model (see Figure 25.15), in the case of a three-day prediction. This site has been selected by the Laboratoire Central des Ponts et Chaussées (Paris, France) as an experimental site to study the impact of a road embankment on an already unstable slope. The superficial layers belong to the upper Oligocene (*chattien*), and consist of marl of which the first 1–10 m constitute the unstable mass. This moving layer is subject to groundwater level that is affected by strong seasonal fluctuations (with 1–5 m amplitudes). As would be expected, sliding velocity fluctuations are related to the position of the groundwater level. This site is heavily instrumented; piezometric readings are taken daily close to the sliding surface with electric pressure transducers. Displacements are derived from angular measurements done with a servo-accelerometer installed in the zone of maximum shear strains. In the present analysis of the Sallèdes Landslide, data from mainly 1992 are used. The total analysed period consists of 365 days, starting February 25th 1992. The input vectors are divided into three sets. The training set contains the first 216 days, the validation set the following 19 days, and the test set the remaining 128 days. One-day prediction was found to be of excellent quality. Figure 25.15 shows a three-day prediction of the velocity (in fact, the angular velocity), and using a recurrent neural network. It can be seen that the quality of prediction is relatively good. Attempts to produce a prediction of more than

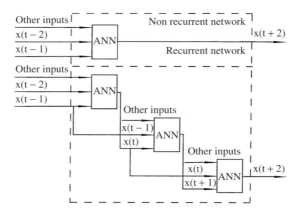

Figure 25.14   Recurrent and non-recurrent network for three-day prediction

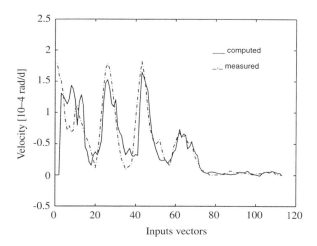

Figure 25.15 Sallèdes landslide: three-day prediction of velocity

three days were also performed, with the following outcome: (i) prediction of pore water pressure as a function of rain amounts remains reliable up to five days; and (ii) prediction of velocity (as a function of rain, pore water pressure and past displacement) deteriorates after three days.

25.4.5   Comments on neural network models

It is clear that as neural network models are based on existing monitored data, the availability of regular data measurements (daily measurements if possible) is essential. In order to predict the behavior of the slope movement more than three days ahead, meteorological estimates of rainfalls should be used.

As the geometry of the landslide keeps changing, the mechanism may change gradually also. As a consequence, it will probably be found more appropriate to revise the weights of the network gradually, by shifting the training set with time.

In cases where the physics and mechanics of the landslide should change completely (major plastification, creation of different kinematics, new shear bands, etc.), the error on the predicted velocity can be used as a warning signal in an automatic alarm system.

A modified version of this type of model could be used to great advantage in connection with deterministic (physical) models presented in Sections 25.2 and 25.3, in particular for the definition of boundary conditions.

25.5   Concluding Remarks

Three families of models for the simulation and prediction of slope movements are presented. Each of them has pros and cons, but they all constitute, despite their level of sophistication, crude approximations of the real behavior of natural slopes. However, they can be used to some extent to predict sliding velocity of slow-moving landslides, as was shown in the various applications.

It can be foreseen that major improvements in these models for the analysis, prevention, and mitigation of natural disasters related to slope movements will be achieved with parallel advances in:

- site investigation (quality of the geotechnical model, *in situ* state of stress);
- site monitoring (in particular, knowledge of the full 3-D kinematics and the essential feature of groundwater seepage);
- material modeling;
- large scale 3-D numerical modeling;
- prediction of boundary conditions (connection with atmosphere physics);
- soft computing (expert system, neural network, artificial intelligence).

This contribution addresses some of these points.

Acknowledgements

The present work is the result of several years of collaboration with several bright people and friends. The author wishes to thank: Professor K. Hutter, University of Darmstadt, his PhD co-advisor (with Professor E. Recordon), for the fruitful work on the different flow models; Professor C. S. Desai for his continuing collaboration in the domain of constitutive modeling and finite elements, in particular for the elasto-viscoplastic model, and for having accepted the stay of the author at the University of Arizona (U. F A.) from 1987 to 1989. The author also wishes to thank Samtani, a former PhD student of the University of Arizona whose thesis focused on the rheology of landslides, the research team at the Soil Mechanics Laboratory of the Swiss Federal Institute of Technology Lausanne (EPFL) and, in particular, Mr F. Mayoraz and Dr T. Cornu for their contributions on neural networks and Mr M. Moreni for the numerical analysis of the Chloewena landslide.

Finally, the support of the Swiss National Science Foundation in the δ_{vp} project (grant no. 82.480.0.97) and the Swiss Federal Highway Administration in the project 'Prediction of landslide with neural network', is greatly appreciated.

References

Desai, C. S. A general basis for yield, failure and potential functions in plasticity. *Int. J. Num. Analy. Meth. in Geomech.*, **4**, (1980) 361–375.

Desai, C. S. Modeling and testing: implementation of numerical models and their application in practice. In: *Numerical Methods and Constitutive Modelling in Geomechanics*, C. S. Desai and G. Gioda, (eds.). Springer-Verlag, Vienna, Austria, (1980).

Desai, C. S. and Fishman, K. L. Plasticity-based constitutive model with associated testing for joints. *Int. J. Rock Mech. and Mining Sci.*, **28**(1), (1991) 15–26.

Desai, C. S. and Siriwardane, H. J. *Constitutive Laws for Engineering Material*. Prentice-Hall, Englewood cliffs, NJ, (1984).

Desai, C. S., Somasundaram, S. and Frantziskonis, G. A hierarchical approach for constitutive modelling of geologic materials. *Int. J. Num. Analy. Meth. in Geomech.*, **10**, (1980) 225–257.

Desai, C. S. and Zhang, D. Viscoplastic model for geologic materials with generalized flow rule. *Int. J. Num. Analy. Meth. in Geomech.* **11**, (1987) 603–620.

Desai, C. S., Samtani, N. and Vulliet, L. Constitutive modeling and analysis of creeping slopes. *J. Geotech. Engrg.*, **121**(1), (1995) 43–56.

Engel, T. Nouvelles méthodes de mesures et d'analyse pour l'étude des mouvements du sol en terrains instables. *Thèse EPFL*, No 601. Ecole polytechnique fédérale de Lausanne.

Hagan, M. T. and Menhaj, M. B. Training feed-forward networks with the Marquadt Algorithm. *IEE Trans. Neural Network*, **5**(6), (1994) 989–993.

Haykin, S. *Neural Network – A comprehensive foundation.* MacMillan, New York, (1994).

Hutter, K. *Theoretical Glaciology: Mathematical approaches to geophysics.* Reidel, Dordrecht, (1983).

Hutter, K. and Vulliet, L. Gravity-driven slow creeping flow of a viscous body at elevated temperatures. *J. Thermal Stresses*, **8**, (1985) 99–138.

Mayoraz, F., Cornu, T. and Vulliet, L. Using neural networks to predict slope movements. *Proc. 7th Int. Symposium on Landslides*, pp 295–300. Trondheim, Norway, 17–21 (June 1996).

Mayoraz, F., Cornu, T., Djukic, D. and Vulliet, L. Neural networks: a tool for the prediction of slope movements. *Proc. 14th Int. Conf. on Soil Mech. and Foundation Engrg.*, Hamburg Vol. 1, (1997) 703–706.

Owen, D. R. and Hinton, E. *Finite Elements in Plasticity: Theory and practice.* Pineridge Press, Swansea, UK, (1980).

Perzyna, P. Fundamental problems in viscoplasticity. *Adv. Appl. Mech.*, **9**, (1966) 243–377.

Pouget, P. and Livet, M. Relations entre la pluviométrie, la piézométrie et les déplacements d'un versant instable (site expérimental de Sallèdes). *Etudes et recherches des laboratoires des Ponts et Chaussées: S. Géotechnique GT 57*; (February 1994).

Samtani, N. *Constitutive modeling and finite element analysis of slowly moving landslides using hierarchical viscoplastic material model. PhD Thesis*, The Univesity of Arizona, Tucson, USA, (1990).

Samtani, N., Desai, C. S. and Vulliet, L. Viscoplastic model of a clay using hierarchical approach. *Proc. 3rd Int. Conf. on Constitutive Laws for Engineering Materials*, Tucson, AZ, January 7–10 (1991) 325–328.

Samtani, N., Desai, C. S. and Vulliet, L. A viscoplastic model for creeping natural slopes. *Proc. 8th Int. Conf. on Computer Methods and Advances in Geomechanics*, Morgantown, West Virginia, Vol. 3, 22–28 (May 1994) 2483–2488.

Samtani, N., Desai, C. S. and Vulliet, L. An interface model to describe viscoplastic behavior. *Int. J. Num. Analy. Meth. in Geomech.*, **20**, (1996) 231–252.

Vulliet, L. *Modélisation des pentes naturelles en mouvement.* Thèse de l'Ecole Polytechnique Fédérale de Lausanne, No 635 pour l'obtention du titre de docteur ès sciences techniques, EPF-Lausanne, (1986a).

Vulliet, L. *Vital1 et Vital2, manuels de l'utilisateur. Rapports internes LMS, EPFL*, Lausanne, (1986b).

Vulliet, L. and Hutter, K. A continuum model for natural slopes in slow movement. *Geotechnique*, **38**(2), (1988a) 199–217.

Vulliet, L. and Hutter, K. Set of constitutive models for soils under slow movement. *J. Geotech. Div.*, **114**(9), (1988b) 1022–1041.

Vulliet, L. and Hutter, K. Viscous-type sliding laws for landslides. *Canadian Geotech. J.*, (25) (1988c) 467–477.

Vulliet, L. and Hutter, K. A multi-layer, multi-sliding surface model for three-dimensional creeping slopes. *Proc. 5th Int. Symp. on Lanslides*, Lausanne, Vol.1, (1988d) 793–797.

Vulliet, L. and Hutter, K. Some constitutive laws for creeping soils and for rate-dependent sliding at interfaces. *Proc. 6th Int. Conf. on Numerical Methods in Geomechanics*, Innsbruck, Austria, (1988e) 495–502.

Vulliet, L. and Desai, C. S. Viscoplasticity and finite elements for landslide analysis. *Proc. XIIth Int. Conf. on Soil Mech. and Found. Eng.*, Vol. 2, Rio de Janeiro, Brazil, (1989) 801–806.

Vulliet, L., Samtani, N. and Desai, C. S. Material parameters for an elasto-viscoplastic law. *Proc. 10th ECSMFE*, Vol. 1, Firenze, Italy, (1991) 281–284.

Vulliet, L. Predicting large displacements of landslides. *Proc. NUMOG V, Numerical models in geomechanics*, Davos, Switzerland, 6–9 (September 1995) 527–532.

Vulliet, L. and Bonnard, Ch. The Chloewena landslide: prediction with a viscous model, *Proc. 7th Int. Symposium on Landslides*, Trondheim, Norway, 17–21 (June 1996) 397–402.

26

Reliability and Stability of Slopes

John T. Christian
Consulting Engineer, Waban, MA, USA

26.1 Introduction

Using relatively crude strength data obtained from a preliminary boring program, an engineer calculates that the factor of safety of a slope is 1.55. The engineer then carries out a program of further investigation to improve the estimates of the soil properties, and finds that the computed factor of safety becomes 1.40. The engineer is much more confident of the latter result because it is based on better data, but how is he or she to convince the regulatory authorities to be satisfied with the smaller value?

A mining company operates a large tailings pond. One day the maintenance crew notices some signs of distress in the embankment that retains the tailings. The consequences of a failure include serious damage to the environment, but stopping the filling of the pond requires closing production and major losses in revenue. The management would like to make a decision based on probabilistic assessment of the alternatives. How is the geotechnical engineer to respond to management's request for the probabilities of the various different outcomes?

A plant is located near a slope in a Karst terrain. From historical records, it is possible to estimate the annual rate of occurrence of landslides in the geologic formation, and the management of the plant is using that estimate to help decide whether to expand the plant and to cut back the slope. The management asks the consulting engineer what effect the earthquake hazard will have on the probability of failure. How can the effect be estimated?

Each of the above cases is loosely based on an actual situation; most engineers can add other examples to the list. As geotechnical engineers are increasingly called on to quantify their confidence or uncertainty, reliability theory, which provides a well established way to attack such problems, has become more important in geotechnical practice. This chapter describes the basis of reliability approaches to geotechnical problems, shows how it can be applied in practice, and discusses some of the current problems associated with its use.

The present discussion concentrates on the stability of slopes, but the methodology can be applied to many other problems in geotechnical engineering. In this context, it should be understood that 'failure' does not necessarily mean a catastrophic collapse, but could be any unsatisfactory performance. Indeed, many agencies now try to avoid using the word 'failure'. Leonards (1982) stated, " 'Failure' is an unacceptable difference between expected and observed performance". The present chapter refers to 'failure' of slopes, but the reader will bear in mind the broader definition of the term.

26.2 Reliability Theory

As the name implies, reliability theory describes analytically how reliable a facility or structure is. Although there are several different ways of expressing the theory, most start by considering the situation illustrated in Figure 26.1. Based on the best estimates of the geometry and soil properties, an engineer has calculated the factors of safety against failure for two slopes. One has an estimate of 1.2, and the other has an estimate of 1.5. However, the engineer is much more confident of the result for the first slope, because the soil properties and the geometry of the strata and failure surfaces are well known. The standard deviation of the factor of safety is estimated to be 0.1. In the other case, the best estimate of the factor of safety is 1.5, but the standard deviation of the factor of safety is 0.5.

Figure 26.1 shows the probability density functions for the reasonable assumption that the factors of safety are normally distributed. The probability of failure in each case is

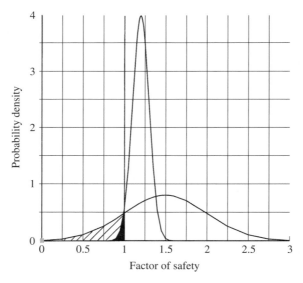

Figure 26.1   Probability density functions for two cases. The case with $E[F] = 1.2$ and $\sigma_F = 0.1$ has a lower total probability of failure than the case with $E[F] = 1.5$ and $\sigma_F = 0.5$

the area under the probability density function to the left of the vertical line at $F = 1.0$. It is clear that the area for the second case is much larger. In other words, the probability of failure is greater for the case with the larger factor of safety.

This result conforms to geotechnical experience. Geotechnical engineers know that the factor of safety by itself does not provide enough information to evaluate the safety of a slope, and that the criteria for a satisfactory value depend upon the degree of confidence that the engineer has in the information that went into the calculation. Reliability theory is one way to express that confidence.

The *reliability index* is represented by β, and defined by the equation

$$\beta = \frac{E[F] - 1.0}{\sigma_F} \tag{26.1}$$

In this equation, $E[F]$ is the estimated value of the factor of safety (usually the value calculated from the best estimates of the parameters), and σ_F is the standard deviation of the factor of safety. In some cases, β is defined in terms of the driving or loading forces and the resisting forces:

$$\beta = \frac{E[R] - E[Q]}{(\sigma_R^2 + \sigma_Q^2)^{1/2}} \tag{26.2}$$

where R and Q stand for the resistance and the load, respectively. Properly applied, the two formulations are equivalent, but computational considerations may dictate using one or the other.

If the factor of safety is normally distributed, the probability of failure, that is the probability that $F \leq 1.0$, is easily computed as 1.0 minus the cumulative distribution function for a normal distribution with mean value of 0.0 and standard deviation of 1.0 evaluated at β. Figure 26.2 shows the probability of failure for a range of values of the reliability index. Similar plots can be created for other distributions; they are close to the values in Figure 26.2 for β less than about 2.

The reliability index provides a measure of how much confidence the engineer can have in the computed value of the factor of safety, and leads to an estimate of the probability of failure. It should be borne in mind that this probability is actually the lower bound on

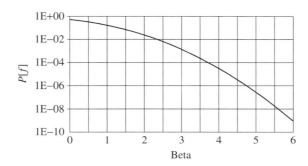

Figure 26.2   Probability of failure as a function of reliability index for normally distributed factor of safety

the probability of failure because it ignores all the factors that have not been considered by the engineer. A great many failures result from the failure to consider a possible mode of behavior rather than from failure to use the correct parameters in design. With that proviso in mind, the question is then how to compute the reliability index.

26.3   Uncertainty in the Parameters

The first problem is how to estimate the uncertainties in the parameters that go into the calculation of F and σ_F. In some cases, all the uncertain parameters are soil properties. In a more general and more realistic case, there should also be uncertainty in the location or size of different elements of the soil profile, in the pore fluid pressure, or in other aspects of the analysis. These uncertainties can be treated in the same way as the uncertainties in the soil properties, except that numerical approximations become necessary as more uncertain parameters are introduced.

The uncertainty of any parameter can be divided into two components: *scatter* and *systematic error*. Scatter represents the random variability; it is a parameter due to inherent spatial variability and errors in measuring the values of the parameter. It can be considered to involve the random variation about the mean. Systematic error is the uncertainty that applies across all locations. It is the uncertainty in the value of the mean itself.

Scatter is further composed of two components: *real spatial variability* and *random testing errors* or *noise*. Real spatial variability is, as the name implies, something that is inherent in the nature of the soil or rock profile. Random testing error, on the other hand, is an artifact of the testing process. Since only the real spatial variability contributes to the uncertainty of the actual facility, it is necessary to separate it from the noise and to eliminate the noise from the estimate of the uncertainty. Christian *et al.* (1994) describe one widely used method for doing this. It starts by finding how the values of the parameter are correlated over distance, a relation that is called the 'autocovariance function'. This is then extrapolated back to a distance of zero; the result should be the local variance of the parameter. Usually, this is found to be less than the measured value of the variance, and the difference must be due to measurement error because the measurement error must be purely local and uncorrelated over distance. When the measurement error is subtracted from the measured variance, the result is the variance due to real spatial variance. This procedure also gives a measure of the correlation between values of the parameter over distance, which is useful in the subsequent procedures. DeGroot and Baecher (1993) have shown that formal methods using maximum likelihood provide superior estimates than traditional curve-fitting approaches.

Systematic error can also be divided into two parts: *statistical error in the mean* and *measurement bias*. Statistical error in the mean arises because even an exact measurement process that takes a finite number of samples will have some error in the estimate of the mean. A well known way to estimate this error is to divide the computed variance in the parameter by the number of points used to evaluate it. This gives a measure of how uncertain the mean value is. Measurement bias depends upon the particular measurements used. It arises because many engineering parameters are evaluated by tests that measure something other than the actual parameter of interest. For example, the field vane test does not measure the shear strength on surfaces likely to be involved in a slope failure, but its results can be converted into meaningful numbers. The errors in that conversion make up the measurement bias.

Thus, the engineer must identify, for each parameter, how the uncertainty is divided between spatial variation and systematic error. Sometimes, this must depend upon judgment; sometimes the analysis dictates how the division must be done. For example, the thickness of a stiff crust may be uncertain, and the uncertainty would be divided into the two categories. On the other hand, the depth to a stiff lower stratum might also be uncertain, but it would be considered entirely a systematic error because it limits the location of the failure surface, and this applies across the entire geometry of the analysis.

The main reason for separating the uncertainty into two categories is that the spatial variables tend to be averaged out over the failure surface, or over the geometry that applies to any other analysis. If L is the significant length of the analysis (for example, the length of the failure surface) and r_0 is the *autocorrelation distance* (the distance over which the spatial variability decreases to $1/e$ of its value at $r = 0$), it has been found that the ratio between the average variance of the parameter to be used in analysis to the computed spatial variance is approximately $2r_0/L$. When the failure surface is large compared to the correlation distance, this has the effect of removing most of the spatial variability from the calculation.

Another factor that should be considered is the error in the analytical model. The model error has two effects. First, the model can be expected to be in error by some average amount or ratio, and this factor can be applied directly to the computed mean value of the estimated factor of safety. Secondly, the uncertainty in the model results, expressed as a variance, can be added to the computed variance of the factor of safety. These effects go in opposite directions if the analytical model is believed to be conservative. The effect on the mean is to raise the value of F, which increases β. The uncertainty increases the variance of F, which decreases β. Christian *et al.* (1994) show one case in which the reliability increases and two in which it decreases because of model uncertainty.

26.4   Computational Procedures

There are several methods of calculating the value of the standard deviation of the factor of safety, σ_F, or equivalently, its variance VAR_F, which is simply the square of σ_F. They include:

- *Direct calculation:* In some cases the equation for the factor of safety or for the margin of safety can be solved analytically to give the variance of the factor of safety. In particular, if the margin or factor of safety can be expressed as the sum of independent random variables, the variance is simply the sum of the variances of each contributing variable. Unfortunately, such cases are rare.

- *Monte Carlo simulation:* Modern computers make it possible to generate a large number of solutions in which each of the variables is allowed to vary randomly according to its prescribed probability distribution. The technique is easy to program on a computer; in many cases, it can be performed using a modern spreadsheet. Its main advantage is simplicity. Its major disadvantage is that it may be necessary to run a very large number of cases to get robust results. However, if the method is used to estimate the mean and the variance of the distribution of F, rather than to find the probability of failure directly, the number of stability calculations is reduced significantly.

- *First Order Second Moment (FOSM) methods:* These techniques are based on using the first terms of a Taylor series approximation for the variance. They give a direct approximation for σ_F and, at the same time, provide direct insight into the relative contributions of the various parameters. The basic method is easy to use. Its major disadvantage is that it gives an approximate solution that may be quite inaccurate.

- *The Hasofer–Lind method:* Hasofer and Lind (1974) developed an improvement to the FOSM method to account for the fact that the derivatives used in the FOSM should be evaluated at a point on the failure surface, i.e. when $F = 1.0$. When the underlying assumptions of the FOSM method are satisfied, this technique may improve the accuracy of the results. However, it does require an iterative procedure that is not intuitively obvious, and when the FOSM assumptions are not reasonable, the technique can actually give results that are less accurate than other techniques. Ang and Tang (1990) give a very clear description of the computational procedures.

- *The point estimate method:* Wolff (1996) has described this approximate technique, which has been widely adopted by the US Army Corps of Engineers (e.g. US Army, 1995). It involves computing the factor of safety using values of each variable that are one standard deviation above or below the expected value, and combining the results to estimate the expected value and standard deviation of the factor of safety. For N variables this requires 2^N computations, which is much less than the number needed for a Monte Carlo simulation. The method is simple but approximate. When the variables are not normally or log-normally distributed, it may not be intuitively obvious how to choose the paris of values to be used. Furthermore, when the number of uncertain variables (N) becomes large, the number of required safety calculations (2^N) becomes intractably large.

The best way to illustrate the use of reliability theory and the various methods of performing the calculations is to consider an example. This is done in the next section.

26.5   An Example Using the Culmann Method

The Culmann method is used to analyse the stability of a slope such as that illustrated in Figure 26.3. This method is limited by its assumption that the surface of sliding is a plane, so other techniques such as the various methods of slices have superseded it

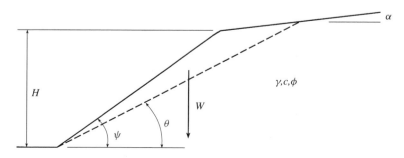

Figure 26.3   Geometry and notation for example Culmann problem

in engineering practice. Nevertheless, it remains valid for cases in which there are pre-existing planes of weakness such as joints. Since it combines some of the complexity of slope stability calculations with computational formulas that are simple enough to be manipulated, it serves to illustrate the application of reliability theory to the problem of slope stability.

With the notation defined in Figure 26.3, the factor of safety for a condition in which water pressures do not apply is found to be

$$F = \frac{c + \dfrac{1}{2}\dfrac{\gamma H}{\sin\psi}\sin(\psi - \theta)\cos\theta\tan\phi}{\dfrac{1}{2}\dfrac{\gamma H}{\sin\psi}\sin(\psi - \theta)\sin\theta} \tag{26.3}$$

In this equation the notation, in addition to that defined in Figure 26.3, is that c and ϕ are the strength parameters and γ is the unit weight of the soil or rock. For purposes of illustration, let it be assumed that the inclination of the failure plane is known – perhaps because the joint set has been well established. This leaves three parameters whose values are uncertain: the two strength parameters and the unit weight.

Two cases will be used for illustration. The height H is fixed at 10 m, ψ is 26°, θ is 20°, and α does not enter into the calculation. The slope of 26° corresponds approximately to a 2:1 slope. It will be assumed that the variables are all normally distributed. Table 26.1 shows the means and the coefficients of variation for the uncertain parameters. (The coefficient of variation, COV, is the ratio between the standard deviation and the mean of a variable.) The COV for the friction component refers to the tangent of ϕ, not to ϕ itself. The COVs for $\tan\phi$ correspond approximately to COVs of 0.1 and 0.2 for ϕ in cases 1 and 2, respectively. Substituting the mean values into Eq. (26.3) gives factors of safety of 1.29 and 2.14 for the two cases. However, it is clear that the values of the parameters are much less well determined in the second case. As the uncertain parameters appear in both the numerator and the denominator of Eq. (26.3) direct calculation if σ_F is not possible.

26.5.1 *Monte Carlo simulation*

Monte Carlo simulation involves generating a large number of sets of values of the parameters using a random number generator modified so the resulting distributions agree with the desired distributions of the parameters. In this case, all the parameters are normally distributed, and the means and standard deviations can be found from Table 26.1. The

Table 26.1   Parameters for example cases

Variable	Units	Case 1		Case 2	
		Mean	COV	Mean	COV
γ	KN/m^3	22	0.1	22	0.2
c	KN/m^2	5	0.1	5	0.4
ϕ	degrees	15		30	
$\tan\phi$		0.267 949	0.07	0.577 350	0.24

simulation can be done on any modern spreadsheet. Two simulations were done for Case 1. The first simulation gave $E[F] = 1.295$, $\sigma_F = 0.1012$, and $\beta = 2.915$. The second simulation gave $E[F] = 1.309$, $\sigma_F = 0.0950$, and $\beta = 3.251$. A single simulation for Case 2 gave $E[F] = 2.137$, $\sigma_F = 0.4985$, and $\beta = 2.280$.

Although the number of sets of values is rather small, these results illustrate several points. First, they show that the reliability index for the first case is larger than for the second case, although the $E[F]$ is much larger in the second case. In other words, the uncertainty in the parameters in the second case makes the slope significantly less reliable. The probabilities of failure corresponding to these values of β are 1.78×10^{-3} and 0.58×10^{-3} for the two simulations in Case 1 and 11.3×10^{-3} for Case 2. Case 2 has a probability of failure that is about one order of magnitude larger than Case 1.

Secondly, different simulations for Case 1 gave different answers, so the user must be aware of the uncertainty that is inherent in Monte Carlo simulation. This can be reduced by taking a larger set of samples in the simulation, but it remains a drawback of Monte Carlo simulation. It should be noted that in this case Monte Carlo simulation was used to estimate $E[F]$ and σ_F, which are established by the middle range of the results. Direct calculation of the probability of failure would require a very large number of samples.

26.5.2   First order second moment

The FOSM method estimates the standard deviation in F by expanding F in a Taylor series. The details can be found in most texts on reliability theory (e.g. Ang and Tang, 1990). The result is, for the case that the variables are not correlated:

$$\sigma_F^2 \approx \sum_{i=1}^{n} \left(\frac{\partial F}{\partial x_i} \right)^2 \sigma_{x_i}^2 \tag{26.4}$$

The derivatives can be evaluated analytically or numerically. One of the advantages of this approach is the possibility of using numerical differences to approximate the derivative in complicated situations for which analytical differentiation may be difficult. Another important advantage is that, for each parameter, the method shows the combined contribution of the derivative and the variance, and thus identifies clearly those variables that have the strongest influence on the uncertainty. Further studies should then concentrate on those variables that have the largest influence.

The FOSM method gives $\beta = 3.117$ for Case 1 and 2.513 for Case 2. The corresponding probabilities of failure are 0.91×10^{-3} and 5.99×10^{-3}, respectively. The relative reliability of the two cases remains the same, but the numbers are somewhat changed from the simulation results.

26.5.3   Hasofer–Lind

The Hasofer–Lind method is a refinement of the FOSM method. It starts from a recognition that the right-hand side of Eq. (26.4) is actually the second term in the Taylor series, and that the first term can be ignored only if the derivatives are evaluated at a point where $F = 1.0$. Finding the nearest such point involves an iteration that will not be described here. However, the results for the two cases are $\beta = 3.436$ and 2.574, respectively. The corresponding probabilities of failure are 0.30×10^{-3} and 5.03×10^{-3}.

26.5.4 Point estimate method

In the point estimate method, the analyst selects two values for each variable. One is at the mean plus one standard deviation, and the other is at the mean minus one standard deviation. Since there are N variables and two points for each, there are 2^N combinations. In the present case $N = 3$, so there are eight combinations, for each of which the factor of safety is calculated. Each of the combinations is weighted equally. Therefore, the estimate of the mean is simply the sum of the results divided by the number of combinations, in this case 8. Similarly, the sum of the squares of each result is also divided by the number of combinations. If the result of each combination is designated F_i, the variance can be estimated by

$$\sigma_F^2 \approx \frac{1}{N} \sum_{i=1}^{N} x_i^2 - \left(\frac{1}{N} \sum_{i=1}^{N} x_i \right)^2 \tag{26.5}$$

The point estimate method yields $\beta = 3.149$ for Case 1 and 2.519 for Case 2. The corresponding probabilities of failure are 0.82×10^{-3} and 5.88×10^{-3}. Again, the relative values of β and the probability are similar, but the numbers differ from those from the other methods.

26.5.5 Summary of numerical results

Table 26.2 summarizes the results for the two cases from the different methods of calculation. The results show clearly that Case 2 is much more reliable than Case 1 by any method of calculation. However, the actual numerical results do differ between the various methods of calculation. In this instance the Hasofer–Lind method, which is intended to be an improvement on the FOSM method, gives the largest values of β and the smallest probabilities of failure.

Table 26.2 Summary of numerical results for example problem

	$E[F]$	σ_F	β	$p[f]$
Case 1				
FOSM	1.294	0.0942	3.117	0.91×10^{-3}
H–L	1.294	0.0725	3.436	0.30×10^{-3}
Point Est.	1.299	0.0950	3.149	0.82×10^{-3}
Monte Carlo	1.295	0.1012	2.915	1.78×10^{-3}
Case 2				
FOSM	2.144	0.4550	2.513	5.99×10^{-3}
H–L	2.144	0.4358	2.574	5.03×10^{-3}
Point Est.	2.167	0.4631	2.519	5.88×10^{-3}
Monte Carlo	2.137	0.4985	2.280	11.30×10^{-3}

Table 26.2 demonstrates that, if one is making comparisons between alternative designs, the comparisons should be made using the same methods of computation. Of course, the same observation could also be made if the comparison were to be made using factors of safety; the different techniques for calculating the factor of safety do not give the same or even comparable results. Another conclusion is that, however the reliability is evaluated, the slope in Case 1, which has a much smaller mean value of the friction angle, is far more reliable and far less likely to fail than the slope in Case 2. Since the aim of calculation should be to establish the relative reliability of a structure, not simply an arbitrarily defined factor of safety, the results in Table 26.2 are more meaningful than a simple calculation of factor of safety.

26.5.6 An exact case

If the unit weight is completely known so that its variance is zero, then all the uncertain variables appear in the numerator of Eq. (26.3). The variance of the sum or difference of two independent variables is simply the sum of the variances of each variable, so this case can be solved exactly. The result for Case 1 is that $\beta = 3.867$, and the probability of failure becomes 0.055×10^{-3}. The FOSM, Hasofer–Lind, and point estimate methods all give the same result. One trial of Monte Carlo simulation gives $\beta = 3.816$, but it is to be expected that the simulation results will show some scatter. For Case 2, when the unit weight is certain, $\beta = 2.592$, and the probability of failure is 4.77×10^{-3}. Again the FOSM, Hasofer–Lind, and point estimate methods give the same result, but Monte Carlo simulation is slightly in error with $\beta = 2.620$. These results how that, as one would expect, when the uncertainty in one parameter is reduced the overall uncertainty is also reduced.

26.6 Application to Earthquake Hazard

An interesting extension of reliability theory arises when one wishes to estimate the effect of seismic hazard on the probability of failure. The details are described by Christian and Urzua (1997), but the basic procedure is as follows. First, historical observation of failures leads to an estimate of the annual probability of failure. This can be converted into value of β, and a reasonably conservative value of σ_F leads to an estimate of the static F.

Equation (26.3) can be extended to include the effects of a horizontal acceleration factor. Algebraic manipulation of the new equation and Eq. (26.3) leads to

$$F^* = \frac{F - Aa_h \tan \phi}{\dfrac{Aa_h}{\tan \theta} + 1} \tag{26.6}$$

where F^* is the dynamic pseudo-static factor of safety, F is the static factor of safety, A is the amplification factor, a_h is the horizontal acceleration in gs, and the other parameters have been identified. If it is assumed that the variance in F^* is the same as in F, the probability of failure (F^* less than 1.0) can be expressed as a function of the acceleration. The probability distribution of the accelerations can be determined from seismic hazard analyses or maps.

The analysis consists then of combining Eq. (26.6) with the annual probability of accelerations at different levels. The result is an annual probability of failure due to the earthquake hazard alone. Results for one area of moderate seismicity are that the annual hazard of failure is increased by a factor of only about 0.1 to 0.15.

26.7 Conclusions

Reliability theory is a powerful tool. It allows the engineer to investigate cases in which the values of parameters are not known exactly, or where the engineer wishes to know the uncertainty associated with the calculations. The results are in a form that can be combined with other estimates of risk and reliability. Because it is seldom possible to compute the uncertainty in the factor of safety directly, several approximate techniques have been developed, and these are illustrated by a simple example. A major task in performing a reliability analysis is determining the uncertainty in the relevant parameters. It is important to distinguish between local uncertainties and systematic uncertainties, and to incorporate averaging effects. Local uncertainties are often largely averaged out, and the analytical methods need to account for this. The theory can also be extended to estimate the contribution of seismic hazards to the overall risk of slope failure.

References

Ang, A. H.-S. and Tang, W. *Probability Concepts in Engineering Planning and Design, Vol II – Decision, Risk, and Reliability*, reprinted from the 1975 edition originally published by John Wiley & Sons, New York, (1990) 333–447.

Christian, J. T., Ladd, C. C. and Baecher, G. B. Reliability applied to slope stability analysis. *J. Geotech. Engrg.*, **120**(12), (1994) 2180–2207.

Christian, J. T. and Urzua, A. A probabilistic evaluation of earthquake induced slope failure. *J. Geotech. Geoenvironmental Engrg.* (submitted 1997).

DeGroot, D. J. and Baecher, G. B. Estimating autocovariance of in-situ soil properties. *J. Geotech. Engrg.*, **191**(1), (1993) 147–166.

Hasofer, A. M. and Lind, N. An exact and invariant first order reliability format. *J. Engrg. Mechanics*, **100**(EM1), (1974) 111–121.

Leonards, G. A. Investigation of failures. *J. Geotech. Engrg. Div.*, **108**(GT2), (1982) 187–246.

Taylor, D. W. *Fundamentals of Soil Mechanics.* Wiley, New York, (1948) 453–455.

US Army. Introduction to Probability and Reliability Methods for Use in Geotechnical Engineering, *Engineer Technical Letter No. 1110-2-547*, Department of the Army, Corps of Engineers, CECW-EG, (30 September 1995).

Wolff, T. F. Probabilistic slope stability in theory and practice. *Uncertainty in the Geologic Environment, ASCE, Proceedings of a Specialty Conference*, Madison, WI. (Also *ASCE Geotechnical Special Publication No. 58*, Vol. I, (1996) 419–433.)

Index